# SUSTAINABLE AGRICULTURAL SYSTEMS

Edited by
Clive A. Edwards, Rattan Lal, Patrick Madden,
Robert H. Miller and Gar House

Soil and Water Conservation Society
7515 Northeast Ankeny Road
Ankeny, Iowa 50021

Manufactured in the United States of America

Library of Congress Catalog Card No. 89-26370

ISBN 0-935734-21-X

$40.00

*Library of Congress Cataloging-in-Publication Data*

Main entry under title:

Sustainable Agricultural Systems / edited by Clive A. Edwards...(et al.).
p. cm.
Includes bibliographical references.
ISBN 0-935734-21-X
1. Sustainable agriculture. 2. Agricultural systems.
I. Edwards, C. A. (Clive Arthur), 1925- .
S494.5.S86S86 1990 89-26370
338.1'62—dc20 CIP

# CONTENTS



## I. An Overview of Sustainable Agriculture

## II. Components of Sustainable Agricultural Systems

## III. The Importance of Integration in Sustainable Farming Systems



## IV. Sustainable Agricultural Systems in the Tropics

## VI. Improved Ecological Impacts of Sustainable Agriculture

# FOREWORD

Early in my married life, I went to India as a staff assistant to the American ambassador for four years. It was there I came to appreciate the degree to which agriculture really is the cornerstone of the economic health and well-being of any country. To live in India from 1963 to 1967, which as many of you know were critical years in the emergence of the green revolution in that country, was a most exciting time. I worked with an ambassador named Chester Bowles, who was a great believer that agriculture came before industrial development. He regularly debated that proposition with Prime Minister Nehru and appealed to a book of ancient wisdom in India that began, "Agriculture is the lynchpin of the social chariot."

From a local standpoint, Ohio is part of those midwestern states that comprise about a quarter of our nation's population, but provide four out of ten of the farmers in our nation. These farmers produce more than 44 percent of our nation's food products. Here in Ohio, which is often thought of as an industrial state, and it is, agriculture, in terms of food production, processing, and marketing, is the state's number one industry. In 1988, we faced the worst drought in many years, perhaps on record, and it has reminded us of just how fragile our hold on this agricultural wealth is.

The history of agriculture in the 20th century is a story of dramatic technological advances that have led to increased production. Indeed, I've often thought that the manufacturing industry could learn a great deal from agriculture about how to increase productivity.

Still, without the recent rainfall, I believe this country would have faced a disaster of unmatched proportions despite all of the advanced technology available to us. The impact of such a disaster on American farmers and consumers would, of course, have been harsh, but it would have been catastrophic for millions of people around the world whose survival still depends on repeated bumper crops of American agricultural products.

Just as the Midwest is the nation's breadbasket, it has, in many respects, been the world's breadbasket. And while we're proud of that fact, I think we ought also to be humbled by that responsibility and to recognize that it is something that needs to change over time. All of this perspective gives, I think, additional meaning to the concept of sustainable agriculture. To me the word "sustainable" has several important aspects, not the least of which implies the future. When we talk about sustainable agriculture, we talk not only about low inputs for optimum production, we are also talk about agriculture with a future, agriculture with a dependable future.

The experience of our last few months here in Ohio and in this part of our country also reminds us of how indivisible we are as a human family. Two years ago, Ohio farmers provided relief for their counterparts in the southeastern United States as they were reeling from a drought there. Many of those farmers had the opportunity to return the favor this year. Ohio farmers were struck by the fact that as we worried about the drought in our own state, our sister province in China, Hubei Province, suffered a similarly severe drought. Our farmers commented on the irony that here we were, sister state and sister province, both confronted by a similar challenge to our agricultural abundance.

The fact is that we live in an increasingly integrated global economy where virtually everything is interconnected, and the days are long since past when we could refer to the U.S. economy as a free-standing entity. In this, the year the earth talked back, we have also begun to understand that we live in a single ecosystem where the greenhouse effect, the destruction of the ozone layer, the impact of acid rain, the consequences of deforestation are shared by all, and make the prospects of sustainable growth the greatest challenge before us. This is not an American problem or a Third World problem. This is a worldwide challenge that we all must begin to meet, and we must meet it as much here in America's heartland as in any other part of the globe.

*Richard F. Celeste*

# PREFACE

Modern agriculture in developed countries currently depend upon high inputs of inorganic fertilizers and synthetic chemicals for pest control and tends towards monoculture of cash crop varieties that require such inputs. These practices have increased overall productivity, but they have also led to overproduction of certain crops in the United States and Europe, which has reduced farmers' profit margins because of the inevitable drop in market prices of crops. Intensive production also has the potential to accelerate wind and water erosion of soils and to result in the contamination of surface water and groundwater.

There is a growing awareness about the need to adopt more sustainable and integrated systems of agricultural production that depend less on chemical and other energy-based inputs. Such systems can often maintain yields, lower the cost of inputs, increase farm profits, and reduce ecological problems. Some developing countries, having used subsistence agricultural practices and subsequently increased yields by adopting higher input methods, are now experiencing greater pest, disease, and weed problems; increased soil erosion; environmental hazards; and economic stress. Yet the need to increase food production in these countries is greater than ever before. Hence, there is an urgent need for both research and education on sustainable farming systems that can increase productivity and profits for farmers without endangering the resource base and polluting the environment.

This same situation exists throughout the world. In the United States, the Department of Agriculture (USDA) made significant new funds available to promote sustainable agriculture in 1988. These funds are expected to remain available and perhaps even increase in the years ahead. The governments of most European countries likewise are paying increased attention to the need for lowering chemical inputs to avoid environmental problems, and these same countries are reviewing means of reducing production overall. There are also major programs in the U.S. Agency for International Development, the World Bank, and the World Resources Institute aimed at promoting the concept of sustainability into the 21st century in Third World countries.

This circumstance, worldwide, makes these proceedings of the International Conference on Sustainable Agricultural Systems that was held at Ohio State University, Columbus, Ohio, in September 1988, and jointly organized by Ohio State University, Pennsylvania State University, and North Carolina State University, extremely timely. The conference was made possible by the generous sponsorship of USAID and three USDA agencies, the Cooperative States Research Service, the Soil Conservation Service, and the Agricultural Research Service, along with the Rodale Institute and the Farm Foundation. We are also grateful to the Soil and Water Conservation Society for the prompt publication of this material.

The conference included 38 formal presentations, which form the contents of this book, and 40 poster presentations, some of which are being published elsewhere. There were also five workshops that addressed critical issues on (1) "Government Policies and Strategies to Promote Sustainable Agriculture," (2) "Innovative Technical Assistance and U.S. Aid Policies to Promote Sustainable Agriculture in Developing Countries," (3) "Research Education and Extension Strategies in Sustainable Agriculture," (4) "The Agrochemical Industry in Relation to Sustainable Agriculture," and (5) "Building on Success: Farmer Participation in Sustainable Agriculture." There was one informal discussion group as well on the "Role of Women in Sustainable Agriculture." Participants in each of these sessions were asked to discuss present policies and strategies and develop recommendations on future research, teaching, and extension activities. The discussion in each case was summarized by a rapporteur and presented to a plenary session of the entire conference for comment and discussion. The conclusions have since been published as a booklet by the Rodale Institute.

*Clive A. Edwards*

# INTRODUCTION

The topic of agricultural sustainability is a high priority in all countries around the earth, whether they are developed or developing in their current economies. Some would say it has been too long in receiving attention. But I prefer to view it as timely and relevant.

There is no more important question before us on this globe today than that of the sustainability of agricultural systems. Desertification, deforestation, and accumulation of chemicals in soils and waters are of increasing concern in many ecosystems and different parts of the world. One can find a growing number of such citations in both scientific and popular publications, to the degree that not only scientists but also the general public are raising serious questions about the current state of affairs and potential alternatives for the future.

It has been authentically reported that some agricultural systems that were once popular have disappeared over time because they could not be sustained for a variety of reasons. Others have been sustained for thousands of years and are still flourishing. Not enough has been done to analyze the differences between those systems and practices that persisted and those that did not. Such analyses might provide insight concerning the present and future.

Many definitions of the term "sustainable agriculture" have been presented, and that is as it should be. We must recognize the varied points of view that enter such a discussion. For me the term sustainable merely adds a long-term dimension to consideration of any agricultural system. It requires studies that are conducted over a long period of time, such as decades, rather than for three or four years. I might point out the Rothamstead plots in England and the Morrow, Jordan, and Sanborn plots in the United States are excellent examples of long-term experiments that have contributed invaluable information. I am convinced that we must initiate new studies with a long time frame that integrate many of the current and innovative farming practices in a variety of ecosystems.

Much has been written recently regarding the environmental and the economic aspects of sustainable agricultural systems. A sustainable system must be both economically profitable and environmentally compatible. As William Ruckelshaus has pointed out, "Unlike railroad tracks, economic development and environmental protection really do converge if you take a long enough view."

*F. E. Hutchinson*

# AN OVERVIEW OF SUSTAINABLE AGRICULTURE

# A HISTORY OF SUSTAINABLE AGRICULTURE

Richard R. Harwood

To provide a conceptual setting for the definition of sustainable agriculture and to show evolutionary trends in its development, two reference points in that evolution are of special importance. These two reference points are not meant to be exclusive but rather to represent a spectrum of thought.

The first reference point should be placed in the early 1980s, with the emergence of the concepts of regenerative agriculture (Rodale, 1983) and the articulation of a sustainable agriculture (Jackson, 1980). The early concept has evolved into a construct of agriculture based on principles of ecological interaction. It is referred to as an ecological definition of sustainability. This concept now forms the philosophical basis for most alternative agricultural groups.

A second reference point is the increased use of the term sustainable, starting in 1987, to refer to a "stable" agriculture in the global sense, involving all facets of agriculture and its interaction with society. It is the "universal" sense that seems to be the object of this book.

## A Framework for Universal Definition

The word sustainable implies steady state. If one sees a steady-state situation, one must look over horizons to some distant goal. A careful reading of development literature reveals as many ideas about direction as there are authors, so consensus on an equilibrium point would be impossible. Lack of understanding; of hard data; or of consensus on resource bases, global climate and its variation, technologies of the future, the role of people in agriculture, and the relationship between people, agriculture, and the environment all make prediction of an end point a futile exercise. Others could argue as well that there may never be an end point or equilibrium but, as with the rest of the universe, a continual process of evolution.

Given the limits of vision, of data, and of the imprecision of a process for arriving at consensus, I suggest using a "framework" definition that can be filled with appropriate detail by country and by desired time frame. A workable definition is *"an agriculture that can evolve indefinitely toward greater human utility, greater efficiency of resource use, and a balance with the environment that is favorable both to humans and to most other species."*

This definition is heavily value-laden, but it is consistent with the parameters of an emerging social and political agenda for agricultural development. It is also very much generic. To understand the process by which it is translated into substance in any national setting, some sense is needed of public agendas, the translation of those agendas into policy, and the roles of agendas and policy in development.

## Evolution of the Sustainability Concept

Agricultural evolution always has been guided by a perception of what should be, sometimes called the model, the goal, or even the ideology. The difference between that goal and agriculture as it exists presently is the development gap. The breadth or all-inclusiveness of the model likewise changes with time. We could analyze at great length the philosophical bases for development, but we will limit ourselves to just a few key concepts that seem most closely related to present sustainability concepts.

Some analysts take us back to origins of current conflict in a Newtonian world view. Rifkin (1980) characteristizes that view by four relevant elements: a mechanical view of nature, a rigid dichotomy between nature and society, a faith in progress, and a consumerist ethic. Others point to the 17th century English philosopher John Locke who wrote on the social goal of efficiency in agriculture by stating, "He that encloses land, and has a greater plenty of the conveniences of life from 10 acres, then he could have from a 100 left to nature, may truly be said to give 90 acres to mankind." Others point to Thomas Jefferson's linkage of agricultural practices with morality. In notes on the State of Virginia in the late 1700s, Jefferson (1984) wrote, "Those who labor in the earth are the chosen people of God, if ever he had a chosen people, whose breasts he has made his peculiar deposit for substantial and genuine virtue. . . ." Corruption of morals in the mass of cultivators is a phenomenon of which no age or nation has furnished an example. The poet and philosopher Emerson expressed a similar belief. Some would say that the thoughts of Newton are reflected closely in the global view of the Battelle Institute (1985) and those of both Jefferson and Emerson in the works of Wendell Berry (1978). Certainly, the conflicts now faced in the articulation of a universal paradigm are in many instances the same as those addressed by writers of two centuries ago.

However, we are not as concerned here with philosophical content as much as with process. At the turn of the 20th century, U.S. agriculture was in the early stages of industrialization. The conflict between an urban "agrarian" lifestyle and what were seen as radical changes being brought on by industrialization was already present (Danborn, 1979). More important, however, were the divisions among and between farmers and the growing community of "land-grant" scientists. Now these divisions have returned to haunt us 80 years later.

In the early 1900s, popular thinking among farmers had led to rejection of the portion of Jeffersonian thought that held individualism to be supreme. Politically, this led to establishment of organizations, such as the Grange. Farmers felt that they should develop and share technological knowledge among themselves. There were two sources of that knowledge. The "systematic agriculturists" looked to the emerging industry as their model. The second group, the "scientific agriculturists," looked to nature as their model, with the objective of rationalizing and formalizing their experiences as "natural historians" (Marcus, 1985). At the same time, land-grant scientists were beginning to have an impact (Rossiter, 1975). It is these philosophies and to these turn-of-the-century farmer-scientist groups that we can trace the roots of much of the current debate on sustainable agriculture.

U.S. agriculture was in a major expansionist mode during the early 1900s. The number of farms reached a peak of 6.8 million in the early 1930s (Hardin, 1988). Mechanization was being adopted rapidly, spurred by rising costs and the scarcity of labor brought on both by area expansion and by competing demands for industry. Technologies were increasing rapidly, as exemplified by the development and widespread adoption of crop hybrids. The land-grant system was a major determinant in the articulation of the development paradigm. During the early 20th century, the concepts of conservation evolved, first giving emphasis to preservation of natural areas. The progressive conservation movement of the early 1900s established the intellectual foundations of the later conservation programs (Batie et al., 1985).

A series of conservative programs reflects these common roots, including the Agricultural Adjustment Act of 1933, the Soil Bank Program of 1956, and the Food Security Act of 1985 (Phipps and Crosson, 1986). These programs addressed both the problems of soil conservation and of growing surplus production through the mechanism of land reserves or set-aside acres (Hardin, 1988; Jeske, 1981). Parallel to these developments and to the spread of new crop technologies and to rapidly expanding mechanization was the spread of "chemical" technologies. The use of industrially produced fertilizers spread rapidly after World War II, and that of pesticide development followed close behind, leading to what Rifkin calls "the age of alchemy" (Rifkin, 1983).

I would like momentarily, however, to review a thought development process of major significance to the concept of sustainability that traces its origin through Malcolms' "scientific agriculturists." At the turn of the 20th century, the concepts of wholism versus reductionism were taking shape (Harwood, 1983). The emergence of thought on wholism, of looking to natural systems as a model, and of the role of farmers in evolving their own systems (all concepts mentioned above) led to what is today generally referred to as "alternative agriculture." Alternative agriculture evolved during the 1900s in a course parallel to that of industrial agriculture, borrowing liberally but selectively from technologies, such as new crop varieties, mechanization, and soil nutrient testing. A review of that evolution helps greatly to understand today's debate. Many of today's alternative agriculturists trace their history back, surprisingly enough, to Darwin.

Charles Darwin spent his later years in England meticulously studying soil floral and faunal activity. His extremely interesting work, *The Formation of Vegetable Mold Through the Action of Worms, With Observations of Their Habits,* documents in great detail the intricate biological balances in the soil (Darwin, 1882).

In the early 1900s, several works focused on the broader, nonsimplistic aspects of agriculture and their complex interrelationships. Elliot (1907) wrote of the complexities of pasture mixes and their importance to soil fertility in rotations. The true classic, however, which stimulated later thinkers of the British and American schools, was King's *Farmers of Forty Centuries* (1911). King described in this book and in his following book, *Soil Management* (King, 1914), the complexity of integration in the then highly productive, traditional systems of Asia. The interrelationships between these systems were the key to the thinking of all agriculturists who followed.

## Biodynamic Agriculture

The first organized and well-defined movement of growers and philosophies was the biodynamic movement, which arose from a series of lectures given by Rudolf Steiner, the founder of anthroposophy, in 1924 (Steiner,

1958). The basic tenets of biodynamic farming include:

■ Sound farming and gardening techniques, no matter whether old or new.

■ Such principles as diversification, recycling, avoiding chemicals, decentralized production and distribution, etc.—ideas held in other biological movements. Since the 1920s, biodynamic farmers have developed the execution of such principles and also reintroduced useful traditional techniques.

■ The specific biodynamic measures and concepts as they evolve from Steiner's spiritual teaching, which mold the method into a consistent whole (Koepf et al., 1976).

The latter point is usually what separates biodynamic practices from the rest of biological agriculture. It includes "the stimulation and regulation of complex life processes by biodynamic preparations for soils, plants, manures" (Koepf et al., 1976). It also includes the consideration of cosmic and terrestrial forces on biological organisms. Biological rhythms are affected by a range of cosmic forces. Although growing evidence, mostly from biomedical research, suggests the occurrence of such effects, their importance in agriculture has not been evaluated.

Early writers on biodynamic agriculture include Pfeiffer (1934, 1943, 1956) and Baker (1940). These publications set forth the arguments for the disruptive effects of concentrated synthetic fertilizer and pesticides, which have been major aims of all biological or organic practitioners through the years. The connections between the biological "health" of the soil and the health of animals and humans associated with it, or using produce from it, were also articulated at this time.

All of these concepts did not originate with the biodynamic school, but they became an integral part of the thinking of Steiner and his followers. More recent summaries of the biodynamic concepts include Koepf (1981), Steiner (1958), Rateaver and Rateaver (1973), Pauli (1967), Koepf et al. (1976), Pank (1976), and Jeavons (1979). Although the biodynamic movement is concentrated in European and Scandinavian countries, a limited number of practitioners, both commercial and home-garden, are found in the United States and Canada.

## Development of "Humus Farming" Concepts

A school of thought evolved both as a part of and in addition to the biodynamic school that focused on the importance of humus in agriculture. This concept provided the foundation for several philosophies of biological agriculture that emerged from the 1930s through the 1960s.

Browne, in 1855, wrote *The Field Book of Manures or the American Muck*

*Book*. Roberts (1907), Fletcher (1907), and Waksman (1936) wrote basic works on humus-oriented soil fertility that, at the time, were considered states-of-the-art in scientific thought.

A major development that not only advanced techniques of compost-making but began to discuss the disruptive effects of concentrated synthetic fertilizers was that of Howard and Wad (1931). This work marked a major point of departure for the humus-farming school. Publication of definitive technical books and applied humus-farming books soon followed (Billington, 1942; Bruce, 1943, 1945a, 1945b; King, 1943; Waksman, 1936).

In 1943, Sir Albert Howard's book, *An Agricultural Testament*, became a new landmark. Not only did it add significantly to the emerging thought on humus farming through its exposition of the Indore method of composting, it restated in positive, modern terms the concept of integrated farming. *An Agricultural Testament* influenced the Soil Society work in England as well as the writing of J. I. Rodale in the United States. Subsequent works by Howard elucidated further the connections between soil and health and clarified the methods to be used in an agriculture based on biological structure rather than on the use of synthetic chemical inputs (Howard, 1945, 1946, 1947).

The humus-farming philosophy reached its peak in the early 1950s with publications by Sykes (1949, 1952, 1959) and Seifert (1952). These works proved to be the mainstay of the organic farming movement that followed. The principles of composting and compost use were well articulated by this time, and considerable research has since been done on the handling of municipal waste, with emphasis on methodologies. Many feel the culmination of agricultural composting studies is the *Rodale Guide to Composting* (Minnich and Hunt, 1979).

## Emergence of the Organic Philosophy

The basic tenets that led to organic, biological, and ecological agriculture and eventually to the regenerative farming movement can be traced to Sir Albert Howard's *An Agricultural Testament* (1943). The ideas of an integrated, decentralized, chemical-free agriculture were advocated by Northburn (1940) in a largely overlooked work. As far as we can tell, he was the first to use the word organic to refer to the entire philosophy and practice.

Graham (1941) and Barlow (1942) exemplified the rethinking of agricultural practices that occurred in the 1940s. Barlow was especially critical of the impacts of agriculture in the early 1940s on soil degradation and reduction in diversity through specialization. The momentum increased significantly with the publication of Lady Eve Balfour's *The Living Soil* (1943). Faulkner's *Plowman's Folly* (1943) was another classic, spurred by the Dust Bowl of the 1930s in the American Great Plains. Faulkner described in forceful terms the biological and human tragedy resulting from misdirected technology. In 1945, J. I. Rodale's *Pay Dirt* became a rallying point that carried the organic movement in America through the difficult 1960s. A lengthy series of books by J. I. Rodale was to follow (1948, 1953, 1954, 1977).

The late 1940s and early 1950s were prolific periods for organic literature. Faulkner (1946, 1947, 1952) was not only a critic of contemporary agriculture but an experienced extension agent and farmer as well. He detailed his own experiences in the regeneration of worn-out soil with organic farming practices.

Louis Bromfield also contributed significantly with his accounts of organic farms on which people, crops, and livestock were intermeshed in a living system (Bromfield, 1946, 1947, 1950, 1955). Bromfield felt strongly that the sensitivity, skill, and dedication required of a good farmer meant that "not everybody can farm" (1950). Several other authors, including Pfeiffer (1947), Cocannouer (1950, 1954, 1958), Hainsworth (1954), Howard (1947), and Widkenden (1949), continued through 1956 to articulate the increasing environmental harm and resource degradation brought about by "modern" farming methods. They repeatedly advocated the holistic approach to agriculture.

As with the earlier notions of Newton, Locke (1980), and Emerson (1904), many of the issues debated and the relationships suggested during the first half of the 20th century have become focal points for discussion in today's debate on sustainability. The concepts of wholeness, of an ecological model, of a fragile relationship with the environment, and a host of farming practices are being reconsidered.

## The 1960s: A Transition Period of Narrow Focus

By the late 1950s, the evolution and spread of industrial technologies had increased exponentially. In the developed nations, the industrial model was widespread. Moves toward crop specialization on the farm, permitted by the availability and low price of fertilizers and pesticides, had accelerated. The increased need for power as farmers grew only one or two crops was met by larger horsepower tractors.

Capital for investment was readily available, perhaps generated in large part by undervalued energy costs. The major problems were agricultural surpluses. For those of us who went through our graduate training in agriculture during the early 1960s, it was a time of scientific euphoria. We were in the post-Sputnik era and very literally on our way to the moon. As scientists, we considered that we were masters of our fate and the fate of humankind. Our technologies and our opinions, spoken from the dias of science and academia, dominated the formulation of the development paradigm of the day. Gone were the traditions of humus farming, of mechanical weed control, and of the need for large portions of our population to be involved in agriculture. Farming was now a business, to be run as efficiently as any other industrial enterprise. Soil conservation seemed to be the only major theme from past decades that remained in the model. We gave it major attention and resources, but our focus was on correction of the problems caused by crop technologies, not on prevention of them. We focused on terraces, levees, and farm ponds to slow and stop the runoff from the bare-soil corn and soybean fields. We were structuring our farms and our technologies according to valid Newtonian principles, applied with full intentions to dominate the earth.

There was little or no debate during those years, in the biological sciences at least, on development direction. The success of current technologies was so overwhelming that it stifled serious debate of alternatives. The alternative farming "schools" were practically nonexistent and certainly in disrepute. I remember clearly graduate school discussions in 1964 about the "crackpot Rachel Carson and her whistling in the wind" against the great benefits of DDT. In looking back on those heady years, I wonder about our arrogance and narrowness of vision. I also wonder, parenthetically, if many of us still remain intellectually in the comfortable era of the early 1960s when we trained. But the results of the narrow focus were far from being entirely negative. The concentration of scientific and development resources during the late 1960s and early 1970s achieved dramatic results.

## The Green Revolution

Agricultural development trends and breakthroughs up to and including the Green Revolution are interestingly summarized by Dahlberg (1979). He gives heavy emphasis to the emerging influence of the foundations and to international development assistance during the 1960s and 1970s as determining the development paradigm of the era. Those working in international development at the time followed the "commandments" according to Moser (1969):

- Research to find and develop new and improved farm (and related) technologies.
- Arranging for the importation and/or domestic production of farm supplies and equipment needed to put the new technology into use.

■ Creating a progressive rural structure, or "organization of the countryside," that provides channels through which goods and information can move easily back and forth between each farm and the total society in which it is located.

■ Creating and maintaining adequate incentives for farmers to increase production.

■ Improving agricultural land.

■ Educating and training technicians to accomplish all of these tasks competently.

In the process, the extension agent was seen as the "advisor, teacher, analyzer, and organizer" (Moser, 1969).

That the approach had significant impact is without question. In spite of massive and unprecedented increases in population since the 1950s and in the face of predictions of (and actual instances of) starvation in Asia, country after country, including India, Bangladesh, China, the Phillippines, Indonesia, and many others, have achieved food self-sufficiency and even food surpluses. The approach has worked best, however, in areas with good soil and water resources where returns to infrastructure development, to technology application, and to inputs have been high. Farm size, interestingly, has not been a factor in responsiveness where population density is high and where agriculture remains the predominant employer. In Asia, at least, mechanization has played only a modest role, limited to a few key technological areas. This latter distinction is significant because it plays a major role in the definition of sustainable agriculture for many, if not most, Third World countries. Those national definitions are now focused on many of the shortcomings of the Green Revolution model: the problems of equity, of rural income, of product diversity, of environmental impact, and of huge neglected areas of poor soil and water resources that must support increasing numbers of people.

## Broadening the Profile for Sustainability

In the late 1960s and early 1970s, several trends or events occurred to spur agricultural development and thinking beyond the new boundaries of the early 1960s' model. The increasing awareness of the impact of modern (industrial) technologies on the environment became clear as we traced pesticides in our food chains. Crop nutrients began to accumulate in streams and in underground aquifiers. Water resources became oversubscribed, and the "spaceship earth" concept was born. An event that shook our consciousness, however, was the energy "shortage" of the early 1970s. For the first time, we became painfully aware that earth's resources were limited.

The analysis by Hill and Cleveland (1981) of the energy cost of exploration and of recovery of oil and gas was reported by Gever et al. (1986):

"In the 1950s, we discovered about 50 barrels of oil for every barrel invested in drilling and pumping. Today, the figure is only five for one. Sometime between 1994 and 2005 that figure will become one for one. In other words, perhaps as early as 1994, it will generally become uneconomical to search for any oil for energy in the United States."

While certainly not everyone shares this view of the short time frame involved, there is little question that business as usual should be questioned. For these and for a broad range of other reasons, agricultural development directions have come under serious debate and analysis. The university-based scientific communities have been joined by a plethora of private "think-tank" and industrial groups all making contributions.

From the alternative agriculture point of view, several scenarios for sustainable agriculture have been articulated. Most include principles evolved during the early 1900s and stress the following (Harwood, 1983):

- The interrelatedness of all parts of a farming system, including the farmer and his family.
- The importance of the many biological balances in the system.
- The need to maximize desired biological relationships in the system and to minimize use of material and practices that disrupt those relationships.

The several modern articulations of world views of "alternative agriculture" include but are not limited to Berry's *The Unsettling of America* (1988) (stressing the importance of human participation from a morality standpoint reminiscent of Jefferson), Walters' *The Case for Eco-Agriculture* (1975), Rodale's *Breaking New Ground* (1983), and Jackson's *New Roots for Agriculture* (1980). These authors all derived their thinking from the alternative agriculture tradition, but they differ markedly in their approaches.

Closely associated with these works are those of the agroecology movement, best known by scientists through the work of Altierri (1987). These authors combine the scientific method of modern ecology with the older concept of the scientific agriculturists of learning from nature. Although the idea seems romantic, it has evolved in an age of realism. Quinney of New Alchemy wrote: "Today, although we have a better understanding of the limits of this concept, nature as inspiration is still powerful and increasingly useful" (Quinney, 1987). Perhaps the most eloquent of these works is Dover and Talbot's *To Feed the Earth: Agro-Ecology for Sustainable Development* (1987).

For those who would attempt to articulate any national sustainable agriculture paradigm, there are several other key readings: *Farmland or Wasteland: A Time to Choose* (Sampson, 1981); *Paying the Price: Pesticide Sub-*

*sidies in Developing Countries* (Repetto, 1985); *Defusing the Toxic Threat* (Postel, 1987); *Ecological Aspects of Development in the Humid Tropics* (National Academy of Science, 1982); *State of the World: A Worldwatch Institute Report on Progress Toward a Sustainable Society* (Brown et al., 1986); *Crop Productivity: Research Imperatives Revisited* (Gibbs and Carlson, 1985); and *Agriculture 2000: A Look at the Future* (Battelle Memorial Institute, Columbia Division, 1985).

## National Agendas for Agricultural Development

A public agenda is an accumulation of issues that attract debate and concern. The contributors include individuals, social groups, institutions, government agencies, and power brokers. Issues achieve agenda status when they receive widespread and continuing public recognition. Public agenda items then receive policy status when they receive sanction in the form of law, funding, or other official pronouncement or action. Present U.S. agricultural development agenda items can be grouped into the following five categories (with examples of frequently heard, specific concerns):

■ *Increase the utility of agriculture.* Maintain adequate production. Provide adequate livelihood (considering equity, stability, safety, lifestyle) for a desired number of participants. Provide food of acceptable quality and diversity (no pesticides, low heavy metals, little fat, good flavor, little processing, few preservatives, no antibiotics, regulated levels of synthetic hormones).

■ *Increase productivity.* Develop more productive biotypes (with pest resistance, tolerance to adverse conditions). Maintain soil organic matter, tilth. Maintain crop diversity. Practice rotations. Use integrated animal/fish/crop/tree systems. Practice nutrient cycling.

■ *Maintain an environment favorable to humans and most of other species.* Protect groundwater from contamination. Reduce or eliminate use of pesticides. Reduce use of synthetic fertilizers. Encourage wildlife maintenance. Recognize animal rights (reduce stress in confinement, provide for a degree of natural activity).

■ *Assure the ability to evolve indefinitely.* Minimize soil loss (from erosion, conversion to nonagricultural use). Stop overdraft of fossil groundwater. Reduce energy use (especially of fossil fuels). Develop better technologies for biological nitrogen fixation. Develop perennial cereals. Maintain existing genetic diversity.

■ *Develop patterns of geographical distribution and scale (macro structure) consistent with national agendas.* Create adequate physical and institutional infrastructure. Develop market channels that respond to market and social needs. Manage corporate activities that may control portions of the agricultural sector. Monitor (or manage) land ownership (land is usually considered to be a quasi-public resource).

Recognition of these points is given or implied in the above definition of sustainable agriculture. Most of the five categories are recognized in a current definition (TAC, 1988): Sustainable agriculture should involve the successful management of resources for agriculture to satisfy changing human needs while maintaining or enhancing the natural resource base and avoiding environmental degradation.

The U.S. Agency for International Development (1987) avoids a specific definition but identifies a long list of parameters that fall into the five suggested categories.

These five categories are purposefully broad to include most possible items. The breadth is the result of historical process, as we shall see below. In most countries where debate is prevalent, the concerns are remarkably similar to those in the United States. The priorities change with resource base, stage of agricultural development, and national politics. The consistency and speed with which particular items reach policy status depends upon the size and influence of the proponent group, the perceived seriousness of the problem, and government responsiveness. Those relationships are little understood, even here in the United States. They are influenced to some extent by prominent events, such as pesticide spills, farm bankruptcies, or major disasters.

The public agenda must be both nation- and time-specific. Its establishment is a people-driven process that differs from country to country. A process of goal-setting, of identifying gaps between existing and desired future states, and, finally, of priority setting and resource allocation completes the process. In most countries, neither farmers nor agricultural scientists are the sole or even the major determinants of what is sustainable. Their roles in technology development are probably their most significant contribution.

## From Concept to Action

Our concept of the multiple dimensions of a sustainable agriculture is more broad today than at any time in history. We are more aware of the potency of technologies, of the fragility of the earth's environment, and of humankind's ability to disrupt it. We have a notion of earth's limited resources. This is appropriate at the threshold of our transition from the age of alchemy to the age of biotechnology.

As we survey our past, it seems that consensus is possible on three major points: (1) Agriculture must be increasingly productive and efficient in resource use, (2) biological processes within agricultural systems must be much more controlled from within (rather than by external inputs of pesticides), and (3) nutrient cycles within the farm must be much more closed.

A less well-recognized point is that crop nutrients must come from management of nutrient flow into and out of the soil organic matter fraction, a "farming of the organic matter" rather than a "farming of the soil nutrient solution." There is ample circumstantial evidence from alternative agriculture on this point but, as yet, little scientific evidence. If we are to learn one central lesson from all of alternative agriculture, I think it would be this one.

We may develop the sustainable model for the United States, but how do we approach sustainable Third World development? No Third World country has so broad a public agenda for development as do the western developed nations, nor do they have such a plethora of well-funded public and private agencies that are contributing to that agenda. Do we take our own agenda with us when we go to the tropics? Certainly, we must not ignore the differences in priorities, which is a primary emphasis in Asia, on rural income and employment as opposed to our own priorities. We must see the drain of wealth from rural areas as a result of inappropriate structure of the agricultural system. But should we impose our own priorities of food safety, of environmental impact, or of human safety in agriculture on developing countries? We have no ready answers, but we must be sensitive and responsive to national agendas in each country in which we work. We must choose our attack points carefully, remembering that progress is most rapid when effort and resources are focused best.

But how about our own agenda for sustainability? As scientists, how do we react to the realization that we no longer dominate the agenda-setting process for agriculture. Do we understand the impact of all five areas of a sustainable profile? If you are a land-grant scientist, how do you relate? The words of Sandra Batie, in my opinion, are wise counsel: "The new agendas of a concerned public should be seen neither as a threat nor as irrelevant to the land-grant tradition but as challenges and opportunities to better serve the needs of society. Land-grant colleges of agriculture must embrace opportunities to assist in identifying and designing solutions which are in our finest tradition of being the 'people's' university" (Batie, 1988).

As we survey the past and then move ahead to determine our future, we should cherish the diversity of thought and experience that provides the "raw material" for evolution of a new paradigm. The implementation of that new model requires new attitudes, new policies, and new technologies.

## REFERENCES


Altierri, M. A. 1987. Agroecology: The scientific basis of alternative agriculture. Westview Press, Boulder, Colorado.

Baker, A. 1940. The laboring earth. Heath Cranton Ltd., London, England. 216 pp.

Balfour, E. B. 1943. The living soil. Faber and Faber, London, England. 246 pp.

Barlow, K. E. 1942. The discipline of peace. Faber and Faber, London, England. 214 pp.

Batie, S. 1988. Agriculture as the problem: New agendas and new opportunities. Virginia Polytechnic Institute Staff Paper 88-6. Blacksburg. 22 pp.

Batie, S. S., L. A. Shabman, and R. A. Kramer. 1985. U.S. agriculture and natural resource policy. *In* The dilemmas of choice. Resources for the Future, Washington, D.C.

Battelle Memorial Institute, Columbia Division. 1985. Agriculture 2000: A look at the future. Battelle Press, Columbus, Ohio.

Berry, W. 1978. The unsettling of America. Sierra Club Books, San Francisco, California.

Billington, F. 1942. Compost for the garden plot or 1,000-acre farm. Faber and Faber, London, England. 88 pp.

Bromfield, L. 1946. Pleasant valley. Harper and Bros., New York, New York.

Bromfield, L. 1947. Malabar farm. Ballantine Books, New York, New York. 470 pp.

Bromfield, L. 1950. Out of the earth. Harper and Bros., New York, New York. 305 pp.

Bromfield, L. 1955. From my experience. Harper and Bros., New York, New York. 355 pp.

Brown, L. R., W. U. Chandler, C. Flavin, C. Pollock, S. Postel, L. Starke, and E. C. Wolf. 1986. State of the world. A Worldwatch Institute Report on progress toward a sustainable society. Worldwatch Institute, Washington, D.C. 263 pp.

Browne, D. 1855. The field book of manures or the American mulch book. C. M. Saxton and Co., New York, New York. 422 pp.

Bruce, M. 1943. From vegetable waste to fertile soil. Faber and Faber, London, England.

Bruce, M. 1945a. Common sense compost making. Faber and Faber, London, England. 93 pp.

Bruce, M. 1945b. Quick return method of compost making. Rodale Press, Inc., Emmaus, Pennsylvania. 92 pp.

Cocannouer, J. 1950. Weeds, guardians of the soil. Devin-Adair, New York, New York. 179 pp.

Cocannouer, J. 1954. Farming with nature. University of Oklahoma Press, Norman. 147 pp.

Cocannouer, J. 1958. Water and the cycle of life. Devin-Adair, New York, New York.

Dahlberg, K. A. 1979. Beyond the green revolution: The ecology and politics of global agricultural development. Plenum, New York, New York.

Danborn, D. B. 1979. The resisted revolution: Urban America and the industrialization of agriculture, 1900-1930. University of Iowa Press, Iowa City. 195 pp.

Darwin, C. 1882. The formation of vegetable mold through the action of worms, with observations of their habits. D. Appleton and Co., New York, New York. 326 pp.

Donaldson, F. 1961. Approach to farming. Faber and Faber, London, England. 248 pp.

Dover, M., and L. M. Talbot. 1987. To feed the earth: Agro-ecology for sustainable development. World Resources Institute, Washington, D.C.

Elliot, R. H. 1907. The Clifton Park System of Farming. Faber and Faber, London, England. 261 pp.

Emerson, R. W. 1904. Farming: The complete works of Ralph Waldo Emerson. Concord Edition, Volume 7. Houghton-Mifflin, Boston, Massachusetts.

Faulkner, E. 1943. Plowman's folly. Grosset and Dunlap, New York, New York. 155 pp.

Faulkner, E. 1946. Uneasy money. University of Oklahoma Press, Norman.

Faulkner, E. 1947. A second look. University of Oklahoma Press, Norman. 193 pp.
Faulkner, E. 1952. Soil development. University of Oklahoma Press, Norman. 232 pp.
Fletcher, S. 1907. Soils: How to handle and improve them. Doubleday, Page and Co., New York, New York. 438 pp.
Gever, J., R. Kaufman, D. Shole, and C. Parosmarty. 1986. Beyond oil: The threat to food and fuel in the coming decades. Complex Systems Research Center, Durham, New Hampshire. 300 pp.
Gibbs, M., and C. Carlson, editors. 1985. Crop productivity: Research imperatives revisited. Proceedings of an international conference held at Boyne Highlands Inn, October 13-15, 1985, and Airlie House, December 11-13, 1985. 304 pp.
Graham, M. 1941. Soil and sense. Faber and Faber, London, England. 274 pp.
Hainsworth, P. 1954. Agriculture, a new approach. Faber and Faber, London, England. 248 pp.
Hardin, L. S. 1988. Thirty years of agriculture: A review of North America. Span. 30(3): 98-101.
Harwood, R. R. 1983. International overview of regenerative agriculture, in resource-efficient farming methods for Tanzania. Rodale Press, Emmaus, Pennsylvania. 24-25 pp.
Howard, A. 1943. An agricultural testament. Oxford University Press, London, England. 253 pp.
Howard, A. 1945. Farming and gardening for health or disease. Faber and Faber, London, England.
Howard, A. 1946. The war in the soil. Rodale Press Inc., Emmaus, Pennsylvania. 96 pp.
Howard, A. 1947. The soil and health: A study of organic agriculture. Devin-Adair, New York, New York. 307 pp.
Howard, A., and Y. D. Wad. 1931. The waste products of agriculture. Oxford University Press, London, England. 167 pp.
Jackson, W. 1980. New roots for agriculture. Friends of the Earth, San Francisco, California.
Jeavons, J. 1979. How to grow more vegetables than you ever thought possible on less land than you can imagine. Ten Speed Press, Palo Alto, California. 159 pp.
Jefferson, T. 1984. Literary classics of the United States. Writings, New York.
Jeske, W. E., editor. 1981. Economics, ethics, ecology: Roots of productive conservation. Soil Conservation Society of America, Ankeny, Iowa.
King, F. C. 1943. The compost gardener. Titus Wilson and Son Ltd., Kendal, England. 90 pp.
King, F. H. 1911. Farmers of forty centuries. Rodale Press Inc., Emmaus, Pennsylvania. 441 pp.
King, F. H. 1914. Soil management. Orange Judd Co., New York, New York. 311 pp.
Koepf, H. H. 1981. The principles and practice of biodynamic agriculture. *In* B. Stonehouse [editor] Biological husbandry. Butterworths, London, England. pp. 237-250.
Koepf, H. H., B. Patterson, and W. Schaumann. 1976. Biodynamic agriculture: An introduction. Anthroposophic Press, Spring Valley, New York. 429 pp.
Locke, J. 1980. Second treatise of government. C. B. MacPherson (editor). Hackett Publishing Co., Indianapolis, Indiana.
Marcus, Alan J. 1985. Agricultural science and the quest for legitimacy. Iowa State University Press, Ames.
Minnich, J., and M. Hunt, editors. 1979. The Rodale guide to composting. Rodale Press Inc., Emmaus, Pennsylvania. 405 pp.
Moser, A. T. 1969. Creating a progressive rural structure: To serve a modern agriculture. Agricultural Development Council, Inc., New York, New York. 172 pp.
National Academy of Science. 1982. Ecological aspects of development in the humid tropics. National Academy Press, Washington, D.C. 297 pp.
Northburn, Lord. 1940. Look to the land. Dent, London, England.
Pank, C. J. 1976. Dirt farmer's dialogue, twelve discussions about biodynamic farming. Biodynamic Press, Sprakers, New York. 133 pp.

Biodynamic Press, Sprakers, New York. 133 pp.
Pauli, F. W. 1967. Soil fertility: A biodynamic approach. Adam Hilger Ltd., London, England. 204 pp.
Pfeiffer, E. 1934. New methods in agriculture and their effects on foodstuffs: The biodynamic method of Rudolf Steiner. Rudolf Steiner Publishing Co., London, England.
Pfeiffer, E. 1943. Biodynamic farming and gardening. Soil fertility renewal and preservation. Anthroposophic Press, New York, New York. 240 pp.
Pfeiffer, E. 1947. The earth's face and human destiny. Rodale Press Inc., Emmaus, Pennsylvania. 184 pp.
Pfeiffer, E. 1956. Biodynamics: Three introductory articles. Biodynamic Farming and Gardening Association, Wyoming, Rhode Island. 40 pp.
Phipps, T. T., and P. R. Crosson. 1986. Agriculture and the environment: An overview in agriculture and the environment. *In* T. T. Phipps, P. R. Crosson, and K. A. Price [editors] Resources for the future. Washington, D.C. pp. 3-29.
Postel, S. 1987. Defusing the toxic threat: Controlling pesticides and industrial waste. Worldwatch Institute, Washington, D.C.
Quinney, J. 1987. Still crazy after 18 years. New Alchemy Quarterly (summer).
Rateaver, B., and G. Rateaver, editors. 1973. A condensation of biodynamic farming and gardening. Rateaver, Pauma Valley, California. 115 pp.
Repetto, R. 1985. Paying the price: Pesticide subsidies in developing countries. World Resources Institute, Washington, D.C. 33 pp.
Rifkin, J. 1980. Entrophy: A new world view. Bantam, New York, New York.
Rifkin, J. 1983. Algeny. Viking Press, New York, New York. 298 pp.
Roberts, I. 1907. The fertility of the land. Macmillan Publishing Co., New York, New York. 415 pp.
Rodale, J. I. 1945. Pay dirt. Rodale Press, Inc., Emmaus, Pennsylvania. 242 pp.
Rodale, J. I. 1948. The organic front. Rodale Press, Inc., Emmaus, Pennsylvania. 199 pp.
Rodale, J. I. 1953. The organic method on the farm. Organic Farmer No. 41. Rodale Press, Inc., Emmaus, Pennsylvania. 128 pp.
Rodale, J. I. 1954. Organic merry-go-round and an organic trip to England. Rodale Press, Inc., Emmaus, Pennsylvania. 95 pp.
Rodale, J. I. 1960. The complete book of composting. Rodale Press, Inc., Emmaus, Pennsylvania. 1,007 pp.
Rodale, J. I. 1977. Organic gardening: How to grow healthy vegetables, fruits, and flowers using nature's own methods. Rodale Press, Inc., Emmaus, Pennsylvania. 224 pp.
Rodale, R. 1983. Breaking new ground: The search for a sustainable agriculture. The Futurist 1(1): 15-20.
Rossiter, M. 1975. The emergence of agricultural science: Justus Liebig and the Americans, 1840-1880. Yale University Press, New Haven, Connecticut.
Sampson, R. N. 1981. Farmland or wasteland: A time to choose. Rodale Press Inc., Emmaus, Pennsylvania. 422 pp.
Seifert, A. 1952. Compost. Faber and Faber, London, England.
Steiner, R. 1958. Agriculture: A course of eight lectures. Biodynamic Agriculture Association, London, England. 175 pp.
Sykes, F. 1949. Humus and the farmer. Rodale Press Inc., Emmaus, Pennsylvania. 393 pp.
Sykes, F. 1951. Food, farming, and the future. Faber and Faber, London, England. 294 pp.
Sykes, F. 1959. Modern humus farming. Rodale Press Inc., Emmaus, Pennsylvania. 270 pp.
Technical Advisory Committee (TAC). 1988. Sustainable agricultural production: Implications for international agricultural research. (AGR/TAC:IAR 87/22). Consultative Group on International Agricultural Research, Washington, D.C. 108 pp.
U.S. Agency for International Development, Office of Agriculture, Bureau for Science and Technology. 1987. Sustainable agriculture. Washington, D.C.
Waksman, S. 1936. Humus: Origin, chemical composition, and importance in nature.

Williams and Wilkins Co., Baltimore, Maryland. 526 pp.
Walters, C., Jr. 1975. The case for eco-agriculture. Acres U.S.A., Raytown, Missouri.
Widkenden, L. 1949. Make friends with your land: A chemist looks at organiculture. Devin-Adair, New York, New York. 132 pp.

# MAKING AGRICULTURE A SUSTAINABLE INDUSTRY

N. C. Brady

There is growing awareness that agricultural systems must provide not only what humanity needs today but what the human family will require a decade or even a century from now. Sustainable agriculture is a topic whose time has come.

## Sustainability: A Question of Increasing Concern

Agricultural sustainability was not a major issue in the 1960s and 1970s because food production resources did not appear threatened by overuse. In the 1960s, prevention of the mass starvation predicted by the doomsayers of that era was the primary concern. The focus of the Green Revolution was to produce large quantities of food, particularly wheat and rice, close-growing crops in which soil erosion was not as serious as with row crops.

Neither was natural resource conservation high on the development agenda in the 1960s and 1970s. Conservation was a concern for the future, whereas the burning issue of that era was how to grow enough food for the current year.

Today, sustainability has become a significant issue in the United States and internationally. A number of scientists and laymen have persistently asserted their concern about conservation and the environment. Some are concerned with the dangers of excessive chemical fertilizer and pesticide use; others focus on the problems of soil and water conservation. The Rodale Institute and the Soil and Water Conservation Society have voiced these various concerns for many years. We can't claim that we did not know about these problems.

Eventually, we began to do something about them. Today, despite our romance with the plow, some form of reduced tillage is used on nearly 40 percent of the cropland in the United States. I would like to think that we were motivated to change because we really became believers. But while conservation elements soon surfaced, this change initially was stimulated by rising fuel costs and the availability of effective herbicides. When the price of gasoline went up, farmers suddenly realized that traveling over those fields with a tractor three, four, and five times was expensive. They began to look at alternatives because it was to their economic advantage to do so. They could control weeds by using herbicides, rather than through numerous tillage operations. From a conservation point of view, we are lucky that these factors stimulated the use of minimum tillage, which not only saves fuel and time, but dramatically reduces soil erosion and runoff from American fields. Now, however, we have begun to be concerned about the long-term effects of herbicide use and runoff.

The emergence of knowledge about the so-called greenhouse effect brought with it the realization that the destruction of tropical forests through slash-and-burn agriculture was a major source of carbon dioxide. I am not talking about American farmers. I am talking about the 300 million people who survive each year by slashing and burning tropical forests and other areas in order to have enough ashes in the soil to raise a few crops. Slash-and-burn agriculture emerged as a global problem, and people began to realize that increased food production must come from existing cropland, not from land that is yet to be brought into cultivation.

At one time we thought that forested land was available and ready to be taken over by agriculture, as if it had no other important function. That mode of thinking has begun to change. Suddenly, we realize that some serious environmental concerns arise when natural vegetation is removed and the land is used for agriculture. We realize that we must seek means of increasing food production other than by expanding the land under cultivation.

And we do need to produce more food. The temporary food surpluses of 1985-1987, especially in the United States, brought into question the wisdom of increased food production and U.S. help to support such production in the developing world. How to increase food production no longer seemed to be an appropriate, primary concern in the United States or overseas.

In fact, the U.S. Agency for International Development (AID) was blamed for the problems of the U.S. agricultural sector because the agency's programs were helping Indians, Indonesians, Filipinos, and others to improve their food production. Africa was not mentioned because it was having a drought. But the United States garnered blame for every other region because it was supplying the technology that made it possible for people to feed themselves and, therefore, to avoid buying agricultural products from the United States.

The drought in the United States during 1988, droughts the previous year in India and Indonesia, a slowdown in Chinese agricultural production, and floods in Bangladesh have revived concern about food production.

## Two Decisive Factors

There are a couple of decisive factors that brought to a head the importance of agricultural sustainability. One of them is the population situation. The other is a realization that most unused land should not be cultivated, that people should not automatically move in and start cutting and burning forests as they have done before, even in the United States when it was first settled.

The population increase we experienced in this country during the "baby boom" from the late 1940s through the early 1960s was nothing compared to what is now happening in the developing world. In 1987, 102 million babies were born on this planet, 95 million of them—better than nine out of ten—in developing countries (Figure 1) (ODC, 1982; Population Reference Bureau, 1986). All of them must be fed, of course.

In the 1960s, even as recently as the 1970s, the assumption was that there was plenty of potentially arable land from which to feed additional human beings. But where is it?

Comparatively little exists in densely populated Asia, already home to almost three-fifths of the human family. There is considerable land not now cultivated in Latin America and Africa, but much of it is infertile or fragile. Neither is there much in Europe, nor in North America where we thought there was plenty, particularly if we take environmental considerations into account (Figure 2) (Revelle, 1976).

Opening forests to cultivation may greatly increase the destruction of these already endangered areas and with them a wealth of unique, irreplaceable biological resources. Assessments by the United Nations Food and Agriculture Organization (FAO) suggest that for most African countries the area of land now cultivated exceeds what is compatible with sustainable agriculture. Soil erosion losses in the next few years are going to be considerable.

Central America also has experienced extreme deforestation. In the last three and one-half decades the dense forest cover of that region has all but disappeared (Figure 3) (USAID Country Environmental Profiles; Nations and Komer, 1983).

## The AID Approach to Agricultural Sustainability

AID officials took a hard look at this world situation and came up with three elements that they think are terribly important to agricultural sus-

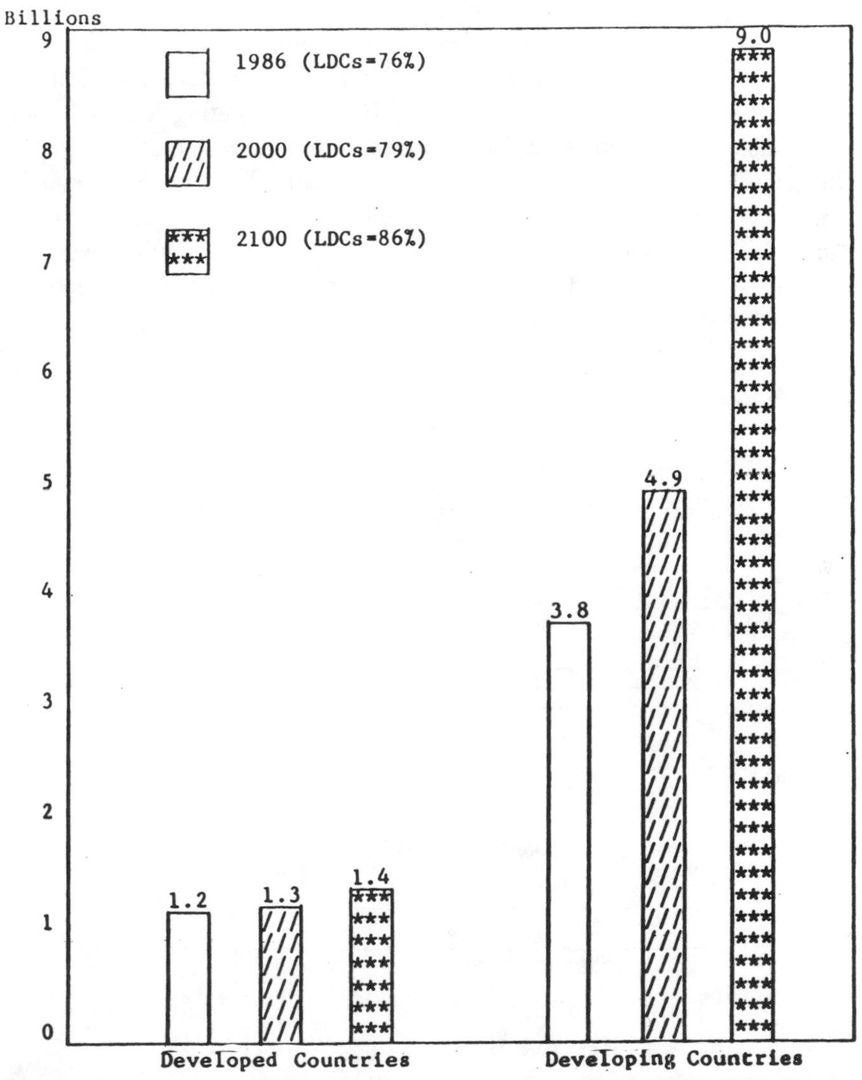


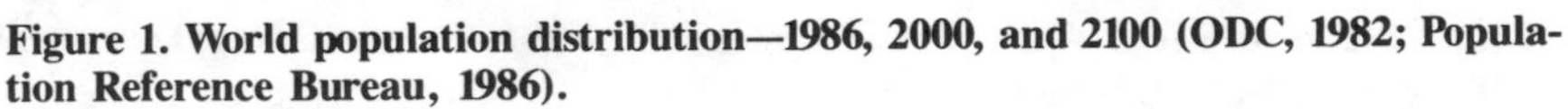


**Figure 1. World population distribution—1986, 2000, and 2100 (ODC, 1982; Population Reference Bureau, 1986).**

tainability. The first element is income generation, particularly among the poor. The second element is expanded food availability and consumption; this means having more food available through increased production and improved marketing. The third element is the conservation and enhancement of natural resources.

***Income Generation.*** Income generation is directly affected by government policies on land tenure, pricing, availability of inputs, access to efficient markets, and other agricultural production incentives. Changing established but inadequate policies can be politically risky for developing country governments, which often makes the policy arena a difficult one in which to offer assistance. However, AID is encouraging African governments to commit themselves to getting economic growth restarted by making their economic systems more open and market-oriented. Farmers are receiving higher prices, access to efficient markets and agricultural inputs, and other incentives to plant and produce more crops while government regulations are reduced. AID particularly encourages country leaders to consider policy alternatives that can increase the productivity and incomes

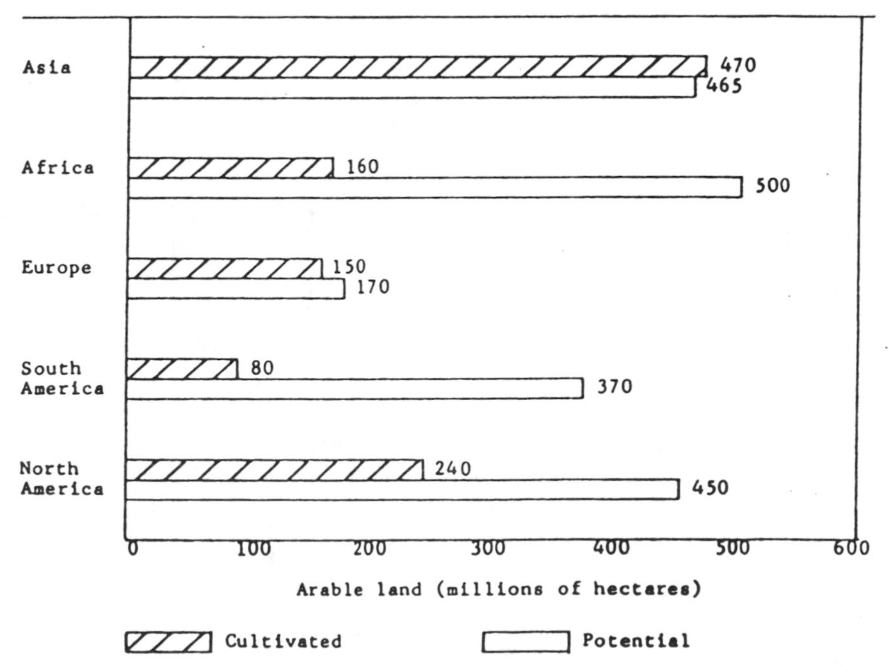


**Figure 2. Estimated cultivated land and potentially arable land on different continents (Revelle, 1976).**

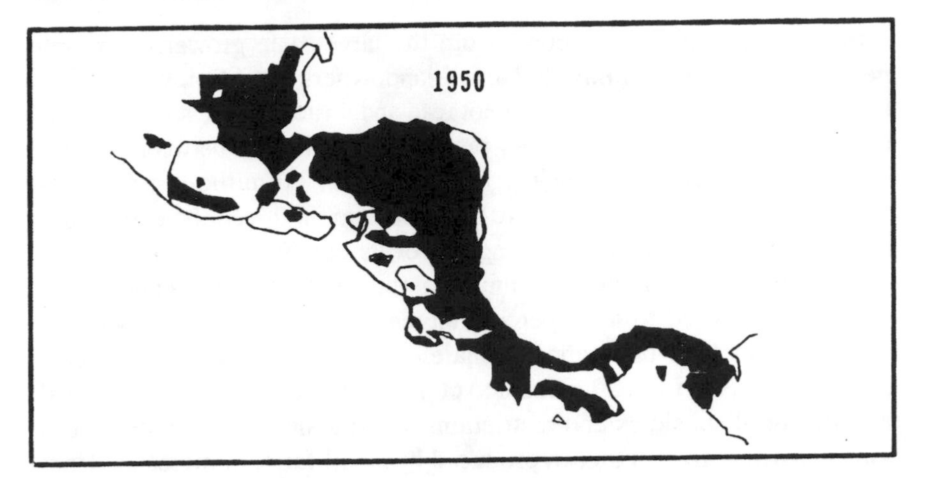


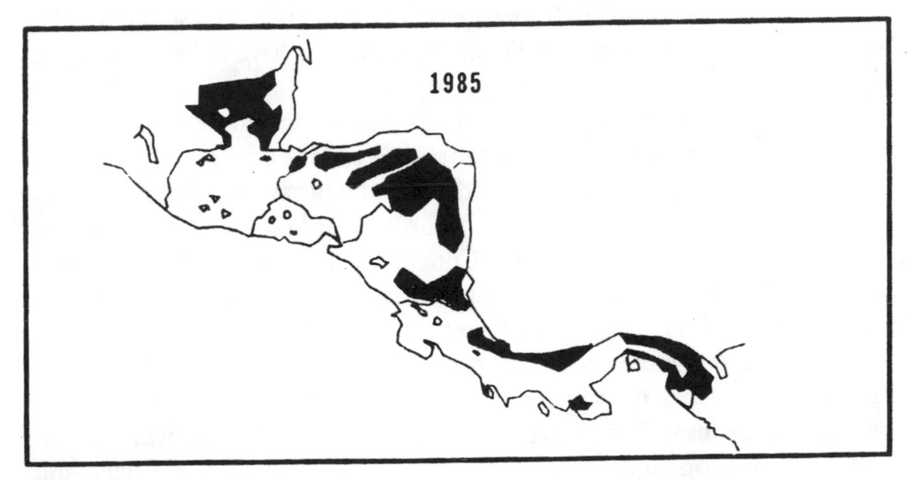


Dense forest cover

**Figure 3. Deforestation in Central America, 1950-1985, excluding coastal mangrove forests and open pine savanna. (USAID country environmental profiles; Nations and Komer, 1983).**

of small-scale, rural farmers.

Between 1980 and 1986, for example, AID officials worked with Zimbabwe through the agency's Agricultural Sector Assistance Program. The agency supported the Government of Zimbabwe's efforts to help small farmers benefit economically from improved technologies, such as new hybrid maize varieties, and associated technologies that they had been researching for some 15 years. The government set stable prices, and according to its game-plan, most of the extra maize from the hybrid varieties and fertilizers was supposed to come from the large-scale growers. Government planners did not think that small landowners would pick up on the new technology because they could not read and write. But the small farmers fooled them. The share of corn they produced tripled, from between 400,000 and 500,000 metric tons annually in the 1970s to 1.6 million metric tons in 1985 (Haykin, 1987). They were the ones who produced the extra food, and Zimbabwe had maize "coming out of its ears."

There are other examples. In Zambia, improved technologies and higher producer prices led to a 65 percent increase in marketed corn between 1983-1984 and 1985-1986, and the share of corn marketed by small-scale growers increased from 20 percent to 60 percent (Haykin, 1987). Removal of agricultural subsidies and restrictions led to a similar improvement in Somalia, where irrigated corn produced by small farmers increased from an annual average of 107,000 metric tons in the 1970s to 280,000 metric tons in 1985 (Haykin, 1987). AID's policy initiatives have helped to achieve similar improvements elsewhere.

This increased on-farm productivity has improved rural incomes, as have innovative, income-generating, nonfarm rural endeavors. To encourage and support entrepreneurial efforts of both rural and urban citizens, the agency and the U.S. Congress are giving priority attention to microenterprise development. In fiscal year 1988, Congress called upon AID to expend at least $50 million on credit, training, and technical assistance for development of very small-scale enterprises.

To provide guidelines for this program, the administrator of AID appointed an advisory committee of nongovernmental individuals with experience in business, banking, private and voluntary aid, management consulting, education, and other areas of expertise essential to successful development of microenterprises. Working with enterprise development experts within AID, the committee produced guidelines on such matters as business size, target entrepreneurs, loan size, number of women-owned firms, access to credit programs, strengthening of local business-promotion institutions, and need for training and technical assistance. The guidelines are now being sent to AID's field missions for review and comment.

Central support projects, like Assistance to Resource Institutions for Enterprise Support (ARIES), which is ongoing, and Growth and Equity Through Microenterprise Investments and Institutions (GEMINI), which is being put together this year, help missions to design, implement, and evaluate microenterprise endeavors and act as a central source of materials and information on research in this critical income-generating area.

Increased income generated by policy changes, improved agricultural technologies, and alternative economic endeavors enable people to improve

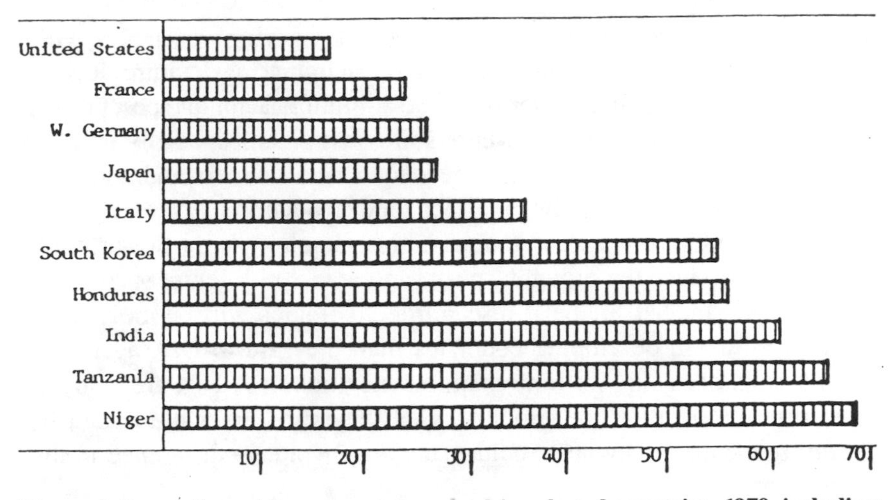


**Figure 4. Proportion of income spent on food in selected countries, 1979, including food, beverages, and tobacco (Mackie, 1983).**

their lives in several ways. Primary education and adequate health care become more readily accessible.

***Food Availability and Consumption.*** Poor people spend a larger portion of their income on food (Figure 4) (Mackie, 1983), and increased income helps them to acquire and consume more food. But even in developing countries where production and incomes have improved, major problems can reoccur.

During a recent conference sponsored by AID's Asia/Near East Bureau, economists voiced a note of caution: "You cannot say that food production is no longer a problem in Asia." There are still many more hungry people in Asia than in Africa, and the newly acquired self-sufficiency of a few countries in that region was undone by a season of drought. Indonesia is beginning to buy more rice to replenish reduced food stocks. India is eating into its surplus wheat and rice.

The battle for food availability in Asia and elsewhere is certainly not over, and there are several things we must do as we continue this fight. First, we must maintain the gains we have already achieved. We must not listen to those who think that research on wheat and rice is no longer essential. Maintenance research is needed to ensure that we do not slip backward, but rather continue the upward surge of the Green Revolution.

Second, we must give greater attention to crops for which production and marketing practices have not yet been researched. Research is needed to improve the component technologies for growing vegetables, edible legumes, and roots and tubers, particularly in upland agriculture. Research is needed for areas where water is the most limiting factor and pest pressures are high. Research is needed where soil constraints, especially toxicities, limit production. And research is needed to better understand how credit, marketing, and public policies affect food production.

Use of modern research techniques, biotechnology in particular, will be critical to produce the drought-, pest-, and acid-soil-tolerant varieties that are needed. Indeed, modern research technologies will be more critical to agriculture in developing countries than they are to U.S. agriculture. If our soils are acid, we lime them. If the soils in Africa are acid, they just plant their crops and produce what they can. An acid-tolerant lima bean, soybean, or cowpea would make a tremendous difference in those environments.

When I was at the International Rice Research Institute (IRRI), every once in a while my colleagues who worked for chemical pesticide companies would accuse me of trying to put them out of business by developing rices that were resistant to the green leafhopper, the brown planthopper, and plant diseases. I told them I was not there to keep them in business or put them out of business. My job was to try to give poor farmers, who had neither the money nor the expertise to use pesticides, access to crop varieties that would produce without excessive pesticide application.

Today, research networks are essential to get the job done, particularly where the sophisticated research techniques of biotechnology are concerned. Collaboration between researchers at U.S. universities and international agricultural research centers is essential. Cooperation between international centers and developing country scientists is essential. And networking between U.S. university and developing country researchers, as is now taking place in our collaborative research support programs (known as CRSPS), is essential.

Through these cooperative endeavors, we are using biotechnology to explore solutions to the most pressing agricultural and environmental problems in developing countries. Examples include:

■ Colorado State University's tissue culture work that produced rice cells tolerant to such constraints as highly acid or saline soil, heat, and drought. Seeds regrown from the tolerant cells are being field-tested in the Philippines at IRRI. Successful seeds will be distributed to national programs and farmers.

■ The University of Hawaii's coordination of work at 20 U.S. universities, in cooperation with several international agricultural research centers, to increase the efficiency of nitrogen-fixing organisms and their production, distribution, and use in almost 30 countries in developing regions.

■ Research by the University of California at Davis on a better vaccine for rinderpest, an acute, highly contagious viral disease of ruminants. The vaccine is being tested at the U.S. Department of Agriculture's facility at Plum Island, New York, in preparation for field trials in Africa.

■ Research at the International Livestock Research and Development Center (ILRAD) in Nairobi, Kenya, on trypanosomiasis, using the best biotechnology has to offer. A vaccine against this dreaded disease would open an area of Africa about a third the size of the United States to cattle production. Today, the farmers in that region, mostly women and children, do their hoeing and other agricultural chores by hand because the threat of trypanosomiasis prevents them from herding cattle.

We at AID are mindful of the many serious issues surrounding the use of biotechnology. This past April, the agency sponsored a five-day conference to strengthen understanding and collaboration between international researchers and U.S. private sector firms involved in agricultural applications on this new technology. Participants agreed that continuing attention must focus on such issues as laboratory safety, intellectual property rights, quality control, public education, and environmental protection.

***Natural Resource Conservation.*** The effect of increased food production on the natural resource base has become a central development concern in recent years. Working with the environmental community, AID has assumed a leadership role in these efforts. Even before I arrived in 1981, the agency was helping countries to identify environmental problems and their major causes. Using environmental profiles, we have already helped about 40 developing countries to determine the degree to which industry, agriculture, waste disposal, and other activities contribute to environmental degradation. Other countries will benefit from similar surveys in the future.

Natural resource conservation must get ongoing attention. The deterioration of tropical forests is reducing biological diversity, leaving land open to erosive forces, and contributing to increased carbon dioxide concentration in the earth's atmosphere.

During my years at AID, I have become aware of the lack of interaction between those concerned with woody species and those working with food crops. U.S. scientists who do research on woody species have had little if any experience in the tropics. Even within our foreign assistance agency, it took almost three years to get a Forestry Fuelwood Research and Development project (F/FRED) started in Asia. The opposition did not come from scientists in the developing countries; it came from those within our own system who did not understand what we were trying to do.

F/FRED research networks are now underway in countries of southern and southeastern Asia and in Kenya, Uganda, Rwanda, and Burundi in East Africa. A third phase is planned for Latin America. In each region, research on multipurpose tree species and on appropriate land and forest resource management for different tropical environments is developed and strengthened. There are trials on woody species just like those that CIMMYT (in Spanish, the Centro Internacional de Mejoramiento de Maiz y Trigo, in English, the International Maize and Wheat Improvement Center) and others started on maize 20 years ago. The goals are the same—to try to find out which species will grow best. The International Center for Research in Agroforestry (ICRAF), in Africa, is identifying and assessing fast-growing, multipurpose woody species that grow well in different soils and climates. These species can be used for reforestation of deforested areas. They can also be the source of fuelwood, animal fodder, and green manure in farming systems.

Alley-cropping systems that incorporate leguminous tree species are viable alternatives to harmful slash-and-burn agriculture. Other alternatives to slash-and-burn agriculture, which include contour strip farming, minimum tillage, and other soil and water conservation practices, are being adapted to help farmers produce more on the same plot of flatland. Soil and water losses can be controlled through the use of these new technologies and management systems. As a result, forests, uplands, and other easily eroded areas, along with their valuable biological flora and fauna, can remain intact.

AID's Soil Management Collaborative Research Support Program, better known as TropSoils—an important program in this effort—is managed by North Carolina State University. A recent TropSoils report cited many soil management remedies for developing countries. The report asserted that "proper management of one acre of permanent cropland in the humid tropics can prevent the need for clearing about five acres of tropical rainforest per year.... The same principle applies to savannas and steeplands [and] the dry lands of Africa's Sahel." Other findings include the contribution of legumes to soil nutrition, the superiority of mulching to tillage in reestablishing vegetation on barren forest soils, and the benefits of "strip rainfall harvesting" to reduce rainfall loss.

Africa's fragile natural resources make much of that continent particularly vulnerable to agricultural constraints and the environmental problems they cause. AID is emphasizing the conservation of natural resources and the need for biological diversity on that continent. Of our Development Fund for Africa (DFA), fully 10 percent ($50.7 million) was targeted to natural resource conservation in fiscal year 1988. A projected 10 percent ($55 million) of the DFA is again targeted to natural resource conservation in fiscal year 1989.

## Four Challenges

I would leave four challenges for U.S. universities and cooperating institutions that are concerned about sustainable agricultural systems. *First* is the challenge to better understand the true nature of agricultural sustainability in developing countries. There are serious weaknesses in the notion that we can simply transfer to the developing countries the technologies and systems we have in the United States. Practices and policies that we have found successful for our mechanized agriculture often fail miserably on small farms where the hoe and manual labor prevail. Likewise, trying to handle environmental problems by merely passing laws against degradation simply will not work. We must better educate ourselves as to the biological, social, economic, and political environments of developing countries.

The *second* challenge is to work with the developing countries in creating and testing improved technologies and systems that will help them increase rural incomes, alleviate hunger, and conserve natural resources. Once again, there are limitations as to what we can transfer directly from U.S. agriculture to developing countries. Technologies and systems must be adapted for the needs and environments of those developing countries using the most modern scientific tools.

*Third* is the challenge to create an economic and social environment that will encourage the adoption of the new technologies and systems. Policies must encourage this adoption, and needed outputs must be available. Universities and private, voluntary organizations have key roles to play in helping to identify needed policy changes and to encourage their implementation.

*Fourth* is the challenge to establish linkages between U.S. and developing country scientists and institutions to expedite cooperation. Such linkages will further the concept that sustainability is an international problem and opportunity. The United States will gain, along with our overseas partners.

The struggle to further the sustainability of agriculture in the developing world has just begun. Resources, like this book, should help us all to better understand how such sustainability can be assured.

### REFERENCES

Belize. 1982. Environmental report on Belize. Compiled for the Agency for International Development by S. Hilty. Arid Lands Information Center, University of Arizona, Tucson.

Costa Rica. 1981. Draft environmental profile of Costa Rica. Compiled for AID by

J. Sillman. Arid Lands Information Center, University of Arizona, Tucson.
El Salvador. 1982. Environmental profile of El Salvador. Compiled for AID by S. Hilty. Arid Lands Information Center, University of Arizona, Tucson.
Guatemala. 1981. An environmental profile of Guatemala. Prepared for AID by the Institute of Ecology. University of Georgia, Athens.
Haykin, Stephen M. 1987. Policy reform programs in Africa: A preliminary assessment of impacts. Office of Development Planning, Bureau for Africa, U.S. Agency for International Development, Washington, D.C.
Honduras. 1981. Environmental profile of Honduras. Prepared for AID by J. Silliman and P. Hazelwood. Arid Lands Information Center, University of Arizona, Tucson.
Leonard, H. Jeffrey. International Institute for Environment and Development. 1987. Natural resources and economic development in Central America: A regional environmental profile. Transaction Books, New Brunswick, New Jersey, and Oxford, England.
Mackie, Arthur B. 1983. The U.S. farmer and world market development. Mimeograph. International Economics Division, U.S. Department of Agriculture. Washington, D.C.
Nations, J. D., and D. I. Komer. 1983. Central America's tropical rainforests: Positive steps for survival. Ambio 12(5): 232-238.
Nicaragua. 1981. Environmental profile of Nicaragua. Prepared for AID by S. Hilty. Arid Lands Information Center, University of Arizona, Tucson.
Overseas Development Council (Roger and Hansen and contributors). 1982. U.S. foreign policy and the Third World: Agenda 1982. Praeger, New York, New York.
Panama. 1980. Panama: State of the Environment and Natural Resources. Agency for International Development, Washington, D.C.
Population Reference Bureau. 1986. World population data sheets. Washington, D.C.
Revelle, R. 1976. The resources available for agriculture. Scientific American 235: 64-168.
U.S. Agency for International Development Country environmental profiles. Washington, D.C.

# INTERNATIONAL GOALS AND THE ROLE OF THE INTERNATIONAL AGRICULTURAL RESEARCH CENTERS

Donald L. Plucknett

Until after World War II, there was little international cooperation in agricultural research. By 1971, however, when the Consultative Group on International Agricultural Research (CGIAR) was formed, the world had seen what could be achieved with international effort. High-yielding wheat varieties, developed in Mexico in the 1950s and 1960s at what was to become the Centro Internacional de Mejoramiento de Maiz y Trigo (CIMMYT), had been transferred to Pakistan and India in the mid-1960s with tremendous success. Both countries were on their way to self-sufficiency in wheat production. In the Philippines at the International Rice Research Institute (IRRI), semidwarf rice varieties were bred and released during the 1960s.

CGIAR began with these two institutions, CIMMYT and IRRI, and with two other international centers launched by the Rockefeller and Ford Foundations in the 1960s—the Centro Internacional de Agricultura Tropical (CIAT), located at Cali, Colombia, and the International Institute of Tropical Agriculture (IITA), located at Ibadan, Nigeria. Over the next eight years, nine additional centers were added to the group. Centro Internacional de la Papa (CIP) had a commodity mandate. Others had mainly a geographical perspective: International Center for Agricultural Research in the Dry Areas (ICARDA); International Crops Research Institute for the Semi-Arid Tropics (ICRISAT); International Livestock Center for Africa (ILCA); International Laboratory for Research on Animal Disease (ILRAD); and West Africa Rice Development Association (WARDA). Three had a policy and/or service orientation: International Board for Plant Genetic Resources (IBPGR); International Food Policy Research Institute (IFPRI); and International Service for National Agricultural Research (ISNAR).

Other international centers or initiatives have been discussed; about a dozen centers now are being reviewed for inclusion in the consultative group.

CGIAR has three cosponsors—the Food and Agriculture Organization of the United Nations (FAO); the United Nations Development Program (UNDP); and the World Bank. It had more than 40 donors in 1988. The group has a Technical Advisory Committee (TAC) with 14 members and a chair. In 1988, CGIAR's total operating budget was $243 million. The 13 centers operate as nearly autonomous institutions. Research program decisions are made at the center level. The board of trustees, director general, and staff of each center plan and carry out the research programs under mandates developed by CGIAR and its TAC. Five-year program and planning reviews are conducted by teams of outside reviewers.

The work of these centers has included developing technology and systems that contribute to sustainable agriculture. A recent TAC paper on priorities and future strategies, however, stressed the need for even greater emphasis on sustainability. CGIAR concurred and requested TAC to consider how to achieve such an objective. A TAC sustainability subcommittee has studied and prepared a document on this subject. Systemwide input and reviews, as well as outside commentary, have contributed to TAC's final paper, which was released in 1988 at CGIAR's mid-term meeting in Berlin.

The paper primarily addresses CGIAR and its role in international agricultural research. It recognizes, however, that "many of the problems identified cannot be solved by CGIAR institutions or through agricultural research alone." While focusing on the role of CGIAR, it emphasizes that national governments and their research and development services must bear the brunt of the problem in the developing world. It also stresses that both developing and industrialized countries face serious problems of agricultural sustainability. Moreover, it urges bilateral and multilateral donors to give high priority to agricultural sustainability in their support of developing-country programs.

## General Goals

From the beginning, CGIAR gave priority to food production—increasing the quantity and improving the quality of food supplies in developing countries. The most recently approved goal statement reads: "Through international agricultural research and related activities, to contribute to increasing sustainable food production in developing countries in such a way that the nutritional level and general economic well-being of low-income people are improved."

This goal statement specifies and focuses on:

■ Developing, not developed countries.

■ Research-related activities, not development or technical assistance.

■ International research, not national or regional.

■ Food and feed, not industrial commodities.

■ Technologies for long-term sustainable production, not those that sacrifice ecological stability for short-term gains in productivity.

■ Improved nutrition and economic well-being of low-income people, not only through increased food production, but also through improved food quality, more equitable distribution, more stable food supplies, and increased purchasing power.

Eight objectives under the complex CGIAR goal provide a framework for the work of the centers in collaboration with national agricultural research systems and other partners in the global system:

■ Managing natural resources for sustainable agriculture.

■ Increasing productivity of essential food crops within improved production systems.

■ Improving productivity and ecological stability of livestock production systems.

■ Improving post-harvest technologies to utilize agricultural products more fully in rural and urban areas.

■ Promoting better human health and economic well-being through improved nutritional quality of foods, more equitable access to foods, expanded economic opportunities, and better management of overall family resources.

■ Ensuring the formulation of rational agricultural and food policies that favor increases in food production and commodity productivity.

■ Strengthening national agricultural research capacities in developing countries to generate, adapt, and use enhanced technologies faster.

■ Integrating efforts within and among centers of CGIAR and its partners in the global system.

The system's involvement in each of these areas varies greatly. In some, such as crop and African livestock production, the centers play a major role. In other areas, the system is primarily a catalyst, stimulating and supporting research at other institutions. In areas such as commodity conversion and utilization the system integrates the work of other leading institutions into the results of center programs.

## Implementing Sustainability

TAC's recently released paper, "Sustainable Agricultural Production: Implications for International Agricultural Research," discusses sustainability as a dynamic concept that allows for changing needs of a steadily increasing global population. "Sustainable agriculture should involve the successful management of resources for agriculture to satisfy changing human needs while maintaining or enhancing the quality of the environment and conserving natural resources."

Sustainability involves the complex interactions of biological, physical, and socioeconomic factors and requires a comprehensive approach to research in order to improve existing systems and develop new ones that are more sustainable.

These biological considerations will be important for future sustainability:

■ Conservation of genetic resources must be continued and strengthened.

■ Yields per unit of area and per unit of time must be substantially increased to meet the needs of rapidly increasing populations.

■ Long-term pest control must be developed through integrated pest management and built-in resistance because intensified production will tend to encourage build-up of pests and break down the effectiveness of pesticides and host-plant resistance.

■ Improved methods for disease and parasite control will also be important to sustain animal production.

■ A balanced production system involving both crops and livestock will be needed to enhance productivity and avoid overgrazing.

These physical factors and constraints are deemed most important:

■ Soil is the most important resource for ensuring sustainability; loss of topsoil through erosion and a reduction in soil fertility by not replacing nutrients both turn a renewable resource into a nonrenewable one.

■ Agriculture is the principal user of water globally; inefficiently using fossil water and overdrafting rechargeable aquifers can result in another renewable resource being eroded.

■ Poor soil and water management in rainfed agriculture can cause severe land degradation.

■ Misuse of agricultural and industrial chemicals can contribute to the accumulation of toxic substances in soil and water.

■ Atmospheric changes brought about by human activities will adversely affect agricultural production.

■ Energy consumption required by high-yielding production systems will probably be justified in the foreseeable future as using a nonrenewable resource, oil, to protect soil from being reduced to a nonrenewable resource.

These socioeconomic and legal constraints also affect long-term sustainable strategies:

■ Weak infrastructure in many developing countries is a major constraint to delivering inputs and transporting farm products.

■ Financial and administrative programs often are biased toward urban consumers.

■ Land tenure systems can discourage farmers from conserving natural resources and investing in future productivity; many countries do not have laws to protect forests and rangelands from indiscriminate exploitation.

To achieve sustainability, the constraints that threaten it must be alleviated, and major efforts must be made in increasing productivity to meet the immediate demands of a growing global population. Although CGIAR resources are small in comparison to total global expenditures on agriculture, the CGIAR system has had a greater influence on the nature of research at other institutions. The donors and other components of the system are in a position to help focus attention on this issue and to encourage governments and relevant institutions to give highest priority to sustainability.

TAC recommends specific strategies for research within the CGIAR system:

*Research with a sustainability perspective.* TAC does not view research related to sustainability as a separate or discrete activity. Concern for sustainability should be reflected in the way research is approached. TAC recommends that research at the centers designed to generate agricultural innovations should be planned and conducted with a sustainability perspective. TAC further suggests that in formulating or revising strategic plans centers include proposals for maintaining a sustainability perspective throughout their programs.

*Balance in research.* Although productivity research includes many aspects of resource management, strengths of the various components of the multidisciplinary approach must be kept under review to ensure an appropriate balance. Plant breeding, for example, can continue to contribute much to sustainability, but it must not dominate center programs so that other approaches are neglected. TAC recommends that centers with commodity mandates review the balance of activities in their productivity research. Sustainability concerns may make it desirable, if not essential, for some centers to give increased attention to research on problems of resource management.

*Short-term and long-term objectives.* If the goal of sustainable agriculture is to meet the changing needs of people, research must clearly consider both short-term and long-term needs. Yet, stability of the environment should not be consciously sacrificed for short-term gains. Each center's aim should be to devise technologies that can meet short-term requirements while at the same time maintaining or enhancing ability to meet long-term needs.

*Levels of input.* The centers should give greater emphasis to research designed to optimize productivity from the use of low levels of purchased inputs, consistent with the requirements of sustainability. The aim should be to promote gradual evolution toward greater productivity from balanced systems, which may require progressively higher levels of purchased inputs to ensure that the requirements of sustainability are met. At all levels, the aim should be to use inputs as effectively as possible. Centers should review the emphasis given to low-input farming in their research programs and increase it where appropriate. They should also review their approaches to research on low-input farming to ensure that sustainability is adequately considered. High levels of industrial inputs can contribute to sustainability, and TAC recommends that high-input production systems and related policy issues be included in CGIAR center research. These centers should avoid duplicating research already being done in industrialized countries, however.

*Sustainability and equity.* TAC reaffirms its earlier recommendation that the centers give greater emphasis to development of techniques applicable in less-endowed regions.

*Improved production systems.* Centers should continue to investigate aspects of more intensive production systems based on sound ecological principles and resource conservation. Where appropriate, this work should include aspects of agroforestry.

*Advances in biotechnology.* Centers involved in productivity research should have the capability to monitor advances in biotechnology and, when appropriate, develop in-house capacity to use techniques that would assist their programs in a cost-effective manner.

*Policy research.* In its study of priorities and future strategies, TAC recommended a significant increase in policy research. TAC reaffirms this recommendation.

*Relations with national agricultural research systems.* Centers should give high priority to strengthening the capacity of national agricultural research systems to incorporate sustainability into their research approach.

*Training.* Centers should give high priority to incorporating sustainability into training programs, adjusting as needed to meet the needs of national agricultural research systems and blending their approaches.

*The role of developing countries.* Whatever help the centers provide to national systems in research and training, success in achieving sustainability will ultimately depend upon the commitment of developing countries themselves.

*Collaboration with institutions outside CGIAR.* In view of the contributions the centers can make to the solution of large-scale and long-term problems of sustainability, the need for effective collaboration is greater now than ever before, not only with national systems, but also among the centers themselves, as well as with institutions outside the CGIAR system. Centers should continue to explore the potential for collaboration with other research institutions, including the private sector, especially on strengthening research related to sustainability.

*Research needs and resource implications.* In view of the serious problems limiting the achievement of sustainability and the urgency for additional research to solve them, centers should review their budget priority for sustainability research and increase it where appropriate. TAC will support the centers in attempts to attract funding for well-conceived, new projects related to sustainability. The international donor community as well as the governments of developing countries have crucial roles to play in emphasizing the need to consider sustainability in allocating future resources and orienting future thrusts.

TAC has characterized sustainability mainly in terms of the dynamics of population growth and resource conservation. The common challenge facing all concerned is to find new ways of removing any impediment to sustainable agricultural production—technical, economic, social, institutional, political, or a combination.

## The Challenges

The resource base for agriculture, unless husbanded carefully and replenished continually, will dwindle in its capacity to produce at levels of global demand. While the initial challenge of averting Malthusian famine has been met, at least for the present, new ones have emerged:

- Globally, can yields be increased to and maintained at their technical and economic potential?
- Can productivity be improved in less-favored areas that have become the last frontier of agricultural expansion?
- Will production technologies maintain soil fertility and other vital resources upon which production depends?

The overuse of favored areas, exploitation of underutilized areas, and excessive irrigation or chemical usage will lead to the same unintended result: a narrower base for future production. The critical difference in outcomes—enhanced and increasing production over the long run or accelerated depletion of resources—depends upon the knowledge base that agricultural researchers develop. This involves describing the upper bounds for genetic potential, finding the most suitable technological package for expressing that potential, determining appropriate environments for its use, and then making compromises among the upper bounds on yield, immediate demands for food, what is socially and economically feasible, and conservation for the future.

TAC believes that the three main challenges for agricultural science today are to:

- Maintain and increase yields in favored areas.
- Bolster productivity in less-endowed areas.
- Employ technologies in both settings that conserve or enhance the resource base.

The availability of physical resources—soil, water, nutrients, and energy—determines whether land is cultivatable or not and whether it is classified as favored or less-endowed and marginal. Favored areas were cultivated first, and efforts to intensify production also occurred in these same areas first. They are the backbone of agricultural production, so ensuring their fertility and maintaining yields there are paramount. On less-endowed areas, subsistence is the norm, climate harsh, resources few, and external inputs inaccessible or prohibitively costly. Agriculture can be sustainable here, however, developed by trial and error over hundreds of years.

Because of the shrinking availability of cultivatable land, gains in production will be from intensification rather than expansion schemes. Intensification will occur on both favorable and less-endowed land. On favorable land, with high-input systems, better management will be especially crucial to achieving higher yields. For less-endowed and marginal land, pressed upon by population, a combination of new cultivars and management practices will be required to achieve both greater productivity and yields. In either case, conserving and enhancing the resource base will be vital.

Underlying each challenge is the need for more data—on the ecological setting; the crop's optimal requirements for water, solar radiation, and nutrients; and its tolerance or resistance to pests and disease—before better germplasm or management practices can be suggested.

## Research Strategies

Strategic research explores the scientific ideas upon which progress in agricultural technology depends. Its fundamental questions are these: Where are the next likely advances in productivity? What strategies should be pursued to achieve those advances? Which problems limit production the most? In Africa and many parts of Asia and Latin America, for example, low levels of nitrogen in the soil and forage plants severely limit plant and animal production. The CGIAR response, shared across many centers, is to work simultaneously on identifying and developing nitrogen-efficient cultivars; exploring the nitrogen-fixing properties of legumes and plant helpers, such as rhizobia and mycorrhizae; ascertaining more efficient levels of fertilizer usage; and using crop residues and green manures as sources of nitrogen fertilizer, as well as experimenting with pasture production systems that use forage legumes and grasses to enhance soil fertility and animal nutrition.

Maintaining soil fertility for most crops mainly requires nitrogen and phosphorus. Efforts to make soil phosphorus more available is the goal of several CGIAR research projects. Phosphorus often is present in the soil but in a relatively unabsorbable form, so plant helpers, such as mycorrhizae, are needed. Unlike rhizobia, which are bacteria, the plant helpers are fungi that invade plant roots, form an associative relationship with the plant, and enhance phosphorus uptake.

***Nitrogen-Efficient Cultivars.*** In resource-poor environments, low-cost, low-input technologies are often seen as the most efficient approach to improving production. CIMMYT's maize program has increased its attempts to identify maize that can be grown with limited nitrogen. At IRRI, significant differences in the ability of 37 lowland rices to support biological nitrogen-fixation (measured by acetylene-reducing activity at heading stage for three consecutive days) suggest that it should be possible to breed rices for high nitrogen-fixing ability. Results show that atmospheric nitrogen was higher in the grain of IR42 than in other varieties. Also seeking nitrogen-efficient cultivars, CIP scientists grew 64 potato varieties on a soil that supplies 60 kilograms of nitrogen per season through mineralization.

***Commercial Fertilizer.*** Through the International Network on Soil Fertility and Sustainable Rice Farming (INSURF, formerly INSFFER), some 50 rice scientists in 22 countries collaborated in research on integrated nutrient management. Sulfur-coated urea, the most expensive nitrogen product, and deep-placed urea supergranules outperformed the best split application of prilled urea at 24 lowland irrigated sites in seven countries. Fifty-one percent less nitrogen was required to obtain a ton-per-hectare yield increase from sulfurcoated urea and 48 percent less from supergranules than from prilled urea. At 31 rainfed sites in nine countries, yield responses also were significantly higher in 33 percent of the trials, with sulfur-coated urea and supergranules requiring 57 percent and 62 percent less nitrogen, respectively, than prilled urea.

Earlier work at IRRI showed that fertilizer nitrogen recovery is only 30 percent or less if fertilizer is broadcast into field water, the most common practice. Incorporating fertilizer into the soil before planting can double its nitrogen efficiency. Soil incorporation of nitrogen is likely to be more important for broadcast-seeded rice than for transplanted rice, according to experiments involving ammonia volatilization. Scientists from IRRI and the Commonwealth Scientific and Industrial Research Organization (CSIRO) of Australia compared total nitrogen loss and ammonia volatilization using simple techniques. Comparing different management and application rates, maximum volatilization occurred when urea was broadcast into floodwater and least when urea was incorporated into soil without standing water.

On calcareous soils of the arid zones of the ICARDA region, separate experiments involving ammonia volatilization of top-dressed urea showed that soil temperature appears to affect ammonia volatilization. With increased soil temperature, the rate of hydrolysis of urea and the rate of reaction leading to ammonia volatilization increase. In northwest Syria, for example, the temperature in February usually is low, and urea appears to be an efficient and cheap source of nitrogen fertilizer. Losses of nitrogen applied at planting of wheat are expected to be only 5 to 10 percent, and only 10 to 20 percent when nitrogen is applied as top-dressing in early February.

CIP scientists, under contract research with the National Agrarian University in Lima, Peru, are investigating basic fertilizer requirements for potatoes in diverse soils and environments. For heat-adapted potato cultivars, mulch is recommended at planting to improve crop emergence and establishment; however, mulch makes it difficult to side-dress a split application of nitrogen fertilizer. CIP experiments conducted during two seasons at separate locations in Peru indicate that a split fertilizer application is not superior. Total fertilization at planting in combination with mulching obviates the need to build soil ridges along the rows and reduces the possible entry of bacterial wilt and other pathogens into the crop. Maximum yield was achieved at 70 days by applying all nitrogen at planting. Similar experiments during the dry season also indicated no benefit to tuber yield by splitting fertilizer applications.

***Fertilizer and Water-Use Efficiency.*** Based on ICARDA's regional network set-up to calibrate soil tests with crop responses in cereals and legumes, it is clear that economically optimal fertilizer use depends upon crop rotation, weed control, soil fertility, previous fertilizer usage, and rainfall. In ICARDA's predominantly dry region, soil deficiencies are widespread—particularly nitrogen and phosphorus. Even in harsh environments, fertilizer can improve water-use efficiency and farmers' profits can increase, as indicated by results with barley fertilizer experiments. In on-farm trials in Syria, the use of 20 kilograms per hectare of nitrogen and 60 kilograms per hectare of phosphate resulted in increased farmer incomes. The increase was enough that after only two seasons of collaboration the Ministry of Agriculture and Agrarian Reform provided agricultural credit for fertilizer in low-rainfall areas (350 millimeters of annual rainfall). In 11 trials harvested, mean yields ranged from 0.9 ton per hectare to 3.3 tons per hectare for grain and between 1.4 tons per hectare and 3.9 tons per hectare for straw.

***Phosphate.*** On marginal land with low pasture productivity and overgrazing by sheep, phosphate fertilizer experiments by ICARDA scientists showed significant improvement in total herbage yield (legumes and grass). Legume seed yield on marginal land increased 27 percent and 61 percent in response to 25 kilograms and 60 kilograms, respectively.

Among the inputs suggested to farmers in ICRISAT's village studies in Niger, phosphorus is being adopted by farmers; a general increase in use, up to 60 kilograms per hectare, has been noted. An improved low-cost production package was tested on a pilot scale in farmer-managed and researcher-managed tests in four villages in Burkina Faso. The package includes runoff-reducing bunds made from field rocks, mechanically built tied ridges, a low dose of chemical fertilizer, and the improved ICRISAT white-sorghum variety ICSV 1002. Yield increases exceeding 157 percent were obtained under complete farmer management. Results of phosphorus research on millet are consistent with earlier findings: a phosphate application of only 30 kilograms per hectare can triple millet yields.

A collaborative ICRISAT/International Fertilizer Development Center (IFDC) study in Niger confirmed that rock phosphate processed from locally available material is as good as water-soluble phosphate. It is also less expensive than imported phosphate fertilizers. Yields at least doubled with 24 kilograms per hectare of phosphate applied, both on the research station and in on-farm trials.

Crop residues and green manures are being used to maintain soil fertility in the semiarid regions of West Africa, where farmers are being forced, due to population pressures, to change from traditional shifting cultivation and fallow systems to continuous cultivation and reduced fallow. In a two-year experiment in Burkina Faso involving IITA and the Semi-Arid Food Grain Research and Development (SAFRAD), six crop residue and four tillage treatments were tested in cowpea production. Crop residue treatments were etablished the first year and tillage methods in the second year. Before planting, all plots received 50 kilograms per hectare of phosphate fertilizer. Because of beneficial effects on physical and chemical properties of the soil, cowpea seed yields were positively associated with the amount of crop residues left in the field, either as in situ mulch on no-till plots or incorporated into the soil in tilled plots. Maize residues retained in situ led to early flowering and maturity. No-till with in situ mulch was as effective as conventional tillage.

Recycling crop residues and growing and incorporating organic manures also can significantly increase nitrogen supply to crops and reduce chemical fertilizer requirements. For rice, the nitrogen accumulation of the water fern *Azolla* (which grows in association with nitrogen-fixing bluegreen algae) and of *Sesbania rostrata* (a fast-growing, stem-nodulating green manure species from West Africa that will grow in standing water) has been confirmed in IRRI experiments conducted since 1985. Growing and incorporating *Azolla* and *Sesbania* before rice increase the soil's nitrogen content and the supply of nitrogen to rice.

Before each rice crop, *Sesbania* was grown for 45 days during the wet season and for 55 to 60 days during the dry season, then incorporated. In *Azolla*-treated plots, *A. microphylla* no. 418 was grown and incorporated three times before rice planting and once at 25 days after transplanting. The inorganic fertilizer treatment received 50 to 60 kilograms of nitrogen per hectare per rice crop. Rice yield and fertilizer efficiency of four croppings showed a significant yield increase over the control, attributed to *Sesbania* and *Azolla*. Both biofertilizers also produced higher rice yields than the inorganic nitrogen fertilizer. Studies of long-term effects on soil fertility will be made after several croppings.

Phosphorus deficiency limits *Azolla* growth and nitrogen fixation, so soil or fertilizer phosphorus is essential. At IRRI, by enriching *Azolla* inoculum with up to 1.1 percent phosphorus, *Azolla* multiplied seven times in the field. Proper application of water-soluble phosphorus allowed *A. microphylla* no. 418 to fix 80 kilograms of nitrogen per hectare in 35 days, the highest rate measured so far. *A. microphylla* is more tolerant of high temperatures than other species and has erect growth that increases biomass production. Incorporating this quantity of *Azolla* supplies as much nitrogen to rice as adding an equivalent amount of urea.

***Legumes.*** Legumes are viewed as a major alternative to expensive and largely unavailable nitrogen fertilizers in conditions of low soil fertility where increased cropping pressures are reducing fallow periods and depleting soil resources. Their importance to smallholders is their nitrogen-fixing and often erosion-halting capacities, at low cost and risk. A major concern to subsistence farmers and other smallholders is the impact of legumes on food crops. In mixed livestock/cropping systems, and in rangeland livestock production, improved protein content of legumes for animal feed is the premium asset.

More than five years of experiments by IITA scientists with two legumes—*Centrosema pubescens* and *Psophocarpus palustris*—demonstrate the effectiveness of tropical legumes in adding organic matter to soil and improving soil properties at low-input levels without a long period of bush fallow. As "live mulch" ground cover in 10 crops of maize, the tropical legumes controlled weeds and contributed nitrogen to the soil, resulting in high yields. No nitrogen was applied. Yields were superior to those in conventional and no-tillage systems at low levels of inorganic nitrogen fertilization. The live mulch also depleted weed seeds in a manner similar to cropland subjected to long bush fallow. After four years of cropping, the organic carbon content of soil in the live mulch plots approached the level in newly cleared tropical forest, while that in no-till and conventional tillage sites remained relatively low.

Data from long-term crop rotation experiments on black soils at the ICRISAT center confirm the good residual effects of grain legumes. Grain yields of rainy season sorghum with no added fertilizer increased from 1,400 kilograms per hectare to 3,400 kilograms per hectare where an intercrop of pigeonpea and cowpea was grown the previous year.

At IRRI, rice and legume intercropping evaluated in upland rice areas showed that yields from the best treatments in many experiments exceeded 5.0 tons per hectare, while the average local upland rice yield is 1.5 tons per hectare. Yields from the first rice crop in the pattern were 4.0 tons per hectare, with high returns above variable costs. Returns were highest for the rice and cowpea intercrop; yields were high and costs were low because fertilizer was not applied.

Additional research is being conducted on dual-purpose legumes. In the Sudan savanna and Sahel of West Africa, cowpea fodder is as important to farmers as the grain because of dry-season shortages of fodder. A dual-purpose variety that produces 600 to 800 kilograms of grain per hectare and retains its foliage through the end of the season is valued because both grain and fodder attract almost equal prices in times of scarcity. IITA scientists in collaboration with program scientists in Nigeria and Niger evaluated medium-maturing cowpeas for this purpose. Preliminary results indicated that TVX 465903E, a multiple resistant variety, gave grain yields of 1.7 tons per hectare and fodder of 4.5 tons per hectare and could be a suitable dual-purpose cultivar. In Niger, ICRISAT/IITA work on early maturing cowpeas, widely intercropped with millet, shows the best dual-purpose line yielded 1,600 kilograms of grain per hectare and 2,070 kilograms of hay per hectare in fewer than 70 days.

***Intercropping.*** In Latin America, major production zones for cassava include poorer, more acid soils, and irrigation is not normally available. About 40 percent of total cassava production occurs in mixed cropping systems with maize, beans, and cowpeas. Technology is generally labor-intensive with little use of fertilizer, herbicides, and pesticides. Use of CIAT's low-input technology has increased yields of local clones in Colombia to 20 tons per hectare, while the national average is 8 tons per hectare. On-farm validation trials have shown that small farmers can readily increase yields 70 percent.

Complementary work by IITA on intercropped groundnut-cassava in Zaire showed that the protein equivalents of cassava and groundnut were higher in intercropping systems than those of only cassava and only groundnut. An intercrop of cassava planted in double rows with groundnut resulted in the highest yields of both crops. Groundnut also fixes nitrogen, reduces soil erosion, and controls serious soil pests, such as nematodes. Groundnut also contains amino acids that reduce the cyanide problems common among people who rely on cassava as a basic staple food.

***Alley Cropping.*** Another form of intercropping, called alley cropping, has been an important research thrust at IITA since 1978 in an attempt to find a permanent, more productive solution to the bush-fallow slash-and-burn cultivation practiced in much of tropical Africa. Alley cropping is an agroforestry system that grows food crops in alleys between hedgerows of trees or shrubs. These trees and shrubs perform the restorative powers of the bush fallow period, which relies on regrowth of deep-rooted trees and shrubs to recycle plant nutrients and build up soil organic matter.

Most recently, IITA evaluated the nitrogen contributions of three species of hedgerow trees (*Flemingia congesta, Cassia siamea* and *Gliricidia sepium*) to a maize crop on a degraded Alfisol in Ibadan. Phosphorus and potassium were broadcast on the plots before planting. *Gliricidia* prunings yielded a greater amount of nitrogen, equivalent to 90 kilograms of nitrogen fertilizer per hectare. For alley cropping, where labor for pruning is a major constraint, preliminary results from *Calliandra,* a fast-growing leguminous tree intercropped with maize and cowpea, suggest that good maize yields result from pruning alone without added fertilizer. In Kagasa, Rwanda, a harsh, semiarid environment with poor soils, yields of maize, bean, and sorghum in alley cropping with leguminous shrubs were as good as yields where shrubs were not grown. No fertilizers were applied in these trials.

In the humid zone of Africa, ILCA's agronomic work with IITA has focused on integrating fodder trees into traditional farming systems and developing fodder production strategies. The aim is to improve livestock production and make cropping practices more stable. Food crops are grown in alleys between lines of leguminous trees (*Gliricidia and Leucaena*). Alley farming after a two-year grazed fallow produced maize yields 60 percent and 100 percent higher, respectively, than maize yields following continuous maize cropping without trees. Soils under alley farming were richer in organic material and major nutrients. The nitrogen content of maize leaves was highest in alley farming following fallow, providing improved crop residue feed for livestock. The number of alley farms under farmer management in Nigeria's humid zone increased from 4 in 1982 to 250 in 1987. The technique is being extended to other countries.

***Forage and Pasture Crops.*** At CIAT, top priorities are to develop grazing systems that recycle nutrients through soil microorganisms and to integrate soil and plant nutrition. Forage legumes increasingly are being introduced into the traditionally grass-based pasture systems that predominate in subhumid and humid ecosystems of Latin America. Legumes in symbiosis with rhizobia are expected to contribute directly to the improved diet of animals in terms of protein (particularly during the dry season) and to improve the yield, quality, and persistence of grasses through enhanced nitrogen availability.

Legumes also are emphasized in ICARDA's research on pastures and forages in rotation with cereals to improve native pastures and animal nutrition, hence, productivity and the effective use of crop by-products. Analysis of a four-year series of trials to test the feasibility of replacing fallow with forage legumes, such as vetch and lathyrus, indicates that forage substantially increases barley's water-use efficiency.

One common question asked by farmers about pastures and forage crops is what effect they will have on a subsequent cereal crop. Based upon ILCA research, *Vicia dasycarpa* and *Trifolium steudneri* produced average yield increases of 72 percent in sorghum grain and 91 percent in maize grain.

Intercropping forage legumes with cereals can produce higher total biomass and protein yields than legumes and cereals grown in pure stands. For example, intercropping the cowpea CII and maize produced 24 percent and 38 percent more dry matter, respectively, than did the two crops grown in pure stands.

## Policy Strategies

Three of the CGIAR centers deal primarily with policy and service to national programs. IBPGR's mandate is to promote the conservation of plant genetic resources through the exploration, collection, characterization, multiplication, evaluation, and storage of crop plants, pastures, fruits, and vegetables and their wild and weedy relatives. The CGIAR supports this work "to ensure that the diversity of germplasm is safely maintained and available for use in programs of research and crop improvement for the long-term benefit of all people." Further, CGIAR encourages all countries to support the unrestricted interchange of germplasm throughout the world and supports relevant research on genetic resources at its own institutions and through collaborative projects with others throughout the world.

During IBPGR's first 10 years, more than 300 collecting missions were undertaken in 88 countries; these missions involved more than 550 collectors. The resulting materials, covering 138 crop species, were stored in gene banks by more than 450 organizations in 91 countries, well over half of them in developing countries. IBPGR has developed a research program that supports germplasm exploration, collection, characterization, and storage work.

IFPRI's mandate is to identify and analyze alternative national and international strategies and policies for meeting the food needs of the developing world.

Current IFPRI research programs include a food data evaluation program in which trends in food consumption are being analyzed, along with the associated pressures of demand on food and feed production and imports and exports. Development, diffusion, and sustainability of production technology (particularly in the area of food grain production in India and 30 other countries) is a long-term study of adoption of modern farm practices and factors affecting input use of fertilizers and irrigation. In global food policy studies, researchers are looking at cash crops versus food crops and their effect on domestic consumption, supply and demand forces influencing substitution of traditional crops in West Africa, the effects of switching from semisubsistence to commercial agriculture, and the social and economic consequences of clearing tropical rain forests in Brazil for agricultural settlement. Development strategy involves the dynamic interaction of agricultural and nonagricultural sectors and focuses upon growth linkages, infrastructural development, and the problems of populations in low-potential regions. It is in this area that IFPRI thinks policy analysis has the potential to contribute most significantly to sustainability.

ISNAR's mandate to assist national agricultural research programs provides an opportunity to encourage research with a sustainability perspective. In the Advisory Services Program, ISNAR teams assess the adequacy of research policy, organization and structure, and management in relation to stated and/or implied national program objectives and needs.

## Future Outlook

With its strategy in place, CGIAR now is examining ways the strategy can be implemented. Most CGIAR centers are conducting research related to sustainability. Much of that work is related to protecting past gains and ensuring broader levels of resistance to diseases and pests—and tolerance to environmental stresses. A continuing dialogue on research related to sustainability is planned.

# SUSTAINABLE AGRICULTURE IN THE UNITED STATES

J. F. Parr, R. I. Papendick,
I. G. Youngberg, and R. E. Meyer

The U.S. Department of Agriculture's Report and Recommendations on Organic Farming (USDA, 1980) cited increasing concern among farmers, environmental groups, and the general public about the adverse effects of the U.S. agricultural production system, particularly the intensive monoculture of cash grains and the extensive and often excessive use of agricultural chemicals, both fertilizers and pesticides. Among the concerns most often expressed to the USDA study team were the following:

- Increased cost of, and dependence on, external inputs of chemicals and energy.
- Continued decline in soil productivity from excessive soil erosion and nutrient runoff losses.
- Contamination of surface and groundwater from fertilizers and pesticides.
- Hazards to human and animal health and to food quality and safety from agricultural chemicals.
- Demise of the family farm and localized marketing systems.

Because of these concerns, questions have been increasingly raised in recent years about the long-term sustainability of the U.S. agricultural production system, which has become so dependent on nonrenewable resources and exploitive of the natural resource base.

The USDA report found that many farmers, in addressing these concerns, had shifted away from conventional (chemical-intensive) farming systems to a less intensive, low-input approach based primarily on sod-based rotations and mixed crop-livestock enterprises. A major conclusion of the report was that these low-input farming systems are environmentally sound, energy-conserving, productive, stable, profitable, and tended toward long-term sustainability.

## The Concept of Low-Input/Sustainable Agriculture

A number of terms and definitions have emerged in recent years that refer to a spectrum of low-chemical, resource- and energy-conserving, and resource-efficient farming methods and technologies. For example, words such as "biological," "ecological," "regenerative," "natural," "biodynamic," "low-input," "low-resource," "agroecological," and "eco-agriculture" are specific terms used by certain spokespersons and groups to refer to various alternative agricultural production technologies and practices that, they feel, are essential to the development of long-term sustainable farming systems. We tend to view the words "organic" and "alternative" as more general terms that appear to embrace a number of the more specific words.

According to Lockeretz (1988), "sustainable agriculture" is a loosely defined term that encompasses a range of strategies for addressing a number of problems that afflict U.S. agriculture and agriculture worldwide. Such problems include loss of soil productivity from excessive erosion and associated plant nutrient losses; surface and groundwater pollution from pesticides, fertilizers, and sediment; impending shortages of nonrenewable resources; and low farm income from depressed commodity prices and high production costs. Furthermore, "sustainable" implies a time dimension and the capacity of a farming system to endure indefinitely (Lockeretz, 1988).

While it is often implied that sustainable agriculture can be attained through the development of long-term, stable, and profitable conservation and production systems, it may be that these systems will have to await the test of time. For example, we may not know whether a particular system has a high, medium, or low level of sustainability for possibly a decade or more. Currently, however, sustainable agriculture has settled in as the ultimate goal. How we achieve this goal will depend upon creative and innovative alternative methods and practices that provide farmers with economically viable and environmentally sound options in their various farming systems.

In 1985, the U.S. Congress passed the Agricultural Productivity Act as part of the Food Security Act, Public Law 99-198 (otherwise known as the 1985 farm bill). This act provided USDA the authority to conduct research and education in alternative agriculture, or, more specifically, on low-input or sustainable farming systems (USDA, 1988). In December 1987, Congress appropriated $3.9 million to implement the research and education programs called for in the Agricultural Productivity Act. The concept that has emerged from this effort is one of low-input/sustainable agriculture, or LISA, which addresses multiple objectives, such as increasing agricultural productivity, conserving energy and natural resources, reducing soil erosion and loss of plant nutrients, increasing farm profits, and developing more stable and sustainable conservation and production systems for U.S. agriculture.

At this point it seems appropriate to offer a definition of low-input farming systems that tends to characterize the farmers who were interviewed during compilation of the USDA organic farming report, as well as those low-input/sustainable farmers who are discussed in a following section.

By way of definition, then, low-input farming systems seek to optimize the management and use of internal production inputs (i.e., on-farm resources) in ways that provide acceptable levels of sustainable crop yields and livestock production and that result in economically profitable returns. This approach emphasizes such cultural and management practices as crop rotations, recycling of animal manures, and conservation tillage to control soil erosion and nutrient losses, and to maintain or enhance soil productivity. Low-input farming systems seek to minimize the use of external production inputs (i.e., off-farm resources), such as purchased fertilizers and pesticides, wherever and whenever feasible and practicable, to lower production costs, to avoid pollution of surface and groundwater, to reduce pesticide residues in food, to reduce a farmer's overall risk, and to increase both short- and long-term farm profitability.

## Misconceptions About Low-Input/Sustainable Farming Systems

Three misconceptions about low-input/sustainable farming systems commonly arise:

■ Low-input/sustainable farming systems represent a return to agriculture that was practiced in the 1930s. This is simply not true. These farmers use modern equipment, certified or hybrid seed, soil and water conservation practices, conservation tillage, and the latest innovations in livestock feeding and handling. They minimize the use, and need for, off-farm purchased inputs of fertilizers and pesticides through sod-based crop rotations, integrated crop/livestock management, and recycling crop residues and animal manures to maintain soil productivity.

■ Low-input farming methods result in low output. On the contrary, many of these farmers insist that their crop yields from low-input systems are equal to or even higher than their more conventional neighbors. Studies have shown that crop yields from low-input farming systems might actually exceed those of conventional cash grain farmers during periods of below average rainfall (USDA, 1980; Lockeretz, 1981). Nevertheless, even if the crop yields from low-input farming systems are lower than from chemical-intensive cash grain production systems, the bottom line is how their yields translate into net returns, which in many cases will be higher with low-input systems.

■ Low-input farmers are really farming at the lower end of the crop response curve. Actually, in many cases the low-input farmer's productivity (i.e., efficiency or output-input ratio) will be high enough to place him or her near the top of the curve. That is, through good management of on-farm resources and crop rotations to provide the necessary levels of plant nutrients and to conserve available soil moisture, he or she may have pushed the crop yield potential up to a maximum level without using any chemical fertilizers or pesticides.

## A Study Tour on Sustainable Agriculture

The USDA/U.S. Agency for International Development Project on Dryland Agriculture (USDA/USAID 1987, 1988) held its fourth annual sustainable agriculture study tour, July 10-16, 1988. The participants included some 35 project managers and program directors from the World Bank, USAID, and agencies representing the USDA, including the Agricultural Research Service (ARS), Soil Conservation Service (SCS), Economic Research Service (ERS), Cooperative State Research Service (CSRS), Extension Service (ES), Office of International Cooperation and Development (OICD), and the Institute for Alternative Agriculture (IAA), a nonprofit organization in Greenbelt, Maryland.

The tour visited farms in Kansas, Nebraska, and Iowa where farmers have developed conservation/production systems that minimize the use of external production inputs, such as chemical fertilizers and pesticides. Through innovative management and on-farm research, they have lowered their production costs, reduced their overall risk (even during the severe drought of 1988), and increased their short-term and long-term productivity, profitability, and sustainability. A brief description of the unique features of the conservation and production practices employed by these farmers is noteworthy.

***Heiniger Dairy Farm, Fairview, Kansas.*** There are a number of unique features about the Cory and Shiela Heiniger farm. The Heinigers have demonstrated that with careful planning and innovative management, over a five-year period, they have been able to develop and sustain a 70-cow herd of registered Holsteins on 160 acres of cropland with few purchased inputs; at the same time they have improved soil productivity. Average annual milk production approaches 20,000 pounds per cow, and the farm operation has become a profitable enterprise. The key to success has been computerized feeding and record-keeping, intensive production of a balance of high quality forage, and reducing the need for chemical fertilizers and pesticides. The cows are maintained in a drylot; all forage is chopped and fed directly or stored as silage in plastic bags for use during the nongrowing season. The crop rotation is alfalfa (four to five years)-corn-wheat-sorghum sudan (double-cropped after wheat). The sorghum sudan is sown no-till into wheat stubble after harvest in June and produces a cutting about every 30 days through the summer. In addition to some agrichemicals, off-farm inputs include the purchase of some bromegrass and alfalfa hay.

The crop rotation provides cover for erosion control, helps to break weed and disease cycles, and adds nitrogen to the soil during the alfalfa sequence. Herbicides are now used only on corn; the rate is about one-half of that recommended on adjacent farms that are more intensively row-cropped. In addition, more than 800 tons of manure are produced annually by the dairy herd, all of which is returned to the fields. Recycling of nutrients and organic matter in the manure and a sod-based rotation has enabled the Heinigers to improve soil productivity, fertility, and tilth, thus reducing the need for chemical fertilizers. Their long-range goal is to eliminate the need for purchased fertilizers and pesticides. The Heinigers do not participate in the USDA farm program.

***Bender Diversified Crop/Livestock Farm, Weeping Water, Nebraska.*** Jim Bender operates a farm with 650 acres of cropland and a 90-cow beef herd on moderately hilly land. His ultimate goal is to sell only beef and soybeans. Other crops, including oats, corn, wheat, sorghum, and turnips, are fed to livestock; some crop stubble is grazed. Bender's main approach to achieving sustainability involves implementation of an effective soil and water conservation program, use of sod-based crop rotations, and a reduction in the need for chemical fertilizers and pesticides in the cropping system. Bender acquired his farm in 1975 in "a poor condition" after many years of intensive row-cropping by previous owners who had used high levels of chemical inputs. In his current state of transition back to crop rotations, he has not used herbicides for eight years or chemical fertilizers for two years.

Effective soil and water conservation measures have been achieved by installing 30 miles of jumbo diversion terraces and 25 miles of grassed waterways. In addition, small grains and hay crops have reduced the percentage of row crops, which has helped to reduce soil erosion and nutrient runoff losses. The broad terraces allow the use of wide field equipment. His key to successful and economical weed control without chemicals is timeliness of cultivation, which is accomplished almost exclusively with a rotary hoe, tine harrow, and cultivator. These tools have a low power requirement and are most effective when weeds are just emerging or otherwise small. This results in a narrow time frame during which effective mechanical cultivation for weed control can and must be accomplished. However, when conditions are optimum, his 90-horsepower tractor pulling a 36-foot rotary hoe can cover about 30 acres per hour. The entire farm can be cultivated in this manner in less than two days. Bender's costs for weed control are as low as $6 per acre. His main weed problems are two perennials, field bindweed and Canada thistle. He is attempting to use patch tilling to eradicate Canada thistle.

Bender has reduced his capital investment by using and maintaining older farm equipment. His father, who is 80 years old, and one person hired for the summer assist him with the farm work. He uses custom planting of some crops, purchases some hay, and hires custom operators for feed grinding and hay stacking to resolve time-labor constraints.

Bender acknowledges that the cattle operation with calving requires much of his personal attention and considerable time is spent building and mending fences and corrals. From his perspective, however, the key to the success of his farming operation is crop diversification and raising cattle, which enhances nutrient and organic matter recycling. Bender participates in the USDA farm program and grows sweetclover for nitrogen and organic matter return on set-aside acres. But he feels the current farm program is an impediment to his goal of establishing a long-term, sustainable agricultural conservation and production system with primary emphasis on sod-based crop rotations, diversification, and integration of crops and livestock.

Bender's primary motive for choosing this particular method of farming, compared with intensive cash grain production, is to protect the environment and soil resource base. He considers his methods to be a more sound, permanent form of agriculture than conventional cash-grain farming. He says that his yields are higher than his more conventional neighbors in years with below normal rainfall and possibly lower in years with above normal rainfall. Wheat yields on his farm have reached 50 bushels per acre; oat yields were 70 bushels per acre in 1988 (a year of below normal rainfall) and 114 bushels per acre last year with higher rainfall. Bender has been working on a special market certification to qualify as a certified organic producer. To date, however, he has been marketing all of his produce through regular commercial channels.

***Akerlund Grain and Livestock Farm, Valley, Nebraska.*** Delmar Akerlund's 760-acre grain and livestock farm emphasizes low-input methods to reduce operation costs and soil improvement practices to maintain crop productivity. Akerlund estimates that his variable costs of production average about $30 per acre across all crops in his diversified rotation. This compares with $65 to $75 per acre for neighboring farms committed to chemical-intensive corn-soybean production. Akerlund made the transition from chemical-intensive, monoculture-based farming to his present system more than 20 years ago because of concern for pesticide effects on his family's health. It took three years, he said, to make the change and to eliminate the residual phytotoxic effects of pesticides in his soil. No pesticides or commercial fertilizers have been used on his cropland since 1967.

A key factor in improving the productivity of Akerlund's soils is the regular addition of paunch manure, which is imported from a packing plant in Omaha, 30 miles away. He pays only hauling costs, which are nominal. Over the years, the manure application has increased the soil organic matter content from 0.5 percent to more than 6 percent in some fields. The basic crop rotation includes oats underseeded with clover, followed by one or two years of corn, a crop of soybeans, and back to oats and clover. Wheat, rye, and alfalfa are also grown, but mainly to break weed, insect, and disease cycles. Cattle and hogs are produced as an additional source of income, not according to a planned schedule, but rather when Akerlund judges economic conditions to be favorable.

The crop yields obtained by Akerlund without irrigation are equivalent to or somewhat higher than those of his neighbors who irrigate and use conventional farming methods. He produces soybean crops of 60 bushels per acre, 125 to 130 bushels of corn per acre, 60 bushels of wheat per acre, 70 bushels of rye per acre, and 90 bushels of oats per acre.

Labor is not a constraint. Akerlund and a single hired man perform all of the farm operations. He places great emphasis on wildlife preservation and has participated in studies on his farm conducted by the U.S. Fish and Wildlife Service. Results have shown that his farming methods have greatly improved bird habitat and populations on his land. Akerlund does not participate in USDA's farm program because he feels it is of no benefit to him. In fact, he feels it would be a constraint to his low-input/sustainable farming system. He strongly believes that such participation would force him back into the monocultural production of feed grain crops and heavy use of chemicals.

***Rosmann Diversified Grain and Livestock Farm, Harlan, Iowa.*** Ron and Maria Rosmann's approach to sustainability emphasizes ecological aspects and seeks to achieve a balanced system through crop diversification and animal production. They avoid monoculture or limited-rotation cropping because of the potential for increased erosion and heavy dependence on the use of pesticides, which they feel can lead to a build-up of insects and weeds that are resistant to chemicals. The basic rotation on their 320-acre farm is corn-soybeans-corn-oats-alfalfa. Soybeans are sometimes overseeded to hairy vetch or rye in September before harvest as a winter cover crop for erosion control and green manure. Hairy vetch provides additional nitrogen from biological fixation.

No herbicides have been used on the farm for six years. Better weed control in soybeans is achieved by late planting to enable more opportunity for preplant weed control. Manure from a farrow-to-finish hog operation and a 50-cow beef herd is allowed to compost in windrows before field application in the spring. The Rosmanns have reduced overall nitrogen fertilizer use by 75 percent; on some fields, only 50 pounds of nitrogen per acre are applied to corn.

The Rosmanns feel their way of farming fulfills the concept of sustainability, and they believe it could serve as a model system for the rural community because it accommodates soil and water conservation, environmental protection, and is easily adapted to smaller farms. In their view, sustainable agriculture gets away from large, capital-intensive monoculture farming and promotes the smaller, traditional family-type farm, which has been one of the real strengths of U.S. Agriculture.

The Rosmanns currently participate in the USDA farm program, though reluctantly, because they are young farmers who inherited a relatively small corn base acreage. This base, they feel, must be maintained for now because of federal farm policy and economic uncertainties.

***Thompson Grain and Livestock Farm, Boone, Iowa.*** The 320-acre Richard and Sharon Thompson farm approaches a self-contained ecosystem. Pork and beef are the only produce sold from the farm. All manures and those crop residues not fed to livestock are returned to the soil. In addition, all of the sewage sludge from the city of Boone (population 12,000), approximately 200 dry tons annually, containing about four percent nitrogen, is applied to the fields. No herbicides and only small amounts of fertilizer are used.

Dick and Sharon Thompson began their unique system of farming in 1967 when they experienced problems with residual herbicide phytotoxicity and an increase in hard-to-control weeds, such as foxtail, which they attributed to the excessive use of nitrogen fertilizer.

The basic crop rotation is corn-soybeans-oats-meadow (three years). The corn is a cross between open-pollinated and a hybrid. The crops are fed to a 50-beef cow and 80-sow livestock operation. The livestock manure, along with the sewage sludge, is stored in a large concrete bunker until it can be applied to the fields in early spring before planting.

An innovative feature of the Thompson's farming methods is their development of a ridge-till system for corn and soybeans without use of herbicides. The crops are planted on ridges formed the previous summer, either by a special operation for corn following meadow, or by the last cultivation of corn for the soybeans. Manure is applied uniformly across the fields as early as possible in the spring. The planting operation removes about an inch of soil, weeds, and trash from the ridge peak where seed and starter fertilizer (if used) are placed. The soil, along with manure, crop residues, and any weed seeds, is moved into the interrow or wheel track area. Early cultivation for weed control is performed with a rotary hoe. Later cultivation is done with an implement that mixes the crop residue and manure with soil in the interrow zone and moves the mixture onto the row, thereby rebuilding the ridge. Enough surface residue is maintained to prevent soil erosion and nutrient runoff.

Most soil fertility needs for maximum crop production are met with manure applications and return of crop residue. However, their soils have become somewhat deficient in potassium, and a starter fertilizer containing potassium is often used, especially for soybeans. The starter also provides nutrients for early crop growth before the added organic materials begin to mineralize. The Thompson's ultimate goal is soil improvement so that all fertility needs of their Webster and Clarion soils are supplied through manure and sludge applications.

The Thompsons also conduct extensive on-farm research on tillage practices for weed control, the ridge-till system for soil erosion control, and crop response to different forms of nitrogen. The latter experiments help to monitor how well nitrogen demands are being met with applications of manure and crop residue. The Thompson's do not participate in the USDA farm program.

Thompson is also past-president of the Practical Farmers of Iowa. This organization is an on-farm research and demonstration network of Iowa farmers who are interested in shifting to low-input/sustainable production systems.

## Relevant Research on Low-Input/Sustainable Agriculture

The sustainable agriculture study tour visited land grant universities and agricultural experiment stations in Kansas, Nebraska, and Iowa where research relevant to low-input/sustainable farming systems is in progress. Some of this research is briefly highlighted here.

### *Kansas State University*

■ *Herbicide movement affected by tillage.* Studies are underway to determine the potential for movement of two commonly used herbicides, atrazine and alachlor, into shallow groundwater (15-foot water table) as affected by tillage practices and methods of herbicide incorporation into soils. Models will be used to interpret and extrapolate the results. Preliminary results for the field experiment, which is being conducted on a coarse-textured soil, showed no evidence of deep movement of the herbicides.

■ *Nitrogen credits for corn following soybeans.* Studies show that the yield of corn following soybeans averaged 50 bushels per acre more than continuous corn when no nitrogen was applied. The yield difference declined as fertilizer nitrogen applications increased. Similarly, the yield of soybeans following corn was four bushels per acre higher than continuous soybeans. The results show that for each bushel of soybeans produced a pound of nitrogen is supplied to the following corn crop.

■ *Long-term crop rotations and tillage interactions.* Long-term studies were initiated in the mid-1970s to determine the effect of conservation tillage and crop rotations on crop yields and soil fertility/productivity. Rotations of soybeans-wheat and soybeans-grain sorghum under three tillage/residue management systems are being compared with these same crops grown in monoculture. Results show that wheat and soybean yields were higher in rotation with sorghum than in monoculture. Experiments are underway to determine the nitrogen credits that farmers can expect from soybeans for the following grain sorghum crop, as well as effects of tillage and residue management. Another highly relevant study being conducted is to determine the feasibility and practicability of substituting legumes for fallow in the wheat-fallow rotation that is common in the dryland regions of the U.S. Great Plains.

■ *Bioregulation of soil fertility.* Studies are being conducted to determine whether the sequencing of certain crops could effectively enhance nutrient cycling in soils. Particular emphasis is being put on soil phosphorus. The system involves "accumulator plants," some of them deep-rooted, that can use phosphorus from "pools" that are not readily available to other plants and from well below the usual soil-root zone. The phosphorus used by the accumulator plants then becomes chemically and positionally available for uptake by "user plants," thereby reducing the need for phosphorus fertilizer inputs.

***University of Nebraska and Agricultural Research Service***

■ *Nitrogen credits for corn from legume cover crops.* Research is being conducted cooperatively by ARS and the University of Nebraska to determine the relative contributions of different nitrogen sources to corn, including soil organic matter, hairy vetch, and nitrogen fertilizers. Additional factors include tillage and crop residues. Hairy vetch has been shown to be sufficiently winter hardy for Nebraska's climate, whereas crimson clover is not. The winter cover crop depleted soil moisture to the extent that germination and emergence of the following corn crop was delayed 12 days during a below normal rainfall year. Thus, winter cover crops may not be practical in low rainfall years in this region.

■ *Long-term crop rotations compared with monoculture corn.* The University of Nebraska initiated a study in 1975 comparing an organic rotation of oats/clover-corn-soybeans-corn with conventional continuous corn. The organic rotation received feedlot manure instead of chemical fertilizers and no pesticides. The continuous corn receives the recommended applications of chemical fertilizers and pesticides. This study has shown that (a) the yield potential of corn in rotation is higher than when grown in monoculture, especially in warmer and drier years; (b) the organic treatments result in increased soil organic matter content, pH, phosphorus, potassium, and soil nitrogen compared to other treatments; and (c) the legume cover crop can use enough moisture in dry years to slow the germination and growth of the following crop.

***Iowa State University, Leopold Center for Sustainable Agriculture, and National Soil Tilth Laboratory***

■ *Improved management for more efficient use of fertilizer nitrogen.* Research is being conducted on the effect of nitrogen fertilizer rates on crop yields, nitrogen losses, energy consumption, and farmers' net profits. Economic analyses will determine the costs to the farmer of both overfertilization and underfertilization. Results have shown that more than 50 percent of the applied nitrogen fertilizer is lost from Iowa farms through leaching, runoff, and denitrification. Nitrogen fertilizer must be applied in accordance with the crop's nitrogen requirement and expected yield and with proper credit for the available nitrogen already in the system (e.g., residual fertilizer nitrogen). Nitrogen fertilizer should also be applied near the time of greatest demand by the crop so as to minimize losses and maximize nitrogen use efficiency. Methods that show promise include improved soil and plant diagnostic tests and giving proper credits for the contribution of residual fertilizer nitrogen in soil and nitrogen fixed by legumes in crop rotations.

■ *Leopold Center for Sustainable Agriculture.* This new institution was established by the Iowa state legislature in 1987 as part of the Iowa Groundwater Protection Act. The mission of this center is to facilitate, coordinate, and support research that promotes the development of low-input, environmentally sound, and sustainable farming systems for U.S. agriculture. The center's research agenda includes such topics as transition to low-input agriculture, nutrient cycling and efficient pesticide use, soil erosion control, soil quality (tilth), ground and surface water quality, preservation and improvement of recreational and wildlife areas, and community stability.

The center will provide research data and technical reports to farmers, policymakers, scientists, regulatory agencies, agriculture-related industries, environmental groups, and the agricultural extension service.

■ *National Soil Tilth Laboratory.* The newly established ARS National Soil Tilth Laboratory on the Iowa State University campus will focus on developing basic knowledge about soil physical, chemical, and biological properties to improve soil structure and enhance soil and water conservation and plant growth. Studies will also be conducted on how management affects soil tilth, with special emphasis on conservation tillage, soil fertility management, and crop rotations. The ultimate goal is to develop management systems that promote good soil tilth for long-term, sustainable farming systems.

## The Role and Influence of Public Policy: Some Considerations

Sustainable farming systems in the United States have developed and continue to perform remarkably well, despite a formidable array of specific policy disincentives and broad institutional and structural constraints. On balance, the dominant thrust of U.S. agricultural technology and of trends in farm structure since World War II (e.g., energy and chemical intensiveness, larger farm units, and the specialization and intensification of production practices and enterprises) has created an industrial form of agriculture. These structural changes adversely affect the use of sustainable farming methods, such as those described earlier (Youngberg and Buttel, 1984). Institutional conditions in agriculture, such as low relative energy prices; inflationary land markets; availability of large-scale, capital-intensive farm technologies; favored access of large farm operators to agricultural credit; various features of the tax code; and commodity-based price support policies, have directly influenced farm adaptation strategies in the United States over most of the past five decades. The expansion of farm size and the shift to highly specialized and intensified management practices throughout U.S. agriculture during this period reflect rational farmer responses to these interactive policy and institutional conditions. The oft-repeated admonition to American farmers to "get bigger or get out" simply reinforces a decision-making pattern deeply embedded in modern American agriculture.

Many of these conditions and constraints were noted by the low-input farmers described earlier. However, the most pervasive policy-related theme revolved around the influence of U.S. commodity policy on the adoption of sod-based rotations, a practice widely acknowledged as being of central importance to the development of stable, low-input farming systems (USDA, 1980).

As a general rule, the specific provisions of U.S. commodity policy place farmers, particularly cash grain farmers wishing to include a hay, small grain, or green manure crop in their rotations, at a distinct disadvantage (Anderson, 1985). Consequently, most low-input farmers either forego participation in these programs or participate marginally and sporadically, depending upon individual farm circumstances. There are a number of specific commodity-based program provisions that account for this behavior.

First, U.S. commodity programs tend to encourage chemical-intensive, monocultural cropping systems by focusing program benefits on a handful of crops (Young and Goldstein, 1987). Corn and other feed grains, wheat, cotton, and soybeans receive roughly three-fourths of all U.S. crop subsidies. These same commodities account for approximately two-thirds of U.S. agrichemical use (Fleming, 1987). Most agricultural economists agree that "the selective largess of U.S. commodity programs directs resources away from nonsupported commodities and toward supported commodities" (Young and Goldstein, 1987).

The manner in which program payments are calculated also reinforces farmer decisions to adopt chemical-intensive, monocultural cropping systems. Three factors, crop acreage bases, crop yields, and target prices, determine payments received by farm program participants. We will use corn to illustrate how these factors influence decision-making by individual farmers. However, the same general conditions exist for the other major program crops.

Under current legislation, each farm is assigned a corn base by the county committee of the Agricultural Stabilization and Conservation Service (ASCS). The size of individual farm bases results from the number of acres planted to corn during a designated historical base period. Thus, the more acres planted to corn during the base period (bases are adjusted periodically), the higher the corn base assigned to a given farm.

The second factor used to calculate farm program payments is the average number of bushels produced per acre on a given farm over a stipulated period of time. Under these arrangements, when subsidy payments are warranted because of low market prices, yields per acre, coupled with the size of the commodity base, bear directly on total payments received by program participants.

Target prices, first introduced in 1973 with passage of the Agriculture and Consumer Protection Act, interact with farm base and crop yield characteristics to determine the total amount of individual farm subsidy payments. Eligible farmers (all producers if no set-asides are required) are guaranteed a per bushel target price. In 1988, for example, the target price for corn was $2.93 per bushel. If average annual market prices over a stipulated period of time fall below the target price, producers receive deficiency payments equal to the difference between the market and target prices. Payment levels are thus tied directly to the size of the crop base on each farm and the number of bushels produced per acre on the base. Total payments are maximized through a combination of large crop bases and high yields per acre.

Taken together, these interrelated program provisions provide powerful, direct incentives for farmers to adopt continuous monocultures (in this case, corn) and to increase their yield per acre. Failure to do so can result in the loss of favorable crop base and yield histories, which can substantially reduce future potential program benefits (Taff and Runge, 1987). From a strictly economic perspective, it is irrational under these conditions for a farmer to introduce long-term, sod-based rotations into his or her cropping system or restrict chemical inputs (i.e., fertilizers and pesticides) on planted acres. Not only do large acreage bases and yield-per-acre crop histories ensure maximum payments under current law, they also add to the value of the farm itself. When farmland is offered for sale, a primary consideration of prospective buyers is the base acreage and established yields of farm program crops. If such bases are currently in effect and acceptably large, the seller can expect a premium price for his land. The common practice of bidding these program characteristics into farm value creates yet another significant economic incentive for adopting high-yield, monocultural, cash-grain production strategies.

The heightened activity in sustainable agricultural research and education programs within the USDA/land grant community is beginning to address the urgent need for reliable and readily available information on low-input farming technologies and systems. Although much work in this area remains to be done, farmer involvement and the emphasis on practical information in these developing programs are particularly relevant and encouraging. Despite these positive trends within the research and education community, it seems clear that future adoption of low-input farming systems will be severely constrained until appropriate adjustments in commodity and several related policy areas can be implemented.

## Research Needs and Priorities

The following research needs should be given high priority by USDA, land grant universities, and nonprofit research organizations to facilitate the development of low-input/sustainable farming systems for U.S. agriculture:

■ Conduct research on low-input/sustainable agriculture systems using a holistic approach. Some research that has been conducted by USDA and the land grant universities may be potentially useful to low-input farming systems. Such systems are undoubtedly complex and involve poorly understood chemical, physical, and biological interactions. However, much of the research conducted to date that relates to low-input, sustainable agriculture is piecemeal and fragmentary. A systems or holistic approach, which may require the development of new methods and technologies, is needed to thoroughly investigate and elucidate these interactions and their relationship to organic recycling, nutrient availability, crop protection, energy conservation, and environmental quality.

■ Assess the economic aspects of low-input/sustainable farming systems. This should be done on a whole farm basis. Such data are absolutely essential because herein is the essence of credibility as to whether low-input/sustainable agricultural systems are economically viable.

■ Determine the reasons for reduced crop yields during transition from conventional to low-input farming systems. Research is needed to determine the underlying causes of yield reductions so that farmers can make this transition in a shorter time and without experiencing undue risk and economic loss.

■ Conduct on-farm research to obtain more relevant data. Scientists should be involved directly with farmers in conducting on-farm research. In conducting on-farm tests, farmers usually go through a sequence of experimentation, assessment, and reevaluation, much of which is based on trial and error. Consequently, the farmer knows what happened, but often he or she does not know why. The research scientist could play a vital role in making this determination.

■ Develop new techniques for control of weeds, insects, and plant diseases using nonchemical methods. Pest control methods using parasites, predator insects, and other biological methods to eradicate unwanted species are vitally needed to further the development of low-input, sustainable farming systems.

■ Determine the nutritional quality of crops and the bioavailability of food nutrients for crops grown in low-input farming systems. As cultural and management practices change and as new cultivars are introduced into low-input farming systems, it will be important to monitor changes in nutritional quality, for example, the vitamin content and kind and amount of fiber, to ensure consumers that dietary standards are being met.

■ Develop improved methods for technology transfer. One of the most effective means of transferring technical information and practical methodology is through organizations like the Practical Farmers of Iowa. This is a network of farmers who have agreed to conduct on-farm research and demonstrations on low-input farming. They meet regularly at each others farms to share information and compare results. The land grant community should be involved and promote the development of such networks.

■ Assess the economic and farm structure implications of widespread adoption of low-input, sustainable agriculture. Accurate and honest assessments are needed to determine what the impact might be on existing agricultural production and marketing systems from low, medium, or extensive adoption of low-input, sustainable agriculture by U.S. farmers.

■ Develop farm program and policy innovations that are compatible with low-input farming systems. Many farmers would like to shift from conventional to low-input systems. As USDA farm program participants, however, they cannot do so without forfeiting a portion of their feed grain base. In future farm legislation, policymakers must allow a greater degree of flexibility for those farmers who would choose to shift toward low-input, sustainable farming systems.

■ Establish and assign proper nitrogen credits for calculating nitrogen fertilizer rates for crops. Agricultural scientists and extension workers must do more to see that farmers determine nitrogen credits from (a) soil residual nitrate, (b) irrigation water, (c) legumes in rotation, and (d) animal manures, green manures, and other organic amendments in calculating nitrogen fertilizer rates. Many Corn Belt farmers still apply heavy applications of nitrogen fertilizer for corn following alfalfa when research has shown that there is already sufficient soil nitrogen available for at least one and probably two consecutive corn crops. This can lead to excess nitrate nitrogen in the soil profile, which is subject to leaching and can cause contamination of groundwater.

### The Ultimate Goal

The ultimate goal of many U.S. farmers is to achieve sustainability in agricultural production systems. The primary objectives of these farmers are to develop farming systems that (a) maintain or improve the natural resource base, (b) protect the environment, (c) ensure profitability, (d) conserve energy, (e) increase productivity, (f) improve food quality and safety, and (g) create a more viable socioeconomic infrastructure for farms and rural communities.

To achieve these objectives, the farmers previously described employ a variety of alternative production and management practices designed to maximize the use of on-farm resources, such as animal manures and legumes, to provide plant nutrients; crop residues, cover crops, and conservation tillage to control soil erosion and runoff; crop rotations to control weeds, insects, and diseases; and energy-conserving tillage systems to save energy and reduce operation costs. By the same token, they seek to minimize their dependence on costly off-farm resources, such as chemical pesticides and fertilizers.

These mixed crop-livestock systems are vital to a sustainable agriculture. It is questionable whether conventional, monocultural feed-grain cropping systems or intensive row-crop production can fulfill the objectives and criteria of sustainability because of their necessary heavy chemical inputs, potential for environmental degradation, and excessive energy costs.

Considerable research is now underway at state agricultural experiment stations, including Kansas, Nebraska, and Iowa, that is highly relevant to the development of low-input, sustainable agricultural systems. Additional research needs and priorities are suggested that would contribute significantly to this effort.

The 1985 Food Security Act is not particularly compatible with, or conducive to, alternative agricultural practices or the development of low-input, sustainable farming systems. Certain aspects of current public agricultural policy contribute to conventional production practices and pose barriers to the wider scale adoption of more sustainable practices and systems. For example, oats and legumes, such as alfalfa, vetch, and sweet clover, are essential for establishing low-input crop rotations. Under current policy, however, planting these crops means that farmers would necessarily forfeit a portion of their feed grain base. Most of them are not willing to do this because of economic considerations and uncertainties about future farm policy legislation. Hopefully, policymakers in future farm legislation will provide a greater degree of flexibility and accommodation for the development of low-input, sustainable agriculture systems for U.S. farmers.

## REFERENCES

Anderson, James E. 1985. Farm programs and alternative agriculture. *In* Proposed 1985 farm bill changes: Taking the bias out of farm policy. Institute for Alternative Agriculture, Washington, D.C.

Fleming, Malcolm H. 1987. Agricultural chemicals in ground water: Preventing contamination by removing barriers against low-input farm management. American Journal of Alternative Agriculture 2: 124-130.

Lockeretz, W. 1988. Open questions in sustainable agriculture. American Journal of Alternative Agriculture 3(4): 174-181.

Lockeretz, W., G. Shearer, and D. Kohl. 1981. Organic farming in the Corn Belt. Science 211: 540-547.

Taff, Steven J., and C. Ford Runge. 1987. Supply control, conservation, and budget restraint: Conflicting instruments in the 1985 Farm Bill. *In* Daniel W. Halbach, C. Ford Runge, and William E. Larson [editors], Making soil and water conservation work: Scientific and policy perspectives. Soil Conservation Society of America, Ankeny, Iowa.

U.S. Department of Agriculture. 1980. Report and recommendations on organic farming. Washington, D.C. 94 pp.

U.S. Department of Agriculture. 1988. Low-input/sustainable agriculture: Research and education program. Washington, D.C. 7 pp.

U.S. Department of Agriculture/U.S. Agency for International Development. 1987. Technology for Soil Moisture Management Project (TSMM) progress report: January 1984 to December 1986. Washington, D.C. 36 pp.

U.S. Department of Agriculture/U.S. Agency for International Development. 1988. Technology for Soil Moisture Management Project (TSMM) annual report: January 1987 to January 1988. Washington, D.C. 22 pp.

Young, Douglas L., and Walter A. Goldstein. 1987. How government farm programs discourage sustainable cropping systems: A U.S. case study. pp. 443-459. *In* Proceedings of the farming systems research symposium. University of Arkansas, Fayetteville, Arkansas.

Youngberg, I. Garth, and Frederick H. Buttel. 1984. Public policy and socio-political factors affecting the future of sustainable farming systems. *In* David F. Bezdicek, et al. [editors] Organic farming: Current technology and its role in a sustainable agriculture. American Society of Agronomy, Madison, Wisconsin.

# SOCIETY'S STAKE IN SUSTAINABLE AGRICULTURE

Charles M. Benbrook

For an agricultural production system to be sustainable in the long-run, the following conditions must be satisfied:

■ Soil resources must not be degraded in quality through the loss of soil structure (i.e., compaction) or through the buildup of salts, selenium, or other toxic elements; nor can topsoil depth be significantly reduced through erosion, thereby reducing water-holding capacity.

■ Available water resources must be managed in a way that assures that crop needs are satisfied, and excessive water has to be removed through drainage or otherwise kept from inundating fields.

■ The biological and ecological integrity of the system must be preserved through management of plant and animal genetic resources, crop pests, nutrient cycles, and animal health. The development of resistance to pesticides must be avoided.

■ The system must be economically viable, returning to producers an acceptable profit.

■ Social expectations and cultural norms must be satisfied, as well as the food and fiber needs of the population.

Farming systems collapse or are forced to change when they become unprofitable to the farmer or when they impose on farm families, neighbors, rural communities, or perhaps even whole nations clearly excessive indirect costs or burdens. Examples of the latter could include a farming region in which nitrate levels in drinking water are found to exceed acceptable levels, triggering steps to manage and calibrate nitrogen fertilizer and manure applications to crop needs more carefully; the falling levels of Mono Lake in California or the Aral Sea in the Soviet Union, caused by excessive withdrawals of water for irrigation from rivers feeding these major lakes; or recurrent flooding from siltation in low-land areas, exacerbated by deforestation and excessive erosion in upland regions.

The world is tragically well-supplied these days with examples of clearly unsustainable farming practices. Drought, floods, deforestation, uncontrollable insect pest outbreaks, erosion, and economic calamity in food production systems threaten at least some farming regions in virtually all countries. In the developing world, many people die as a result, millions suffer hunger and disease, and economic development is thwarted because scarce capital and foreign aid resources are diverted to more immediate needs.

## The Contemporary "Sustainability" Record

What are the benefits of more widespread adoption of sustainable agriculture in the United States? To answer this question requires some sense of where American agriculture is now relative to sustainable agriculture and how it might change as progress is made in adopting cropping patterns and management systems more consistent with the principal features of sustainability previously outlined.

First, some general points regarding the sustainability of American agriculture at the present time. The economic score card is well-known and clearly not good in several key respects: More than 60,000 commercial farmers have lost their operations in the 1980s; the average capital asset value of farmland and machinery dropped about 40 percent from 1981 to 1986 and has now recovered perhaps one-quarter of its lost value; and the stabilization of net farm income within politically acceptable bounds has required truly massive federal expenditures, coupled with an unprecedented degree of government involvement in agricultural planting and investment decisions.

Fortunately, the economic health of American agriculture is clearly improving, but its overall contribution to national economic activity and long-term prospects are suspect. It will always be sustained economically in some fashion because everyone wishes to continue eating, but how and at what scale and cost remains uncertain. Somewhat more than one-fourth of U.S. cropland is devloted currently to export commodities. Will this country's commitment to export competitiveness wane as domestic, social, and environmental concerns gain prominence as policy goals?

Second, the problems faced by American agriculture that could undermine sustainability differ greatly by region, both in degree and character. Many, if not most, agricultural systems currently practiced in the United States are likely to remain sustainable for many years to come, albeit perhaps at higher cost to the treasury and natural environment than desirable. Adjustments in cropping practices and technology surely will be needed for most farms to remain sustainable and competitive, but such adjustments will be made.

Third, environmental concerns are the most volatile element in the sustainability equation today in the United States. Without doubt, nitrate levels in drinking water in many major farming regions have reached worrisome levels, concern grows about pesticide residues in food, the Endangered Species Act looms as an unfilled promise to some and a nightmare to others, and evidence mounts of tragic impacts on wildlife in some areas from toxic chemicals and minerals. These environmental problems and challenges are real, although often quite localized. The extent to which society demands a vigorous, swift remedial response to these problems will evolve from a series of political decisions. In general, stricter and more aggressive environmental laws and standards will hasten the need for change in farming systems.

Fourth, conventional farming practices in many major farming regions today are sustained economically, at least in part, by high levels of government support. Economic challenges in the 1980s already have had a fundamental and in many cases a profound impact on the attitudes and practices of American farmers. In many regions, driven by economic necessity, movement is at least underway toward more sustainable production systems, and farmers everywhere are seeking options in the form of new technologies, tools to support management decision-making, and changes in policy. As a result, the American agricultural research and extension system faces new challenges and expectations of unprecedented complexity and, in many cases, urgency.

The ongoing debate over commodity policy also demonstrates that certain key features of current farm policy are now viewed as politically unsustainable. A consensus is emerging on the need to reorient the incentives within commodity price support programs from maximum production of a single specified crop to maximum profit from more diverse cropping systems closely tied and responsive to market conditions.

Fifth, sustainable farming systems and practices will be adopted when and only if they offer farmers a convincing opportunity to earn higher profits than from other systems, taking into account all existing government program options and policies that have impacts on farm-level economic decision-making.

Sixth, new technologies, particularly biotechnology, could shift the economic viability of sustainable production systems dramatically by providing genetic and biological solutions to long-standing pest control problems, by increasing levels of nitrogen fixation, and through novel animal husbandry practices.

Finally, we also must remember that current systems often are used because they were used the year before; are proven; reduce short-term risk; and require modest public and private investments in education, training, management skills, and equipment. In contrast, the transition to sustainable systems will impose typically both economic and managerial costs on a farm operation, increase near-term risk, and require some capital investment. It must also overcome social pressures accompanying neighborly dialogue at local cafes.

## Features of Sustainable Systems

Before reviewing the benefits and costs of adopting sustainable production methods, the important changes in agronomic and pest control practices that will result as progress is made toward sustainable agriculture must be identified. The changes noted below, to become truly widespread, would invariably take a decade or more to evolve.

***Crop Rotations.*** In row-crop systems, continuous cropping patterns generally would be replaced by multiyear rotations, including field crops, legumes, and forage crops. Soybeans would play a key role in common rotations. Small grains also would be included in rotations, either as principal crops or cover crops to assist in establishing forage stands.

Continuous cropping is a common practice today, affecting perhaps two-thirds of U.S. cropland in most years. Following a truly full-scale transition to sustainable agriculture, continuous cropping would occur on perhaps less than one in five acres.

Given the large acreages idled most years in the United States, some 75 million acres in 1988, it would be possible to adopt more diverse multiyear rotations while actually increasing, if need be, total levels of production of major crops. One essential ingredient that would allow such a change would be a series of reforms in commodity price support program rules. Set-aside requirements and paid land diversions must be phased out as incentives for rotations on all cropland base acres are phased in. It is intriguing and significant that just such a set of policy changes lie at the heart of current decoupling proposals before the U.S. Congress and innovative conservation legislation authored by Senator Robert Dole (S. 2045).

***Nutrient Management.*** A much more careful job can and must be done in estimating nutrient needs. Then, fertilizers, manure, organic materials, and legumes could be managed more carefully to maximize the efficiency of plant uptake of available nutrients. Manure would be managed and applied with the goal of preserving a higher percentage of potentially available nitrogen. Unnecessary levels of fertilizer applications, regrettably common today in American agriculture, would become a much less frequent occurrence.

***Weed Control.*** Rotations and tillage would be relied upon more heavily to gain an acceptable level of weed control. Rotary-hoeing of newly emerging row crops would become much more commonplace, followed by two or more cultivations. Herbicides, when used, would be applied in bands around growing crops, and the much safer, more selective modern materials would be used more often, taking into account possible carryover problems in rotations.

***Insect Control.*** American agriculture faces a wide array of insect pests, only some of which routinely require insecticide treatments. Most potential insect pests are managed effectively through a variety of cultural and biological controls. Nonetheless, it is increasingly difficult for farmers to deal with certain pest-crop combinations in some regions. Resistance to registered pesticides is a widespread phenomenon and an extremely serious threat to sustainability in hundreds of specific regions for at least one major class of pest. Moreover, the rate at which insects develop resistance and cross-resistance is often high and accelerating, and the rate of introduction of efficacious new classes of insecticides is low and clearly becoming even slower. In major fruit and vegetable regions, resistance to pesticides and economic and environmental problems arising as a direct result will be in many cases "the straw that breaks the camel's back."

***Livestock Production.*** Trends toward large livestock operations and high levels of livestock concentration in limited geographic areas would reverse. A significant portion of livestock production now concentrated in just a few regions, particularly beef and dairy cattle, would move back into row-cropping regions dominated by crop farms. Such a change is integral to progress toward sustainable agriculture because there must be a profitable use of forages grown in rotations; manure will be needed to supply part of plant nutrient needs; and a more dispersed pattern of livestock production is a necessary step to reduce the often severe regional water quality problems that result when the available local supply of manure exceeds by severalfold the capacity of cropland to use the nutrients in the manure efficiently.

In most regions and cropping systems, the changes noted above will be evolutionary, nudged along by economic forces and the development of effective new technologies. In a few locations for some crops, the magnitude of the changes necessary to make these adjustments and the gaps in viable technological options will make the process of change much more traumatic. In such cases, the pace of change probably will lag until government policy intervenes in one way or another to alter the economics of choice among systems, hasten the emergence of new technological options, or simply forbid by some sort of regulation certain unacceptable farming practices.

It is also important to note that these adjustments must typically proceed hand-in-hand. The practices common to sustainable systems, such as rotations, mechanical weed control, integrated pest management, cover-cropping, and use of manure, will work only if incorporated into a well-designed, integrated farming system. Some progress can be made on a piece-meal basis, such as doing a better job of managing nitrogen or selecting genetically resistant varieties. But the sort of major shifts outlined above will occur only over many years and in the presence of substantially different economic incentives and opportunities.

## Consequences, Costs, and Benefits

The diversity of American agriculture makes it difficult to characterize briefly the consequences of a successful transition to sustainable agriculture. Nonetheless, I will try, assuming that the transition to sustainable agriculture leads to diversified crop-livestock farms producing some combination of row crops, small grains, and forages. I will focus on how these farms might contribute to achieve three major goals: resource conservation, water quality, and enhancing profitability.

***Stewardship of Natural Resources.*** New policies adopted in the Food Security Act of 1985 have brought about tremendous progress in controlling excessive soil erosion. Although the administration relaxed certain key provisions of the law markedly, major progress in reducing erosion will still be made over the next decade, even if crop prices recover and planted acreage expands (assuming no further major policy changes are made).

Sustainable agricultural practices on diversified farms would greatly facilitate further progress in erosion control by fostering rotations, cover crops, and efficient use of organic materials, such as crop residues and manure. These practices will reduce erosion rates by at least a factor of three from rates commonly expected with conventional tillage and continuous cropping. Such practices will prove adequate to reduce erosion rates to or below tolerable levels on at least 90 percent of U.S. cultivated cropland.

Indeed, erosion is the only major natural resource challenge facing American agriculture that is a recurrent threat to sustainability, at least on most of the cultivated cropland base, and that could be largely overcome. Moreover, this accomplishment would require, at current and even somewhat higher levels of production, no major technological breakthroughs, nor would it impose unbearable costs.

The major benefit to farmers from controlling erosion will be that economic input costs will fall as moisture and nutrients are kept on the land. Soil structure will improve, increasing water-holding capacity and, hence, yields because some drought periods are experienced almost every year.

The cost of such practices to farmers will not be great and generally can be spread over a variety of agronomic and ecological benefits, such as nutrient and moisture retention, weed and pest control, improved soil structure, and higher yields. An exception could be more steeply sloping land that will require terracing or other costly structural practices to keep erosion in check, if such lands are cropped on a continuous basis.

The principal benefits to society from erosion control will be cleaner water by virtue of less sedimentation in streams, rivers, lakes, and other water bodies. Wildlife habitat will improve and new recreational opportunities will emerge. Over time, the economic value of these benefits, while difficult to quantify, surely will exceed several billion dollars each year.

***Improving Water Quality.*** By far the most significant benefits in the United States that will evolve as progress is made toward sustainable agriculture will flow from improved water quality. Without near-term changes in the efficiency of nutrient and pesticide use, extremely costly remedial and regulatory actions undoubtedly will emerge in the 1990s as a necessary step to achieve state and federal environmental objectives.

Foregoing the need to incur such costs will be among the most dramatic near-term benefits to farmers and society from a heightened rate of progress in adopting sustainable agricultural production systems. Over the long-run, the opportunity to sustain high levels of agricultural production without any serious degradation of surface water or groundwater will be of enormous significance to the U.S. economy, the farm community, and the world as a whole, which will probably grow more reliant, at least periodically, on U.S. agricultural exports.

The need to reduce nitrate levels in drinking water already has triggered aggressive government actions in several states. More states are sure to follow and perhaps even the U.S. Environmental Protection Agency. It is fortunate for American agriculture that the types of best management practices recommended (perhaps soon to be required) to protect water quality from excessive levels of nutrients often are similar, if not identical, to practices recommended to reduce per unit production costs and reliance on pesticides.

***Economic Performance.*** Despite shrinking surpluses and rising prices triggered by the severe drought of 1988, American agriculture still faces a serious long-term competitive challenge in world commodity markets. Despite the admittedly volatile nature of contemporary, multilateral trade negotiations, the only sustainable way to meet the challenge of the international marketplace is to reduce per unit production and marketing costs faster and more consistently than the country's major competitors.

This outcome can be achieved in four basic ways: increasing yields without raising costs; reducing capital asset values and/or returns to labor and management; reducing costs with little or no loss in yields; or reducing transportation, handling, and marketing costs. The first three of these paths to lower costs have each contributed in the 1980s to the dramatic progress made by many farmers in reducing per unit costs. To make further progress, however, more fundamental changes in policy, production systems, and technology will be essential.

Such changes must be compatible with and, indeed, a cause of widespread adoption of sustainable agriculture systems and production methods. Four of the most significant common paths to lower costs or greater income will be the following:

- Expand production of specific crops in low-cost regions while phasing out production in regions facing the highest costs. Important commodity program changes must occur before this critical adjustment can proceed.
- Reduce unrealistically high yield goals in certain regions.
- Use all cropland for its most profitable and sustainable economic use each year: crops, trees, forages, or wildlife habitat. Idling productive lands should become a thing of the past because it imposes such high costs on farmers, the economy, and the treasury. (This is why many agricultural policy leaders are seeking a more effective, affordable way to stabilize production and farm income.)
- Maintain a growing crop on land for more months each season, thereby more thoroughly and efficiently capturing the solar energy falling on farmland. Widespread use of cover crops and forages in rotations will greatly facilitate achieving this goal. (Continuous corn production in the United States fully uses available sunlight for only about three months each year.)

The benefits to farmers and society of reducing production costs are self-evident. Despite heroic efforts by most farmers to survive the 1980s and reduce costs, the overall economic record of American agriculture has been dismal throughout this decade. The nation has suffered an enormous loss of wealth, and the infrastructure of agriculture has deteriorated and been grossly underutilized. Jobs have been lost, and public and private investments in research, technology, and natural resource stewardship have been postponed or languished less than fully utilized. The human toll brought on by foreclosures, forced sales, suicides, drought, and stress within families and communities will remain a deep scar for generations.

## Society's Stake

Society has an enormous stake in fostering progress toward profitable, environmentally stable farming systems. The two most dramatic, near-term benefits from such progress will involve, first, improved economic performance and job creation, made possible largely by progress in reducing per unit production costs and exploiting new opportunities in both domestic and international markets, and, second, reduced movement of agrichemicals into surface water and groundwater.

Despite growing public concern about agriculture's impact on the environment and increasingly comprehensive and aggressive regulatory policies, incessant economic pressures to cut costs probably will remain the dominant agent of change in most of American agriculture. Regulatory actions may force rapid change in some production systems in isolated cases. Because of the many linkages between environmental and economic performance, such farms or farming practices likely will be economically tenuous at best when the need arises for restrictive regulatory actions.

There is little reason to hope that progress toward sustainable agriculture will occur until such systems are viewed by farmers as more profitable and practical. Relatively few farmers to date have adopted such systems because continuous cropping patterns with conventional management practices are generally familiar, easier to manage, less labor intensive, and sufficiently profitable. Unprecedented economic pressures to change farming systems emerged in the 1980s, but were substantially alleviated by a large increase in government program payments. Looking ahead, fiscal pressures and competing policy objectives may result in markedly lower government expenditures for farm subsidies. As a result, necessity, that mother of invention, may work its will, and sustainable agriculture can come of age.

# SUSTAINABILITY: AN OPPORTUNITY FOR LEADERSHIP

Robert Rodale

Why are we all here at a conference on sustainable agriculture? If you say that we are here because Ohio State University and other sponsors invited us to come, you would be right only superficially. I believe that this meeting is happening now because nature, that creative and controlling force in the universe, has sent us all a message. Therefore, nature, to my way of thinking, is the primary source of the invitations to this meeting.

Of course, nature does not speak to us directly and certainly does not write letters of invitation. But with the use of a little imagination, I can visualize nature saying something like this: "I have been around a long time, and will be here for many millions more years. And I will continue to make resources available for the use of all living creatures and plants. But you human beings are becoming very numerous and are taking more from me all the time without giving back things that I can recycle and purify. So it is time that you start using my resources in a different way. Be more gentle, and don't try to dominate me. Especially, learn to farm your land more sustainably. Get together with people from all over the world at this conference and who read the proceedings and start work on new ideas."

So that is why we are at this conference. We responded to the thought that it is time to learn to work with nature in better ways. But that brings a question to my mind. Why is this just an agricultural meeting? How is it that all of us here are farmers, farm researchers, or have a direct connection in some way to the land? Isn't it true that nature is the creative and controlling force in all of the universe, not just the farm universe.

Farm people work and live close to nature, so we heard and responded to this conference first because our income goes up when the way we do our work fits the requirements that nature gives us. Moreover, our bank accounts usually get smaller when our farming methods fall out of step with the resources nature provides us. In other words, we have a relationship of close accountability with nature. If we plow a hillside the day before a heavy rain, nature allows our soil to wash away. But if we find a way to keep that hillside in grass, nature helps our soil get richer, whether it rains or not.

Protection of our soil from erosion is just the beginning of our close relationship with nature. We can make soil better as well as protect it by understanding and using nature's nutrient cycles. Often, the best way to do that is to mimic nature's own pressures to have a diverse range of plants live in the best agricultural regions. So we, like our ancestors, can move away from crop rotation and techniques like interplanting only at our peril. The alternative is mounting bills for fertilizers and other agrochemicals.

Pest control offers a similar challenge. We can do much to prevent disease and insect attack by using diversity, rotations, and many other old and new techniques common to sustainable agriculture. Those farmers who ignore such possibilities will be forced to suffer the consequences, which include the labor and cost of a more extensive pest control program.

Of course, city people are also accountable to nature, just as everything on Earth is. However, people in cities live at an arm's length relationship to nature. For example, they need water every day from the well, but they don't know where the well is, and they have little knowledge of how deep the water is in it. They get their food from stores, which, incidentally, most farmers do also these days. But at least farmers know that food is not actually made in the store in some kind of magic, industrial way unconnected with nature.

City people draw their resources from many places in nature. For instance, if they get food from one set of farms that are wearing out because of bad farming practices, they start buying food from other farms that still have good soil left. We should not blame city people for doing those things because that is just the way of city life and cities are here to stay. They are, in fact, getting bigger while our farm sizes and population decline. Hence, the people in cities are a potent force in our society, socially, politically, and economically.

However, we need to be aware that city people are just beginning to think and do things to try to make their way of life more sustainable, which is a change from the past. Historically, city people have treated nature as a dumping ground, rather than as a creative and controlling force in the universe, which is the dictionary definition of nature. They took what they needed from farms and from the rural environment, used the materials, then turned them into waste products and threw them into the nearest convenient river, bay, ocean, or hole in the ground.

City people also tend to fill their minds with social, political, or economic theories that have little basis in nature's realities. I maintain that they have never tried to create a meaningful and whole concept of life that is rooted firmly in an understanding of how this world really works. As a result, city people have come up with thoughts that we rural people instinctively believe are crazy.

An example is the concept that manufacturing doesn't matter and that machines operating on their own will soon take care of making the things we need. So the future business of people will be to perform services for one another, such as operating insurance schemes, banks, and running entertainments of various kinds.

My instincts tell me that that kind of thinking is wrong, and I have said that on occasion to city people. But until recently, they paid little attention because, for the most part, they have had faith that their conception of nature will allow them to keep enjoying not only all the blessings of the past but many new ones that they can dream up for the future. They have thought that science fiction is a good predictor of future science fact.

Now, that almost blind faith in the power of people to dominate nature with clever schemes is hitting the wall, to use a phrase from my friends who run marathon races. (And life in fact is a marathon race of the longest imaginable kind.) The most important wall city people are hitting is what I like to call the trash wall. They are running out of holes in the ground to dump the wastes they produce from the materials farmers and factories produce. This is a more serious problem in certain cities than in others, but in New York and Philadelphia, two cities near to where I live, it's about as serious as it can possibly get. Soon, in the New York area, the only hole left for dumping will be the nearby Atlantic Ocean. But if that is used, people will have to stop swimming in it. So they are in fact very close to hitting the proverbial wall.

What does a trash wall look like? In New York it is called Fresh Kills landfill, perhaps that is an apt name. Once it was a marsh, home to blue crabs and other wildlife, but today it is a high mound that soon will be the highest point on the whole eastern seaboard. Here is a description of Fresh Kills today, written by Douglas Martin of the *New York Times* and published in that newspaper on September 10, 1988:

"It is grim almost beyond description. The stench recalls the festering, viscous bottom of a filthy garbage can. Hills where stones and gravel have been spread over generations of refuse are eerily bare. Over newly arrived garbage, tens of thousands of gulls swarm crazily about, shrieking delight over their revolting fare. Here, at Fresh Kills, is where things, all things, end up."

City people are not like those gulls, because they don't see where their trash and garbage end up. However, at least here they are reading about it. And I hope they are getting ready to think about what to do next when places like Fresh Kills come to a point at the top and can no longer be expanded.

What this situation does is to make urban people very receptive to the idea of sustainability and to the urgent need to learn to think and live in more sustainable ways. They are now at least beginning to change their minds about their relationship with natural resources, such as air, water, soil, and the many diverse forms of life itself. They are ready to change, I think, but don't yet know how. They lack vision of a better future, and no person or group has yet come forward to lead them effectively.

## A Great Opportunity for Agriculture

My central message is that we who believe in sustainable agriculture have the capacity to provide that leadership. Those of us who have taken at least a few steps up the learning curve of sustainable agriculture now have the opportunity to think beyond just the farm field and the animal pen. We can now take our knowledge about sustainable ways to produce food and begin talking with urban people as well as our farming and scientific colleagues. We can start toward the making of a shared vision of a sustainable world.

The first step in doing that is to know our history. Agriculture has a long and honored history that is much longer than the history of cities and of the factories of the industrialized world. We who cultivate the land go back 10,000 years or more into prehistory. Cities and the urban culture come later, and it was farmers who made cities possible. Without our food, the concept of a city in people's minds couldn't even form.

The importance of the whole history of agriculture is that important elements of sustainable systems can be found in it from the earliest times. I am not saying that we should go back and reform agriculture in its ancient pattern. Our mission is not to move back to some primitive former way of working. But we do need to look carefully and pluck from history the elements of sustainability that have been demonstrated by centuries of practical use.

The best way to begin doing that is to read the book *Farmers of Forty Centuries* by Franklin H. King. That's how I began my career in this field

39 years ago, and it is still a good starting point today. Certainly, the publication of King's book is the most clear marking point I know of the sustainability movement itself.

Franklin King was a soils professor at the University of Wisconsin and later chief of the Division of Soil Management in the U.S. Department of Agriculture. He was also something of a fighter, who fought much of his professional life for the view that soil and farming systems had a regenerative capacity that would ultimately have to be used to ensure their permanence. His opponents were the advocates of the view that mined and manufactured fertilizers would have to become the primary basis for agricultural productivity.

King lost those early battles, but he left behind his book describing in detail the farm systems that he knew were sustainable. Those of the oriental farmers, the tillers of rice paddy and the savers of all things organic to be returned to the land.

Later, thinkers took King's insights and built from them the organic farming idea and the more recent sustainability movement. In a linguistic sense at the very least, the real pioneer of sustainability was Lady Eve Balfour, a leader in agriculture who is now 90 years old. She first used the phrase sustainable agriculture in the late 1970s. From that act of leadership came the movement of alternative agriculture from being a permanent alternative to an idea and a method destined to have broad appeal and application. These were two of the pioneers; there were many more. However, I don't want this to be a discussion about history as much as a discussion of the future, so I will move to the present.

Where are we now on the sustainability leadership curve? There has been much progress in the last 10 years. We have taken the first steps to create sustainable methods of farming. To me, the most clear and useful of those steps is the understanding that agriculture has two separate resource streams. One is internal and one external. The internal stream of resources never leaves the farm. It is there when a farm is born and when it dies. Of course, the internal resources of farms can become weaker or stronger depending on how they are used and how they are regenerated. But they are permanent and, therefore, sustainable.

External inputs to farms are also important. They are the things we buy to keep production going, such as energy, fertilizers, pesticides, and farming information, to name a few. Those external things are not sustainable because we can't be sure they will always be available at prices we can afford, and perhaps some will eventually not be available at all.

Figure 1 is one way to illustrate the difference between internal resources and external inputs into farming. I have worked with a number of people in various places on the crafting of this diagram. It may not yet be perfect, but at least it shows in a fairly precise way how everything we need to farm can be found internal to the farm system. And it also lets us visualize how these internal building blocks of sustainability are countered by matching external inputs, which are not sustainable (Figure 1).

Viewing agriculture "through" this diagram lets us see that the basic challenge of sustainability is to make better use of our internal resources. In effect, to move the dividing line between the two sets of resources more toward the right, so that the area occupied by internal resources becomes larger and more important. We need to do that by either minimizing the

| | *Internal Soil* | *External Hydroponic Medium* |
|---|---|---|
| Sun | Main source of energy | Energy used as "catalyst" for conversion of fossil energy |
| Water | Mainly rain and small irrigation schemes | Increased use of large dams and centralized water distribution systems |
| Nitrogen | Collected from air and recycled | Primarily from synthetic fertilizer |
| Minerals | Released from soil reserves and recycled | Mined, processed, and imported |
| Weed and pest control | Biological and mechanical | With pesticides |
| Energy | Some generated and collected on farm | Dependence on fossil fuel |
| Seed | Some produced on-farm | All purchased |
| Management decisions | By farmer and community | Some provided by suppliers of inputs |
| Animals | Produced synergistically on farm | Feed lot production at separate locations |
| Cropping system | Rotations and diversity enhance value of all of above components | Monocropping |
| Varieties of plants | Thrive with lower moisture and fertility | Need high input levels to thrive |
| Labor | Most work done by the family living on the farm | Most work done by hired labor |
| Capital | Initial source is family and community; any accumulation of wealth is reinvested locally | Initial source is external indebtedness or equity, and any accumulation flows mainly to outside investments |

**Figure 1. Internal/external chart.**

amount of external inputs we use, by regenerating internal resources more effectively, or preferably by doing both.

The same analytical approach can be used by cities. Urban areas have internal resources and also need external inputs, just as farms need fertilizer, energy, and other purchased resources. What are the internal resources of cities? Of course, they are different from those of a farm. The nitrogen in the air is much less important to a city than it is to a farm, to cite just one example. In cities, the internal resources are primarily the skilled and productive people, organizations of all kinds, buildings, and the infrastructure, such as streets, pipes, bridges, parks, and so forth.

And what about the food inputs of cities? I am focusing on food imports, to keep this discussion as close as possible to farming. And I will ask the question this way, in order to still further link this discussion of cities to the agricultural theme of this meeting:

How can the sources of the food imports of a city be changed to increase the welfare of the farmers living and working in the nearby region? And what effect would that kind of change have on the sustainability of those farms?

Those are big and important questions. They are also questions that a research group at Rodale Press, known as the Cornucopia Project, worked on for a number of years. They produced many useful studies of the agricultural potential of regions near cities. And they created market development techniques for farmers that are now in use in several areas.

When farms near cities are sustainable and regenerative, so are the cities. That was the central theme of the Cornucopia Project, which was a successful effort of leadership. Today, some states and urban regions are using the Cornucopia Project reports for policy development. Much more work of that kind is needed if the full potential of the movement toward sustainability is to be realized.

## Highlighting the Accomplishments of Sustainable Agriculture

The next step for all of us who aspire to a broader role of leadership should be to learn to think more deeply about what sustainability in fact is. We need to know more about sustainability than just ways to reduce input costs and improve soil. We've got to develop the skills of standing back from farming in a conceptual sense, learning to see what sustainability is over the full horizon of the whole world—the urban world as well as the farm field. When you do that, you may see some interesting things. For example, you can see that sustainability is not basically a method; rather, it is a question about permanence.

If that disappoints you, it shouldn't. I have learned over the years that questions are extremely valuable and often more important than answers. If you know the right question to ask, the rest of the research and development effort is simplified tremendously. That is not to say that answers come automatically, but at least with the right question you are heading in the best direction.

I will give you one example of the power of the concept of sustainability as a question. Think for a moment about the debate between the believers in the possibility of permanent growth and the advocates of limits to growth; the followers of Julian Simon on the one hand and Dennis Meadows on the other. They have been debating at scientific meetings for years without any clear resolution of the argument. Some policymakers have been mesmerized by that argument, waiting for a clear answer before moving on to the business of managing the future.

Now, when we ask the question of the sustainability of systems, we can at least suggest to policymakers a way to move around the growth-no-growth stalemate. We can educate them about the value of permanence of internal resources. We can show how resources can be expanded and regenerated, that renewable resources are not fixed at some definite upper limit of strength. We can get them moving toward sustainable solutions, using the best of both internal and external streams of resources.

I realize that that is a somewhat complicated concept, one that deserves more than a few paragraphs in a general and somewhat introductory paper like this. But even that fact illustrates well the great value of sustainability as a question. Once asked, it forces us to look at old problems and even difficult arguments in new ways.

## Regeneration, A Particularly Useful Answer to Sustainability

There are now in use a number of different labels for the general technique of sustainable agriculture. Quite recently, a consensus has developed that the acronym LISA (low-input, sustainable agriculture) is the preferred term of researchers and agricultural policymakers. It will be interesting to see if LISA stands the test of time and becomes a widely used label.

Regenerative agriculture is the term we prefer to use at the Rodale Institute and the Rodale Research Center. In 1981, we made the decision to use the term regenerative, instead of sustainable, to label the question of agriculture we planned to develop. There were a number of reasons why we took that step, but space doesn't permit a detailed description here of that decision process. I will, therefore, focus on only two of the reasons:

- In our opinion, enhanced regeneration of renewable resources is essential to the achievement of a sustainable form of agriculture. Other techniques will also contribute to sustainability, but regeneration is the component of sustainability that we felt was most important. Our research has confirmed that most soils and farms do have considerable underused regenerative capacity.

■ We felt that the concept of regeneration would be relevant to many economic sectors and social concerns. It could be the main answer to the sustainability question of agriculture, but it could also become part of the language of renewal, reconstruction, and permanence of urban people as well. Therefore, we felt that regenerative agriculture could nurture new ideas for general social leadership, as well as solve farm problems.

A further reason that we use the word regeneration is most important to capitalizing on the present opportunity for leadership facing agriculture. There is opening before farm people now a unique window of opportunity to lead the people who live in towns and cities. Urban dwellers soon will know that they must have a more sustainable basis for their lives. But right now they do not even know where to look for it. We have the chance to show them the value of regenerative capacity. We can work with nonfarm people to ask the sustainability question and to explore all possible answers.

We in agriculture have never before had such an enticing opportunity to lead. Remember that *Farmers of Forty Centuries* was published 80 years ago. That gives a big head start at least on the conceptualization of a sustainable environment. Moreover, we now have 10 years of intensive work on sustainable agricultural systems to build on. That gives us the position on the learning curve to be able to start speaking like leaders to others about sustainability.

I realize that this is an ambitious concept to present to people at a conference convened solely to work on challenges facing agriculture. I know we have much work to do in our own field of agriculture and that those problems are probably as serious as any facing the cities. So I can understand if some of you in this audience would prefer to think only about agronomy, or soil science, or plant genetics, or agricultural economics.

"Let people in cities solve their own problems," you might feel like saying, but that would be a big mistake. The prestige of agricultural scientists is quite low today, compared with the respect and rewards society accords to other scientists. I would even say that we in agriculture are almost at the bottom of the heap. At the top are physicists, molecular biologists, and certain medical specialists. They are the ones who get most of the big offices, laboratories, equipment, and large pay checks. Occasionally they are even treated as celebrities.

It was not always that way. Think of Washington and Jefferson, our country's founding fathers who were farmers. Most leaders of society either knew farming well or were naturalists and generalists in the best sense. Those who could lead others toward better production on the land were held in highest esteem.

We cannot and should not try to return to the past, but it is obvious that agriculture is even more important to the world's welfare today, in a total sense, than it was 200 years ago. Agricultural leaders should think about that often and should see in the question of sustainability, and the answer of regeneration, the opportunity to be of much wider service to the world.

Will our sustainable and regenerative agricultural community of today be able to seize this opportunity? There is at least a chance that can happen because I am sure that not everyone working on sustainability will choose to think that broadly. But we at the Rodale Research Center are continuing the effort to make our facility a base where those who aspire to become such leaders can find inspiration.

# COMPONENTS OF SUSTAINABLE AGRICULTURAL SYSTEMS

# SOIL NUTRIENT MANAGEMENT IN THE UNITED STATES

Larry D. King

Nutrient cycling is the key to nutrient management in sustainable agricultural systems. Cycling can be viewed at several levels. On a field level in a natural system nutrients move from soil into plants and are returned to the soil via residue as plants die (Figure 1). Most of the nutrients are conserved in the cycle, but inputs from the atmosphere and losses due to erosion, leaching, denitrification, and ammonia volatilization must be considered. Agricultural systems differ from natural systems because nutrients are removed from the cycle in the harvested product (Figure 2). If the agricultural system is to continue, these nutrients must be replaced. In conventional agricultural systems nutrients are replenished with commercial fertilizer.

The nutrient cycle can be expanded to include an entire farm (Figure 3). On a crop/livestock farm nutrients are removed from the fields and leave the farm either in harvested crops or in animal products. A large fraction of the nutrients consumed by animals do not leave the cycle because they are returned to the soil in manure. Nutrients lost from the system are replenished with fertilizer and purchased feed.

The cycle can be expanded further to include nutrient cycling in a region (Figure 4). Harvested crops, animals, and animal products leave the farm and are processed before being sold to consumers in the city. Most of the nutrients in these products end up in landfills or in surface water rather than being cycled back to the farm. In a few cases by-products from food processing may be returned to farms near the processor. More prevalent is the recycling of nutrients in sewage sludge. However, although many cities apply sludge to agricultural land and some apply sewage effluent to land, only a small fraction of the total cropland in a region is affected. Consequently, nutrients lost from the cycle are replaced mainly with commercial fertilizer.

To make an agricultural system more sustainable, losses from leaching, erosion, denitrification, and ammonia volatilization must be minimized while maximizing nitrogen input via biological nitrogen fixation; utilization of nutrients currently present in the soil; and, where practical, recycling of nutrients from off-farm sources.

## Reducing Losses

***Manure.*** Use of manure for crop production is an ancient practice and one of the most obvious methods of recycling nutrients because most of the nutrients entering an animal via feed are excreted in the manure. Unfortunately, a considerable fraction of these nutrients are lost from the nu-

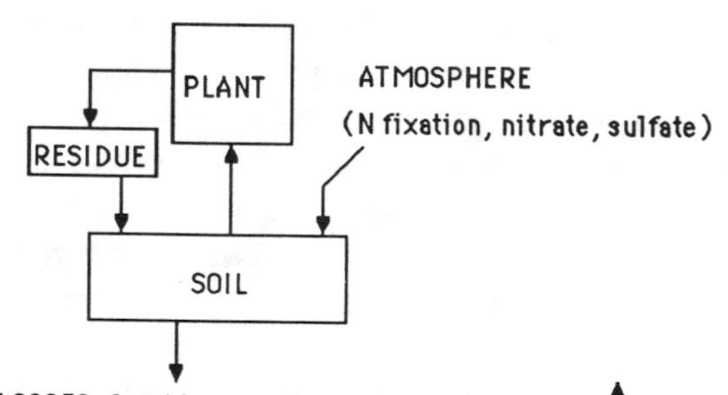


**Figure 1. Nutrient cycling in natural systems.**

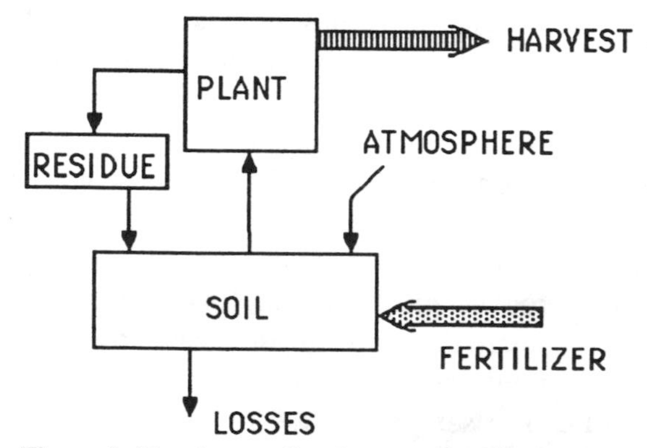


**Figure 2. Nutrient cycling in an agricultural system.**

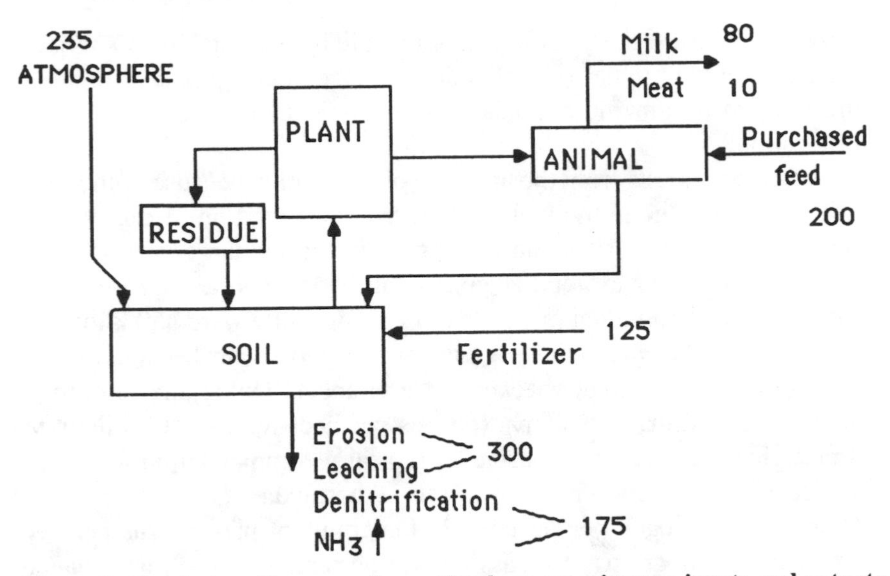


**Figure 3.** Nutrient cycling on a farm. Numbers are nitrogen inputs and outputs (kilograms per cow per year) from a Connecticut dairy farm (Frink, 1969).

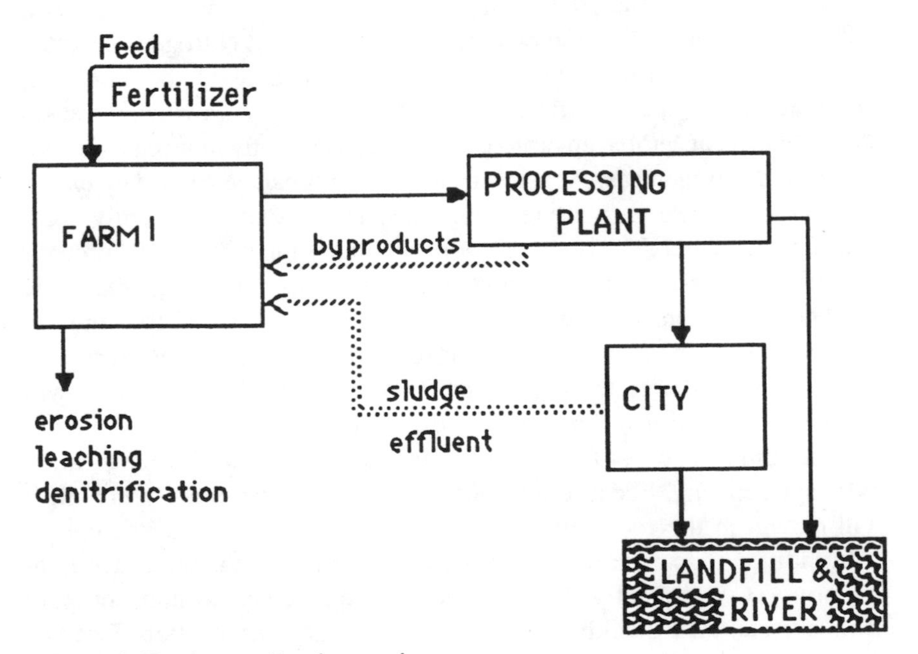


**Figure 4.** Nutrient cycling in a region.

trient cycle. Although direct return of manure by grazing animals relieves the farmer of the problem of collection, storage, and application to crops, direct return by grazing animals is an inefficient mechanism of pasture fertilization.

Inefficiency results from the high rate of application because of the small area covered by the individual excreta, the concentration of excreta near watering and feeding sites and along fences, and the large portion of the pasture not receiving excreta. Petersen et al. (1956) reported nutrient application rates for individual excreta from grazing cattle were 850 kilograms of nitrogen per hectare, 170 kilograms of phosphorus per hectare, and 410 kilograms of potassium per hectare for feces and 450 kilograms of nitrogen per hectare, 7 kilograms of phosphorus per hectare, and 400 kilograms of potassium per hectare for urine. A typical recommendation for annual application of commercial fertilizer to a bermudagrass pasture is 270 kilograms of nitrogen per hectare, 27 kilograms of phosphorus per hectare, and 100 kilograms of potassium per hectare. Because plants cannot use these high rates of excreta-applied nutrients efficiently, much of the nitrogen and potassium is carried below the root zone and lost from the nutrient cycle in the pasture.

Nitrogen loss is particularly high because it is lost via several mechanisms. Much of the nitrogen in feces and urine is in the ammonia form or is quickly converted to ammonia. Because most of the excreta remains on the soil surface, the potential for ammonia volatilization is high. Nitrogen converted to nitrate can be lost from the cycle by being leached below the root zone, or it may be lost by denitrification. Denitrification occurs when nitrate is subjected to an anaerobic environment and is subsequently reduced to nitrous oxides or dinitrogen gas. Anaerobic environments can be caused by excessive soil moisture or, as with fecal deposits, a large source of readily available organic matter. The organic matter stimulates rapid microbial growth, oxygen concentration is reduced via microbial respiration, and anaerobic conditions develop. In feces deposits, nitrogen is converted to nitrate in the aerobic surface layers of the deposit. If this nitrate diffuses into anaerobic zones in the center of the deposit or at the deposit-soil interface, it will be denitrified and lost from the cycle (Figure 5).

Potassium can be lost from the cycle by leaching if soils have low cation exchange capacity. Because phosphorus does not move readily in soil, it will remain in the root zone and, thus, not be lost from the cycle.

In addition to excreta being deposited at high rates, it is deposited nonuniformly. Consequently, much of the pasture receives no nutrient input from excreta. At a stocking rate of 2.5 dairy cattle per hectare, Petersen et al. (1956) estimated that 10 years would be required for 95 percent of

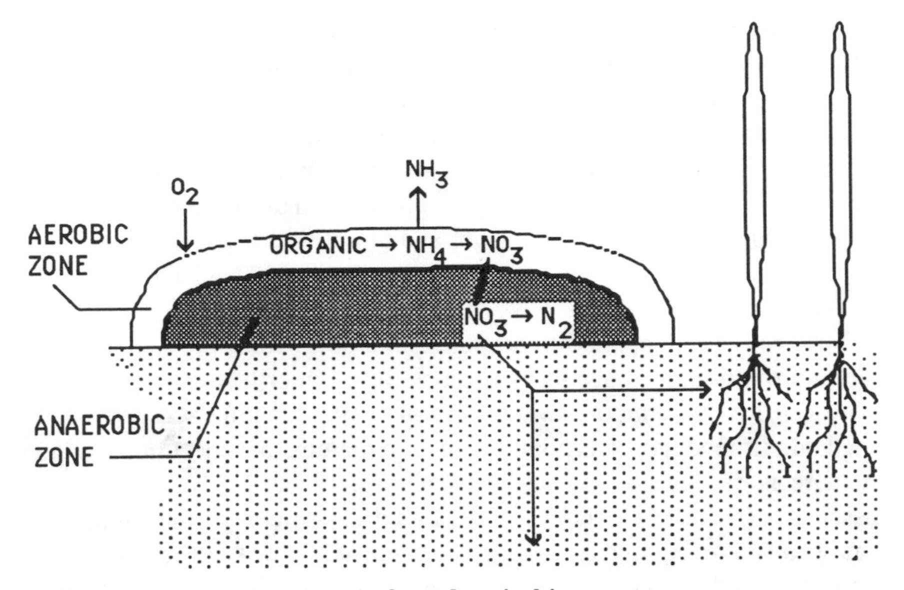


**Figure 5. Pathways of nitrogen in feces deposited in a pasture.**

the pasture to be covered by at least one excretion (feces or urine). The practice of intensive rotational grazing may improve nutrient use efficiency in pastures because excreta distribution is more uniform as a result of the higher stocking density.

When animals are confined, manure can be managed to use the nutrients more efficiently. Management strategy should be to minimize the loss of nutrients during collection, storage, and application and to apply manure uniformly to maximize nutrient use by crops. As in a pasture system, nitrogen is the nutrient that is most readily lost from manure in confinement. Therefore, manure should be collected as quickly as possible and properly stored or applied to fields to reduce ammonia volatilization. Safley and associates (1986) reported a 23 percent loss of nitrogen from dairy manure from the time of defecation in the barnlot until removal from storage (stored as liquid in above-ground tanks or earthen lagoons). Most of the loss occurred during the 24-hour period between barnlot cleanings. Nitrogen loss during storage was negligible. Potassium loss from defecation to removal was 10 percent, probably from loss of urine in the barnlot. Phosphorus loss was essentially zero. Muck and Richards (1983) reported 40 to 60 percent losses of total nitrogen in free-stall dairy barns.

In confined animal systems in which large quantities of water are used for removing and transporting manure, lagoons are often used to store the dilute manure. The use of mechanical aeration in lagoons to reduce odor by keeping the surface layer aerobic results in up to an 80 percent loss of nitrogen by ammonia volatilization and by nitrification-denitrification as a result of the aerobic surface layer and anaerobic lower layer (Barker et al., 1980). Composting is sometimes advocated as a means of stabilizing manure prior to applying it to fields, but a big disadvantage of composting is the loss of nitrogen during the process and the low plant availability of the remaining nitrogen. Castellanos and Pratt (1981) found that availability of nitrogen in composted dairy manure averaged only half of that in fresh manure.

Manure may be applied to cropland as a solid from dry storage, as a slurry from liquid storage tanks/lagoons, or as an effluent from a lagoon. Nitrogen loss by ammonia volatilization is likely from each application method. Immediate incorporation of dry manure or injection of liquid manure is essential to reduce ammonia volatilization. Beauchamp and associates (1982) reported an average loss of 29 percent of the ammonium nitrogen (48 kilograms of nitrogen per hectare) via volatilization over a six-day period from surface-applied liquid dairy manure. Lauer and associates (1976) reported ammonium-nitrogen losses of 55 to 75 percent from surface-applied solid dairy manure (mean loss of 100 kilograms of nitrogen per hectare). Although injection of liquid manure reduces ammonia loss, it does increase the potential for denitrification. The aerobic zone around the injection trench and the anaerobic zone in the trench result in the same nitrification-denitrification sequence outlined for manure in a pasture (Figure 5). Westerman et al. (1983) found a 15 percent loss of nitrogen due to ammonia volatilization during the application of lagoon effluent to cropland via spray irrigation.

After techniques have been developed to minimize nitrogen loss during application, rates should be used to supply adequate but not excessive nutrients to crops. One example of excess application is the practice of applying high rates of manure to fields near the manure storage facility to minimize the time and distance involved in manure application. In a study of three dairy farms in the Piedmont of North Carolina, nitrate-nitrogen in the upper two meters of soil was found to range from 200 to 700 kilograms of nitrogen per hectare (L.D. King, unpublished data). Most of this nitrate was below the root zone and thus unavailable to crops. Efficiency of nutrient use also can be improved by applying manure close to the time crops will need the nutrients; for example, fall application for spring-planted crops usually results in significant nutrient loss from the cycle.

Refeeding animal manure is a "shortcut" in the normal nutrient cycle. Crinkenberger and Goode (1978) reported that broiler litter (feces, bedding, feathers) mixed at a rate of 30 percent by weight with corn and en-

**Table 1. Nutrient losses from runoff and erosion (Albouts, et al., 1978; Kilmer et al., 1974; Miller and Krusekopf, 1932; Olness et al., 1975).**

| *Location* | *Crop* | *Site* | *Annual Loss* Nitrogen | *Annual Loss* Phosphorus |
|---|---|---|---|---|
| | | | — kg ha — | |
| Iowa | Corn | 2-18% slope, contour-cropped | 57 | 2* |
| | | Terraced | 5 | <1* |
| Missouri | Corn-wheat-clover | — | 29 | 9 |
| North Carolina | Grass | 35-40% slope | 3 | <1 |
| Oklahoma | Cotton | <0.5% slope | 6 | 5† |
| | Wheat | <0.5% slope | 4 | 3† |
| | Alfalfa | <0.5% slope | 3 | 2† |
| | Pasture | 3% slope | | |
| | | continuous grazing | 8 | 5† |
| | | rotational grazing | 1 | 1† |

*Total phosphorus.
†$NaHCO_3$-extractable phosphorus.

siled produces a feed that provides adequate crude protein, calcium, and phosphorus for most beef cattle. Stacking poultry litter, allowing it to heat, and then adding 20 percent ground corn results in a good wintering ration for beef cattle.

***Erosion.*** While the mechanisms of nutrient losses from manure are subtle, one can readily observe runoff water and sediment leaving a field during a rainstorm or dust being carried away by high winds and realize that nutrients are being removed by these losses. The magnitude of loss is affected by such factors as cropping system, conservation measures, and slope (Table 1). Thus, much of this potential loss can be reduced through management. Obviously, the magnitude of nutrient loss is directly related to the amount of soil loss, but that magnitude is also a function of the nutrient content of the soil. From a crop production standpoint, the importance of the nutrient loss depends less on the amount of nutrient loss than on how available the nutrients would have been to crops if the nutrients had remained on site. For example, in some erosion studies the total amount of potassium lost via erosion is reported (Olness et al., 1975), but in others an estimate of plant-available potassium, that is, the amount extractable with a soil test extractant, is reported (Alberts et al., 1978). Therefore, in interpreting nutrient losses by erosion the decision must be made as to whether the concern is the immediate effect on nutrient cycling (loss of "available" nutrients) or the long-term effect (loss of total nutrients).

Because erosion studies like those shown in table 1 are very expensive to conduct, it is not feasible to measure erosion and nutrient loss for all possible situations. However, nutrient loss by erosion from any site can be estimated by first using the universal soil loss equation (USDA, 1976) to calculate soil loss and then using either total nutrient content of the soil or some measure of available nutrients to estimate the nutrient loss from erosion. This procedure overestimates loss from most fields because it does not consider redeposition of sediment within a field. Thus, it should be considered a worse-case estimate for most sites.

***Denitrification.*** As mentioned above, additions of organic matter can cause the anaerobic conditions required for denitrification. Examples of nitrogen loss by denitrification resulting from application of organic wastes include loss of more than half of the nitrogen in fermentation wastes injected into soil in field studies (Rice et al., 1988), 20 percent loss of nitrogen with manure in greenhouse studies (Guenzi et al., 1978), 30 to 60 percent loss of nitrogen with industrial waste (King and Vick, Jr., 1978), and 20 percent loss of nitrogen with sewage sludge (King, 1972) in laboratory incubation studies. This effect of organic matter poses a dilemma because use of manures, other organic by-products, and green manure crops is important in agricultural sustainability.

Soil moisture status can sometimes be modified by management, such as subsurface drainage or irrigation scheduling to minimize saturated conditions. But in many situations little can be done to control soil moisture. One management technique is to minimize the period that nitrate is subject to denitrification; that is, nitrogen sources should be applied as closely as possible to the period the crop will need the nitrogen. Thus, fall application of manure for a spring crop should be avoided. To reduce the potential for denitrification during the winter, cover crops can be used to accumulate residual nitrate, store it during the winter, and then release the nitrogen for use by subsequent crops the following spring.

***Leaching.*** The magnitude of nutrient loss via leaching is controlled by such factors as rainfall distribution; type of crop; the specific nutrient and its concentration in the soil; and soil properties, such as permeability, cation exchange capacity, and aluminum, iron, and calcium content. The numerous possible combinations of these factors result in a wide range of leaching losses. The effect of rainfall is most pronounced in the southeastern United States where leaching losses are greatest (Figure 6). The potential for leaching losses decreases as one moves north or west. Nitrogen is the nutrient most susceptible to leaching because the nitrate form is negatively charged and is not retained appreciably in soil. In contrast, phosphorus

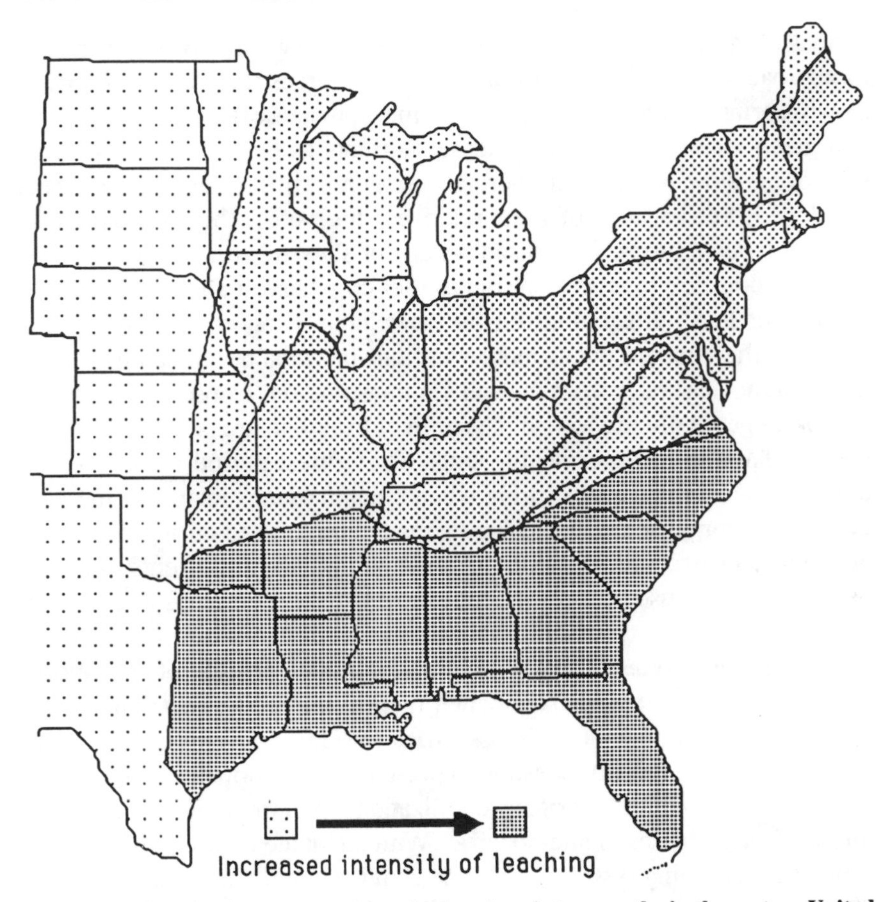


**Figure 6. Relative intensity of leaching during the winter months in the eastern United States (Nelson and Uhland, 1955).**

reacts with aluminum and iron compounds in acid soils and with calcium in alkaline soils to form insoluble compounds; hence, leaching losses are negligible except in very sandy soils. Movement of potasium ($K^+$), calcium ($Ca^{++}$), and magnesium ($Mg^{++}$) cations is influenced by the cation exchange capacity of the soil. Extremes in losses are represented by a loss of two kilograms of potassium per hectare per year from a silt loam soil in Illinois and a loss of 140 kilograms (out of a total of 150 kilograms of total potassium present) from a coarse sandy soil in a greenhouse study in Florida (Tisdale and Nelson, 1966).

An obvious method of reducing leaching losses is to supply nutrients to a crop at a rate equal to the rate of crop uptake. One effort to achieve this balance is to apply commercial fertilizer several times during the growing season rather than applying it all prior to planting the crop. Split applications of fertilizer are common on very permeable soils. Another method to approach this balance is to use nutrient sources that release nutrients slowly during the growing season. The most obvious examples are manures and green manure crops that release nutrients as they decompose in the soil. Some slow-release commercial nitrogen fertilizer sources have been developed, but because of their relatively high cost, they are used only with high-value crops. One disadvantage of slow-release materials is that they may continue to release nutrients after crop harvest and, thus, increase the risk of loss by leaching during the winter (King et al., 1977).

During the summer, plants reduce leaching potential by taking up nutrients and by removing water from the soil via transpiration. Consequently, leaching losses generally are low in the summer except in areas with high summer rainfall and very permeable soils. The critical period for leaching is in the late winter and early spring when precipitation is relatively high and evapotranspiration is low. Leaching losses can be reduced during this period by using winter cover crops to accumulate and store nutrients and later release them for use by subsequent crops as the cover crops decompose.

***Summary of Losses.*** The key to increasing sustainability of agricultural systems is to develop management practices to reduce nutrient losses. Because the magnitude of each loss varies with location, topography, cropping system, etc., it is not possible to present a table showing typical losses from erosion, leaching, nitrogen volatilization, and crop removal. The magnitude of these losses is site-specific. With adequate data on climate, soil properties, cropping systems, etc., one could estimate the potential for loss via the various mechanism mentioned above. For example, Gambrell et al. (1975) estimated nitrogen losses in the Coastal Plain of North Carolina (Table 2). The greatest nitrogen removal from the cycle was in the harvested grain. But leaching losses on the well-drained soil and denitrification losses on the poorly drained soil were appreciable. Erosion losses measured may be exaggerated above what actually would be lost from a field because measurements were made on small plots. On nearly level fields like the experimental sites, significant redeposition of sediment can occur within the field.

## Nutrient Inputs

Once management practices have been adopted to reduce nutrient losses to the extent practical, external sources of nutrients must be used to offset the uncontrollable losses and loss due to crop harvest. However, before external inputs are considered, the present fertility status of the soil should

**Table 2. Nitrogen budget for continuous corn grown in the Coastal Plain of North Carolina (Gambrell et al., 1975).**

| | *Nitrogen Budget by Soil Drainage Class* | |
|---|---|---|
| | *Well Drained (Aquic Paleudult)* | *Poorly Drained (Typic Umbraquult)* |
| | *kg ha*$^{-1}$yr$^{-1}$ | |
| Inputs | | |
| Fertilizer nitrogen | 160 | 196 |
| Losses | | |
| Harvested grain | 92 | 92 |
| Runoff/erosion* | 22 | 29 |
| Leaching | 46 | 16 |
| Denitrification | 0 | 60 |

*Measurement of erosion from small plots probably overestimates losses from nearly level fields because much of the sediment may be redeposited in the field.

be determined because fields that have received fertilizer for a long time often contain phosphorus and potassium concentrations that far exceed the critical concentration required for maximum crop production (Cramer et al., 1985). Two common reasons for these high concentrations are (a) continued use of a fertilizer program developed for a soil that was originally infertile and not adjusting rates down as unused nutrients accumulated and (b) recommended fertilizer rates by some soil testing laboratories that are higher than needed for maximum yield.

Soil test data from 1950 and 1987 in North Carolina were analyzed to determine the extent of excess phosphorus and potassium in the Coastal Plain soils used for corn production (Figure 7). In 1950, 70 percent of the soils sampled were within a range that a response to phosphorus fertilizer would not be expected. Thirty-seven years later, about the same percentage of the soils (68 percent) would not be expected to respond to fertilizer phosphorus. Data for potassium for the two time periods were quite different. In 1950, only 13 percent of the soils were in ranges such that response to potassium fertilizer would not be expected. By 1987, the percentage of potentially nonresponsive soils had increased to 56 percent.

Similar data from Ohio for 1987 showed less reserve fertility than in the North Carolina soils (Figure 8). For a yield goal of 9,400 kilograms of corn grain per hectare, the extractable phosphorus concentration above which no phosphorus fertilizer would be recommended is 95 kilograms per hectare. The corresponding potassium concentration is dependent on cation exchange capacity and is 550 kilograms per hectare for a cation exchange capacity of approximately 15 milliequivilants per 100 grams. On this basis, 30 percent of the soils would not require phosphorus; only 5 percent would not require potassium.

This reserve of nutrients can be used as the sole source of nutrients for crop production for several years. McCollum (1987) reported on a 30-year study of phosphorus on a Portsmouth soil (fine-loamy, siliceous, thermic Typic Umbraquults) in the Coastal Plain of North Carolina. Phosphorus

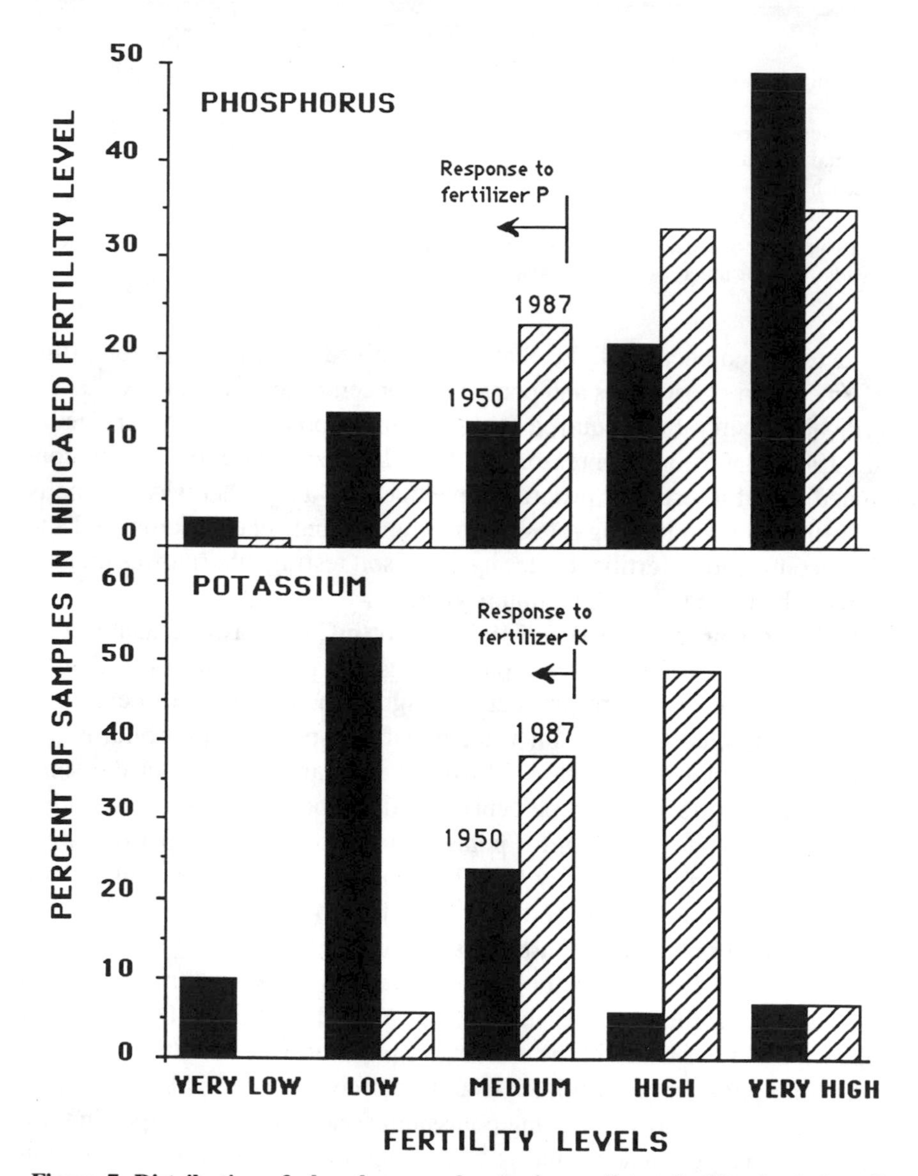


**Figure 7. Distribution of phosphorus and potassium soil test fertility levels in soils from corn fields in the Coastal Plain of North Carolina in 1950, 3,731 samples (Welch and Nelson, 1951), and 1987, 40,400 samples, (Tucker, 1988). Extractant in 1950 was 0.05N HC1; in 1987, Mehlich 3 (Mehlich, 1984).**

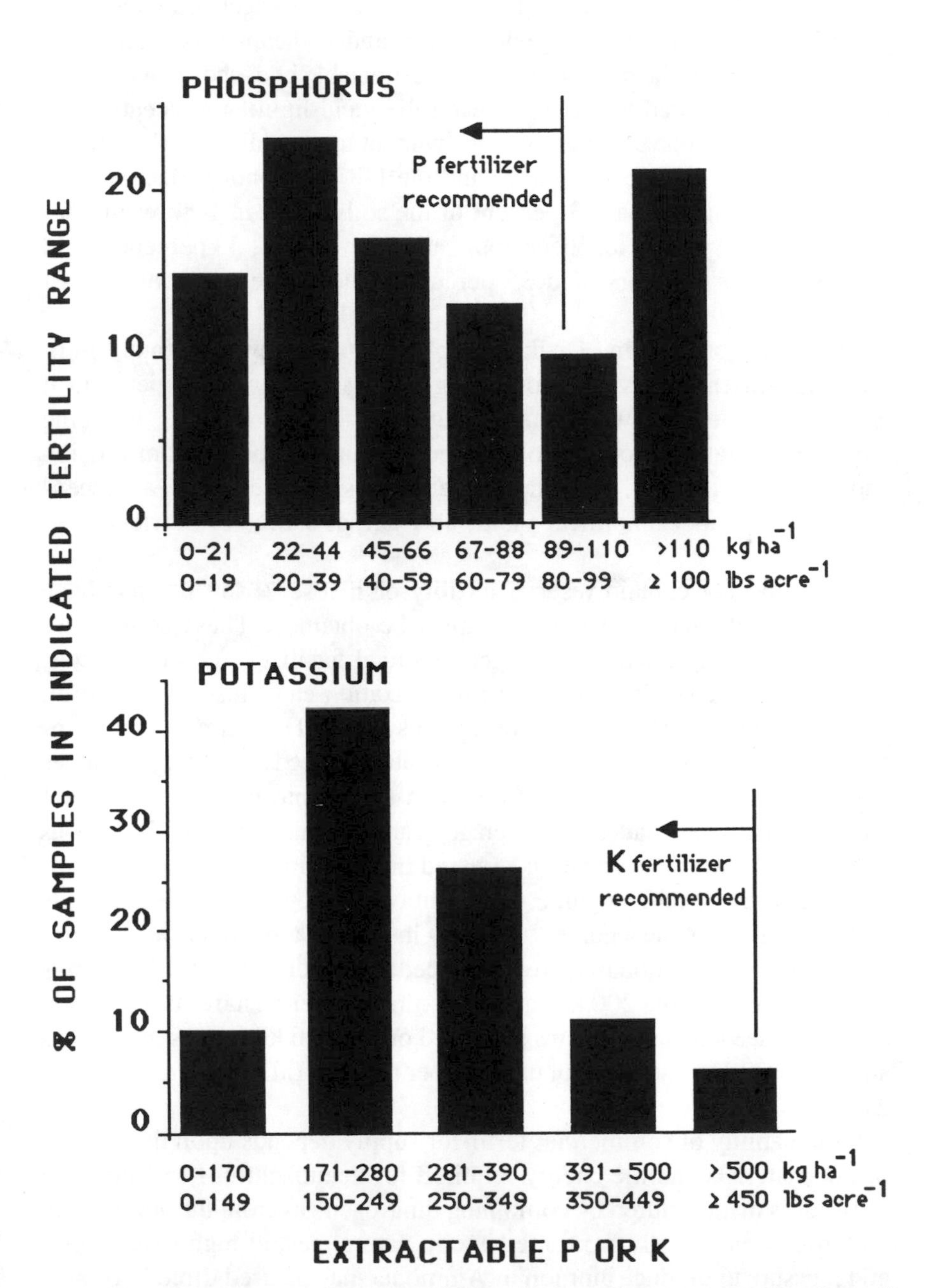


**Figure 8. Distribution of soil test phosphorus and potassium concentrations from all agronomic samples (29,620) analyzed by the Ohio State University Research-Extension Analytical Laboratory, Wooster, Ohio, in 1987 (Watson 1987).**

was added over an eight-year period to achieve very high fertility levels. Then the decline in extractable phosphorus and the effect of extractable-phosphorus concentration on yield of corn and soybeans was determined for 22 years. From the data on decline of extractable phosphorus with time, McCollum calculated the time to reach the yield-limiting concentration, that is, the period one could grow crops without using additional phosphorus fertilizer. Using this decline curve and the 1987 phosphorus data (Figure 6), one can estimate that 23 percent of the soils tested in 1987 would not respond to phosphorus fertilizer application for 3 years, 33 percent would not respond for 10 years, and 35 percent would not respond for at least 12 years.

Suspending or reducing fertilizer application for a few years and "mining" the nutrient reserves is not a long-term, sustainable practice, but it is a viable option that will reduce input costs for several years. Bringing these concentrations down also will reduce nutrient losses from erosion and leaching. However, once the nutrient status has declined to a concentration that limits economical yield, an external source of nutrients will be required.

If soils do not contain reserve fertility or if reserve fertility has been used, external sources of nutrients must be obtained. These sources include biologically fixed nitrogen, commercial fertilizer, recycled wastes, and purchased feed. Biological nitrogen fixation can supply much or all of the nitrogen required in some cropping systems. For example, soybeans and forage legumes can supply all of their nitrogen needs and forage legumes in rotations or a legume cover crops can supply nitrogen to subsequent crops. Table 3 shows data on the average quantity of nitrogen fixed by various legumes. The word "average" is stressed because, in addition to being influenced by the specific legume, the quantity of nitrogen fixed also depends upon the growth of the legume. In a study in North Carolina, crimson clover planted in a conventionally prepared seedbed produced 7,200 kilograms of biomass containing 200 kilograms of nitrogen per hectare. But crimson clover overseeded into soybeans produced only 4,500 kilograms of biomass and contained 130 kilograms of nitrogen per hectare (L.D. King, unpublished data).

Sustainability of commercial fertilizer supply depends upon the reserve of raw materials and the energy required for manufacture. Most nitrogen fertilizer is manufactured by combining dinitrogen gas from the atmosphere with hydrogen, usually from methane in natural gas, at high temperature and pressure to produce ammonia. Ammonia may be used directly or processed to produce other types of nitrogen fertilizer, such as urea, ammonium nitrate, etc. The energy consumed in production of fertilizer nitrogen is

**Table 3. Average fixation of N by legumes (Tisdaleaid Nelean, 1966).**

| *Legume* | *Nitrogen Fixed* *(kg $ha^{-1}$)* |
|---|---|
| Alfalfa | 217 |
| Ladino clover | 200 |
| Sweet clover | 133 |
| Red clover | 128 |
| Kudzu | 120 |
| White clover | 115 |
| Cowpeas | 100 |
| Lespedeza | 95 |
| Vetch | 90 |
| Peas | 72 |
| Soybeans | 65 |
| Winter peas | 56 |
| Beans | 45 |
| Peanuts | 44 |

18,000 kilocalories per kilogram of nitrogen (Pimentel et al., 1973), so the availability of nitrogen fertilizer depends mainly upon energy supply. Sustainability of phosphorus and potassium fertilizer supplies is dependent upon world reserves of rock phosphate and potassium chloride deposits. The life expectancy of these raw materials is influenced by the current rate of use, the projected annual rate of increase in use, and the increased cost of mining as the easily mined reserves are depleted. Known and estimated phosphate rock reserves may last 650 years if use continues at the 1981 annual rate (Council for Agricultural Science and Technology, 1988). However, basing these calculations on the 1981 rate with a projected 3.6 percent annual increase, these reserves would last only 88 years. Similarly, potassium reserves are estimated to last 3,640 years if used at 1974 rates, but only 107 years if the annual growth rate is 5 percent. Thus, the increase (or decrease) in nutrient use in the future will have a dramatic effect upon the life of the reserves. The energy required to produce phosphorus and potassium fertilizer is much less than that required in nitrogen fertilizer production: 3,000 kilocalories per kilogram of phosphorus and 2,300 kilocalories per kilograms of potassium (Pimentel et al., 1973). Thus, their availability and cost are less affected by energy availability and cost than in the case of nitrogen.

Agricultural use of municipal and food processing wastes is a method of returning nutrients to the farm cycle (Figure 4). Table 4 shows median concentrations of nutrients and heavy metals in samples of sludge and other wastes from the southern United States. A large fraction of these wastes are being applied to agricultural land, and that practice is increasing. How ever, the overall impact of these nutrients on U.S. agriculture is not great. The Council for Agricultural Science and Technology (1976) estimated that it would require about 0.5 percent of U.S. cropland to receive all municipal sewage sludge expected to be produced in the U.S. in 1985 at rates to supply 112 kilograms available nitrogen per hectare. Assuming a phosphorus concentration of 1.6 percent and that the sludge applied supplied 25 kilograms of total phosphorus per hectare, then about 4 percent of the U.S. cropland would be required. These estimates show that sludges cannot supply a large fraction of the national need for nutrients. But they are a valuable source of nutrients for cropland within economical hauling distance of wastewater treatment plants.

When interest in sludge use increased in the 1960s, the question of potential heavy metal toxicities to crops, animals, and humans was raised. Concern about heavy metals has decreased since that time for several reasons. Industrial pretreatment ordinances have eliminated much of the metal input from industries. Research conducted during the last 20 years has shown that heavy metal accumulation in plants is minimal if sludges with very high concentrations of metals are avoided, if sludges are applied at rates to supply only the nutrient needs of the crops, if crops that accumulate metals are avoided (e.g., tobacco), and if soil pH is maintained at 6.0 or greater.

One other external source of nutrients is purchased animal feed. The importance of this source would be lowest on a beef cattle farm, approximately 10 percent purchased feed; intermediate on a dairy farm, about 40 percent; and highest on poultry and swine farms, 90-100 percent. In fact, on many poultry and swine farms this high nutrient input causes an en-

**Table 4. Median nutrient and heavy metal concentrations in wastes from the southern United States (King, 1986).**

| | *Concentration by Waste Type* | | | |
|---|---|---|---|---|
| | *Municipal* | *Textile* | *Fermentation* | *Wood Processing* |
| | *% of dry weight* | | | |
| Nitrogen | 2.6 | 2.8 | 3.5 | 0.4 |
| Phosphorus | 1.6 | 0.9 | 0.2 | 0.1 |
| Potassium | 0.2 | 0.2 | 0.1 | 0.1 |
| | *mg kg⁻¹* of dry weight | | | |
| Lead | 335 | 135 | 6 | 36 |
| Zinc | 1,750 | 940 | 40 | 73 |
| Copper | 475 | 416 | 13 | 58 |
| Nickel | 37 | 40 | 18 | 60 |
| Cadium | 11 | 4 | <1 | <1 |
| Cnromium | 380 | 1,820 | 10 | 30 |

vironmental problem because inadequate cropland is available on which to apply the manure.

## Summary

From a plant nutrient standpoint, the key to sustainable agriculture is nutrient cycling. In an agricultural system, one large loss from the cycle is the harvested crop. This loss is minimized if crops are fed to animals and only animal products leave the farm. Management practices can be used to minimize some other losses. Once these losses are minimized, inputs must be obtained to offset the losses. Commercial fertilizer is now the main source of nutrients. However, past inputs in the form of residual fertility may offset the need for new inputs for a short or extended period. Legumes in rotations or as cover crops can supply some or all of the nitrogen required for subsequent crops. Some nutrients leaving in the harvested crop or animal product may be recycled via sewage sludge or food processing wastes.

REFERENCES

Alberts, E. E., G. E. Schuman, and R. E. Burwell. 1978. Seasonal runoff losses of nitrogen and phosphorus from Missouri Valley loess watersheds. Journal of Environmental Quality 7: 203-208.

Barker, J. C., F. J. Humenik, M. R. Overcash, R. Phillips, G. D. Wetherill. 1980. Performance of aerated lagoon—land treatment systems for swine manure and chick hatchery wastes. *In* Livestock wastes: A renewable resource. Proceedings, Fourth International Symposium on Livestock Wastes. American Society of Agricultural Engineers, St. Joseph, Michigan. pp. 127-220.

Beauchamp, E. G., G. E. Kidd, and G. Thurtell. 1982. Ammonia volatilization from liquid dairy cattle manure in the field. Canadian Journal of Soil Science 62: 11-19.

Castellanos, J. Z., and P. F. Pratt. 1981. Mineralization of manure nitrogen—correlation with laboratory indexes. Soil Science Society of American Journal 45: 354-357.

Council for Agricultural Science and Technology. 1976. Application of sewage sludge to cropland: Appraisal of potential hazard of the heavy metals to plants and animals. Report No. 114. Ames, Iowa.

Council for Agricultural Science and Technology. 1988. Long-term viability of U.S. agriculture. Report No. 114. Ames, Iowa.

Cramer, C., G. DeVault, M. Brusko, F. Zahradnik, and L. J. Ayers, editors. 1985. The Farmer's Fertilizer Handbook. Regenerative Agriculture Association, Emmaus, Pennsylvania. 176 pp.

Crickenberger, R. G., and L. Goode. 1978. Guidelines for feeding broiler litter to beef cattle. 5-7-78-2M, AG-61. North Carolina Agricultural Extension Service, Raleigh.

Frink, C. R. 1969. Frontiers of plant science. Connecticut Agricultural Experiment Station, New Haven.

Guenzi, W. D., W. E. Beard, F. S. Watanabe, S. R. Olsen, and L. K. Porter. 1978. Nitrification and denitrification in cattle manure-amended soil. Journal of Environmental Quality 7: 196-202.

Gambrell, R. P., J. W. Gilliam, and S. B. Weed. 1975. Nitrogen losses from soils of the North Carolina Coastal Plain. Journal of Environment Quality 4: 317-323.

Kilmer, V. J., J. W. Gilliam, J. F. Lutz, R. T. Joyce, and C. D. Eklund. 1974. Nutrient losses from fertilized grassed watersheds in western North Carolina. Journal of Environmental Quality 3: 214-219.

King, L. D. 1972. Mineralization and gaseous loss of nitrogen in soil-applied sewage sludge. Journal of Environmental Quality 2: 356-358.

King, L. D. 1986. Introduction to the agricultural use of municipal and industrial sludge. *In* L. D. King [editor] Agricultural Use of Municipal and Industrial Sludges in the Southern United States. Southern Cooperative Series Bulletin 314. North Carolina State University, Raleigh.

King, L. D., A. J. Leyshon, and L. R. Webber. 1977. Application of municipal refuse and liquid sewage sludge to agricultural land: II Lysimeter study. Journal of Environmental Quality 6: 67-71.

King, L. D., and R. L. Vick, Jr. 1978. Mineralization of nitrogen in fermentation residue from citric acid production. Journal of Environmental Quality 3: 315-318.

Lauer, D. A., D. R. Bouldin, and S. D. Klausner. 1976. Ammonia volatilization from dairy manure spread on the soil surface. Journal of Environmental Quality 5: 134-141.

Mehlich, A. 1984. Mehlich 3 soil test extractant: A modification of Mehlich 2 extractant. Communications on Soil Science Plant Analysis 15: 1,409-1,416.

McCollum, R. E. 1987. The build-up and decline in soil phosphorus: Thirty-year trends on Portsmouth soil. Soil Science Society of North Carolina Proceedings 30: 28-64.

Miller, M. F., and H. H. Krusekopf. 1932. The influence of systems of cropping and methods of culture on surface runoff and soil erosion. Research Bulletin 177. Missouri Agriculture Experiment Station, Columbia.

Muck, R. E., and B. K. Richards. 1983. Losses of manurial nitrogen in free-stall barns. Agricultural Wastes 7: 65-79.

Nelson, L. B., and R. E. Uhland. 1955. Factors that influence loss of fall applied fertilizers and their probable importance in different sections of the United States. Soil Science Society of America Proceedings 19: 492-496.

Olness, A., S. J. Smith, E. D. Rhoades, and R. G. Menzel. 1975. Nutrient and sediment discharge from agricultural watersheds in Oklahoma. Journal of Environmental Quality 4: 331-336.

Petersen, R. G., W. W. Woodhouse, Jr., and H. L. Lucas. 1956. The distribution of excreta by freely grazing cattle and its effect on pasture fertility: I. Excreta distribution. Agronomy Journal 48: 440-444.

Pimentel, D., L. E. Hurd, A. C. Bellotti, M. J. Forster, I. N. Oka, O. D. Sholes, and R. J. Whitman. 1973. Food production and the energy crisis. Science 182: 443-449.

Rice, C. W., P. E. Sierzega, J. M. Tiedje, and L. W. Jacobs. 1988. Stimulated denitrification in the microenvironment of a biodegradable organic waste injected into soil. Soil Science Society of America Journal 52: 102-108.

Safley, L. M., Jr., P. W. Westerman, and J. C. Barker. 1986. Fresh dairy manure characteristics and barnlot nutrient losses. Agricultural Wastes 17: 203-215.

Tisdale, S. L., and W. L. Nelson. 1966. Soil fertility and fertilizers. The Macmillan Company, New York, New York.

Tucker, M. R. 1988. Soil test summary from 7/1/86-6/30/87. Agronomic Division, North Carolina Department of Agriculture, Raleigh.

U.S. Department of Agriculture, Soil Conservation Service. 1976. The universal soil loss equation with factor values for North Carolina. Raleigh, North Carolina.

Watson, M. E. 1987. Soil test summary-1987. Ohio Agricultural Research and Development Center, Wooster.

Welch, C. D., and W. L. Nelson. 1951. Fertility status of North Carolina Soils. North Carolina Department of Agriculture, Raleigh.

Westerman, P. W., L. D. King, J. C. Burns, and M. R. Overcash. 1983. Swine manure and lagoon effluent applied to fescue. EPA-600/S2-83-078. Robert S. Kerr Environmental Research Laboratory, U.S. Environmental Protection Agency, Ada, Oklahoma.

# CROP ROTATIONS IN SUSTAINABLE PRODUCTION SYSTEMS

C. A. Francis and M. D. Clegg

Crop rotations and biological diversity long have been cornerstones of successful, traditional agricultural production systems. Rotations received concentrated interest during the first half of this century. Some research continued through the past several decades, yet the introduction of relatively inexpensive nitrogen after World War II provided an economically attractive alternative to farmers and a focus for university research and extension. Many producers today equate soil fertility with rates of applied fertilizer. It is important to explore further the potentials of rotation effects as they contribute to sustainable agricultural production system.

Pest management can be enhanced by rotations of different crop species, one important component of "integrated pest management." Rotation of insecticides and herbicides represents one strategy, even in continuous single-crop culture, to reduce the probability of pest resistance to chemicals. Successful pest management, especially of insects and weeds, can be enhanced by a process termed the "biological structuring" of systems (Francis et al., 1986). This is the conscious choice of crop sequences and management practices to take advantage of biological processes and their interactions with climate and with imposed cultural practices in the production system.

Complicating the current use of rotations are a number of economic and policy factors that override their biological advantages. Many farmers who would prefer rotations and diversity are locked into participating in payment programs that preclude broader farming options because of short-term economic necessity.

## Early Research in Rotations and Crop Diversity

A symposium at the annual meeting of the American Society of Agronomy in Washington, D.C., in November 1924 was titled, "Symposium—The Legume Problem." Specific articles included an economic overview of legumes, their use in several regions of the United States, the benefits of annual legumes and sweetclover, and the potentials for encouraging legume use through extension programs. Clearly, there has been a dilemma for farmers—between the long-term soil-building potential of legumes in a rotation and short-term economic returns from the most profitable cash grain crop.

Another dimension of crop rotations that should be explored is the biological diversity introduced with a sequence of dissimilar crops. Diversity, in a temporal or spatial dimension or both, has long been a part of cropping systems in most parts of the world. Continuous planting of the same crop, whether cotton in the South or maize in the Midwest of the United States, has been cited as the ultimate homogeneity in agriculture. This monoculture pattern leads to biological problems: more insects, plant diseases, weeds, and reduced fertility. It is valuable to explore the biological advantages of rotations over time, as compared to the historical experience with diversity in the field in a given season. We also need to determine whether continuous culture of a crop mixture could provide some of the advantages of crop rotations.

Oakley (1925) traced the use of rotations and legumes before recorded history. He associated the advance of civilization with that of agriculture and coupled the use of legumes to improve soil fertility with that advance. Simkhovitch (1913) credited grasses and legumes for the development of Europe: "The introduction of grass seed and clovers marked the end of the Dark Ages of agriculture." On the North American continent, native legumes were given credit for much of the productivity of prairie soils (Warren, pre-1925).

Legume acreage expanded rapidly during the early years of this century, with campaigns such as "Ten Acres of Legumes on Every Farm" and "Lime and Legumes." Yet even in 1924, Oakley (1925) and others in the Washington symposium questioned the economic impact in the short run of planting too many acres of legumes if there were not a ready market for hay or seed. Advantages of integration of crop/livestock enterprises were readily recognized, as stated by Kenney (1925): "...the agronomy field worker who cannot intelligently interpret the interlocking of his crop program with the livestock needs of his territory is not worth his hire. He merely serves to clog the wheels of progress."

Likewise, benefits of diversity have been known to farmers from the early days of crop domestication. The first cultivators grew mixtures of desirable plant species near their dwellings and along paths leading to water sources, natural food-gathering sites, or hunting areas (Plucknett and Smith, 1986). Perhaps because many of their plants grew spontaneously in mixtures at first, in garbage dumps near the camps where a few seeds were discarded, or perhaps because mixtures more closely resembled the natural ecosystem with which they were familiar, these first farmers grew intercrops with great diversity. Whether they observed the improved fertility or the potential sustainability of mixed cropping systems is not known. We do know that today's subsistence farmers on each continent plant multiple crop systems for a number of biological, economic, and social reasons (Francis, 1986). These complex systems, especially at low levels of technology, are remarkably stable.

The history of rotations and legume use in Nebraska can be cited as an example of what has happened over the past 50 years (Walters, 1987), and a parallel story could be told for each of the other northern states in the U.S. The primary legumes that had been in use for soil improvement in rotation with maize had been alfalfa (*Medicago sativa*) and sweetclover through the 1930s. F. L. Duley et al. (1953) conducted a wide range of field rotation trials and screening projects with specific alternative legumes and tillage systems over 20 years. Results were published in a number of experiment station bulletins (for example, Duley et al., 1953), annual reports, and journals.

Nationally, the use of legumes in rotations declined dramatically after relatively inexpensive nitrogen fertilizers were introduced. Legume seed production, not including alfalfa, dropped from 123 million kilograms (1959) to 22 million kilograms (1979), while total nitrogen fertilizer applied increased from 2.4 million tons to 9.5 million tons during the same period (Power and Doran, 1984; Power, 1989).

Many rotation experiments were discontinued in the 1950s, but A. D. Flowerday, W. W. Sahs, and M. D. Clegg initiated trials in Nebraska to test new crop varieties and hybrids and alternative fertility sources in the 1960s and 1970s (Fahad et al., 1982). Yields and returns (Sahs and Lesoing, 1985) and economic risk (Helmers et al., 1986) have been evaluated in these long-term trials. Several of the rotation trials at the Agricultural Research and Development Center near Mead, Nebraska, are managed by an interdisciplinary team including agronomists, economists, and animal scientists. Similar rotation trials are being carried out at other district stations around the state. This illustrates the current interest and importance given to rotation studies.

## Rotation Effects in Crops

Crop rotations do not have to contain a legume, although most frequently these are the dominant types of rotations. Examples of rotations and their effects follow.

***Legumes.*** There is general agreement that legumes contribute nitrogen to a succeeding crop (Heichel, 1987; Power, 1989). Use of legumes was a broadly recommended procedure for improving soil fertility prior to the manufacture of relatively cheap chemical fertilizers. Two types of legumes can be used to improve soil fertility, primarily through nitrogen contributions: annual seed legumes and perennial forages used as green manure crops.

***Seed Legumes.*** Brodie (1908) reported that soil fertility could be improved using cowpeas (*Vigna sinensis*). Cotton yields doubled and maize yields increased from 0.96 to 2.3 megagrams per hectare following these legumes. Maize yields (Table 1) were increased by 16 to 17 percent when grown after soybeans (*Glycine max*) compared to continuous maize (Robinson, 1966; Higgs et al., 1976; Randall, 1981; Hesterman et al., 1986). Early researchers concluded that rotation effects were due primarily to increased nitrogen availability after the soybean; more detailed research has shown this to be a primary factor. Nitrogen fixation by soybeans may vary from 57 to 94 kilograms per hectare per year (Evans and Barber, 1977).

The nitrogen-equivalent contribution of soybeans, as measured by applying different nitrogen rates to continuous maize and to soybean-maize rotations, is illustrated in table 2. The contribution appears to vary from about

**Table 1. Maize yields (kg ha$^{-1}$) following soybeans compared to continuous cropping of maize with no additional nitrogen fertilizer.**

| *Year(s)* | *Maize Yield* | | *Reference* |
|---|---|---|---|
| | *Following Maize* | *Following Soybeans* | |
| | *(kg ha$^{-1}$)* | | |
| 1962 | 1,483 | 4,089 | (*44*) |
| 1967-1984 (8 years) | 5,259 | 8,412 | (*28*) |
| 1980 | 4,450 | 6,890 | (*41*) |
| 1982-1983 (2 years) | 3,100 | 3,600 | (*26*) |

**Table 2. Maize yields (kg ha$^{-1}$) following soybeans compared to continuous cropping of maize at different nitrogen levels.**

| *Year(s)* | *System* | *Nitrogen Level* | | | | | | *Reference* |
|---|---|---|---|---|---|---|---|---|
| | | *kg ha$^{-1}$* | | | | | | |
| | | *0* | *75* | *150* | *300* | | | |
| 1967-1974 | Continuous | 5,259 | 7,646 | 8,405 | 7,807 | | | (*28*) |
| | After soybeans | 8,412 | 8,835 | 8,979 | 8,504 | | | |
| | | *0* | *55* | *110* | *165* | *180* | *225* | |
| 1980 | Continuous | 4,450 | 5,800 | 6,460 | 6,850 | 7,360 | 7,530 | (*41*) |
| | After soybeans | 6,890 | 8,140 | 8,700 | 8,970 | 9,370 | 9,370 | |
| | | *0* | *55* | *110* | *165* | *220* | | |
| 1982-1983 | Continuous | 3,100 | 4,100 | 6,300 | 6,500 | 6,300 | | (*26*) |
| | After soybeans | 3,600 | 5,600 | 7,600 | 8,600 | 8,800 | | |

**Table 3. Sorghum yields (kg ha$^{-1}$) in continuous culture and following soybeans at different nitrogen levels in Nebraska.**

| | *Sorghum Yields* | | |
|---|---|---|---|
| *Nitrogen Fertilizer* | *Continuous Soybeans* | *Sorghum After Soybeans* | *Reference* |
| | *(2-Year Average)* | | |
| *(kg ha$^{-1}$)* | *(kg ha$^{-1}$)* | | |
| 0 | 2,560 | 4,740 | (22) |
| 56 | 4,840 | 5,710 | |
| 112 | 6,150 | 6,480 | |
| 168 | 6,720 | 6,200 | |
| | *(6-Year Average)* | | |
| 0 | 4,890 | 6,120 | (*46*) |
| Manure (15 ton/ha) | 6,040 | 6,490 | |
| 90 | 5,990 | 6,340 | |

50 to near 150 kilograms of nitrogen per hectare, although these responses are confounded with other effects of the rotation. Maximum maize grain yields following soybeans were not achieved even at the highest nitrogen levels in continuous maize.

The effect of soybeans on a subsequent crop of sorghum (*Sorghum bicolor*) is very similar to the effect on maize. Sorghum yield increases when grown after soybeans have been widely reported (Robinson, 1966; Marty and Hilaire, 1979; Clegg, 1982; Gakale and Clegg, 1987; Roder et al., 1989). The response to nitrogen fertilizer in continuous sorghum compared to sorghum following soybeans is shown in table 3. Nitrogen equivalent from the soybeans was from 30 to 75 kilograms of nitrogen per hectare. These nitrogen contributions from soybeans are widely reported (Power, 1989).

Yields of grain legumes also may increase as a result of rotation with cereals. Soybean yields in Nebraska from 1981 to 1987 were higher following grain sorghum than when following soybeans, and these were significant (.05 level) in five of seven years. Comparing rotation soybeans with continuous soybeans, yields were 2,780 versus 2,570 kilograms per hectare with zero nitrogen, 2,790 versus 2,540 kilograms per hectare with manure, and 2,820 versus 2,620 kilograms per hectare with 45 kilograms of nitrogen per hectare (Roder et al., 1989). Thus, rotations may benefit both the cereal and the legume in a crop rotation.

***Perennial Forages.*** Maize yields after sweetclover are consistently greater than maize yields when fertilized with nitrogen. In a six-year experiment in Indiana, maize after sweetclover produced 5,952 kilograms per hectare, while maize with 94 kilograms of nitrogen per hectare produced 5,506 kilograms per hectare (Anonymous, 1958). Grain sorghum increased yields over four seasons when grown after sweetclover, with the greatest effect in the first year (Adams, 1974). In another study, maize yields were higher following two years of alfalfa than in continuous maize, and this was consistent with zero to 180 kilograms of nitrogen per hectare (Adams et al., 1970). Yields were even higher after three years of alfalfa.

Cotton yields were increased 20 percent following alfalfa compared to continuous cotton (Turner et al., 1972). Baldock and Musgrave (1980) concluded that two years of alfalfa contributed 135 kilograms of nitrogen per hectare to a maize crop and the nitrogen amounts from fertilizer, manure, or rotation were additive. Adams et al. (1970) concluded that the major contribution of vetch (*Vicia* sp.) to maize was nitrogen, since additional nitrogen did not increase yields.

In summary, both seed and green-manure legumes may improve soil fertility and otherwise increase soil productivity for a following crop. Green-manure legumes have the advantage of contributing more organic matter to the nitrogen pool for two to three years.

***Nonlegumes.*** Rotations of nonleguminous crops also may result in increased yields of the succeeding crop. Robinson (1966) reported that sorghum yielded 36 percent more when grown after maize than following sorghum, 3,164 versus 2,335 kilograms per hectare. Maize yielded 45 percent more when grown after sorghum than following maize, 3,229 versus 2,225 kilograms per hectare.

Maize produced higher yields when grown after rye (*Secale cereale*) than in continuous maize (Adams et al., 1970), but 45 kilograms of nitrogen per hectare was needed for maximum yields. This suggests that nitrogen is maintained at a higher level in rotations than in continuous cropping

**Table 4. Maize yields (kg ha$^{-1}$) of five hybrids grown continuously, in rotation with other hybrids, and in rotation with soybeans (27).***

| *Maize Hybrid* | *Hybrid Follows Itself* | *Hybrid Follows Other Hybrid* | *Hybrid Follows Soybeans* |
|---|---|---|---|
| MO17×873 | 7,969 | 8,466 | 9,138 |
| Pioneer 3780 | 7,055 | 7,942 | 9,010 |
| W64A×Oh43 | 7,566 | 7,646 | 8,842 |
| W64A×W117 | 7,700 | 7,451 | 7,633 |
| A632×A619 | 7,902 | 7,633 | 8,775 |

*Average of two locations and two years.

with corn, as observed by Bartholomew et al. (1957) in a maize-oats (*Avena* sp.) rotation. A 12-percent yield increase in cotton occurred when cotton followed maize (Turner et al., 1972).

Rotation of cultivars (hybrids) within a species may result in a yield increase. Hicks and Peterson (1981) showed a significant yield increase in maize, with rotation in three of five hybrids tested (Table 4). They suggested that the effect may be due to favorable allelopathic differences between certain hybrids. Yields of all five hybrids were greater in rotation with soybeans.

***Soil and Other Effects.*** Soil aggregation, bulk density, microbial biomass, and water infiltration and extraction are influenced by rotations. Soil aggregates were reduced by the soybean crop (Spurgeon and Grisson, 1965; Fahad et al., 1982). Bulk density was the highest in continuous soybeans and continuous sorghum at high nitrogen levels (Santos and Clegg, 1986). Soil aggregation and bulk density probably account for the reduced water infiltration associated with continuous soybeans (Fahad et al., 1982) and added nitrogen to sorghum (Turner et al., 1972); there is increased water infiltration (Adams et al., 1970) and reduced compaction (Bolton et al., 1982) in rotated crops. Increases in soybean yields in rotations may be due to system-related water differences early in the season (Roder et al., 1989).

Organic matter was higher in rotation plots when sorghum was included in the cropping system (Mannan, 1962). Wilson and Wilson (1928) reported that sorghum root residues were easier than maize root residues to oxidize by soil microorganisms. This process would increase the soil microbial population and tend to deplete the soil of available nitrogen. Roder et al. (1988) reported higher microbial biomass in rotations of soybeans (plus 9 percent) and sorghum (plus 10 percent) than in their respective continuous monocultures. Shorter growth, later maturity, and yellower foliage of plants grown after sorghum may mean nitrogen is less available (Robinson, 1966).

Increased nitrogen in the soil following seed legumes depends upon legume yields and other factors. For example, nitrogen fixed by the soybean is about 61 percent of the nitrogen requirement for an average soybean crop in Nebraska, creating a 39 percent net loss after soybeans. Power et al. (1986) reported that nitrogen of soybean residues is mineralized the first year and is available for crop use. Gakale and Clegg (1987) reported that the soil nitrogen in the profile was about 60 kilograms per hectare after soybeans, enough to support a sorghum crop. In the same study, Lohry et al. (1987) concluded that the higher amount of nitrogen removed by sorghum in the rotation was due to increased availability from the previous soybean crop. Thus, the nitrogen cycling from the legume to the cereal is short in duration.

Soil erosion can be a problem with crop rotations. Soil loss depends upon the amount of residue left on the field surface, previous crop, and tillage method, as well as slope, rainfall, and soil type. The average loss of soil was 40 percent greater following soybeans than following maize (Moldenhauer and Wischmeier, 1969). Over an 18-year period, the soil loss was 45 percent greater for a maize-soybean rotation than for continuous corn (Van Doren et al., 1984). In contrast, a sorghum-sweet clover rotation resulted in less soil loss than with continuous sorghum (Adams, 1974). Some of this soil loss can be prevented by contour stripcropping and establishing the rotation within a field rather than between fields.

## Pest Management in Crop Rotations

Conventional wisdom among farmers is that crop rotations reduce the incidence of insects, plant pathogens, nematodes, and weeds. Research has shown that this is accomplished by breaking the reproductive cycles of these species because different pests generally are found on or with different crops. The case of weeds is somewhat more complicated because some weeds may occur with different crops in the sequence, and the seed of some species may last for many years in the soil. In addition to the variation in natural occurrence of pests with dissimilar crops, there are different pesticides available to control these problems. Pesticide alternation helps to avoid buildup of pests that are tolerant to a specific chemical product and reduces the probability of pests developing resistance to particular chemical formulations.

The greater the differences between crops in a rotation sequence, the better cultural control of pests can be expected. Rotation of summer annuals with winter annuals, perennial crops with annual crops, legumes with cereals, long-season with short-season crops, and in the tropics wet-season with dry-season crops are examples. This is a central part of integrated pest management, along with genetic resistance in crops and limited applications of pesticides where needed. These principles lead to practical application of rotation to reduce pest incidence and severity and contribute to a sustainable agriculture.

***Insect Management in Rotations.*** Crop rotation is one of the most effective cultural practices to provide control against insects that have specific host ranges and relatively short migration distances (Ware, 1980). Maize rootworm (*Diabrotica* sp.) in the north central United States can be controlled readily in most fields by rotation of maize with soybeans. Corn rootworm at economically important levels generally occurs in less than one percent of maize fields that were in soybeans the previous year and in less than one-third of fields in maize the previous year (Baxendale, 1985). A more economically desirable and environmentally sustainable management strategy is to scout for these insects each year and use charts to determine the probability of economically important infestations for the next year.

In Colombia, maize fall armyworm (*Spodoptera frugiperda*) and bean leaf beetle (*Diabrotica* sp.) were fewer in numbers in intercrop systems of the two crops, compared to their monoculture counterparts in the same trials (Altieri et al., 1977).

These are two examples of the importance of rotations and diversity in controlling insects. More research and information are needed on life cycles, cultural control alternatives, and economic levels of infestation on the principal insect pests of the most important crops.

***Plant Disease Management in Rotations.*** Occurrence of economically significant levels of plant pathogens can be influenced by cropping sequence. Brown leaf spot (*Pleiochaeta setosa*) reached 63 percent infection levels in continuous lupin (*Lupinus* sp.) cultivation in Australia, as compared with only 18 percent in a lupin-wheat rotation (Reeves et al., 1984). Likewise, the incidence of fungus diseases in wheat was negligible in rotation, while infection levels caused up to 36 percent deadheads in the third year of continuous wheat. Root rot (several species) in field peas (*Pisum sativum*) in Wisconsin was more likely when this legume was included two or more times in an eight-year period, compared to fields with a history of other crops (Temp and Hagedorn, 1967). Thus, the type of pathogen and soil, the length of time between crops of a given susceptible crop species, and the availability of other forms of control, such as genetic resistance or tolerance, influence the value of rotations for control of plant pathogens. Cropping systems that incorporate rotation of appropriate species to control plant pathogens generally are more environmentally benign than alternatives that use chemical biocides for control.

***Nematode Management in Rotations.*** Literature on the effects of crop species and rotations on nematode populations is providing the technical basis for practical management decisions on nematode control. Okra (*Hibiscus esculentus* L.) or cucumber (*Cucumus* sp.) monoculture increased larval populations of one nematode, while chili pepper (*Capsicum annuum* L.) practically obliterated the nematodes according to one report (Birat, 1969). Different population levels of four *Pratylenchus* sp. occurred in fields cropped to maize, soybeans, oats (*Avena sativa*), wheat, and forages, indicating that crop rotations could influence the potential severity of damage from any one nematode species (Ferris, 1967). Thus, diversity and crop rotation influence nematode populations in the soil. A management strategy to minimize economic damage must consider alternatives to applying chemical nematocides.

***Weed Species Management in Rotations.*** Weed populations are especially sensitive to changes in crop species and herbicides used from one season to the next. As outlined above, the rotation of summer crops with winter crops is useful because it provides an opportunity to control both summer weeds and winter weeds. Rotation of a perennial with an annual crop also gives some cultural control of weeds not adapted to both systems. Combining cultural management with inexpensive chemical control methods can be especially effective (Ennis et al., 1962). Weed seed populations also can be influenced by which crop is in the field in a given season (Dotzenko et al., 1969), and this will affect the cost and effectiveness of weed control strategies in the following season. Tillage for weed control gives some indication of higher yields of soybeans, compared to the herbicide alternative under conditions in the eastern Great Plains (Burnside and Moomaw, 1984). Rotations of crops away from maize and grain sorghum have been effective in controlling shattercane (*Sorghum bicolor*), a difficult weed to control in cereals in Nebraska. Although rotation and tillage can provide adequate weed control in some seasons, there are other modifications in the system that may lead to even better cultural control; these modifications include ridge tillage, multiple years in hay crops, or allelopathic crop species (Rice, 1974).

These are some of the potentials for cultural and biological control of major pest species in crops. Many practices can substitute for chemical pesticides, or be used in conjunction with reduced levels or less frequent applications of chemicals. The potential results include a more rational use of management for sustainable production systems that have less effect on groundwater and on favorable insect species in the environment.

## Biological Structuring of Systems

Strategies using rotations have been used for incorporating diversity into cropping systems, for providing crop nutrients, and for managing pests in the field. The actual mechanisms that function in the plant and animal interactions on a farm could be called the biological structuring of a system (Harwood, 1985). Harwood described the need for efficient transfer of energy and growth factors among crops and niches within a system in order to maintain sustained yields. If this type of efficiency is not achieved, high levels of productivity can be sustained only with high and continuous applications of inputs based on fossil fuels, i.e., fertilizers and pesticides. Examples of these high-input systems are temperate region monocultures of maize, wheat, and rice.

Cropping systems that can be sustained with a greater dependence upon internal, renewable resources (Francis and King, 1988) are based on a deeper understanding of the biological and natural environment and on the complex interactions among components in a cropping sequence. Efficient biological structuring depends upon these interactions and interdependencies among crops and other biotic factors. Many of the most intimate interactions occur among crops present in a field at the same time or those that overlap or follow each other (Figure 1). These complex interactions could be called the "progressive biological sequencing" in a field (Francis et al., 1986), the sum total of the linear and cyclical changes that occur in the field environment as a result of cropping activities and the soil modifications that occur as a result of the crops and their management.

Because any single field or enterprise does not operate in isolation from other activities—crop or animal—on the farm, it is important to conceptualize how these primary interactions function across fields or pastures.

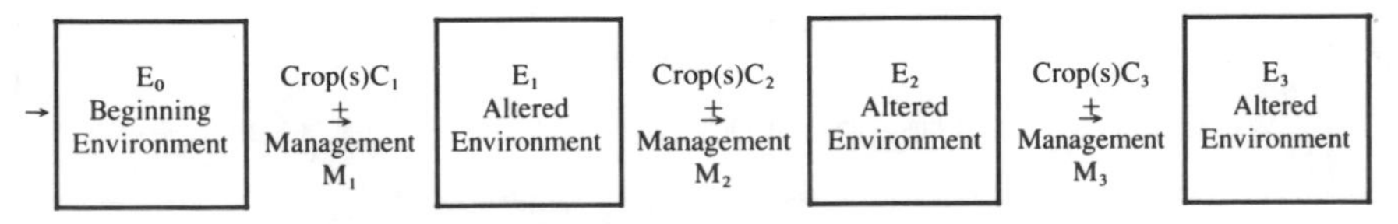


**Figure 1. Conceptual pattern of dynamic cyclical and linear changes in one field crop environment as a result of successive crops and management decisions. (Francis, C. A., R. R. Harwood, and J. F. Parr, 1986).**

Examples are the harvest of maize silage from one field, then fed to beef cattle, and the manure spread on another field. The management decisions about input distribution among fields, which cultural operations to accomplish each day, and the long-term planning for the entire farm could be called the "integrative farm structuring" of that farm shown in figure 2 (Francis et al., 1986). Proper structuring can lead to rational distribution of resources, including labor among enterprises, and a sustainable food supply and income for the farm family, as well as an environmentally sound set of practices that can help to build rather than destroy soil productivity.

## Economic and Policy Aspects of Crop Rotations

Conflicts between long-term soil fertility and system sustainability and short-term economic viability are not new. Oakley (1925), in a 1924 symposium on legumes and rotations, described the value of legumes in crop rotations and said, "This enthusiasm...has frequently resulted in campaigns for increased legume production...(the) farmer who is endeavoring to wrest a living from a stubborn soil is by force of circumstances frequently compelled to follow expedient courses without much regard for the needs of posterity. He must meet his interest payments and other pressing obligations, therefore, he cannot indulge in all the features of farming that promise ultimately to be profitable to him."

Today's management environment is further complicated by the greater

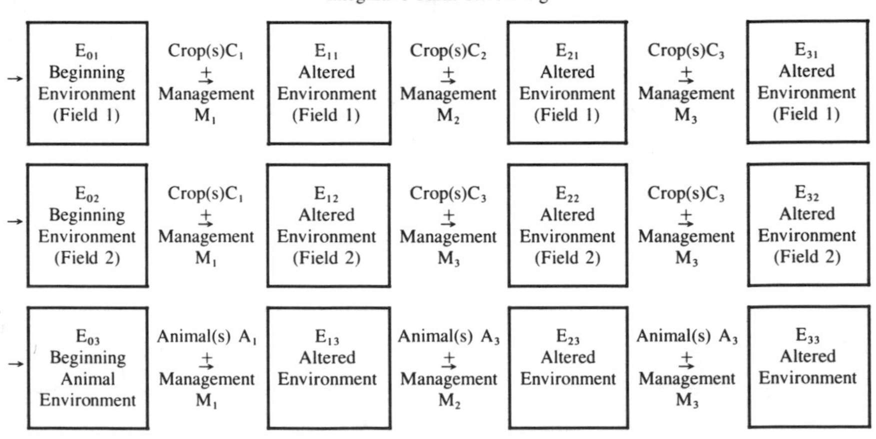


**Figure 2. Conceptual pattern of intersections and integrations of primary crop and animal enterprises on a resouce-efficient farm (Francis, C. A., R. R. Harwood, and J. F. Parr, 1986).**

specialization of many cash grain farmers, the lack of immediate markets for hay and animals to put on pasture, and the broad participation by farmers in government price support and supplemental payment programs for feed grains. These programs require the individual producer to maintain a certain feed grain acreage based on historical production, and this participation is viewed as the only route to survival in the current volatile market and cost/price situation. In fact, current land values and tenant arrangements frequently are tied to the base acreage, often overriding any long-term fertility or environmental concerns that the producer may have for a given field or farm. These policy and program constraints restrict the options available to the farmer who is serious about improving the productivity of a given field's soil resource or considering a rotation that reduces the feed grain acreage.

## Future Research on Rotations for Sustainable Agriculture

A number of key issues remain on the research agenda for rotations in the temperate zone. First is an in-depth study of rotation effects. There is a confirmed contribution of legumes to soil fertility and a proven effect of this fertility on subsequent cereal crops in the sequence. This legume contribution is often in excess of 50 kilograms per hectare, depending upon legume species, total biomass and harvested portion of the legume, and the nature of the crop sequence. Many other factors operating in the rotation of crops are poorly understood. Examples include how maize-sorghum rotations affect yields in the sequence and the specific biological mechanisms in allelopathy and soil microflora interactions. This information will help the researcher design crop sequences to take better advantage of rotation effects. To date there has been more success in identifying and quantifying these effects in a number of rotations than in determining the precise causes of the effects. This is a research priority for the future, and one which will make rotations more useful and applicable in more situations for a sustainable agriculture.

### REFERENCES

Adams, J. E. 1974. Residual effects of crop rotations on water intake, soil loss, and sorghum yield. Agronomy Journal 66: 299-304.

Adams, W. E., H. D. Morris, and R. N. Dawson. 1970. Effect of cropping systems and nitrogen levels on corn (*Zea mays*) yields in the southern Piedmont Region. Agronomy Journal 62: 655-659.

Altieri, M. A., C. A. Francis, J. D. Doll, and A. van Schoonhaven. 1977. A review of insect prevalence in maize (*Zea mays* L.) and bean (*Phaseolus vulgaris* L.) polycultural systems. Field Crops Research 1: 33-49.

Anonymous. 1958. Higher crop yields from improved soils. Purdue University Agricultural Experiment Station Mimeo AY 11c. p. 4.

Baldock, J. O., and R. B. Musgrave. 1980. Manure and mineral fertilizer effects in continuous and rotational crop sequences in central New York. Agronomy Journal 72: 511-518.

Bartholomew, W. V., W. D. Shrader, and A. J. Englehorn. 1957. Nitrogen changes attending various crop rotations on Clarion-Webster soils in Iowa. Agronomy Journal 49: 415-418.

Baxendale, F. P. 1985. First year corn: Does it pay to treat? Crop Focus '85. pp. 200-203. *In* Sustainable agriculture...wise and profitable use of our resources in Nebraska. Department of Agronomy, Cooperative Extension Service, University of Nebraska, Lincoln.

Birat, R.B.S. 1969. Effect of momoculture on the population of Meloidogyne javanica. Nematologica 15: 154-155.

Bolton, E. F., V. A. Dirks, and M. M. McDonnell. 1982. The effect of drainage, rotation and fertilizer on corn yield, plant height, leaf nutrient composition and physical properties of Brookston Clay soil in southwestern Ontario. Canada Journal of Soil Science 62: 297-309.

Brodie, D. A. 1908. Building up a run-down cotton plantation. Farmers Bulletin 326. U.S. Department of Agriculture, Washington, D.C. p. 22.

Burnside, O. C., and R. S. Moomaw. 1984. Influence of weed control treatments on soybean cultivars in an oat-soybean rotation. Agronomy Journal 76: 887-890.

Clegg, M. D. 1982. Effect of soybean on yield and nitrogen response of subsequent sorghum crops in eastern Nebraska. Field Crops Research 5: 233-239.

Dotzenko, A. D., M. Ozkan, and K. R. Storer. 1969. Influence of crop sequence, nitrogen fertilizer and herbicides on weed seed populations in sugar beet fields. Agronomy Journal 61: 34-37.

Duley, F. L., J. C. Russel, T. H. Goodding, and R. L. Fox. 1953. Soil conservation and management on sandy farm land in northeast Nebraska. Nebraska Agricultural Experiment Station Bulletin 420.

Ennis, W. B., W. C. Shaw, L. L. Danielson, D. L. Klingman, and F. L. Timmons. 1962. Impact of chemical weed control on farm management practices. Advances in Agronomy 15: 161-210.

Evans, H. J., and L. E. Barber. 1977. Biological nitrogen fixation for food and fiber production. Science 197: 332-339.

Fahad, A. A., L. N. Mielke, A. D. Flowerday, and D. Swartzendruber. 1982. Soil physical properties as affected by soybean and other cropping sequences. Soil Science Society of America Journal 46: 377-381.

Ferris, V. R. 1967. Population dynamics of nematodes in fields planted to soybeans and crops grown in rotation with soybeans. I. The genus Pratylenchus (Nematoda:Tylenchida). Journal of Economic Entomology 60: 405-410.

Francis, C. A. (Editor). 1986. Multiple cropping systems. Macmillan Publishing Co., New York. 383 pp.

Francis, C. A., and J. W. King. 1988. Cropping systems based on farm-derived, renewable resources. Agricultural Systems 27: 67-75.

Francis, C. A., R. R. Harwood, and J. F. Parr. 1986. The potential for regenerative agriculture in the developing world. American Journal of Alternative Agriculture 1: 65-74.

Gakale, L. P., and M. D. Clegg. 1987. Nitrogen from soybean for dryland sorghum. Agronomy Journal 779: 1,057-1,061.

Harwood, R. R. 1985. The integration efficiencies of cropping systems. *In* T. C. Edens, C. Fridgen, and S. L. Battenfield [editors] Sustainable agriculture and integrated farming systems. Michigan State University Press, East Lansing, Michigan.

Heichel, G. H. 1987. Legumes as a source of nitrogen in conservation tillage systems. pp. 29-34. *In* J. F. Power [editor] The role of legumes in conservation tillage systems. Soil Conservation Society of America, Ankeny, Iowa. pp. 29-34.

Helmers, G. A., M. R. Langemeier, and J. Atwood. 1986. An economic analysis of alternative cropping systems for east-central Nebraska. American Journal of Alternative Agriculture 1(4): 153-158.

Hesterman, O. B., C. C. Shaeffer, D. K. Barnes, W. E. Lueschen, and J. H. Ford. 1986. Alfalfa dry matter and N production and fertilizer N response in legume-corn rotations. Agronomy Journal 78: 19-23.

Hicks, D. R., and R. H. Peterson. 1981. Effect of corn variety and soybean rotation on corn yield. pp. 84-93. *In* Proceedings, 36th Annual Corn and Sorghum Research Conference, Washington, D.C.

Higgs, R. L., W. H. Paulson, J. W. Pendleton, A. E. Peterson, J. A. Jackobs, and W. D. Shrader. 1976. Crop rotations and nitrogen: Crop sequence comparisons on soils of the driftless area of southwestern Wisconsin, 1967-1974. Research Division of College of Agriculture and Life Science, University of Wisconsin-Madison Research Bulletin R2761.

Kenney, R. 1925. The utilization of legumes in the rotation. B. In the Middle West. Journal of American Society of Agronomy 17: 389-394.

Lohry, R. D., M. D. Clegg, G. E. Varvel, and J. S. Schepers. 1987. Fertilizer use efficiency in a sorghum-soybean rotation. *In* J. F. Power [editor] The role of legumes in conservation tillage systems. Proceedings, National Conference, University of Georgia, Athens, April 27-29, 1987. Soil Conservation Society of America, Ankeny, Iowa.

Mannan, M. A. 1962. Organic matter, nitrogen, and carbon:nitrogen ratio of soils as affected by crops and cropping systems. Soil Science 93: 83-86.

Marty, J. R., and A. Hilaire. 1979. Etudes preliminaires des effets del'insertion du soju dans les rotations cerealieres. Ann. Agronomy 30(2): 191-211.

Moldenhauer, W. C., and W. H. Wischmeier. 1969. Soybeans in corn-soybean rotation permit erosion but put the blame on corn. Conservation Crop Soils 21-20.

Oakley, R. A. 1925. The economics of increased legume production. Journal of American Society of Agronomy 17: 373-389.

Plucknett, D. L., and N.J.H. Smith. 1986. Historical perspectives on multiple cropping. pp. 20-39. *In* C. A. Francis [editor] Multiple cropping systems. Macmillan Publishing Co., New York.

Power, J. F. (editor). 1987. The role of legumes in conservation tillage systems. Proceedings, National Conference, University of Georgia, Athens. Soil Conservation Society of America, Ankeny, Iowa.

Power, J. F. 1989. Legumes and crop rotations. *In* C. A. Francis, C. Flora, and L. King [editors] Sustainable agriculture in temperate zones. J. Wiley & Sons, New York (in press).

Power, J. F., and J. W. Doran. 1984. Nitrogen use in organic farming. pp. 585-598. *In* R. D. Hauch [editor] Nitrogen in crop production. American Society of Agronomy, Madison, Wisconsin. pp. 585-598.

Power, J. F., J. W. Doran, and W. W. Wilhelm. 1986. Uptake of nitrogen from soil, fertilizer, and crop residues by no-till corn and soybeans. Soil Science Society of America Journal 50: 137-142.

Randall, G. W. 1981. Rotation nitrogen study, Waseu, 1980. Soil Series 109. Agricultural Experiment Station Miscellaneous Publication 2 (revised). Department of Soil Science, University of Minnesota, St. Paul. pp. 115-116.

Reeves, T. G., A. Ellington, and H. D. Brooke. 1984. Effects of lupin-wheat rotations on soil fertility, crop disease and crop yields. Australian Journal of Experimental Agricultural Animal Husbandry 24: 595-600.

Rice, E. L. 1974. Allelopathy. Academic Press, New York.

Robinson, R. G. 1966. Sunflower-soybean and grain sorghum-soybean rotations versus monoculture. Agronomy Journal 58: 475-477.

Roder, W., S. C. Mason, M. D. Clegg, and K. R. Kniep. 1988. Plant and microbial responses to sorghum-soybean cropping systems and fertility management. Soil Science Society of America Journal 52 (in press).

Roder, W., S. C. Mason, M. D. Clegg, and K. R. Kniep. 1989. Yield-soil water relationships in sorghum-soybean cropping systems with different fertilizer regimes. Agronomy Journal 81 (in press).

Sahs, W. W., and G. Lesoing. 1985. Crop rotations and manure versus agricultural chemicals in dryland grain production. Journal of Soil and Water Conservation 40: 511-516.

Santos, J.R.A., and M. D. Clegg. 1986. Evaluating the long-term effects of sorghum legume rotation. Philippine Council for Agriculture Resources Research and Development Proceedings. Sorghum Research and Development in the Philippines. Los Banos, Laguna, Philippines, PCARRD. Book Series No. 25:58-65.

Simkhovitch, V. G. 1913. Hay and history. Political Science Quarterly 28: 385-403.

Spurgeon, W. I., and P. H. Grissom. 1965. Influence of cropping systems on soil properties and crop production. Mississippi State University Agricultural Experiment Station Bulletin 710. p. 20.

Temp, M. V., and D. J. Hagedorn. 1967. Influence of cropping practice on Aphanomyces root rot potential of Wisconsin pea fields. Phytopathology 57: 667-670.

Turner, J. H., E. G. Smith, R. H. Garber, W. A. Williams, and H. Yamada. 1972. Influence of certain rotations upon cotton production in the San Joaquin Valley. Agronomy Journal 64: 543-546.

Van Doren, Jr., D. M., W. C. Moldenhauer, and G. B. Triplet, Jr. 1984. Influence of long-term tillage and crop rotation on water erosion. Soil Science Society of America Journal 48: 636-640.

Walters, D. T. 1987. Early studies on the use of legumes for conservation tillage in Nebraska. pp. 9-10. *In* J. F. Power [editor] The role of legumes in conservation tillage systems. Soil Conservation Society of America, Ankeny, Iowa.

Ware, G. W. 1980. Complete guide to pest control, with and without chemicals. Thomson Publications, Fresno, California.

Warren, J. A. pre-1925. Unpublished field report. Office of Farm Management, Bureau of Agriculture Economics, U.S. Department of Agriculture, Washington, D.C.

Wilson, B. D., and J. K. Wilson. 1928. Relation of sorghum roots to certain biological processes. Journal of Americac Society of Agronomy 28: 247-254.

# ECOLOGICAL AND AGRONOMIC CHARACTERISTICS OF INNOVATIVE CROPPING SYSTEMS

Benjamin R. Stinner and John M. Blair

Sustainable agricultural systems are those that rely on lower inputs of energy and chemicals to achieve long-term productivity and environmental compatibility (Poincelot, 1987). Stinner and House (1988) and others (Conway, 1987; Hart, 1986) have argued that better and more informed management, and specifically management of ecological interactions and processes, will be required to replace high inputs in sustainable systems. High-input agriculture has attracted criticism for overproduction (Altieri, 1987), lack of economic stability (Ehrenfeld, 1987), and environmental degradation (Hallberg, 1986). Attention to these issues has been focused on both temperate and tropical agroecosystems (Dover and Talbot, 1987). Dialogue, conferences, and publications generated in the past few years suggest strongly that agriculture is undergoing a rapid transition or even revolution.

During the past 40 to 50 years, in the United States and Europe, agriculture has changed from farming systems using relatively low amounts of energy and chemical subsidies to those requiring high inputs of energy derived from fossil fuel, chemical fertilizers, and pesticides (Edwards, 1988). Later, in the 1960s and 1970s, the Green Revolution program exported this high-technology, high-chemical-input agriculture to developing countries, mainly in the tropics (Dahlberg, 1979). These high-input systems and higher yielding crop varieties dramatically increased production per unit of land area in temperate and tropical regions (Brady, 1974; Lal, 1987). High-input management remained profitable throughout the 1970s. But now, for a variety of complex economic and ecological reasons, many argue that the maximum yield concept cannot provide long-term sustainability and that cropping systems will have to change.

Sustainable agriculture differs from conventional, high-input agriculture in that it emphasizes long-term yield stability with minimal environmental impact—in contrast to focusing more on short-term goals, such as maximum yields. The goals of sustainable agriculture cannot be met by simply lowering inputs. New and innovative cropping systems also will have to be designed and adopted.

Major ecological and agronomic characteristics of conventional and innovative sustainable cropping systems are compared in table 1. Sustainable systems share some similarities with more mature and less disturbed plant communities undergoing ecological succession—with the distinction that agroecosystems export large quantities of energy and nutrients as yield and, therefore, require some input subsidies (Hendrix, 1988). These innovative systems incorporate some of these later successional properties, which should confer sustainability characteristics and contribute to minimizing the need for external subsidies.

High-input systems, based typically upon large-scale annual monocultures, are by their design and management very susceptible, for example, to nutrient losses and pest outbreaks. Stinner and House (1988) have pointed out that although ecological phenomena occur in all agroecosystems, high inputs tend to override or mask ecological processes. Sustainable, innovative systems should be based upon ecological principles rather than on chemical and energy inputs. Invariably, sustainable, lower-input agroecosystems are more complex than high-input systems; thus, their management will need to be more sophisticated.

## Multiple Cropping

Multiple cropping—using the same field to produce two or more crops a year—is generally believed to be the oldest form of organized agriculture, and it remains one of the most common practices employed by "traditional" farmers in tropical regions. In Africa, for example, 98 percent of all cowpeas, a major leguminous crop, are grown in combination with other crops (Dover and Talbot, 1987). Most native farming systems in the tropics are polycultural, as were earlier traditional systems in temperate zones (Plucknett and Smith, 1986). Multiple cropping systems are much less common in developed countries than they once were.

The term multiple cropping actually encompasses a variety of cropping systems, whose commonality is a diversification of crops in time and/or space. (For a more complete discussion of the terminology associated with multiple-cropping systems, see Francis, 1986). One example is sequential

**Table 1. Comparison of ecological characteristics between conventional and innovative systems.**

| | *Conventional* | *Innovative-Sustainable* |
|---|---|---|
| Fossil fuel energy | High | Low |
| Labor/management | Low? | High?, more complex |
| Fertilizer | Inorganic | Organic |
| Tillage (soil disturbance) | Lower | Lower |
| Crop diversity | Low | High |
| Life history characteristics | Annual | Mixed/more permanent |
| Pests | Less stable/ chemical control | More stable/biological and cultural control |
| Nutrient cycling | Open/pulsed emphasis on physical/ chemical control | Closed emphasis on biological control |
| Animal integration | Low | High |
| Importance of decomposer processes | Low | High |

cropping, or crop rotation, in which two or more crops are grown sequentially in the same field. In this case, the diversification of crops is in time only. In intercropping, two or more crops are grown simultaneously in the same field. In intercropped systems there is some degree of overlap among crops so that crops are diversified both in time and space. Intercropped systems include mixed intercropping, where two or more crops are grown without a distinct row arrangement; row intercropping, where at least one crop is planted in rows; strip intercropping, where two or more crops are planted in strips wide enough to allow for independent cultivation but narrow enough to interact with one another ecologically; and relay intercropping, where a second crop is planted into a first crop before harvest so that there is some overlap in the life cycles of the two crops. Variations in relay intercropping include overseeding (e.g., legumes into previously established crops) and "living mulch" systems (e.g., corn planted into alfalfa).

The most important characteristic of multiple-cropping systems is increased diversity, both in terms of habitat structure and species. In this respect, multiple-cropping systems resemble natural plant communities more closely than do conventional high-input monoculture systems where considerable effort is expended to minimize diversity. Some forms of intercropping, such as mixed intercropping and relay intercropping, increase what Whittaker (1960) termed alpha, or within-habitat diversity. Other practices, such as strip intercropping in relatively wide rows, increase beta, or between-habitat, diversity. Many tropical intercropped agroecosystems appear to be very stable, and it is tempting to suggest that increased diversity imparts increased stability to multiple-cropped systems. However, the relationship between diversity and stability is complex (Odum, 1983), and the design of stable agricultural systems must consider more than simply increasing diversity (Dover and Talbot, 1987). Still, there are apparent advantages of carefully designed increased diversity in agricultural systems. One very important advantage of multiple-cropping is decreased risk of total crop failure. By utilizing more than one crop, often with different resource requirements, multiple-cropping systems minimize the risk of failure due to pest problems, weather, price fluctuations, and so forth. Studies at ICRISAT in India have shown that the probabilities of economic failure over a range of income levels are consistently less for intercropping than for monoculture farming (Rao and Willey, 1980). Increased spatial and plant species diversity also has important implications for increasing resource utilization efficiency, internal nutrient cycling, and biological control processes.

Intercropped systems often have a land equivalent ratio (LER), a measure of the relative amount of land planted in monoculture that would be required to produce the same yield, greater than one. For example, Ahmed and Rao (1982) examined intercropped maize-soybean systems grown under a range of nitrogen inputs at 14 locations in seven countries and found LERs greater than one in all but a single case. They also observed a greater monetary return from the intercropped systems, compared to producing either crop in monoculture. A frequently cited reason for the success of many traditional intercropped systems is increased resource utilization, both in space and time (Gliessman, 1985; Trenbath, 1986). Because crop species vary in their resource requirements, including light, water, and nutrients, a properly designed multiple-cropping system should utilize available resources more efficiently than a monoculture system. For example, a more complex canopy structure may increase total light utilization (Marshall and Wiley, 1983). Intercropped plants with different water and nutrient requirements and different rooting patterns may utilize available resources more fully. A secondary advantage of greater resource utilization by multiple-crop species is reduced resource availability to weeds.

In addition to using available resources more efficiently, intercropped agroecosystems should be more efficient in internal recycling of nutrients and less "leaky" than conventional monocultures. Combinations of crops that differ in the timing of nutrient uptake may minimize the potential for nutrient losses through leaching. Plants with deeper rooting patterns may transport nutrients from lower soil horizons to the surface for use by other crops that would not otherwise be able to access them (Harwood, 1984). Apart from increased crop yields, total biomass production often is greater in intercropped systems (Gliessman and Amador, 1980). We observed significantly greater total biomass production (17,500 versus 11,200 kilograms per hectare) from a corn-alfalfa intercropped system in Ohio, as compared with corn grown in monoculture (Stinner et al., unpublished data). Above-ground corn plant biomass also had higher concentrations (1.7 versus 1.1 percent) of nitrogen in the intercropped system. The additional noncrop biomass can be reincorporated directly into the soil or, in the case of alfalfa, used as feed for livestock whose manure can be used as fertilizer.

The use of legumes is an important aspect of many multiple-cropping systems and is particularly relevant to the development of low-input sustainable agricultural systems. Nitrogen fixed by legumes may be utilized by other nonlegume crops in both legume/nonlegume intercropping systems and legume/nonlegume rotations. An example of the former is the traditional corn/bean/squash intercropped system used in Mexico, where corn yields increased as much as 50 percent relative to corn grown in monoculture (Gliessman, 1985). Rotations of legumes and cereals, such as soybean and corn, can greatly reduce or eliminate the need for external nitrogen inputs (Harwood, 1984; Francis and Clegg, this volume).

Increased crop diversity often results in decreased insect pest problems. Reduced pest populations can result from a number of factors that increase the abundance of natural predators and parasites and interfere with a pest species' host-finding abilities. Increased diversity of habitats and potential prey species in multiple-cropping systems may support increased populations of generalized predators. For example, Brust et al. (1986a) found greater densities of predacious soil macroarthropods and increased predation on larval lepidopterous pests from corn intercropped in an alfalfa-orchard grass mixture (Figure 1). Recent studies in Ohio indicate that beneficial parasitic wasps (Hymenoptera: Chalcidoidea) are more abundant in soybean-corn narrow-strip intercropping systems than in monoculture soybeans (Tonhasca, unpublished data). Increased crop diversity may also impede host-finding by pest species by interferring with chemical and visual stimuli, and by creating barriers of nonhost species that may restrict movement. In a survey addressing the impact of crop diversification on insect pest and natural enemy dynamics, Andow (1983) reported that diversifying cropping systems decreased abundance of 58 species of pest herbivorous insects while increasing only three species. Monophagous insect herbivores were affected more in annual than perennial systems.

Decoy or trap crops can be used to control nematode pests in multiple-cropping systems. Decoy crops are nonhost crops designed to activate larval nematodes in the absence of a host that would allow them to continue their development. Altieri and Liebman (1986) list over a dozen decoy crops used to control 10 species of nematodes in a variety of crops. Trap crops are host plants that are planted to attract pests, such as nematodes, and then destroyed before the nematodes can reproduce.

Multiple-cropping systems also have the potential to reduce weed density by limiting effectively the availability of resources to weed species. Some level of controlled weed growth may be desirable in certain systems (Chacon and Gliessman, 1982). Weeds may further reduce nutrient losses from the system and also increase structural heterogeneity, and thus impact on insect pest populations. Diversification of crops also may reduce incidence of plant loss to disease (Altieri and Liebman, 1986).

Many of the same benefits apply to sequential multiple-cropping systems. The influence of crop rotation on certain diseases and insect pests is a familiar example. In addition to plant pathogens, crop rotations also may affect populations of beneficial plant-growth-promoting rhizobacteria and deleterious (though not normally considered pathogenic) rhizosphere micro-

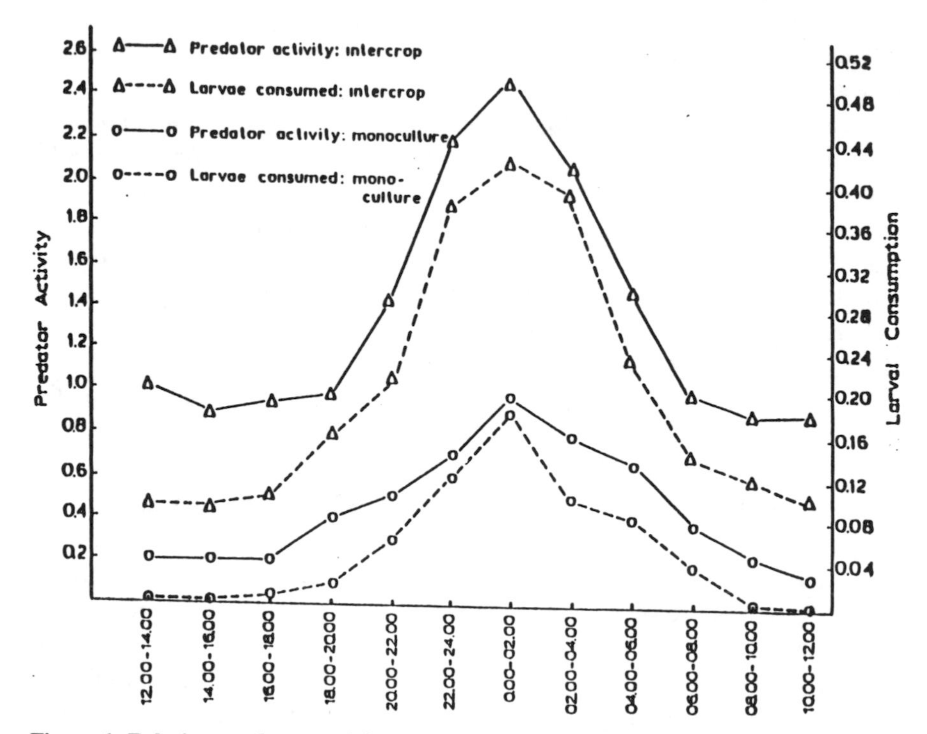


**Figure 1. Relative predator activity (mean number of attacks on tethered larvae per 2-hour interval) and larval predation (mean number of tethered larvae consumed per stake per 2-hour interval) over a 24-hour period in corn intercropped in an alfalfa-orchardgrass mixture and monoculture corn (Brust et al., 1986a).**

organisms, which may cause considerable reductions in plant yield (Schippers et al., 1987). The effects of intercropping on plant-growth-promoting rhizobacteria and deleterious microorganisms remain to be investigated.

***Potential for Multiple Cropping in U.S. and European Agriculture.*** Multiple-cropping systems that appear most suitable for adaptation to large-scale agriculture in the United States and Europe, other than sequential cropping and rotations, are forms of stripcropping and relay intercropping. Stripcropping may be the easiest to adopt since, by definition, the strips are wide enough to allow independent cultivation, eliminating the need for special farm machinery. One option is to alternate strips of a legume, such as alfalfa, with strips of a nonlegume, such as corn. The two crops can be rotated annually so that the nonlegume receives a nitrogen benefit and soil erosion is reduced by the presence of the legume, especially if the ribbons are oriented perpendicular to slopes. This type of biculture, with the use of a perennial legume such as alfalfa, maintains year-round vegetative cover on the landscape, and stabilizes both hydrologic and soil characteristics as well as populations of pest and beneficial organisms. The stripcropping also reduces potential competition among crops, compared to living mulch alfalfa-corn systems. One problem with stripcropping is that soil pests and diseases may carry over in spite of crop rotation among strips if strips are not precisely placed. Alternatively, we have found that with narrow-strip, or ribbon, intercropping of corn and alfalfa, populations and activity of soil-dwelling predatory arthropods were enhanced by the presence of the alfalfa compared to monoculture corn (Stinner et al., unpublished data).

Variations in relay cropping, such as living mulch systems and overseeding of legumes into previously existing crops, also appear to have promise. However, there are several obstacles to increased adoption of intercropped agroecosystems by large-scale agriculture. One is the lack of crop varieties suitable for use in intercropped systems. Most crop varieties used in large-scale agriculture for developed countries have been selected to produce high yields in response to high-input regimes. Other traits, such as resource use efficiency, competitiveness, shading tolerance, and plant architecture are important considerations in developing varieties for intercropped agroecosystems (Smith and Francis, 1986). Another major factor in the adoption of multiple-cropping systems is agricultural infrastructure, including governmental policies that may affect a farmer's decision on cropping practices, regardless of agronomic or ecological considerations. However, the movement toward low-input sustainable agricultural systems seems to be linked to some form of multiple-cropping systems.

## Minimum Tillage

Minimum, or conservation, tillage is defined as a crop planting system that leaves 30 percent or more of crop residues on the soil surface. Conservation tillage systems have been gaining acceptance in the United States and elsewhere, largely because these systems can lower farm expenses, reduce runoff and soil erosion, aid in soil organic matter conservation, and increase soil moisture retention.

Minimum tillage practices have significant ecological, as well as agronomic and economic, impacts that promote their use in innovative cropping systems. Minimum tillage agroecosystems, by reducing soil disturbance, promote the development of a more complex decomposition subsystem that enhances soil system stability and efficiency of internal nutrient cycling (House et al., 1984; Stinner et al., 1984; Hendrix et al., 1986). By increasing the complexity of interactions among plants, invertebrates, and microorganisms, minimum tillage systems resemble natural ecosystems more closely than do conventional agricultural systems. Increased biotic regulation of processes, such as decomposition and nutrient release, under minimum tillage regimes should result in more efficient, less environmentally degrading, and more sustainable agroecosystems (Stinner and Stinner, 1989).

Some of the biological, physical, and chemical characteristics of minimum tillage systems are indicated in figure 2. Soil structure in minimum and no-tillage agroecosystems is characterized by increased stratification, with

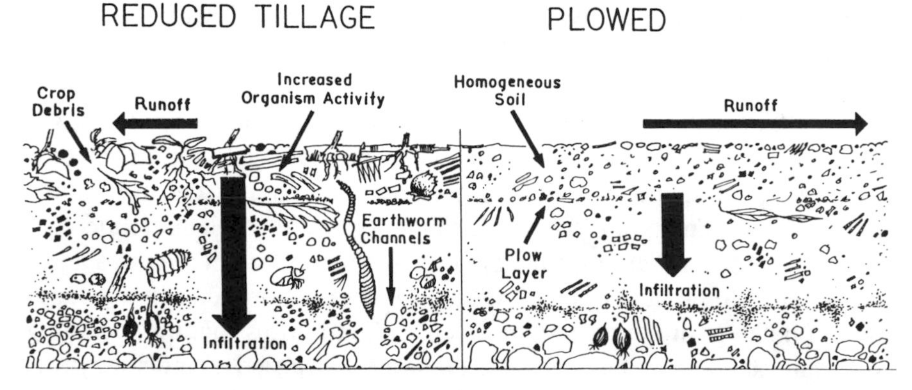


**Figure 2. A comparison of biological, physical, and chemical properties of minimum tillage and plowed soils (from Stinner and Stinner, 1989).**

higher concentrations of organic matter and nutrients near the soil surface. A gradient of decomposition products, from fresh residues at the surface to humified organic matter at deeper horizons, develops under continuous minimum and no-tillage systems. No-tillage agroecosystems also tend to conserve soil organic matter and nutrients more than conventionally tilled systems (Blevins et al., 1983; Hendrix et al., 1986; Stinner et al., 1988). This conservation may be related, in part, to tillage-induced changes in the decomposer community (Holland and Coleman, 1987). For example, there is evidence that no-tillage systems favor a fungal-based detritus food web, while conventional tillage favors a bacterial-based food web (Hendrix et al., 1986; Holland and Coleman, 1987).

Tillage practices affect the distribution and abundance of pest and nonpest soil arthropods (Blumberg and Crossley, 1983; House and Stinner, 1983; Hendrix et al., this volume). Detrimental effects of tillage on particular groups of organisms have been abundantly documented. For example, increased tillage has been shown to reduce populations of beneficial soil microarthropods (Hendrix et al., 1986) and earthworms (Edwards and Lofty, 1977). Minimum tillage systems tend to have a more abundant decomposer community and, by reducing the frequency and intensity of soil disturbance, also allow for greater stability of the soil biota. Another important feature of minimum tillage systems is enhanced soil predator abundance and activity. Studies of the soil macroarthropod community in conventionally plowed and no-till corn agroecosystems in Ohio indicated lower numbers of herbivores and higher numbers of predators in the no-till system than in the plowed (Stinner et al., 1986). Predator activity and incidence of predation was also greater in no-tillage corn systems (Brust et al., 1986b).

A major criticism of reduced tillage agroecosystems is their reliance on high inputs of herbicides for weed control. In order to reconcile minimum tillage with sustainable agriculture's goals of less reliance on nonrenewable inputs and less negative environmental impacts, it is necessary to use lower input minimum tillage techniques with alternative methods of weed control. One way is to combine minimum tillage with innovative multiple-cropping systems, such as incorporation of legume cover crops through overseeding, sod-based rotations, sod strip intercropping, and various living mulch systems. Vrabel et al. (1980) commented on the feasibility of using legume cover crops in vegetable systems to reduce the use of chemical herbicides. Overseeding cover crops into rows of grain crops also may provide some degree of weed control as well as controlling soil erosion and contributing to nitrogen inputs. An example is the use of no-tillage agriculture with legume intercropping using living mulch systems where corn is planted into alfalfa (Elkins et al., 1982). In this case, the alfalfa provides ground cover and weed control as well as reducing the need for inputs of exogenous fertilizer. To minimize intercrop competition that could potentially reduce yields, the choice of crop varieties and timing of planting are important considerations in these cropping systems. Current crop varieties bred for use in high-input systems may not be the most suitable for use in innovative minimum tillage systems. Again, we stress the need for developing crop varieties appropriate for use in lower input innovative cropping systems.

## Agroforestry

Agroforestry, including trees in cropping systems, is a very ancient and still widespread practice in many regions of the world (Farrell, 1987). Agroforestry probably was the original form of agriculture in areas that were naturally forested (Bishop, 1983). Agroforestry systems are receiving renewed attention, particularly in tropical regions, because of their value in achieving the goals of sustainable agriculture. Agroforestry is not only compatible with production of annual and perennial herbaceous crops, but also integrates well with animal agriculture (Bishop, 1983).

Ecologically and agronomically, agroforestry can accomplish a great deal that other cropping systems cannot. Overall, agroforestry stabilizes cropping systems. Figure 3 illustrates a generalized type of agroforestry where trees form successive rows of shelterbelts. The trees provide permanent above- and below-ground structure to the cropping system. In this way, water and wind movement is reduced and consequent soil erosion losses can be decreased dramatically (Vergara, 1982). Trees in agricultural landscapes modify microclimatic conditions by decreasing temperature extremes. The trees also intercept air moisture and redistribute it to the soil (Farrell, 1987).

Agroforestry can impact significantly upon soil fertility. McGuahey (1986), working in Africa, reported that Acacia trees grown in agroforestry systems with annual crops of millet and sorghum sustained yields for 15 to 20 years compared to only three to five years when annual crops were grown alone. The same study also reported that the Acacia trees significantly increased nutrient concentrations in the upper soil strata. Nitrogen, potassium, and calcium concentrations were increased 186, 76, and 22 percent, respectively. Organic matter increased from 40 to 269 percent, and cation exchange capacity from 50 to 120 percent. In Mexico, Farrell (1984) reported that phosphorus, nitrogen, calcium, and carbon all increased significantly in surface soil under capulin (*Prunus*) trees compared to soil located away from the trees. In the Majjia Valley of Africa, grain yields in agroforestry systems with neem trees as windbreaks were 40 percent higher than in open

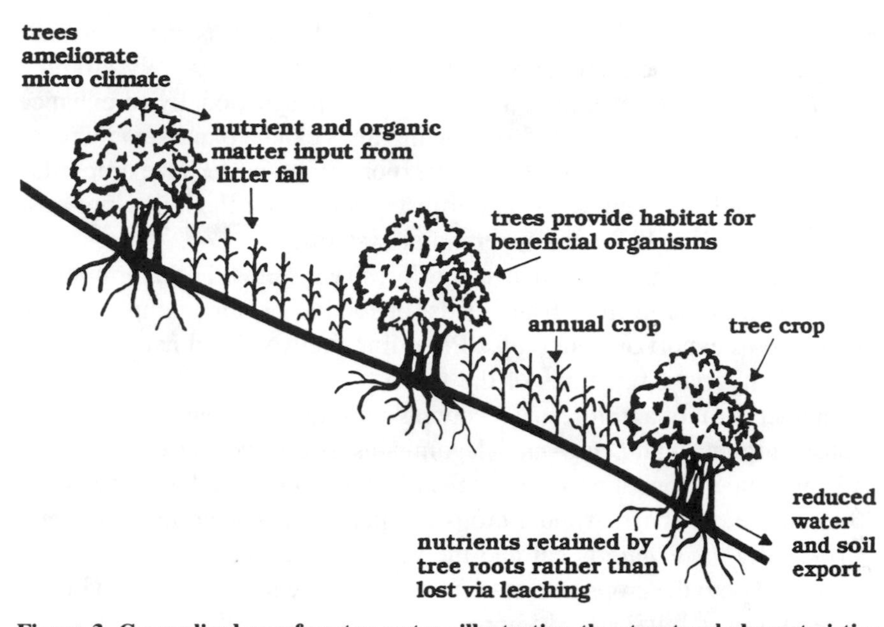


**Figure 3. Generalized agroforestry system illustrating the structural characteristics and ecological processes derived from introducing trees into cropping systems (Vergara, 1982).**

areas.[1] Furthermore, utilizing leguminous trees and shrubs, such as *Acacia* and *Leucaena,* provides a mechanism for nitrogen acquisition via biological fixation.

Agroforestry systems can exert appreciable influences on populations of invertebrates and microorganisms and, consequently, the roles that these organisms play in agroecosystems. For example, research has indicated that there is higher insect diversity and activity in and adjacent to shelterbelts than away from the trees (Lewis, 1969; Pollard et al., 1971). Evidence also suggests that forested areas in agricultural landscapes serve as reservoirs for predators and parasites (Pollard, 1971; Lundholm and Stackerud, 1980). Formann and Baudry (1984) indicated that forested strips of vegetation in agroecosystems intercept more solar energy than do adjacent crop fields. In early spring, this causes increased biological uptake of nutrients in shelterbelt areas, indicating that the shelterbelts play an important role in nutrient conservation during periods when nutrient demand by herbaceous crops is minimal.

Agroforestry can be based on either annual or perennial herbaceous crops, pastoral systems for animals, or any combinations of these systems. De-

[1]Dennison, S. "The Majjia Valley Windbreak Evaluation Study: An Examination of Progress and Analyses to Date. Unpublished report for CARE, February 1986.

pending on socioeconomic and cultural needs, the systems can be managed to favor one function over another. For example, in severely erodible regions, the density of trees can be increased proportionally to enhance the trees' protective role. Or, fast-growing trees can be planted to increase firewood supplies. The ratio of cash or export versus sustenance crops also may be varied. In many ways, agroforestry presents a flexibility not possible in annual or herbaceous perennial systems.

Trees support many functions in agroforestry systems: Food for humans or livestock in the form of fruit, seeds, or fodder, or, for example, coffee and cacao as export or cash crops. Providing fuelwood and natural fences also are important functions of the trees.

Another form of agroforestry receiving considerable attention in Nigeria is alley cropping with *Leucaena*, a leguminous shrub, and maize or sorghum as intercropped annual grain. *Leucaena* is planted in parallel hedges from 10 to 20 meters apart. Annual crops are planted between the *Leucaena*. Periodically, *Leucaena* is cut and the nitrogen-rich leaves and stems are distributed onto the crops as a nutrient subsidy. The system has been highly successful in stabilizing and improving soil characteristics, and it has the potential to increase and stabilize populations of beneficial predatory and parasitic insects (Lal, 1981; Plucknett and Smith, 1986).

Still another approach to agroforestry developed for the tropics by Robert Hart (1986) incorporates ecological succession, where an agroecosystem is designed and managed to parallel or mimic natural ecosystem development. The cropping system is initiated by planting annual, rapidly growing species, such as beans or maize, followed by progressively more long-lived and woody plants, such as coffee and avocado, and finally, large trees such as coconut and cacao. Some native farming systems in the tropics have followed this pattern for thousands of years. In this type of agroforestry, each stage of succession modifies the environment, resulting in favorable conditions for the next stage. In these systems, trees also provide cash crops of bananas, coffee, and avocadoes. Nutritionally, these agroforestry systems yield 31 percent more protein, 54 percent more calories, and 62 percent more carbohydrates than monocultures (Behmel and Neumann, 1985).

Agroforestry has provided one of the most important means of reclaiming degraded lands, especially in upland areas. In Nyabisindu, Rwanda, agroforestry played an invaluable role in restoring denuded and severely eroded hill land that had been subjected to excessive shifting cultivation and chemical agriculture. An agroforestry system combining herbaceous crops, animals, and trees was used to stabilize this degrading ecosystem. Trees created the permanent structure that stabilized the soil and hydrologic budgets. Diverse crops emphasizing perennials were employed to reduce risks from pests and weather variablity. Fertility depended on organic sources from animal manures and plant materials.

***Potential for Agroforestry in U.S. and European Agriculture.*** Extant agroforestry systems are primarily a tropical phenomenon. Yet the principles demonstrated by agroforestry should be applicable to designing sustainable, innovative cropping systems for temperate regions (Douglas and Hart, 1978). Essentially, this is taking an inverse approach to the Green Revolution and asking what temperate, large-scale mechanized agriculture can learn from sustainable tropical agriculture. During the last 20 years in the United States, especially in the Midwest, agricultural landscapes have lost many fencerows, shelterbelts, windbreaks, woodlots, and other trees and shrub vegetation. This loss of perennial vegetation was motivated, in part, by governmental policies and economic conditions during the 1960s and 1970s, when land values and commodity prices were high (Batie and Healy, 1980). The shelterbelt vegetation had been invaluable in stabilizing soil and water resources (Schroeder, 1988), a protective value that has declined across agricultural landscapes in the United States.

A number of tree and shrub species could be adapted readily to agroforestry practices. In Ohio, for example, we are experimenting with a species of legume called the Siberian pea shrub or *Caragana arborescens*. This legume attains a height of 3 to 6 meters, depending upon cultivar type. *Caragana* has a brushy, dense growth form that provides excellent habitat for wildlife as well as windbreaks (PFRA, 1986). We have planted the *Caragana* in parallel rows 15 meters apart, with corn or wheat grown between the trees. Certain cultivars of black locust (*Psuedoacacia robinia*), or honey locust (*Gleditsia tricanthos*) are legume species that can be planted in similar patterns.

Alternatively, the protective value of trees could be incorporated into an agricultural landscape by simply permitting defined areas—narrow strips of land, etc.—to follow natural successions within crop fields. Planting species of trees with specific economic value would increase the productivity and economic value of agroecosystems. Even though planting trees would decrease the total amount of acreage in herbaceous crops, the benefits accrued in environmental stability and long-term economic value of the trees should more than compensate. As in the tropics, the ratio of trees to associated crops could be varied to meet the specific needs of different regions and economic conditions and, in this way, provide greater flexibility than with typical annual monoculture systems.

An innovative approach to agroforestry in temperate regions would be to parallel Hart's successional model developed for the tropics. In the temperate version, the system could begin with annual crops of maize, soybeans, or wheat and then progress to orchard or nut tree crops as a polyculture system. The system could be modified to include animals or interspersed pasture. Here again, the complex nature of these systems is an advantage in that it allows for a high degree of flexibility in management options as buffers against economic and environmental vagaries.

## Integrating Plant and Animal Systems

Most traditional farming systems incorporate some form of animal husbandry. In arid regions, animal agriculture tends to be practiced as pastoral systems, and in humid regions as intensive agriculture in conjunction with grain or other herbaceous crops. During the past 40 to 50 years, large-scale, high-input plant agriculture has moved away from integration with animal husbandry. Integrating crop systems with animals can help accomplish economic and environmental sustainability. For example, animal products increase economic diversity and provide alternative pathways for nutrient cycling processes.

Most farms combining animal husbandry and crop systems are of small to medium size. For example, a typical family-managed dairy farm in the northeastern United States is 50 to 100 hectares. In fact, most farms in the midwestern United States that are practicing lower input agriculture incorporate animal husbandry into their cropping systems.

We are working on two farms in Ohio where low-input practices have been in place for 20 years. One farm, owned by the Spray family in Mount Vernon, raises beef cattle in conjunction with a 220-hectare rotational cropping system of maize, soybeans, legume forage, and small grain. The Spray family practices nonchemical farming, utilizing the legumes and animal manure as fertility sources, and conservation tillage practices with innovative, shallow cultivation techniques for weed control. Another farm, owned by the Hartzler family in Smithville, is centered around a dairy operation. The farm uses a rotation of maize and alfalfa, most of which is fed to cattle. The Hartzlers do not use inorganic fertilizer or pesticides for weed control, and they also rely on cultivations.

In a recent study of soil chemical and biological characteristics on the Spray farm, compared with those of an adjacent farm using chemical inputs of fertilizer and pesticides, we found soil organic matter to be almost twice as high (4.11 versus 2.67 percent). Densities of soil invertebrates (Acarina and Collembola, key animals in decomposition) were four times higher on the Spray farm than on the adjacent farm (Blair and Zaborski, unnpublished data).

The Chinese have devised animal-based cropping systems that extend plant-animal agriculture to used detritus food chains (Chen, 1989). In these systems, rice is the major field crop. When the grain is harvested, the straw is composted along with animal manure in a biogas digester that produces methane that is used for cooking and light. Sludge from the digester is used for mushroom production; after mushrooms are harvested, the remaining organic residues are returned to the rice fields as an organic fertilizer. This system is extremely efficient in energy use and in nutrient cycling.

An essential component of animal- and crop-integrated agriculture is the cycling of manure through plant-soil systems. Many studies have been devoted to investigating the role of manure in cropping systems and, in particular, comparing effects of manure versus inorganic fertilizer on plant growth and soil fertility. Long-term studies have indicated that crop yields obtained with manure can be comparable to or greater than those obtained using inorganic fertilizer (Baldock and Murgrove, 1980; Welch, 1979). There often is a greater positive crop response to manure than expected from nutrient content alone, suggesting that factors other than macronutrient content have beneficial influences on plant growth.

An important issue in lower-input farming is the relative merits of maintaining soil fertility and plant production with manure and nonmanure organic sources, such as cover crops and rotation with legumes. Nutrients supplied via organic matter may be released more slowly to plants than nutrients derived from inorganic fertilizer. In a four-year study in California, Pratt et al. (1976) found that 40 to 50 percent of the original nitrogen contained in farmyard manure was mineralized during the first year, 20 percent the second, and 5 percent the third year. Experiments at Rothamsted Experiment Station in England demonstrated that manure applications over 117 years were more effective at raising levels of soil phosphorus than superphosphate fertilizer, when each was applied at equal phosphorus levels (Olsen and Barber, 1977). Brogen (1981) reported that after 25 years of treatment with mineral fertilizer, green manure, and animal manure, the nitrogen stored in soil humus in the green manure and animal manure systems was two and three times, respectively, those in the inorganically fertilized system. In the same study animal and green manure systems mineralized 2.1 and 2.5 times more nitrogen per year, respectively, than the inorganically fertilized treatment. Clearly, integration of animal husbandry into lower input systems can improve long-term soil fertility.

## Conclusions

Multiple cropping, minimum tillage, agroforestry, and integration with animal husbandry all have the potential to sustain productivity using lower chemical and energy inputs. Although most of these practices have historical bases in some very old, traditional farming systems, the challenge toward innovation derives from understanding the ecological basis for their success and applying these principles to new systems in modern agriculture. In this way, sustainable systems are not a return to the past, but a synergistic blending of new and old concepts and technology to address the needs of the future.

Design and management of innovative, sustainable agricultural systems should be based on ecological principles. With this approach, interactions among agroecosystem components and processes are the central theme, as opposed to the more singular approach of trying to maximize any one parameter, such as grain yields or milk production. Perhaps the most important barriers to adoption of sustainable agricultural systems are not biological or technological, but sociological and political.

## REFERENCES


Ahmed, S., and M. R. Rao. 1982. Performance of maize-soybean intercrop combinations in the tropics: Results of a multi-location study. Field Crops Research 5: 147-161.

Alteiri, M. A. 1987. Agroecology. The scientific basis of alternative agriculture. Westview Press, Boulder, Colorado.

Altieri, M. A., and M. Liebman. 1986. Insect, weed and plant disease management in multiple cropping systems. *In* C. A. Francis [editor] Multiple cropping systems. Macmillan Publishing Co., New York.

Andow, D. 1983. Effect of agricultural diversity on insect populations. *In* W. Lockeretz [editor] Environmentally sound agriculture. Praeger, New York.

Baldock, J. O., and R. B. Musgrave. 1980. Manure and mineral fertilizer effects in continuous and rotational crop sequences. Agronomy Journal 72: 511-518.

Batie, S. S., and R. G. Healy. 1980. The future of American agriculture as a strategic resource. The Conservation Foundation, Washington, D.C.

Bishop, J. P. 1983. Tropical forest sheep on legume forage/fuelwood fallows. Agroforestry Systems 1: 79-84.

Blevins, R. L., G. W. Thomas, M. S. Smith, W. W. Frye, and P. L. Cornelius. 1983. Changes in soil properties after 10 years non-tilled and conventionally tilled corn. Soil and Tillage Research 3: 135-146.

Blumberg, A. Y., and D. A. Crossley, Jr. 1983. Comparison of soil surface arthropod populations in conventional tillage, no-tillage and old field systems. Agro-Ecosystems 8: 247-253.

Brady, N. C. 1974. The nature and properties of soils. Macmillan Publishing Co., New York.

Brogen, J. C. 1981. Nitrogen losses and surface runoff from land spreading of manures. Nifholff and Junk, The Netherlands.

Brust, G. E., B. R. Stinner, and D. A. McCartney. 1986a. Predation by soil inhabiting arthropods in intercropped and monoculture agroecosystems. Agriculture, Ecosystems and Environment 18: 145-154.

Brust, G. E., B. R. Stinner, and D. A. McCartney. 1986b. Predator activity and predation in corn agroecosystems. Evironmental Entomology 15: 1,017-1,021.

Chacon, J. C., and S. R. Gliessman. 1982. Use of the "non-weed" concept in traditional tropical agroecosystems of south-eastern Mexico. Agro-Ecosystems 8: 1-11.

Chen, Rongjun. 1989. Energy and flow through detritus food chains. *In* M. G. Paoletti [editor] Agricultural ecology and environmental issues. Elsevier, New York. In press.

Conway, G. R. 1987. The properties of agroecosystems. Agricultural Systems 24: 95-117.
Dahlberg, K. A. 1979. Beyond the Green Revolution: The ecology and politics of global agricultural development. Plenum Press, New York.
Douglas, Sholto J., and R. A. Hart. 1978. Forest farming. Rodale Press, Emmaus, Pennsylvania. 199 pp.
Dover, D. J., and L. M. Talbot. 1987. To feed the earth: Agro-ecology for sustainable development. World Resources Institute, New York.
Edwards, C. A. 1988. The concept of integrated systems in lower input/sustainable agriculture. American Journal of Alternative Agriculture 2: 148-152.
Edwards, C. A., and J. R. Lofty. 1977. Biology of earthworms. Chapman and Hall, London.
Elkins, D. M., R. Marashi, and B. McVay. 1982. No-till corn production in chemically suppressed grass and legume sods. Progress Report on No-till Corn Production in Legume and Non-Legume Sods. Southern Illinois University, Carbondale.
Ehrenfeld, D. 1987. Implementing the transition to a sustainable agriculture. Bulletin of the Ecological Society of America. 68: 5-8.
Farrell, J. G. 1987. Agroforestry systems. *In* Agroecology: The scientific basis of alternative agriculture. Westview Press, Boulder, Colorado.
Formann, R. T., and J. Baudry. 1984. Hedgerows and hedgerow networks in landscape ecology. Environmental Management 8: 499-510.
Francis, C. A. 1986. Introduction: Distribution and importance of multiple cropping. *In* C. A. Francis [editor] Multiple cropping systems. Macmillan Publishing Co., New York.
Gliessman, S. R. 1985. Economic and ecological factors in designing and managing sustainable agroecosystems. *In* T. C. Edens, C. Fridgen, and S. L. Battenfield [editors] Sustainable agriculture and integrated farming systems. Michigan State University Press, East Lansing.
Gliessman, S. R., and M. F. Amador. 1980. Ecological aspects of production in traditional agroecosystems in the humid lowland tropics of Mexico. *In* J. I. Furtado [editor] Tropical ecology and development. International Society of Tropical Ecology, Kuala Lumpr.
Hallberg, G. R. 1986. From hoes to herbicides: Agriculture and groundwater quality. Journal of Soil and Water Conservation 41: 257-364.
Hart, R. D. 1986. Ecological framework for multiple cropping research. *In* C. A. Francis [editor] Multiple cropping systems. Macmillan Publishing Co., New York.
Harwood, R. R. 1984. Organic farming at the Rodale Research Center. *In* Organic farming: Current technology and its role in a sustainable agriculture. American Society of Agronomy Special Publication 46.
Hendrix, P. F. 1988. Strategies for research and management in reduced input agroecosystems. American Journal of Alternative Agriculture 2: 166-172.
Hendrix, P. F., R. W. Parmelee, D. A. Crossley, Jr., D. C. Coleman, E. P. Odum, and P. F. Groffman. 1986. Detritus food webs in conventional and no-tillage agroecosystems. Bioscience 36: 374-380.
Holland, E. A., and D. C. Coleman. 1987. Litter placement on microbial and organic matter dynamics in an agroecosystem. Ecology 68: 425-433.
House, G. J., and B. R. Stinner. 1983. Arthropods in no-tillage soybean agroecosystems: Community composition and ecosystem interactions. Environmental Management 6: 23-28.
House, G. J., B. R. Stinner, D. A. Crossley, Jr., and E. P. Odum. 1984. Nitrogen cycling in conventional and no-tillage agroecosystems: Analysis of pathways and processes. Journal of Applied Ecology 21: 991-1,012.
Lal, R. 1981. Managing the soils of Sub-Saharan Africa. Science 236: 1,069-1,076.
Lewis, T. 1969. The diversity of the insect fauna in a hedgerow and neighboring fields. Journal of Applied Ecology 6: 453-459.
Lundholm, B., and Stackerud, M. 1980. Environmental protection and biological forms of control of pest organisms. Swedish Commission for Research on Natural Resources

and the Swedish Products Control Board, Stockholm, Sweden. 171 pp.
Marshall, B., and R. W. Wiley. 1983. Radiation interception and growth in an intercrop of pearl millet/groundnut. Field Crops Research 7: 141-160.
McGuahey, M. 1986. Impact of forestry initiatives in the Sahel. Chemonics International. Washington, D.C.
Odum, E. P. 1983. Basic ecology. Saunders College Publishing, New York.
Olsen, S. R., and S. A. Barber. 1977. Effect of waste application on soil phosphorus and potassium. *In* Soils for management of organic wastes and waste waters. American Society of Agronomy, Madison, Wisconsin.
Plucknett, D. L., and N.J.H. Smith. 1986. Historical perspectives on multiple cropping. *In* C. A. Francis [editor] Multiple cropping systems. Macmillan Publishing Co., New York.
Poincelot, R. P. 1987. Toward a more sustainable agriculture. AVI Publishing Co., Inc., Westport, Connecticut.
Pollard, E. 1971. Hedges. VI. Habitat diversity and crop pests: A study of Brevicoryne brassicae and its syrphid predators. Journal of Applied Ecology 8: 751-780.
Prairie Farm Rehabilitation Administraton (PFRA). 1986. Shelterbelt species. Agriculture Canada, Indian Head, Saskatchewan.
Pratt, P. F., S. Davis, and R. G. Sharpless. 1976. A four-year field trial with animal manures. I. Nitrogen balances and yields. II. Mineralization of nitrogen. Hilgardia 44: 99-125.
Rao, M. R., and R. W. Wiley. 1980. Evaluation of yield stability in intercropping: Studies on sorghum/pigeon pea. Experimental Agriculture 16: 29-40.
Schippers, B., A. W. Bakker, and P.A.H.M. Bakker. 1987. Interactions of deleterious and beneficial rhizosphere microorganisms and the effects of cropping practices. Annual Review of Phytopathology 25: 339-358.
Schroeder, W. R. 1988. Planting and establishment of shelterbelts in humid severe-winter regions. Agriculture, Ecosystems and Environment 22: 441-463.
Shrader, W. D. 1975. Organic farming—pro and con. Iowa State University, Ames.
Smith, M. E., and C. A. Francis. 1986. Breeding for multiple cropping systems. *In* C. A. Francis [editor] Multiple cropping systems. Macmillan Publishing Co., New York.
Stinner, B. R., D. A. Crossley, Jr., E. P. Odum, and R. L. Todd. 1984. Nutrient budgets and internal cycling of N, P, K, Ca and Mg in conventional tillage, no-tillage, and old field ecosystems on the Georgia piedmont. Ecology 65: 354-369.
Stinner, B. R., and D. H. Stinner. 1989. Plant-animal interactions in agricultural systems. *In* W. G. Abrahamson [editor] Plant-animal interactions. McGraw-Hill Book Co., New York.
Stinner, B. R., and G. J. House. 1988. The role of ecology in lower input, sustainable agriculture. American Journal of Alternative Agriculture 2:145-147.
Stinner, B. R., G. J. House, J. K. Pechmann, D. E. Scott, and D. A. Crossley, Jr. 1988. Phosphorous and cation dynamics of component and processes in conventional and no-tillage soybean agroecosystems. Agriculture, Ecosystems and Environment 20: 81-100.
Stinner, B. R., H. R. Krueger, and D. A. McCartney. 1986. Insecticide and tillage effects on pest and non-pest arthropods in corn agroecosystems. Agriculture, Ecosystems and Environment 15: 11-21.
Trenbath, B. R. 1986. Resource use by intercrops. *In* C. A. Francis [editor] Multiple cropping systems. Macmillan Publishing Co., New York.
Vergara, N. T. 1982. New directions in agroforestry: The potential of tropical legume trees. Environmental and Policy Institute, Honolulu, Hawaii.
Vrabel, T. E., P. L. Minotti, and R. D. Sweet. 1980. Seeded legumes as living mulches in sweet corn. Paper 769. Department of Vegetable Crops, Cornell University, Ithaca, New York.
Welch, L. F. 1979. Nitrogen use and behavior in crop production. Illinois Agricultural Experiment Station Bulletin 761. Champaign, Illinois.
Whittaker, R. H. 1960. Vegetation of the Siskiyou Mountains, Oregon and California. Ecological Monographs 30: 279-338.

# BIOTECHNOLOGY AND CROP BREEDING FOR SUSTAINABLE AGRICULTURE

Holly Hauptli, David Katz,
Bruce R. Thomas, and Robert M. Goodman

Crop plants have been manipulated genetically to suit mankind's purposes since agriculture began. Dramatic changes in morphology, physiology, and pest and disease resistance have resulted in what are now the major crop plants. Domestication and breeding achievements over the first eight millennia of agriculture have been quite impressive. Among them are the control (and in some cases elimination) of seed-dispersal mechanisms; increase in size of seeds, fruit, or other harvested structures (harvest index); elimination of naturally occurring toxins; alterations in seed dormancy; and the selection of polyploid variants (Harlan, 1975). Many new species were created during this period, some of which became totally dependent upon man for survival (Beadl, 1980). It has not been easy to duplicate results such as those that occurred with this kind of selection in one person's working lifetime. There are no modern examples of complete domestication of a wild plant, and there is but one example of a new, stable, intergeneric hybrid crop, Triticale (Larter, 1974).

Genetic manipulation of crop species has become more sophisticated in the last century. A primary ingredient for plant improvement by breeding is genetic variability. Before the advent of recombinant DNA technology, the sources of variability were derived from hybridization and mutagenesis. The successful use of plant breeding to produce improved crop varieties has depended upon assembly of new useful combinations within the available variability. Progress in modern crop improvement can be attributed partly to previous advances in broadening the base of available genetic variability. Much progress has also been the result of increasingly more powerful or efficient methods of utilizing that variability (Anonymous, 1981).

Plant breeders seeking sources of new traits have been limited to the genetic variability found in a particular crop's germplasm, or in germplasm of closely related, cross-compatible species. The breeding time required for development of a new variety is long, requiring several to many iterations of the life cycle of the crop species being bred. Additional plant generations are then necessary for field evaluation. Both of these conventional breeding requirements, limited germplasm resource and long generation times, have influenced the genetic complexion of today's high-yielding crop varieties.

The economic and biological constraints affecting the strategies chosen by plant breeders have resulted in modern crop varieties that (a) are usually homozygous, or uniform within a cultivated population; (b) derive their pest and disease resistances from single genes; and (c) have high yield potential. Modern plant breeding has produced crop varieties that are very productive when combined with an intensive input management regime. Extensive use is made of mechanization, chemical biocides and fertilizers to replace labor, and monocultures. Modern crop varieties show strong nitrophilic growth, responding dramatically to available nitrogen typically provided as inorganic fertilizer. Height has been decreased in cereal grains, making plants sturdier and increasing harvest the index (Austin et al., 1980). The development of hybrid maize to exploit the phenomenon of heterosis has been responsible for at least 20 percent of the dramatic genetic gain in yield seen by the introduction of hybrid seed technology over the last 50 years of maize improvement in the United States (Borlaug, 1983). The success of the maize example has encouraged breeders to develop hybrid technology in other crops. Mechanization has required uniform fruit and seed ripening. These changes in crop plant characteristics prove the utility of genetic manipulation in producing plants that are responsive to the agricultural practices of the mid-twentieth century. These practices in turn were made possible by shifts in the relative cost and availability of labor and energy and were responses to the demands of a rapidly increasing world population.

Modern agriculture is a very recent development, when considered in the context of evolution or even human history. It is best considered as an experiment still in progress. Contrasts with the agriculture that has fed humankind for most of its history are dramatic. The land race varieties of major food crops that were grown for centuries in subsistence farming agroecosystems were genetically diverse, carried polygenic disease and pest-resistances, and were environmentally stable (Wilkes and Wilkes, 1972). By comparison with today's varieties, however, they were low-yielding. Farms were small, labor-intensive, and characterized by a mix of species, both plant and animal. The agriculture of primitive humankind, or even of the early decades of "industrialized" agriculture in the late nineteenth century, was less damaging to the environment than today's agriculture may prove to be. It is indisputable that modern agricultural practices are among the many factors that threaten the long-term stability of the earth's environment. Agricultural practices must be adjusted to serve the long-term need for a sustainable agriculture.

## Defining Sustainable Agriculture

"Sustainable agriculture" has emerged in the last 10 years as the most agreed-upon term to describe the varied field of agricultural practices that differ from conventional concepts of modern agricultural production (Bidwell, 1986; World Bank, 1981). The most prevalent definition of sustainable agriculture is one that is "ecologically sound, economically viable, socially just and humane" (Gips, 1987). Most definitions of sustainable agriculture share two key elements: Use of farm chemicals is minimized, especially pesticides and soluble fertilizers, and the farm system is viewed as a whole when making management decisions, even though specific decisions do not appear to have impact outside the area of use or application.

The focus and specific definitions of the subcategories of sustainable agriculture and its practices are many. Some practitioners choose to focus on the ecological aspects of the farming system, calling it organic (USDA, 1980), biological (Friend, 1981; Hodges, 1978), ecological, natural, or alternative (Crosson and Ostrow, 1988). Others focus more narrowly on resource dynamics of the agroecosystem, calling it low-input, or resource-efficient agriculture. Some emphasize the social and ecological aspects (Berry, 1977), while others refer to a specific set of practices or management concepts combined with an ecological/social overview, such as biodynamics (Koepf et al., 1976) and permaculture (Mollison, 1988).

Sustainable agriculture is a system of farming based on the principle that agriculture is, first and foremost, a biological process (Hodges, 1978; Katz, 1984). In practice, this means that sustainable agriculture attempts to mimic the key characteristics of a natural ecosystem, while still maximizing the yield of one or more components. It strives to build complexity into the agroecosystem, to cycle nutrients more efficiently, and to maintain the primacy of the sun as the main source of energy driving the system. The management focus in sustainable agriculture is on long-term optimization of the system as a whole rather than on short-term exploitation. The farmer and/or researcher must select strategies that balance the need for high yields each year with the longer-term biological requirements that contribute to ecological stability. This requires a sophisticated approach that emphasizes stewardship. It also requires an understanding of the internal agroecosystem relationships, especially population dynamics and nutrient monitoring.

In sustainable agriculture, soil is seen as a living system; it is managed to maximize diversity and the well-being of its organisms. Soil chemistry is carefully monitored, and nutrients are supplied in stable forms. Soluble fertilizers are applied carefully to limit their uncontrolled mobility in the ecosystem. Care is taken to ensure minimal soil physical deterioration. Tillage systems are selected to minimize impacts such as erosion, compaction, and oxidation. Soil is kept under continuous cover to the extent feasible.

Complexity and diversity are built into sustainable agriculture at every opportunity. This requires maintenance of a wide range of plant types and habitats on the farm. It also implies use of a sophisticated understanding of population dynamics to manipulate host/pest/predator relationships in the ecosystem, without causing major disruptions requiring other kinds of intervention to manage. Pesticides are used with caution and in such a way as to avoid any type of contamination or disruption of the ecosystem. When biocides are employed, they must be specific to targeted organisms and meet criteria of low mammalian toxicity, limited persistence, and low environmental mobility.

To achieve a sustainable agriculture that embodies these ecological values, national agricultural policy and the economic climate must encourage or mandate the use of practices consistent with these values. Many longer-term benefits of sustainable agriculture, such as reduced damage to soils and to water quality, will not be reflected in the short-term economic calculations of farmers unless these practices also provide the possibility of short-term economic success. If sustainable agriculture is to be achieved, national farm programs and international policy must provide a foundation that will ensure the short-term economic viability of farmers. Today's intense pressures of a burgeoning world population (Brown, 1988; Burki, 1982), the vagaries of massive climatic shifts, and mounting environmental and health concerns directed to the agricultural sector cannot be ignored (Edens and Koenig, 1980). A proper response to these pressures requires that the world's farmers maintain high productivity while avoiding strategies that involve inefficient use of inputs or allow unacceptable levels of environmental degradation (Reed, 1982).

## Engineering for Input-Intensive Agriculture

Today's modern mechanically harvested tomato crop represents an outstanding example of how genetic technology has been used to develop plants specifically adapted for use with an input-intensive style of agriculture. The biology of the species, the history of its domestication, and the efforts of modern breeders have all contributed to the realization of this goal.

The tomato and its wild relatives belong to the genus *Lycopersicon*. The botanical center of origin for the genus is in the Andean region of South America. Domestication of the cultivated tomato, *Lycopersicon esculentum*, is thought to have occurred in what is now Mexico (Jenkins, 1948). Shortly after the Spanish conquest of this region in 1519, the tomato was introduced into Europe. A number of publications from the mid-1500s described the cultivation and consumption of tomatoes in Europe (McCue, 1952). European cultivars returned to the New World by the introduction of tomatoes to North America in the 1700s (Rick, 1978).

The ancient genetic variability within the cultivated tomato species was reduced progressively since the time of its initial domestication. Variability was reduced by "founder effects" as the tomato was brought from South America to Mexico, Europe, and finally North America. Today, in South America, the genetic variability of the cultivated tomato is maintained by insect-medicated cross-pollination between cultivars and the wild tomato species that grow locally as weeds (Rick, 1958). This opportunity for natural gene flow does not occur in Europe and North America, where the tomato is cultivated in isolation from natural pollinating insects and wild tomato species (Rick, 1949). Thus, restoration of the genetic variability that was lost due to genetic drift or to selection by the early breeders was unlikely. The reduction of genetic variability within the modern tomato cultivars is reflected in the greatly reduced isoenzyme polymorphism of modern temperate cultivars relative to that of the cultivars of South America (Rick and Fobes, 1975).

Genetic uniformity in tomato cultivars has been increased intentionally for a variety of reasons. Genetic uniformity within a cultivar together with the practices of certified seed production help to ensure reproducible agronomic performance in successive years. Genetic uniformity among varieties is promoted by the sharing of germplasm resources among breeders; the common goal of development of varieties that are adapted to the contemporary agronomic practices; and the shared needs of tomato processors, who produce a rather uniform set of processed tomato food products.

The biology of the tomato has contributed to the ease with which this genetic uniformity was achieved. Modern tomato cultivars are completely self-pollinated in most areas where they are grown, by virtue of their shortened style (Rick, 1978) and the absence of natural insect pollinators; a high degree of homozygosity is maintained, unless there is intentional cross-pollination. Tomato is a diploid species, thus polyploidy is eliminated as a potential mechanism for maintaining genetic variability within an individual homozygous plant.

Development of a tomato variety suitable for mechanical harvesting was initiated by Jack Hanna in the early 1940s. By 1948, Hanna began working with Coby Lorenzen, the agricultural engineer who headed the project to develop the mechanical harvesting machine (deJanvry et al., 1980; Dickman and Brienes, 1978). Early breeding objectives in the development of the mechanical harvest system concerned fruit ripening and growth habit characteristics. Uniform fruit ripening time was sought because the harvester destroys the plant during harvest, preventing maturation and later collection of immature fruits. Compact plant growth habit was important to facilitate lifting of the plant onto the harvester for separation of fruit from the stems. Firm fruits were required to withstand rough treatment by the mechanical harvester. Diseases became more damaging to the plants modified for these changes in growth habit; therefore, genetic resistance was required as well.

The recessive allele of the *L. esculentum* gene self-pruning *(sp)* (Yeager, 1927) has a major role in the simultaneous ripening and compact, determinate growth habit of VF145, the first processing tomato variety that was widely grown for mechanical harvest. No single major gene controlling fruit firmness has been identified in VF145 or the firmer varieties that have followed. Hanna and the breeders who followed him selected a number of minor genes that change the toughness of the skin, the internal architecture of the fruit (watery locule versus fleshy pericarp), and the overall size and shape of the fruit.

Disease resistance was critical to the success of VF145, because simultaneous fruit development was a stress that increased the susceptibility of the plants to *Verticillium* and *Fusarium* wilt diseases. The resistance gene I initially provided resistance to *Fusarium*. This gene was introgressed into the cultivated tomato from the related wild species *Lycopersicon pimpinellifolium.* A second race of the pathogen arose promptly in response to the widespread use of resistant varieties on increasingly larger acreages. Many of the later varieties have the I-2 gene, which provides resistance to that second race of the pathogen. The I-2 gene also was found in *L. pimellifolium*. A third race of *Fusarium* was first observed in 1983. No single gene giving resistance to race 3 has been found, but a multigene resistance has been identified (Stevens and Rick, 1986).

The situation is quite similar with respect to *Verticillium* resistance. VF145 has the gene Ve for resistance to *Verticillium*, which was introgressed from *L. esculentum var. cerasiforme* (Schnaible et al., 1951). In 1962, a second race of *Verticillium* was reported for the first time. No resistance to race

2 has yet been identified, but some tomato genotypes with a good level of tolerance have been found. Work is underway to incorporate this trait into modern varieties (Stevens and Rick, 1986).

The success of the mechanical harvester led to further increases in the genetic uniformity of tomatoes grown for processing. Within the United States, the growing of processing tomatoes has been concentrated largely in the central valley of California (greater than 220,000 acres; greater than 85 percent U.S. processing tomato yield). Here, Hanna's VF145 was the single predominant processing tomato variety for over 10 years. A succession of many newer varieties have since replaced VF145, but the genetic uniformity of the crop remains high because the newer varieties continue to share a large proportion of their ancestry with VF145.

Disease resistance genes provide examples of how wild tomatoes have been used increasingly to broaden the germplasm base for development of improved tomato cultivars (Rick, 1973). The processing tomato varieties used commercially in any given year remain relatively uniform because any new gene or allele that is economically successful is quickly incorporated into varieties of all the competing seed companies. Thus, new genes, rather than increasing the genetic diversity of the modern cultivars, are normally used in a manner that gives only an upgraded state of genetic uniformity.

In summary, the history of the development of processing tomatoes illustrates how modern plant breeding has tended to reduce genetic variability as a crop is genetically modified to fit a particular agricultural management style. The range of genetic variability found in primitive tomato cultivars has been distilled to yield a relatively narrow breeding germplasm base and homogeneous varieties, each bearing similar sets of traits, many of which are controlled by single genes. Genes already present within the genus *Lycopersicon* have been recombined, via cross-pollination and selective breeding, with those traits necessary for mechanical harvest (single genes as well as polygenic traits). Of the many improvements that have been made in modern tomato cultivars, traits that would decrease reliance on the use of chemicals have not been among them. Large amounts of a variety of different chemicals (pesticides, herbicides, and synthetic fertilizers) are used in production of processing tomatoes.

## Engineering for Sustainability

Tomato cultivars other than those designed for modern, chemically intensive agriculture will be necessary for the development of a more sustainable tomato industry. The biological principles needed to achieve a successful switch to sustainable agriculture are analogous to those that underlay the stability of land race varieties under subsistence production conditions. Cultivar characteristics—such as lowered reliance on synthetic organic chemicals, increased genetic diversity by a wider use of polygenic traits, and tolerance rather than absolute resistance to crop pests—would facilitate the establishment of sustainable agricultural production techniques on a large scale. Tolerance, instead of resistance, is desirable as a way of decreasing selection pressure in pest populations to prevent selection for resistant pest biotypes. These characteristics will have to be incorporated into cultivars, while maintaining high yield potential, so that it will be economical for farmers to adopt sustainable agricultural techniques. The history of the development of the modern tomato cultivars indicates that genetic manipulation is a powerful tool that can be used to modify plants to fit the requirements of a management system. Can this tool be successfully applied to address the agenda of sustainable agriculture? What contribution might modern biotechnology make to advance this agenda?

Progress in breeding for mechanically harvested tomatoes came through the introgression of desirable genes from tomato variants and wild species. It has not always been easy to transfer what is inherited as a single gene from one species to another without also transferring extra, undesirable DNA. It is now possible to isolate, characterize, and modify genes at the molecular level. Recombinant DNA technology allows engineering and transfer of genes between species that are not sexually compatible. These new tools can contribute both to creating variability and to its effective use in managing a breeding program.

***Weed Control.*** The first genes of agricultural interest to be tested using the new technology were genes conferring tolerance to herbicides. Early attention was focussed on the herbicide N-phosphonomethyl glycine (glyphosate), a potent inhibitor of the shikimate pathway leading to the synthesis of aromatic amino acids in bacteria and plants. Two independent research groups set out to genetically engineer resistance to this herbicide in the early 1980s. One group concentrated on engineering a mutant bacterial gene for expression in plants (Comai et al., 1985; Comai et al., 1988). The other group pursued cloning and expression of a plant gene (Shah et al., 1986). Both were successful and had glyphosate-tolerant crop plants in field trials in 1987 and 1988.

Contrary to the claims of some critics of biotechnology, some herbicide tolerances may result in lowered overall usage of herbicidal chemicals (Benbrook and Moses, 1986; Goodman, 1987). Glyphosate tolerance in tomatoes grown for processing is a case in point. Herbicides currently play a major role in processing tomato culture because weed control is crucial to achieving high yields at all stages of tomato culture. Early in the season, competition with weeds can cause yield reductions and delay harvest. At harvest, weeds can hinder mechanical harvesting. Current practices with processing tomatoes in the major U.S. production area, California's central valley, include at least one preplant and preemergent application as well as a "lay by" herbicide application next to the plant row. As many as nine different chemicals have been recommended for spray and soil incorporation (Sims et al., 1979), and typical practice involves the use of at least three of these chemicals on each acre. With the use of a glyphosate-tolerant tomato, post-emergence applications of the herbicide would economically control weeds without harming the tomato crop. The herbicide has a very wide phytotoxicity spectrum but low mammalian toxicity, a relatively short environmental half-life, and is systemic. This could result in a significant decrease in overall herbicide usage and a decrease in toxicity from all the chemicals that are applied, because glyphosate is much less toxic than many other chemicals recommended for use with tomatoes. Fewer applications mean directly lowered overhead costs in time and chemical applied. Less immediately obvious, but no less important, are the benefits that would come in the form of reduced soil compaction from fewer passes over the field with heavy equipment, and less energy expended on soil incorporation of herbicide.

***Insect Resistance.*** A second area of research that has received considerable interest has been the engineering of crop plants for resistance to insect predation. For many years the very specific insecticidal activity of naturally occurring strains of *Bacillus thurengiensis* (Bt) has been known. Preparations of the bacterial spores containing an insecticidal toxin have been formulated and sold for control of lepidopteran pests for at least 20 years. Several groups have reported the expression of the Bt toxin in transgenic plants, including tomato (Fischhoff et al., 1987; Vaeck et al., 1987), and have found insect control in laboratory bioassays as well as in field tests.

Several strategies have been proposed to address the possibility of evolution in pest populations after exposure to plants expressing Bt toxin. Several factors may deter such pest evolution. There are a number of Bt toxin genes, and the range of susceptible insect species is somewhat different for each (Dulmage, 1981). The concurrent use of more than one engineered Bt toxin gene, each with a different toxicity profile, may further reduce the possibility of pest evolution. Industrial and academic scientists are aware of evolution of resistance by pests and have formed a worldwide working group to research the potential for insect evolution and to formulate guidelines for responsible deployment of Bt toxin genes (Marrone, 1988). In addition to the complete resistance promised by tandem gene transfer as mentioned earlier, three other strategies are proposed for Bt toxin gene transfer for tolerance to insect damage. Using genetic engineering techniques, the expression of toxin genes could be limited in overall level or to particular tissues to yield a partial insect kill. Mixtures of transgenic and nontransgenic seed could be developed as multilines (Gould, 1988).

***Disease Resistance.*** A third area where biotechnology can contribute to sustainable agriculture is disease resistance. An impressive example of a disease tolerant phenotype resulted from engineering genes for virus coat protein. Expression of the genes conferred a high degree of tolerance to engineered plants by delaying the onset of disease (Powell Abel et al., 1986) by a margin that may be economically significant to the farmer. In some cases, transgenic plants are resistant to infection by the virus (Cuozzo et al., 1988). The effect is virus-specific; for example, plants expressing the capsid protein of tobacco mosaic virus are resistant to tobacco mosaic virus but are not resistant to tobacco ringspot virus. This phenomenon has been demonstrated in laboratory experiments for at least six plant virus families, and field trials with transgenic tomato plants tolerant to tobacco mosaic virus (TMV) are underway.

Another strategy also has been demonstrated to reduce the damage in crops caused by viral diseases. At least two plant virus groups have satellite RNAs associated with them. Satellite RNAs are small RNA molecules that are completely dependent for their replication and transmission on a helper virus (Francki, 1985). When a virus infection includes a satellite RNA, the symptoms of virus infection often are reduced. DNA copies of satellite RNA sequences expressed in transgenic plants confer this ability to attenuate the symptoms of a virus infection just as a naturally occurring satellite/virus coinfection (Gerlach et al., 1987; Harrison et al., 1987).

There is no example yet of the molecular isolation of a naturally occurring plant disease resistance gene, and the natural biological mechanism of resistance conferred by such genes is unknown. It is possible that examples of genetically engineered disease resistance phenotypes described here use the same mechanism as the naturally occurring resistance genes. In any case, both examples of engineered virus resistances represent novel sources of variability that effectively widen the range of genetic variation available for plant improvement. Although no chemical can impede viral infection directly, the use of plant genotypes resistant or tolerant to a virus would obviate the use of pesticides for the control of viral vectors, such as thrips, aphids, beetles, and white flies.

***Fruit quality.*** Tomato fruit quality and ripening also have been modified using genetic engineering techniques. The enzyme polygalacturonase (PG) is expressed in tomato during fruit ripening and causes the softening of tomato fruit by the hydrolysis of polygalacturonic acid in the cell walls and middle lamellae of fruit. A PG gene has been constructed containing a cauliflower mosaic virus 35S promoter and a full-length copy of the PG gene in reverse ("antisense") orientation. Functional PG enzyme is reduced dramatically during fruit ripening in transgenic plants. This result is due to duplex formation by the constitutively expressed antisense transcript with normally occurring PG sense mRNA. The evidence suggests that this occurs in the nucleus before normal sense RNA can be translated to protein (Sheehy et al., 1988; Smith et al., 1988). Production of lycopene, the red pigment expressed in ripe tomatoes, is not inhibited in the antisense PG plants.

If fruit from transgenic plants bearing the antisense PG gene do not soften immediately after ripening, as do nontransgenic plants, this kind of modification could have profound effects for breeding of both processing and fresh market tomatoes. Fruits that do not become soft immediately after ripening would allow a more complete processing tomato yield, as the first fruits to ripen could be "stored" on the plant until all of the fruits are ripe. Fresh market tomatoes expressing the PG antisense gene could be allowed to ripen on the vine, instead of being picked while still green and "ripened" with applications of ethylene after transport as they are today. This could reduce storage and shipping costs and could improve consumer acceptance.

***Quantitative Genetic Traits.*** A final example of the contribution that recombinant DNA and its associated technologies is making to plants and agriculture is the use of molecular markers in breeding programs (Beckmann and Soller, 1986). The DNA sequences of genes in different individuals within a species or from closely related, sexually compatible species can differ in subtle ways. These differences can be revealed as variations in the pattern observed when total DNA isolated, for example, from individual progeny of a cross, is cut with restriction endonucleases and the resulting fragments separated on the basis of length on an agarose gel. This technology can be useful in managing breeding programs (Helentjaris et al., 1985; Soller and Beckmann, 1983) and in identification and manipulation of quantitative traits (Nienhuis et al., 1987).

Although the tomato genome has been mapped with mutant genes and isozyme polymorphisms, these markers are few in number relative to restriction fragment length polymorphism (RFLP) markers, and they have left much genetic territory still uncharted. RFLP mapping has increased dramatically the detail in the existing tomato genome map. Quantitative traits, each composed of many linked genes with small effects, have been distinguished collectively as Mendelian factors. The tomato RFLP markers associated with such polygenic characters as fruit mass, soluble solids, and fruit pH will simplify the transfer of these traits from exotic species into horticulturally important varieties (Paterson et al., 1988). Undoubtedly, this technology will be applied to other complex characters, such as horizontal disease resistance, to facilitate the breeding of these complex traits in tomato (Rick, 1973; Young and Miller, 1988).

## Conclusions and Future Prospects

The genetically engineered improvements outlined here are all quite recent. No new tomato variety has yet been commercialized that incorporates even one of these genes, although successful field trials have been completed with several transgenic varieties. Considerable commercial interest has been expressed in all of them. Field trials are being conducted to assess the efficacy of proposed plant varieties and to identify environmental issues that may need to be addressed before these varieties can be released for general use. Progress has been made much more quickly in recent years than was predicted. Choices for research and development projects involving plant genetic engineering indicate a trend toward the creation of new genes and varieties that are compatible with the goals of establishing sustainable agricultural management systems.

The development of genetic engineering techniques and the potential for rapid commercialization have raised concerns from both the scientific community and the general public with regard to the potential dangers of genetically engineered organisms. A major area of concern is the possibility that the biotechnology industry will produce new crop cultivars tailor-made to require the use of specific chemicals during the production cycle, thereby increasing the overall use of chemicals in agriculture. This concern is heightened by the well-publicized involvement of large agrochemical companies in plant biotechnology, and the close connections that have emerged between large chemical companies and some biotechnology firms. The conjecture is that such partnerships and the products they create will contribute to the current cycle of consolidation and vertical integration in the food system.

Companies seeking to profit from plant genetic engineering advances like those overviewed here, would not be able to sell the farmer seed if required a higher overhead cost of production by the requirement of additional chemicals. The few examples of potential new genetically engineered varieties will command a higher usage. The seed of genetically engineered varieties will command a higher price if the grower's overhead costs are significantly reduced. Growers will be able to afford to pay more for seed if there will be a corresponding savings elsewhere, in the form of lowered input and energy costs, resulting in lower net overhead costs. Saving the farmer money by reducing overhead costs can be consistent with addition of traits that would enhance sustainable agricultural management systems. While some degree of unforeseeable risk must be recognized, the current regulatory environment is rigorous (for example, USDA, 1987), and all sectors of the American public are becoming more involved and vociferous in the determination of agricultural and environmental policy. This rigorous regulatory climate has even caused some outside observers to voice the fear that advances in the application of biotechnology may be stifled by excessive regulation (McCormick, 1987; Young and Miller, 1988). Potential risks are extremely unlikely to occur and would be more than offset by the benefits genetic engineering offers that can contribute to developing a more sustainable agriculture.

## REFERENCES

Anonymous. 1981. The manipulation of genetic systems in plant breeding. Philosophical Transactions, Royal Society of London B 292: 399-609.

Austin, R. B., J. Bingham, R. D. Blackwell, L. T. Evans, M. A. Ford, C. L. Morgan, and M. Taylor. 1980. Genetic improvements in winter wheat yields since 1900 and associated physiological changes. Journal of Agricultural Science Cambridge 94: 675-689.

Beadl, G. W. 1980. The ancestry of corn. Science of America 242: 112-119.

Beckmann, J. S., and M. Soller. 1986. Restriction fragment length polymorphisms in plant genetic improvement. Oxford Survey of Plant Molecular and Cell Biology 3: 196-250.

Benbrook, C., and P. B. Moses. 1986. Engineering crops to resist herbicides. Technology Review (November-December): 55-79.

Berry, W. 1977. The unsettling of America—culture and agriculture. Sierra Club Books. San Francisco, California.

Bidwell, 1986. Where do we stand on sustainable agriculture? Journal of Soil and Water Conservation 41(5): 317-320.

Bingham, J. 1981. The achievements of conventional plant breeding. Philosophical Transactions, Royal Society of London B. 292: 441-455.

Borlaug, N. E. 1983. Contributions of conventional plant breeding to food production. Science 219: 689-693.

Brown, Lester. 1988. The growing grain gap. World Watch 1:5, 10-19.

Burki, Shahid Javed. 1982. Feeding a growing world. Cornell Executive, Cornell University, Ithaca, New York. Spring 1982. pp. 33-34.

Comai, L., D. Facciotti, W. R. Hiatt, G. A. Thompson, R. E. Rose, and D. M. Stalker. 1985. Expression of aroA in plants confers partial tolerance to glyphosate. Nature 317: 741-744.

Comai, L., N. Larson-Kelly, J. Kiser, C.J.D. Mau, A. R. Pokalsky, C. K. Shewmaker, K. McBride, A. Jones, and D. M. Stalker. 1988. Chloroplast transport of a RuBP carboxylase small subunit-EPSP synthase chimeric protein requires part of the mature small subunit in addition to the transit peptide. Journal of Biological Chemicals (in press).

Crosson, P. R., and J. E. Ostrow. 1988. Alternative agriculture: Sorting out its environmental

benefits. Resources (summer): 14-16.
Cuozzo, M., K. M. O'Connell, W. Kaniewski, R-X Fang, N-H Chua, and N. E. Turner. 1988. Viral protection in transgenic tobacco plants expressing the cucumber mosaic virus coat protein or its antisense RNA. Bio/Technology 6: 549-557.
deJanvry, A. P. LeVeen, and D. Runsten. 1980. Mechanization in California agriculture: The case of canning tomatoes. Unpublished study. Department of Agriculture and Research Economics, University of California, Berkeley.
Dickman, A. I., and M. Brienes. 1978. Interviews with persons involved in the development of the mechanical tomato harvester, the compatible processing tomato and the new agricultural systems that evolved. The Oral History Center, Shields Library, University of California, Davis.
Dulmage, H. T. 1981. Insecticidal activity of isolates of Bacillus thurengiensis and their potential for pest control. *In* H. D. Burges [editor] Microbial control of pests and plant diseases 1970-1980. Academic Press, New York.
Edens, T. C., and H. E. Koenig. 1980. Agroecosystem management in a resource-limited world. Bioscience 30: 697-701.
Fischhoff, D. A., K. S. Bowdish, F. J. Perlak, P. G. Marrone, S. M. McCormick, J. G. Niedermeyer, D. A. Dean, K. Kusano-Kretzmer, E. J. Mayer, D. E. Rochester, S. G. Rogers, and R. T. Fraley. 1987. Insect tolerant transgenic tomato plants. Bio/Technology 5: 807-813.
Francki, R.I.B. 1985. Plant virus satellites. Annual Review Microbiology 39: 151-174.
Friend, G. 1981. Sustainable food systems—the potential for biological agriculture. Unpublished paper given at the Annual Conference of the Agricultural Institute of Canada, St. Catherines, Ontario, August 10, 1981.
Gerlach, W. L., D. Llewellyn, and J. Haseloff. 1987. Construction of a plant disease resistance gene from the satellite RNA of tobacco ringspot virus. Nature 328: 802-805.
Gips, T. 1987. Breaking the pesticide habit: Alternatives to twelve hazardous pesticides. International Alliance for Sustainable Agriculture, Minneapolis, Minnesota.
Goodman, R. M. 1987. Future potential, problems, and practicalities of herbicide-tolerant crops from genetic engineering. Weed Science 35: 28-31.
Goodman, R. M., H. Hauptli, A. Crossway, and V. C. Knauf. 1987. Gene transfer in crop improvement. Science, New York 236: 48-54.
Gould, F. 1988. Evolutionary biology and genetically engineered crops. BioScience 38(1): 26-33.
Harlan, J. R. 1975. Crops and man. American Society of Agronomy, Madison, Wisconsin.
Harrison, B. D., M. A. Mayo, and D. C. Baulcombe. 1987. Virus resistance in transgenic plants that express cucumber mosaic virus satellite RNA. Nature 328: 799-802.
Helentjaris, T., G. King, M. Slocum, C. Siedenstrang, and S. Wegman. 1985. Restriction fragment polymorphisms as probes for plant diversity and their development as tools for applied plant breeding. Plant Molecular Biology 5: 109-118.
Hodges, R. D. 1978. The case for biological agriculture. Ecology Quarterly 10: 187-203.
Houck, C. M., D. Shintani, and V. C. Knauf. 1988. Agrobacterium as a gene transfer agent for plants. Frontiers of Applied Microbiology (in press).
Jenkins, J. A. 1948. The origin of the cultivated tomato. Economic Botany 2: 379-392.
Katz, D. 1984. Sustainable agriculture: Basic guidelines for future farming. *In* Context, (winter): 37-39.
Klee, H., R. Horsch, and S. Rogers. 1987. Agrobacterium-mediated plant transformation and its further applications to plant biology. Annual Review of Plant Physiology 38: 467-486.
Koepf, Herbert H. et al. 1976. Bio-dynamic agriculture: An introduction. Anthroposophic Press, Spring Valley, New York.
Larter, E. N. 1976. A review of the historical development of triticale. pp. 35-52. *In* Tsen [editor] Triticale: First man-made cereal. American Association of Cereal Chemists.

St. Paul, Minnesota.
Marrone, P. 1988. Minutes of the first meeting of the Bt resistance working group. Monsanto Agricultural Company, St. Louis, Missouri.
McCormick, D. 1987. Evening the odds. Bio/Technology 5: 999d.
McCue, G. A. 1952. The history of the use of the tomato: An annotated bibliography. Annals of the Missouri Botanical Garden 39: 289-348.
Mollison, Bill. 1988. Permaculture: A design manual. Tagari Publications, Tyalgum, Australia.
Nienhuis, J., T. Helentjaris, M. Slocum, B. Ruggero, and A. Schaefer. 1987. Restriction fragment length polymorphism analysis of loci associated with insect resistance in tomato. Crop Science 27: 797-803.
Paterson, A. H., E. S. Lander, J. D. Hewitt, S. Peterson, S. E. Lincoln, and S. D. Tanksley. 1988. Resolution of quantitative traits into Mendelian factors by using a complete linkage map of restriction fragment length polymorphisms. Nature 335: 721-726.
Powell Abel, P., R. S. Nelson, B. De., N. Hoffman, S. G. Rogers, R. T. Fraley, and R. N. Beachy. 1986. Delay of disease development in transgenic plants that express the tobacco mosaic virus coat protein gene. Science 232: 738-743.
Reed, J. F. 1982. A changing agriculture and our role in it. Crops and Soils 34: 5.
Rick, C. M. 1949. Rates of natural cross-pollination of tomatoes in various locations in California as measured by the fruits and seeds set on male-sterile plants. Proceedings, American Society of Horticulture Science 54: 237-252.
Rick, C. M. 1958. The role of natural hybridization in the derivation of cultivated tomatoes of western South America. Economic Botany 12: 346-367.
Rick, C. M. 1973. Potential genetic resources in tomato species: Clues from observations in native habitats. pp. 255-268. *In* A. M. Srb [editor] Genes, enzymes, and populations. Plenum, New York.
Rick, C. M. 1978. The tomato. Scientific American 239: 76-87.
Rick, C. M., and J. F. Fobes. 1975. Allozyme variation in the cultivated tomato and closely related species. Torrey Botanical Club Bulletin 102: 376-384.
Schnaible, L. O. S. Cannon, and V. Waddcups. 1951. Inheritance of resistance to vertillium wilt in a tomato cross. Phytopathology 41: 986-990.
Shah, D. M., R. B. Horsch, J. J. Klee, G. M. Kishore, J. A. Winter, M. E. Turner, C. M. Hironaka, P. R. Sanders, C. S. Gasser, S. Aykent, N. R. Siegel, S. G. Rogers, and R. T. Fraley. 1986. Engineering herbicide tolerance in transgenic plants. Science 233: 478-481.
Sheehy, R. E., M. Kramer, and W. R. Hiatt. 1988. Reduction of polygalacturonase activity in tomato fruit by antisense RNA. Proceedings, National Academy Science USA (in press).
Sims, W. L., M. P. Zobel, D. M. May, R. J. Mullen, and P. P. Osterli. 1979. Mechanized growing and harvesting of processing tomatoes. Leaflet 2686. Division of Agricultural Sciences, University of California, Davis.
Smith, C.J.S., C. F. Watson, J. Ray, C. R. Bird, P. C. Morris, W. Schuch, and D. Grierson. 1988. Antisense RNA inhibition of polygalacturonase gene expression in transgenic tomatoes. Nature 334: 724-726.
Soller, M., and J. S. Beckmann. 1983. Genetic polymorphisms in varietal identification and genetic improvement. Theoretical and Applied Genetics 67: 25-33.
Stevens, M. A., and C. M. Rick. 1986. Genetics and breeding. pp. 35-109. *In* J. G. Atherton and J. Rudich [editors] The tomato crop: A scientific basis for improvement. Chapman and Hall, London and New York.
U.S. Department of Agriculture. 1980. Report and recommendations on organic farming. USDA, Washington, D.C.
U.S. Department of Agriculture, Animal and Plant Health Inspection Service. 1987. 7CFR Part 340. Fed. Reg. Vol. 52, no. 115. Tuesday, June 16, 1987; Final Rule.
Vaeck, M. A. Reynaerts, H. Hofte, S. Jansens, M. De Beuckeleer, C. Dean, M. Zabeau,

M. Van Montagu, and J. Leemans. 1987. Transgenic plants protected from insect attack. Nature 328: 33-37.
Wilkes, G. H., and S. Wilkes. 1972. The green revolution. Environment 14: 32-39.
World Bank. 1981. Principles and technologies of sustainable agriculture in tropical areas: A preliminary assessment. The World Bank, Washington, D.C.
Yeager, A. F. 1927. Determinate growth in the tomato. Journal of Heredity 18: 263-265.
Young, E. F., and H. I. Miller. 1988. "Old" biotechnology to "new" biotechnology: Continuum or disjunction? Communication for Advancement of Biotechnology. Proceedings, Biosymposium, Bio Fair Tokyo 88, October 19-22, 1988, Tokyo, Japan. pp. 63-85.
Young, N. D., D. Zamir, M. W. Ganal, and S. D. Tanksley. 1988. Use of isogenic lines and simultaneous probing to identify DNA markers tightly linked to the Tm-2a gene in tomato. Genetics 120: 579-585.

# PEST MANAGEMENT IN SUSTAINABLE AGRICULTURAL SYSTEMS

John M. Luna and Garfield J. House

Until 1962, when Rachael Carson's *Silent Spring* exploded the balloon of environmental complacency, the philosophy and practice of agricultural pest control was centered almost entirely on the use of pesticides. Until *Silent Spring*, only a relatively small handful of entomologists and environmentalists had sounded alarms that pesticides were something less than the perfect solution to pest control problems. Just three years earlier, a landmark paper had appeared in the journal *Hilgardia*. The paper, written by four California entomologists, marked the beginning of the end of a euphoric era of pesticides. Entitled, "The Integrated Control Concept," this paper set forth a new pest control paradigm that sought to integrate economics and ecology (Stern et al., 1959). The authors called for an approach to pest control based on an understanding of pest and crop ecology and "integration" of a variety of biological, cultural, as well as chemical controls into an ecologically and economically sound pest management strategy. Pesticides were to be used, but only as necessary, based on population monitoring and "economic thresholds." In the United States, this approach soon became known as integrated pest management (IPM), while in Europe it was more commonly called integrated control.

In a very brief time, the IPM paradigm—understand the ecology of the system, maximize natural and cultural controls, and use pesticides only as a last resort—became an accepted philosophy in academia, with many agricultural colleges offering undergraduate and graduate programs in integrated pest management. In less than 20 years, a dramatic shift in consciousness had occurred: from that of conceiving pests as organisms to be controlled to that of perceiving pests as members of communities within agroecosystems that need to be managed. Along with this evolution of consciousness, new technologies have emerged to enhance the development of ecologically based pest management systems: pest population sampling and pheromone monitoring systems, rubidium marking techniques for studying insect dispersal, microbial-based pesticides, ridge-till cultivation equipment, computer simulation modelling, expert systems, and many others.

## Defining Integrated Pest Management

Like the term sustainable agriculture, the acronym IPM is widely used under various contexts. For the purpose of this discussion, we define integrated pest management as:

"A strategy of pest containment that seeks to maximize the effectiveness of biological and cultural control factors, utilizing chemical controls only as needed and with a minimum of environmental disturbance."

This definition clearly emphasizes the importance of nonchemical control tactics and the need to minimize pesticide use. Inherent in the IPM approach is the "integration" of control tactics into comprehensive management strategies that are economically viable and ecologically sound. In many situations, the IPM approach has yielded pest control systems that satisfy both economic as well as environmental objectives. Frequently, however, compromises must be made between economic and environmental considerations.

Optimizing pesticide use through the use of scouting and economic thresholds has been the most common approach in the development of IPM programs. Individual fields are "scouted" on regular intervals during the growing season by farmers or by specially trained individuals employed by farmers. Pest densities are estimated using standardized sampling procedures, and data are compared with "economic thresholds" to determine whether pesticide applications are economically justified. Stern et al. (1959) defined the economic threshold as "the density at which control measures should be initiated to prevent an increasing pest population from reaching the economic injury level." The economic injury level was defined as "the lowest population density that will cause economic damage." Economic damage, in turn, was defined as "the amount of injury which will justify the cost of artificial control measures." Another way of looking at the economic threshold concept is by plotting the theoretical net revenue derived from pesticide-treated and untreated fields against pest density (Figure 1). The economic threshold, or action point, becomes the x-coordinate of the interception of the net revenue functions.

The ability to accurately sample pest population and predict whether

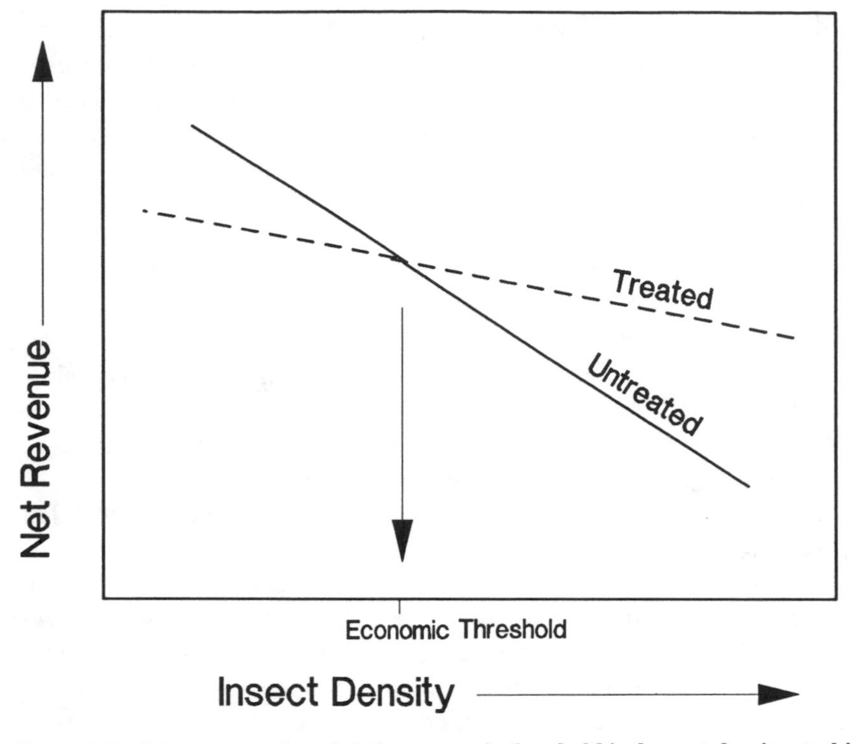


**Figure 1. In this conceptual model, the economic threshold is the pest density at which the net revenue derived from spraying a pesticide equals the net revenue resulting from not applying a pesticide.**

a pesticide application is economically justified is technically quite difficult to attain. The economic threshold concept involves site-specific prediction of pest and natural enemy population dynamics, the relationship between the pest infestation and crop yield or quality, the amount of physical damage that can be prevented by the control measure, costs of control measures and potential crop value, and the costs associated with failure to control the reproduction of a pest (Pedigo et al., 1986). Obviously, the economic threshold is a dynamic variable dependent on many site-specific factors, as well as interacting biological, environmental, and economic components.

## Cultural Control

The use of sampling and economic thresholds is aimed primarily at optimization of pesticide use. Other important components of the IPM approach are cultural and biological controls. Cultural control is the reduction of pest damage or abundance through the purposeful manipulation of the environment. Although cultural control is often associated only with mechanical operations, such as tillage or burning, it involves many aspects of crop and soil management, including crop rotations, time of planting and harvesting, trap cropping, and cropping system diversification (Metcalf and Luckman, 1975). Cultural control practices may inflict direct mortality on pest species, but these practices more often are oriented at prevention of pest outbreaks rather than control of an existing pest infestation. Although some cultural control practices, such as crop rotation, are general practices that affect a variety of pests, many cultural controls are relatively pest-specific and require a thorough knowledge of the life history and habits of the target pest. The most vulnerable stage or stages of the pest's life cycle must be determined and modifications made in cultural practices to prevent attack, kill the pest, or reduce its rate of reproduction.

Because cultural control is aimed primarily at prevention and reduction of pest outbreaks, the results of these practices often are unseen and difficult to quantify. When cultural control practices are integrated easily with other cultural practices, they are usually readily adopted by growers. However, when these practices require significant modification in farming practices, the advantages and disadvantages must be weighed carefully. Although cultural practices alone may not give completely satisfactory pest control, they are important in minimizing pest injury and should be included in any integrated control program.

## Crop Rotation

Crop rotation systems offer numerous advantages in soil structure, fertility and erosion management, as well as aiding in control of various pest species. Crop rotation for pest management consists of a planting pattern alternating susceptible and nonsusceptible crops. The interval necessary between susceptible crops depends upon the length of the pest's life cycle, reproductive potential, degree of specificity, and dispersal characteristics of the target pest. This approach is most useful for fairly immobile, soil-dwelling pest species, such as rootworms or *Diabrotica* sp., and also those pests with a restricted host range or a life cycle of one year or more, such as wireworms or Elateridae. The value of crop rotation is limited in control of highly mobile insects, pathogen spores, or air-borne weed seeds that move readily from field to field.

In the Southeast, crop rotation from soybeans to nonleguminous crops commonly is used to reduce root-knot nematode severity in soybeans (Bailey et al., 1978). In Virginia, gray leaf spot of corn (*Cercospora zeae-maydis*), et al., 1978). In Virginia, gray leaf spot of corn (*Cercospora zeae-maydis*), a fungal disease, is becoming an increasingly serious problem in no-till and reduced tillage production. Rotation of leaf spot infested fields out of corn for at least one year has been recommended as one method to reduce gray leaf spot severity (Stromberg, 1986).

Crop rotation is useful for cultural control of many insect pests, including the western corn rootworm, white-fringed beetle, and wireworms. Adults of the western corn rootworm, *Diabrotica virgifera*, lay eggs in cornfields in late summer. These eggs overwinter and hatch the following spring. If corn is planted in the field, rootworm larvae feed and develop on the corn roots. Because rootworm larvae feed only on corn and the adults only lay eggs in corn fields, rotation of corn with any other crop provides effective cultural control of the rootworm.

## Planting and Harvest Dates

Planting and harvesting dates of some crops can be altered to reduce or to avoid pest damage. A classic example is the delayed planting of wheat as a means of controlling the Hessian fly, *Mayetiola destructor.* In Virginia soybean production, early-planted corn and soybeans are far less susceptible to corn earworm damage than are the late-planted crops. Late-planted corn is also more susceptible to European corn borer damage. In establishment of alfalfa, spring-planted stands are far more susceptible to potato leafhopper damage during the establishment phase than are fall-planted stands. Because potato leafhopper populations usually decline in mid-August, late summer and fall plantings avoid this pest. However, fall plantings can be severely damaged by the fungal pathogen *Sclerotinia* sp., particularly in no-till plantings. Because this disease is linked to the presence of legumes, no-till planting in the fall into sods that contain clovers or other legumes should be avoided whenever possible (VanScoyoc and Stromberg, 1984).

Early harvest of alfalfa can often be used to control both alfalfa weevil and the potato leafhopper (Luna, 1987). The harvesting process, particularly where the hay is chopped for haylage, destroys most weevil larvae and is an effective control where crop growth and weather conditions are favorable. This approach is not foolproof, however, because under certain conditions enough larvae will survive the harvest to damage the next alfalfa regrowth. Thus, when early harvest is used for weevil control, the regrowth must be examined carefully for damage to new growing tips. Early harvest is also used to minimize potato leafhopper damage. Because leafhopper feeding causes internal vascular clogging of the alfalfa plant, harvesting removes this damaged tissue and allows the crop to regrow. Harvesting also kills most immature leafhoppers.

## Biological Control

One of the most successful, nonchemical approaches to pest management is that of biological control. Numerous organisms exist in nature that feed upon, or infect insect pests, pathogens, and weeds. Collectively, these organisms provide a significant level of "natural control," in many cases preventing many insect species from ever reaching the status of pests. The importance of natural control is demonstrated when natural enemy populations are destroyed by insecticides and a previously unimportant insect suddenly escapes from natural control and becomes a major pest. The cabbage looper, *Trichoplusia ni* (Hubner), is an example of this phenomenon. In parts of the San Joaquin Valley in California, the cabbage looper is found commonly in and around cotton fields, but it seldom becomes a pest. However, a catastrophic event, such as treatment of the field with a broad-spectrum insecticide, may destroy the complex of predators that regulates looper abundance; the cabbage looper population then may explode (Reynolds et al., 1975).

Biological control can be defined as the manipulation of parasites, predators, and pathogens to maintain pest populations below economically injurious levels. Natural control is generally defined as the action of these natural enemies without human intervention. A highly successful means of pest control, biocontrol has been used for nearly 100 years since the introduction of the vedalia beetle to control cottony cushion scale on California citrus in 1888. Since this beginning, commercial pest control success has been achieved using biological control methods in at least 253 projects around the world (Van Driesch and Ferro, 1987). According to an analysis by Paul DeBach (1975), each dollar invested in biological control in California has resulted in a $30 increase in net return through reduced crop damage and chemical control costs.

In addition to the benefits of classical biological control involving the importation of natural enemies of a pest, other forms of biological control also have been used successfully to control a large number of pest species in various crops. Augmentations of natural enemies through rearing and mass release have been used extensively for pest control, including release of predatory phytoseid mites for control of spider mites in strawberries (Oatman, 1968), mass release of a coccinellid predator (*Stethorus picipes Casey*) for control of the avocado brown mite (McMurtry et al., 1969), control of bollworm and budworm in cotton through release of green lacewing larvae (Ridgeway and Jones, 1969), and by release of *Trichogramma* wasps (Lingren, 1970).

Periodic releases of natural enemies also are used extensively for pest control in greenhouses, particularly in Europe. Several species of entomophagous arthopods are used, but use of *Encarsia formosa* for control of greenhouse whitefly and predatory mites for spider mite control is common (Hussey, 1985; van Lenteren and Woets, 1988). Releases of natural enemies also have been used to control houseflies around livestock operations. Recent work by Craig Turner and Lorraine Kohler, at Virginia Polytechnic Institute and State University, has shown periodic releases of the housefly predator, *Ophyra aenescens*, into deep-pit layer houses to be effective in reducing housefly population levels below the nuisance threshold (C. Turner, unpublished data). Reviews of the use of augmentation of natural enemies for biological control are provided by Stinner (1977) and by King et al. (1985).

Recent research using entomogenous nematodes in the genera *Steinernema* (*Neoaplectana*) and *Heterorhabditis* has shown promise for control of several insect pests, particularly those living in the soil (Morris, 1985; Kard et al., 1988). About 250 insect species encompassing 10 orders have been found to be susceptible to *Steinernema feltiae* (Poinar, 1979). Recent work on encapsulation of nematodes with calcium alginate by Kaya and Nelson (1985) has shown considerable promise for increasing survivorship of entomogenous nematodes used in biocontrol programs. In addition to active research programs in the United States, research on biological control using entomogenous nematodes is being conducted in Argentina, Australia, Canada, China, England, France, Italy, New Zealand, and the Soviet Union (Gaugler, 1981).

Yet, in spite of the many successes and potential of biological control, funding commitments from the U.S. Department of Agriculture (USDA) to research and extension efforts in biological control continue to decline. Funding for biological control is estimated at less than 20 percent of the funding allocated for IPM programs (Hoy and Herzog, 1985). Pushed into the background by chemical pesticides since the 1950s, biological control funding is currently losing more ground to the perceived prospects of biotechnology.

Van Driesche and Ferro (1987) warn that the benefits of classical biological control may be lost in the "biotechnology stampede." According to these authors, "Like other control strategies, pest controls based on genetically-modified organisms are likely to be of value in specific cases, but will not be a panacea. Yet in the rush by the USDA and the Entomology Departments in Land Grant Universities to create Biotechnology Centers, support for classical biological control has seriously diminished. Positions are being lost. Classical biological control specialists are being reassigned to other duties. Laboratory budgets and staffing are being drastically cut. The truism that what you get out depends on what you put in is certainly true for biological control. We cannot simultaneously allow biological control's infrastructure (positions, agency focus, funding) to be eliminated and also expect to reduce U.S. agriculture's dependence on chemical pesticides. Biotechnology has yet to control its first pest; classical biological control has worked for 100 years and yet its potential has barely been touched. Clearly, one of the best ways to control pests without using pesticides is to use biological control methods."

## Biotechnology and Pest Management

Crop breeding and selection has been an important element in the evolution of agriculture since prehistoric times. Development and use of pest-resistant varieties has had a tremendous impact on global agricultural production. Recent advances in genetic engineering have raised hopes of greatly accelerating classical crop breeding efforts, as well as incorporating new resistance mechanisms. Of particular significance to pest management is the ability of genetic engineering to overcome the narrow taxonomic restrictions on the sources of genes that can be transferred to a given crop. The successful transfer of a toxin-producing gene from a bacterium (*Bacillus thuringiensis*) to tobacco (Goodman et al., 1987) may auger a new era in pest control technologies.

Applying evolutionary theory to the use of genetically engineered pest resistance, Gould (1988) has warned of potential problems of greatly accelerated adaptation by the target pest organisms. According to Gould, "One of the tenets of IPM involves using the ecologically least disruptive tactic that can limit economic loss. The problem here is that highly resistant cultivars will cause the same selection pressure for pest adaptation whether the density of the pest is high or low. Some pests that outbreak sporadically may adapt to a widely planted resistant cultivar before the resistance factor has ever been useful in reducing economic losses due to the pest."

Looking to future applications of genetic engineering for pest resistance, we must not forget the lesson of the southern corn blight epiphytotic in 1970. At that time, almost all major U.S. maize varieties had a common source of susceptibility to the southern corn leaf blight, and in 1970 corn blight caused a greater economic loss on a single crop in a single year than any similar agent known in the history of agriculture (Sill, 1982). This disaster resulted from a breeding program targeted at high yields, but with a very narrow genetic base.

Genetically engineered pest resistance in crops could offer powerful economic and environmental incentives for large-scale adoption by growers, again setting the stage for large-scale disasters when resistance is overcome by the target pest species. Historically, monogenic (single gene) breeding efforts for pest resistance have produced resistance to crown rust of oats, Hessian fly of wheat, powdery mildew of barley, *Cladosporium* leaf mold, tobacco mosaic virus on tomato, golden nematode, and late blight of potato. But in each case, the life of the developed resistant varieties has been relatively short (Day, 1972).

Genetic engineering is still in its infancy and the expectations for it are great. A number of useful products will undoubtedly result from the enormous resources being applied to this subject. However, more lasting progress in ecologically sound pest management and sustainable agriculture will result from agroecological research focused on redesigning the structure and operation of agricultural ecosystems.

## Redefining IPM for Sustainable Agricultural Systems

Extension Service-sponsored IPM programs save farmers more than $500 million annually and significantly reduce pesticide use (Rajotte et al., 1987). This $500 million return results from a meager annual investment of $10 million by USDA in research and Extension IPM programs. American farmers spend an additional $22 million annually for IPM services and information through private consultants and grower-financed programs operated by the Cooperative Extension Service. IPM programs have resulted in dramatic decreases in pesticide use in several crops. For example, from 1971 (when IPM programs were initiated) to 1982, insecticide usage in cotton decreased from 73.4 million pounds of active ingredients to 16.9 million pounds, with a 46 percent decrease in total acreage treated with insecticides (Frisbie and Adkisson, 1985). Similar reductions were realized for grain, sorghum, and peanuts.

But, in spite of these obvious successes and the apparent benefits of IPM, only an estimated 8 percent of cropland (11 million hectares) is enrolled currently in 30 state IPM programs supervised by Cooperative Extension in the United States (Mueller, 1988). Insecticide usage has increased on corn and soybeans, with 15 percent and 60 percent increases, respectively, in acres receiving insecticides (Frisbie and Adkisson, 1985). More than 337 million kilograms of pesticides are used on croplands annually in the United States alone. Of this total, 195 million kilograms of herbicides are used.

Classical IPM, centered primarily around the use of intensive monitoring and judicious application of needed pesticides, may have taken us about as far as it can in terms of pesticide reduction within conventional, high-input farming systems. How does IPM fit in the development of low external input, sustainable agricultural systems?

Pest monitoring and economic thresholds are the cornerstones of the IPM concept. But in real-world IPM programs what are called "economic thresholds" are actually "action thresholds" that are generalized over large areas and are developed on limited research data. Development of truly dynamic economic threshold models involves extremely expensive, long-term research requiring extensive replication under various climatic and soil conditions. Unfortunately, short-term funding and the extensive use of personnel with a two- to three-year time commitment to field work, usually results in fairly limited data sets for implementing Extension IPM programs. Once a sampling program and economic threshold for a particular pest becomes adopted by a state, there is often adoption of that decision-making system by many other states' IPM programs without independent validation of the economic utility of the system in each geographic area. Thus, an enormous gap exists between the action thresholds used in most IPM programs and the theoretical dynamic economic threshold incorporating site-specific parameters described above (Posten et al., 1983).

In spite of the inherent inaccuracy assocated with a lack of long-term research data, IPM programs based on sampling and economic thresholds are nevertheless useful during very high or very low pest population levels. But when pest densities are at intermediate levels (usually those that approach the economic threshold), considerable uncertainty prevails. Uncertainty, or the probability of making a pest control error, can be reduced by increasing the level of complexity of the decision-making model, incorporating more site-specific parameters, and obtaining more field data. In other words, the information input must be increased. This hypothetical relationship between error probability and information input is shown in figure 2a. However, with increasing model complexity and the need for data gathering are the associated costs (Figure 2b). When the increasing cost of information, for example, scouting, exactly equals the cost of chemical control, the grower no longer can afford to spend money on information to make a correct decision and will usually apply a chemical pesticide.

It is critical that the farmer using scouting and economic thresholds is convinced that these methods aid in making the correct pest management decision. With the year's income frequently riding on a single pest control decision, buying insurance by applying a pesticide is a safe decision. A

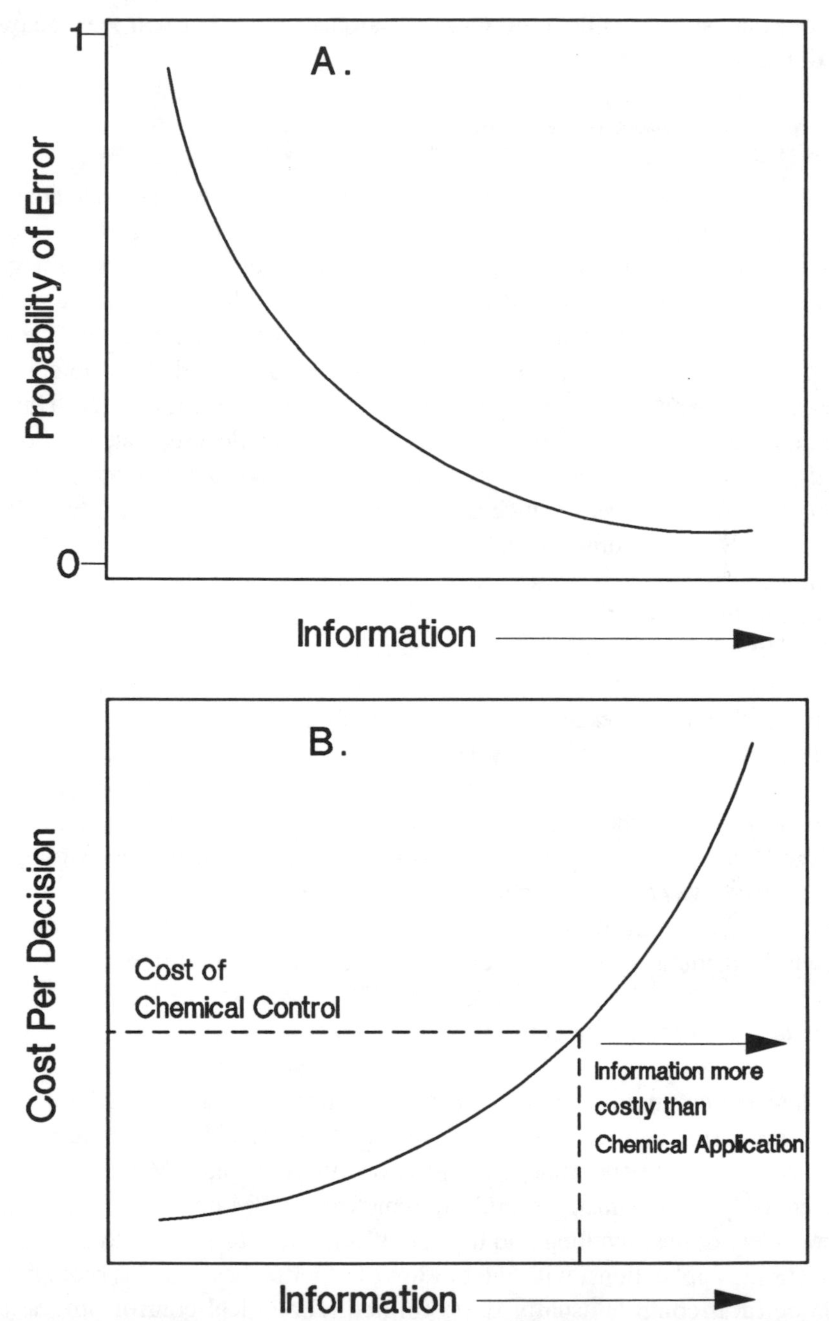


**Figure 2. In these conceptual models, as the quantity and quality of information increases, the probability of error associated with a particular pest management decision decreases (A). However, the cost per decision increases rapidly as information increases (B).**

farmer may spend a little more money than necessary, but will feel secure in doing so.

## Economic Analyses of IPM and Off-Farm Costs

The economics of pesticide use and IPM are based only on private costs, that is, production costs actually borne by the farmer. The farmer does not currently pay the external, off-farm costs of environmental damage caused by overuse of pesticides or of pesticide regulation, as in the case of fines for pollution in the industrial sector of the U.S. economy. These costs, such as pesticide pollution, pesticide regulation, and clean-up costs, are substantial and are currently being borne by taxpayers in general. Thus, the microeconomic benefits associated with pesticide use, that is, those to the farming sector, are grossly inflated. Although the driving motivation of IPM has been to reduce environmental contamination through reduced pesticide use, unfortunately, many IPM programs increase and justify pesticide use through economic analyses that ignore the total costs of pesticide use.

IPM programs also have been centered around single commodities—soybeans, cotton, apples, etc.—rather than being centered around whole farming systems, for example, a mixed cropping/livestock system or a fruit and vegetable truck-farming system. Focus by most applied agricultural scientists has been narrowed to the specialty crops for which they are responsible. Thus, when there is interdisciplinary cooperation among agricultural scientists, soybean agronomists work with soybean entomologists and soybean weed-control specialists. We certainly need interdisciplinary cooperation in research and extension, but individuals must begin looking at whole farming systems rather than specialty commodities.

## Pest or Pesticide Management?

The IPM paradigm is predicated on an understanding of the ecology of the crop/pest system and an integration of control tactics into comprehensive pest management strategies. However, in practice, IPM has been reduced to "pesticide management" in many cases. IPM programs often consist merely of pest scouting and the use of economic thresholds to optimize insecticide applications, with integration of other tactics, such as biological and cultural controls usually quite limited. Biological control programs and IPM programs often are commonly developed independently from each other. The full potential of integrated pest management is far from realized.

Clearly, the IPM paradigm has been extremely useful in reducing insecticide use. A consciousness of using pesticides only when necessary has coevolved with the technology of pest monitoring and yield loss prediction. But, in some ways, this simple IPM paradigm is analogous to the soil-testing paradigm: "Take soil tests and only apply as much fertilizer as the soil tests call for." This paradigm has been useful in reducing overuse of fertilizers, but as the 1982 series in *The New Farm* magazine on "Testing, Testing, Testing..." (DeVault, 1982) revealed, there is considerable variation among soil-testing laboratories concerning fertilizer recommendations. For example, some laboratories recommended 109 kilograms of nitrogen fertilizer, while others recommended none.

Merely relying on soil tests for fertilizer recommendations does not address a more fundamental issue in sustainable agriculture: How do we reduce the overall need for fertilizers in the first place? Crop rotations and the use of legumes to provide biologically fixed nitrogen have been shown to reduce the need for nitrogen fertilizers. Similarly, with IPM, scouting and using pesticides only when threshold levels are reached reduces their use. However, the issue of how we reduce our dependence on pesticides is avoided. Increased emphasis needs to be placed on cultural and biological control practices that reduce pest populations, prevent pest outbreaks, and reduce the need for chemical pesticides.

## Need for Farming Systems Design

Of paramount importance is the redesign of farming systems to accomplish the multiple objectives of sustainable agriculture: more profitable production of healthy livestock and crops while minimizing soil erosion and environmental contamination. From its inception, the IPM paradigm has called for the "system approach" to understanding the complex interactions occurring within agroecosystems. As Edwards (1988) has pointed out, cultural practices, particularly the use of agricultural chemicals, frequently have impacts on a wide array of nontarget organisms. A few examples of these nontarget impacts include:

■ Stimulation of aphid outbreaks in cole crops through the use of nitrogen fertilizers (van Emden, 1966).

■ Insecticides increasing weed populations by killing the natural enemies of the weeds (Smith, 1982).

■ Use of carbofuran insecticide increasing the growth of crabgrass and other grassy weeds (D. Wolf, unpublished data).

■ Fungicides killing soil fungi that exert natural control over nematode populations (Kerry, 1988).

■ Insecticides and fungicides reducing earthworm populations, hence, lowering soil fertility and water infiltration rates (Edwards and Lofty, 1977).

Clearly, there is a fundamental need to study the ecology of agricultural systems. But rather than continuing to study conventional high-input systems in an effort to manage these systems more efficiently, sustainable agriculture places a renewed emphasis on designing and developing new agroecosystems that maximize beneficial ecological processes and minimize expensive off-farm inputs. Historically, IPM programs have been designed to function within the confines of an established, high-intensive, high-input agricultural system. Success under such narrow and therefore precarious circumstances is difficult. Under such systems, pest control is viewed as one of many farm operations. Biological activity beyond crop photosynthesis and plant nutrient uptake is ignored or suppressed, for example.

Thus, sustainable agriculture is more than methodology; rather, a different philosophical approach is required. Although sustainable agricultural systems are associated with lower inputs of fossil fuel-based chemicals, they require increased knowledge about and management of ecological processes. Thus, sustainable systems could be described better as involving low material input and high information input (Figure 3) (Stinner and House,

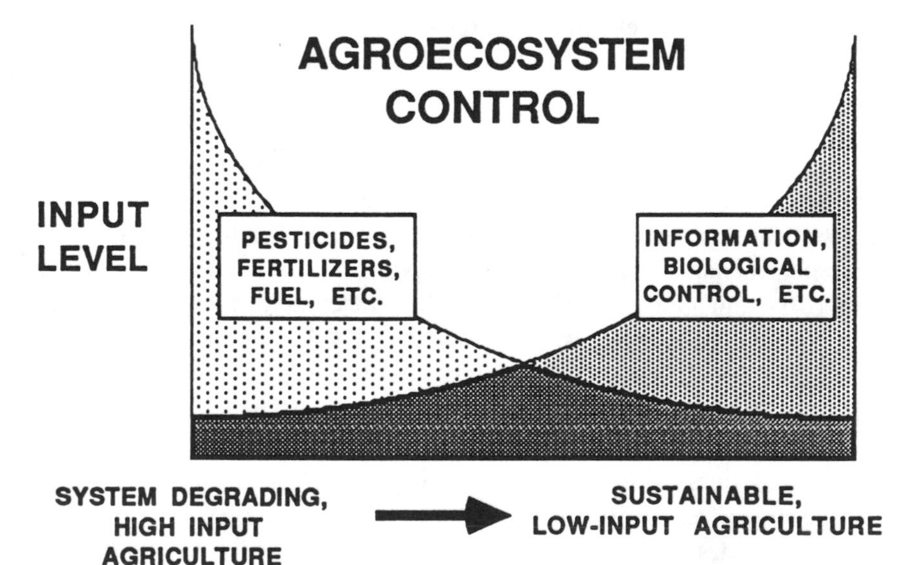


**Figure 3. Sustainable agricultural systems are conceptualized here as being low in material input—pesticides, inorganic fertilizers, etc.—and high in information input—applied ecological knowledge of the system. High chemical input practices conceal and depreciate the importance of ecological processes occurring in agricultural systems. However, as pesticides, fertilizers, and other inputs are reduced, greater knowledge of the interactions occurring in agroecosystems is required for success (Stinner and House, 1988).**

1988). High chemical input practices mask the ecological processes occurring in agricultural systems. For example, adding large quantities of inorganic fertilizer to cropping systems allows a farmer to ignore or pay little attention to nutrient cycling processes that involve diverse groups of soil organisms and the associated mechanisms affecting nutrient loss from soil systems, while still obtaining high yields. Similarly, applying herbicides to reduce weeds eliminates that component from an agricultural system, so aspects of interspecific associations among plants, such as competition and mutualism, become irrelevant. Stinner and House (1988) suggest that as agricultural chemical inputs are lowered (Figure 3), there is a concomitant increase in the need for fundamental understanding and management of ecological processes in agricultural systems. In other words, we need to substitute knowledge and management of ecological processes for large energy and chemical subsidies.

With the current groundswell of interest in low-input, sustainable agriculture, we are at a historical juncture. There is an increasing willingness to question "business as usual" concerning how we grow and market crops. We have the potential to create an agriculture than can reduce the dependency on agricultural chemical inputs dramatically by substituting information-intensive, ecologically compatible alternatives. To realize this potential, we must have vision and the commitment to explore alternative approaches to agricultural production. Sustainable agriculture is not merely a collection of discrete farming practices, but is a vision of the future. We believe that sustainable agriculture has the potential to achieve a common ground between environmental and economic goals. But success ultimately depends upon our willingness to accept and implement a new philosophy of agricultural production.

## REFERENCES

Bailey, B. A., E. B. Whitty, D. H. Teem, F. A. Johnson, R. A. Dunn, T. A. Kucharek, and R. P. Cromwell. 1978. Soybean production guide. Circular 277E. Florida Cooperative Extension Service, Gainesville. 16 pp.

Chang, N. T., R. R. Wiseman, R. E. Lynch, and D. H. Habeck. Influence of N fertilizer on the resistance of selected grasses to fall armyworm larvae. Journal of Agricultural Entomology 2: 137-146.

Coulson, R. N., and M. C. Saunders. 1987. Computer-assisted decision-making as applied to entomology. Annual Review of Entomology 32: 415-438.

Day, P. R. 1972. Crop resistance to pests and pathogens. pp. 257-272. *In* Pest control strategies for the future. Division of Biology Agriculture. National Research Council, National Academy of Science, Washington, D.C.

DeBach, P. 1975. Biological control by natural enemies. Cambridge University Press, Cambridge, England.

DeVault, G. 1982. Testing...testing: Nitrogen, the never-never land of N. The New Farm 4:29-41.

Doutt, R. L. 1964. The historical development of biological control. pp. 21-42. *In* Paul DeBach [editor] Biological control of insect pests and weeds. Chapman and Hall, London, England.

Edwards, C. A. 1988. The concept of integrated systems in lower input/sustainable agriculture. American Journal of Alternative Agriculture 2: 148-152.

Edwards, C. A., and J. R. Lofty. 1977. The biology of earthworms (second edition). Chapman and Hall, London, England. 333 pp.

Frisbie, R. E., and P. L. Adkisson. 1985. IPM: Definitions and current status in U.S. agriculture. pp. 41-49. *In* M. A. Hoy and D. C. Herzog [editors] Biological control in agricultural IPM systems. Academic Press, New York, New York.

Gaugler, R. 1981. Biological control potential of Neoaplectanid nematodes. Journal of Nematology 13: 241-249.

Goodman, R. M., H. Hauptli, A. Crossway, and V. C. Knauf. 1987. Gene transfer in crop improvement. Science 236: 48-54.

Gould, F. 1988. Evolutionary biology and genetically engineered crops. BioScience 38: 26-33.

Hoy, M. A., and D. C. Herzog. 1985. Preface: Biological control in agricultural IPM systems. *In* M. A. Hoy and D. C. Herzog [editors] Biological control in agricultural IPM systems. Academic Press, New York, New York.

Hussey, N. W. 1985. History of biological control in protected culture. pp. 11-22. *In* W. Hussey and N. Scopes [editors] Biological pest control: The glasshouse experience. Cornell University Press, Ithaca, New York.

Kard, B. M., F. P. Hain, and W. M. Brooks. 1988. Field suppression of three white grub species (Coleoptera: scarabaeidae) by the entomogenous nematodes Steinernema feltiae and Heterorhabditis heliothidis. Journal of Economic Entomology 81: 1,033-1,039.

Kaya, H. K., and C. E. Nelson. 1985. Encapsulation of steinernematid and heterorhabditid nematodes with calcium alginate: A new approach for insect control and other applications. Environmental Entomology 14: 572-574.

Kerry, B. 1988. Fungal parasites of nematodes and their role in controlling nematode populations. *In* C. A. Edwards, B. R. Stinner, and D. H. Stinner [editors] Interactions between soil organisms. Elsevier Science Publishers, Amsterdam, The Netherlands.

King, E. G., K. R. Hopper, and J. E. Powell. 1985. Analysis of systems for biological control of crop arthropod pests in the U.S. by augmentation of predators and parasites. pp. 201-227. *In* M. A. Hoy and D. C. Herzog [editors] Biological control in agricultural IPM systems. Academic Press, New York, New York.

Lingren, P. D. 1970. Biological control—can it be effectively used in cotton production today? pp. 236-240. *In* Proceedings, Second Annual Texas Conference on Insect, Plant Disease, Weed and Bush Control. Texas A&M University, College Station.

Luna, J. M. 1987. Forage crop pest management. *In* Pest management guide for forages and pastures. Publication 456-014. Virginia Cooperative Extension Service, Blacksburg.

Luna, J. M. 1988. Influence of soil fertility practices on agricultural pests. pp. 589-600. *In* P. Allen and D. Van Dusen [editors] Global perspectives on agroecology and sustainable agricultural systems. Agroecology Program, University of California, Santa Cruz.

McMurtry, J. A., H. G. Johnson, and G. T. Scriven. 1969. Experiments to determine effects of mass releases of Stethorus picipes on the level of infestation of the avocado brown mite. Journal of Economic Entomology 62: 1,216-1,221.

Metcalf, R. L., and W. H. Luckman, editors. 1975. Introduction to insect pest management. John Wiley and Sons, New York, New York.

Morris, O. N. 1985. Susceptibility of 31 species of agricultural insect pests to the entomogenous nematodes Steinermema feltiae and Heterorhabditis bacteriophora. Canadian Entomology 117: 401-407.

Mueller, W. 1988. IPM: A wise discipline. So why hasn't it caught on? Agricultural Chemical Age. 32: 6-7, 10, 22-23.

Mumford, J. D., and G. A. Norton. 1984. Economics of decision making in pest management. Annual Review of Entomology 29: 157-174.

Oatman, E. R., J. A. McMurtry, and V. Voth. 1968. Suppression of the two-spotted spider mite on strawberry with mass releases of Phytoseiulus persimilis. Journal of Economic Entomology 61: 1,517-1,521.

Pedigo, L. P., S. H. Hutchins, and L. G. Higley. 1986. Economic injury levels in theory and practice. Annual Review of Entomology 31: 341-368.

Poinar, G. O. 1979. Nematodes for biological control of insects CRC. Boco Raton, Florida.

Poston, F. L., L. P. Pedigo, and S. M. Welch. 1983. Economic injury levels: Reality and practicality. Bulletin Entomology Society of America 29: 49-53.

Rajotte, E. G., G. W. Norton, R. F. Kazmierczak, M. T. Lambur, and W. A. Allen. 1987. The national evaluation of extension's integrated pest management (IPM) programs. Publication 491-010. Virginia Cooperative Extension Service, Blacksburg.

Randall, G. W. 1987. Ridge planting: A system for the future. Crops and Soils 39: 9-11.

Reynolds, H. T., P. L. Adkisson, and R. F. Smith. 1975. Cotton insect pest management. pp. 379-442. *In* R. L. Metcalf and W. H. Luckman [editors] Introduction to insect pest management. John Wiley and Sons, New York, New York.

Ridgeway, R. L., and S. L. Jones. 1969. Inundative releases of Chrysopa carnea for control of Heliothis on cotton. Journal of Economic Entomology 62: 177-180.

Rodriquez, J. G. 1958. The comparative NPK nutrition of Panonychus ulmi (Koch) and Tetranychus telarius (L.) on apple trees. Journal of Economic Entomology 51: 369-373.

Sill, W. H. 1982. Plant protection: An integrated interdisciplinary approach. Iowa State University Press, Ames. 297 pp.

Smith, R. J., Jr. 1982. Integration of microbial herbicides with existing pest management programs. pp. 189-206. *In* R. Charudattan and H. L. Walker [editors] Biological control of weeds with plant pathogens. John Wiley and Sons, New York, New York.

Stern, V. M., R. F. Smith, R. van den Bosch, and K. Hagen. 1959. The integrated control concept. Hilgardia 29: 81-101.

Stinner, B. R., and G. J. House. 1988. Role of ecology in lower-input, sustainable agriculture: An introduction. American Journal of Alternative Agriculture 2: 146-147.

Stinner, R. E. 1977. Efficacy of inundative releases. Annual Review of Entomology 22: 515-531.

Stromberg, E. 1986. Gray leaf spot disease of corn. Publication 450-072. Virginia Cooperative Extension Service, Blacksburg.

Van Driesche, R. G., and D. N. Ferro. 1987. Will the benefits of classical biological control be lost in the "biotechnology stampede?" American Journal of Alternative Agriculture 2: 50, 96.

van Emden, H. F. 1966. Studies on the relations of insect and host plant. III. A comparison of the reproduction of Brevicoryne brassicae and Myzus persicae (Hemiptera: Aphididae) on brussels sprout plants supplied with different rates of nitrogen and potassium. Entomological Experiments Applied 9: 444-460.

van Lenteren, J. C., and J. Woets. 1988. Biological and integrated pest control in greenhouses. Annual Review Entomology 33: 239-269.

VanScoyoc, S. E., and E. L. Stromberg. 1984. Results of the 1984 Sclerotinia crown and stem rot survey. Virginia Cooperative Extension Service Plant Protection Newsletter (2): 7-13.

# EVOLVING STRATEGIES FOR MANAGING WEEDS

Emilie E. Regnier and Rhonda R. Janke

Management of weeds in agricultural fields with herbicides is under scrutiny because of off-target herbicide effects, primarily ground and surface water contamination (Schweizer, 1988). There is a flux of research on alternative methods of controlling weeds as well as on new concepts underlying weed control. Prior to 1944, when 2,4-D was introduced as a herbicide (Ross and Lembi, 1985), weeds were controlled through the integrated use of crop rotation, competitive or "smother" crops, and tillage (Runnels and Schaffner, 1931). With increasing use of herbicides, these methods have declined in importance. Herbicides have replaced tillage, allowing reduced tillage systems to become possible, while present economics disfavor crop rotation (Stonehouse et al., 1988).

Herbicides constituted 85 percent of total U.S. pesticide use in 1988; 226, 112, and 37 million pounds were used in corn, soybean, and cotton production, respectively (Anonymous, 1988a). Herbicide costs in winter cereals in the United Kingdom exceeded 20 percent of total variable costs in crop production, or 6 percent of gross output in terms of grain yield (Sim, 1987). Effectiveness, labor and time savings, and ability to reduce tillage have led to reliance on herbicides for weed control. Over-reliance may be risky; abrupt removal of a herbicide from the market due to toxicological or environmental problems could leave producers unprepared for controlling weeds by other means. In developing countries, while herbicides may be effective, herbicide technology may be inappropriate because of different cropping systems, particularly the use of intercropping; lack of application equipment, knowledge, or literacy; or cost (Mercado, 1987).

While reduced or nonchemical weed control alternatives are being sought, adoption of conservation tillage is increasing rapidly in the United States. Projections of the total planted cropland in conservation tillage in the year 2010 range from 63 to 82 percent (Schertz, 1988). Alternative weed control methods are needed that are compatible with reduced tillage.

## Mechanical Weed Control

Mechanical cultivation for weed control in crops probably is the oldest method used except for hand-weeding. Changes have come about over the centuries as equipment and power available for field crop production has increased, from hand hoes in prehistoric and historic times (Anonymous, 1897), to plows and other horse-drawn implements beginning in the 17th and 18th centuries, to tractor-drawn implements in the 20th century (Timmons, 1970). Hand-weeding still is the most widely used method in underdeveloped countries today (Mercado, 1987).

Recent developments in mechanical weed control are primarily in the area of plant ecology—increasing knowledge of how plants respond to tillage at different points in their life cycle. Application of this knowledge allows more effective and efficient use of available tillage tools in combination with other weed-control practices. Equipment continues to be improved; recent innovations include the development of row-crop cultivators designed to work in high residue (conservation tillage) systems and improvement of flame or thermal control of weeds.

Tillage control of weeds can be divided into four categories, based on the point in the crop's life cycle at which tillage is employed. These include (1) primary, (2) secondary, (3) selective, and (4) fallow-season repeated tillage. Primary tillage is the initial breaking of soil in the fall or spring, prior to planting, with the moldboard, chisel, or disk plow. Chisel and disk plowing are forms of reduced tillage in which the soil is not completely inverted as it is with moldboard plowing (conventional tillage); some crop residues are left on the soil surface. Secondary tillage involves additional seedbed preparation and/or weed control operations with lighter equipment, such as disk or springtooth harrows and field cultivators. Selective or postplant tillage involves broadcast tools, such as rotary hoes or light harrows used at crop planting and shortly after crop emergence, and interrow tillage with row-crop cultivators used after crop emergence. Fallow-season repeated tillage is used to control perennial weeds or to reduce weed seed content in the soil.

Shallow tillage, rather than deep tillage, is recommended to prevent burying and thus preserving weed seed and to avoid infesting the upper inches of soil with dormant but viable seed from lower depths (Runnels and Schaffner, 1931). Fall primary tillage is recommended to allow more time for secondary tillage operations in the spring. The tools used in primary tillage affect weed spectra and herbicide efficacy. Changes in tillage regimes have caused at least as many changes in weed species over the years as herbicides and other factors (Haas and Streibig, 1982). Moldboard plowing partially suppresses perennial weeds, while chisel and disk plowing are less effective (Ross and Lembi, 1985). Reduced tillage thus results in a shift toward perennial weeds (Koskinen and McWhorter, 1986) and robust winter annual weeds, such as horseweed (Brown and Whitwell, 1988), that can survive infrequent tillage. Small-seeded annual grasses and broadleaves (foxtails, fall panicum, pigweeds, and common lambsquarter) are also found in increased densities in reduced tillage systems, while large-seeded broadleaf weeds, such as velvetleaf, decrease (Koskinen and McWhorter, 1986; Buhler and Daniel, 1988). The reason for this species shift may have to do with a requirement for deeper burial for germination of large-seeded weeds than for small-seeded weeds.

Crop residues can alter herbicide performance by intercepting and retaining herbicides, thereby necessitating higher herbicide rates. A 1980 survey (Duffy and Hanthorn, 1984) reported greater herbicide use and increased herbicide costs in reduced versus conventional corn and soybean tillage systems in the Midwest, mid-South, and southwestern United States. Herbicide costs and inadequate weed control, particularly of perennial weeds, were listed as major problems in conservation tillage.

Repeated secondary tillage has traditionally been recommended for control of winter and summer annual weeds (Runnels and Schaffner, 1981; Hardy, 1931; Brown and Whitwell, 1988). The practice is common among organic farmers (McLeod and Swezey, 1980). The purpose of repeated shallow tillage is to stimulate germination of weed seed in the upper soil by exposing it to light, abrasion, warmer temperatures, or oxygen and also to bring it to the soil surface where it is susceptible to decay and predation. Each flush of weeds following a tillage operation is killed by the subsequent tillage operation. This tillage technique results in loss of germinable weed seed from the upper soil. Based on his experiments, Roberts (1970) estimated that 30 to 60 percent of the seeds of a given weed species can be eliminated annually from the soil by tillage. Others have found that tillage results in greater annual loss of weed seeds, wild oats seed, for example, declined 80 percent annually due to tillage (Wilson, 1978), and barnyardgrass seed was eliminated from the soil by two years of tillage (Standifer, 1979). Crop planting is delayed as long as possible in fields known to contain a large number of seeds in order to allow more tillage operations.

Many weeds have patterns of emergence that peak in April and May (Ogg and Dawson, 1984). A delayed planting date (after peak emergence) thus may have weed control benefits whether or not weed germination is stimulated by secondary tillage. Shallow tillage increased the cumulative (overall) emergence of common lambsquarter, redroot pigweed, and black nightshade, but germination of Russian thistle, cutleaf nightshade, or hairy nightshade did not increase (Ogg and Dawson, 1984). This may explain why Robinson and Dunham (1956) found no advantage in weed control from repeated versus a single secondary tillage operation for soybeans, although soybeans planted in late June were less weedy than soybeans planted earlier. Stimulation of weed germination by tillage depends in part on environmental conditions, such as rainfall and temperature, and on seed physiological factors, such as dormancy. Where conditions are not suitable, tillage may be ineffective in inducing germination.

Selective or post-plant tillage is the practice usually associated with mechanical weed control, although it is only one aspect of mechanical control. There are two phases of post-plant tillage. The first is immediately after planting until approximately one week after crop emergence. During this period, shallow cultivation tools, such as the rotary hoe (Anonymous, 1988b) and light harrows can be used to control weeds in the "white" stage (roots elongated, no true leaves with little green showing). The tools control both the within-row and between-row weeds because they are used to "blind cultivate" (not attempting to stay between the crop rows). These tools can be used on large-seeded crops planted one to two inches or more deep or on crops with well-established root systems, such as winter grains at spring time or established alfalfa.

Once the crops have emerged to the point that injury is likely with broadcast tools, a row-crop cultivator can be used. The cultivator tool bar can be equipped with one or more shovels, sweeps, shanks, flexible and rigid tines, disk hillers, and rotary cultivators (Buckingham, 1976). These can be of an infinite number of sizes, shapes, and spacings; adjustment to specific field and crop conditions is an art. In addition, various types of stationary and rolling shields can be attached to the cultivator to minimize crop injury during early cultivations.

The length of season available to cultivate is limited by the height of the crop with respect to the tool bar of the cultivator and by the potential for damage to crop roots growing near the soil surface. Late-season cultivations probably are not necessary if competitive crop cultivars are grown, and weed control for most crops is required only for the first four to eight weeks following planting (Ross and Lembi, 1986; Stoller et al., 1986). Row spacing influences this critical weed-free period. Row spacings of 106, 79, and 53 cm in cotton required 14, 10, and 6 weeks of weed-free maintenance, respectively, for maximum yield (Rogers et al., 1976). In cultivation experiments with cotton over a five-year period, Buchanan and Hiltbold (1977) found that deep and shallow cultivation at weekly intervals for the full season, for half of the season, and three times (factorial design) did not result in significantly different yields of cotton. Yields did not differ from four different herbicide regimes except in three cases where yields were lower with the herbicide. They noted no beneficial effect of cultivation beyond that of weed control.

Fallow-season repeated tillage has been used historically to control perennial weeds if infestations are too severe to manage otherwise (Runnels and Schaffner, 1931). The underground vegetative structures of perennial weeds are "starved" through a system of repeated shoot removal by tillage or mowing, which eventually results in eradicating the weed from the field. Repeated tillage and mowing, when performed before flowers open, also prevent seed reproduction. Control of severe perennial weeds, such as quackgrass, Johnsongrass, and Canada thistle, involves sequences tailored for each weed of a full- or partial-season fallow at a specified depth depending on the rooting depth of the weed, followed by smother or competitive crops, and cultivatable crops (Runnels and Schaffner, 1931). One or more seasons of fallow are required to fully eradicate a perennial weed by tillage alone, with tillage repeated every three weeks or more frequently (Timmons and Bruns, 1951; Bakke et al., 1944). During this time, the land is out of production, and the soil left in a disturbed state. Recommendations to farmers include leaving the surface rough if possible, or with residues left on the surface. Fallow-season repeated tillage also is recommended to reduce soil weed seed numbers in fields severely infested with annual weeds.

An example of an innovative tillage system, first investigated in the 1950s, is ridge tillage. No primary tillage is used. At planting, the soil, crop residue, weeds, and weed seeds are scraped off the top of ridges formed the previous season and dumped between rows; this creates a weed- and residue-free seedbed on the top of the ridges (Forcella and Lindstrom, 1988). Subsequent cultivations remove the weeds and soil from between the rows, rebuilding the ridges; a herbicide band or rotary hoeing is required to control weeds in the row. This system may be advantageous in that weed seeds are removed from proximity to the crop at the time the crop is planted (Forcella and Lindstrom, 1988), which may help to reduce weed competition with the crop during crop emergence and early growth. Further, ridge tillage offers potential as a system in which weed control may be provided by interrow cultivation within a reduced tillage system. More research is needed to determine the feasibility of weed control in ridge-till with interrow tillage and with or without herbicide bands.

Flame-weeding, also known as thermal control, is another option for preplant weed control; it can be combined with a stale seedbed technique to achieve a weed-free field at the time of crop emergence. Flame-weeders have been investigated for weed control in onions, carrots (Desvauz and Ott, 1988), and corn (Geier and Vogtmann, 1988) in Europe. Flame-weeding involves heating but not burning weeds. Disadvantages are that it is at least as energy intensive as current mechanical and chemical control, and timing is critical. Advantages include good control of spring weeds stimulated to germinate by secondary tillage operations, followed by no soil disturbance to further germinate weed seed and no residual effect (as with herbicides) that may limit relay crop, cover crop, and rotational crop options later in the season.

Few comparisons of the costs of mechanical and chemical weed control are available. This probably reflects that weed control costs vary from farm to farm and from a few dollars per acre for a single herbicide to $20 to $40 per acre for a no-till herbicide mixture. Nonmonetary costs and tradeoffs between chemical and mechanical control from the farmer's point of view, such as convenience, timing, and flexibility, are aptly compared in an extension bulletin published by the University of Nebraska (Bender, 1987).

Potential for soil erosion is a disadvantage of using tillage for weed control. Tillage implements are needed that maintain soil cover by residue while effectively controlling weeds. Erosion is caused primarily by lack of crop residues on the soil surface and also pulverization of the soil (Griffith et al., 1986), both of which are increased by tillage. However, systems based on tillage for weed control tend to include longer crop rotations consisting of row crops, winter grains, and perennial forage crops. Thus, effects of tillage on the soil may be diluted over time or partially compensated by the inclusion of perennial crops.

Herbicide use makes effective weed control possible with monocultures and short crop rotations. In these systems, row crops, such as corn or soybeans, are preferred—crops that are more prone to soil erosion than perennial sods or legumes. It would be interesting to compare such tillage and crop rotation-based systems to herbicide-based weed control systems while keeping primary tillage constant.

## Weed Control Prior to Herbicide Introduction

Weed control, as practiced before the introduction of herbicides and still practiced today by some farmers, consisted of crop rotation, smother crops, and tillage (Runnels and Schaffner, 1931). Weed management was not considered achievable without the use of all three of these methods in an integrated fashion.

A three- to five-year crop rotation is used to disrupt weed life cycles and prevent buildup of adapted weeds as well as to provide different environments in which other weed control methods (e.g., tillage or mowing) can be employed. Smother crops are competitive enough by themselves when seeded in narrow rows to suppress weeds and are used to suppress annual and particularly perennial weeds. Smother crops include highly competitive annual crops, such as cowpeas, buckwheat, rye, forage soybeans, Sudangrass, or millet. Others are competitive biennial or perennial crops, such as sweet clover and alfalfa. Clipping of a perennial crop, such as alfalfa, increases the effectiveness of this crop as a smother crop by periodically removing shoots of perennial weeds, which further weakens the weeds, and by suppressing seed reproduction in perennials and annuals.

The overriding emphasis in weed control using tillage with crop rotation and smother crops is prevention of reproduction by sexual or asexual means and depletion of the soil seedbank. Spot hand-weeding is used to prevent seed production by low densities of highly competitive or perennial weeds. These weed control methods are routine or "fixed" (King et al., 1986) in the sense that standard control methods are implemented without considering the actual weed populations and their various abilities to reduce yield. The concept of an economic threshold—the minimum weed density and/or duration of weed interference at which the cost of control equals the potential loss in yield caused by the uncontrolled weeds (Stoller et al., 1986)—was developed after herbicides were introduced. This concept has stimulated interest in rational or flexible weed control (King et al., 1986), where herbicides are applied only when weed populations exceed or are predicted to exceed economic threshold levels. Fixed weed management strategies emphasize prevention of weed emergence and infestation for the long term, while flexible weed management strategies emphasize limited use of control measures for economic gains.

## Cultural Weed Control

Cultural control methods include the development of competitive and/or allelopathic cultivars, mulching with allelopathic crop residues, and intercropping systems to supplement crop competition or extend it over a longer period of time during the growing season.

***Competitive and/or Allelopathic Crops.*** Crops differ in their competitiveness with weeds; van Heemst (1985) rated 25 different crops for their competitiveness with weeds based on the mean percentage yield reductions caused by weeds and ranked wheat at the top, as least affected by weeds; soybeans placed fourth, corn seventh, and onions were at the bottom. The most obvious reason for the differential ranking: differences in rate and amount of canopy development, based on the crop's morphology, spacing at which it is typically planted, and, possibly, its life cycle (fewer weeds may be present during winter for a winter grain crop than during the spring and summer).

Within a crop species, different cultivars also have been found to differ in competitiveness. The Amsoy 71 soybean cultivar was more competitive with both early and late emerging weeds than was the Beeson variety (Burnside, 1979). Yield reductions for Bragg and Semmes, two soybean varieties of intermediate maturity date, were less when competing with Johnsongrass and cocklebur, respectively, than four other varieties with different maturity dates (McWhorter and Hartwig, 1972). Yield reductions among 10 soybean varieties grown with a mixed stand of tall waterhemp and green foxtail ranged from 3 to 20 percent (Burnside, 1972). Similarly, differences in competitiveness of cultivars have been observed for cotton (Andries et al., 1974; Chandler and Merredith, 1983), corn (Staniforth, 1961), grain sorghum (Guneyl et al., 1969), snap beans (unpublished data from William and Warren), winter wheat (Challaiah et al., 1986; Ramsel and Wicks, 1988), and rice (Smith, 1988).

Information is lacking on growth characteristics that contribute to cultivar competitiveness. Forcella (1987) found that the competitiveness of eight soybean varieties was associated with greater leaf-area expansion and branch production during the first month of growth, but not with maturity group, height, or habit. Ramsel and Wicks (1988) reported that greater competitiveness in wheat cultivars was associated with greater above-ground dry wheat weight and yield, but not with greater light interception or height. Other authors (Appleby et al., 1976; Challaiah et al., 1986) found competitiveness of wheat cultivars was associated with greater height. Challaiah and associates (1986) reported that number of tillers, crop canopy, and height all were positively correlated with weed suppression, and height correlated best. Downy brome dry weight was reduced 41 to 44 percent more by the most competitive cultivar than the least competitive cultivar. They noted, however, that the most competitive cultivar also had relatively poor yields. Smith (1988) found that tall rice cultivars with greater dry weight and long, lax leaves competed with weeds better than short-statured and semidwarf cultivars, early maturing cultivars, and cultivars with erect leaves. Competitiveness of cotton cultivars of different maturity heights, in another study, did not differ (Bridges and Chandler, 1988). In general, a more dense crop canopy, as influenced by canopy architecture, leaf angle, leaf shape, or leaf size, improves crop competitiveness.

Interference of crops with weeds also may involve the production of inhibitory allelopathic substances by the living roots and/or shoots of crops. Cultivars differing in production of allelochemicals have been found for oats, sunflower, cucumber, and sweet potato (Fay and Duke, 1977; Leather, 1983; Lockerman and Putnam, 1981; Harrison and Peterson, 1986). Cultivars with greater allelochemical production suppressed weeds better than those with low production. Selection of a more competitive or allelopathic cultivar may improve weed control and permit reduced use of herbicide or tillage. Forcella (1987) found that the most competitive soybean varieties required only one half as much herbicide for good weed control and maximum yields as the least competitive varieties. A reservation with breeding crops for increased competitiveness or allelopathic potential is that characteristics contributing to greater overall interference with weeds may also result in reduction of yield or yield quality (Duke and Lydon, 1987). In some regions, small reductions in yield may be accepted if weed control practices can be reduced substantially as a result. In regions such as the Phillipines, however, producers are interested primarily in high-yielding crop varieties; acceptance of lower yielding yet competitive varieties is likely to be low (Mercado, 1987).

Crops planted in narrow row spacings suppress weed growth more than when planted in wide row spacings (Felton, 1976; Bendixen, 1988; Teasdale and Frank, 1983). For example, weed weight in soybeans planted in 50-cm rows was less than one-third of weed weight in soybeans planted in 100-cm rows (Felton, 1976). Fischer and Miles (1973) developed a theoretical model of weed and crop growth as a function of planting pattern; they varied the distance between row crops and between crop plants within rows while keeping the total number of crop and weed plants per unit surface area constant. The model predicted that when crops and weeds emerge at the same time and have equal growth rates the most suppressive crop spacing pattern is an equilateral triangular lattice with crop plants equidistant from each other. Square lattices are nearly as suppressive, but suppression declines as the lattice becomes more rectangular, that is, distance between rows increases while distance between crop plants within the rows decreases. Increased crop density has little added effect on weed suppression in a given planting pattern.

***Allelopathic Crop Residues.*** Mulch crops, like rye, traditionally have been planted to compete with weeds, cover the soil in the winter, and improve soil tilth when they are plowed or disked in the spring. Recent innovations in planting equipment have encouraged research on no-till planting into standing green mulches or mulches that have been killed with herbicides or by mowing. Several mulch crops have been investigated for their use in field and high-value crops (Smeda and Putnam, 1988). In addition to the physical weed suppression from the mulch, many plant species, including crops, contain allelochemicals that suppress weeds and other plants. Extensive research has examined this phenomenon over the last 30 years in both natural and agricultural systems. Several monographs have been published (National Academy of Sciences, 1971; Putnam and Tang, 1986; Rice, 1974; and Silverstein and Simeone, 1983). Some of the research on allelopathy has been motivated by a search for biodegradable herbicidal compounds.

Much of the current published research focuses on small grains, such as rye, wheat, oats, and barley (Barnes and Putnam, 1983; Eckert, 1988; Leibl and Worsham, 1983; Mitchell and Teel, 1977; Moschler et al., 1967; Putnam and DeFrank, 1983; Putnam et al., 1983). Considerable research has focused on identification and characterization of allelochemical properties of rye (Barnes and Putnam, 1986; Shilling, 1983). Work on legumes has included hairy vetch (Teasdale, 1988), crimson clover (Mitchell and Teel, 1977; Moschler et al., 1967; Peele et al., 1946), and subterranean clover (Enache and Ilnicki, 1988; Else and Ilnicki, 1988).

Most of this work demonstrates significant weed suppression by all of the above-mentioned mulch crops; in many cases there is strong evidence of an allelochemical effect of the mulch crops in addition to physical effects. Most of these studies involved the use of a herbicide to kill the mulch crops, so they are less useful as pioneers toward the development of no-herbicide, no-till cropping systems.

***Intercropping.*** Intercropping of two harvested crops is used extensively in the tropics to maximize land use and to ensure against crop failure (Mercado, 1987). Nonchemical control measures may be especially useful in these systems because herbicides that can be used in an intercrop system without injury to one of the crops or to a rotational crop are few.

Research on intercropping for weed control consists currently of investigations on relay intercropping of soybeans into wheat or other winter grain and the use of a low-growing plant cover (living mulch) to control weeds in row crops, such as corn or soybeans. Both of these practices are in the preliminary phase of research with respect to weed control. Both are based on the concept of increasing crop competition to control weeds by using competition provided by an intercrop. Intercropping methods such as these have the potential to provide weed control in reduced tillage systems.

***Relay Intercropping.*** Relay intercropping involves planting no-till soybeans into standing green wheat or other winter grain and subsequently harvesting both wheat and soybeans. The system has been investigated with respect to obtaining maximum combined wheat and soybean yields by selecting appropriate wheat and soybean varieties, appropriate row spacing of both crops, and appropriate soybean planting date and wheat harvest date (Reinbott et al., 1987; Jeffers and Triplett, 1979). Generally, yields of wheat and soybeans are reduced compared to monocropping of either crop, but soybean yields are higher than when doublecropped after wheat, and the combined yields of wheat and soybeans offset the yield reductions in the crops, which can range from 16 to 43 percent (Reinbott et al., 1987). A major benefit of relay intercropping is the provision of a winter cover crop to prevent soil erosion and to suppress spring-germinating weeds, possibly providing some weed control for soybeans. Often, small grains are effective in suppressing weeds by themselves and do not require weed control for maximum yields.

The possibility of using relay intercropping as a means to reduce herbicide use in soybeans by suppressing early weed growth has been investigated recently. Two years of research at the Rodale Research Center (Peters et al., 1988) revealed that weed pressure was moderate to low in relay-intercropped soybeans as a result of competition by wheat, and the researchers found that weeds had a slight adverse effect on soybean yield in only one of the two study years. Another study (Prostko and Ilnicki, 1988), conducted for two years in New Jersey with a barley-soybean relay-intercropping system, showed that the intercropped system with no herbicide provided 70 to 93 percent control of fall panicum, redroot pigweed, and giant ragweed and resulted in soybean yields comparable to the intercrop system with herbicides. Weed control with herbicides was equal to or slightly better than control with no herbicides (weed control provided by wheat and soybeans alone).

More studies need to be conducted, including long-term experiments, to determine the effects on weed species and densities of relay-intercropped soybeans with and without herbicide use. The system appears attractive because it conserves soil, provides economic return for two crops per year instead of one, and may provide extra suppression of weeds by continuous crop cover, which may reduce the need for herbicide application, depending on resultant weed populations. The relay-intercropping system has potential to include more competitive and/or allelopathic winter grains to increase weed suppression. Increased interference of winter grains may affect soybean growth, however. Potential problems with this system include shifts toward summer annual weeds able to germinate at cool temperatures in the early spring (common lambsquarter, giant and common ragweed) and able to tolerate shading by wheat. Such weeds would have a head start on soybeans and compete strongly with seedling soybeans. Another potential problem is the shift to perennial weeds over a period of time because tillage in this system is reduced.

***Living Mulches.*** An alternative method of weed control in conventional and reduced tillage systems that may also reduce dependence on herbicides is the intercropping of a low-growing cover crop, referred to as a "live" or "living mulch" (Lal, 1975), to provide weed control in a summer annual crop, such as corn or soybeans. The living mulch must establish itself readily, covering the ground rapidly so as to smother weeds but not compete with the main crop.

Living mulches have been of interest to researchers for soil conservation and weed control. Live plant covers as companion crops were investigated in the 1950s in the Midwest (Burwell, 1956; Kurtz et al., 1952; Robinson and Dunham, 1954) and later in the 1970s and 1980s in Nigeria, and the northeastern and north central United States (Akobundu, 1980; Berkowitz et al., 1986; Hartwig, 1988; Vrabel et al., 1980, 1983; Lake and Harvey, 1985; Regnier and Stoller, 1981; Regnier and Stoller, 1987; Warnes, 1985). Most research has involved seeding corn or soybeans at a given row spacing into a preestablished, relatively low-growing winter grain, legume, or grass sod subjected to partial or complete suppression by herbicide treatments, rototilling, mowing, or a combination of methods. The majority of studies (Berkowitz et al., 1986; Burwell, 1956; Hartwig, 1988; Lake and Harvey, 1985; Regnier and Stoller, 1981; Vrabel et al., 1982; Warnes, 1985) showed that the species selected as living mulches do not suppress weeds selectively but also suppress the crop and must be managed carefully to reduce their competition with the crop. In studies where yield reductions were not observed (Robinson and Dunham, 1954; Vrabel et al., 1980), the living mulch was established at the same time as the crop, rather than preestablished, and planted in bands between crop rows, rather than broadcast, or planted with the crop seeded in narrow rather than wide rows.

The interference of living mulches with crops, which appears to involve competition for water and/or allelopathy (Barnes and Putnam, 1986; Berkowitz et al., 1986; Lake and Harvey, 1986), poses the greatest problem in developing living mulches for weed control. Results of several studies (Enache and Ilnicki, 1988; Lake and Harvey, 1986; Regnier and Stoller, 1987; Vrabel et al., 1980) suggest that living mulch competition with crops may be reduced while maintaining adequate weed suppression by (a) selection of less competitive mulches (finding the right mulch), (b) partial suppression of preestablished mulches with herbicide bands, and (c) use of narrow rows of the crop in combination with planting the living mulch at the same time as the crop. Selection of a competitive crop variety also may improve crop yield in this system.

A recent two-year study in New Jersey (Enache and Ilnicki, 1988) demonstrated the success of subterranean clover as a living mulch in corn. The clover, a winter annual planted in the fall, provides soil cover during the erosive winter and spring months. It completes its life cycle in the spring, thereby not competing with the corn during the summer. Yields of corn planted in subterranean clover with no additional weed control or suppression of the clover were comparable to yields of corn with no mulch and conventional weed control (herbicides), and a higher net profit was obtained with the clover. Subterranean clover provided good to excellent control of seven broadleaf annual weeds, but acceptable to little control of fall panicum.

A winter annual species seems to have a promising life cycle for a living mulch. More research is needed in this area before these systems are practicable. For areas where low-growing winter annual species are not available or where water during the spring is limiting, living mulch systems will continue to involve suppression of the mulch with herbicides or other management techniques to reduce competition with the row crop while providing weed control. Potential problems with living mulches for weed control over an extended period of time include shifts toward weeds with the ability to compete with the living mulch, particularly perennial weeds, and potential for the living mulch to harbor crop diseases and pests, such as rodents, nematodes, and insects, or their natural enemies.

Both relay-intercropping and living mulches offer a potential for alternative weed control in reduced tillage systems. Yet these systems, like chemical weed control in reduced or no-tillage systems, probably will encounter increased perennial weed pressure. Research is needed to evaluate the degree of weed control required in relay-intercropping systems and to develop practicable living mulch systems as well as to evaluate the long-term effects of both these systems on weed populations.

## Biological Weed Control

There are two different approaches in biological control of weeds: classical biological control and mass-exposure or inundative biological control. The former approach is used for control of perennial weeds, usually introduced, that are widespread in perennial crops, such as pasture or rangeland, or in wetlands and other noncrop land. A biological agent is introduced in small quantities and builds up sufficient numbers over a period of years to keep the weed at economic threshold levels. The inundative approach, developed more recently, is applicable to annual crops with annual weed problems. A biological agent, usually a fungal agent, is released in quantities sufficient to control the annual weed before it causes a reduction in crop yields. Because this approach is similar to herbicide application, these biological agents are called "mycoherbicides." Research in both these areas has and will continue to provide innovative, alternative means of controlling weeds. In general, biological agents are highly specific, and biological control must be combined with other control measures to control all of the weeds in a given crop.

***Classical Biological Control.*** Biological agents have given effective control of perennial weeds in the past; examples are the leaf-eating insect *Chrysolina quadrigemina* (Suffr.), which controlled Klamath weed (*Hypericum perforatum* L.) (St. Johnswort) in California, and the boring insect *Cactoblastic cactorum* (Berg.), which controlled prickly pear (*Opuntia stricta* Haw.) in Australia (Ross and Lembi, 1985). Several other organisms show promise as classical biological control agents for perennial weeds. Among these are *Sphacelothecu holci,* a fungal agent (loose smut fungus) that inhibits seed production, for the control of Johnsongrass, a severe perennial weed (Massion and Lindow, 1986). Since Johnsongrass is widely spread by seed in addition to its rhizomes, this control may be very useful. Other possibilities are the stem and root mining beetle, *Oberea erythrocephaia*, for the control of leafy spurge, a perennial weed in western rangeland (Rees et al., 1986). Adults feed externally on stems and leaves; thus, the main effect of the insect is reduced vigor of the weed the following year. The rhinocyllus weevil (*Rhinocyllus conicus* Froelich, a seed-eating weevil) shows potential for the control of musk thistle (*Carduus nutans* L) and was released in 1969 in several areas of Montana (Rees, 1986). Three insects—a seed chalcid, rose hip borer, and cane borer—as well as a disease (rose rosette) show potential for control of multiflora rose, a perennial shrub in midwestern pastures (Hindal and Wong, 1988; Mays and Kok, 1988). A fungal agent, *Sclerotinia schlerotiorum,* reportedly has potential for the control of Canada thistle (Brosten and Sands, 1986). An indigenous rust fungus, *Puccinia canaliculata,* also shows potential for the control of yellow nutsedge (Phatak et al., 1987).

Other research in this area involves modelling the growth of perennial weeds and then using sensitivity analysis to determine the most vulnerable time in their life cycle, that is, the time at which attack by a biological agent would most curtail reproduction and population growth by the weed

(Maxwell et al., 1988). A different approach involves developing methods to contain grazing animals economically for grazing control of perennials. Recently, a method was devised using electric shock collars to contain goats for the control of leafy spurge (McElligott and Fay, 1988).

The control of perennial weeds by biological agents that by themselves are not sufficiently effective can be enhanced by combination with other methods, such as timely herbicide application and/or mowing, and growth with a highly competitive crop, such as subterranean clover (Diamond et al., 1988; Lee, 1986).

***Mycoherbicides.*** Mycoherbicides have considerable potential for biological control of weeds because they can be applied in the same way as herbicides for control of annual weeds in annual crops. Success has been achieved with the marketed mycoherbicides Collego™ [*Colletotrichum gloesporioides* (Penz.) Sacc. f. sp. *aeschynomene*] for the control of northern jointvetch in rice and soybeans in Arkansas (Smith, 1986) and DeVine™ [*Phytophthora palmivora* MWV pathotype (P.p.)] for the control of stranglervine in Florida citrus orchards (Ridings, 1986).

Potential mycoherbicides include *Colletotrichum coccodes* for control of eastern black nightshade (Anderson and Walker, 1985), a weed that severely interferes with soybean harvesting; *Fusarium lateritium* for control of prickly sida and velvetleaf (Boyette and Walker, 1985); *Colletotrichum gloeosporiodes* (Penz.) Sacc. f. sp. *malvae* for control of round-leaved mallow and velvetleaf (Mortensen, 1988); *Alternaria cassiae* for control of sicklepod (Walker and Boyette, 1985); and *A. crass* for control of jimsonweed (Quimby et al., 1988). Most of these weeds occur in soybeans.

Mycoherbicides, like classical biological control agents, have the advantage of no environmental contamination or toxicity to humans; they are also highly selective. From a marketing standpoint, however, their selectivity can be a disadvantage because each agent controls only one weed species, and crop fields typically are infested by several different weed species. The problem of a limited spectrum can be solved if the organism can be combined with another control agent or with chemical herbicides (Templeton, 1986). Other problems with mycoherbicides are the requirement for a long dew period for infection to occur after application (Quimby et al., 1988), difficulty of commercially producing and formulating the organism while maintaining its viability (Bowers, 1986; Kenney, 1986), and susceptibility of the organisms to fungicides used to control crop pathogens (Khodayari and Smith, 1988). The requirement for a long dew period (frequently standing water on the leaf is required for sufficient infection to occur) appears to be lessened with some potential mycoherbicides by combining them with postemergence herbicides (Quimby et al., 1988), which retard evaporation of the suspension from the leaf.

### Herbicides from Natural Compounds

Natural compounds from plants (allelochemicals) or from microorganisms (microbial toxins) can be used as bioherbicides or as a source of new herbicide chemistry. The compounds may be produced by the organisms and then applied to weeds as herbicides. This differs from biological control in that the product of the organism is applied rather than the organism itself. Alternatively, the chemistry of the naturally produced toxins may be synthetically modified or used as a model for a synthetically produced herbicide.

Advantages of developing herbicides from natural compounds include the potential to discover novel chemicals, potential to reduce investments in synthetic chemistry, and greater likelihood that the compounds will be biodegradable (Duke and Lydon, 1987). The potential for toxicity of the compounds to humans and other animals is not necessarily less than that of synthetic herbicides.

Allelochemicals are secondary metabolites partially bred out of crop plants because of undesirable qualities, such as bad taste or reduced yield. Most allelochemicals are relatively nonphytotoxic compared with synthetic herbicides, sometimes antagonistic in combinations, and may be inactivated by soil (Duke and Lydon, 1987). They often are toxic to the parent plant; thus, breeding crop plants for production of high levels of allelochemicals may be counterproductive. Other allelochemicals, such as a-terthienyl and hypericin, which are patented herbicides, act as phototoxins and absorb visible or ultraviolet radiation to become active (Duke and Lydon, 1987). When activated, the compounds combine with DNA or produce singlet oxygen, which destroys membranes. These compounds are effective at low doses, but they are also toxic to almost all living organisms. Examples of other allelochemicals with potential as herbicides are coumarins (common in grasses, legumes, and citrus); juglone (found in walnut trees); secondary compounds from the terpenoid pathway, such as 1,8-cineole (found in sage species); and artemisinin (found in annual wormwood) (Duke and Lydon, 1987).

A ubiquitous plant constituent, delta-aminolevulinic acid (ALA), an amino acid (also referred to by the popular press as a "laser" herbicide) has been investigated recently for potential as a herbicide (Duke and Lydon, 1987). ALA is used with 2,2′-dipyidyl, a nonnatural compound, to activate the chlorophyll biosynthetic pathway. The compounds cause massive accumulation of chlorophyll precursors that act similarly to phototoxins, creating singlet oxygen upon exposure to light. The long dark period required after application of ALA for sufficient build-up of precursors, its high cost, potential toxic effects of the 2,2'-dipyridyl, and lack of proven efficacy under field conditions limit development of this compound.

Microbially produced toxins are regarded more favorably for potential as herbicides than allelochemicals (Duke and Lydon, 1987). They are more selective and effective at low rates than plant-produced phytotoxins and, compared with use of the actual pathogens, are easier to formulate, less likely to spread disease to nontarget species, and less dependent on environmental conditions. However, the living agents have the advantage of being self-perpetuating. Microbial toxins may be obtained by fermentation and used in their natural state, subjected to synthetic modification, or their chemistry used as a basis for a synthetic herbicide.

Anisomycin and bialaphos are products of *Streptomyces* strains. They are the first microbial metabolites to be used both directly and indirectly as commercial herbicides. Anisomycin is the chemical basis for a synthetic herbicide for rice ("NK-049") in Japan, and bialaphos is used directly as a herbicide in Japan (Duke and Lydon, 1987). Bialaphos is relatively nonselective, effective on monocot and dicot weeds as well as perennial weeds, has low mammalian toxicity, and a relatively short soil half-life (20-30 days). In plants it is metabolized to the active form, phosphinothricin. This form is being developed synthetically as the herbicide glufosinate. Other microbial toxins under investigation are herbicidin and herbimycin, from *S. saganonensis,* developed as rice herbicides for monocots and dicots, and tentoxin, from *Alternaria alternata,* which is active on several species, in particular Johnsongrass in corn and broadleaf weeds in soybeans.

Limiting factors in the development of microbially produced toxins as herbicides are the low yields produced by fermentation and difficulty and expense of synthesis due to the complexity of the structures. Commercialization of tentoxin, for example, is limited by both factors. Biotechnology can be used to alter the genome of the fermenting organisms and may improve the rate of production of these compounds. Alternatively, research on the efficacy and environmental safety of similar but more simple synthetic compounds, based on the chemistry of the natural compounds, may aid in development of natural compounds as herbicides.

## New Herbicide Technology

Improvements in herbicide technology to reduce environmental contamination also are underway. Research in this area probably will give results faster than in other areas because research funding is readily available and the technology base will require little modification (i.e., the way in which herbicides are applied). Because herbicides are widely accepted, changes in herbicide technology also will be readily accepted. Improvements in herbicide technology consist of more effective herbicides applied at lower rates, reduced mammalian toxicity of herbicides, improved application equipment, and controlled release technology (Schweizer, 1988).

The trend in new herbicide chemistry has been toward compounds, such as the imidazolinones, mono- and diphenylethers, and sulfonylureas, that are effective at much lower rates (g/ha instead of kg/ha) and have very low mammalian toxicity. These two characteristics should contribute to reduced risk of environmental contamination (Schweizer, 1988). Some of these compounds are very active and persistent, however, and have a strong potential to injure nontarget crops, such as crops planted in a rotation.

Research on improved application equipment includes application of low volumes of spray solution (1-50 L/ha) with low herbicide rates and distribution of uniform droplets with no drift (Schweizer, 1988). Techniques investigated are controlled-droplet applicators, air curtain sprayers, spinning disk sprayers, electrostatic sprayers, rope-wick applicators, and recirculating sprayers. Controlled release technology is another research area in herbicide application that involves encapsulation, polymerized herbicides, copolymers and crop-seed coatings to release soil-applied herbicides gradually, thus reducing herbicide runoff, leaching, volatility, and potentially reducing application rates while increasing crop selectivity. Advances in application technology should reduce overapplication and drift, thereby reducing environmental contamination.

## Bioengineering Contributions

Genetic engineering can contribute to alternative weed control in two ways: through development of herbicide resistance and improved biosynthesis of microbial toxins. Resistance to herbicides is expected to be helpful by allowing greater use of nonleachable and/or rapidly degraded herbicides that are not currently selective for a given crop. Resistance to glyphosate, a nonselective, nonleachable herbicide with low mammalian toxicity and no soil activity, has been introduced by genetic engineering from bacteria into plants (Comai et al., 1985). Resistance to several imidazolinones in a corn line has been introduced (Schweizer, 1988), and resistance to some of the sulfonylurea herbicides also has been accomplished through alteration of acetohydroxyacid synthase enzyme, which interacts with the herbicide to produce the phytotoxic effect. A potential disadvantage of engineering for resistance in crops is increased selection for resistance in weeds because of increased selection pressure.

## Weed Thresholds and Modeling

Much research has been conducted on the density and duration effects of weeds on yield loss (Stoller et al., 1986; Aldrich, 1988). Yield losses increase exponentially with weed density, but low densities may result in no statistically detectable yield losses (Aldrich, 1988). Weeds also vary greatly in their ability to reduce yield (Stoller et al., 1986); for example, common cocklebur is approximately twice as competitive with soybeans as jimsonweed or velvetleaf. A crop need not be weed-free for its entire growing season for maximum yields; research shows that weed removal beyond a critical weed-free period, typically from four to eight weeks after crop emergence, does not improve yield (Stoller et al., 1986). Crops can tolerate competition with weeds for a limited amount of time after crop emergence; generally, weeds allowed to emerge with the crop but removed within four to six weeks after crop emergence do not cause yield losses. The effects of weed densities, the length of the critical weed-free period, and the time after crop emergence of weed competition tolerated by the crop all depend on the crop and weed species, planting arrangement (wide or narrow rows), and environmental conditions. This information is of use in herbicide application because it indicates how long a soil-applied herbicide must remain active after application to provide sufficient weed control and/or when after crop emergence a postemergence herbicide must be applied to avert yield losses. These time thresholds also could be useful for timing of selective tillage. Density and duration thresholds have led to the development of the concept of flexible weed control. With flexible weed control, measures are based on knowledge of actual or potential weed densities and their economic thresholds, rather than implemented on a routine or fixed basis (King et al., 1986).

Models have been developed to aid weed management decision-making by predicting net returns from a weed control measure, based on measured or predicted densities of weed species, crop yield losses associated with the weed species, costs and effectiveness of herbicides, and anticipated value of the crop (Marshall, 1987; Sim, 1987; King et al., 1986; Cousens et al., 1985; Wilson and Wright, 1987; Heitefuss et al., 1987; Koch, 1988, Lapham, 1987). The value of these models lies in the potential to apply herbicides on a prescribed basis and thus potentially apply less herbicide, reducing both cost and the potential for environmental contamination. Models are easily developed for use with postemergence compounds because weed densities and species can be measured readily with a quadrat or other sampling device. Models for use with preemergence compounds rely on predictions of weed populations that will emerge, based on sampling of weed populations in the previous crop, samples of seed content in the soil, as well as data on germinability of seed in the soil, or rate of seed production by weeds present in the previous crop. Models that combine a fixed weed control, such as the application of a soil-applied herbicide, with flexible postemergence control, applied if and when needed, also have been developed (King et al., 1986). Important issues that are just now being addressed are the impact of subthreshold weeds on weed infestation in future crops, the cost of gathering the data to run the model, and the cost of running the model.

Few models have been tested. A model developed for Colorado (King et al., 1986) showed that flexible strategies required 64 percent less herbicide to control weeds in corn and gave annualized net returns of $83 per hectare more than fixed strategies. Similarly, a study in Germany revealed that 20 to 50 percent of fixed weed control measures were uneconomical (Heitefuss et al., 1987). Marshall and associates (1987) found that major weed problems did not occur when thresholds were used to make spray decisions. In considering the long-term effects of weed control strategies in relation to the use of a model, Wilson and Wright (1987) found that for cleavers, a weed in wheat with extreme competitiveness and high potential rate of increase when present at low populations, a threshold was not practical in taking into account costs and efficiency of control, seed production, and seed deaths. The long-term threshold for cleavers was well below one plant per 20 square meters. Such a low threshold was impossible to monitor, so essentially a zero threshold for cleavers was necessary. Lapham (1987) developed a long-term threshold for yellow nutsedge in a tobacco-grass-ley rotation in Zimbabwe (one year tobacco, three years of grass-ley); the threshold was less than 1.5 yellow nutsedge tubers per square meter, an extremely low number. Cousens and associates (1985) found that economic thresholds calculated for annualized returns were four times greater than thresholds developed for longer periods, such as 10 years.

The length of time to sample a field for weed densities or soil seed content and the accuracy of estimates from quadrat surveys are major concerns in the use of models. A study in the United Kingdom (Marshall, 1987) revealed that a sampling intensity of 6 points per hectare produced an error of up to 100 percent for relatively rare species and 30 percent for abundant species. This sampling intensity was the most practical, requiring 15 min per hectare, but the author still considered it impractical for a farmer. With these concerns and others, such as the uncertainties of herbicide performance due to the unpredictable environment, future economics, narrow window for herbicide application, unknown potential for subthreshold weeds to increase and generate unmanageable weed populations in the future, etc., Cousens and associates (1985) suggested that thresholds be low and calculated with a large safety factor. The researchers said that calculated low thresholds may, in fact, be comparable to aesthetic thresholds that farmers routinely use in judging the necessity of weed removal in their fields. Mortimer (1987) emphasized the need to predict the long-term effects of weed management strategies and combine prediction modeling with economic risk analysis. For producers who rely on tillage for control of weeds, it is probable that a more preventative philosophy is more appropriate than a flexible philosophy because of the environmental constraints imposed on tillage. There always is the danger that a field cannot be cultivated because of wet weather. In this case, previous preventative measures ensuring low weed seed content in the soil might prevent a disastrous weed population from occurring.

## Integrated Weed Management

Integrated weed management is a popular term for the combination of weed control practices, thus reducing dependence on any one type of weed control. Cultural, mechanical, biological, and chemical control measures may be combined. The combination of various methods is especially important for control of perennial weeds or particularly prolific or competitive annual weeds that are generally inadequately controlled by any one method (Lee, 1986; Glaze, 1988; Bridges and Walker, 1987; Charudattan, 1986). The application of integrated weed management also includes knowledge of past weed populations in fields, competitive crop cultivars, improved crop and soil management practices, regular monitoring for annual and perennial weeds, hand-weeding, spot-treating, and appropriate selection of herbicides (Schweizer, 1988).

Control of yellow nutsedge, for example, requires a program of crop rotation that includes crops with rapid canopy development, preplant tillage to stimulate tuber germination and promote tuber decay by bringing tubers to the soil surface, high plant populations of competitive crops to cause shading, and cultivation plus herbicides during the growing season to keep populations at manageable levels (Glaze, 1987). All components of this program are considered necessary for control of moderate to severe nutsedge infestations. Cultivation is considered particularly necessary because nutsedge proliferates under systems of reduced tillage with herbicides. A combination of methods also was considered essential for the control of rush skeletonweed, a perennial weed, such as the combination of a competitive crop, subterranean clover, and the rust fungus *Puccinia chondrillina* Bubak and Syd (Lee, 1986).

Management of sicklepod in the South requires crop rotations; tillage management, including cultivation; reduced row spacings; and appropriate herbicide selection and applications (Bridges and Walker, 1987). Lack of rotation in particular allows sicklepod seed numbers to increase and reduces the profitability of continuous soybean production. Minimum-till planting of soybeans into a mulch of small grain residue effectively reduces sicklepod competition in comparison with conventional till-planting of the soybeans.

A recent example of the value of integrated management is the combination of *Urophora cardui* L. (Diptera:Tephritidae), which reduces flower production but is not effective alone in reducing Canada thistle populations, with a spring and mid-season mowing, plus application of 2,4-D as well as dicamba or clopyralid in Canada (Diamond et al., 1988).

Other combinations of weed control methods include mycoherbicides combined with synthetic herbicides to provide control of a broader spectrum of weeds than with the mycoherbicide alone (Templeton, 1986) or to improve the efficacy of the mycoherbicide (Quimby et al., 1988; Smith, 1986). Obviously, cultural methods that improve crop germination, giving it a head start, are vital. Crop rotation can be combined with surface tillage during that part of the rotation in which row crops are grown, and hay crops in the rotation that are grazed or clipped frequently contribute greatly to weed control (Ross and Lembi, 1985). Cultivation can be combined with herbicide bands on the crop rows in either a conventional tillage system or a ridge-till system. The ridge-till system offers a way to cultivate yet reduce tillage.

Another area currently receiving attention is use of reduced broadcast herbicide rates. This has been used particularly in combination with narrow row spacings of soybeans or peanuts (Bendixen, 1988; DeFelice et al., 1988; Baldwin et al., 1988, Cardina et al., 1987) or with more competitive soybean varieties at narrow row spacings (Forcella, 1987). In several studies, the increased competitiveness of the crop when planted in narrow rows (drilled) versus wide rows, or when a more competitive variety is used, resulted in equivalent weed control with lower herbicide rates (Bendixen, 1988; Forcella, 1987; Cardina et al., 1987). Bendixen (1988) found that Johnsongrass biomass in soybeans planted in 25-cm rows was one-fourth that in soybeans planted in 76-cm rows, and higher herbicide rates were required in wide-row soybeans to obtain control equivalent to that obtained in narrow-row soybeans. In the Phillipines, half-rates of herbicides combined with hand-weeding or cultivation frequently are used for economic reasons (Mercado, 1987).

## The Quest Continues

Alternative weed control involves combining the traditional cultural practices of crop rotation, the use of smother crops and competitive crop varieties, and tillage with newer chemical, biological, and cultural technologies. New cultural practices include relay cropping, either of one crop into another crop, or a crop into a mulch crop, as with living mulches. Research points to the feasibility of using dead mulches, both grasses and legumes, for weed control in no-till and reduced-till systems, with or without minimal use of herbicide. New tillage systems, such as ridge tillage and other systems that allow for mechanical cultivation in high residue systems, open up more options for farmers, as does flame weeding. Classical biological control agents, mycoherbicides, biologically derived herbicides (phytotoxic compounds from plants or microbes), and other means that limit the synthetic compounds released into the environment have potential to help reduce environmental problems caused by current herbicide use patterns.

The use of an integrated pest management philosophy, along with newly developed crop-weed competition models and the availability of postemergence herbicides, gives farmers flexibility to move away from a fixed or preventative management style with the purely cultural (no herbicide) or purely chemical control methods. It gives them a means to predict the impacts of various weed management strategies over a given time period. Strategies that combine a fixed component, such as a preemergence herbicide or sequence of secondary tillage operations, with a flexible component, such as a postemergence herbicide or interrow cultivation, may not be as economical as purely flexible strategies (King et al., 1986). Yet they may be more acceptable to producers in that at least some portion of the potential weed population is controlled on a routine basis.

Development of alternative weed control methods for reduced tillage systems will continue to be important because tillage or hand-weeding still is the major practicable alternative to weed control with herbicides. Alternative reduced or nonchemical methods for intercropping systems may be of special interest in tropical and underdeveloped countries where herbicides for these systems are either unavailable or unaffordable, yet an alternative to hand cultivation is desired. With sound crop-weed management practices and integration of new cultural practices and new herbicide technologies with traditional preventative practices, both acceptable levels of weed control and improved environmental quality can be achieved.

## REFERENCES

Akobundu, O. 1980. Live mulch: A new approach to weed control and crop production in the tropics. Proceedings, 1980 British Crop Protection Conference-Weeds 2: 377-382.

Aldrich, R. J. 1987. Predicting crop yield reductions from weeds. Weed Technology 1: 199-206.

Andersen, R. N., and H. L. Walker. 1985. Colletotrichum coccodes: A pathogen of eastern black nightshade (*Solaunum ptycanthum*). Weed Science 33: 902-905.

Andries, J. A., A. G. Douglas, and A. W. Cole. 1974. Herbicide, leaf type, and row spacing response in cotton. Weed Science 22: 496-499.

Anonymous. 1897. Twenty-five most harmful weeds. Yearbook of Agriculture. U.S. Department of Agriculture, Washington, D.C. pp. 641-644.

Anonymous. 1988a. U.S pesticide use still climbing. Pesticides and You 8(4): 4.

Anonymous. 1988b. Rotation, rotary hoe important on Schroeder farm. Ohio Ecological Food and Farm Association Newsletter 8(4): 4.

Appleby, A. P., P. D. Olson, and D. R. Colert. 1976. Winter wheat yield reduction from interference by Italian ryegrass. Agronomy Journal 68: 463-466.

Bakke, A. L., W. G. Gaessler, L. M. Pultz, and S. C. Salmon. 1944. Relation of cultivation to depletion of root reserves in european bindweed at different soil horizons. Journal of Agricultural Research 69: 137-147.

Baldwin, F. L., L. Oliver, and T. Tripp. 1988. Arkansas' experiences with reduced-rate herbicide recommendations. Abstract, Weed Science Society of America 28: 45.

Barnes, J. P., and A. R. Putnam. 1983. Rye residues contribute weed suppression in no-tillage cropping systems. Journal of Chemical Ecology 9: 1,045-1,057.

Barnes, J. P., and A. R. Putnam. 1986. Evidence for allelopathy by residues and aqueous extracts of rye (*Secale cereale*). Weed Science 34: 384-390.

Bender, J. 1987. Convenience of alternative weed control methods in row crops. Cooperative Extension Service, University of Nebraska, Lincoln.

Bendixen, L. E. 1988. Soybean (*Glycine max*) competition helps herbicides control Johnsongrass (*Sorghum halepense*). Weed Technology 2: 46-48.

Berkowitz, A. R., H. C. Wien, and B. F. Chabot. 1986. Competition for water between beans and interplanted perennial grass cover crops. Agronomy Abstracts 86: 107.

Bowers, R. C. 1986. Commercialization of Collego™—an industrialist's view. Weed Science 34 (supplement 1): 24-25.

Boyette, C. D., and H. L. Walker. 1985. Evaluation of Fusarium lateritium as a biological herbicide for controlling velvetleaf (*Abutilon theophrasti*) and prickly sida (*Sida spinosa*). Weed Science 34: 106-109.

Bridges, D. C., and J. M. Chandler. 1988. Influence of cultivar height on competitiveness of cotton (*Gossypium hirsutum*) with Johnsongrass (*Sorghum halepense*). Weed Science 36: 616-620.

Bridges, D. C., and R. H. Walker. 1987. Economics of sicklepod (*Cassia obtusifolia*) management. Weed Science 35: 594-598.

Brosten, B. S., and D. C. Sands. 1986. Tield trials of Sclerotinia sclerotiorum to control Canada thistle (*Cirsium arvense*). Weed Science 34: 377-380.

Brown, S. M., and T. Whitwell. 1988. Influence of tillage on horseweed, *Conyza canadensis*. Weed Technology 2: 269-270.

Buchanan, G. A., and A. E. Hiltbold. 1977. Response of cotton to cultivation. Weed Science 25: 132-134.

Buckingham, F. 1976. Fundamentals of machine operation: Tillage. Deere and Company, Moline, Illinois.

Buhler, D. D., and T. C. Daniel. 1988. Influence of tillage systems on giant foxtail, *Setaria faberi*, and velvetleaf, *Abutilon theophrasti* density and control in corn, *Zea mays*.

Weed Science 36: 642-647.
Burnside, O. C. 1972. Tolerance of soybean cultivars to weed competition and herbicides. Weed Science 20: 294-297.
Burnside, O. C. 1979. Soybean (*Glycine max*) growth as affected by weed removal, cultivar, and row spacing. Weed Science 27: 562-565.
Burwell, R. W. 1956. Skip-row planting of corn. Agronomy Facts, C-12. University of Illinois College of Agriculture, Urbana.
Cardina, J., Au. C. Mixon, and G. R. Wehtje. 1987. Low-cost weed control systems for close-row peanuts (*Arachis hypogaea*). Weed Science 35: 700-703.
Challaiah, O. C. Burnside, G. A. Wicks, and V. A. Johnson. 1986. Competition between winter wheat (*Triticum aestivum*) cultivars and downy brome (*Bromus tectorum*). Weed Science 34: 689-693.
Chandler, J. M., and W. R. Merredith, Jr. 1983. Yields of three cotton (*Gossypium hirsutum*) cultivars as influenced by spurred anoda (*Anoda cristata*) competition. Weed Science 31: 303-307.
Chardattan, R. 1986. Integrated control of waterhyacinth (*Eichhornia crassipes*) with a pathogen, insects, and herbicides. Weed Science 34 (supplement 1): 26-30.
Comai, L., D. Facciotti, D. M. Stalker, G. A. Thompson, and W. R. Hiatt. 1985. Expression in plants of a bacterial gene coding for glyposate resistance. *In* M. Zaitlin, P. Day, and A. Hallaender [editors] Biotechnology in plant sciences—relevance to agriculture in the eighties. Academic Press, New York, New York.
Cousens, R. B., J. Wilson, and G. W. Cussans. 1985. To spray or not to spray: The theory behind the practice. Proceedings, British Crop Protection Conference, Weeds 7A-1: 671-678.
DeFelice, M. S., W. B. Brown, R. J. Aldrich, B. D. Sims, D. Judy, and D. Guethle. 1988. Weed control in soybeans with below-label rates of postemergence herbicides. Abstract, Weed Science Society of America 28: 17.
Desvaux, R., and P. Ott. 1988. Introduction of thermic weed control in southeastern France. pp. 479-482. *In* P. Allen and D. Van Dusen [editors] Proceedings, Sixth International Science Conference.
Diamond, J. F., M. G. Sampson, and A. K. Watson. 1988. Integrated management of Canada thistle [*Cirsium arvense* (L.) Scop.] in pastures in eastern Canada. Abstract, Weed Science Society of America 28: 50.
Duffy, M., and M. Santhorn. 1984. Returns to corn and soybean tillage practices. Agricultural Economics Report 408. U.S. Department of Agriculture, Washington, D.C. 14 pp.
Duke, S. O., and J. Lydon. 1987. Herbicides from natural compounds. Weed Technology 1: 122-128.
Eckert, D. J. 1988. Rye cover crops for no-tillage corn and soybean production. Journal of Production Agriculture 1: 207-210.
Else, M. J., and R. D. Ilnicki. 1988. Allelopathic properties of subterranean clover. Proceedings, C. E. Weed Science Society 42: 65.
Enache, A., and R. D. Ilnicki. 1988. Subterranean clover: A new approach to weed control. Proceedings, Northeast Weed Science Society 42: 34.
Fay, P. K., and W. B. Duke. 1977. An assessment of the allelopathic potential in Avena germplasm. Weed Science 25: 224-228.
Felton, W. L. 1976. The influence of row spacing and plant population on the effect of weed competition in soybeans. Australian Journal of Agricultural and Animal Husbandry 16: 926-931.
Fischer, R. A., and R. E. Miles. 1973. The role of spatial pattern in the competition between crop plants and weeds. A theoretical analysis. Mathematical Biosciences 18: 335-350.
Forcella, F. 1987. Characteristics associated with highly competitive soybeans. Agronomy

Abstracts 87: 111.
Forcella, F., and M. J. Lindstrom. 1988. Movement and germination of weed seeds in ridge-till crop production systems. Weed Science 36: 56-59.
Geier, B., and H. Vogtmann. 1988. Weed control without herbicides in corn crops. pp. 483-486. *In* P. Allen and D. Van Dusen [editors] Proceedings, Sixth International Conference.
Glaze, N. 1987. Cultural and mechanical manipulation of Cyperus spp. Weed Technology 1: 82-83.
Griffith, D. R., J. V. Mannering, and J. E. Box. 1986. Soil and moisture management with reduced tillage. *In* Milton A. Sprague and Glover B. Triplett [editors] No-tillage and surface-tillage agriculture. John Wiley and Sons, New York, New York.
Guneyli, E., O. C. Burnside, and P. T. Nordquist. 1969. Influence of seedling characteristics on weed competition ability of sorghum hybrids and inbred lines. Crop Science 9: 713-716.
Haas, H., and J. C. Streibig. 1982. Changing patterns of weed distribution as a result of herbicide use and other agronomic factors. *In* H. M. LeBaron and J. Gressel [editors] Herbicide resistance in plants. John Wiley and Sons, New York, New York.
Hardy, E. A. 1931. Machinery for weed control. Agricultural Engineering 12(10): 369-373.
Harrison, H. F., Jr., and J. K. Peterson. 1986. Allelopathic effects of sweet potatoes (*Ipomoea batatas*) on yellow nutsedge (*Cyperus esculentus*) and alfalfa (*Medicago sativa*). Weed Science 34: 623-637.
Hartwig, N. L. 1988. Ten years of corn yields with and without a crownvetch (*Coronilla varia* L.) as a living mulch. Abstract, Weed Science Society of America 28: 19.
Heitefuss, R., B. Gerowitt, and W. Wahmhoss. 1987. Development and implementation of weed economic thresholds in the F. R. Germany. Proceedings, British Crop Protection Conference-Weeds. 10A-1: 1,025-1,035.
Hindal, D. F., and S. M. Wong. 1988. Potential biocontrol of multiflora rose, Rosa multiflora. Weed Technology 2: 122-131.
Jeffers, D. L., and G. B. Triplett, Jr. 1979. Management needed for relay intercropping soybeans and wheat. Ohio Report 58: 67-69.
Kenney, 1986. Devine®—the way it was developed—an industrialist's view. Weed Science 34 (supplement 1): 15-16.
Khodayari, K., and R. J. Smith, Jr. 1988. A mycoherbicide integrated with fungicides in rice, Oryza sativa. Weed Technology 2: 282-285.
King, R. P., D. W. Lybecker, E. E. Schweizer, and R. L. Zimdahl. 1986. Bioeconomic modeling to simulate weed control strategies for continuous corn (*Zea mays*). Weed Science 34: 972-979.
Koch, W. 1988. Weed science in Germany. Weed Technology 2: 388-395.
Koskinen, W. C., and C. G. McWhorter. 1986. Weed control in conservation tillage. Journal of Soil and Water Conservation 41(6): 365-370.
Kurtz, T., S. W. Melsted, R. H. Bray, and H. L. Breland. 1952. Further trials with intercropping of corn in established sods. Soil Science Society of America Proceedings 16: 282-285.
Lake, G. G., and R. H. Harvey. 1985. Corn production using alfalfa sod as a living mulch. North Central Weed Control Conference 42: 136-137.
Lal, R. 1975. Role of mulching techniques in tropical soil and water management. Technical Bulletin 1. International Institute of Tropical Agriculture, Ibadan, Nigeria.
Lapham, J. 1987. Population of dynamics and competitive effects of *Cyperus esculentus* (yellow nutsedge)—prediction of cost-effective control strategies. Proceedings, British Crop Protection Conference-Weeds. 10A-3: 1,043-1,050.
Leather, G. R. 1983. Sunflowers (*Heilianthus annuus*) are allelopathic to weeds. Weed Science 31: 37-42.
Lee, G. A. 1986. Integrated control of rush skeletonweed (*Chondrilla juncea*) in the western U.S. Weed Science 34 (supplement 1): 2-6.

Liebl, R. A., and A. D. Worsham. 1983. Tillage and mulch effects on morning glory (*Ipomeoea* spp.) and certain other weed species. Proceedings, Southern Weed Science Society 36: 405-414.

Lockerman, R. H., and A. R. Putnam. 1981. Mechanism for differential interference among cucumber (*Cucumis sativus* L.) accessions. Botanical Gazzette 142: 427-430.

Marshall, D.J.P. 1987. Using decision thresholds for the control of grass and broad-leaved weeds at the Boxworth E.H.F. Proceedings, British Crop Protection Conference-Weeds. 10A-5: 1,059-1,066.

Massion, C. L., and S. E. Lindow. 1986. Effects of Shacelotheca holci infection on morphology and competitiveness of Johnsongrass (*Sorghum halepense*). Weed Science 34: 883-888.

Maxwell, B. Dl, M. V. Wilson, and S. R. Radosevich. 1988. Population modeling approach for evaluating leafy spurge (*Euphorbia esula*) development and control. Weed Technology 2: 132-138.

Mays, W. T., and Loke-tuck Kok. 1988. Seed wasp on multiflora rose, *Rosa multiflora*, in Virginia. Weed Technology 2: 265-268.

McElligott, V. T., and P. K. Fay. 1988. Containment of grazing goats with electric shock collars. Abstract, Weed Science Society of America 28: 85.

McLeod, E. J., and S. L. Swezey. 1980. Survey of weed problems and management technologies. Research Leaflet Series, October 1980. U.S. Appropriate Technology Program, University of California, Davis. 10 pp.

McWhorter, C. G., and E. E. Hartwig. 1972. Competition of Johnsongrass and cocklebur with six soybean varieties. Weed Science 20: 56-59.

Mercado, B. 1987. Future role of weed science in international agriculture. Weed Technology 1: 107-111.

Mitchell, W. H., and M. R. Teel. 1977. Winter-annual cover crops for no-tillage corn production. Agronomy Journal 69: 569-573.

Mortensen, K. 1988. The potential of an endemic fungus, *Colletotrichum gloesporioides*, for biological control of round-leaved mallow (*Malva pusilla*) and velvetleaf (*Abutilon theophrasti*). Weed Science 36: 473-478.

Mortimer, A. M. 1987. The population ecology of weeds—implications for integrated weed management, forecasting and conservation. Proceedings, British Crop Protection Conference-Weeds. 9A-1: 935-943.

Moschler, W. W., G. M. Shear, D. L. Hallock, R. D. Sears, and G. D. Jones. 1967. Winter cover crops for sod-planted corn: Their selection and management. Agronomy Journal 59: 547-551.

National Academy of Sciences. 1971. Biochemical interactions among plants. Washington, D.C.

Ogg, A. G., Jr., and J. H. Dawson. 1984. Time of emergence of eight weed species. Weed Science 32: 327-335.

Peele, T. C., G. B. Nutt, and O. W. Beale. 1946. Utilization of plant residues as mulches in the production of corn and oats. Soil Science Society of America Proceedings 11: 356-360.

Peters, S. E., R. W. Andrews, and R. R. Janke. 1988. Comparing soybean intercropped with small grains to a soybean monoculture within a cropping systems experiment (in review).

Phatak, S. C., M. B. Callaway, and C. S. Vavrina. 1987. Biological control and its integration in weed management systems for purple and yellow nutsedge (*Cyperus rotundus* and *C. esculentus*). Weed Technology 1: 84-91.

Prostko, E., and R. D. Ilnicki. 1988. Residual weed control in intercropped soybeans. Proceedings, Northeast Weed Science Society 42: 22-23.

Putnam, A. R., and Chung-Shih Tang, editors. 1986. The science of allelopathy. John Wiley and Sons, New York, New York.

Putnam, A. R., and J. DeFrank. 1983. Use of phytotoxic plant residues for selective weed control. Crop Protection 2: 173-181.

Putnam, A. R., J. DeFrank, and J. P. Barnes. 1983. Exploitation of allelopathy for weed control in annual and perennial cropping systems. Journal of Chemical Ecology 9: 1,001-1,010.

Quimby, P. C., Jr., F. E. Fulgham, C. D. Boyette, R. E. Hoagland, and W. J. Connick. 1988. New formulations/nozzles boost efficacy of pathogens for weed control. Abstract, Weed Science Society of America 28: 52.

Ramsel, R. E., and G. A. Wicks. 1988. Use of winter wheat (*Triticum aestivum*) cultivars and herbicides in aiding weed control in an ecofallow corn (*Zea mays*) rotation. Weed Science 36: 394-398.

Rees, N. E. 1986. Two species of musk thistle (*Carduus* spp.) as hosts of *Rhinocyllus conicus*. Weed Science 34: 241-242.

Rees, N. E., R. W. Pemberton, A. Rizza, and P. Pecora. 1986. First recovery of *Oberea erythrocephala* on the leafy spurge complex in the United States. Weed Science 34: 395-397.

Regnier, E. E., and E. W. Stoller. 1981. A living mulch for weed control in soybeans. Proceedings, North Central Weed Control Conference 36: 24-25.

Regnier, E. E., and E. W. Stoller. 1987. Cropping methods affecting the performance of living mulches in soybeans. Proceedings, North Central Weed Control Conference 42: 36-37.

Reinbott, T. M., Z. R. Helsel, D. G. Helsel, M. R. Gebbhardt, and H. C. Minor. 1987. Intercropping soybean into standing green wheat. Agronomy Journal 79: 886-891.

Rice, E. L. 1974. Allelopathy. Academic Press, New York, New York.

Ridings, W. H. 1986. Biological control of stranglervine in citrus—a researcher's view. Weed Science 34 (supplement 1): 31-32.

Roberts, H. A. 1970. Viable weed seeds in cultivated soils. *In* National Vegetable Research Station Annual Report 1969. Wellesbourne, Warwick, England.

Robinson, R. G., and R. S. Dunham. 1954. Companion crops for weed control in soybeans. Agronomy Journal 46: 273-281.

Robinson, R. G., and R. S. Dunham. 1956. Pre-planting tillage for weed control in soybeans. Agronomy Journal 48: 493-495.

Rogers, N. K., G. A. Buchanan, and W. C. Johnson. 1976. Influence of row spacing on weed competition with cotton. Weed Science 24: 410-413.

Ross, M. A., and C. A. Lembi. 1985. Applied weed science. Burgess Publishing Company, Minneapolis, Minnesota. 340 pp.

Runnels, H. A., and J. H. Schaffner. 1931. Manual of Ohio weeds. Bulletin No. 475. Ohio Agricultural Experiment Station, Wooster.

Schertz, D. L. 1988. Conservation tillage: An analysis of acreage projections in the United States. Journal of Soil and Water Conservation 43: 256-258.

Schweizer, E. E. 1988. New technological developments to reduce groundwater contamination by herbicides. Weed Technology 2: 223-227.

Shilling, D. G. 1983. The suppression of certain weed species by rye (*Secale cereale* L.) mulch. Isolation, characterization and identification of water-soluble phytotoxins from rye. Ph. D. thesis. North Carolina State University, Raleigh.

Silverstein, R. M., and J. B. Simeone. 1983. Special Issue: North American Symposium on Allelopathy. Journal of Chemical Ecology 9(8): 935-1,292.

Sim, L. C. 1987. The value and practicality of using weed thresholds in the field. Proceedings, British Crop Protection Conference-Weeds. 10A-6: 1,067-1,071.

Smeda, R. J., and A. R. Putnam. 1988. Cover crop suppression of weeds and influence on strawberry yields. Hortscience 23: 132-134.

Smith, R. J., Jr. 1986. Biological control of northern jointvetch (*Aeschynomene virginica*) in rice (*Oryza sativa*) and soybeans (*Glycine max*)—a researcher's view. Weed Science

34 (supplement 1): 17-23.

Smith, R. J., Jr. 1988. Weed thresholds in southern U.S. rice, *Oryza sativa*. Weed Technology 2: 232-241.

Standifer, L. C. 1979. Some effects of cropping systems on soil weed seed populations. Abstract 160. Proceedings, 32nd Annual Meeting, Southern Weed Science Society.

Staniforth, D. W. 1961. Response of corn hybrids to yellow foxtail competition. Weeds 9: 132-136.

Stoller, E. W., S. K. Harrison, L. M. Wax, E. E. Regnier, and E. D. Nafziger. 1986. Weed interference in soybeans (*Glycine max*). Weed Science 3: 155-182.

Stonehouse, D. P., B. D. Kay, J. K. Baffoe, and D. L. Johnston-Drury. 1988. Economic choices of crop sequences on cash-cropping farms with alternative crop yield trends. Journal of Soil and Water Conservation 43: 262-270.

Teasdale, J. R. 1988. Weed suppression by hairy vetch residue. Proceedings, Northeast Weed Science Society 42: 73.

Teasdale, J. R., and J. R. Frank. 1983. Effect of row spacing on weed competition with snap beans (*Phaseolus vulgaris*). Weed Science 31: 81-85.

Templeton, G. E. 1986. Mycoherbicide research at the University of Arkansas—past, present, and future. Weed Science 34 (Supplement 1): 35-37.

Timmons, F. L. 1970. A history of weed control in the United States and Canada. Weed Science 18: 294-307.

Timmons, F. L., and V. F. Bruns. 1951. Frequency and depth of shoot-cutting in eradication of certain creeping perennial weeds. Agronomy Journal 43: 371-375

van Heemst, H.D.J. 1985. The influence of weed competition on crop yield. Agricultural Systems 18: 81-93.

Vrabel, T. E., P. L. Minotti, and R. D. Sweet. 1980. Seeded legumes as living mulches in sweet corn. Proceedings, Northeast Weed Science Society 34: 171-175.

Vrabel, T. E., P. L. Minotti, and R. D. Sweet. 1983. Regulating competition from white clover in a sweet corn living mulch system. Abstract, Weed Science Society of America 23: 4.

Walker, H. L., and C. D. Boyette. 1985. Biocontrol of sicklepod (*Cassia obtusifolia*) in soybeans (*Glycine max*) with *Alternaria cassiae*. Weed Science 33: 212-215.

Warnes, D. D. 1985. Use of winter rye to help in weed control in narrow row soybeans. North Central Weed Control Conference 42: 230-231.

Wilson, B. J. 1978. The long-term decline of a population of *Avena fatua* with different cultivations associated with spring barley cropping. Weed Research 18(1): 52-54.

Wilson, B. J., and K. J. Wright. 1987. Variability in the growth of cleavers (*Galium aparine*) and their effect on wheat yield. Proceedings, British Crop Protection Conference-Weeds. 10A-4: 1,051-1,057.

# CONSERVATION TILLAGE IN SUSTAINABLE AGRICULTURE

R. Lal, D. J. Eckert, N. R. Fausey, and W. M. Edwards

Sustainable agriculture implies profitable farming on a continuous basis while preserving the natural resource base. It is the most economic-cum-efficient harnessing of solar energy in the form of agricultural products without degrading soil productivity or environmental quality. Sustainable agriculture, however, must look beyond production economics. Linkages are needed among production economics, ecological stability, and environmental quality. In this context, sustainable agriculture is not synonymous with "low-input," "organic," or "alternative" agriculture. In some cases, low inputs may sustain profitable and environmentally sound farming. In others, it may not (Figures 1-2). Addition of organic amendments may enhance soil productivity, but may not eliminate the need for balanced fertilizer because large quantities are required for economic returns. Pest incidence may be reduced through crop management, but chemical pest control may be more efficient and cost-effective.

World population has been growing exponentially since 1650; the present growth rate is about 2.1 percent per year. The present population is 5.1 billion people, and 230,000 humans are born each day. But the population is distributed unequally. A high proportion lives in developing countries—76 percent in 1986. This proportion will increase to 79 percent in the year 2000 and 86 percent in 2100 (U.S. Department of Census, 1983; Population Reference Bureau, 1986). Most farmers in developing countries use low-input agricultural practices. Impoverished soils, low levels of purchased inputs, and low returns have been responsible for a low standard of living and widespread malnutrition. This is especially true of the shifting cultivators and subsistence farmers of the tropics and subtropics.

More than 97 percent of the world's food is produced on land. A large proportion is produced by technology that eliminates soil and environmental constraints to agricultural production through energy-based inputs. It is the high energy flux of agricultural ecosystems that has broken yield barriers in the so-called "modern agriculture." If the concept of sustainable agriculture is to be widely accepted, it must have a double-edged strategy. The energy flux must be substantially increased to get the shifting cultivator out of subsistence farming, but energy use efficiency must be increased to render intensive modern agriculture more profitable (Figure 1). The "system" organization of extensive shifting cultivation must be improved by increasing the energy influx to counteract "entropy." On the other hand, "profit" maximization for intensive modern agriculture must be achieved by regulating overall energy input and reducing losses. Important causes of losses are erosion, leaching, and volatilization for nutrients and pesticides. In both scenarios, the optimization of energy influx is crucial to sustainability and environmental quality.

Intensification and adoption of modern agricultural techniques are necessary prerequisites to feeding the earth's expected 10.5 billion inhabitants by the year 2110. An alternate agriculture based on a level of low input similar to that of the shifting cultivator and causing a reduction in production is not a solution. Energy-efficient technology is not necessarily

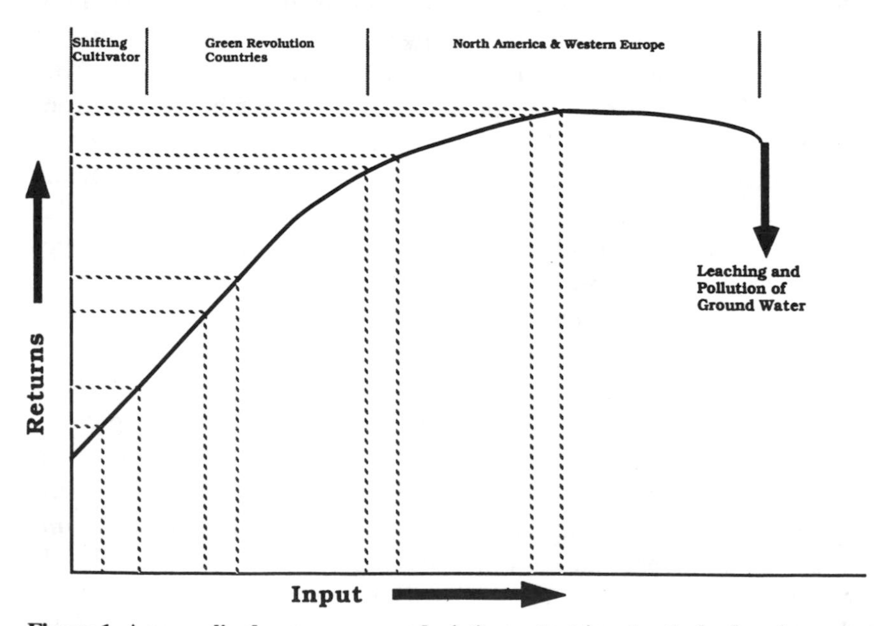


**Figure 1. A generalized response curve depicting output-input ratio for farming systems ranging from subsistence shifting cultivator to large-scale commercial agriculture.**

**Table 1. Energy efficiency (outinput ratio) for crop production in different production systems.**

| Crop | Energy Efficiency | |
|---|---|---|
| | *Less Developed Countries* | *Developed Countries* |
| Wheat | 6.25 (India) | 2.0 (United States) |
| Rice | 7.7 (Philippines) | 1.35 (United States) |
| Corn | 16.7 (Mexico) | 2.7 (United States) |
| Sorghum | 20.0 (Sudan) | 3.1 (France) |

Source: Le Pape and Merica (1983)

**Table 2. Food energy output per man-hour of farm labor.**

| *System* | *Crops* | *Output (MJ/Man-Hour)* |
|---|---|---|
| Subsistence tropics | Rice | 11-19 |
| | Maize, millet, sweet potato | 25-30 |
| Semi-industrial (Green Revolution) | Rice | 40 |
| | Maize | 23-48 |
| Industrial crops | Rice (United States) | 2,800 |
| | Cereals (United Kingdom) | 3,040 |
| | Maize (United States) | 3,800 |

Source: Leach (1976)

a high-production technology. Tables 1 and 2 indicate that subsistence agriculture is more energy-efficient than intensive agriculture. However, widespread adoption of the most energy-efficient and low-input agriculture would also lead to the lower nutritional and living standards now experienced by shifting cultivators and resource-poor farmers of less-developed countries. Adoption of low-input agriculture at the level of subsistence farming is also a nonsolution.

Furthermore, the level of input required to obtain an economic response depends on soil properties and cropping systems. For some fertile soils (soil A in Figure 2), high yields are obtained even with no input. In contrast, soil C does not respond to any level of input. There are vast areas of marginal soils of category C that should not be cultivated. A large proportion of arable land falls under category B and respond to input. The level of input should be judiciously managed for soils of category B.

## Tillage and Sustainability

Tillage and seedbed preparation can be a major contribution to energy influx in a crop production system (Table 3). The diesel fuel consumption for plow-based conventional tillage systems ranges from 60 to 80 liters per hectare (Rask and Forster, 1977). Plow-based systems also increase soil erosion on undulating and sloping cropland. Dealing with sediment in surface waters costs the United States from $4 billion to $16 billion annually (in 1985 dollars) (Crosson and Ostrov, 1988). Cropland erosion is respon-

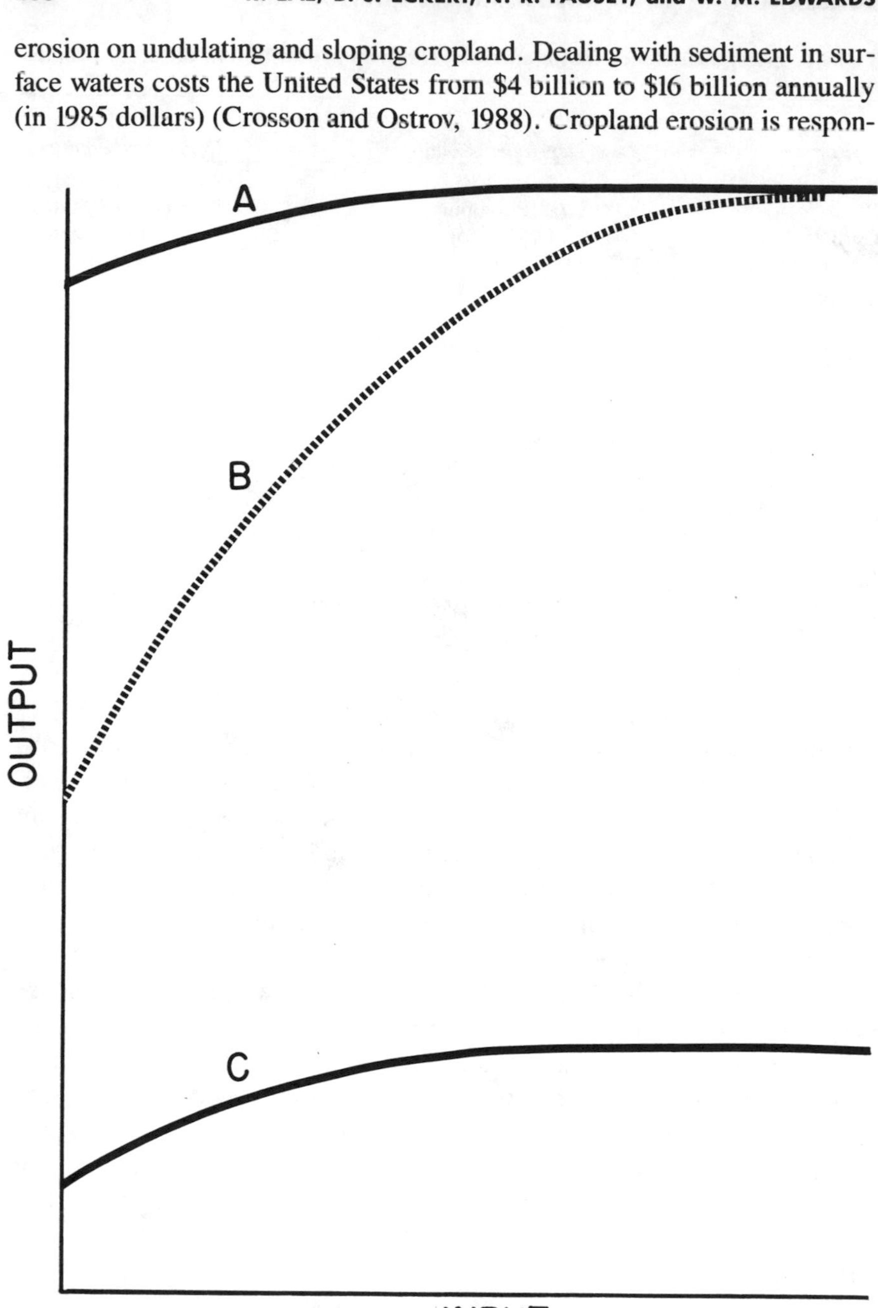


**Figure 2. A generalized response function depicting output-input for three soils with varying level of antecedent fertility—A: highly productive soil, B: soils of medium fertility that respond to input, C: marginal and relatively infertile soil that do not respond to input.**

**Table 3. Estimated fuel requirements for various field crop operations in Indiana.**

| *Field Operation* | *Fuel (L/ha)* |
|---|---|
| Moldboard plowing | 21.6 |
| Chisel plowing | 14.3 |
| Heavy tandem disking | 12.2 |
| Standard tandem disking | 9.7 |
| Field cultivation | 9.7 |
| Row crop planting | 10.3 |
| Grain drilling | 10.3 |
| Rotary hoeing | 7.3 |
| Row crop cultivation | 10.3 |
| Knifedown $NH_3$ application | 13.0 |

Source: Stout (1984), FEA/USDA (1977)

sible for about one-third of this damage. Reductions in frequency and intensity of tillage reduce the erosion risks and energy input. In many cases, however, reduction in tillage also reduces yield and profit.

Reduced tillage systems generally are termed "conservation tillage"—defined currently as any system that leaves at least 30 percent of the previous crop residue on the soil surface after planting. The various types include minimum tillage, chisel plowing, plow-plant, ridge tillage, and no-tillage. When successfully applied, conservation tillage may reduce energy consumption and control erosion. Similar to the management of other energy-based inputs, the goal is to apply the right amount of tillage needed to optimize yield, while decreasing energy input and erosion potential.

## Conservation Tillage and Energy Efficiency

The number and type of required field operations differ among tillage systems (Tables 4 and 5). In the United States, primary and secondary tillage operations consume some 10-12 trillion Kcals a year (Stout, 1984). In comparison, energy used in production of fertilizers applied is about 160-170 trillion Kcals. Thus a reduction in frequency and intensity of tillage would reduce a small proportion of energy input.

Conservation tillage systems have many potential energy-saving benefits (Triplett and Van Doren, 1977; Sprague and Triplett, 1986). If properly done, conservation tillage improves soil structure (Lal, 1983). Tilling a well-structured soil (chisel or disk plowing) requires less mechanical power and hence less fuel than a soil with massive and degraded structure. Energy-saving benefits of conservation tillage are (a) less fuel consumption due to reduced field operations, (b) less power required due to better soil structure, (c) less time and labor required, (d) possibility of double cropping, and (e) lower investment in farm machinery. Yet some activities associated with conservation tillage require more energy. Exampes are (a) weed control with herbicides, (b) control of high incidence of insects and pathogens, (c) addition of fertilizers, and (d) higher seeding rates and special seeding equipment.

Actual energy needs for a conservation tillage system thus depends on the balance among the factors listed above. In Kentucky, Frye and Phillips (1981) observed that savings in fuel by adopting no-till may be as much as 40 percent (Table 6). The saving in total energy by switching from con-

**Table 4. Farm operations required in four tillage systems in the Corn Belt of the United States.**

| | | *Conservation Tillage* | | |
|---|---|---|---|---|
| *Season* | *Conventional tillage* | *Minimum-tillage* | *Ridge-tillage* | *No-tillage* |
| Fall/ Winter | Harvest, moldboard plow, (apply fertilizer).* | Harvest, chisel plow, (apply fertilizer). | Harvest, (apply fertilizer). | Harvest, apply fertilizer). |
| Spring/ Summer | Disc harrow, apply herbicides, plant, (apply fertilizer), (cultivate). | Disc, spray herbicides, plant, (apply fertilizers). | Plant, spray herbicides, (apply fertilizer), cultivate, and form ridges. | Plant, spray herbicides, (apply fertilizer). |

*Operations in parentheses are either not performed by all operators or are not dependent on season.

**Table 5. Farm operations required in three farm systems in the tropics.**

| | | *Intensive Systems* | | |
|---|---|---|---|---|
| *Season* | *Shifting Cultivation* | *Conventional Tillage* | *No-tillage* | *Alley Cropping* |
| Dry | Harvesting, land clearing, burning, making mounds and ridges. | Harvesting, plowing at the rain end. | Harvesting | — |
| Monsoon | Planting, weeding, staking, harvesting. | Plowing, harrowing, spraying herbicides, applying fertilizer, planting, cultivation, harvesting. | Mowing, spraying herbicides, applying fertilizer. | Pruning, plowing, harrowing, applying fertilizer, planting, cultivating, harvesting. |

**Table 6. Energy requirements for several field operations in different tillage systems.**

| *Operation/ Input* | *Diesel Fuel (L/ha)* Conventional Tillage | Chisel Plow | Disk | No-tillage |
|---|---|---|---|---|
| Moldboard plow | 17.2 | — | — | — |
| Chisel plow | — | 10.5 | — | — |
| Disk | 5.9 | 5.9 | 5.9 | — |
| Apply herbicide and disk second time | 6.8 | 6.8 | 6.8 | — |
| Spray herbicides | — | — | — | 1.2 |
| Plant | 4.0 | 4.0 | 4.0 | 4.7 |
| Cultivate (once) | 3.9 | 3.9 | 3.9 | — |
| Herbicides | 16.4 | 18.8 | 21.0 | 26.9 |
| Machinery and repair | 17.4 | 15.1 | 11.7 | 5.6 |
| Total | 71.6 | 65.0 | 53.4 | 38.4 |

Source: Frye and Phillips (1981), Poincelot (1986).

**Table 7. Energy/kg required to produce active ingredients of various herbicides.**

| *Herbicide* | *MJ/kg* |
|---|---|
| Atrazine | 238 |
| Alachlor | 396 |
| 2, 4-D | 203 |
| Paraquat | 414 |
| Trifluralin | 374 |

Source: Southwell and Rothwell (1977); Stout (1984).

ventional tillage to conservation tillage may, however, only be about 3 percent for reduced tillage (Fluck and Baird, 1980) and 7 percent for no-tillage (Phillips, et al., 1980). The adoption of legume-based rotations may lead to greater energy saving (Heichel, 1978), through improvements in soil nitrogen. The saving in energy by conservation tillage depends on the degree of weed infestation and its control. Energy value of most herbicides is high (Table 7). The effects of reduced tillage on saving energy comes from the possibility of high production or high production efficiency. In some instances, weed control through herbicides is more cost- and energy-effective than mechanical measures of manual or motorized operations (Table 8).

In the tropics, traditional farming systems are labor-intensive and use none or few purchased inputs. Energy requirement per man-day is estimated at 300 to 600 MegaJoules (Fluck and Baird, 1980). Land clearing and weeding use as much as 80 percent of the labor consumed. Energy requirements for intensive conventional and no-till systems are high and similar to those of the Corn Belt region in the United States. The alley-cropping system is an intermediate system, a combination of both shifting cultivation and intensive farming. It is labor-intensive and requires additional input for plowing, mowing, and spraying herbicides. The fertilizer needs may be somewhat reduced (Kang et al., 1981).

Saving of energy is not a principal advantage of conservation tillage systems for intensive row cropping in either temperate or tropical climates. Mechanized operations reduce human drudgery by eliminating many back-breaking jobs. Substantial energy saving can be achieved by:

- Using no-till planting where possible.
- Using chisel plowing or disking instead of moldboard plowing.
- Avoiding primary tillage deeper than 8 inches (20 cm).
- Reducing the number of secondary tillage operations.
- Avoiding secondary tillage deeper than 4 inches (10 cm).
- Combining various operations; e.g., pulling a spring-tooth harrow behind a disk, or applying herbicides along with secondary tillage.
- Traveling in the highest gear practical, and throttling down.
- Maintaining proper drive-wheel slippage.
- Ensuring that tillage equipment is properly adjusted.
- Matching implement size to tractor size.

## Erosion Control

Soil erosion from nonfederal land in the United States is estimated at 5 billion tons per year by water and 1.5 billion tons per year by wind (USDA, 1987). The relative magnitude of soil erosion for different land use systems shown in table 9 indicates an average rate of 10 tons per hectare per year from croplands. Economic and environmental effects of accelerated erosion are related to reduced yield due to loss of topsoil, available water and nutrient reserves on-site, and siltation and water pollution off-site. Reduction in crop yield due to erosion may be 5 to 10 percent for very fertile soils in the United States (Larson et al., 1983) and 20 to 30 percent for each 2.54 centimeters of soil loss for shallow soils of the tropics (Lal, 1987a).

Conservation tillage systems can be extremely effective in reducing soil

**Table 8. Energy relationships in weed control in six experiments on corn.**

| *Method of Weed Control* | *Energy Input For Weed Control (MJ/ha)* | *Yield of Corn* | | *Net Energy Gain Due to Weed Control (MJ/ja)* |
|---|---|---|---|---|
| | | *Kg/ha* | *MJ/ha* | |
| None | 0 | 3,395 | 56,300 | — |
| Cultivation | 580 | 5,090 | 84,400 | 27,520 |
| Herbicides | 390 | 5,660 | 93,800 | 37,100 |
| Hand labor | 340 | 5,780 | 95,900 | 39,270 |

Source: Nalewaja (1974); Stout (1984).

**Table 9. Sources and rates of soil erosion in the United States.**

| Erosion | Land Use | Percentage of Total | Annual Average Rate (m tons/ha) | Extreme Rate | | |
|---|---|---|---|---|---|---|
| | | | | Highest | Percentage of Land | Percentage of Erosion |
| Water | Cropland | 38 | 10.3 | 66.0 | 2 | 25 |
| | Rangeland | 8 | 6.2 | 11.0 | 12 | 57 |
| | Pastureland | 3 | 5.7 | 11.0 | 11 | 50 |
| | Forests | 29 | 2.6 | — | — | — |
| | Grazed forests | — | 9.2 | — | — | — |
| Wind | Rangeland | 45 | 4.4 | 30.8 | 3 | 31 |
| | Cropland | 55 | 11.7 | 30.8 | 9 | 53 |

Source: OTA (1982); Poincelot (1986).

**Table 10. Runoff and sediment yield from plowed and no-till watershed growing corn at Coshocton, Ohio.**

| Treatment | Slope (%) | Rainfall (cm) | Runoff (cm) | Sediment Yield (t/ha) |
|---|---|---|---|---|
| Plowed, clean tilled, sloping rows | 6.6 | 13.97 | 11.18 | 50.7 |
| Plowed, clean tilled, sloping rows | 5.8 | 13.97 | 5.84 | 7.2 |
| No-till, contour rows | 20.7 | 12.88 | 6.35 | 0.07 |

Source: Harrold and Edwards (1972).

**Table 11. Effect of no-till system on runoff and soil erosion from corn for a tropical Alfisol in western Nigeria.**

| Slope (%) | Runoff (mm) | | Soil Erosion (t/ha) | |
|---|---|---|---|---|
| | No-till | Plowed | No-till | Plowed |
| 1 | 11.4 | 55.1 | 0.0 | 1.2 |
| 5 | 11.8 | 158.7 | 0.2 | 8.2 |
| 10 | 20.3 | 52.4 | 0.1 | 4.4 |
| 15 | 21.0 | 89.9 | 0.1 | 23.6 |
| Mean | 16.1 | 89.0 | 0.1 | 9.4 |

Source: Lal (1976).

erosion. Many studies have shown that conservation tillage can reduce erosion by as much as 90 percent (Crosson, 1981; Mannering, 1979; Langdale et al., 1978). Data from Coshocton, Ohio (Table 10) and Ibadan, Nigeria (Table 11) indicate that conservation tillage can be extremely effective in controlling soil erosion.

The choice of conservation tillage for most effective erosion control and water conservation is influenced by soil, drainage, cropping system, and resources available. The effectiveness of a reduced tillage system in controlling erosion depends on (a) the surface area covered by mulch, (b) the area disturbed by tillage, (c) soil compaction, (d) surface crusting, and (e) degree of worm activity and status of biopores. For dense, compacted soil, some tillage may be necessary to allow root proliferation, but some residue cover should be maintained. Ridge tillage and basin tillage or tied-ridges are effective in soil drainage (Stewart et al, 1981; Fausey, 1984; Eckert, 1987a, and Lal, 1987b). No-till or reduced tillage is relatively more effective for erosion control on coarse-textured and well-drained soils than on heavy-textured and poorly drained soils (Edwards and Amerman, 1984). Slot mulching or vertical mulching sometimes is recommended for heavy-textured soils (Saxton et al., 1981; Parr and Papendick, 1983).

## Crop Yields

Crop response to tillage systems is hard to predict. Agronomic yield depends on a range of associated practices such as drainage method; planting date; variety selection; cropping geometry; plant population; form of fertilizer, and time and mode of its application; pest control; cropping system; and type of equipment. Above all, crop growth and yield in relation to tillage are greatly influenced by antecedent soil properties and climate. Important soil factors are texture, internal drainage, soil compaction, soil temperature and moisture regimes, and fertility status.

***Soil Drainage.*** No-till and reduced tillage systems produce satisfactory yields on well-drained soils. Yield reduction is commonly observed whenever no-till is used on somewhat poorly drained soils (Table 12). Date of

**Table 12. Tillage × drainage interactions for corn grain yield (t/ha) on a Clermont soil.**

| | | *Distance From Drain (m)* | | | | | |
|---|---|---|---|---|---|---|---|
| *Tillage* | *Year* | *0-3* | *3-6* | *6-9* | *9-12* | *12-15* | *Average* |
| Beds | 1978 | 9.88 | 9.52 | 9.00 | — | 9.25 | 9.41 |
| | 1979 | 9.52 | 9.33 | 9.52 | 9.06 | 8.80 | 9.24 |
| | 1980 | 10.57 | 9.98 | 9.66 | 9.83 | 9.73 | 9.25 |
| | Average | 9.99 | 9.61 | 9.39 | 9.44 | 9.26 | |
| Chisel | 1978 | 9.29 | 8.64 | 8.43 | — | 7.66 | 8.50 |
| | 1979 | 9.90 | 9.81 | 9.23 | 9.40 | 9.10 | 9.49 |
| | 1980 | 11.33 | 10.96 | 10.36 | 10.18 | 9.58 | 10.48 |
| | Average | 10.17 | 9.80 | 9.34 | 9.79 | 8.78 | |
| No-till | 1978 | 9.30 | 8.50 | 8.03 | — | 8.38 | 8.55 |
| | 1979 | 9.95 | 9.42 | 9.25 | 9.64 | 9.23 | 9.50 |
| | 1980 | 9.45 | 6.62 | 6.17 | 6.55 | 7.47 | 7.25 |
| | Average | 9.57 | 8.18 | 7.82 | 8.10 | 8.36 | |

Source: Unpublished data, N. R. Fausey.

**Table 13. Effects of tillage and drainage on corn and soybean yields (t/ha) following oats in a lake-bed soil in Ohio.**

| | *Corn* | | *Soybean* | |
|---|---|---|---|---|
| *Tillage* | *Tile* | *No Tile* | *Tile* | *No Tile* |
| Moldboard plow | | | | |
| 18-cm row | — | — | 5.1 | 4.3 |
| 76-cm row | 10.4 | 8.3 | 4.7 | 3.6 |
| Ridge (76-cm row) | | | | |
| Slot plant | 11.2 | 9.9 | 4.7 | 3.7 |
| Till plant | 11.4 | 9.7 | 4.7 | 3.6 |
| LSD (0.05) | NS | 0.8 | 0.2 | 0.4 |

Source: Eckert (1987).

**Table 14. Effects of tillage and drainage on soil temperature in early spring in Ohio (unpublished data of N. Fausey).**

| *Distance From* | *Soil temperature (C°) at 5 cm depth* | | | |
|---|---|---|---|---|
| *Drain (m)* | *Ridges* | *Beds* | *No-till* | *Plow* |
| 0 | 8.5/1.0 | 7.2/2.0 | 7.3/2.8 | 7.9/1.9 |
| 9 | 8.1/1.3 | 7.3/2.5 | 7.2/3.0 | 7.7/2.1 |
| 18 | 7.7/1.7 | 7.5/2.3 | 7.3/3.0 | 7.6/2.4 |
| 27 | 7.8/1.7 | 7.4/2.7 | 7.2/3.0 | 7.2/2.5 |

Maximum/minimum measured on April 3, 1984.

planting is also delayed on wet soils (Eckert, 1984). Cereals sown direct exhibit symptoms of nutrient deficiency. Fertilizer needs for no-till usually are different from those of plow-till (Eckert et al., 1986; Eckert, 1987b). In northern Indiana, Griffith et al. (1982) observed that conventional tillage produced 10 percent more yield than no-tillage on a poorly drained soil. In contrast, no-tillage produced 8 percent more yield than conventional tillage on a well-drained soil.

Ridge planting is an effective way to improve drainage. On a lake-bed soil (*Mollic ochraqualf*) in Ohio, Eckert (1987a) observed that yields of corn planted on ridges generally were equal to or greater than those of corn planted following fall plowing (Table 13). For a Clermont soil, Fausey (1984) reported that in poorly drained conditions raised beds and chisel plowing produced higher corn yields than no-till (Table 12). In northern latitudes, slow warming and cooler temperatures on poorly drained soil can inhibit germination, causing stunted initial growth. Tillage-induced temperature differences in the seed zone can be 2 to 5 degrees C (Table 14).

***Soil Structure and Compaction.*** No-till systems do not perform well on compacted soils and on those with massive structure. Alleviating soil compaction is essetnial to crop growth and yield. A one-year experiment conducted on a lake-bed soil in Ohio indicated that oat yield in no-till was suppressed in comparison with that of plowed treatment. Yield suppression was greater in compacted than uncompacted soils (Table 15).

Soil compaction can be alleviated by mechanical or biological means. In contrast to subsoiling and chiseling, biological means of alleviating soil compaction have slow but long-lasting effects. Growing leguminous and grass covers improve soil physical properties and enhance soil fertility (Wilson et al., 1982). Rotation experiments in Ohio by Van Doren et al. (1976) showed that no-tillage can reduce yields on corn grown in monoculture. Table 16 shows that growing corn in rotation improved yields. Rotation-induced improvements in yields were more significant for poorly drained Hoytville clay than for well-drained Wooster silt loam. An experiment conducted on a root-restrictive Alfisol in Western Nigeria indicated substantial structural improvements by growing a deep-rooted legume. Growing tap-rooted pigeon-pea (*Cajanus cajan*) for two years lowered soil bulk density, increased macroporosity, and improved root penetration by the following maize crop (Table 17). In addition to improving soil structure, legumes also enhance soil fertility (Frye et al., 1983; McCown et al., 1985; Lal, 1987b) and suppress weeds (Palada et al., 1983). Cover crops also enhance population and activity of soil fauna such as earthworms. Stable

**Table 15. Oat yield (Mg/ha) on a lake-bed soil in northwest Ohio as influenced by tillage method and soil compaction level (unpublished data of R. Lal).**

| | *Single Axle Load (Mg) for Harvesting Traffic* | | |
|---|---|---|---|
| *Tillage Method* | *0* | *10* | *20* |
| No-till | 1.97 | 1.89 | 1.16 |
| Chisel till | 2.40 | 2.22 | 1.71 |
| Plow till | 2.41 | 2.17 | 2.08 |
| LSD (.10) | | | |
| Tillage (T) | 0.45 | | |
| Compaction (C) | 0.45 | | |
| T × C | 0.77 | | |

**Table 16. Rotation and tillage effects on corn grain yield on two soils in Ohio.**

| | | *Corn Grain Yield (Mg/ha)* | | |
|---|---|---|---|---|
| *Soil* | *Rotation* | *No-till* | *Plow-till* | *Probability* |
| Wooster silt loam | Continuous corn | 9,400 | 8,420 | ? 0.001 |
| | Corn-soybean | 9,480 | 8,720 | 0.03 |
| | Corn-oats-hay | 10,450 | 9,720 | 0.01 |
| Hoytville clay | Continuous corn | 6,820 | 8,000 | ? 0.001 |
| | Corn-soybean | 7,920 | 8,260 | ND |
| | Corn-oats-hay | 8,180 | 8,390 | ND |

Source: Van Dohen et al. (1976).

**Table 17. Effects of cropping systems on soil bulk density and yield of maize on a tropical Alfisol.**

| | *Soil Bulk Density (Mg/m³)* | | *Grain Yield (Mg/ha)* | |
|---|---|---|---|---|
| *Cropping System* | *29 Days After Sowing* | *90 Days After Sowing* | *First Season* | *Second Season* |
| Continuous maize | 1.41 a | 1.39 a | 2.0 a | 2.2 a |
| Pigeon pea-maize | 1.36 b | 1.33 b | 2.8 b | 2.7 b |

Source: Hubugalle and Lal (1986).

and continuous biopores created by burrowing activity of soil fauna facilitate water and air movement and deep root penetration of the following crop. Crop covers must, however, be properly suppressed prior to no-till seeding of a grain crop. If not, live mulches can cause a severe yield reduction (Eckert, 1988). Low crop stand and severe competition for water and nutrients can be responsible for reduced yield in live-mulch systems.

***Climate and Drought.*** Climate is an important determinant of crop response to tillage methods. In general, conservation tillage systems perform poorly in wet springs and better in moderately dry conditions. The effects of climate can be drastically altered by soil properties and management skills. For some soils with a high plant-available water capacity, modest variations in rainfall pattern (total amount and its distribution) do not have drastic effects on crop yield. Data from Coshocton, Ohio, indicate relatively little effect of rainfall amount on corn yield by no-till or plow-till method of seedbed preparation (Figure 3). In other soils, the seed drill passing over a clayey soil in a wet spring may create a smeared groove that remains open even after the seed is dropped. Soil near the groove is somewhat compacted (Figure 4). Seed germination is satisfactory if rains are favorable and occur frequently, but seedling mortality is high and crop stand poor if a wet period is followed by a prolonged dry season. This was the case in Ohio in 1988. Stand establishment also depends on operator skill and equipment adjustment. Soil temperature also is more favorable in a well-drained, coarse-textured soil than in a poorly drained, clayey soil. Microclimate, pore-size distribution, soil strength, and biophysical environments within the seed zone are vital to crop establishment and subsequent growth and yield.

Meso- and macro-climate also affect crop response to tillage. In well-drained soils, yield stability with conservation tillage often is better than with conventional tillage. The trend is reversed in poorly drained soils. Table 18 compares long-term yield averages for three tillage methods at two Ohio sites. The standard deviation of the mean was more for no-till than for chisel-till or plow-till. Long-term yield averages for no-till treatment were more significantly correlated with crop stand than for chisel or plow-till treatments. Complete crop failure at the Wooster site was observed in 3 out of 25 years for no-till. At the South Charleston site, complete crop failure in one out of 25 years was observed for no-till only.

## Tropical vs. Temperature Climates

There are subtle differences in tropical temperature regions (Figure 4) that must be considered while assessing the applicability of conserva-

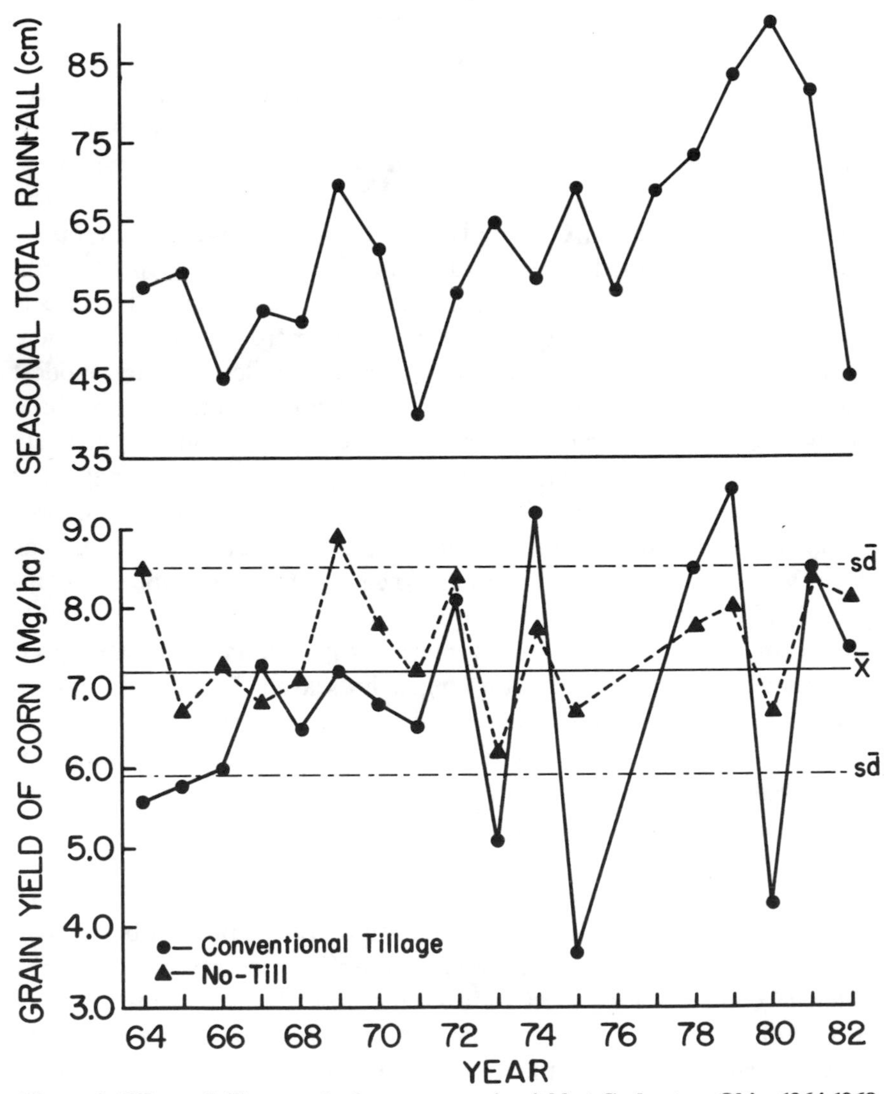


**Figure 3. Effects of tillage methods on corn grain yield at Coshocton, Ohio, 1964-1968.**

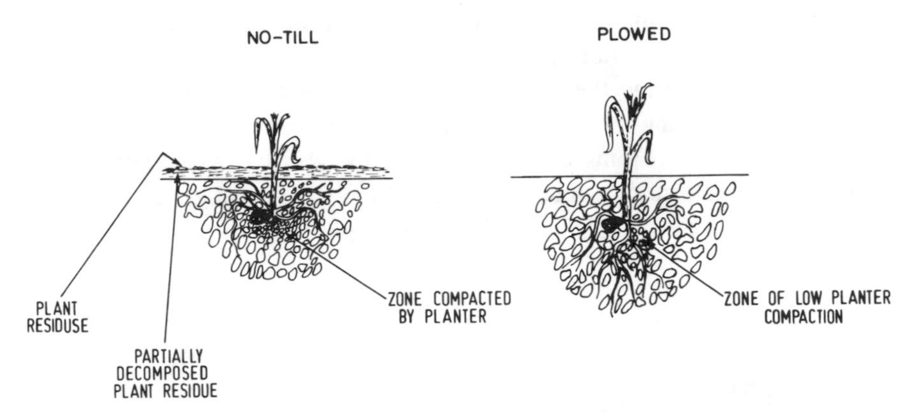


**Figure 4. Soil conditions in the seedling zone for no-till and plowing systems of seedbed preparation.**

tion tillage systems. The major difference is the physical and climate environments in the early spring. In temperate regions, soil is cold and wet in early spring and gets warm and relatively dry as the season changes to summer. In contrast, soil is hot and dry in early spring in the tropics and gets cooler and moist as the monsoons begin. While the presence of crop residue mulch may be a disadvantage for a soil with suboptimal temperature in the temperate zone, it is a definite advantage for a soil in the tropics with supraoptimal temperature regime. Presence of crop residue mulch on a wet and cool soil can cause prevalence of anaerobic conditions that may hinder seed germination and seedling establishment. Soils in the northern latitudes undergo repeated cycles of freezing and thawing and have a natural structure-improving system conducive to formation of crumb structure and favorable tilth. There is no such mechanism for soils of the tropics. On the contrary, soils in the wet/dry tropics experience ultra-desiccation, are hard-set and develop a massive structure during the prolonged dry season. There are also differences in the net radiation received. In the temperate zone, net radiation received is low in early spring and increases as the season progresses into summer. Net radiation received in the tropics

**Table 18. Long-term corn yield average (Mg/ha) and standard deviation for three tillage methods at two sites in Ohio (yield record from 1962-1987).**

| *Tillage* | *Wooster Silt Loam (Wooster)* | *Crosby Silt Loam (South Charleston)* |
|---|---|---|
| No-till | $7.23 \pm 3.09$ | $7.31 \pm 2.50$ |
| Chisel | $7.06 \pm 2.50$ | $8.06 \pm 1.79$ |
| Plow till | $6.85 \pm 2.36$ | $7.98 \pm 1.91$ |

is higher at the beginning of the monsoon than during the rainy season. Grain yields in the equatorial climate are limited by the low level of radiation, especially during the reproductive stages of crop growth.

Strongly interacting with the climate are differences in soil characteristics. All other factors remaining the same, soils in the wet/dry tropics (subhumid and semiarid) are more susceptible to crusting, hard-setting, and compaction than soils in the northern latitudes. This implies that mechanical loosen-

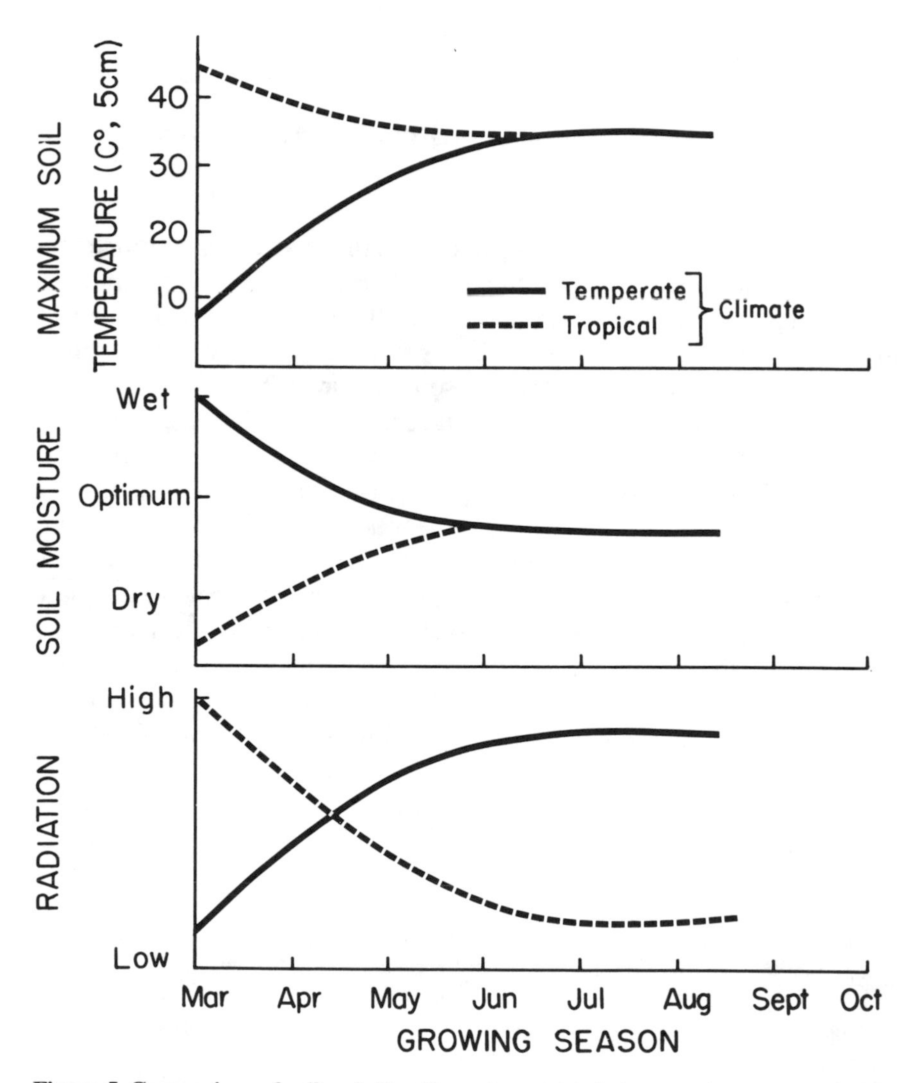


**Figure 5. Comparison of soil and climatic environments in temperate zone versus tropical climates.**

ing based on plow-till systems may be needed more for soils of the wet/dry tropics than for well-drained soils in the temperate regions. Above all, accelerated soil erosion and its adverse effects on soil fertility and crop productivity are usually more severe for soils of the tropics than temperate regions (Lal, 1987b).

## Soil Guide to Tillage Methods

The choice of the most appropriate type of conservation tillage depends on many important soil factors: texture, structure, erodibility, susceptibility to crusting and compaction, slope length, slope gradient, slope aspect and shape, effective rooting depth, plant-available water and nutrient reserves, and internal drainage. There also are socioeconomic considerations. Because there are a wide range of soils even within a single farm unit, it is difficult to recommend an optimum tillage system for all soils.

Ohio possesses a wide range of soils, and the success of a conservation tillage system depends on an interaction between inherent soil properties, drainage improvement, and crop rotation. Tillage requirements for principal soils in Ohio were discussed by Triplett et al. (1973) and are briefly described here:

■ *Soils Developed from High-lime Lake Sediments.* Two types of soils typified by this condition are Paulding and Hoytville series. Paulding is among the most difficult soils to manage in Ohio. With silty clay to clay texture, these soils have extreme wetness problems and occur on level landscapes. These soils do not respond to tile drainage and must be surface drained for reasonable production potential. Soybeans and wheat are major crops because soils often are too wet to prepare for corn early in spring. They often are fall plowed, but more progressive farmers perform only minimum tillage (to 10 centimeters deep). No-till and ridging are not favored because of wetness and soil hardening. These soils have low erosion potential.

Hoytville is similar to Paulding but responds well to tile drainage. It is a very important agricultural soil. Corn-soybean and corn-soybean-wheat rotations predominate. Most farmers fall plow. Severe phytophthora root-rot problems in soybeans are aggravated by reducing tillage. In general, conservation tillage must be accompanied by drainage improvements and crop rotation. In monoculture corn or soybeans, yields decline as tillage intensity declines. Fall plowing predominates, although more farmers are chiseling. Many farmers perform only shallow tillage in soybean residue. No-tillage has generally produced favorable yields if managed intensively. Ridge planting has performed well on these soils, and it may seem to have overcome much of the enhanced rotation effect associated with no-till and ridges observed in 1988. A minimum tillage system, such as chiseling or disking, may provide most consistent crop performance.

■ *Soils Developed from High-Lime Glacial Drift.* These are the most widespread and productive agricultural soils in Ohio. Most fields have more than one soil type and a gradient of drainage characteristics. Rolling topography results in slight to severe erosion problems, depending on location. These soils should be tiled for optimum production potential. The rotation interaction with tillage is not as important as on lake-plain soils but is still considerable. More poorly drained soils (Kokomo) do need rotation, and performance of tillage systems is fairly consistent, resembling Hoytville but at higher average yield levels. On better drained soils (Crosby, Blount) rotation is less important but performance of no-till is very erratic from year to year; average yields are close to those obtained by plowing. Limited experience with ridge planting indicates that yields are similar to no-till. Most farmers fall plow or fall chisel because spring wetness can be a problem. Chiseling seems to be the most consistent conservation tillage system from an agronomic standpoint. Severe crusting and hardening potential exists on bare fields, particularly in dry years. Severe to no apparent problems were observed with no-till in 1988, depending on location and rainfall distribution.

■ *Soils Developed from Low-Lime Glacial Drift.* Many naturally very well-drained soils are ideally suited to no-till. Erosion hazard is very high if plowed because most fields are sloping. No-till responses are best on these soils because of naturally good drainage and tendency to crust after spring rains. Ridge planting is considered impractical because most farmers rotate corn with hay. There were rather severe problems with no-till in 1988, mostly crop growth. The rotation interactions with tillage in corn are not very strong, probably because yields are boosted more by water conservation by cornstalk residue than they are hurt by monoculture. This region has seen the greatest adoption of no-till, probably because of the consistently significant yield increases and the fact that much successful no-till research was conducted at Wooster on similar soils (Figure 6).

■ *Soils Developed from Sandstone and Shale.* This region is characterized by steep slopes and relatively unproductive soils. Good response to no-till is observed on deeper soils with good drainage. No-till corn production is limited by Johnsongrass infestation in southern areas of the region. The region seems to be gradually returning to forest because of a relatively low potential for intensive row-crop production.

The most desirable type of tillage in the tropics likewise depends on soil type, cropping systems, and logistic support. A soil suitability guide for

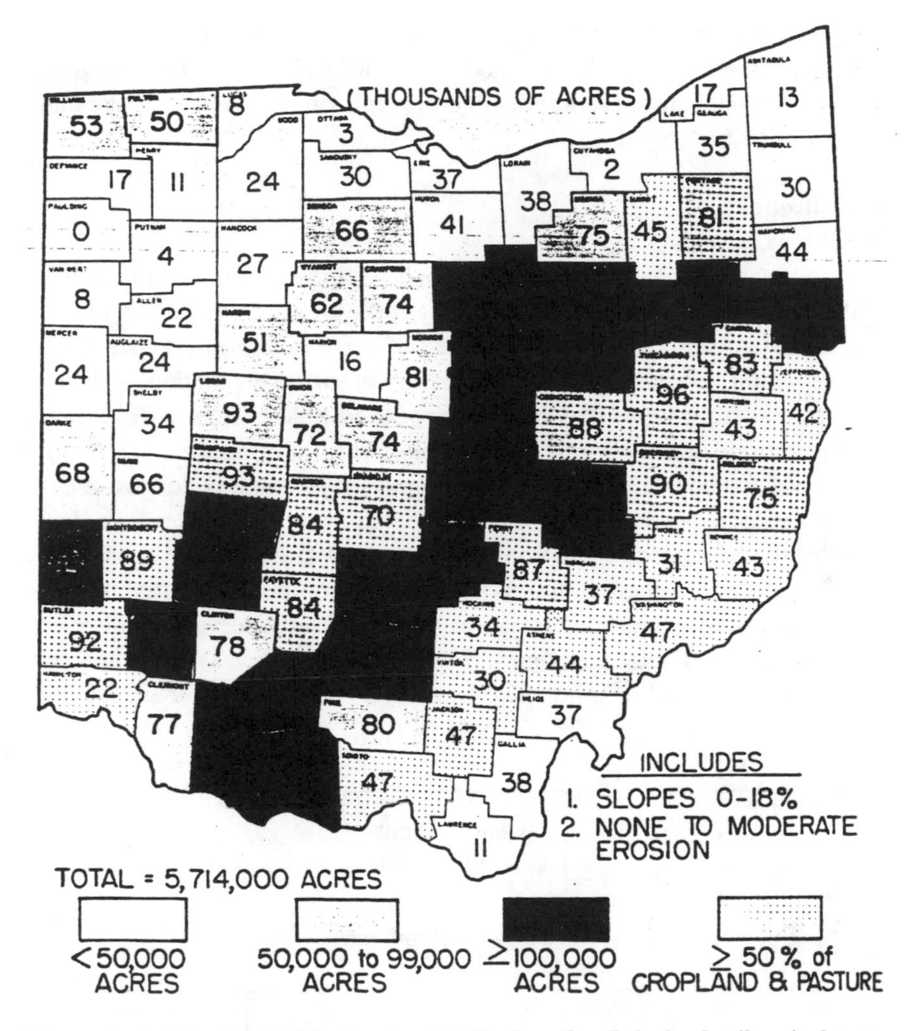


**Figure 6. A soil map of Ohio showing distribution of well-drained soils suited to no-till farming.**

conservation tillage for soils of the tropics was proposed by Lal (1985). A rating system was developed to assess the suitability of the type of conservation tillage for different soils. Soil and climatic properties included in the rating system are erosivity, erodibility, soil loss tolerance, compaction, soil temperature regime, available water-holding capacity, cation exchange capacity, and soil organic matter content. Also included is the quantity of crop residue mulch on the soil surface at seeding. In the humid and subhumid tropics with soils of coarse texture in the surface horizon, no-tillage can be successfully applied for production of upland row crops. In the semiarid regions with fine-textured soils, some form of mechanical loosening of crusted and compacted soils is necessary. The frequency and type of mechanical operation desired depends on soil characteristics and the crops to be grown.

## Conclusions

Some form of conservation tillage can be applied on a wide range of soils and ecological regions by adopting the systems approach—a holistic approach that considers all factors that affect production (D'Itri, 1985). Conservation tillage requires a special set of cultural practices that may be different from those needed for plow-based tillage (Figure 7). Careful consideration should be given to the set of cultural practices specifically developed for conservation tillage. Conservation tillage is not just a concept but a package of cultural practices that are specifically developed and adopted to conserve soil and water resources, sustain high and satisfactory returns, minimize degradation of soil and environment, and maintain the resource base. The close interrelationship between conservation tillage and cultural practices shown in figure 5 requires a high level of management skills and high level of inputs. Conservation tillage can be made an inte-

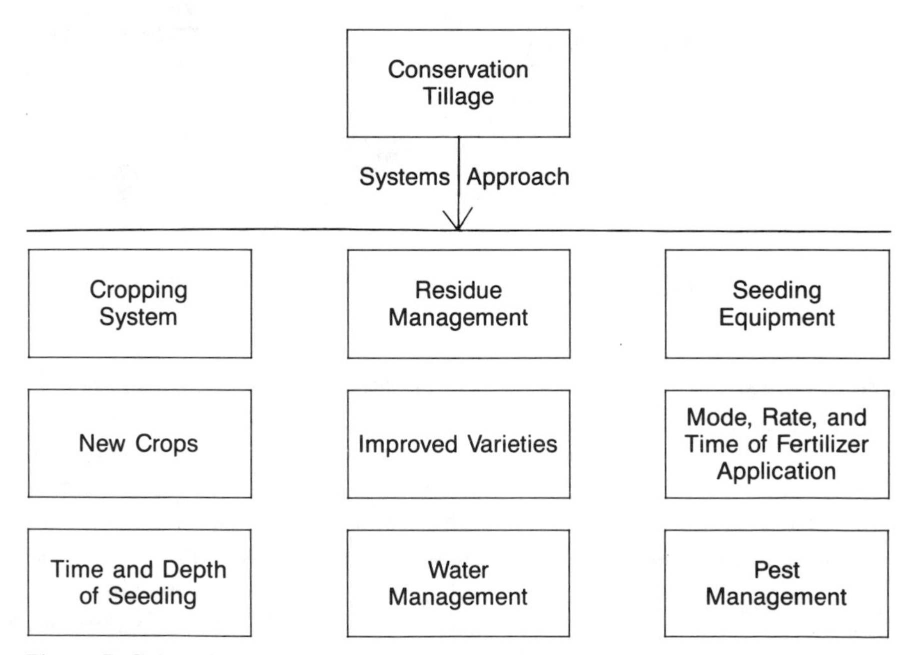


**Figure 7. Cultural practices needed for successful adoption of conservation tillage.**

gral part of sustainable agricultural systems through practically oriented, multidisciplinary research. It is an approach to soil surface management that emphasizes use and improvement of the natural resource rather than exploitation and mining of its productivity for quick economic return.

## REFERENCES

Crosson, P. 1981. Conservation and conventional tillage: A comparative assessment. Soil Conservation Society of America, Ankeny, Iowa.

Crosson, P. R., and J. E. Ostrov. 1988. Alternate agriculture: Sorting out its environmental effects. Resources 92: 13-16.

D'Itri, F. M. 1985. A systems approach to conservation tillage. Lewis Publishers, Inc., Chelesha, Montana.

Eckert, D. J. 1984. Tillage system × planting date interactions in corn production. Agronomy Journal 76: 580-582.

Eckert, D. J. 1987a. Evaluation of ridge planting systems on a poorly drained lake plain soil. Journal of Soil and Water Conservation 42: 208-210.

Eckert, D. J. 1987b. UAN management practices for no-tillage corn production. Journal of Fertilizer Issues 4: 13-18.

Eckert, D. J. 1988. Rye cover crops for no-tillage corn and soybean production. Journal of Production Agriculture (in press).

Eckert, D. J., W. A. Dick, and J. W. Johnson. 1986. Response of no-tillage corn and soybean residues to several nitrogen fertilizer sources. Agronomy Journal 78: 231-235.

Edwards, W. M., and C. R. Amerman. 1984. Sub-soil characteristics influence hydrologic response to no-tillage. Transactions, American Society of Agricultural Engineers 27: 1,055-1,058.

Fausey, N. R. 1984. Drainage-tillage interaction on Clermont soil. Transactions, American Society of Agricultural Engineers 27: 403-406.

FEA/USDA. 1977. A guide to energy savings for the field crop producer. Washington, D.C., GPO.

Fluck, R. C., and C. D. Baird. 1980. Agricultural energetics. AVI Publishing Co., Westport, Connecticut.

Frye, W. W., J. H. Herbek, and R. L. Blevins. 1983. Legume cover crops in production of no-tillage corn. pp. 179-192. *In* W. Lockeretz (editor) Environmentally sound agriculture. Praeger Special Studies, New York, New York.

Frye, W. W., and S. H. Phillips. 1981. How to grow crops with less energy. *In* J. Hayes (editor) Cutting energy costs (The 1980 Yearbook). U.S. Department of Agriculture, Washington, D.C.

Green, M. B. 1978. Eating oil. Westview Press, Boulder, Colorado.

Griffith, D. R., J. V. Mannering, D. B. Mengel, S. D. Parsons, T. T. Bauman, D. H. Scott, C. R. Edwards, F. T. Turpin, and D. H. Doster. 1982. A guide to no-till planting after corn or soybean. Cooperative Extension Service Publication ID-154. Purdue University, W. Lafayette, Indiana.

Harrold, L. L., and W. M. Edwards. 1972. A severe rainstorm test of no-till corn. Journal of Soil and Water Conservation 27: 30.

Heichel, G. H. 1978. Stabilizing agricultural needs: Role of forages, rotations, and nitrogen fixation. Journal of Soil and Water Conservation 33: 279-282.

Hulugalle, N. R., and R. Lal. 1986. Root growth of maize in a compacted gravelly tropical Alfisol as affected by rotation with a woody perennial. Field Crops Research 13: 33-44.

Kang, B. T., G. F. Wilson, and L. Sipkens. 1981. Alley cropping maize and Leucaena in southern Nigeria. Plant and Soil 63: 165-179.

Lal, R. 1976. Soil erosion on Alfisols in western Nigeria. I. Effects of slope, crop rotation and residue management. Geoderma 16: 363-375.

Lal, R. 1983. No-till farming. International Institute of Tropical Agriculture, Monograph 3.

Lal, R. 1985. A soil suitability guide for different tillage systems in the tropics. Soil and Tillage Research 5: 179-196.

Lal, R. 1986. Soil surface management in the tropics for intensive land use and high and sustained production. Advances in Soil Science 5: 1-105.

Lal, R. 1987a. Effects of soil erosion on crop productivity. Chemical Residue Company, Critical Reviews in Plant Science 5: 303-368.

Lal, R. 1987b. Managing soils on sub-Saharan Africa. Science 236: 1,069-1,076.

Langdale, G. W., A. Barnett, and J. E. Box. 1978. Conservation tillage systems and their control of water erosion in the southern Piedmont. Proceedings of the First Annual Southeastern No-Till Systems Conference, Georgia Experiment Station, Watkinsville, Georgia. Special Publication 5: 20-29.

Larson, W. E., F. J. Pierce, and R. H. Dowdy. 1983. The threat of soil erosion to long-term crop production. Science 219: 456.

Le Pape, Y., and J. R. Mercier. 1983. Energy and agricultural production in Europe and the Third World. pp. 104-123. *In* D. Knorr (editor) Sustainable food systems. Ellis Harwood Ltd., Chichester, United Kingdom.

Leach, G. 1976. Energy and food production. IPC Science & Technology Press, Ltd., United Kingdom.

Mannering, J. V. 1979. Conservation tillage to maintain soil productivity and improve water quality. Cooperative Extension Service Publication AY-222, Purdue University, W. Lafayette, Indiana.

McCown, R. L., R. K. Jones, and D.C.I. Peake. 1985. Evaluation of a no-till tropical legume lay farming strategy. pp. 450-472. *In* R. C. Muchow (editor) Agro-research for Australia's semi-arid tropics. University of Queensland. Press, Australia.

Nalewaja, J. D. 1974. Energy requirements for various weed control practices. Proceedings of North Central Weed Control Conference 29: 19-23.

OTA. 1982. Impacts of technology on U.S. cropland and rangeland productivity. Office of Technology Assessment, Washington, D.C.

Palada, M.C.S. Ganser, R. Hofstetter, B. Volak, and M. Culik. 1983. Association of inter-seeded legume cover crops and annual row crops in year-round cropping systems. pp. 181-213. *In* W. Lackeretz (editor) Environmentally sound agriculture. Praeger Special Studies, New York, New York.

Parr, J. F., and R. J. Pappendick. 1983. Strategies for improving soil productivity in developing countries with organic wastes. *In* In W. Lockeritz (editor) Environmentally sound agriculture. Praeger Publishers, New York, New York.

Phillips, R. E., R. L. Blevings, G. W. Thomas, W. W. Frye, and S. H. Phillips. 1980. No-tillage agriculture. Science 208: 1,108-1,113.

Poincelot, R. P. 1986. Toward a more sustainable agriculture. AVI Publishing Co., Westport, Connecticut.

Population Reference Bureau. 1986. International Union for Conservation of Nature and Natural Resources. Population Bulletin.

Rask, N., and D. Lynn Forster. 1977. Corn tillage systems: Will energy costs determine the choice? pp. 289-299. *In* William Lockeretz (editor) Agriculture and energy. Academic Press, New York, New York.

Saxton, K. E., D. K. McCool, and R. J. Papendick. 1981. Slot mulch for runoff and erosion control. Journal of Soil and Water Conservation 36: 44-47.

Southwell, P. H., and T. M. Rothwell. 1977. Analysis of output/input energy ratios of food production in Ontario, Contract Serial No. OSW76-00048. School of Engineering, University of Guelph, Ontario.

Sprague, M. A., and G. B. Triplett, Jr. 1986. No-tillage and surface tillage agriculture:

The tillage revolution. John Wiley & Sons, New York, New York.

Stewart, B. A., D. A. Dusek, and J. T. Musick. 1981. A management system for the conjunctive use of rainfall and limited irrigation of graded furrows. Soil Science Society of America Journal 45: 413-419.

Stout, B. A. 1984. Energy use and management in agriculture. Breton Publishers, North Scituate, Massachusetts.

Triplett, G. B., Jr., and D. M. Van Doren, Jr. 1977. Agriculture without tillage. Science of America 236: 28-33.

Triplett, G. B., Jr., D. M. Van Doren, Jr., and S. W. Bone. 1973. An evaluation of Ohio soils in relation to no-tillage corn production. OARDC Research Bulletin 1068. Wooster, Ohio. 20 pp.

U.S. Department of Agriculture. 1987. The Second RCA Appraisal. Washington, D.C.

U.S. Department of Census. 1983. World Population. Washington, D.C.

Van Doren, D. M., Jr., G. B. Triplett, Jr., and J. E. Henry. 1976. Influence of long-term tillage, crop rotation and soil type combinations on corn yield. Soil Science Society of America Journal 40: 100-105.

Wilson, G. F., R. Lal, and B. N. Okigbo. 1982. Effects of cover crops on soil structure and on yield of subsequent arable crops grown under strip tillage on an eroded Alfisol. Soil and Tillage Research 2: 237-250.

# PASTURE MANAGEMENT

Bill Murphy

The pasture resource in the United States has been mismanaged, wasted, and ignored—probably because the nation had too much land available, too few animals to graze it well, and no pressing economic need to use the land more efficiently. Today, however, with a shrinking agricultural land base and farm financial problems the United States is being forced to take a much closer look at all available resources.

Pasture is a tremendous resource. In the Northeast alone there are 10 million acres of permanent pastureland, which were perceived until recently to be practically worthless. In fact, it has been called marginal land, because most of it cannot be tilled and planted to crops due to soil and site limitations. Another 4 million acres of pastures rotated with other crops are used at a level far below their potential because of defective grazing management (Northeast Research Program Steering Committee, 1976).

Incorporating well-managed pastures into farm feeding programs can reduce production costs and increase farming profitability and sustainability.

## Grazing Management Methods

Pasture is a forage crop capable of intercepting and storing large amounts of solar energy and, consequently, supporting high levels of livestock production at low cost if managed properly. In the United States we are just beginning to learn how to manage pasture for higher solar energy interception. We are realizing that animals are the tools for managing pasture and marketing its forage, not ends in themselves. It follows that pasture needs to be managed in ways that result in as high-quality, dependable, and uniform supply of forage as possible, at as low a cost as possible. We also need to select livestock that do well on forage that can be produced economically, rather than attempting to feed animals to a preselected genetic potential regardless of cost.

The problem then is: How to manage the pasture resource to achieve that high-quality, dependable, and uniform supply of forage over as long a grazing season as possible? (Pastures in the United States are almost exclusively complex mixed swards based on legume-fixed nitrogen. Growing conditions and plant growth rates are highly variable. Consequently, these pastures have little in common with the pure perennial ryegrass nitrogen-fertilized swards growing under relatively uniform conditions in Europe.)

There are two basic methods of grazing management on farms and ranches: continuous, sometimes called set stocking, and rotational, which has many names.

***Continuous Grazing.*** Continuous grazing usually involves trying to match a set number of livestock with pasture growth within the same area during the entire grazing season. Plants are continually exposed to grazing. Since the number of stock carried on U.S. farms using continuous grazing tends to be conservative, continuously grazed pastures tend to be overgrazed in early spring, mid-, and late summer, and undergrazed in late spring, early summer, and autumn due to variation in plant growth rate during the season. When excess forage is available, animals selectively graze, and repeatedly graze more palatable plants, leaving the rest to mature, flower, set seed, and multiply. During times of excess forage, individual animal productivity can be high, but production per unit area usually is low. This is because much of the forage remains uneaten and competes with grazed plants for sunlight, water, and nutrients. Grasses tend to disappear and broadleaf weeds come into the overgrazed areas. The undergrazed patches become rank and even less palatable, clovers are shaded out, soil nitrogen levels drop, and total production eventually falls.

Selective grazing occurs when the stocking density is too low to use all of the forage that is produced. Under continuous grazing, selection can be avoided only by increasing the stocking density, and by adjusting the stocking density continually during the grazing season as plant growth rate and herbage accumulation vary. In practice, adjustments of stocking density to forage availability are not made frequently enough or at all during the season, with the result that continuously grazed pastures in the United States generally are weedy, unproductive messes.

***Rotational Grazing.*** The term "rotational grazing" is an especially poor one in the United States because it can mean so many different things, most of which have been associated with defective grazing management and consequent failure to use pastures well. It can mean two 20-acre paddocks grazed two weeks on, two weeks off, by 20 cows. It can mean moving 10 cows every Monday afternoon in sequence among four 5-acre paddocks. This kind of "management" is convenient for farmers and researchers, but has nothing at all to do with plant growth rate and forage availability.

True rotational grazing that takes into account both the needs of pasture plants and grazing animals was defined by Andre Voisin (1959, 1960). Voisin called his method "rational grazing" because with it pasture forage is rationed out according to animals' needs (just as feed is rationed out in confinement feeding), while protecting the plants from overgrazing. Unfortunately, instead of using Voisin's term for his management method, several different terms are being used for it, including intensive rotational grazing, intensive grazing management, short duration grazing, Savory grazing, controlled grazing management, and Voisin grazing management.

Two of the terms, short duration grazing and Savory grazing, have been discredited for rangeland management in the U.S. West. A modified version of Voisin's grazing method, planned grazing management, must be used on rangelands in brittle environments. Planned grazing management involves planning for specific recovery periods between grazings, monitoring regrowth of severely grazed plants, and using effects of herds to break up uneaten plants and soil crust, which quickens plant decomposition and allows water penetration and seedling establishment (Savory, 1988).

"Mob stocking" is similar to Voisin grazing management; large numbers of animals are concentrated in small paddocks for short periods. Mob stocking is used mainly to clean up coarse, fibrous, rank forage left by poor utilization, rather than for animal production. It is used to improve pastureland, and quickly transforms low-producing, overgrown areas into high-producing ones. When mob stocking is used to improve overgrown pastureland, only nonproducing animals should be used, and care should be taken so that the animals are not overly stressed (Smetham, 1973).

This multiplicity of terms is confusing. All refer to the same method, but all must be defined when used because they imply different methods. To describe the method here, I will use Voisin grazing management.

## Voisin Grazing Management

Voisin grazing management in essence gives pasture plants a chance to photosynthesize and replenish food reserves. The Voisin method con-

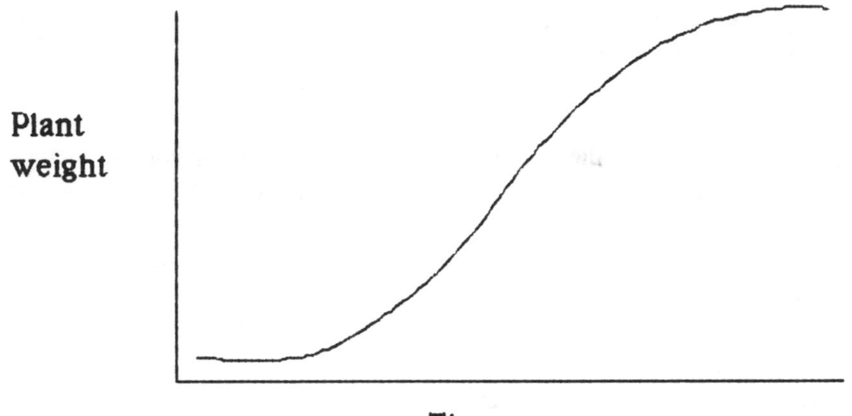


**Figure 1. Plant growth over time (adapted from Voisin, 1959).**

trols what and when livestock eat, by dividing pastures into small areas (paddocks) and rotating animals through them. It rations out pasture forage according to the needs of livestock, rests the plants according to their needs, and keeps forage waste to a minimum. [Much of what follows originally appeared in "Greener Pastures on Your Side of the Fence" (Murphy, 1987); used with permission of the publisher.]

Key parts of Voisin's method concern rest periods between grazings and the length of time that animals are in a paddock.

***The Pasture Plant.*** Pasture plants must be able to regrow after they have been grazed. When plants are cut, either by a mower or animal, very little leaf surface is left. Adequate food reserves must exist in remaining plant parts to form enough leaf surface so that photosynthesis can function normally again. Photosynthesis supplies energy for continued regrowth of the plant and storage of more food reserves. If plants are cut or grazed before enough reserves are stored, regrowth will be retarded or will not occur at all. Pasture plants must have sufficient leaves for variable periods during the growing season.

***Plant Regrowth Curve.*** The regrowth curve (Figure 1) of plants is S-shaped and has three stages: (1) early period of slow growth, (2) middle period of rapid growth, and (3) final period of slow growth. In the first growth of spring or after being grazed or cut off at any time in the season, plants have a limited leaf surface and grow slowly. When enough leaf surface has developed from food reserves to intercept large amounts of sunlight, rapid growth follows. When enough food reserves are stored and shading increases, growth slows down again.

***Rest Periods.*** Besides the variation of the plant growth curve, plant growth rate also differs within the season. One of the main rules of Voisin's method is that rest periods between grazings must vary according to changes in plant growth rates, which reflect changes in growing conditions. Generally, this means that rest periods must lengthen as plant growth rate slows during the season.

Assume that plant growth rate in May through June is twice as fast as it is in August through September, and July is a transition time between the two. This means that rest periods between grazings must be twice as long in August-September as they are in May-June. Of course, plant growth rates differ within regions and prevailing climatic conditions in any season, but these are good average estimates for humid-temperate regions, such as in the northeastern and north central United States.

Productivity of plants and the amount of forage available to animals entering a paddock equal the daily amount of plant regrowth per acre that accumulated since the last time the paddock was grazed. In the northeastern and north central United States, for example, optimum rest periods between grazings are about 18 days in May-June and lengthen to about 36 days by August-September (Figure 2). In these regions, regrowth accumu-

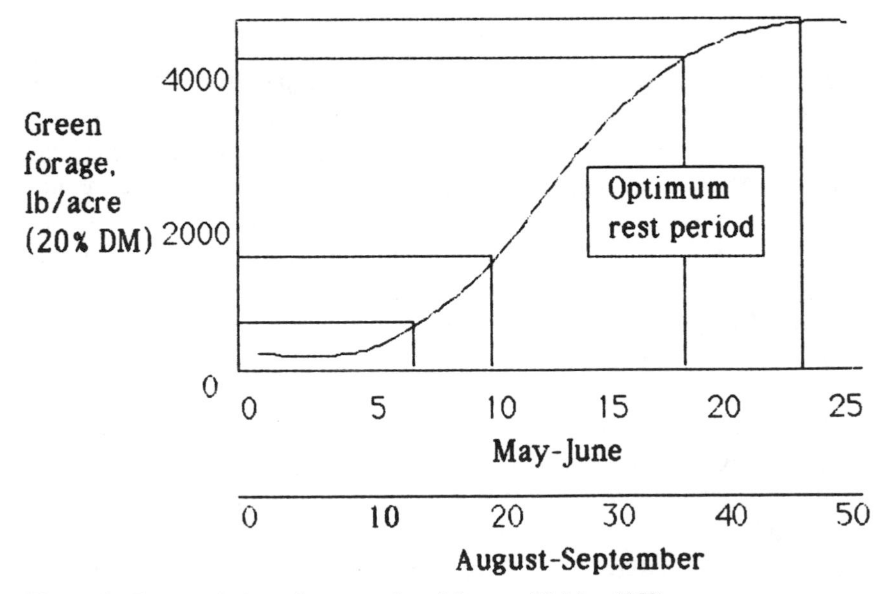


**Figure 2. Rest periods and accumulated forage (Voisin, 1959).**

lates about 4,200 pounds of green forage (20 percent dry matter, average) per acre during optimum rest periods. Forage availability decreases to one-third, or only about 1,400 pounds of green forage per acre, with a rest period of one-half of the optimum. If rest periods are shortened even more, available forage drops to one-tenth, or about 430 pounds per acre. This short rest period corresponds to what happens when pastures are grazed off every time they grow tall enough to be grasped by animals' mouths, or about every 6 days in May-June and 12 days in August-September. If rest periods are longer than the optimum, forage accumulation increases, but the increase is due mainly to more fiber, which lowers the forage feeding value.

These are guidelines for beginning to use the method. With experience, adjustments can be made to better suit local conditions. For example, in the Champlain Valley of Vermont, the following rest period schedule works well: 12 to 15 days in late April to early May, 18 days by May 31, 24 days by July 1, 30 days by August 1, 36 days by September 1, and 42 days by October 1. If a month is unusually hot and dry, more rest may be needed in that month and the following one. If conditions are more favorable than usual for plant growth, less rest may be needed. Plant growth is the best indicator of conditions for the plants. With adequate rest between grazings, pastures may be grazed for a much longer part of the year than would otherwise be possible.

Don't be concerned if rest periods on your farm are not precisely these amounts. The amount of rest needed between grazings will differ with locations. Do become concerned, however, if rest periods shorten by even 12 hours in any rotation, because it may indicate that plant growth has slowed. For example, around June 20 you notice that rest periods, which had been 22 days, begin to shorten by 12 hours in each paddock. Carefully check the growth of plants in paddocks that the animals will go into next. If the plants haven't regrown enough (i.e., 4 to 6 inches tall), you must either increase the pasture area and number of paddocks available for grazing, remove all animals from the pasture and feed them elsewhere, or feed hay or green chop to them in a paddock or paddocks until rest periods are adequate again.

If paddocks are not rested enough, animals move through the rotation faster, due to decreasing amounts of available forage, when the movement should be slowing down because of slowing plant growth rate. Suddenly, the plants become exhausted, stop growing, and there is no more forage to graze.

How do you achieve rest periods that are twice as long in autumn as they are in spring? Three practical ways for doing this are:

- About half of the pasture area must be set aside from grazing in the spring and either hayed or ensiled, because too much forage is produced in May-June for a set number of animals to consume. This means that only about one-half of the total pasture area is grazed in May, June, and part of July. Because using the Voisin method at least doubles or triples plant productivity, you will have to machine-harvest forage from areas where there never has been an excess before. If your pastures are mainly on rough land, you should set aside areas where you can use machinery to harvest the excess forage. You must be prepared for the increase in forage production that will occur; otherwise, the pasture won't be grazed properly and its full potential won't be realized. After the excess forage has been harvested, the areas should be rested for about 25 days, divided into paddocks, and included in the rotation. This increases the area available for grazing and automatically lengthens the rest periods to about what is needed for the remainder of the season.

■ Graze twice as many animals on the total pasture area in May, June, and the first part of July as in the rest of the season. The decrease in animal numbers carried during the second half of the season lengthens rest periods to about what is needed. This is the only way to graze it well if the pasture is all rough land that can't be harvested with machinery. This means that about half of the animals will either have to be fed elsewhere after July 15, or they will have to be sold.

■ A compromise way of keeping pasture forage under control during times of excessive growth on land where machine harvesting isn't possible is not to graze as low as usual. This will result in patchy grazing, but can be minimized by grazing two kinds of animals (e.g., sheep and cows) on the same land, either simultaneously or one after the other.

***Periods of Stay and Occupation.*** The length of time that each group of animals is in a paddock per rotation is called the period of stay. The total time that all groups of animals occupy a paddock in any one rotation is called the period of occupation (also sometimes called the grazing period). If only one group of animals is grazing, the period of stay equals the period of occupation. If two groups graze, their total periods of stay equal the period of occupation.

When plants are grazed off, they can regrow tall enough to be grazed again after about six days in May-June and 12 days in August-September in the northeastern or north central United States. Although most plants take 12 days to regrow to grazing height in August-September, some (e.g., orchardgrass) continue to regrow tall enough to be grazed after only about six days. So, periods of occupation should never exceed six days, to prevent grazing of regrowth in the same rotation, and they really should be two days or less for best results.

Periods of stay for any one group of animals should not be longer than three days, giving total occupation periods of six days for two groups of animals. This is because the longer animals are in a paddock, the less palatable the remaining forage becomes, and the more time and energy they spend searching for desirable feed. Periods of stay of two days or less if animals are grazed as one group, and one day or less for each of two groups, giving total occupation periods of two days or less, are better than longer periods of stay and occupation.

In practice, the shorter the periods of stay and occupation, the more optimum plant and animal production will be. Milking, growing, and fattening animals should not be in a paddock for longer than two days per rotation any time during the season. This keeps them on a consistently high level of nutrition. Milking cows, goats, and sheep produce most if they are given a fresh paddock after every milking. Not only is the forage of higher quality and grazed more uniformly than with less frequent moves to fresh paddocks, but milking animals let their milk down better, anticipating that they are going to a fresh paddock as soon as they are milked. Growing and fattening animals, such as lambs and beef animals, also gain weight most rapidly if they are moved to a fresh paddock every 12 or 24 hours.

Paddocks must be small enough so that all forage in each paddock is grazed completely and uniformly within each occupation period. Occupation periods may need to change because plant growing conditions vary during the season, and the amount of forage available also changes. Lengthening occupation periods indicate that excess forage is available, and paddocks may need to be subdivided or removed from the rotation and machine-harvested, so that the pasture continues to be well grazed. Shortening occupation periods indicate that more paddocks and pasture area are needed.

For example, if animals don't eat enough to keep up with the rapidly growing forage in May and June, remove more paddocks from the rotation and cut them for hay or silage. Suppose that in the first rotation of the season animals occupy paddocks for two days, eating all of the forage available in each paddock within the two days. After they have grazed six or seven paddocks, move them back to the first paddock that was grazed and start the second rotation. Leave the rest of the pasture area for machine harvesting. (Return animals to the first paddock only if the plants in it have regrown to about 4 inches tall and have a fully developed green color. Never allow animals to graze forage that is so young that it is still yellow. It is bad for the animals and may result in the plants growing poorly during the rest of the season.)

***Dividing the Animals.*** The simplest way to graze is to have all the animals in one group. For best results, paddocks should be small enough so that all of the forage is eaten in two days or less. This keeps all of the animals on a medium-to-high level of nutrition. Although only one source of water is needed when animals graze as one group, ideally drinking water should be provided in all paddocks so that animals and their manure remain in the paddocks they are grazing. Also, if drinking water is always readily available in paddocks, animals don't waste energy walking to and from the source of water.

A more efficient way is to divide animals into two groups (leader/follower) according to their production levels and nutritional needs at different physiological states. This allows pasture-feeding value to be closely matched with animal needs. The groups of animals can be handled in two ways:

- Each group can be grazed on separate pasture areas. Animals with the highest nutritional requirements (milking cows, goats, sheep, growing lambs, and beef cattle) are grazed on the best pasture available. Other animals (dry cows and heifers, dry does, dry ewes) are grazed elsewhere on lower quality pasture that is being improved. This has the advantage that only one source of drinking water is needed for each group, although it always is best to provide water in all paddocks. The disadvantage is that more paddocks must be built, requiring more fencing materials.
- Both groups graze within the same pasture area at different times. Animals having the highest nutritional needs are turned into a paddock first, which allows them to eat the best forage quickly. They should not be left in a paddock for more than two days, because after that time they have to work too hard to meet their nutritional needs (12- to 24-hour periods of stay are best). After the first group is removed from a paddock, the second group follows to clean up the remaining forage, which has a lower feeding value than the forage that was grazed first. Paddocks must be small enough so that the combined periods of stay of the two groups are less than six days (two days or less is best). When two groups of animals graze the same area, all paddocks must have drinking water available to them. This is because at least one of the groups must be locked in its paddock to keep the groups separate.

***Number of Paddocks.*** The number of paddocks needed depends on the rest periods between grazings, period(s) of stay or occupation of the animals in each paddock in each rotation, and the number of animal groups graz-

**Table 1. Number of paddocks needed for a 36-day rest period between grazings.**

| *Period of Stay For One Group, Days* | *Total Number of Paddocks Needed For* | |
|---|---|---|
| | *One Group* | *Two Groups* |
| ½ | 73 | 74 |
| 1 | 37 | 38 |
| 2 | 19 | 20 |
| 3 | 13 | 14 |

ing (Table 1). Since shorter periods of occupation favor higher plant and animal yields, the more paddocks there are, the more productive the pasture will tend to be. In deciding how many paddocks to build, consider the topography of the pastureland, the pasture plant botanical composition and its potential yielding ability, the maximum rest periods needed in your area, the livestock, fencing costs, and financial constraints.

First, estimate the rest period likely to be needed during the time of slowest plant growth in your region's grazing season. This may range from less than 36 to more than 100 days, depending on conditions. Remember, these are the total numbers of paddocks needed when pasture plants grow the slowest. During times of fastest growth, only about one-half as many paddocks are needed, since the remainder of the pasture area will be set aside for machine harvesting.

Second, estimate how long the periods of occupation will be and decide whether to use one or two groups of animals.

Third, use this equation to calculate the number of paddocks needed: rest period divided by occupation period plus number of animal groups equals number of paddocks needed.

The more paddocks that can be formed, the better. Don't attempt to use Voisin's method with fewer than 10 paddocks.

***Paddock Size.*** After deciding how many paddocks to have, divide the total pasture area by the number of paddocks to get the average area of each paddock. Paddocks don't have to be equal in area, but they should produce more or less similar amounts of forage to facilitate moving animals on a regular, convenient schedule.

With electric fencing you can relocate some fences or subdivide more if necessary. For example, as your pasture becomes more productive and the animals can't eat everything within the occupation period that you want to use (e.g., 12 hours), divide the paddocks in half, in thirds, or whatever it takes to reduce the amount of available forage. A stocking density (number of animals confined to a certain area) increases, there is more competition among the animals for feed and less selective grazing. Under heavy stocking density, even high-producing animals will graze uniformly and close to the ground, and do well on it.

Paddock sizes must be adjusted according to the intensity of management desired. Paddocks usually should be smaller than two acres, depending on pasture productivity and numbers and sizes of animals. You will have to experiment, starting with about two-acre paddocks for cattle or horses, and one-fourth- to one-half-acre paddocks for sheep, goats, pigs, or poultry. For example:

■ On one Vermont farm, two-acre paddocks carry 70 milking Holsteins for three days in May-June, and they still will not be grazed down properly. If that same farmer wanted to provide a fresh paddock to the cows after each milking (every 12 hours), paddocks would have to be only one-sixth as large (there are six 12-hour periods in three days), or one-sixth × two acres equals one-third acre. That's about one-third of an acre in each paddock for 70 Holstein cows to graze 12 hours—on that soil, on that farm.

■ On good pastures in northern Vermont, one-eighth-acre (or less) paddocks carry about 100 growing lambs for 24 hours.

■ New Zealand farmers routinely and successfully graze 200 milking Fresians in one-acre paddocks for 12 hours on very well-developed pasture with a dense plant population.

Paddock size is not nearly as important as providing the required amounts of rest between grazings. Having more paddocks is better than having fewer, smaller paddocks are better than larger ones, shorter occupation periods are better than longer ones, and adequate rest is essential.

***Pasture Forage Quality.*** Besides increasing pasture forage productivity, applying Voisin grazing management results in high forage quality throughout the grazing season. Table 2 shows the average analyses of forage sampled on six northern Vermont dairy farms each time cows were about to enter a paddock.

Incorporating well-managed pastures into farm feeding programs can reduce production costs and increase farming profitability and sustainability.

**Table 2. Average analyses (dry weight basis) of forage from permanent pastures grazed with Voisin management on six Vermont dairy farms from May 1 to October 1, 1984.**

| *DM* | *CP* | *AP* | *ADF* | *TDN* | *ME* | *NEL* |
|---|---|---|---|---|---|---|
| % | | | | | *Mcal/lb* | |
| 23 | 22 | 21 | 28 | 69.4 | 1.14 | .72 |

## REFERENCES

Murphy, B. 1987. Greener pastures on your side of the fence. Arriba Publishers, Colchester, Vermont.

Northeast Research Program Steering Committee. 1976. Forage crops.

Savory, A. 1988. Holistic resource management. Island Press, Washington, D.C.

Smetham, M. L. 1973. Grazing management. *In* R.H.M. Langer [editor] Pastures and pasture plants. A.H. & A.W. Reed, Wellington, New Zealand.

Voisin, A. 1959. Grass productivity. Philosophical Library, New York, New York.

Voisin, A. 1960. Better grassland sward. Crosby Lockwood & Son, London, England.

# ROLE OF ANIMALS IN SUSTAINABLE AGRICULTURE

Charles F. Parker

Farming systems that are ecologically, biologically, and socioeconomically sound not only involve animals but also are dependent upon their integration with other farm practices. Livestock production is the most important value-added industry in the United States. Animal production characteristics can be exploited to significantly complement agroecosystems throughout the world.

Animals have provided food, clothing, draft, and transportation throughout civilization. Their unique ability to use noncompetitive, renewable resources in the production of quality protein that can be stored and transported *in vivo* remains important to human prosperity in most areas of the world.

Animal involvement in early farming systems dates from the beginning of historical records. Burke (1978) described an eighth century Eastern European, two-field crop rotation system where animals grazed fallow land to provide fertilizing benefits from manure. Transition to a three-field rotation, including a legume combined with cereals, increased output dramatically. Cultural effects of livestock traditions on food, clothing, literature, music, and art have been researched by Willham (1985).

Of a million animal species in the world, only 33 have been domesticated. Ten are major livestock species. Including poultry, an estimated 13 billion domestic animals contribute to world food and fiber needs (FAO, 1987).

## Plant and Animal Complementarity

Integration of plant and animal resources to achieve optimal biomass output within a given ecological and socioeconomic setting should be the ultimate goal for sustainable farming systems. The science of integrated biological functions is very complex, with expectations that certain interrelationships will be important in the derivation of optimal output solutions. Favorable interactions between components should enhance complementarity and synergistic responses, result in improved efficiency of production, and strengthen the economic viability of integrated agricultural systems.

***Enterprise Diversification.*** A recent report by the Council on Agricultural Science and Technology (CAST) (1988) indicated the best strategy for economic viability is flexibility within agricultural systems for food and fiber production. The report outlined that enterprise flexibility can be achieved through reduced input costs and increased diversification of operations. Integrated agroecosystems should provide a greater stabilizing effect against short-run fluctuations in commodity prices. An example is the price buffer or value-added effect livestock provide for the United States corn crop. At present, 60 percent of the corn crop is marketed through livestock products, with the balance divided equally between human foods and export markets.

Perhaps the most limiting resource for sustainable agricultural systems is the managerial ability necessary to develop and maintain an optimal level of enterprise diversification. Monocultural cropping systems are more commonly employed and generally less complex than mixed or integrated systems. Understanding and managing interactions among agroecosystem components provide challenge and opportunity to enhance output for integrated production systems.

## Animal-Forage Integration

One of the most important biological relationships in the world is that between herbivores and forages. The solar-energy-based ligno-cellulosic material from plants has been an important agricultural product since it was first consumed by animals and assimilated into products for human use.

Forages are produced on more than half the land area of the United States. The animal value-added impact on forages generates approximately 30 percent of the total economic value created by United States agriculture (CAST, 1986). This integrated animal food-plant fiber system is the most important agricultural enterprise in the United States. However, the relative significance of the animal-forage relationship is not well recognized by the world agricultural community (Parker, 1982). Greater appreciation is needed of the potential of the animal complement to significantly augment agro-ecosystems.

***Forage Farming.*** Understanding the basics of biological functions and their interactions is fundamental to the establishment of efficient animal-forage production systems. In general, animals are opportunistic creatures relative to the solar-energy-soil-derived plant biomass. Yet they are highly synergistic in their abilities to assimilate quality products for human use, recycle nutrients, and enhance the environment for improved forage production. Economically, the success of forage farming is directly dependent upon animals for the production of value-added sources of income. Solar energy maximization, nutrient cycling, utilizing noncompetitive renewable resources, soil-water conservation, lower capital investment, and enterprise flexibility are all highly favorable characteristics of forage farming with animals. Exploiting the unique production characteristics of plants and animals provides an opportunity for achieving high sustainability in agriculture.

***Animal-Forage Systems.*** Matching the biological characteristics of plants and animals for optimum biomass production and utilization is basic to the management of efficient animal-forage farming systems. Adaptability of plants and animals to the ecological setting establishes the primary resource base for production system development. Research on this subject has been termed as "agroecology" by Knezek and associates (1988). Knowledge of the nutritional needs of plants and animals is basic for efficient systems management. Understanding plant root dynamics is important for sustaining plant communities that are healthy and productive. The nutrient requirements of most food animal species constitutes the major animal production expense. Therefore, systems that provide an economical and available supply of feedstuffs consistent with animal production needs have been the most sustainable.

A major factor affecting forage biomass and quality is the seasonal effect as influenced by temperature and rainfall (VanKeuren, 1976). Approximately 60 percent of the total annual forage yield is produced during the first three months of the growing season in temperate zones. Obviously, seasonality of forage production is an important influence on choice of methods of harvesting and utilizing forages in an animal-forage production system. Seasonal forage availability has also impacted the adaptability and production characteristics of animal species. For example, seasonal forage supply is believed to be the primary cause for the seasonal breeding behavior of sheep and goats located in temperate zones. The young of these species are generally born during the season of highest forage availability, therefore enhancing the likelihood of their viability and species survivability.

Another important biological consideration in animal-forage systems management is the variation in nutrient requirements among animal species and classes within species. Production systems range from extensive management for fiber production by animals fed near maintenance to intensive management for meat and milk production from rapid growing and lactating animals. Diversity among plant and animal species and stage of production among groups within species provide considerable opportunity for enhancing the viability of sustainable agroecosystems.

***Multiple Animal Cropping.*** This concept is presented to further exploit animal variation to efficiently utilize nutrient sources that are variable in terms of location, kind, quality, and quantity. Because of the innate behavior and diet preferences among animal species and the large nutrient requirement variation among homogeneous groups within species, multiple animal cropping has potential for significantly improving biological and economic efficiency of forage utilization.

The advantages of multispecies grazing of livestock, including wildlife, were reported in a conference proceedings (Baker and Jones, 1985). For certain diverse agroecosystems common grazing practices have improved yield of animal products by as much as 90 percent (Cook, 1985).

Seasonal grazing of market animal groups, such as stocker animals and sequential intensively controlled grazing of animals with varying nutritional requirements, are examples of multiple animal cropping groups. The availability of animal groups needed on a seasonal basis to optimize forage carrying capacity will likely originate from different areas or regions as a part of an integrated food production chain. This animal-forage management approach should improve forage utilization and increase diversification of production and total resource output.

The optimal mix of different animal species in a common grazing system is likely to require a relatively higher level of technical knowledge and management skill than commonly practiced. Animal groups within species can have large differences in their nutritional requirements based on varying stages of physiological development and production. Proper management of such groups provides additional benefits to improve biological efficiency through multiple animal cropping. Within-species multiple animal/cropping systems are more likely adaptable for agroecosystems where the use of forage monocultures are common. Multiple animal/cropping systems need further evaluation as an animal-plant equilibrium strategy for improving total efficiency of forage utilization.

***Electrified Fencing.*** During the past 20 years, one of the major technological advances in animal-forage farming has been the development of electric fencing technology (Parker, 1982). Controlled intensive grazing is now feasible due to the economical and labor saving aspects of fencing animals with electrified fences. This advancement in grazing management has improved biological output significantly by increasing plant solarization, photosynthesis, production, and animal efficiency. These favorable biological responses reflect the importance of controlling the stage of plant maturity and amount of root reserve on biomass production. Animals respond to electric fencing more as a psychological than a physical barrier. This method of fencing also has reduced animal wastage due to predation losses, especially among small ruminant populations.

## Mixed Cropping of Plants and Animals

An increase in the integration of plant and animal cropping systems seems likely. The advantages of complementarity and synergism of enterprises for increased efficiency of output and the buffering effect among enterprises are recognized as major strengths of integrated agroecosystems to sustain agricultural production (CAST, 1988). Pond and associates (1980), in looking to the future role of animals in meeting human food needs, predicted that agricultural production systems will place increasing emphasis on forages and recognized that herbivores have special advantages for efficient utilization of forage resources.

Harwood (1982) reported on the value of nitrogen-fixing leguminous forages in crop rotation systems for accumulation of soil nutrients to enhance production of crops with high nitrogen requirements. Additionally, forages are valuable as cover crops to reduce soil and water losses. Redirecting land use and renewed cropping systems to conserve the resource base and stabilize production capacity are expected to increase the availability of higher quality forages, especially legumes. High-quality forages are directly related to animal performance, so that legumes are of greater nutrient value for high performance than grasses, even at equal values of digestibility and intake (Waldo and Jorgensen, 1981). Anticipated agronomic changes should enhance the overall importance of the animal component in sustainable agricultural systems.

***Crop Residue By-Product Materials.*** Animal utilization of crop residues and low-quality cereal grains is significant and provides an economic stabilizer for grain production. Crop residues are a major source of feedstuffs for ruminants. Some 789 million tons of crop residues produced in the United

States were in excess of that needed to prevent serious soil erosion and of sufficient yield to justify harvesting for animals (Lechtenberg et al., 1980). Corn stover provides a majority of the useable crop residue. This feed resource creates additional potential for livestock production, especially in areas where grain production is a major enterprise. The potential value for underutilized by-product materials, such as animal feedstuffs, has been reviewed (NRC, 1983).

***Manuring.*** Animal manuring is an important process for cycling of nutrients to maintain or improve soil fertility, especially in those intensively cropped locations where chemical fertilizers are limited. A major portion of important plant nutrients ingested by ruminants is returned to the soil via feces and urine. Mott (1974) reported that of the plant nitrogen and minerals consumed by grazing, lactating cows and finishing lambs, 75 and 95 percent of the nitrogen and 90 and 96 percent of the minerals were returned to the soil. Because of this high level of nutrient cycling, animal-forage grazing systems are among the most efficient for maintaining soil fertility Animals can be managed to have a significant role in the renovation of marginal land areas, especially where topographical features limit the use of mechanization. For example, herbivores on maintenance level of performance can be used as biological carriers for the transfer and distribution of hard forage seeds in the establishment of new seedings.

***Agroforestry.*** Cropping trees and livestock can be a complementary and sustainable production enterprise. Livestock grazing as a silvicultural tool provides a biological alternative that has economical and ecological advantages (Doescher et al., 1987). Effective grazing management to eliminate the use of herbicides for the control of competing vegetation in clearcut forestry areas and young tree plantations would not only be cost effective but have beneficial effects on soil and water conservation.

Control of animal grazing behavior is imperative to the success of removing competing vegetation without harming tree plantations. Period of grazing and animal conditioning are key management factors for effective grazing control. Learning behavior studies have shown that young herbivores can be influenced to develop preferences for or aversions to specific plants (Provenza and Balph, 1988). Flexibility to manipulate dietary preferences should enhance the importance of animals in silviculture.

***Biological Weed Control.*** The control of weeds and noxious plants is possible through the use of various animal species and grazing systems. Grazing animals can be intensively managed as gleaners or biological scrubbers" to control many species of undesirable plants.

In some plant communities, mixed species grazing can be valuable to reduce the hazard of grazing poisonous plants. Certain plants are highly toxic to some species yet well tolerated by others. Leafy spurge (*Euphorbia esula 1.*) is a highly undesirable perennial that is spreading rapidly in the north central United States and southern Canada. Leafy spurge is harmful to grazing cattle yet nontoxic and palatable to sheep. Therefore, sheep can be used as an effective biological control agent for leafy spurge (Landgraf et al., 1984). Selected herbivores offer an efficient, low-input and ecologically compatible alternative for weed control in mixed farming systems.

## Areas for Further Study

Sustainable agriculture is a very complex, management-intensive interdisciplinary issue. Recognition of the finite nature of land, water, and fossil fuel energy and the need to more completely utilize renewable resources, especially solar energy, is basic for achieving sustainability in agriculture production. Regenerative agriculture, as defined by Rodale (1988), should improve the natural resource base with a strong reliance on renewable internal resources and a relatively low dependence on external inputs. Integrating plant and animal resources to exploit complementarity and synergistic relationships should improve flexibility and enhance the economic viability of agroecosystems regardless of location or socioeconomical structure. Mixed- and multiple-cropping management to optimize biological diversity can further improve differential and total utilization efficiencies. Conceptual models involving plant and animal components are needed to study interrelationships and to identify areas that need further investigation to derive improved alternative production systems. A need for animal integration into sustainable agricultural systems is apparent.

## REFERENCES

Baker, F. H., and R. J. Jones. 1985. Proceedings of a conference on multi-species grazing. Winrock International, Morrilton, Arkansas.

Burke, J. 1978. Connections. Little, Brown and Company, Boston, Massachusetts.

Cook, C. W. 1985. Biological efficiency from rangelands through management strategies. Proceedings on Multispecies Grazing. Winrock International, Morrilton, Arkansas.

Council on Agricultural Science and Technology. 1986. Forages: Resources for the future. CAST Report No. 108. Ames, Iowa.

Council on Agricultural Science and Technology. 1988. Long-term viability of U.S. agriculture. CAST Report No. 114.

Doescher, P. S., S. D. Tesch, and M. A. Castro. 1987. Livestock grazing: A silvicultural tool for plantation establishment. Journal of Forestry 10: 2,937.

Food and Agricultural Organization, United Nations. 1987. 1987 FAO production yearbook. FAO statistics series Vol. 14, No. 76. Rome, Italy.

Harwood, R. R. 1982. Application of organic principles to small farms. Research for small farms. Miscellaneous Publication No. 1422. U. S. Department of Agriculture, Washington, D.C.

Knezek, B. D., O. B. Hesterman, and L. Wink. 1988. Exploring a new vision of agriculture. National Forum. Renewable Resources. 68(3): 1,013.

Landgraf, B. K., P. K. Fay, and K. M. Harsted. 1984. Utilization of leafy spurge (*Euphorbia esula*) by sheep. Weed Science 32-348.

Lechtenberg, V. C., R. M. Peart, S. B. Barber, W. E. Tyner, and O. C. Doering, III. 1980. Potential for fuel from agriculture. Proceedings, 1980 Forage and Grassland Conference, Louisville, Kentucky.

Mott, G. O. 1974. Nutrient recycling in pastures. *In* D. A. Mays [editor] Forage Fertilization. American Society of Agronomy, Madison, Wisconsin.

N.R.C. 1983. Under-utilized resources as animal feedstuffs. Committee on Animal Nutrition, National Academy Press, Washington, D.C.

Parker, C. F. 1982. Increased forage utilization for efficient lamb and wool production. Proceedings, Seventh Annual Forage Day, Minn. Forage and Grassland Council.

Pond, W. G., R. A. Merkel, L. D. McGilliard. 1980. Animal agriculture research to meet human needs in the 21st century. Westview Press, Inc., Boulder, Colorado.

Provenza, F. D., and D. F. Balph. 1988. Development of dietary choice in livestock on rangelands and its implications for management. Journal of Animal Science 66: 2,356.

Rodale, R. 1988. Agriculture systems: The importance of sustainability. National Forum. Renewable Resources, 68(3): 26.

VanKeuren, R. W. 1976. Hill land improvement in the eastern United States. Proceedings, International Symposium on Hill Lands, Morgantown, West Virginia.

Waldo, D. R., and N. A. Jorgensen. 1981. Forages for high animal production: Nutritional factors and effects of conservation. Journal of Dairy Science 64: 1,207.

Willham, R. C. 1985. The legacy of the stockman. Iowa State University and Winrock International, Ames.

# THE IMPORTANCE OF INTEGRATION IN SUSTAINABLE FARMING SYSTEMS

# THE IMPORTANCE OF INTEGRATION IN SUSTAINABLE AGRICULTURAL SYSTEMS

Clive A. Edwards

Crop yields in developed countries have increased dramatically since World War II. Traditionally, farming methods depended upon and maintained the soil's inherent fertility by recycling the nutrients in organic matter. Over the last 40 years, new high-yielding crop varieties have developed. However, high yields depend upon high-energy inputs in the form of inorganic fertilizers and high inputs of synthetic pesticides to combat increased pest disease and weed problems resulting from monoculture or rotations involving only two crops.

The current use of fertilizers and pesticides (Figures 1 and 2) is predicted to continue to increase almost exponentially (Edwards, 1987) unless there are fundamental changes in the philosophy that crop yields should continue to increase, irrespective of the plight of the small farmer and environmental deterioration.

High-input practices have led to overproduction of certain crops in many developed countries in recent years. The inevitable results have been a fall in commodity prices and poorer farm incomes. Moreover, the efficiency of production has not kept pace with the increase in energy needed to produce the chemicals upon which they depend. From 1970 to 1978, U.S. farmers used 50 percent more energy to produce 30 percent more crops (Buttel et al., 1986). Moreover, high inputs are inefficient in energy terms. For every calorie of food currently produced in the United States, three calories are required in production and seven calories are needed for processing, distribution, and preparation (Papendick, 1987). These intensive cropping practices and heavy use of chemicals have created a variety of economical, environmental, and ecological problems. The most important environ-

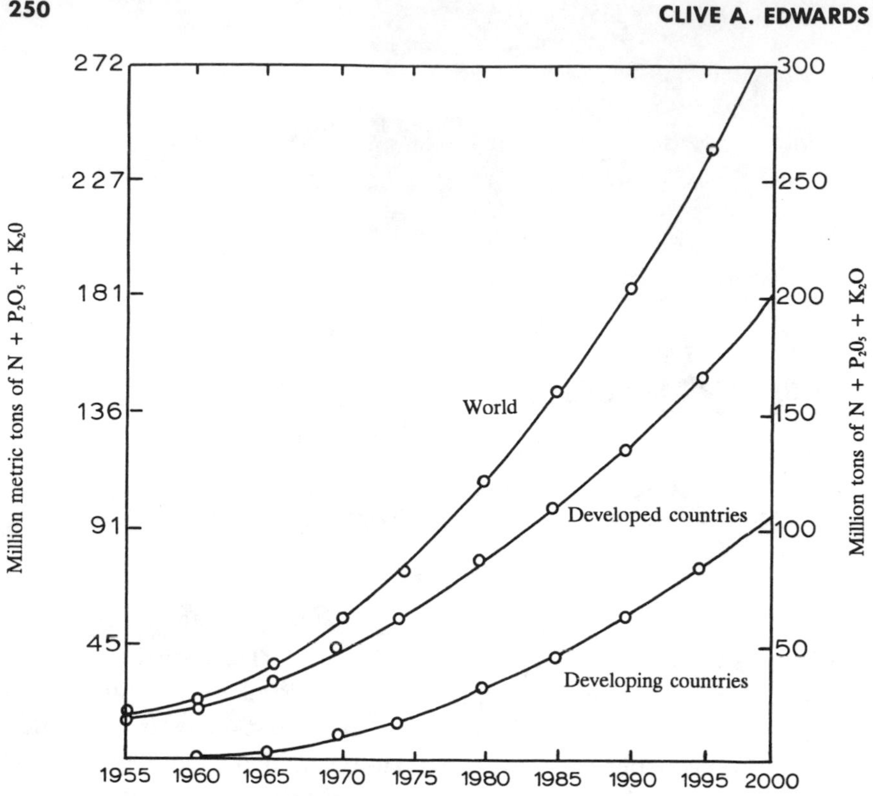


**Figure 1. World fertilizer consumption, 1955-1974 (actual) and 1975-2000 (estimated) (Edwards, 1985).**

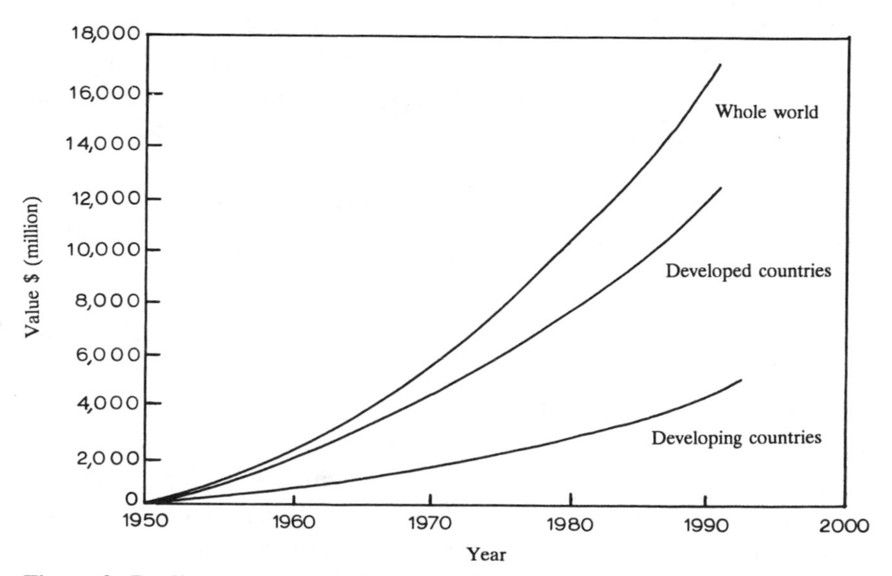


**Figure 2. Predicted world pesticide use (Edwards 1986).**

mental effects are (a) soil erosion, (b) pollution of groundwater and surface water with agricultural chemicals (Edwards, 1987a, 1987b, 1988), (c) destruction and disturbance of wildlife habitats (Jenkins, 1987; Papendick, et al. 1986), and (d) various adverse effects on rural landscapes (Lowrance and Groffman, 1988). The seriousness of these problems can be illustrated by data showing that one-third of the topsoil on U.S. agricultural land has been lost over the past 200 years. One-fourth of the 421 million crop acres currently suffer serious soil losses at rates well above those that permit sustainable crop production (Papendick, 1987). In addition to these serious environmental problems, frequent pesticide use has caused the development of resistant strains of pests and diseases, resulting in a need for even more pesticides and increased costs (Pimentel and Andow, 1984). Moreover, energy-based agrichemicals have become increasingly expensive, causing severe economic pressure on farmers in developed countries as a result of overproduction and falling prices. Thus, many farmers in the United States are tending to reduce their use of these inputs.

Economical and environmental problems associated with higher chemical inputs have also occurred in developing countries. In the 1960s, food production increased dramatically through the Green Revolution, which was based on high-yielding varieties of wheat and rice that responded to high inputs of nitrogenous fertilizers and irrigation. However, fertilizer efficiency is reduced in the tropics because of rapid leaching of nitrogen and a greater degree of phosphorus fixation. Many tropical soils also have poor structures and are much more susceptible to erosion when continually cropped. At the same time, a higher incidence of pests and diseases occurred, because of shorter crop rotations or monoculture. This led to much greater use of pesticides that, in turn, created new pest and disease problems because of the eradication of natural enemies and increased dependence upon chemicals. Hazards to humans are also involved because the hot, humid conditions in tropical countries discourage protective clothing, and the relatively poor education of the farmers often causes environmental hazards through poor methods of application, washing of equipment in water systems used for other purposes, and disposal of pesticide containers.

For more than a decade there has been a growing movement, which originated in developed countries, to find ways of reducing chemicals and other energy-based inputs, such as cultivations, fertilizers, and pesticides (Edens et al., 1985; Buttel et al., 1986; Wagstaff, 1987; Buckwell and Smith, 1986; Lockeretz et al., 1984; Klepper et al., 1977; Youngberg, 1984). Greater economic returns to a farmer can be attained when the use of fewer inputs is associated with little or no reductions in yields, thereby resulting in improved farm profitability. Fewer cultivations and more crop rotations, increased ground cover, and innovative cultural and cropping practices can decrease soil erosion considerably. Lower inputs of pesticides and fertilizers result in greatly reduced contamination of surface water and groundwater and minimization of other environmental impacts. Although developing countries have different problems and will have to continue to depend upon inorganic nutrient sources for some crops and soils, many of their problems are similar to those in developed countries, and solutions will differ mainly in degree and emphasis.

## Major Inputs into Farming Systems

The production of a crop involves sowing seed at an appropriate rate and time with several key inputs. The main inputs are some degree of soil cultivation; provision of plant nutrients by some means of fertilization; methods of crop protection against pests, diseases, and weeds; and suitable crop rotations to maximize productivity (Figure 3). Central to this pattern is farm economics that encompass all other inputs, such as land, labor, buildings, machines, chemicals, and seed, balanced against profits from yields and other economic factors, such as market prices, exports, and subsidies. A farming system is not just a simple sum of all of its components but rather a complex system with intricate interactions. The concept of the central position of farm economics differs markedly from the perception of many agricultural scientists who usually assume that their own specialty, such as pest control, nutrient supply, or cultivation, is the central and most important component. In this context, farm economics mainly deal with microeconomics at the farm level, but also include macroeconomics of farm prices, subsidies, and the cost of environmental pollution.

Farmers and agricultural scientists rarely consider how the amounts of

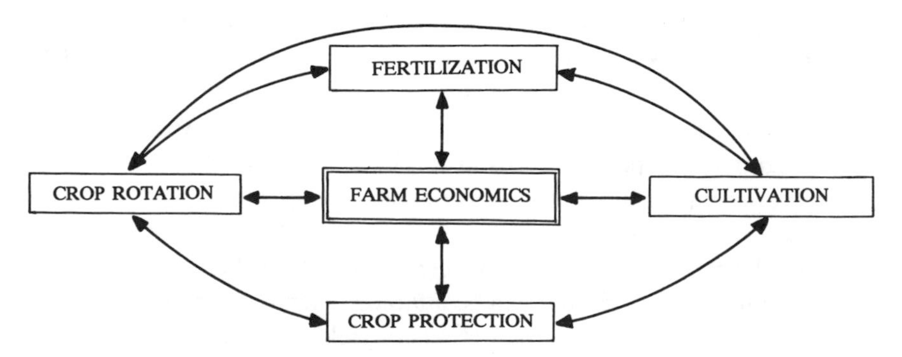


**Figure 3. Interactions among farm inputs.**

fertilizer they use affect pests, diseases, or weeds. In the same way, the impact of cultivation on pest diseases and weed problems is not usually a factor in deciding the type of cultivation a farmer uses. Even in the use of pesticides where integrated pest management systems have been developed, it is rare for any account to be taken of the impact of herbicides on pests and diseases, of insecticides on diseases, or of fungicides on pests.

In conventional "higher-input" farming, large yields can often be obtained without any appreciable attention given to the interactions between various inputs. For example, if heavy fertilizer use renders a crop more susceptible to pests and diseases through production of lush, soft growth, this can be compensated for by adding more pesticides. The decline in natural pest and disease control and consequent increased pest and disease incidence caused by herbicides through loss of foliar and habitat diversity is compensated for by increased use of insecticides and fungicides. Any affect of pesticides on earthworms and other soil organisms that promote organic matter turnover, nutrient cycling, and soil fertility is covered by increased nutrients from the additional inorganic fertilizers used. When chemical inputs are lowered, it is imperative to learn what effects these inputs have on each other in much more detail. Farming systems that use fewer chemicals implicitly require a much better understanding of the interactions between and among inputs in agroecosystems.

## Components of Sustainable Agricultural Systems

***Fertilizers.*** At lower input levels, the increased use of inorganic fertilizers has dramatic effects on crop yields. But as the amount of fertilizer applied increases, the growth and yield response of the crop diminishes exponentially and eventually levels off (Figure 4).

At a certain point, the cost of the fertilizer equals the value of the crop yield increase. It is important to use considerably less inorganic fertilizers than this. Reductions in inorganic fertilizers can be compensated for by using crop rotations, particularly those involving legumes as a source of nitrogen and other nutrients, and using animal manures where available (Sahs and Lesoing, 1985). Other practices that can minimize fertilizer use include regular soil analyses to assess actual fertilizer needs, growing crop varieties that have lower nutrient needs, and placing inorganic fertilizers in the crop row where they have maximum benefit to the crop but do not contribute to weed growth.

There may be great potential to reduce the need for inorganic fertilizers even more as new research results are found. Research that might achieve this includes investigating the potential for increasing biological nitrogen fixation in crops other than legumes by genetic engineering; scheduling treatment with incremental additions of nutrients through the growing season; and using alternative forms of organic matter from urban and industrial sources, which currently cause disposal problems, to supply nutrients.

***Pesticides.*** Pesticides are often used as recommended by chemical dealers or on an insurance basis. Many of the applications used may be unnecessary and/or economically unsound (Pimentel and Andow, 1984). The amounts used could be reduced substantially and a range of alternative methods of pest control used. For instance, insecticide use can be reduced and compensated for or replaced by integrated pest management techniques in which rotations and use of resistant varieties, economic thresholds, pest forecasting, and biological and cultural pest control all play a part (Lisansky, 1981). All of these must be integrated into farming systems for pest and disease management while taking account of their side-effects on other aspects of crop production (Edwards et al., 1988).

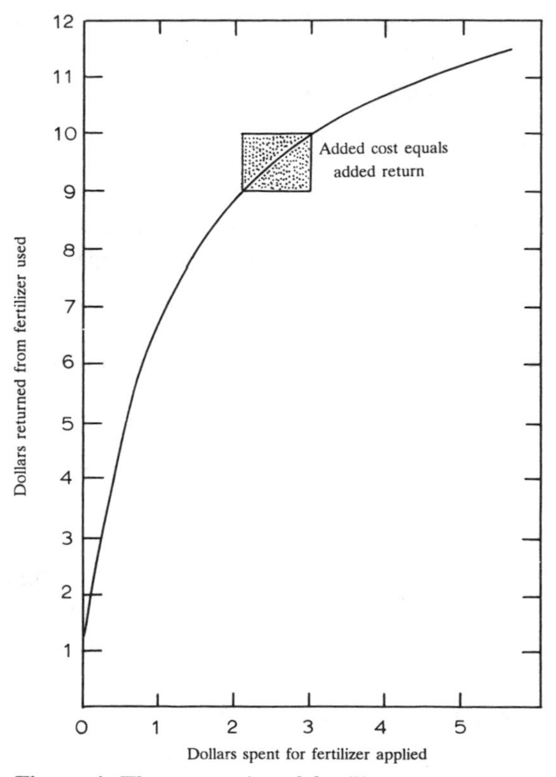


**Figure 4. The economics of fertilizer use.**

In addition, the use of insecticides can be minimized or replaced by other techniques that involve:

■ Minimal use of insecticides based on methods of forecasting pest incidence.

■ Better insecticide placement and formulations, thereby using smaller amounts with improved effectiveness.

■ More crop rotations to avoid carryover of pests from one season to the next and gradual build up of population.

■ Appropriate cultivations that minimize pest attack. The form these cultivars should take depends upon the pest involved.

■ Timing of crop sowing to avoid pest attack.

■ Adoption of controlled weed growth practices as compared with total weed suppression so as to encourage natural enemies of pests.

■ Use of biological insecticides based on insect pathogens that are effective without environmental impacts.

■ Use of nematodes that attack insects to control them. Many nematode varieties have considerable potential but are not yet available on an extensive commercial basis.

■ Release of parasites and predators of pests.

■ Use of pheromones, other allelochemicals, or repellents to keep pests away from crops.

■ Release of sterile male insects to abort reproduction of pests where appropriate.

■ Use of crop varieties resistant to pest attack.

■ Use of crop varieties with toxins implanted into their tissues by genetic engineering.

■ Encouragement of natural predators by maintaining biological diversity among plants and in soil systems.

■ Use of trap crops that promote pest emergence when the main crop is not available.

■ Innovative cultural techniques, such as stripcropping, intercropping, etc., that increase diversity of habitat, flora, and fauna.

Fungicide use can be minimized by:

■ Use of minimal amounts of fungicide based on disease forecasting methods.

■ Use of crop rotations to minimize disease attack.

■ Better application techniques for fungicides using small amounts and better placement.

■ Timing of crop sowing to avoid the disease incidence period or climatic periods favorable to the disease.

■ Use of disease antagonists. A number of microorganisms inhibit the growth of plant pathogens.
■ Use of crop varieties that are tolerant or resistant to disease.

Herbicide applications can be replaced by:
■ Use of mechanical weed control. This can be associated with row spacing to facilitate such cultivations.
■ Use of rotations to avoid volunteer of seedlings from previous crops.
■ Cover cropping to minimize weed seed germination.
■ Use of live mulches to provide soil cover and inhibit seed germination.
■ Use of mycoherbicides. These have been identified and can be produced by genetic engineering techniques.
■ Release of pests of weeds. These have been used successfully against a number of weed species.

***Cultivations.*** Traditionally, land in developed countries has been cultivated annually to a depth of 9 to 12 inches (22.5-30 cm) with the surface soil completely inverted by moldboard plows. This involves a high consumption of energy to pull the plow, particularly in difficult and compacted soils. For the last 30 years there has been a progressive trend toward fewer cultivations with corresponding reductions in energy inputs. This has culminated into a complete absence of cultivation and seeding into the previous crop using special tillage implements, usually after a herbicide application.

Techniques that reduce the number of cultivations required, compared to deep-plowing, include:
■ Shallow plowing to a depth of 6 inches (15 cm) or less.
■ Chisel plowing, which does not invert the soil.
■ Deep subsoiling, which lifts the soil but does not invert it.
■ Ridge tillage.
■ Shallow-tine, soil loosening.
■ Harrowing to create a seed bed.
■ No-till (direct-drilling).

All of these techniques tend to create conditions that reduce soil erosion and create a more natural soil structure, which improves both drainage and water retention and favors biological and natural techniques of pest and disease control because there is less disturbance of the soil ecosystem.

## Additional Components of Sustainable Agricultural Systems

In low-input systems of crop production a number of component techniques in addition to the main inputs are used. These include:

***Rotations.*** In most developed countries there has been a trend in farming over the last 40 years toward monoculture or cropping with only two annually alternating crops. When chemical fertilizers and pesticides are reduced, it usually becomes essential to increase the use of crop rotations to provide nutrients, if possible, through legumes and to lessen pest and disease attack by minimizing infectious carryover from one season to the next.

***Innovative Cultural Techniques.*** As chemical inputs in cropping systems are lowered, there becomes an increasing need for cultural techniques. Possible cultural techniques include:

- Systems of strip intercropping using two crops, with strips normally involving one pass of a tractor and its implements.
- Interrow crop techniques where alternate rows of two crops are sown.
- Undersowing with a legume or other crop.
- Use of varietal or species mixtures to create greater crop diversity.
- Use of trap crops, which may or may not have any commercial value but attract pests away than the main crop.
- Double-row cropping to facilitate weed control by allowing passage of cultivation implements.

***Machinery Inputs.*** Most agricultural machinery used now is developed for farming practices that use large amounts of chemicals. As inorganic chemical inputs are reduced, new machinery is needed for better mechanical weed control. Typical machinery needs include:

- Lighter machinery that causes less soil compaction.
- Machinery for placing fertilizers in the crop row.
- Pesticide placement equipment that applies the chemical where it is required to kill the pest.
- Weed control machinery for a variety of cropping patterns.
- Subsoiling equipment to open up the soil without any inversion.

***Organic Matter Inputs.*** Traditionally, animal manures were the main source of soil nutrients and soil fertility, making crop and animal production interdependent (Figure 5). In developed countries today, animal and crop farming occur together only on smaller farms. Diversified farming is much more common in developing countries. Thus, manurial inputs into crop production in developed countries are relatively low. The use of animals to consume crop residue is of only minor importance because these residues are not always palatable. Sustainable systems should consider increasing the association between crop and animal production. Moreover, there is a wide range of urban and industrial waste organic materials that are used little in agriculture but hold considerable potential as sources of crop nutrients. The organic inputs that could compensate for reduced inorganic chemicals include:

- Animal manures, mainly from cattle, poultry, and hogs.
- Sewage sludge or cake that can be applied as a spray, injected liquid, or solid.
- Domestic lawn clippings and leaf material that can be composted.
- Paper pulp waste that can be sprayed or applied as a dewatered solid.
- Waste from the potato industry, either as liquid washing or solid peelings.
- Brewery wastes consisting largely of yeasts.
- Domestic vegetable and other organic wastes.

***Crop Breeding.*** New crop varieties that respond to high levels of nitrogen are a major reason for the increased crop yields produced currently in developed countries. However, the crop varieties in developing countries have been designed to respond to and produce good yields with fewer inorganic fertilizers because large amounts of these chemicals are either not available or too costly. These two systems may have something to teach each other about developing sustainable agricultural systems.

Traditionally, crop breeding has involved selection of favorable plant traits, crossing to produce new varieties, and building up seed stocks. This can now be expedited by genetic engineering (Figure 6) to develop crops that respond to lower inputs of fertilizers without major decreases in yields and are highly resistant to pests and diseases (Edwards, 1988). With this new

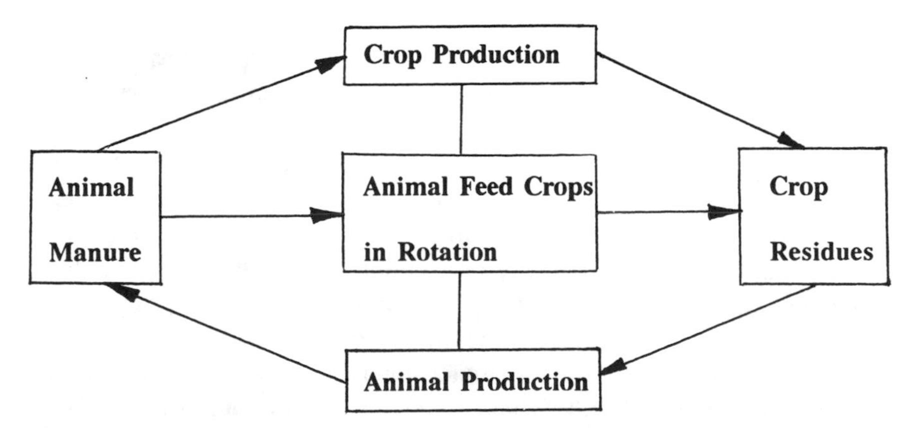


**Figure 5. Integration of crop and animal production.**

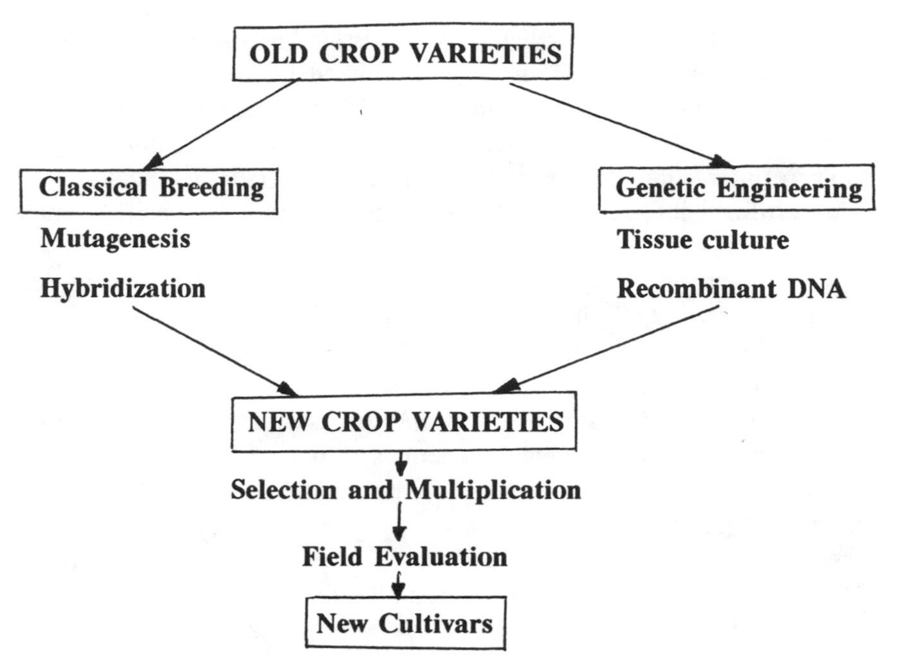


**Figure 6. Development of new crop varieties.**

ability crop breeding has great potential for sustainable agricultural research. These potentials include:

- Breeding plant varieties that respond to fewer inputs of chemical fertilizers.
- Breeding plant varieties that are resistant to pest and disease attack.
- Implanting insect toxins into crop plants to provide pest control.
- Developing crops with disease antagonism that are less affected by pathogens.
- Breeding crops that are resistant to low levels of herbicides when they were previously susceptible.

## Integration of Components

Sustainable agricultural systems depend upon suitable manipulation of the previously mentioned components and on a better awareness of how these components can reduce chemical inputs. These systems also depend upon a much better understanding of how the major and other components interact with each other. In other words, lower input agriculture is more system-oriented and, consequently, management-intensive.

Some interactions among components of agricultural systems are understood, and others can be predicted from existing knowledge. But many remain poorly understood. There is a need to identify the relative importance of all of these interactions in overall crop production. Figure 7 summarizes some of these interactions in rather simplistic fashion.

These interactions and others that are more speculative include:

■ Fertilizers influence the growth of weeds as well as crops (Moomaw, 1987).

■ Fertilizers can increase disease incidence, for example, cereal leaf disease (Jenkyn, 1976; Jenkyn and Finney, 1981).

■ Fertilizers can increase pest attack, for example, aphids on wheat (Kowalski and Visser, 1979).

■ Organic matter can reduce pest and disease incidence by increasing species diversity in favor of natural enemies (Altieri, 1985; Edwards, 1988).

■ Organic matter can promote populations of fungi that control nematodes (Kerry, 1988).

■ Organic matter can adsorb and inactivate pesticides (Edwards, 1966).

■ Organic matter can provide alternative food for marginal pests and

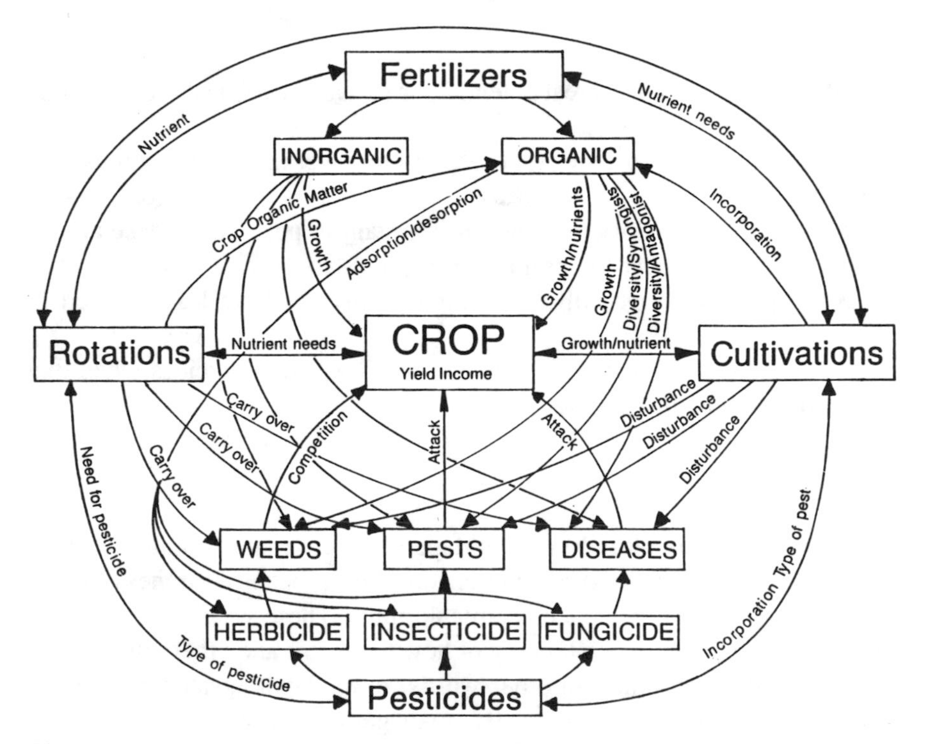


**Figure 7. Interactions between inputs (Edwards, 1987).**

decrease their severity (Edwards, 1979).

■ Cultivations can increase or decrease the incidence of pests or diseases (Edwards, 1975).

■ Cultivations affect the incidence of weeds either mechanically or by burying weed seeds (Klein et al., 1987).

■ Cultivations can affect the amount of fertilizer needed (Follett et al., 1981).

■ Cultivations bring pesticides into contact with the pest, thereby increasing their effectiveness (Edwards, 1966).

■ Cultivations incorporate organic matter into soil where it decomposes more rapidly (Follett et al., 1987).

■ Herbicides can influence the severity of pest and disease attack by removing alternative weed hosts or by reducing the availability of natural enemies (Altieri, 1987).

■ Pesticides can affect soil organisms that break down organic matter and release nutrients (Edwards, 1983).

■ Insecticides can reduce the incidence of viruses and diseases by killing the vectors of these organisms (Edwards and Heath, 1964).

■ Insecticides can increase weed populations by killing the natural enemies of weeds (Smith, 1982).

■ Insecticides kill natural enemies of pests and thereby increase pest incidence or create new pests (Edwards, 1973).

■ Fungicides can kill soil fungi that exert considerable natural control over insect or nematode populations (Kerry, 1988).

■ Fungicides can reduce populations of beneficial soil microorganisms (decomposers and antagonists) as well as those of pathogens (Thompson and Edwards, 1974).

■ Pesticides can deplete earthworm populations and, hence, lower soil fertility (Edwards and Lofty, 1977).

■ Rotations reduce the incidence of most pests and diseases dramatically by interrupting the carryover of organisms from crop to crop (Dabbert and Madden, 1986).

■ Rotations provide crop nutrients, particularly when they include legumes (Follett et al., 1981).

This list of interactions is far from complete. Indeed, there is little doubt that some are still unknown, and many others have not yet been fully documented.

Clearly, as in integrated pest management, a great deal of research is necessary to identify and understand such interactions and to be able to predict how such interactions affect sustainable agriculture systems. Most of this research must be interdisciplinary. There is an urgent need for developing well-designed, holistic agricultural systems that maximize the benefits of the more important interactions among the main components of the systems. There have been relatively few examples of such farming systems to date (El Titi, 1986; Vereijken, 1985). Computer-based, farmer-operated management systems are being developed for many farming systems and have considerable potential for sustainable agricultural systems.

There is little doubt about the ecological, environmental, and economic attraction of lower input farming systems, particularly in developed countries. Such systems would minimize soil erosion and storm runoff as well as avoid contamination of groundwater and surface water. To achieve these ends and still increase farm profitability and the sustainability of the industry, an intensive research program along the lines recommended here is certainly justified and urgently needed. The problems in developing countries are similar in principle but differ considerably in emphasis. Much more research is needed to develop sustainable systems in the tropics.

## REFERENCES

Altieri, M. A. 1985. Diversification of agricultural landscapes—a vital element for pest control in sustainable agriculture. pp. 124-136. *In* T. C. Edens, C. Fridge, and S. L. Battenfield [editors] Sustainable agriculture and integrated farming systems. Michigan State University Press, East Lansing.

Altieri, M. A. 1987. Agroecology, the scientific basis of alternative agriculture. Westview Press, Boulder, Colorado.

Buttel, F. H., G. W. Gillespie, Jr., R. Janke, B. Caldwell, and M. Sarrantonio. 1986. Reduced-input agricultural systems: Rationale and prospects. American Journal of Alternative Agriculture 1(2): 58-64.

Dabbert, S., and P. Madden. 1986. The transition to organic agriculture: A multiyear simulation model of a Pennsylvania farm. American Journal of Alternative Agriculture 1(3): 107-114.

Edens, T. C., C. Fridgen, and S. L. Battenfield, editors. 1985. Sustainable agriculture and integrated farming systems. Michigan State University Press, East Lansing.

Edwards, C. A. 1966. Pesticide residues in soils. Residue Reviews 13: 83-132.

Edwards, C. A. 1973a. Environmental pollution by pesticides. Plenum Publication Co., London, England, and New York, New York.

Edwards, C. A. 1973b. Persistent pesticides in the environment. C.R.C. Press, Cleveland, Ohio.

Edwards, C. A. 1975. Effects of direct drilling on the soil fauna. Outlook on Agriculture 8: 243-244.

Edwards, C. A. 1979. The influence of the soil fauna on crop seed establishment. pp. 38-42. Proceedings, Conference on Crop, Seed, and Soil Environment. ADAS.

Edwards, C. A. 1983. Assessment of the effects of pesticides on soil organic matter breakdown and soil respiration. *In* biological processes and soil fertility. Transactions of the International Society of Soil Science and British Society of Soil Science 65.

Edwards, C. A. 1987. The concept of integrated systems in lower input sustainable agriculture. American Journal of Alternative Agriculture 2(4): 148-152.

Edwards, C. A. 1988a. The importance of integration in lower input agricultural systems. *In* M. G. Maoletti, B. R. Stinner, and G. J. House [editors] Agricultural ecology and

Environment. Elsevier Publishers, Amsterdam, The Netherlands (in press).

Edwards, C. A. 1988b. The application of genetic engineering to pest control. Barcelona Publications Parasitis, Geneva, Switzerland. 1-20.

Edwards, C. A., B. R. Stinner, and N. Creamer. 1988. Pest and disease management in integrated lower input/sustainable agricultural systems. pp. 1,009-1,016. *In* Proceedings, British Crop Protection Conference, Pests and diseases, Volume 3. B.C.P.C. Press, Surrey, England.

Edwards, C. A., and G. W. Heath. 1964. The principles of agricultural entomology. Chapman and Hall, London, England.

Edwards, C. A., and J. R. Lofty. 1977. The biology of earthworms. Chapman and Hall, London, England.

El Titi, A. 1986. Management of cereal pests and diseases in integrated farming systems. pp. 147-155. *In* Proceedings, British Crop Protection Conference, Pests and Diseases. B.C.P.C. Press, Surrey, England.

Jenkins, D. 1984. Agriculture and the environment. *In* Proceedings, ITE Symposium No. 13. Lavenham Press, Suffolk, England.

Jenkyn, J. F. 1976. Nitrogen and leaf diseases of spring barley. pp. 119-128. *In* Proceedings, 12th Colloquium of the International Potash Institute. Worblaufen, Bern, Switzerland.

Jenkyn, J. F., and M. E. Finney. 1981. Fertilizers, fungicides and saving date. pp. 179-188. *In* J. F. Jenkyn and R. T. Plumb [editors] Strategies for the control of cereal diseases. Blackwell, Oxford, England.

Kerry, B. 1988. Fungal parasites of nematodes and their role in controlling nematode populations. *In* C. A. Edwards, B. R. Stinner, and D. H. Stinner [editors] Interactions between soil organisms. Elsevier, Amsterdam, The Netherlands (in press).

Klein, R. N. G. A. Wicks, and R. S. Moomaw. 1987. Cultural practices for weed control-ridge planting. pp. 190-191. *In* Sustainable Agriculture: Wise and profitable use of our resources in Nebraska. Nebraska Cooperative Extension Service, Lincoln.

Klepper, R., W. Lockeretz, B. Commoner, M. Gertler, S. Fast, D. O'Leary, and R. Blobaum. 1977. Economic performance and energy intensiveness of organic and conventional farms in the Corn Belt: A preliminary comparison. American Journal of Agricultural Economics 59: 1-12.

Kowalski, R., and P. E. Visser. 1979. Nitrogen in a crop-pest interaction: cereal aphids. *In* J. A. Lee, S. McNeill, and I. H. Rorison [editors] Nitrogen as an ecological factor. British Ecological Society, Oxford, England.

Lisansky, S. G. 1981. Biological pest control. pp. 117-129. *In* B. Stonehoouse [editor] Biological husbandry, a scientific approach to organic farming. Butterworth Press, London, England.

Lockeretz, W., G. Shearer, D. H. Kohl, and R. W. Klepper. 1984. Comparison of organic and conventional farming in the Corn Belt. pp. 37-48. *In* D. F. Bezdicek et al. [editors] Organic farming: Current technology and its role in a sustainable agriculture. American Society of Agronomy, Madison, Wisconsin. pp. 37-48.

Lowrance, R., and P. M. Groffman. 1987. Impacts of low and high input agriculture on landscape structure and function. American Journal of Alternative Agriculture (4): 175-183.

Mannering, J. V., and C. R. Fenster. 1983. What is conservation tillage? Journal of Soil and Water Conservation 38(3): 141-143.

Moomaw, R. 1987. Low cost weed control. pp. 170-172. *In* Sustainable agriculture: Wise and profitable use of our resources in Nebraska. Nebraska Cooperative Extension Service, Lincoln.

Papendick, R. I. 1987. Why consider alternative production systems. American Journal of Alternative Agriculture 2(2): 83-86.

Papendick, R. I., L. F. Elliott, and R. B. Dahlgren. 1986. Environmental consequences of modern production agriculture: How can alternative agriculture address these issues and concerns? American Journal of Alternative Agriculture 1: 3-10.

Pimental, D., and D. A. Andow. 1984. Pest management and pesticide impacts. Insect Science and Application 5: 141-149.

Plucknett, D. L., and N.J.H. Smith. 1986. Sustaining agricultural yields. BioScience 36: 40-45.

Sahs, W. W., and G. Lesoing. 1985. Crop rotations and manure versus agricultural chemicals in dryland grain production. Journal of Soil and Water Conservation 40: 511-516.

Smith, R. J., Jr. 1982. Integration of microbial herbicides with existing pest management programs. pp. 189-206. *In* R. Charudattan and H. L. Walker [editors] Biological control of weeds with plant pathogens. John Wiley and Sons, New York, New York.

Thompson, A. R., and C. A. Edwards. 1974. Effects of pesticides on nontarget organisms in freshwater and in soil. American Society of Soil Science Special Publication 8, Chapter 13. Soil Science Society of America, Madison, Wisconsin.

Verijken, P.. 1985. Development of farming systems in Nagele: Preliminary results and prospects. pp. 124-136. *In* T. C. Edens, C. Fridgen, and S. L. Batterfield [editors] Sustainable agriculture and integrated farming systems. Michigan State University Press, East Lansing.

Wagstaff, H. 1987. Husbandry methods and farm systems in industrialized countries which use lower levels of external inputs: A review. Agriculture, Ecosystems and Environment 19: 1-27.

Youngberg, G. 1984. Alternative agriculture in the United States: Ideology, politics and prospects. pp. 107-135. *In* D. Knorr and T. R. Watkins [editors] Alterations in food production. Van Nostrand Reinhold Company, New York, New York.

# INTEGRATED FARMING SYSTEM OF LAUTENBACH: A PRACTICAL CONTRIBUTION TOWARD SUSTAINABLE AGRICULTURE IN EUROPE

A. El Titi and H. Landes

The Lautenbach project started in the fall of 1978 at the private estate of Lautenbach in southwestern Germany. Major objectives included long-term multidisciplinary studies to evaluate the economic and ecological effects of a low-input farming system under current market conditions. The project was a direct response to the limited success in introducing integrated pest management (IPM) in European agriculture (Diercks, 1983; Vereijken et al., 1986). Despite knowledge of the biology, life tables, population dynamics, crop injury thresholds, and control techniques of the major target species, however, farmers still find IPM instructions hard to follow. From the farmer's viewpoint, studying control aspects of single pests, mostly in complete isolation from other husbandry measures, is illogical and complicates adoption of IPM. Moreover, researchers and farmers interpret the terms "integrated" and "IPM" in many ways, often colored by their own objectives.

IPM is used to justify complete chemical control programs of commercial interests as well as pesticide-free practices of organic farming (Gutierrez, 1986). In reality, it is neither. The concept takes elements of natural regulation into consideration and incorporates them in the best possible way with farming practices. Components of natural pest regulation, for example, are the background for any integrated approach. The specific biotic and abiotic factors on a growing site are known to regulate the turnover of organic materials, nutrient cycles, affecting both nitrogen fixation and leaching. In comparable ways they significantly influence the population dynamics of pest species and those of their antagonists as well as growth patterns of vegetation, soil-water balance, and soil erosion. The specific ecosystem components can determine the success or failure of the whole farming business (Pimentel, 1982). These factors deeply involved in crop production should not be precluded nor ignored. Maintaining soil fertility at high levels, for example, by recycling of nutritional elements or improving the potential of the beneficial agents, builds the intimate strategies of integrated farming. These also are the fundamental features of sustainable agriculture (Andow, 1983; Pimentel, 1986; Risch et al., 1983). Integration implies fitting farming measures into ecological production processes. The philosophy of integration depends upon making the best possible use of all natural resources before investing fossil energy inputs.

## The Lautenbach Concept

The Lautenbach project is conceived to reach specified goals in reducing inputs, maintaining income, and improving ecological stability. The concept is divided into two main parts.

***Part A: Preparatory Research.*** This includes mainly mono- and bifactorial experiments dealing with the evaluation of single husbandry methods and techniques for use in integrated farming systems (IFS). Studies include effects of different soil tillage regimes on the edaphon and weed seedbanks; optimum nitrogen supply in cereal and sugar beets; effects of undersowing on pest incidence; options for mechanical weed control; and effects of hedgerows on pests, diseases, and yields. Many universities are participating in this research. Responses of the farm manager to the methods and suggestions are considered an important part of judging the practicability of particular practices.

***Part B: Comparisons on a Farm Scale.*** This part is aimed at determining the long-term ecological and economic effects of two farming systems, integrated (IFS) and conventional (CFS). In the integrated farming system, a package of selected husbandry measures—selected from experience gained in part "A" or elsewhere—was implemented on six field plots. This alternative farming system is compared with a current conventional farming system applied to adjacent plots on the same fields. CFS corresponds with intensive farming practices in the region of Lautenbach.

## The Role of Integration in Arable Farming

A farming system commonly is established on the results of agricultural research. These include the results of plant breeding, soil science, microbiology, chemistry, and agroecological studies. Some examples are discussed on how such ecological knowledge can be exploited in integrated farming, identifying the practical management and operational problems that may arise from their adoption.

***Seedlings Pest, Weed Control, and Nitrogen Supply in Sugar Beet.*** Sugar beet is the major crop for sugar production in Europe. The crop reflects most clearly the technological changes in European agriculture of the past few decades (Norton, 1988). As development proceeds and particular techniques or practices are adopted, farmers become locked into a particular development path leading to particular crop protection responses. In current sugar beet growing practices, nitrogen and herbicides are applied during seedbed preparation or at a presowing stage. Modern technologies have led to more pest susceptible crops. To reduce the labor involved in thinning by hoeing, lower seed densities are chosen to obtain final crop stands. Because spacing is wider, potential damage from seedling pests is much greater. Full-insurance insecticide treatments become necessary. Most of the herbivorous arthropods concerned, however, are not specific sugar beet pests (Brown, 1982; Ulber, 1980). They are attracted strongly to sugar beet seedlings because no alternative food plants (weeds, for example, *Onychiurus armatus* or *Blaniulus guttulatus*) are available. The presence of emerging weeds can help to decrease pest attack and reduce seedling damage (Ulber, 1980; Klimm, 1985). Hence, it is wise to control weeds at the postemergence stage. Delayed weed removal, up to four weeks after emergence (period threshold), has reduced the number of missed sugar beet seedlings significantly without yield losses (El Titi, 1986c, 1986d; Hack, 1981; Scott et al., 1979).

An additional function of weeds at the early crop stage (uncovered soils) is preventing runoff of rain water on the slopes (erosion control) and providing shelter for soil surface fauna (Bosch, 1987; Grosse-Wichtrup, 1984). The weed management concept, which takes the ecological functions of the weed flora into account, should replace the current clean field philosophy (Koch, 1979). One of the major difficulties affecting post-emergence weed control is the growth pattern of weeds. Noncrop plants, nitrophile species in particular, make better or faster use of fertilizers (applied at the presowing stage). The weed biomass can reach seriously damaging levels, aggravating weed control operations. Tractor hoeing becomes risky, and higher herbicide dosages are necessary. Alternating the common pattern of fertilizer distribution, in order to favor the crop and prejudice weeds, can help in limiting this problem. The application of nitrogenous fertilizers as an ammonium nitrate-urea solution sprayed in the plant rows, leaving the interrow spaces unfertilized, has reduced weed biomass about one-third (Figure 1). This fertilizing regime saves 40 percent of the total nitrogen recommended and reduces nitrate leaching.

To make use of the ecological functions of the wildflora on sugar beet fields, delayed weed removal helps to reduce soil pest attacks, thereby saving routine insecticide treatments. The risk of uncontrolled weed growth can be limited when mineral fertilizers are applied only to the plant rows.

Like many other sugar beet growers, the farm manager of Lautenbach already has adopted this more cost-effective postemergence weed control regime. The overall spraying of postemergence herbicides, or a combination between band spraying and hoeing, has become a widespread practice. Improvements are recorded on the section of fertilizers. Mineral nitrogen is added now at much lower rates than 10 years ago.

***Antagonistic Agents and Soil Tillage.*** A vast array of natural biological control agents are known to occur on and beyond the soil surface of arable fields (Edwards, 1984; Hokkanen and Holopainen, 1986; Kickuth, 1984; Paul, 1986; Wallwork, 1976). They play a most important role within agro-

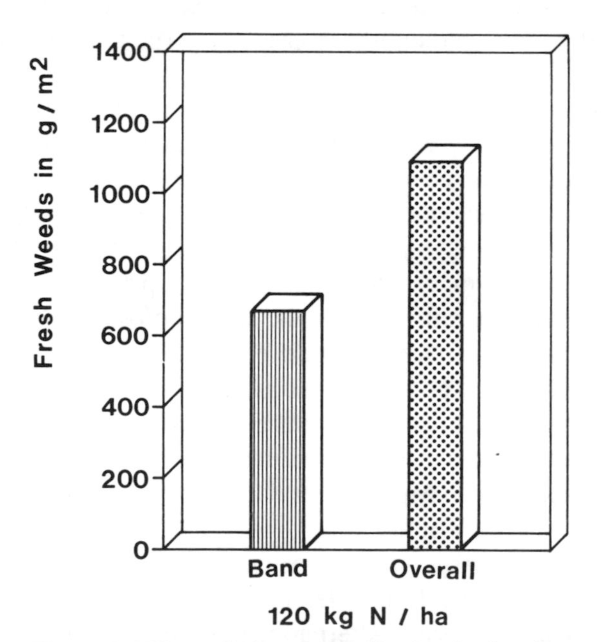


**Figure 1. Effects of nitrogen application in the plant rows of sugar beets on the biomass of weeds, compared with the overall application (average of three years' estimates).**

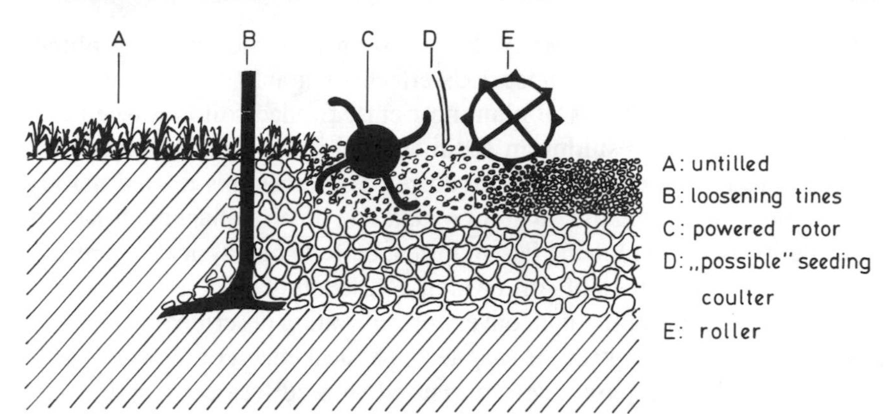

**Figure 2. Schematic design of nonconversion tillage used in the integrated farming system at Lautenbach.**

ecosystems. As predators or parasites they have detrimental effects on the population dynamics of pathogens and pest species. Through these functions they are involved in crop production procedures. Predatory mites (*Gasmasina, Mesostigmata*) are one of the most common groups within the pedoecosystem of arable soils (Inserra and Davis, 1983; Karg, 1983; Karg and Grosse, 1983). More than 60 species have been recorded at Lautenbach, indicating a high diversity. Like many other soil-inhabiting arthropods, gamasids are extremely well adapted to specific microhabitats. Species in deeper soil layers are extremely dependent upon high humidity in the soil pores. They feed mainly on nematodes occurring in the minute soil pores. Conventional tillage regimes of plowing (turning soil layers almost upside down) suppress these predators, causing serious disturbances within the ecosystem.

Field trials at Lautenbach to compare the effects of plowing with those of a nonconversion cultivation (Figure 2) provided good evidence for this hypothesis (El Titi, 1984a; Gottfriedsen, 1987). Tine soil loosening, in combination with shallow incorporation of organic residues into topsoil (5-10 cm), was much less damaging to gamasid mites than plowing. Other soil fauna responded in similar ways to this moderate tillage regime. These include earthworms, *Collembola,* and hymenopterous parasites (Loring et al., 1981). The tillage implement used should allow operational combinations so that seedbed preparation and drilling can be done in one pass. This contributes to a more cost-effective use of farm machinery; another advantage is minimal soil compaction. Integration means in this case the adjustment of soil tillage to suit the natural demands of beneficial organisms and reduce the probability of pest outbreaks.

***Erosion and Tillage Intensity.*** Soil erosion is a widespread problem throughout the world, causing serious effects on arable soils. It results in thinning of topsoil and loss of plant nutrients. Eroded soil is washed into rivers and channels, resulting in sediment and inducing additional labor costs for cleaning operations (social costs). Soil erosion is related mainly to reductions in soil organic matter content and other management factors affecting it. Intensive plowing has significant effects on physical, chemical, and biological properties of soil (Arden-Clarke and Hodges, 1987; Dorn, 1983). Comparisons of the nonconversion cultivation described before with plowing indicate the range of the long-term physical effects of plowing (Table 1). The soil physical parameters measured showed less compaction and higher permeability, water-holding capacity, and aggregate stability of the unplowed fields. These indicate a significant reduction in soil erodibility on unplowed fields after a moderate rain event (Figure 3). Adjustment of the tillage regime can help in reducing erosion and optimizing the soil water/soil pore ratio. Integration of soil cultivations means constructing a fundamental basis for sustainable agriculture.

***Landscape Management/Pest Attack.*** The removal of field boundaries, thereby enlarging the size of field units, has occurred in order to use modern farm machinery. The preclusion of "vegetation islands" has taken place without sufficient knowledge of the ecological functions involved. Hedgerows, shelterbelts, and field-margin vegetation are not merely barriers reducing wind velocity (Arden-Clarke and Hodges, 1987). They also are natural habitats for many different components of agroecosystems (Mader et al., 1986; Schroeder, 1988; Stachow, 1987; Zwoelfer et al., 1984). Mammals, birds, reptiles, spiders, and insects commonly use such vegetation for food, shelter, overwintering, or breeding. Hedgerow inhabitants include a wide

**Table 1. Effects of nonconversion cultivation, compared with plowing, on some soil physical properties at Lautenbach (Dorn, 1983).**

| | *Tillage Regime* | |
|---|---|---|
| *Parameter* | *Nonconversion* | *Plow* |
| Hydraulic conductivity | 26.3 m/d | 22.8 m/d |
| Aggregate size distribution | 2.7 mm | 2.5 mm |
| Aggregate stability | 0.4 mm | 0.7 mm |
| Bulk density | 1.3 g/ccm | 1.4 g/ccm |
| Porosity | 51.8 ccm/ccm | 47.4 ccm/ccm |
| Moisture characteristic | no difference | no difference |
| Wilting point | no difference | no difference |
| Available water percentage | 20.0% | 18.0% |
| Air capacity | 18.0% | 15.0% |

**Figure 3. Gullies 6 to 10 cm deep, caused by slight rain events on a plowed field at Lautenbach, illustrate soil erosion on the slopes.**

range of parasites, predators as well as herbivorous species. Some of the latter can feed on crop plants, causing serious injury. Hence, the majority of herbivores living there are not target species (Bosch, 1986; Marxen-Drewes, 1987). On the contrary, they are hosts or prey of many polyphagous antagonists (parasites and predators) able to control pest populations. The agricultural biocenosis of adjacent fields can be affected by such regulation elements.

Studies on the effects of hedgerows on pest/antagonist relationships were initiated at Lautenbach in 1978. Fifteen different shrubs and trees were chosen according to their plant hygienic aspects, excluding the alternative host species of major pests. The results obtained over seven years indicated an improvement of the entomofauna in the hedge-plantations (Bosch, 1986;

Schaefer, 1984; Zwoelfer et al., 1984). Most evident increases were observed among the hynenopterous parasites, mirid bugs, ground beetles, and spiders. The incidence of attack on adjacent crops indicated significant effects on some potential pest species—*Pegomyia betae* and *Aphis fabae* on sugar beets, and *Sitobion avenae, Metopolophium dirhodum* on wheat. Similar effects of field-edge vegetation on the natural enemies of insect pests also were recorded (Gaudchau, 1981; Klinger, 1984). Phacelia (*Phacelia tanectifolia benth*) and mustard (*Sinapis alba L.*) had enhanced the activity of carabid beetles, staphylinids, and hover flies.

Field surroundings can be manipulated. Noncrop vegetation of defined composition can be an important resource for various biological control agents. Establishment of new hedge plantations on the monocultural landscapes can offer significant options for many animal groups to survive, thereby increasing diversity. This will help prevent pest species from uncontrolled population increases.

These examples illustrate the wide range of integration possibilities that can enhance the natural regulatory components in the field and so reduce crop susceptibility to pest and diseases.

## Comparison of Integrated and Conventional Farming Systems.

***Design and Experimental Layout.*** The Lautenbach estate is a private farm of 245 hectares. The major crops are cereals (winter and spring wheat, spring barley, and oats), sugar beets, and legumes (peas, *Phaeseolus* beans or fababeans), mainly grown for seed production (Steiner et al., 1986). Livestock is limited to 200 to 300 pigs, indicating a high degree of specialization. Blackgrass (*Alopecurus mysuroides*), oat cyst nematode (*Heterodera avenae*), and subterranean Collembola (*Onychiurus armatus*) were the major pest problems on the farm during the preexperimental years. Fertilizer inputs were increasing. The first step was the establishment of a crop rotation that could fit both integration requirements and the economic interests of the farm. Major goals of the new rotation concept were to suppress the target species, ensure a balanced input/output of organic amendments, and enhance beneficial organisms.

The previous four-year (four fields) rotation was replaced by a six-year rotation, and the arable area was divided into six field units. On each field two plots of the same size (four or eight hectares) (Figure 4) were marked and used for comparison of farming systems. The single plot pairs were located on comparable soil types of the same topography. Every six plots represent a small farm of 36 hectares within Lautenbach.

Six single fields were available to implement the integrated farming

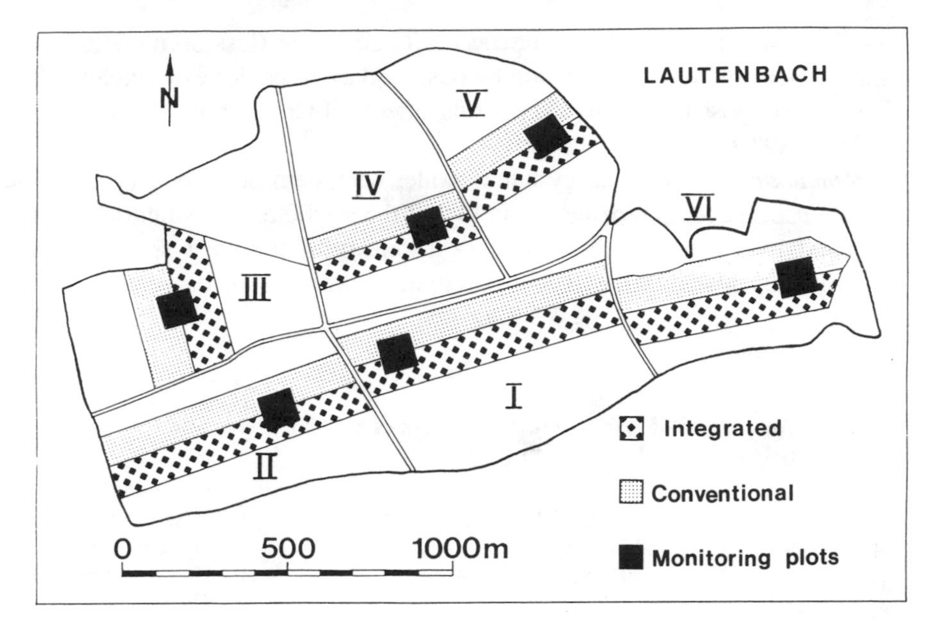


**Figure 4. Design and layout of field plots for integrated and conventional farming systems at Lautenbach with reference to the monitoring plots.**

system. On the adjacent six plots, the conventional farming system was applied. All husbandry operations were recorded for each of both systems according to standard bookkeeping rules. For monitoring the ecological impacts, a number of bioindicators were used. These included earthworms, soil surface fauna, euedaphic mites, Collembola, nematodes, and cellulose decomposition rate. For these purposes a subplot of one hectare, half on IFS and half on CFS, was set up on each of the six field pairs.

The IFS can be characterized by low input of agrochemicals and enhancement of natural regulatory components by considering ecological elements in production processes. In contrast, the CFS relies on a high input of fertilizers and pesticides, disregarding the ecological regulation. Major differences between IFS and CFS are summarized in table 2.

***Results.*** Studies over 10 years indicate significant effects of the low-input, integrated farming system on pest and disease, weeds, nutrients, farm economics, pesticide input, soil physics, and bioindicators.

*Pests and Diseases.* No soil insecticides were used on the integrated sugar beet fields after 1978. Even so, seedling emergence, plant establishment, and yields were 5 to 40 percent higher in the IFS than in the CFS (Figure

5). This was due mainly to improved soil structure (less compaction of soil surface) and lower infestation by pests and diseases, for example, black leg disease, possibly related to grazing effects of Collembola (El Titi and Richter, 1987; Ulber, 1984).

*Nematodes.* Besides oat cyst nematode, beet stem nematode (*Ditylenchus dipsaci*) occurs repeatedly on two of the six fields at Lautenbach (El Titi and Ipach, 1988). The population density of this pest species was significantly lower in the IFS fields than in the CFS fields.

**Table 2. Comparison of main cultural measures used in integrated and conventional farming systems in the Lautenbach project.**

| | *Integrated Farming System* | *Conventional Farming System* |
|---|---|---|
| Crop rotation | 60% cereals, 25% sugar beets, 15% legumes | 60% cereals, 25% sugar beets, 15% legumes |
| Soil tillage | Tine loosening and rotary incorporation —nonconversion | Plowing |
| Sowing | | |
| Sugar beets | 45 cm interrow/20 cm seed space | 45 cm interrow/20 cm seed space |
| Cereals | Double-row 6 cm within, 24 cm between | Drilling 15 cm |
| Faba-beans | 45 cm interrow, 5 cm seed space | Drilling 30 cm |
| Cultivars | Same variety | Same variety |
| Fertilization | | |
| $Ca/K_2O/P_2O_5$ | According to soil chemical analysis | According to soil chemical analysis |
| Nitrogen | According to N-min., but reduced (25%) | According to N-min., optimal supply (recommended dose) |
| Control | | |
| Weeds | Mechanical/herbicides | Herbicides |
| Diseases | Fungicides at high incidence | Fungicides |
| Pests | Insecticides only at high thresholds | Insecticides |
| Hedgerows, Shelterbelts | Included | Not considered |
| Field margins | Native flora accepted | Native flora mowed |
| Fieldedge-attractants | Included | Not considered |

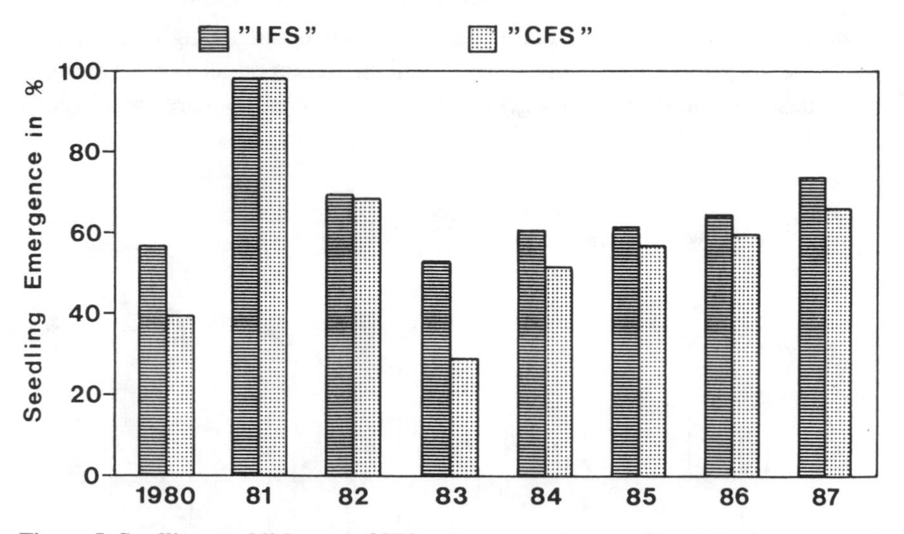


**Figure 5. Seedling establishment of IFS-grown sugar beets at Lautenbach, compared with that of the CFS-grown beets, between 1978-1987.**

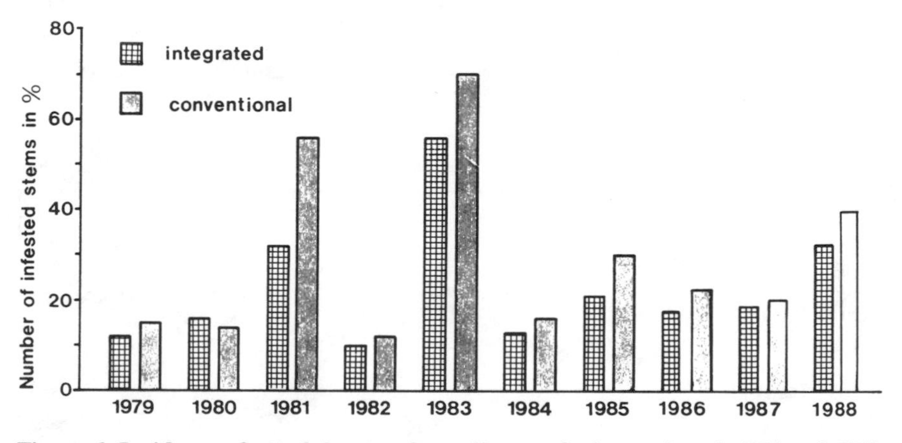


**Figure 6. Incidence of attack by stem base disease of winter wheat in IFS and CFS fields at Lautenbach, 1979-1987.**

*Cereal Diseases.* The incidence of stem base disease (caused by different fungus pathogens) indicated lower infestation levels in the IFS fields (Figure 6). These assessments were made before applying any fungicides. Similar effects were recorded on powdery mildew (*Erysiphe graminis*) and in some cases on brown rust disease (*Puccinia recondita*). Yet a three to nine percent higher infestation was recorded on IFS for leaf and ear blotch disease (*Septoria nodurum*) and *Fusarium* spp. (El Titi, 1984b; El Titi, 1986b).

*Weeds.* Both annual and perennial weeds occurred in higher numbers on the IFS fields. Analysis of the vertical distribution pattern of weed seeds indicates significant changes (Wahl and Hurle, 1988). More weed seeds were left in the topsoil on the unplowed IFS plots. Weed emergence on

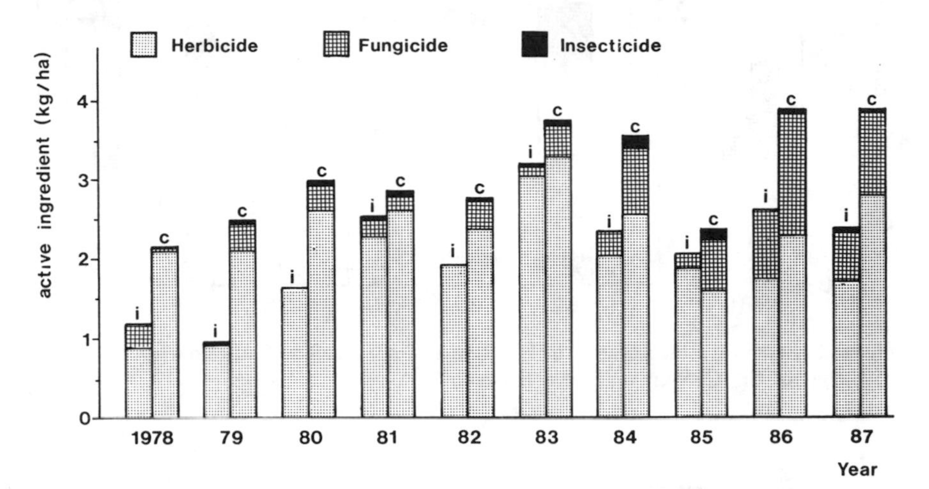


**Figure 7. Annual average of pesticide consumption in IFS and CFS fields at Lautenbach, 1978-1987.**

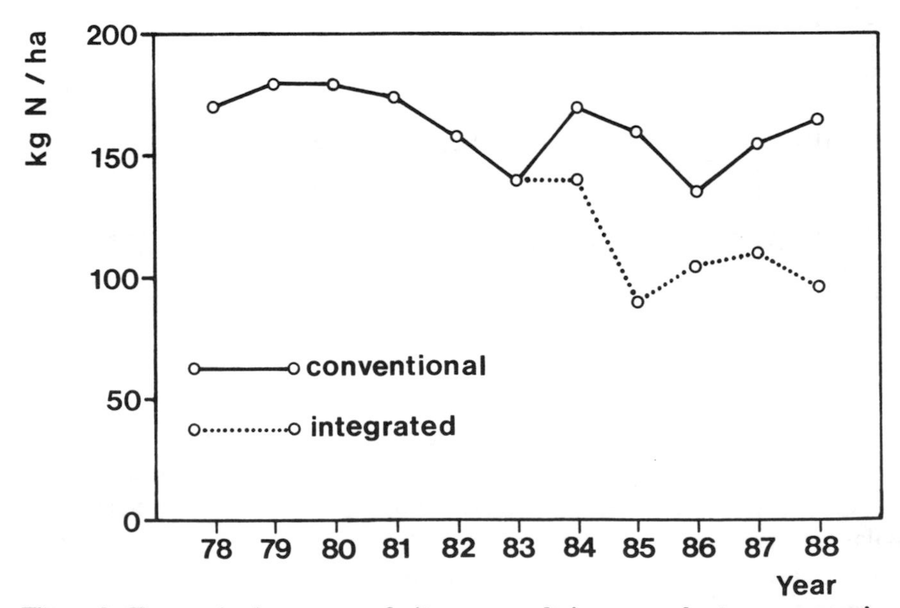


**Figure 8. Changes in the pattern of nitrogen supply in a sugar beet crop grown in IFS and CFS fields.**

**Table 3. The total average of fertilizer input, NPK, in integrated and conventional farming systems at Lautenbach (mean of all fields over 10 years).**

| | *Integrated Farming System* | | | *Conventional Farming System* | | |
|---|---|---|---|---|---|---|
| *Crop* | *N* | *P* | *K* | *N* | *P* | *K* |
| Winter wheat | 116 | 95 | 187 | 155 | 98 | 187 |
| Spring wheat | 129 | 21 | 113 | 175 | 39 | 113 |
| Sugar beets | 144 | 54 | 120 | 162 | 61 | 120 |
| Legumes | 16 | 86 | 177 | 20 | 86 | 194 |

these fields tended to be one to two weeks earlier. Despite higher weed incidence in IFS fields, neither additional treatments nor higher dosages or costs were necessary (El Titi, 1988).

*Consumption of Agrochemicals.* When to apply pesticides depends upon regular monitoring of diseases, pests, and weeds in both farming systems. Results show a significant reduction in pesticide use in IFS (Figure 7). On average, one-third of the total pesticides was saved in the IFS. Lower incidence of pest attack, higher tolerance thresholds, or both made such a reduction in pesticide use in the IFS possible. For all crops and fields, 1,885 kilograms active ingredient per hectare were applied in the IFS and 2,453 kilograms active ingredient per hectare in the CFS.

For more than 75 percent of the total pesticide used, herbicides dominated over all other compounds. Row application and additives use (El Titi, 1988) allowed some herbicide saving. Corresponding efforts with fungicides were mainly unsuccessful in the first rotation. A high reduction was achieved in use of insecticides. Except for a single aphicide treatment in fababeans with reduced dosage (pirmicarb 60-100 g/ha; the commonly recommended dosage was 300 g/ha), no insecticides or nematicides were used in the IFS.

All chemicals used in the sustainable system of Lautenbach were chosen according to their specific environmental side-effects (Hardy and Stanley, 1984). Phosphorus insecticides were not used at all.

Use of mineral fertilizers was reduced sharply in the IFS. Supervised supply of nitrogenous fertilizers in wheat and sugar beet (Table 3, Figure 8) led to a significant reduction in total nitrogen use. Similar results have been achieved in the last two years for phosphorus and are being adopted by the farmer on the whole farm. As a side effect of the reduced nitrogen inputs, less nitrogen was lost by leaching. The mean of the nitrate fraction in the soil profile of the IFS was 30 percent lower than the corresponding CFS level. The key factor seems to be the green manure (e.g., fodder radish) sown in the stubble field as catch crops. Figure 9 shows nitrate estimates in 1987 under both farming systems, with and without a catch crop.

Management of nutrients by including a green manure crop is a realistic

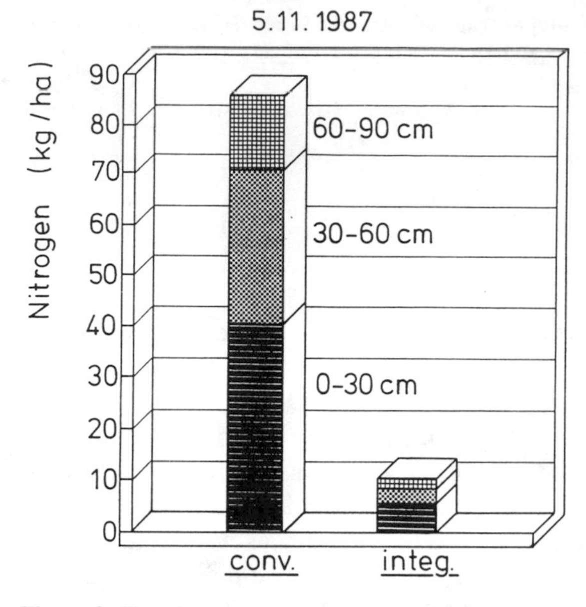


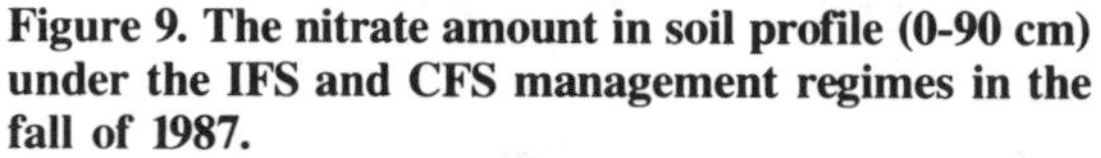

**Figure 9. The nitrate amount in soil profile (0-90 cm) under the IFS and CFS management regimes in the fall of 1987.**

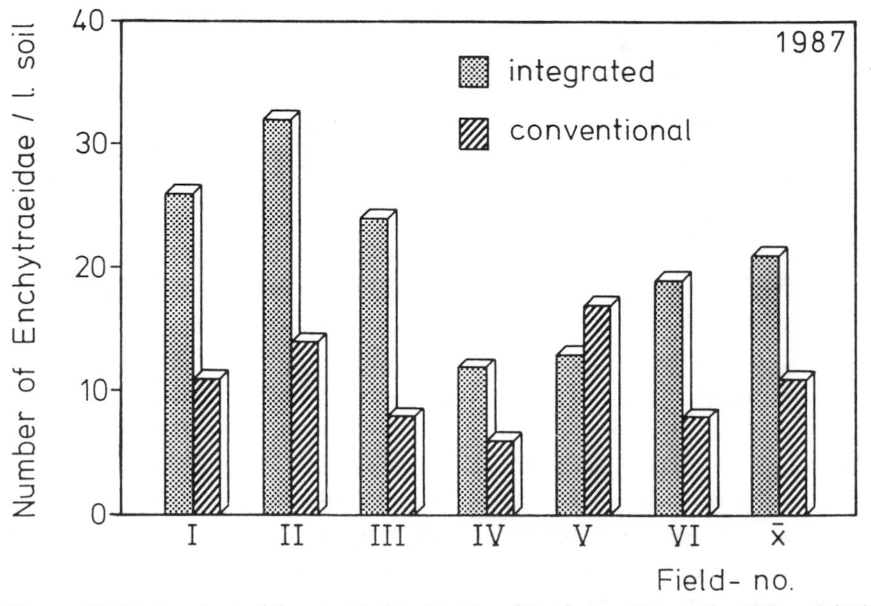


**Figure 10. Comparison of the population density of Enchytraeidae under IFS and CFS regimes at Lautenbach, illustrated by results from 1987.**

method to prevent nitrate losses. For groundwater protection, farmers in the State of Baden-Wurttemberg, Federal Republic of Germany, must, by law, maintain the water-soluble nitrate fraction in their soils below the limit of 45 kilograms per hectare in the fall. Catch cropping is a practical way to manipulate nitrogen leaching in an environmentally sound way.

*Effects on the Agroecosystem.* Changes within the agroecosystems of both farming approaches were recorded by regular monitoring of bioindicators and processes on fixed plots. For these estimates, only known methods and techniques were used. Sampling designs, techniques, sampling frequencies, timing, locations, and distribution patterns were completely comparable in both farming systems. Sampling range was more or less a compromise between the range of confidence and the labor available. Among the results:

■ *Earthworms.* Biomass and number of both adults and juveniles (using the formalin method) were up to six times higher in IFS than in CFS. *Lumbricus terrestris* was the dominate species on all fields (El Titi and Ipach, 1988; Vereijken et al., 1986).

■ Enchytraeidae. The numbers of *Enchytraeidae* extracted indicated higher population densities in IFS compared to CFS. Records of 1986 estimates (Figure 10) show the responses of these annelids to the farming system.

■ *Soil surface fauna.* All assessments were carried out using the pitfall technique. Most of the catches were identified to species level, some only to higher taxa. During the first rotation, total pitfall catches and the number of species were higher in the IFS for carabids, epigeal Collembola, mites, and spiders (*Areneae*), but not consistently for staphylidids (Grosse-Wichtrup, 1984; Paul, 1987). In cases of higher densities on the conventional fields, only a few species were dominant (Drischilo and Erwin, 1982; Speight and Lawton, 1976). In the second rotation (1984 onward) fewer differences were recorded. Species composition of *Carabidae* and *Staphylinidae* showed an unexpectedly high degree of similarity in both farming systems (HS/Eveness). Catches of *Areneae, Diptera,* and *Hymenoptera* completely confirmed the results of the first rotation. In some cases, for example, *Carabus auratus,* the population density increase was extremely high on the IFS.

■ *Soil mites* (Acarina) *and Collembola.* Numbers of these invertebrates were determined after extraction of animals from soil cores taken from all study fields three times a year. Large-scale variations in populations occurred from year to year, from crop to crop, and by sampling time. The subterranean gamasids and Collembola were consistently more prevalent in the IFS soils. This included both degree of diversity (Hs, Eviness, etc.)

and population density, (El Titi, 1986a; Gottfriedsen, 1987) as illustrated, for instance, in 1985 (Figure 11). Subterranean Collembola responded in a similar way. More *Isotomidae* and *Onychiuridae*, the most dominant groups, also were extracted from IFS soils.

■ *Soil microbial activity.* Cellulose decomposition rate was used to assess the breakdown of organic residues in both farming systems (House and Stinner, 1987). Filterpaper of known weight was placed into polyester bags of fine mesh able to exclude microarthropods and earthworms. Ten bags per system and field were buried in five centimeters of topsoil for six to eight weeks. The percentage weight losses were considered as a parameter for describing microbial activity (Table 4). The results obtained over all experimental fields in the course of these studies indicate a slightly higher (not significant) microbial activity in the IFS from the third year onward (Table 4). There was a wide range of variation within the estimates of the single farming systems. The variance within treatments was obviously higher than between treatments. This might be due to weather effects, which can overlap possible differences between the systems.

On the other hand, chemical control costs were reduced an average of 36 percent and variable machinery costs an average of 6.7 percent. As a

ELTON - Pyramid ACARI III

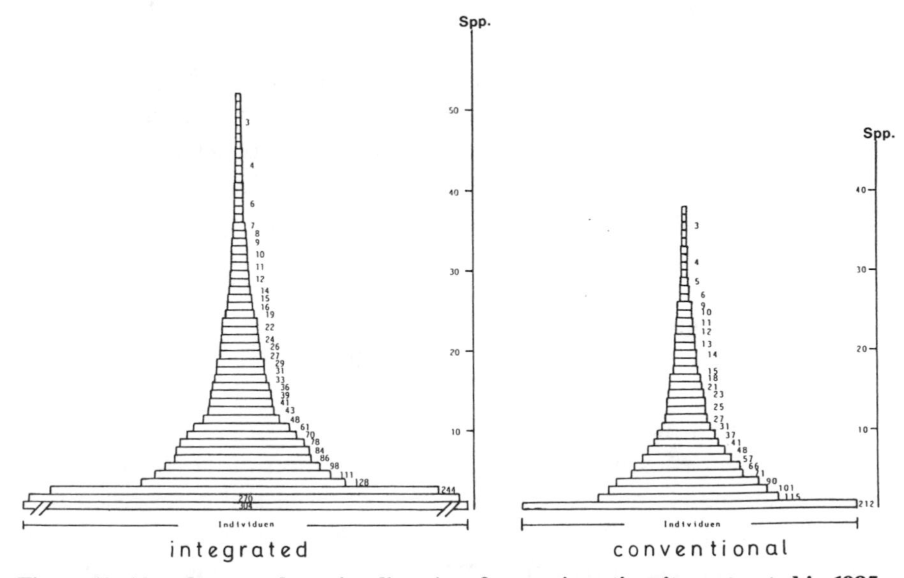


**Figure 11. Abundance and species diversity of mesostigmatic mites extracted in 1985 from integrated and conventionally farmed soils, illustrated by Elton Pyramid (Gottfriedsen, 1987).**

**Table 4. Cellulose decomposition rate in integrated and conventional farming systems at Lautenbach.**

| *Year* | *Integrated Farming System* | *Conventional Farming System* | *Deviation of Systems in Percent* |
|---|---|---|---|
| 1979 | 26.48 | 28.76 | −2.28 |
| 1980 | 54.79 | 55.64 | −0.85 |
| 1981 | 46.65 | 42.48 | +4.17 |
| 1982 | 14.86 | 12.96 | +1.90 |
| 1983 | 43.61 | 34.11 | +9.50 |
| 1984 | 24.56 | 23.27 | +1.29 |
| 1986 | 47.68 | 43.97 | +3.71 |
| Mean | 36.94 | 34.45 | +2.49 |

SED (P=5%) = 18.25

**Table 5. The major farm economic parameters of the integrated farming system at Lautenbach and their percentage deviations from the corresponding figures for the conventional farming system over nine years (Zeddies et al., 1986).**

| *Parameter* | *Deviation of Integrated from Conventional (%)* |
|---|---|
| Yield | − 0,80 |
| Labor | − 2,80 |
| Variable machinery costs | − 6,72 |
| Pesticides | −36,19 |
| Gross margin | 3,52 |

final result of the economic analysis, the gross margin of the IFS was found to be slightly higher (not significant) than the CFS (Figures 12 and 13). A brief summary of the economic results is listed in table 5.

## Farmers Can Adjust

Current agriculture in developed countries is facing serious problems (Diercks, 1983; Vereijken et al., 1986). Declining farm income (despite rising yields), dependence on fossil energy inputs, overproduction, environmental pollution, endangerment of wildlife, and soil erosion have become significant results of intensive land use. A new orientation in agricultural policy is now more necessary than ever before. Therefore, methods of the present agriculture should be reevaluated in the light of up-to-date scientific standards (Seibert, 1985). These will ultimately include impacts on farm economics, on the environment, as well as on the nonagricultural society. There is no realistic option except adoption of farming systems with lower inputs. Improved agricultural systems will be those able to achieve the highest possible degree of self-support. More cost-effective use of farm machinery, producing at least a part of the needed nutrients (e.g., nitrogen by including legumes in the rotation), more natural control of pests and diseases (by enhancement of native biological control agents), and reducing subsoil compaction by growing deep-rooted plant species are examples of how natural components can help in reducing inputs. Various agroecosystem components could be used as production elements in the farming enterprise. A new kind of knowledge is required, however, depending mainly upon the synthesis of the single elements into a completely integrated system.

The integration of the production components characterizes the present research needs. Studies of this kind require a whole-farm scale because of interactions existing between the various farming elements. The project at Lautenbach was the first trial of this kind, at least in West Germany, to study such possibilities. The results obtained showed that farmers can

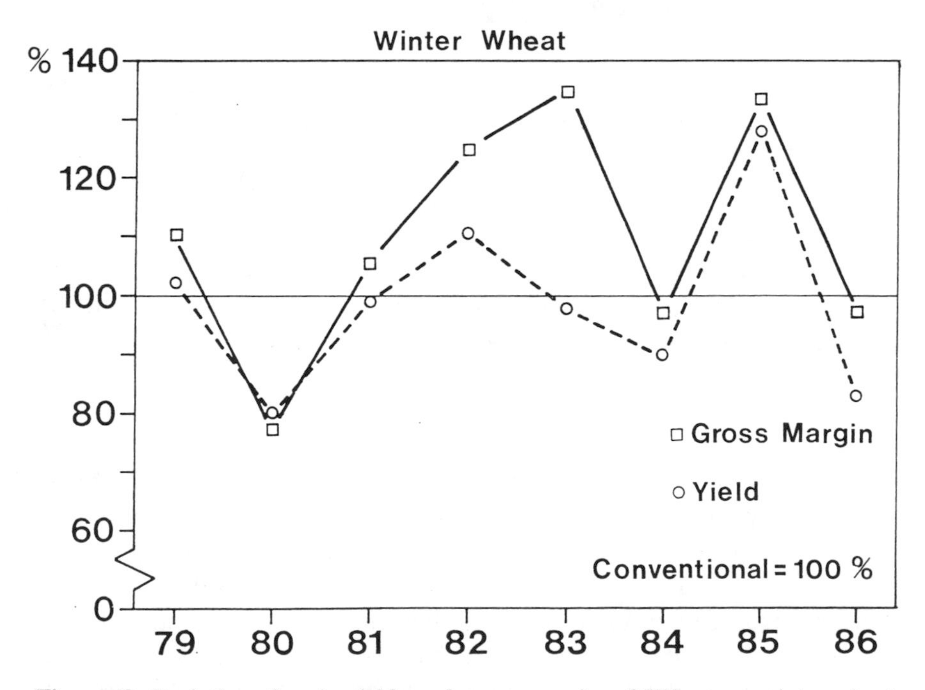


**Figure 12. Deviation of grain yields and gross margins of IFS-grown winter wheat from those grown under CFS, 1979-1986 (conventional system equals 100 percent).**

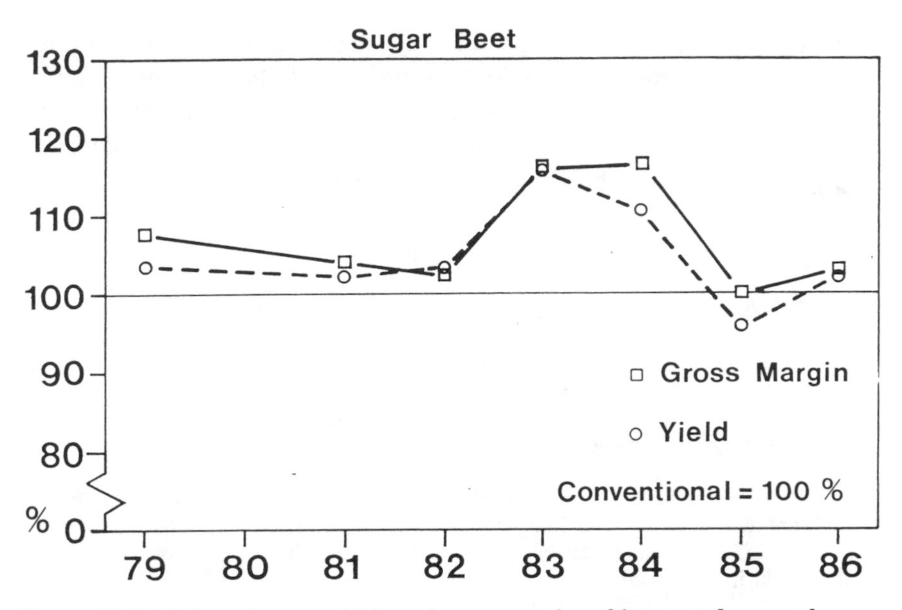


**Figure 13. Deviation of sugar yields and gross margins of integrated cropped sugar beets from those of conventionally grown beets, 1979-1986 (conventional system equals 100 percent).**

adjust their management concepts to fit the goals of integrated farming. They can do this if they are convinced that the techniques recommended are practical. Furthermore, there were significant improvements within the agroecosystem when IFS was implemented, and farm income did not deteriorate within a 10-year period. The deliberate enhancement of predators and parasites obviously contributed to a decrease in the probability of pest outbreaks. It has had proven effects on some potential pest species, such as plant parasitic nematodes. Reducing fungicide treatments and nitrogen supply did not reduce farm income. On the contrary, there was a slight increase (not significant). The concept of nutrient management (N-fixation) through growing legumes or minimizing nutrient losses through catch crops has generated a considerable gain of nutrients. This helped to cover at least a part of the necessary nutrient supply.

The achieved level of self-support at Lautenbach is not the final option. Many other possibilities and options could have been included if the farm had not produced seeds. However, there were some considerable improvements in soil structure, which cannot be calculated in monetary units. The increased populations of earthworms in IFS soils demonstrate the presence of a network of holes and borrows, indicating higher water infiltration potential. Soil workability became better because of the improved soil aggregate structure. The profits of ecological improvements can have great macroeconomical value. Risk analysis would help in understanding the role of sustainable agriculture. The Lautenbach project did not give all the answers to questions, but it has provided significant evidence on the worth of IFS for the farmer, society, and the environment.

## REFERENCES

Andow, D. 1983. Effect of agricultural diversity on insect population. pp. 91-115. *In* W. Lockeretz [editor] Environmentally sound agriculture. Praeger Publisher, New York.

Anonymous. 1977. An approach towards integrated agricultural production through integrated plant protection. Bulletin SROP/WPRS 4.

Arden-Clarke, C., and R. D. Hodges. 1987. The environmental effects of conventional and organic/biological farming systems. I. Soil erosion, with special reference to Britain. Biological Agriculture and Horticulture 4: 309-357.

Bosch, J. 1986. Wirkungen von Feldhecken auf die Arthropoden-fauna und die Ertrage angrenzender Ackerflachen. Mitt. Biol. Bundesanstalt. Heft 232, 308.

Bosch, J. 1987. Der EinfluB einiger dominanter Ackerunkrauter auf Nutz- und Schadarthropoden in Zuckerruben. Z. PflKrankh. und PflSchutz. 94: 398-408.

Brown, R. A. 1982. The ecology of soil-inhabiting pests of sugar-beet, with special references to *Onychiurus armatus*. Ph. D. thesis. University of Newcastle upon Tyne.

Diercks, R. 1983. Alternativen im Landbau. Eugen Ulmer Verlag Stuttgart.

Dorn, L. 1983. Der EinfluB von "konventioneller" und "alternativen" Bodenbewirtschaftung auf einige physikalische Kenn-groBen des Bodens. Dipl. Arbeit Universitat Hohenheim.

Drischilo, W., and T. L. Erwin. 1982. Responses in abundance and diversity of cornfield carabid communities in differences in farm-practices. Ecology 63: 900-904.

Edwards, C. A. 1984. Changes in agricultural practice and their impacts on soil organisms. Agriculture and Environment. pp. 56-66. *In* Proceedings, ITE Symposium No. 13. Institute of Terrestrial Ecology, Cambridge.

El Titi, A. 1984a. Auswirkungen der Bodenbearbeitungsart auf die edaphischen Raubmilben (Mesostigmata: Acarina). Pedobiologia 27: 79-88.

El Titi, A. 1984b. Influence of cultural factors and practices on integrated pest management systems in cereal crops. EG/IOBC/EPPO-conference on integrated crop protection, October 1984, Brussels.

El Titi, A. 1986a. Environmental manipulation detrimental to pest. Parasitis '86. Palexpo-Geneva, Switzerland.

El Titi, A. 1986b. Management of cereal pests and diseases in integrated farming systems. British Crop Protection Conference—Pest and Diseases. Brithton, November 1986.

El Titi, A. 1986c. Unkrautkonkurrenz im Zuckerrubenanbau und ihre praktische Ausnutzung. Z. PflKrankh und PflSchutz 93:136-145.

El Titi, A. 1986d. Zum okonomischen Nutzen von Ackerunkrautern im integrierten Pflanzenschutz, dargestellt am Zuckerrubenanbau. pp. 209-216. *In* Proceedings, Symposium on Economic Weed Control. European Weed Research Society.

El Titi, A. 1988. Angepasste Bestelltechnik: eine Voraussetzung zur Reduzierung des Herbizidaufwandes, Beispiel: Ackerbohnen. Z. PflKrankh. und PflSchutz, Sonderdruck XI: 219-224.

El Titi, A., and J. Richter. 1987. Integrierter Pflanzenschutz im Ackerbau: Das Lautenbach-Projekt III. Schadlinge und Krankheiten 1979-1983. Z.PFL. Krankh. und Pfl. Schutz. 94:1-13.

El Titi, A., and U. Ipach. 1988. Soil fauna in sustainable agriculture: Results of an integrated farming system at Lautenbach/FRG. *In* Proceedings, International Symposium on Agricultural, Ecology and Environment. Padova, April 5-7.

Gaudchau, M. 1981. Zum EinfluB von Blutenpflanzen in intensiv bewirtschafteten Getreidebestanden auf die Abundanz und Effizienz naturlicher Feinde von Getreideblattlausen. Mitt. Dtsch. Ges. allg. angewandte Ent. 3: 312-315.

Gottfriedsen, R. 1987. Angewandt-zoologische Untersuchungen zum integrierten Pflanzenschutz im Projekt Lautenbach. Dipl. Arbeit. Universitat Tubingen.

Grosse-Wichtrup, L. 1984. Populationsdynamik von Getreide-blattlausen und ihren Antagonisten in Winterweizen mit Untersaaten: Eine Untersuchung zum integrierten Pflanzenschutz im Lautenbach-Projekt. Dissertation. Universitat Tubingen.

Gutierrez, A. P. 1986. Systems analysis in integrated pest management. pp. 71-82. *In* V. Delucchi [editor] Protection integree: Quo vadis?

Hack, H. 1981. Vergleich der chemischen Unkrautbekampfung in Zuckerruben mit einer mechanischen Sauberhaltung zu unterschiedlichen Zeitabstanden. Z.PflKrankh. und PflSchutz, Sonderheft IX: 355-363.

Hardy, A. R., and P. J. Stanley. 1984. The impact of the commercial agricultural use of organophorus and carbamate on British wildlife. pp. 72-80. *In* Agriculture and the environment. Institute of Terrestrial Ecology, Cambridge.

Hokkanen, H., and J. K. Holopainen. 1986. Carabid species and activity densities in biologically and conventionally managed cabbage fields. Journal of Applied Entomology 102: 353-363.

House, G. J., and R. E. Stinner. 1987. Decomposition of plant residues in no-tillage agroecosystems: Influence of litterbag mesh size and soil arthropods. Pedobiologia 30: 352-360.

Inserra, R. N., and D. W. Davis. 1983. Hyposaspis ur. aculeifer: a mite predacious on root-knot and cyst nematodes. Journal of Nematology 15(2): 324-325.

Karg, W. 1983. Verbreitung und Bedeutung von Raubmilben der Cohors Gamasina als Antagonisten von Nematoden. Pedobiologia 25: 419-432.

Karg, W., and E. Grosse. 1983. Raubmilben als Antagonisten von Nematoden. Nachrichtenblatt Pflanzenschutz DDR 33: 103-109.

Kickuth, R. 1984. Die okologische Landwirtschaft. Wissenschaftliche und praktische Erfahrungen einer zukunftsorientierten Nahrungsmittelproduktion. Alternative Konzepte 40. Schriftenreihe der Georg Michael Pfaff Gedachtnisstiftung und Verlag C. F. Muller GmbH Karlsruhe. pp. 207.

Klimm, B. 1985. Analyse des Diplopodenbestandes in der Feldflur von Lautenbach. Untersuchungen im Rahmen eines Projektes zum integrierten Pflanzenschutz. Dipl.Arbeit. Universitat Tubingen.

Klinger, K. 1984. Auswirkungen eingesater Randstreifen an einem WinterweizenFeld auf die Raubarthropodenfauna auf den Getreideblattlausbefall. Journal of Applied Entomology 104: 47-58.

Koch, W. 1979. Die Unkrautbekampfung aus dem Blickwinkel des Integrierten Pflanzenschutzes. pp. 235-243. *In* Proceedings, International Symposium of IOBC/WPRS on Integrated Control in Agriculture and Forestry. IOBC Publication.

Loring, S. J., R. J. Snider, and L. S. Robertson. 1981. The effect of three tillage practices on Collembola and Acarina populations. Pedobiologia 22: 172-184.

Mader, H. J., R. Kluppel, and H. Overmeyer. 1986. Experimente zum Biotopverbundsystemtierokologische Untersuchungen an einer Anpflanzung. Schriftenreihe fur Landschaftspflege und Naturschutz. Heft 27.

Marxen-Drewes, H. 1987. Kulturpflanzenentwicklung, Ertragsstruktur, Segetalflora und Arthropodenbesiedlung intensiv bewirtschafteter Acker im EinfluBbereich von Wallhecken. Schriftenreihe. Inst. f. Wasserwirtschaft und Landschaftsokologie der Christian-Albrechts-Universitat Kiel, Heft 6.

Norton, G. A. 1988. Changing problems and opportunities for the adoption of integrated crop protection in cereals and associated crops. pp. 33-42. *In* Proceedings, CEC/IOBC/

EPPO International Joint Conference.

Paul, W. D. 1986. Vergleich der epigaischen Bodenfauna bei wendender bzw. nicht wendender Grundbodenbearbeitung. Mitt. Biol. Bundesanstalt 232, 290.

Paul, W. D. 1987. Okologische Auswirkungen von MaBnahmen des integrierten Pflanzenschutzes auf die epigaische Bodenfauna. Untersuchungen im Rahmen des "Lautenbach-Projektes" (1985-1987). Institut's report.

Pimentel, D. 1982. Environmental aspects of pest management. *In* Proceedings Chemistry and World Food Supplies: The New Frontiers. Pergamon Press, Oxford, England.

Pimentel, D. 1986. Sustainable agriculture: Vital ecological approaches. pp. 85-93. *In* P. Ehrensaft and F. Knelman [editors] The Right to Food: Technology, Policy and Third World Agriculture.

Risch, S. J., D. Andow, and M. A. Altieri. 1983. Agroecosystem diversity and pest control: Data, tentative conclusions, and new research directions. Forum: Environmental Entomology 12(3): 625-629.

Schafer, A. 1984. Neugepflanzte Hecken als Refugien fur Blattlause und ihre Pradatoren. Z. ang. Ent. 98: 200-206.

Schroeder, H. 1988. Primarproduktion von Geholzpflanzen in Wallhecken von Schlehen-Haseltyp, Bedeutung solcher Hecken fur Vogel und Arthropoden sowie einige Pflanzennahrstoffbeziehungen zum angrenzenden intensiv bewirtschafteten Feld. Schriftenreihef. Wasserwirtschaft und Landschaftsokologie der Christian-Albrecht-Universitat Kiel. Heft 7.

Scott, R. K., S. J. Wilcockson, and F. R. Moisey. 1979. The effect of time of weed removal on growth and yield of sugar beet. Journal of Agricultural Science, Cambridge, 93: 693-709.

Seibert, O. 1985. Bedingungen, Formen und Auswirkungen einer Landbewirtschaftung die auf besondere Umweltanforderungen Rucksicht nimmt. Untersuchung im Auftrag der Kommission der Europaischen Gemeinschaften. Inst. f. landliche Strukturforschung an der Johann Wolfgang Goethe-Universitat/Frankfurt.

Speight, M. R., and J. H. Lawton. 1976. The influence of weedcover on the mortality imposed on artificial prey by predatory ground bettle in cereal fields. Oecologia 23: 211-223.

Stachow, U. 1987. Aktivitaten von Laufkafern (Carabidae, Col.) in einem intensiv wirtschaftenden Ackerbaubetrieb unter Berucksichtigung des Einflusses von Wallhecken. Schriftenreihe. Inst. f. Wasserwirtschaft und Landschaftsoklogie der Christian Albrecht Universitat Kiel. Heft 5.

Steiner, H., A. El Titi, and J. Bosch. 1986. Integrierter Pflanzenschutz im Ackerbau: Das Lautenbach-Projekt I. Das Versuchsprogramm. Z.PflKrankh.u. PflSchutz 93: 1-18.

Ulber, B. 1980. Untersuchungen zur Nahrungswahl von Onychiurus fimatus Gisin (Onychiuridae, Collembola), einem Aufgangsschadling der Zuckerrube. Z. angew. Entomology. 90(4): 333-346.

Ulber, B. 1984. Interrelations between soil-inhabiting Collembola and pathogenic soil fungi in sugar beet. XVII International Congress of Entomology, Abstract volume. IOBC Publication.

Vereijken, P., C. A. Edwards, A. El Titi, A. Fougeroux, and M. Way. 1986. Study group management of arable farming systems for integrated crop protection. Bulletin SROP IX/2.

Wahl, S. A. 1988. EinfluB langjahriger pflanzenbaulicher MaBnahmen auf die Verunkrautung-Ergebnisse aus dem Lautenbach-Projekt. Z. PflKrankh.u. PflSchutz, Sonderheft XI: 109-119.

Wallwork, J. A. 1976. The distribution and diversity of soil fauna. Academic Press, London, England.

Zeddies, J., G. Jung, and A. El Titi. 1986. Integrierter Pflanzenschutz im Ackerbau-Das Lautenbach-Projekt II. Okonomische Auswirkungen. Z. PflKrankh.u. PflSchutz 93: 449-461.

Zwolfer, H., G. Bauer, G. Hensinger, and D. Stechmann. 1984. Die tierokologische Bedeutung und Bewertung von Hecken. Berichte der Akademie fur Naturschutz und Landschaftspflege, Laufen/Sulzbach. Beiheft 3, Teil 2.

# RESEARCH ON INTEGRATED ARABLE FARMING AND ORGANIC MIXED FARMING IN THE NETHERLANDS

P. Vereijken

As doubt grows about the sustainability of modern agriculture, interest increases in alternative systems of production. Many new research activities have been started as a result, especially in plant production. In Europe a working group of the International Organization for Biological Control (IOBC) is trying to develop integrated arable farming systems inspired by the aims and methods of integrated pest management (IPM) (Vereijken and Royle, 1989). The two oldest projects are the Lautenbach experimental farm near Stuttgart, West Germany (El Titi, 1989) and the Nagele experimental farm in The Netherlands. Work at the latter study site is the subject here.

## Nagele Experimental Farm

Research at this national experimental farm for the development and comparison of alternative agricultural systems started in 1979. The farm is situated near the village of Nagele in the Northeast Polder, which consists of heavy sandy marine clay (24 percent lutum) three to four meters below sea level. The farm is 72 hectares. Three farming systems have been studied: organic, integrated, and conventional. These systems are run on a commercial basis by one manager and four co-workers.

The organic farm is managed according to the biodynamic method, which is one of the organic systems practiced most in western Europe to date. It is a mixed farm of 22 hectares, with 20 dairy cows and a 10-year rotation, including 50 percent fodder crops. Its main objective is to be self-supporting in fertilizers and fodder. No pesticides are allowed.

The conventional and the integrated farms are concerned exclusively with arable farming. They are each 17 hectares and have the same four-year rotation. The conventional farm, which serves as a reference, seeks to maximize financial returns. The integrated farm should produce a satisfactory financial return, but is also aimed at minimal inputs of fertilizers, pesticides, and machinery to avoid pollution of the environment and save nonrenewable resources. So it may be characterized as an intermediate system.

The research on the farms has three objectives: (a) development of the organic mixed farm and the integrated arable farm in theory and practice, (b) evaluation of the results of the systems based on their specific aims, and (c) comparison of the results of the experimental systems with those of the conventional reference system.

The aim is not to choose between development or comparison of systems but to consider them both as necessary. The experimental systems have to be developed fully before they can be judged on their feasibility and viability in comparison with conventional agriculture. In a previous paper, the initial results of farming and research were presented relating to animal husbandry, crop growth and yield, soil cultivation and weed control, pest and disease control, quality of products, farm economics, and effects on nature and the environment (Vereijken, 1985). One of the most crucial questions in organic farming, that of how to maintain soil fertility, was treated separately (Vereijken, 1986). Herein, the latest research results are evaluated, with special emphasis on development of farm management, inputs of fertilizers and pesticides, and economic results. Based on these results, the perspectives of the two alternative farming systems can be discussed.

## Farming Methods and Techniques

***Crop Rotation.*** An appropriate crop rotation can be very effective in controlling pests, diseases, and weeds and in maintaining soil fertility. In conventional agriculture, the chances for a good rotation have been strongly reduced because most farm holdings in The Netherlands are small and farmers have to grow high-yielding crops in an intensive way, facing increasing production costs and decreasing returns for their products.

For this reason the integrated system had the same crop rotation as the conventional: potatoes-variable-sugar beets-winter wheat. The crop choice for the variable-year crop field depended upon the market situation. Since 1985, peas were grown on half of the field and onions and carrots on a quarter each. A longer rotation would have offered a better barrier against soil-borne pests and diseases, but it also would have been less profitable than the current four-year rotation.

In contrast, the mixed character of the organic system offers excellent opportunities for a diversified and sound rotation. Perennial pastures with grass and clover suppress weeds, restore soil structure, and increase the organic matter and nitrogen content of the soil. Moreover, a high proportion of grassland in the rotation reduces the cropping frequencies of the marketable crops, such as potatoes and cereals. As a result, the pressure of soil-borne pests and diseases is minimized. Until 1985, fodder cereals, such as winter barley and oats, were also part of the rotation. However, their low gross margins had a negative impact upon the economic results of the organic farm. Therefore, they have been replaced by high-yielding crops, such as pea, onion, and carrot. Consequently, a limited amount of supplementary feed has been purchased since 1986. At the moment, the rotation is potato-winter wheat-half carrot, half fodder beet-pea-two-year mowing pasture (alfalfa, red clover, English rye grass)-onion-winter wheat-three-year pasture (white clover/grass mixture). This crop sequence was based largely on alternating positive and negative influences on the structure of the soil and its nitrogen reserves.

***Fertilization and Crop Protection.*** As is usual in Dutch arable farming, fertilization on the conventional farm was mainly of a mineral nature. Organic manure, preferably solid chicken manure, was applied only to the wheat stubble land to supply organic matter. On the integrated farm, fertilization was mainly organic; mineral fertilizers were used only as a complement. In this system, crops were moderately supplied with nitrogen to avoid abundant leaf development and, as a result, high disease susceptibility. Liquid chicken manure was applied right before the sowing of sugar beet and the planting of potatoes; it was plowed under immediately to achieve a maximum nitrogen effect.

In conventional agriculture, green manure is applied to improve the soil structure. On the integrated and organic farms, green manure crops also were grown to fix the nitrate that had been left behind by the main crop or that had mineralized after harvest. Thus, green manure crops served to prevent nitrate leaching.

On the organic farm, only organic manure from the same farm was used. Clover was the main source of nitrogen in the farm cycle. After being consumed as protein by dairy cattle, nitrogen was collected in the stable manure. Together with the other nutrients, nitrogen then was distributed over the various crops as required. Because products are sold off the farm, soil reserves of phosphorus and potassium were depleted gradually (Vereijken, 1986). This was compensated for by purchasing straw and roughage (from natural areas) and some concentrates (partly from conventional origin).

In conventional agriculture, crop protection is chiefly of a chemical nature. On the integrated farm, however, pesticides were used only as a last resort. Chemicals that are known to be highly toxic, persistent, or mobile were avoided. Weeds, diseases, and pests were controlled mainly by using resistant varieties, lowering of the nitrogen dressing, mechanical weed control, use of appropriate sowing times and sowing distances, etc. (Vereijken, 1989a). On the organic farm, ample rotation was indispensable for the prevention of weeds, pests, and diseases because chemical control was prohibited. In both experimental systems, some loss in yield caused by weeds, pests, and diseases was accepted if it was compensated forby savings of expenses.

Cropping systems based on these principles cannot be perfect. Regular observations and reports on management and crop reactions are needed to track imperfections. Ideas from outside the experimental farm (practices, extension, and research) also can improve cropping programs. The fundamental choice of natural practices on the organic farm often called for unusual and risky cropping measures. If successful, they could be introduced on the integrated farm, too. Thus, the biological system serves as a source of inspiration and a pioneer for the integrated system (Vereijken, 1989b).

## Results of Farming and Research

Economics and the environment represent the two main criteria for the social acceptability of the three production systems. The inputs of fertilizers and pesticides were important indicators of environmental impact. The economic viability was indicated especially by net surplus and labor returns. Because of considerable changes in the management of the systems since 1984, only the latest results are presented (1985-1987).

Total returns on the organic farm appear to be considerably higher because of the high premiums on standard product prices (Table 1). Marketable organic crops clearly have higher returns than grassland and fodder crops. However, the total production cost was much higher than on the conventional and integrated arable farms, especially in labor, buildings, and cattle/fodder, which renders by far the lowest net surplus. Inspite of this, returns to labor on the organic farm were highest, although insufficient compared to other professional groups. The integrated farm hardly differed from the conventional farm in total returns and total operation costs. However, the integrated farm gave considerable savings of expenses in fertilizers and pesticides. As a result, the integrated farm achieved a 480-guilders-per-hectare higher net revenue. The three farms were hardly different in intensity of soil use (standard holding units per hectare). Labor productivity

**Table 1. Average farm economic results of the conventional, integrated, and organic farming systems, Nagele experimental farm, 1985-1987.**

| | *Economic Results (Dutch Guilders/hectare)* | | |
|---|---|---|---|
| | *Conventional* | *Integrated* | *Organic* |
| 1. Returns from marketable crops | 6,190 | 6,250 | 12,370 |
| 2. Returns from grassland and fodder crops | — | — | 8,630 |
| 3. Total returns | 6,190 | 6,250 | 10,500 |
| 4. Labor cost* | 2,310 | 2,280 | 5,800 |
| 5. Contract work | 1,020 | 1,020 | 1,290 |
| 6. Equipment and machinery | 1,560 | 1,630 | 2,310 |
| 7. Total operation cost (4 to 6) | 4,890 | 4,930 | 9,400 |
| 8. Land and buildings | 1,290 | 1,290 | 2,630 |
| 9. Cattle and fodder | — | — | 1,820 |
| 10. Fertilizers | 450 | 290 | — |
| 11. Seeds | 690 | 790 | 470 |
| 12. Pesticides | 690 | 260 | — |
| 13. Other cost | 610 | 610 | 900 |
| 14. Total cost (7 to 13) | 8,620 | 8,170 | 15,220 |
| 15. Net surplus (3 minus 4) | −2,430 | −1,920 | −4,720 |
| 16. Labor returns (15 plus 4) | −120 | 360 | 1,080 |
| Technical and economic data | | | |
| 17. Marketable crops (ha) | 17 | 17 | 10.7 |
| 18. Grassland + fodder crops (ha) | — | — | 11.4 |
| 19. Livestock units | — | — | 21.8 |
| 20. Number of labor units | 0.7 | 0.7 | 1.7 |
| 21. Standard holding units (SHU) per ha | 6.2 | 6.2 | 5.5 |
| 22. SHU per labor unit | 149 | 152 | 68 |

*27 guilders/hour was the normal gross reward for the farmer's own labor in Dutch agriculture during 1985-1987.

on the organic farm, however, was less than half of that on the two other farms (standard holding units per labor unit).

On the integrated farm, an important shift has taken place from mineral to organic fertilization (Table 2). Compared to the inputs on the conventional farm, total inputs of potassium and nitrogen were less and the total input of phosphorus was equal. On the organic farm, a large quantity of potassium was brought into circulation by fodder crops and cows. However, phosphorus and nitrogen fertilization here was by far the lowest. Nitrogen availability was clearly the main limiting factor for production on the organic farm, as evidenced by yield comparisons between experimental plots in the pastures with and without clovers (van der Meer and Hofman, 1988). From these results, it has been concluded that biological nitrogen fixation was the main source of nitrogen input in the organic system (Table 3). This table also shows that a deficit on the nutrient balance of phosphorus and potassium caused by sale of products was compensated for by purchase

**Table 2. Fertilization and nitrate-nitrogen content of the drainage water in the three systems averaged for 1985-1987.**

| | *Conventional* | *Integrated* | *Organic** |
|---|---|---|---|
| | *kilograms/hectare* | | |
| Potassium as fertilizer | 135 | 50 | — |
| Potassium in organic manure | 75 | 80 | 155 |
| Total | 210 | 130 | 155 |
| Phosphorus as fertilizer | 10 | — | — |
| Phosphorus in organic matter | 40 | 55 | 20 |
| Total | 50 | 55 | 20 |
| Nitrogen as fertilizer | 135 | 55 | — |
| Nitrogen in organic matter | 80 | 125 | 115 |
| Total | 215 | 180 | 115 |
| | *mg/l* $NO_3-N$ | | |
| Nitrate-N in drainwater | 11.2 | 9.8 | 4.3 |

**Table 3. Balance sheet of nitrogen, phosphorus, potassium nutrients on the organic farm in 1986.**

| | *Nitrogen* | *Phosphorus* | *Potassium* |
|---|---|---|---|
| | *kg/ha* | | |
| Output | | | |
| 84 tons milk with 33.8% protein | 455 | 75 | 125 |
| 4.5 tons fresh weight of cows and calves | 115 | 35 | 75 |
| 4 hectares cereals, 5.5 t ha | 330 | 66 | 85 |
| 1 hectare potatoes, 4.5 t ha | 80 | 25 | 155 |
| 1 hectare onions, 50 t ha | 30 | 20 | 145 |
| 1 hectare carrots, 60 t ha | 55 | 15 | 180 |
| 2 hectares peas, 3.5 t ha | 245 | 25 | 35 |
| 1 hectare chichory, 30 t ha | 25 | 5 | 80 |
| 1 hectare cabbage, 70 t ha | 80 | 40 | 170 |
| Total output over 22 hectares by sale of products | 1,515 | 306 | 1,050 |
| Output by sale of products | 69 | 14 | 48 |
| Input | | | |
| 0.5 ton concentrates | 15 | 3 | 8 |
| 2.1 tons roughage (natural areas) | 42 | 6 | 45 |
| 1.3 tons straw (natural areas) | 13 | 2 | 4 |
| Wet and dry deposition (air pollution) | 35 | 1 | 4 |
| Biological nitrogen fixation | 80 | — | — |
| Total input | 185 | 12 | 61 |
| Natural losses + mutations in soil reserves (input −output)* | 116 | −2 | 13 |

*Nitrogen losses = $NH_3$ − volatilzation + $NO_3$ − denitrifaction + $NO_3$ − leaching. $NO_3$ − leaching = ± 10 kg/ha/yr; potassium losses = potassium leaching = ± 20 kg/ha/yr.

**Table 4. Chemical control in the conventional and integrated systems, 1985-1987.**

| | *Average Number of Treatments per Field* | | *Input of Active Ingredients (kg/ha)* | |
|---|---|---|---|---|
| | *Conventional* | *Integrated* | *Conventional* | *Integrated* |
| Herbicides | 2.2 | 1.1 | 4.5 | 1.7 |
| Fungicides | 4.2 | 2.2 | 5.0 | 2.7 |
| Insecticides | 1.5 | 0.3 | 0.6 | 0.1 |
| Growth regulators | 0.3 | 0.0 | 0.3 | 0.1 |
| Subtotal | 8.2 | 3.6 | 10.4 | 4.6 |
| Nematicides* | 0.3 | 0.0 | 42.7 | 0.0 |
| Total | 8.5 | 3.6 | 56.7 | 4.6 |

*Soil fumigation against potato cyst eelworms.

of feed. Although this deficit existed from 1979 until 1986, the phosphorus and potassium status of the soil is still sufficient, according to conventional standards.

On the organic farm, relatively little nitrate is leached, as shown by the analysis of the average drain water contents (Table 2). Nitrate leaching on the integrated farm remained below the conventional, notwithstanding its principally organic form of nitrogen supply. Apparently, the resulting higher nitrogen mineralization after harvest was recovered successfully by green manure crops. Until now, only the organic farm could meet the standards of the Dutch Ministry of Environment for shallow waters (10 milligrams of nitrate-N per liter). In fact, the drain water from the organic farm was so clean that it can also reach the European Economic Community guidelines for the maximum admissable nitrate content of drinking water (5.6 milligrams of nitrate-N per liter = 25 milligrams of nitrate per liter).

On the conventional farm, 8.5 pesticide treatments per field were applied; only 3.6 were applied per field on the integrated farm (Table 4). If the use of chemical means per year are expressed in kilograms per hectare active ingredient, differences are still greater, that is, 10.4 versus 4.6 and even 56.7 versus 4.6 if routine fumigation of the soil against potato cyst eelworm on the conventional farm is included. When soil fumigation was introduced on the conventional reference farm, as most farmers did at the time, we decided to grow eelworm-resistant potato varieties on the integrated farm.

## Experimental Introduction of Integrated Agriculture

From the experimental results, we have concluded that drastic reduction of the usage of fertilizers and pesticides by means of integrated farm management is attractive from an environmental point of view. The resulting cost reductions also may offer sufficient compensation for lower yields and may bring higher profits. As increasing costs of production and especially decreasing prices of agricultural products put profits under pressure, it becomes attractive to convert to integrated management. Considering the saturation of markets and growing restrictions by environmental legislation, research on integrated farming should be extended by experimental introduction of the system into practice. This latter move would imply the testing of the prototype system developed at Nagele by experienced and commercial arable farmers to attain technically and economically feasible farming scenarios. Undoubtedly, this will also lead to the improvement and broadening of the current integrated cropping programs, promoted by the wide variety of practices in attitude and skill of farmers, nature and size of holdings, soil types, crop rotations, and other factors. Finally, a general strategy for the development and introduction of integrated farming systems is presented (Table 5).

## Perspectives on Organic Farming

The net output of the organic mixed farm has increased steadily since 1985 when low-profit fodder crops were replaced by high-profit vegetables and milk production was raised to a higher level through supplementary purchase of concentrates. Consequently, an acceptable income can be expected in the next few years. To achieve this, it is important that premiums of 50 percent on milk and meat and 100 percent on grain and vegetables be obtained for the organic products compared to the conventional market to make up for the higher investments in capital and labor. This need for high premiums, however, appears to be too high of a threshold for the majority of farmers and consumers until now.

This does not mean that organic farming is doomed to play a marginal role. Several developments are occurring that offer new opportunities for a more radical organic approach. In areas with sensitive ecological characteristics and also in water collection areas, organic farming may play an important role because of its minimum introduction of nutrients and its rejection of chemical pest control measures. Therefore, organic farming in these areas deserves financial support from public funds. Finally, an increasing demand on the European market for organic products is occurring, inspired by growing concerns for mankind and the environment and for the well-being of animals. Sooner or later this may lead to a breakthrough for organic farming into the conventional practices of farm production, trade, and consumption.

**Table 5. Strategy for the development of integrated arable farming systems and their introduction in practice.**

1. Research institutes develop and test the components for integrated farming systems:
   - Varieties with broad resistance and good production.
   - Biological, physical, and chemical methods of crop protection.
   - Methods for the maintenance of soil fertility.
   - Efficient cropping systems with emphasis on quality.
   - Equipment, machines, and buildings for a technically optimum management.
   - Ways of investment with maximum returns of soil, labor, and capital.

2. Experimental stations coordinate the composition and testing of experimental systems on regional experimental farms:
   - Experimental farms on representative locations in specific growing areas. For example in the Netherlands: Nagele in the central clay district (1979), Veendam in the peaty sand district (1986), and Vredepeel in the light sand district (1988).

3. Research and extension introduce and test the experimental systems on a small scale:
   - Regional formation of pioneer groups of farmers for planned conversion from conventional to integrated farming (Dutch central clay district 1990).
   - Technical, economic, and environmental progress has to be monitored and evaluated.
   - Major input/output relations have to be optimized and generally usable cropping and farming scenarios have to be developed.

4. Extension and education introduce integrated production systems on a large scale:
   - Manuals and courses for extension specialists/and teachers.
   - Adaptation of subject matter in agricultural schools.
   - Courses and study clubs for farmers.
   - Appropriate cropping manuals and view data.

## REFERENCES

El Titi, A., and H. Landes. 1989. The integrated farming system of Lautenbach: A practical contribution toward sustainable agriculture in Europe. pp. 265-286. *In* Clive A. Edwards, Rattan Lal, Patrick Madden, Robert H. Miller, and Gartbuse [editors] Sustainable agricultural systems. Soil and Water Conservation Society, Ankeny, Iowa.

van der Meer, H. G., and T. Baan Hofman. 1988. Contribution of legumes to yield and nitrogen economy of leys on a biodynamic farm. *In* P. Planquaert and R. Haggar [editors] Legumes in farming systems. Proceedings, Workshop on Legumes in Farming Systems, May 25-27, 1988, Boigneville, France. Gluwer A.C. Publishers, London, England.

Vereijken, P. 1985. Alternative farming systems in Nagele: Preliminary results and prospects. pp. 124-135. *In* Thomas C. Edens, Cynthia Fridgen, and Susan L. Battenfield [editors] Sustainable agriculture and integrated farming systems. Michigan State University Press, East Lansing.

Vereijken, P. 1986. Maintenance of soil fertility on the biodynamic farm in Nagele. pp. 23-30. *In* Hartmut Vogtman, Engelhard Boehncke, and Inka Fricke [editors] The importance of biological agriculture in a world of diminishing resources. Proceedings, Fifth IFOAM International Scientific Conference at the University of Kassel (Germany). Witzenhausen: Verlagsgruppe Weiland, Happ, Burkhard, West Germany.

Vereijken, P. 1989a. From integrated control to integrated farming, an experimental approach. Agriculture, Ecosystems and Environment 26: 37-63.

Vereijken, P., 1989b. Experimental systems of integrated and organic wheat production.

Agricultural Systems 30(2): 187-197.
Vereijken, P. and D. J. Royle, editors. 1989. Current status of integrated farming systems research. Report of the working group of integrated arable farming. Bulletin 1989/XII/5. International Organization for Biological Control, West Palearctic Regional Section (IOBC/WPRS). Wageningen, The Netherlands.

# SUSTAINABLE AGRICULTURE SYSTEMS IN THE TROPICS

# ECOLOGICAL AGRICULTURE IN CHINA

Shi ming Luo and Chun ru Han

After a study of agriculture in China, Korea, and Japan in the early 1900s, F. H. King (1911) wrote *Farmers of Forty Centuries*. Since then, new forms of agricultural practice, such as organic farming (Oelhat, 1978), biodynamic agriculture (Koepf et al., 1976), natural farming (Masonabu, 1978), ecological agriculture (Worthington, 1981), and biological agriculture, have emerged in the West (Boeringa, 1980). Many methods proposed in these alternative agricultural systems are not strange to Chinese farmers. They are similar to the methods practiced in their homeland for generations. However, Chinese farmers today are using herbicides, plastic sheets, pelleted feed, tractors, hybrid seeds, and so on.

At this stage, an ecological agricultural concept, different from the Western concept, was proposed in China. An introduction to the historical background and present situation is important in understanding why the concept of ecological agriculture was proposed and how it relates to traditional Chinese agriculture and the modernization of Chinese agriculture.

The origin of agriculture in China can be traced back more than 7,000 years (Cheng, 1978). The development of that agriculture has been seriously shaped by nature and society. Those most widely practiced and recorded in ancient Chinese agricultural literature were ecologically reasonable and sustainable (Dang, 1988; Fan, 1985; Xia, 1979).

## The Challenge to Chinese Agriculture

In the last century, especially in the past 30 years, the basic condition of China's agriculture has changed greatly. The population has increased exponentially from about 400 million before 1900 to 1,041 million in 1985 (Figure 1). Representing about 22.2 percent of the world's population, the Chinese rely on 9 percent of the world's arable land. Worldwide, an average of 3.47 people have one hectare of arable land. In China, each hectare of arable land must support 7.5 people, which is a heavy burden.

Great efforts were made to increase cropland in the 1960s and early 1970s by terracing hilly areas, plowing grassland in semiarid areas, or enclosing muddy sea beaches along the coast. However, the rate of increase in cropland could not match the loss of cropland for nonagricultural usage and the increase in population. From 1957 to 1977, 26.7 million hectares of farmland were lost. The average farmland per person declined 1.9 percent each year, from 0.18 hectare in 1949 to about 0.13 hectare in 1983. Before the late 1970s, the ecosystem concept had not been established widely in China. Most efforts to increase cropland neglected the ecological consequences. This intensified the problems of soil erosion and desertification.

Today, of China's 9.6 million square kilometers of land, about 10.3 percent is arable land, 33.0 percent is grassland, 12.0 percent is forest, 2.8 percent is inland water surface, 0.2 percent is seabeach, and the rest is desert or built-up areas. The eroded area has increased from 1.16 million square kilometers in the 1950s to 1.50 million square kilometers in the 1980s. More than 50 million tons of topsoil are lost each year. About 7 million hectares of cultivated land have saline-alkali soils. About 30.3 percent of

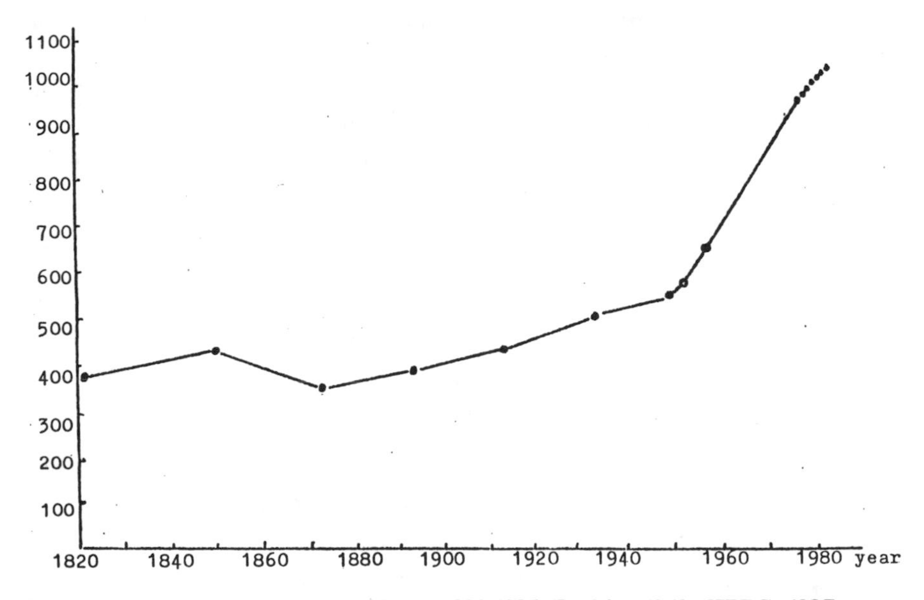


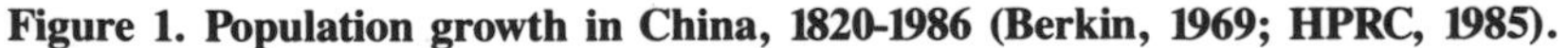

**Figure 1. Population growth in China, 1820-1986 (Berkin, 1969; HPRC, 1985).**

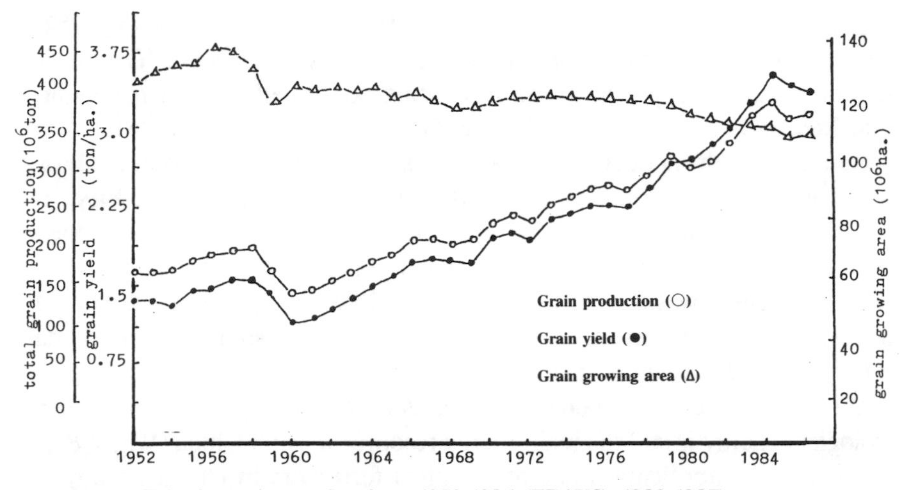


**Figure 2. China's grain production, 1952-1986 (EBAYC, 1980-1987).**

the grassland is overgrazed and degenerating. Desertification reaches 17 million hectares; 29.4 percent of those deserts formed in the past 50 years. Only about 0.8 percent of the land is potentially arable (Agricultural Regionalization Committee of China, 1987). Total production increases in the past 30 years were due mainly to yield increases. For example, total grain production in 1985 was 2.87 times that in 1950, while the grain yield in 1985 was 2.82 times that in 1950. During this period, grain crop areas decreased from 114.41 million hectares in 1950 to 108.84 million hectares in 1985 (Figure 2).

Since the 1950s, the Chinese have paid great attention to increasing crop yield by increasing inputs and improving varieties and cultural methods. For a long period, agricultural modernization in China meant "mechanization, electrification, chemicalization, and adequate irrigation." Obviously, it was a model adapted from developed countries. The subsidized energy input of China's agriculture has increased exponentially (Figure 3). From 1952 to 1982, large- and middle-sized tractors increased from 1,307 to 812,000, hand tractors increased from zero to 1,671,000, the rural electricity supply increased from 50 million kilowatt hours to 3,969 million kilowatt hours, and chemical fertilizer supplies increased from 0.3 million tons to 68.1 million tons. Pesticide production increased from 1,920 tons to 456,900 tons. The area irrigated by water pump increased from 0.3 million hectares to 25.1 million hectares (Editorial Board of Agricultural Yearbook of China, 1980-1987).

As a result of these efforts, food production did increase faster than the population growth. The total population in 1985 was 1.81 times that in 1952, while total grain production in 1985 was 2.31 times that in 1952. The nutrition of Chinese people has been improving continually (Human Population Research Center of the Chinese Academy of Science, 1985).

Can these high-input strategies be successful in the future? After analysis of the general situation, two main difficulties can be identified. First, it is difficult for China to reach the high-input levels common in developed countries. In China, national production per person in 1983 was only $230.6 U.S., whereas it was $2,655.1 in the Soviet Union, $13,887.0 in the United States, and $8,973.1 in Japan. Energy consumption per person in China is only about 9 percent of that in the United States. Although about 10 percent of the commercial energy in China is used in the agricultural sector, biological energy is still the main energy resource in rural China (Wu, 1983).

Second, the application rate of chemical fertilizers in China already is rather high (Table 1). Diminishing returns of input increases can be found in China's production records (Table 2). It is certain that the yield response to further fertilizer increases will not be so marked.

Because it is unrealistic to adopt a fully industrialized model in China, what should be the strategy to use the limited resources to improve the

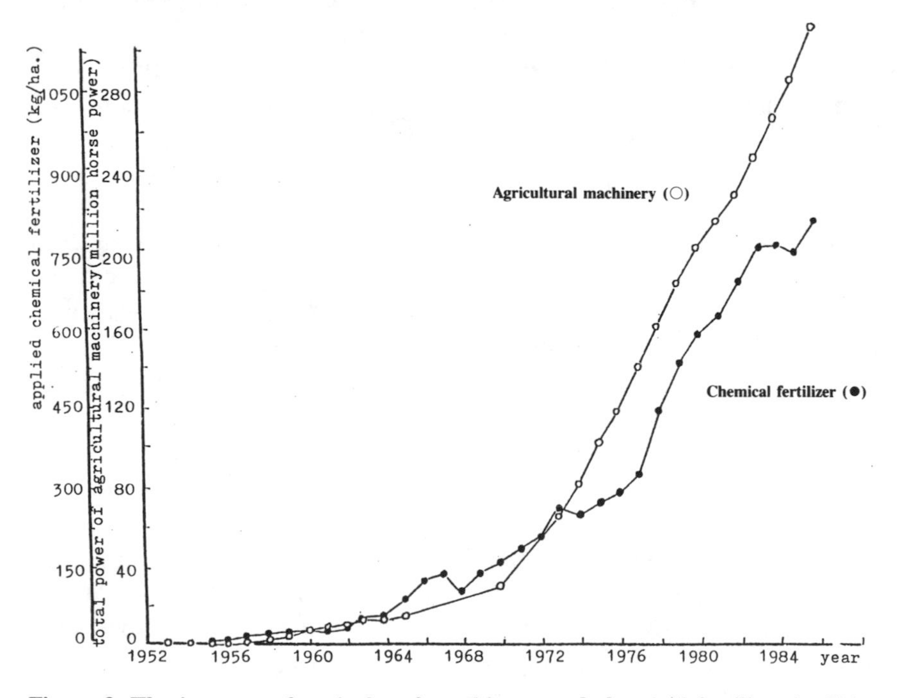


**Figure 3. The increase of agricultural machinery and chemical fertilizer in China, 1952-1986 (EBAYC, 1980-1987).**

**Table 1. Consumption of fertilizer per hectare of arable land and permanent crop in 1983 (FAO).**

| | *World* | *China* | *United States* | *Japan* |
|---|---|---|---|---|
| | *Kg/ha* | | | |
| Nitrogen (N) | 45.5 | 135.6 | 53.2 | 146.1 |
| Phosphate ($P_2O_5$) | 22.3 | 37.0 | 23.5 | 159.4 |
| Postash ($K_2O$) | 17.3 | 8.0 | 27.2 | 131.5 |
| Total | 85.1 | 180.6 | 103.5 | 437.0 |

**Table 2. Relation of chemical fertilizer input and grain yield in China (EBAYC, 1980-1987).**

| *Year* | *(A) Grain Yield (kg/ha)* | *(B) Chemical Fertilizer (kg/ha)* | *(C) Ratio of A/B* |
|---|---|---|---|
| 1952 | 1,320.0 | 3.0 | 440.0 |
| 1957 | 1,462.5 | 15.7 | 93.2 |
| 1960 | 1,170.0 | 30.0 | 39.0 |
| 1966 | 1,770.0 | 122.3 | 14.5 |
| 1970 | 2,010.0 | 156.8 | 12.8 |
| 1975 | 2,347.5 | 266.3 | 8.8 |
| 1979 | 2,782.5 | 527.3 | 5.3 |
| 1985 | 3,480.0 | 735.9 | 4.7 |

life of the Chinese people? The concept of agricultural modernization again became a hot topic in the late 1970s and early 1980s.

## Ecological Agriculture—A Developing Concept in China

In 1981, Ye Xan Ji proposed the concept of ecological agriculture as a development strategy for China's agriculture. Later, he defined ecological agriculture as an ecological and economical system with multipurpose, multifunction, multicomponent, and multilevel organization. It has optimum composition, rational structure, coordinated relationships, and balanced development. It is different from traditional Chinese agriculture, but possesses profound Chinese characteristics (Ye, 1985). Shi (1986) and Yang (1983) proposed the main characters of ecological agriculture as follows:

- Ecological agriculture requires production practices that observe the basic principles of ecology. Agricultural production should adapt itself to the environment and resources. It should maintain the balance of ecosystems.
- Ecological agriculture is a well-planned, optimum system with good social effects, economic effects, and ecological effects.
- Ecological agriculture is highly effective, knowledge intensive, and does not pollute the ecosystem (Shi, 1983; Yang, 1983; Shi, 1986).

There has been a lot of discussion about Chinese ecoagriculture in recent years. Although there is not a generally agreed-upon definition yet, the main trends are obvious.

*Ecological agriculture is considered as a long-term development trend of agriculture.* Ecological agriculture combines good practices of traditional Chinese agriculture with the latest science and technology to create a sustainable agriculture for the future. It tries hard to avoid the resource and environmental problems caused by mechanization and is suitable to the specific conditions of China (Figure 4).

*The concept of ecological agriculture expands the narrow views formed in traditional subsistance agriculture in China.* The concept of food supplies means not only grains but also eggs, meat, milk, fish, fruit, and vegetables. Agriculture means not only crop production but also forestry, fishery, animal husbandary, and other side businesses. Land resources mean not only arable land but also grassland, mountains, seabeaches, and inland water surfaces. For agricultural production, field operations are important, but preproduction service and postharvest processing and marketing are also important. To handle the complications of agricultural production, it is important to observe principles of ecology and economics.

*Ecological agriculture concerns not only economics but also the sustainability of society and environment* (Luo, et al., 1987). The minimum criteria for ecological agriculture include:

■ *Ecologically renewable.* The regenerative rate of renewable resources is greater than the rate of their consumption. The rate of identifying stock of nonrenewable resources or their substitutes is greater than the rate of their consumption. The rate of loss of pollutants is greater than the input rates of them.

■ *Economically survivable.* Under the current economic environment, the income of the system during a production cycle is greater than total cost.

■ *Socially compatible.* Basic needs of society, such as nutrition and em-

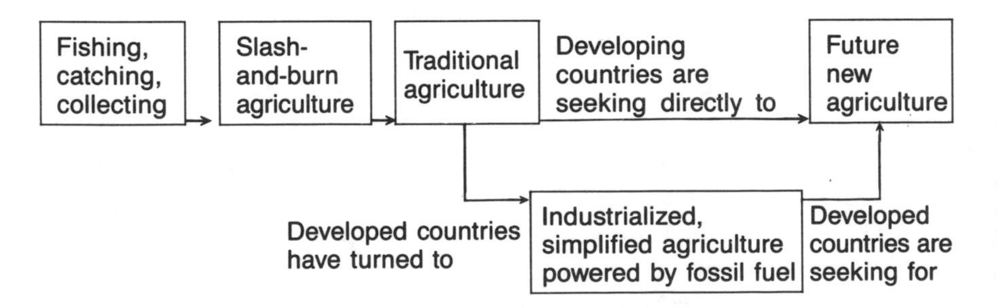


**Figure 4. The development path of agriculture.**

**Table 3. The design of ecological agriculture.**

1. *Qualitative structural design*
   - 1.1 Vertical structure design. Tree, shrub, and grass interplanting; altitude adaptation of species; plant community design; multilayer of fish.
   - 1.2 Horizontal structure design. Species selection for different habitat, density design.
   - 1.3 Time sequence design. Seasonal field structure, such as rotation, relay cropping, etc.; long-term and short-term structural design.
2. *Quantitative functional design*
   - 2.1 Material flow design. Main nutritional balance of crop, animal, and soil; soil erosion control; organic matter balance.
   - 2.2 Energy flow design. Food chain design, primary production design.
   - 2.3 Monetary flow design. Financial balance, optimum scale.
3. *Useful method for design*
   Evaluation, prediction, and optimization by experience; primary mathematics; computer simulation; mathematical programming; fussy mathematics; gray systems theory; etc.

ployment, must be seriously concerned by the system.

*Ecological agriculture emphasizes the relationship between components within the system and the relationship between agroecosystems with their natural and social environments.* On the one hand, systems analysis and system design is a suitable approach to have an optimum system with multiple-purposes (Table 3). On the other hand, special techniques to link different components within and between systems have to be developed and adopted in ecological agriculture (Table 4).

The concept of ecological agriculture in China is somewhat different than the current concept of ecological agriculture in Western countries. First, most ideas about ecological agriculture in China do not totally avoid subsidized energy inputs, such as chemical fertilizer. However, they also emphasize the nutrient cycling process, biological usage of by-products, and integrative pest management. Fertilizer inputs are considered necessary to maintain high yields from limited land resources. Also important is pushing agroecosystems from a low equilibrium stage to a higher equilibrium level. Inputs of chemical fertilizer are the first step to get more organic matter and large-scale material cycling within the system.

Second, ecological agriculture is not confined to only small scales or to the farm level. Ecological agriculture can be practiced in farmyards, on small subsistance farms, or on large-scale commercial farms. According to the organization level, such terms as ecological agricultural household, ecoagricultural village, and ecoagricultural farm usually are used.

In April 1987, a national scientific conference on ecological agriculture was held in Guang Dong Province. Chinese scientists suggested that development of ecological agriculture should be adopted as a national policy for sustainable agricultural development. Scientists in the fields of agronomy, geology, environmental science, mathematics, and computer science are becoming more interested in national or local ecoagricultural projects. Such journals as *Ecological Economics, Rural Eco-Environment, Agricultural Modernization,* and *Acta Ecologica Sinica* have been involved in the exchange of ideas and research results on ecoagriculture. Model farms of ecoagriculture have been set up in many places, such as Beijing, Nanjing, Jiejang, Jang su, Guang dong, An hui, Si chuan, and Shang hai. According to the specific system, ecological agriculture is also called agroforestery, vertical agriculture, and new energy village in different places.

In 1984, "The Decision of Environmental Protection" by the State Council pointed out that "government organizations of environmental protection should cooperate with other organizations to extend ecological agriculture actively in order to prevent contamination and destruction of agricultural environments" (Luo et al., 1987). In May 1987, a National Symposium of Ecological Agriculture, sponsored by the Ministry of Agriculture and other four scientific societies, was held in An hui. After the exchange of ideas and experiences, the Ministry of Agriculture decided to set up typical model farms in different places to get reliable data and make suitable policy decisions (Zhang, 1987).

## Examples of Ecological Agriculture in South China

South China has a humid tropical and subtropical monsoon climate. The rainy season lasts from March to October. Typhoons invade the South China

**Table 4. Examples of ecoagricultural techniques.**

| |
|---|
| Biogas digester |
| High efficiency stove |
| Edible mushroom production using crop and animal by-products |
| Earthworm raising |
| Fly pupae production |
| Chicken waste used as swine feed |
| Rice field fishery |
| Multilayer fish culture |
| Raising of natural enemies of pests |
| Biological method of erosion control |
| Windbreak building |
| Firewood production |
| Agroforestry techniques |
| Biological wastewater treatment |
| Intercropping |
| Solar heater |

Coast two to five times a year from June to November. Average yearly temperatures are 20°C to 28°C. Annual rainfall ranges from 1,500 to 2,000 millimeters.

The most important zonal soils are latosols; in Guang Dong Province, 69 percent of the soils are latosols. These soils have low bases and silica content, with rapid mineralization of organic matter. In a natural ecosystem in tropical and subtropical areas, most nutrients are stored in living organisms or their remains. Soils formed from granite and violet sandy shales are especially vulnerable to erosion by heavy rainfall if the vegetation is removed.

The most important nonzonal soil is paddy soil; in Guang Dong Province, 14 percent of the soils are paddy soils. They are formed in a long-term practice of rice culture in the area.

There are about 6,600 plant species, 130 vertebrate species, and 1,550 fish species. Many insect species can increase to the extent that they influence crop and livestock production in humid and hot environments if the natural balance is altered.

One of the main characteristics of agroecosystems in South China is that the scale of material and energy flow-through is larger than in corresponding agroecosystems in North China. Hence, the turnover time of energy and material of the systems is shorter.

Under these circumstances, several traditional and modern structures of agroecosystems, which are sustainable in tropical and subtropical environment, are adopted widely in South China.

***Multilayer Perennial Plant Communities in Mountainous and Hilly Areas.*** These are artificial plant communities simulating seasonal rainforest communities in the area. There are several advantages to these communities:

*They are highly effective in erosion control and soil improvement.* If mountainous or hilly areas are used for annual crop production, the soil is exposed directly to the sun and rainfall for a long period during a year. If only one tree species is used, it takes many years before a complete canopy forms. The soil protection is incomplete. The multilayer perennial communities not only form a vegetative cover very quickly but also have complete protective mechanisms to check erosion.

Xiao Liang is a hilly area in the southern coastal part of Dian Bai County, Guang Dong Province. The latosols are formed from granite. Because of deforestation, the area used to be a "red desert" with serious erosion. A comparative study of bare soil; soil covered by species of *Eucalyptus excuta;* and soil covered by multilayers of trees, shrubs, and grasses was conducted (Table 5). The multispecies site included *Aphanamixis polystachya,*

*Chukrasia tabularis, Aguilaria sinensis, Albizia odoratissima, Acacia auriculaeformis,* and another 330 species of plants. The erosion was reduced greatly on this multispecies site. Moisture, biomass, litter depth, and animal numbers in the soil on the multispecies site were similar to the natural forest. The combination of biological methods of multilayer perennial communities, together with engineering methods of contour terraces, contour ditches, and sand dams, has changed Xiao Liang into a "green ocean" (Research Station of Ding Hu Shan Forest, 1984).

*They have positive effects on improving production.* Rubber tree and tea complexes in tropical China benefit both rubber production and tea production. Rubber trees are planted in plots 10 to 15 meters by 2.5 meters. Tea is planted between two rows of rubber. This complex community reduced erosion. The soil organic matter and main nutrients under the complex community generally were higher than those under either single rubber tree communities or tea communities. The production of rubber in a rubber and tea complex is higher than in a monorubber community. Under the shade of the rubber trees, the chemical composition of tea is generally improved. The total income from a rubber-tea complex is 12,000 to 15,000 yuan per hectare, while it is only 7,500 to 9,000 yuan per hectare in the monorubber community (Feng, 1985).

There are many other types of multilayer perennial communities in South China, such as rubber-pepper, rubber-banana, pine-stylo, eucalyptus-molass grass, and lichi-tea.

**Table 5. Comparative research results of bare site, monospecies tree site, multispecies site, and secondary natural forest of about 100 years, in Xiao Liang, Guang Dong Province, China (RSDHSF, 1984).**

| | *Bare Site* | *Monospecies Site* | *Multispecies Site* | *Secondary Natural Forest* |
|---|---|---|---|---|
| Erosion rate (kg/ha/yr) | 26,901 | 6,210 | 3 | — |
| Runoff rate ($m^3$/ha/yr) | 1,902.6 | 3,902.85 | 6.29 | — |
| Litter depth (cm) | 0 | 1.0-2.0 | 2.3-5.0 | 3.3 |
| Average soil temperature (°C) | | | | |
| 0 cm | 14.9 | 14.4 | 11.5 | — |
| 20 cm | 23.7 | 22.9 | 22.9 | — |
| Average air | | | | |
| humidity (mb) | 24.2 | 24.8 | 25.2 | — |
| relative humdity (%) | 83 | 85 | 87 | — |
| Moisture content of soil (%) | 5.0-13.2 | 10.7-16.1 | 12-16.9 | 21-29.4 |
| Biomass of soil (g/$m^2$) | 1.30 | 35.66 | 72.18 | 68.01 |
| Soil animal (number/$m^2$) | 15 | 25 | 31 | 27 |

**Table 6. Composition of runoff water from rice field and adjacent peanut field (Luo, 1987).**

| *Composition* | *Rice Field in Late Tillering Stage* | *Peanut Field in Early Flowering Stage* |
|---|---|---|
| Solid material (mg/1) | 11.25 | 2,443.25 |
| | *ppm* | |
| Amonian nitrogen | 1.64 | 11.77 |
| Total nitrogen | 3.96 | 15.65 |
| Rapidly available potasium | 0.71 | 12.22 |
| Total potasium | 0.94 | 24.27 |
| Rapidly available phosphorus | 0.00 | 0.23 |
| Total phosphorus | 0.14 | 2.41 |

*Sampling site: Sui Kang Tean, Guang Li township, Gao Yao County, Guang Dong Province, China. Sampling time: September 3, 1983, during a 58.2 mm rainfall.

***Rice-based Rotation System in Flat Areas.*** Rice (*Oriza sativa*) appeared in China's agriculture at least 6,900 years ago, during the early period of recorded history. Rice was next to millet and wheat and was the third major crop in China. The importance of rice increased continuously. From the Sui Dynasty (581 to 618 A.D.) and the Tang Dynasty (618 to 907 A.D.), rice production exceeded millet and became the second major crop in China. Since the Song Dynasty (960 to 1368 A.D.), rice has become the most important crop. Today, rice remains the most important cereal crop in China, especially in humid tropical and subtropical areas.

Several characteristics of rice-based rotation systems make them more sustainable than other forms of field usage. The water layer of the rice field has a protective and diluting effect to reduce erosion of soil and nutrients during the rainy season (Table 6).

Mineralizational rates of soil organic matter can be reduced in rice fields. A soil organic matter experiments in Ping-sha, set up 15 years ago in the lower reaches of the Pearl River, showed that organic matter was 3.2 percent in a continuous rice field, while it was only 2.7 percent in a continuous sugarcane field (Qai et al., 1982). The mineralizational rate of organic matter in double rice rotation systems ranged from 11.4 to 15.9 percent; in single rice rotation systems, which were covered with water in much shorter time, the rate ranged from 18.7 to 20.9 percent (Soil and Fertilizer Research Laboratory of Guangdong Agricultural Research Institute, 1984).

Nitrogen fixation was high in rice fields. Eighty-one species of blue green algae in rice soil were identified in An Lu County, Hu Bei Province. The nitrogen-fixing activities of five species reached 109.2 to 326.3 nano Mole $C_2H_2$ per second per milligram dry weight of algae. *Azolla* and many species of bacteria in paddy fields also have a nitrogen-fixing ability (Liu, 1984).

Research in the Philippines (Yoshida et al., 1973) showed that rice plants doubled the nitrogen-fixation activity of paddy soils. Among different species of Graminae, rice has the highest potential to increase the nitrogen-fixing ability of soil.

Rice has a wide adaptation in the rotation system. Few weed species and soil-borne pests can survive in both submerged and dry conditions. Rotation of rice with an upland crop, such as sugarcane, sweet potatoes, soybeans, wheat, vegetables, jute, or tobacco, is common in South China. Such rotations provide flexibility for farmers to adapt to changes in social, economic, and natural environments.

***Fish Pond-Raised Field System.*** This system is used mainly in the Pearl River Delta of Guang Dong Province. Farmers dig fish ponds in lowland areas, water-logged areas, or alluvial plains. The fish ponds range from 0.1 to 1 hectare with an average of 0.2 hectare. They are about 1.5 to 2 meters deep. The mud from the fish pond is used to raise the field surrounding the pond. The ratio of fish pond area to raised field area is mostly within 2:6 to 6:4. In this way, formerly unproductive areas become ideal for both fish and crop production. In the fish pond, common carp (*Cyprinus carpio*), silver carp (*Carassius auratus*), big head carp (*Aristichthys nobilis*), grass carp (*Ctenopharyngodon idella*), and dace (*Cirrhina molitorella*) are common. On raised fields, mulberry used to be the most common crop. Other crops include sugarcane, elephant grass, bananas, lichi, flowers, and vegetables. The fish pond-raised field systems are distributed mainly in 110,000 hectares of plain area south of Guang Zhou City.

The earliest record of fish pond-raised field systems in the Pearl River Delta appeared in the ninth century, but remained unimportant until the fourteenth century. From the middle of the Ming Dynasty (1530), fish pond-raised field systems expanded quickly because the demand for silk in international markets increased. At the beginning of this century, the total fish pond-raised field area was more than 67,000 hectares (Zong, 1987).

Components of fish pond-raised field systems have a close relationship. In traditional silk worm-mulberry-fish pond systems, for example, silk worm waste, which includes mulberry leaves and silk worm droppings, are used as feed for fish ponds. Mud formed from soil particles from field erosion, fish waste, and unused feed is raised to cover the field two to three times a year to control weed growth and to provide nutrients for the mulberry. Mulberry leaves are collected for silk worms. In the fish pond, grass carp live in the middle layer of water and feed on grass. They like clear water, but the waste from grass carp could easily destroy the quality of water. Common carp and dace consume the debris from the middle and upper

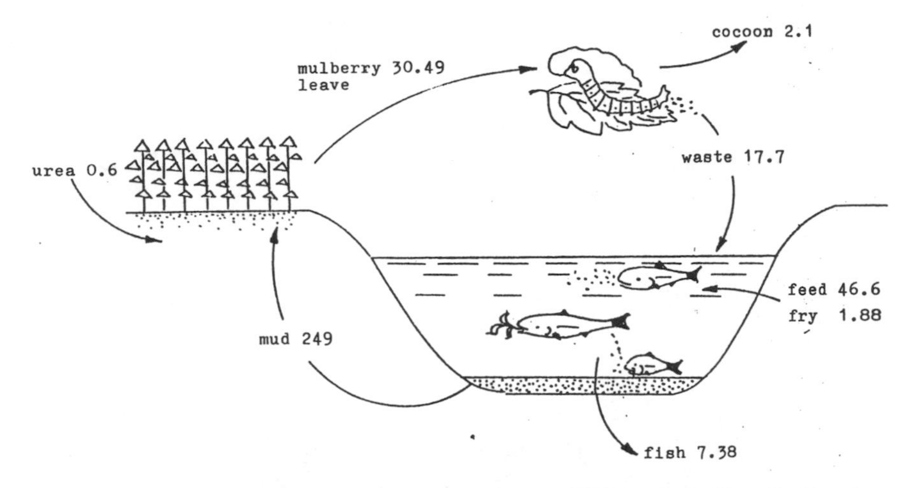


**Figure 5. Relation in a typical fish pond and mulberry field system in Shun De County, Guang Dong, China, 1981 (tons fresh weight/hectare) (JRGSN, 1985).**

layers. These omnivores live in the lower layers and tolerate a low oxygen content. Nutrients dissolved from the feed and waste and leached from the field stimulate the growth of phytoplankton and zooplankton, which are consumed by the upper-layer fishes (Figure 5).

Research by Zong et al. (1987) on typical mulberry-field pond systems showed that fish pond-raised field systems conserve nutrients and soil particles effectively in the humid tropical and subtropical environments. The pond mud increased 13.2 centimeters per year. It contained 249 tons of dry matter, 1,021.5 kilograms of nitrogen, and 4,605 kilograms of carbon per hectare. One hectare of the fish pond received 633 kilograms of nitrogen and 3,597 kilograms of carbon from the runoff and leaching of the field each year (Zong et al, 1987).

From 1978 to 1986, freshwater (Figures 6 and 7) fish-raising areas in Guang Dong Province increased from 197,350 hectares to 276,136 hectares. Most of these were in the form of fish pond-raised field systems. Because of relatively low returns for labor in silk worm production, most of the fields in fish pond-raised field systems have changed from mulberry to sugarcane or elephant grass in the past decade. Many farmers raise ducks and pigs along the edge of the fish ponds. This structure benefits both fish and livestock production. Experience shows that about 3,000 ducks or 45 pigs per hectare of water surface are suitable.

***Deep Ditch and High Bed System.*** This traditional system is also used mainly in the alluvial plains of the Pearl River Delta. To plant vegetables,

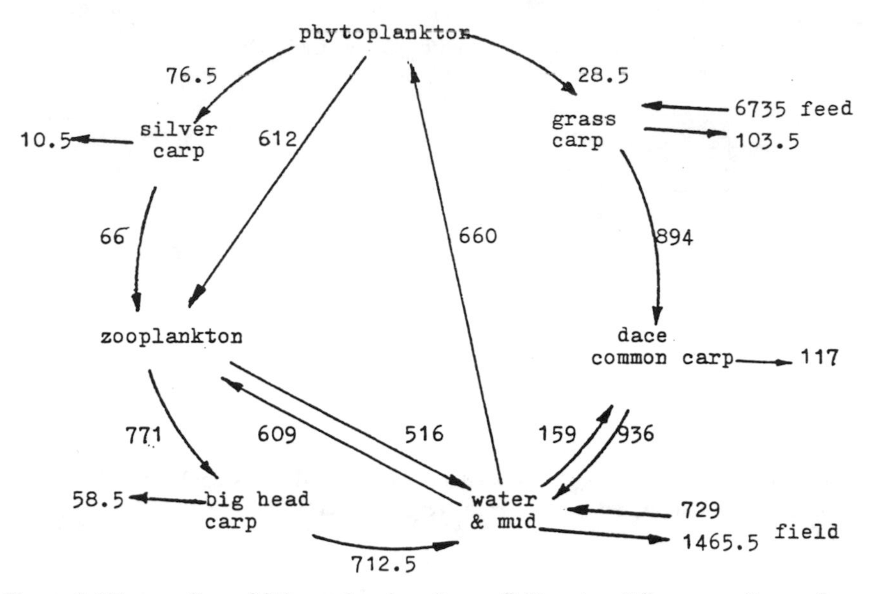


**Figure 6. Nitrogen flow of fish pond and mulberry field system (kilograms nitrogen/hectare/year) (Zong, 1987).**

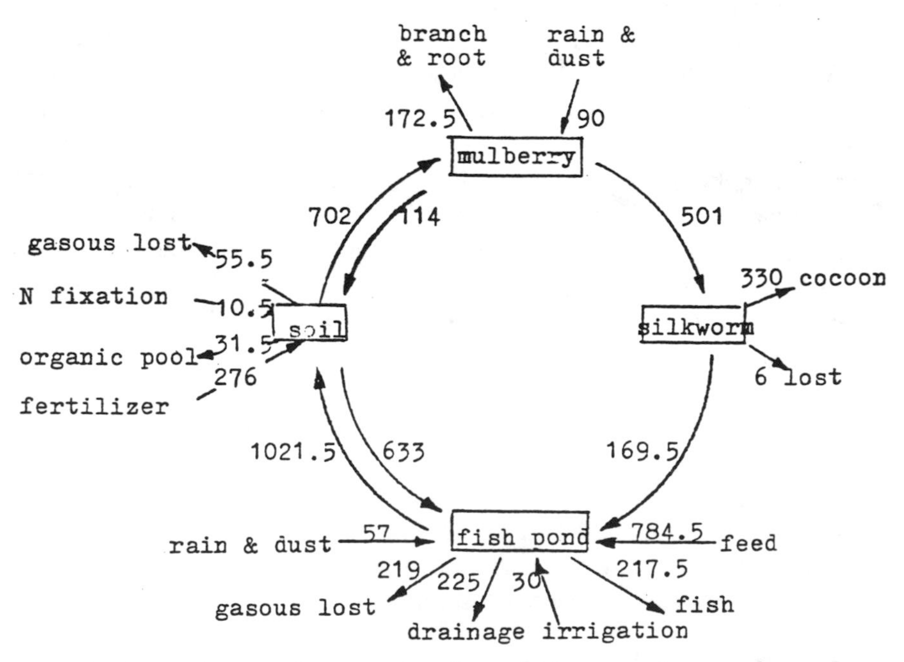


**Figure 7. Nitrogen flow of fish pond and mulberry field system (kilograms nitrogen/hectare/year) (Zong, 1987).**

upland crops, or fruit trees in high water table areas, farmers dig 0.6-meter- to 1.5-meter-deep ditches and make 1.2-meter- to 7-meter-wide beds so the water stays in the ditches. The structure of deep ditch-high bed systems is rather stable and usually maintained for 5 to 10 years or more. The ditches can be used for the production of rice, water hyacinth, snails, or fish; used as a small transportation channel with a special boat; or covered by melon and pea leaves on a bamboo frame overhead. In field beds, long-term fruits, such as lichi, longan, carambola; middle-term fruits, such as orange and banana; and horticultural crops, such as potato, cabbage, and flowers, often are planted (Figure 8). Intercropping is common in these systems. For example, farmers plant soybeans, peanuts, or vegetables around the young orange trees before they fruit. There are about 48,000 hectares of deep ditch-high bed systems in the Pearl River Delta (Lin, 1988).

Comparative research by Lin (1988) showed that deep ditch-high bed systems can lower the water table and promote root growth of the plants (Figures 9 and 10). Deep ditch systems help to reduce soil erosion and to preserve nutrients and organic matter. Sensitivity tests of the nitrogen flow model of deep ditch-high bed systems for orange production showed that if no sedimentation and dilution effect exists in the ditch the cycling index of the system would decrease from 29.19 percent to 5.86 percent. In the present economic environment, the profitability of deep ditch-high bed systems is generally higher than that of rice production. In the past 10 years, 31,500 hectares of fields in the Pearl River Delta have changed to fruit production mainly by deep ditch-high bed methods (Lin, 1988).

***Shelterbelt and Windbreak Networks Along the Coast.*** To prevent crops from damage by typhoons in the rainy season and cold spells in early spring and late fall, shelterbelt systems began to be set up in the late 1950s along the southern coast of China. Today, there are about 140,000 hectares of coastal fields protected by these systems in Guang Dong Province alone.

The main tree species in shelterbelts along the sandy beach is *Casuarina iquisetifolia* Linn. In coastal plains, the main species used in the main belt of windbreak networks, which are perpendicular to the main wind direction, are different than those in the auxiliary belt, which are parallel with the main wind direction. *Metasequcia glyptostroboides* is especially suitable as a main belt species in high water table areas. Other main belt species include *Casuarina equisetifolia* Linn., *Taxiduym distichum* (L.) Rich., *Glyptostrobus pensilis* (L.) Benth, *Melia azedarach* Linn., *Leucaena leucocephala* cr. Saluador, *Acacia mangium* Willd, *Acacia acurriculas* formis Cunn., and some bamboo species. Many fruit tree species and other economic species can be used for the auxiliary belt. *Lichi chinensis,*

*Euphoria longan, Psidium guagava, and Livistona chinesis* R.Br. usually are used. The distance between two auxiliary belts ranges from 200 to 250 meters. The protective arca of a belt is 15 to 20 trees high behind the wind direction and 1 to 3 trees high in front of the wind direction. Typhoon damage was reduced greatly within these networks. The damage caused by cold spells was also reduced. Investigations in 1979 and 1981 showed that the filling percentage of rice grains in protected areas was 10 to 20

**Figure 8. Typical deep ditch-high bed system in South China. Upper on bed: banana intercropped with vegetable surgarcane intercropped with melon; in ditch, fish, snail, water hyacinth. Middle on bed: vegetable, fruit tree (orange, lichi); in ditch, rice. Lower on bed: vegetable; above ditch, melon, pea.**

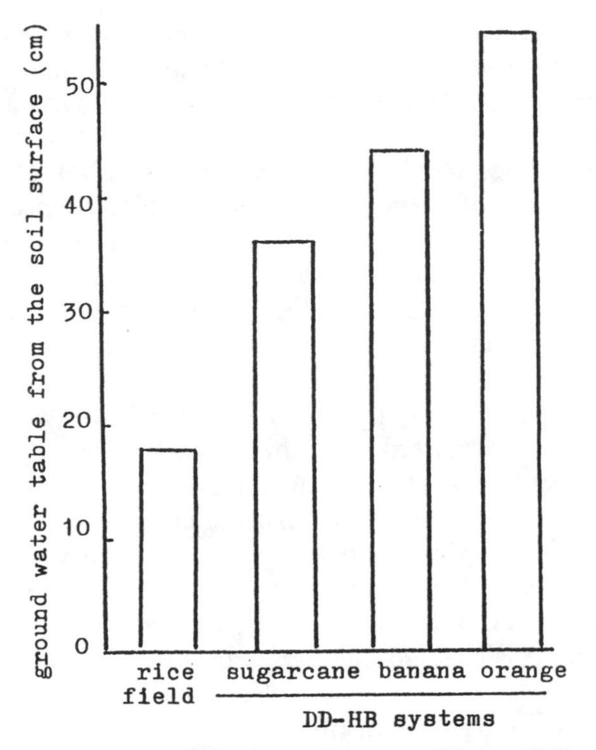


**Figure 9. Depth of water table in different field ecosystems, May 30, 1987, Zhong Shan Guang Dong, China (Lin, 1988).**

percent higher than in unprotected areas after the attack by cool waves in late fall. The thousand seed weight was 0.4 to 3.4 grams heavier and the rice yield was 24.4 percent higher. The rotting rate of rice seedlings in the cold spell in late February to early March of 1980 and 1982 declined 30 to 40 percent because of the windbreak. The introduction of K-strategy species in the form of forest belts has increased system stability. During the attack of cold waves in spring or fall, the air temperature within the network during the day was 0.3 °C to 1.5 °C higher and soil temperatures were 1°C to 2 °C higher than the temperature outside the network (Figure 11).

The direct economic return from shelter belts are high also. In Li Le Township, Xin Hui County, there are 3,300 hectares of coastal fields that are cut into 1,350 sections by 497 kilometers of shelterbelt. The shelterbelt system is formed by 220,000 *Casuarina equisetifolia* Linn., 60,000 *Glyptostrobus pensilis,* 1,200,000 *Liristona chinesis,* 70,000 bamboo, and 18,000 fruit trees of lichi, guagava, and banana. The annual income from the belt from such produce as bamboo, palm leaf, fruits, and wood was more than

1 million Chinese yuan, about 10 percent of total agricultural production. The shelterbelt systems are expanding in the coastal plain (Joint Research Group of Shelterbelt Network in Crop Field of Guangdong, 1985).

***Agricultural By-product Usage by Various Detritus Food Chain Organisms.*** Three models of by-product usage by detritus food chain organisms, such as earthworms, edible mushrooms, flies, and methane-generating bacteria were investigated by Cheng Rong jun (1986) in Feng Xuang County and Xin Ling County of Guang Dong Province (Figures 12, 13, and 14). Energy flow, material flow, and monetary flow analyses of these models show some advantages:

■ *More energy and material can be channeled to economic products.* In the crop/livestock/mushroom/earthworm model (Figure 14), for example, 3.93 percent energy, 14.11 percent nitrogen, 14.99 percent phosphorus, and 1.95 percent potassium in raw material is turned to mushrooms; 6.6 percent energy, 24.1 percent nitrogen, 9.67 percent phosphorus, and 13.4 percent potassium is channeled to earthworms. In the crop livestock/biogas tank/mushroom model (Figure 13), 46.06 percent of the energy from the input by-product is changed into biogas output.

■ *The self-purification ability of agriculture is raised.* The fermentation process in a biogas digester can kill most of the parasites in animal wastes. Odor can be eliminated. The decomposted by-products from a biogas digester, earthworm pits, mushroom houses, and fly cages are ready to be used for fish pond or crop fields. The environmental quality can be improved.

■ *The cycling rate of material also is high.* A return index, which is the material output to the field, divided by the material input from livestock

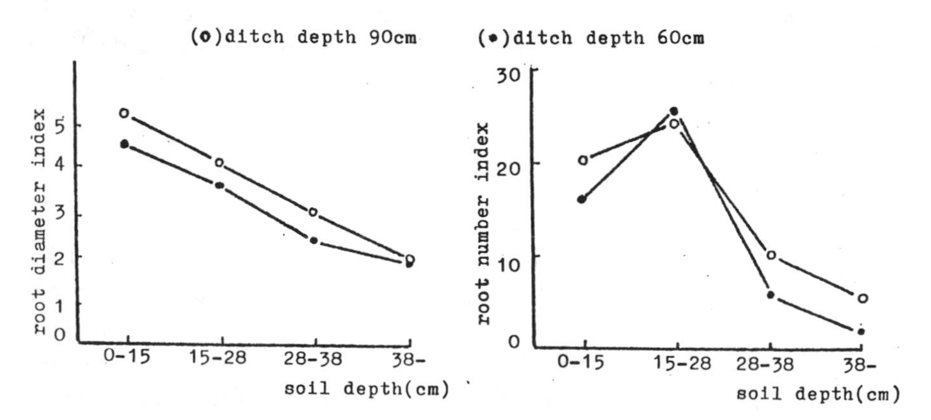


**Figure 10. The influence of ditch depth on the development of root system of banana in deep ditch-high bed systems, Guang Dong, China (Lin, 1988).**

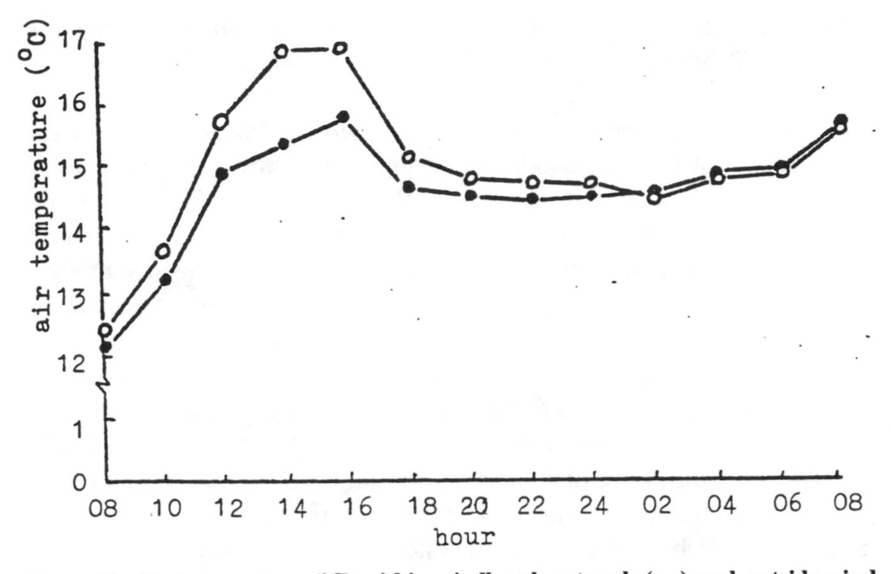


**Figure 11. Air temperature (°C) within windbreak network (o--) and outside windbreak network (•--) in March 15-16, 1980, Da Sha Farm, Dou Men, Guang Dong, China (JRGSH, 1985).**

and crops, times 100 percent, is used to show the efficiency in material cycling in the system.

In a crop/livestock/biogas tank/mushroom model, for example, the return indexes of nitrogen, phosphorous and potasium were 85 percent, 75 percent, and 44 percent, respectively. In a crop/livestock/mushroom/earthworm model, they were 39 percent, 72 percent, and 28 percent, respectively.

■ *Subsidized commercial energy input is small.* No subsidized commercial energy is required in earthworm, fly pupae, and biogas production. In mushroom production of a crop/livestock/biogas tank/mushroom model, only 18.35 percent of the total energy input is from commercial energy, while in a crop/livestock/mushroom/earthworm model, 12.84 percent is from commercial energy.

■ *Labor return is high.* Average income per labor day in 1984 in Feng Xun and Xin Lin Counties ranged from 1.50 to 2.00 Yuan. The labor return on production of the five detritus food chain organisms in the same period was much higher than average, ranging from 4.72 Yuan in fly raising to 5.85 Yuan in biogas production (Chen, 1976).

The biogas and mushroom production developed rather quickly. According to statistics in 1982, there were nearly 5 billion family-size biogas digesters and 38,000 large-size biogas digesters operating in China. These produced 1.028 billion cubic meters of biogas each year. The annual increase rate was 8.8 percent (Editorial Board of Agricultural Yearbook of China, 1980-1987).

***Relationship Between System Structure and Environmental Gradient.*** Investigation of typical counties in Guang Dong Province found that the main influences of the natural and social environment on agriculture can be divided into four groups (Feng, 1985):

- *Topographic gradient.* Statistics show that there are positive relationships between the percentage of area covered by forest and percentage of mountainous and hilly areas in a county, between the percentage of herbivores in livestock and the percentage of grassland areas in the county, between percentages of ducks and geese in poultry and percentages of water surface area in the county, and between percentages of fishery production in total agricultural production and percentages of water surface area in the county. Multilayer perennial plant communities, paddy rice-based rotation systems, fish pond-raised field systems, deep ditch-high bed systems, and shelterbelt networks are suitable in different positions in the watershed, respectively.
- *Climatic gradient.* The main tropical crops, such as rubber, coffee, and oil palm, are grown in counties with an average yearly temperature greater than 22 °C and an average temperature in January greater than 15 °C. Water buffalo are used more in warmer areas with paddy fields, while cattle are raised more in cooler areas with upland fields. The regression equation:

$$y = 103.67 - 1.0523x_1 - 0.65x_2$$

was significant at the 5 percent level, where y represents the ratio of cattle/(cattle + water buffalo) in the county, $x_1$ is the percentage of paddy fields in arable land, and $x_2$ is the average temperature in January (°C).

Acajou and mango are grown more in semidry areas. Lichi grows well only in the central part of Guang Dong Province, where no serious frost exists but winter temperatures are low enough to stimulate the differentiation of flower buds.

- *Population gradients.* In Guang Dong Province, the greater the population density, the greater is the arable land used for stable food production and more crops are planted in a year. With high human population densities, sweet potatoes and wheat become important crops in the county. Wheat is especially adapted to the winter period in South China. Sweet potato can be used on poorer upland fields without irrigation to produce food all year around. Furthermore, viticula of sweet potato, which is about 50 percent of the edible root weight, is good feed for swine. The results

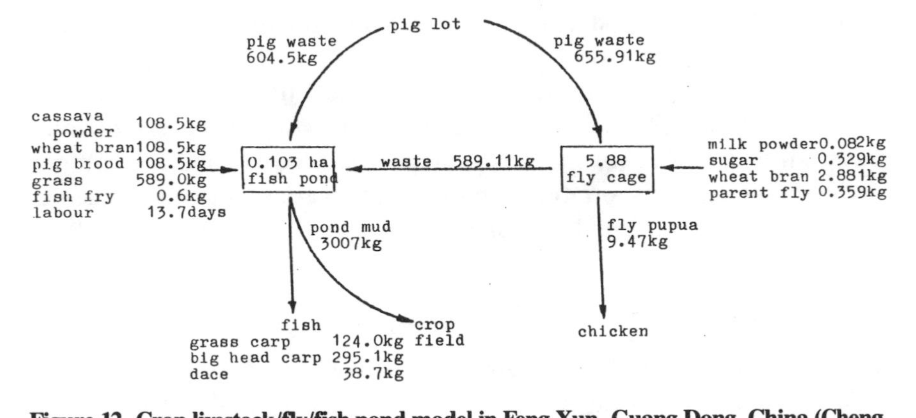


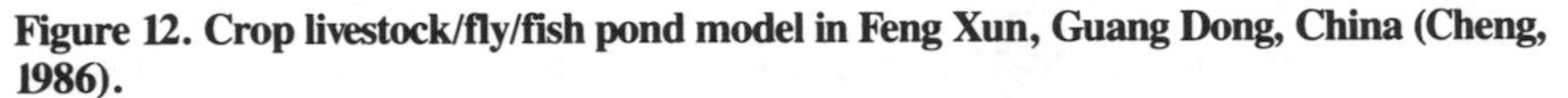
**Figure 12. Crop livestock/fly/fish pond model in Feng Xun, Guang Dong, China (Cheng, 1986).**

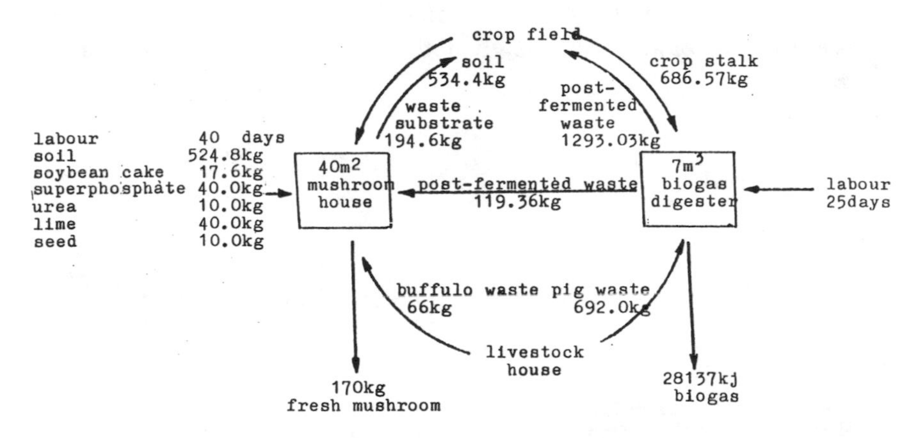

**Figure 13. Crop livestock/biogas digester/mushroom model in Feng Xun, Guang Dong, China (Cheng, 1986).**

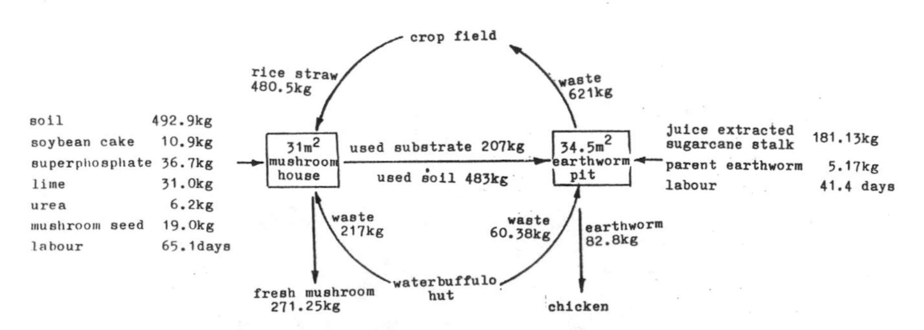

**Figure 14. Crop livestock/mushroom/earthworm model in Xing Ling, Guang Dong, China (Cheng, 1986).**

show that the basic food energy requirement, rather than net income, received a higher priority among farmers in the early 1980s.

■ *Economic gradient.* In developing countries, transportation exerts a major limiting factor in rural economic development. To express the integrated influence of major markets of industrial products and agricultural products and the transportation between these markets and the county in question, relative strength index of city influence (RSI) is used as follows:

$$RSI_i = \sum_{j=1}^{n} \frac{P_j}{D_{ij}}$$

where $RSI_i$ is the RSI of the ith county, $P_j$ is the population of the jth city ($10^4$ person), and $D_{ij}$ is the distance from the ith county to the jth city (km).

The statistics of typical counties of Guang Dong Province show that the larger the county RSI is the greater are the external inputs, such as fertilizer, pesticide, and electricity in agriculture; the higher are the yields of rice, sugarcane, and fish; the greater are the number of pigs, poultry,

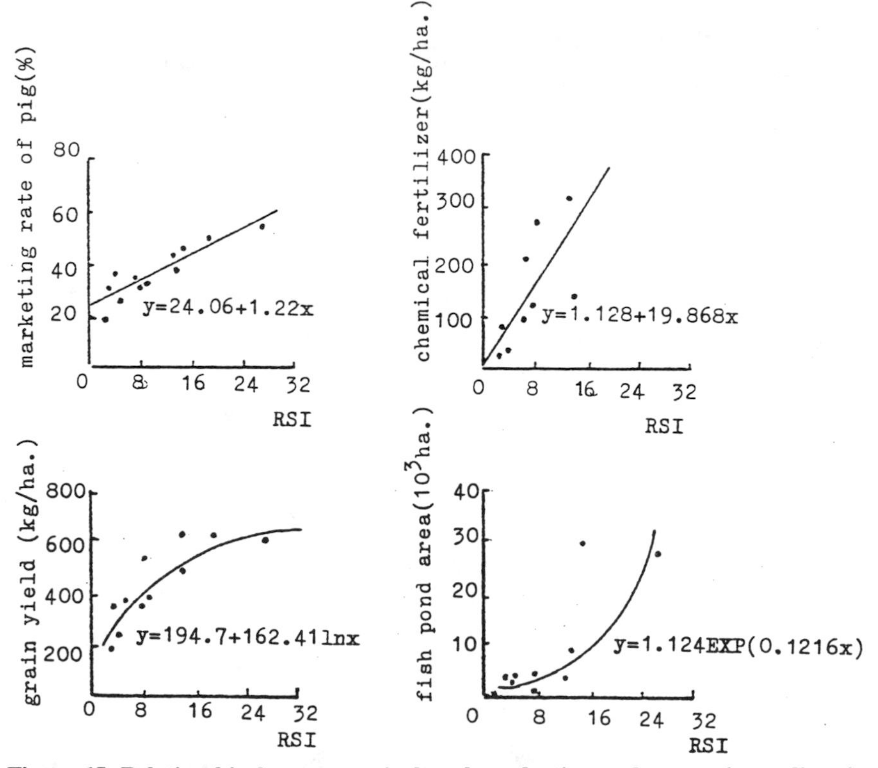


**Figure 15. Relationship between agricultural production and economic gradient in typical counties of Guang Dong, China (Luo, 1986).**

and fish produced and provided to the market (Figure 15).

A well-designed ecoagricultural system must be compatible with its natural and social environment. Hence, the structure and function of an optimum ecoagricultural system must be changed from place to place and from time

## REFERENCES

Agricultural Regionalization Committee of China. 1987. Highlight of agricultural resources and regionalization in China. Survey and Drawing Press and Industrial and Commerce Press, Beijing.

Balandreau, J. P., et al. 1976. Asymbiotic N2 fixation in paddy soils. *In* William E. Newoon [editor] Proceedings, First International Symposium on Nitrogen Fixation (volume 2). Washington State University Press, Pullman.

Berkins, D. H. 1969. Agricultural development in China 1368-1968. Edinburg University Press, Edinburg, England.

Boeringa, R., editor. 1980. Alternative methods of agriculture. Elsevier Scientific Publishing Co., Amsterdam, The Netherlands.

Cheng, Defu. 1988. The long history of ecological agriculture in China. Ecoeconomics 2: 45-50.

Cheng, Rong jun. 1986. Study on energy and material flow in detritus food chain. M.S. thesis. South China Agricultural University, Guang Zhou, China.

Cheng, Wen hua. 1978. Concise illustrations of agricultural science and technology in ancient Chinese history. Agricultural Press, Beijing.

Dang, Defu. 1988. The long history of ecological agriculture in China. Ecoeconomics 2: 45-50.

Deng, Honghai. 1986. New revolution of agriculture—from industrialization to biologicalization. Guangxi People's Press, Nan Ning, China.

Editorial Board of Agricultural Yearbook of China. 1980-1987. Agricultural Yearbook of China. Agricultural Press. Beijing.

Fan Ke. 1985. History of dialectical thinking in China (Early Qin Dynasty). People's Press, Beijing.

Fang, Shengzhi, et al., Shi Hansheng, editor. 1979. Selected agricultural literature of Xi Han Dynasty. Agricultural Press, Beijing.

Feng, Yaozhong. 1985. Research and extension of rubber and tea community. Yunnan Botany Research Institute, Chinese Academic of Science and State Farm Bureau of Hainan, Haikou, China.

Food and Agricultural Organization. 1985. FAO fertilizer yearbook 1984 (volume 134). Rome, Italy.

Han, Chunru. 1985. Energy structure and efficiency of agricultural ecosystems. Rural Ecology and Environment 3: 6-8.

Han, Chunru. 1987. Preliminary discussion on development of ecological agriculture in China. China Environmental Science 7(4): 1-8.

Hu, Zhemin. 1985. Research on agroecosystem arrangement of rice growing area in Gao Yao County. pp. 262-371. *In* Research reports on highly effective subtropical agroecosystems. Agroecology Research Centre of South China Agricultural and University, Guang Zhou, China.

Human Population Research Center of the Chinese Academy of Science. 1985. Yearbook of Chinese Population (1985). The Chinese Academy of Social Science Press, Beijing.

Joint Research Group of Shelterbelt Network in Crop Field of Guang Dong, 1985. Reports on shelterbelt network in crop field of Pearl River Delta. Scientific and Technology Division of Forestry Department, Guang Dong Province, Guang Zhou, China.

King, F. H. 1911. Farmers of forty centuries. Rodale Press, Emmaus, Pennsylvania.

Koepf, H. H., et al. 1976. Bio-dynamic agriculture: An introduction. The Anthropsophic Press, Spring Valley, New York.
Lin, Ri jian. 1988. Research on the structure and function of high bed-low ditch ecosystem in Pearl River Delta. M.S. thesis. South China Agricultural University, Guang Zhou, China.
Liu, Yong ding. 1984. Research on composition and vertical distribution of blue alga community in paddy soil. Aquatic Product Research Institute of Wu Han, Wu Han, China.
Luo, Shi ming. 1986. Relationship between structure of county agroecosystems and relative strength index of city influence. Journal of South China Agricultural University 2: 22-32.
Luo, Shi ming, Chen Yu Wa, and Yan Fu. 1987. Agricultural ecology. Hu nan Scientific and Technical Press, Chang Sha, China.
Masonabu Fukuoka. 1978. The one-straw revolution: An introduction to natural farming. Rodale Press, Emmaus, Pennsylvania.
Oelhat, Robert C. 1978. Organic agriculture. Aelunheld, Osmon & Company, USA.
Qai, Huimin et al. 1982. Preliminary study on the characteristics of soil organic matter in paddy soil of Pearl River Delta. pp. 25-30. *In* Research reports on soil and fertilizer science and technology (1979-1981). Soil and Fertilizer Research Laboratory of Guang Dong Agricultural Research Institute, Guang Zhou, China.
Research Station of Ding hu Shan Forest. 1984. Research reports on tropical and subtropical forest ecosystem research (second volume). Scientific Popularization Press, Guang Zhou, China.
Shi, Shan. 1983. Ecological problem and new phase of agricultural development. Red Flag 17: 20-24.
Shi, Shan. 1986. Ecological agriculture and agricultural engineering. Research on Agricultural Modernization 1: 1-16.
Soil and Fertilizer Research Laboratory of Guang Dong Agricultural Research Institute. 1984. Research on rotation systems of paddy field. SFRL of GARI, Guang Zhou, China.
Worthington, M. K. 1981. Ecological agriculture: What it is and how it works. Agriculture and Environment 6: 349-381.
Wu, Ziang jin. 1983. Agricultural production and energy resources. South China Agricultural College Press, Guang Zhou, China.
Xia, Wei yin. 1979. Interpretation of "Lu Shi Chun Qiu: Shan nong" and other three ancient agricultural literature. Agricultural Press, Beijing.
Yang, Tin xiu. 1983. On "flying bird" strategy of rural economic development. Research on Agricultural Modernization 10: 1-16.
Ye, Xan ji. 1985. Ecological agriculture: Another Green Revolution in China. Science and Technology of South West China Agricultural University 4: 3-15.
Yoshida, T. et al. 1973. Nitrogen fixing activity in upland and flooded rice fields. Soil Science Society of America Proceedings 37: 42-46.
Yu, Rong lian. 1985. Establishment of ecological agriculture is a necessary step for agricultural modernization. Agricultural Archaeology 1: 9-19.
Yun, Zheng ming. 1985. Application of engineering principles of agroecological structure. Editorial board of Agricultural Modernization Research, The Chinese Academy of Science, Beijing.
Zhang, Xi xi. 1987. Symposium of ecological agriculture held in Fu Yan City, An Hui Province. Research on Agricultural Modernization 4: 64.
Zong, Gong pu et al. 1987. Research on fish pond and field system of Pearl River Delta. Science Press, Beijing.

# SUSTAINABLE AGRICULTURAL SYSTEMS IN TROPICAL AFRICA

Bede N. Okigbo

A sustainable agricultural production system is a dynamically stable and continuous production system that achieves a level of productivity satisfying prevailing needs and is adapted continuously to meet future pressing demands for increasing the carrying capacity of the resource base. Sustainability can be achieved only when resources, inputs, and technologies involved are within the capabilities of the farmer to own, hire, maintain, and orchestrate with increasing efficiency to achieve desired levels of productivity in perpetuity, without adverse effects on the resource base and environmental quality.

Agricultural production is a bioeconomic activity with complex implications related to the physicochemical, biological, technological, managerial, and socioeconomic elements. These elements interact in such a way as to determine the location-specific adaptations of various farming systems to different locations, thereby satisfying objectives with minimum adverse consequences. In any situation, a unique set of these elements interacts at a particular time to ensure sustainability. Sustainability can be achieved only with knowledge and understanding of:

- Physicochemical factors, such as soils, climate, moisture, radiation, daylength, etc., and the way they change and interact.
- Biological elements of the production system in terms of crops and/or animals in relation to pathogens, weeds, pests, and beneficial organisms.
- Changing and appropriate technologies available to the farmer.
- Cultural, educational, experiential, and overall decision-making background of the farmer.
- Economic and ecological soundness, cost-effectiveness, and otherwise

appropriateness of various operations that are necessary to satisfy objectives under prevailing infrastructure, institutional, and policy environment(s).

In short, a sustainable agricultural production system is one in which a farmer is equipped continuously with the capabilities for managing the factors of production (land, labor, and capital) with increasing efficiency to satisfy the ever-changing needs and circumstances while maintaining or enhancing the condition of the natural resource base.

What is required to ensure sustainability in Africa must be different, even though certain objectives of ensuring sustainability may be the same worldwide. Science and technology can carry us only so far toward sustainability when there are no adequate policies and measures to curb population growth and satisfy multiuse demands and human activities that interact with each other in causing environmental degradation, loss of biological diversity, and decline of overall quality of life and general well-being.

A major problem in sustainable agricultural production in tropical Africa is that Africa is a continent in crisis (Food and Agriculture Organization, 1986c). This crisis stems from the inability to produce enough food; per-capita food production is lower than two decades ago. The following developments have adversely affected sustainability:

■ Rapid population growth averaging 3.1 percent annually, the highest in the world, vis-a-vis 1.2 percent in food production.

■ Seriously degrading environmental conditions caused by increasing intensity of cultivation, overgrazing, uncontrolled burning of vegetation, deforestation, urbanization, industrialization, and other pressures of modernization.

■ Increasing reliance on imports of food and inputs for attaining food security under very adverse economic conditions of escalating debt burdens, resulting from petroleum and food import bills, loans for development programs, and rapid decline in commodity prices and foreign exchange earnings.

■ Endemic political instability causing inconsistencies and deficiencies in strategies and development programs in addition to an inappropriate policy environment for increased productivity in agriculture.

All these developments have rendered the once ecologically sound and economically viable traditional agricultural production systems increasingly outmoded and unable to satisfy demands for food and other products.

In the present century, the tractor is for agriculture what the steam engine was for the industrial revolution. Mechanical energy, allied with discoveries of chemical and biological research, is initiating an agricultural revolution that calls for profound changes in structure in those countries that want to keep up with the modern world. Attempts at horizontal transfer to Africa of tractorization and various elements of the agricultural revolution of Europe and North America have met with more failure than success. Even before these new technologies could bring the agricultural revolution into Africa, a major concern had developed, starting in the 1950s, about the adverse effects of the mechanical and chemical inputs used to boost agricultural production. Consequently, developed countries now face the problem of how to achieve sustainable agricultural production systems through modifying "modern" agricultural production by scaling down the amount and cost of energy and other inputs. Africa faces the problem of how to develop more productive and sustainable agricultural production systems by carefully upgrading the technologies and the amount and cost of inputs used in the outmoded traditional production systems.

## The African Environment

Tropical Africa extends from latitude 20° North to 26° South. Its widest east to west extension from Rus Hafun (50° 50′ East) and Cape Verde (17° 32′ West), is about 78,200 kilometers, and its north to south extension is about 5,200 kilometers (Pritchard, 1979). Thus, the area of tropical Africa is about 22 million square kilometers, or 72.6 percent of the total area of the continent. It is 10 percent larger than the area of North America, and half as large as Asia. The land mass consists of a series of plateaus with a narrow coastal plain of varying width. More than 40 percent of the area is lowland below 600 meters in altitude. There are isolated highlands over 990 meters in altitude, including Africa's highest mountain, Mount Kilimanjaro (5,895 meters) in Tanzania; Mount Kenya (5,199 meters) in Kenya; Mount Ras Dashan (4,620 meters) in Ethiopia; and Mount Camaroon (4,070 meters) in Camaroon. A line running northwest from Luanda on the coast of Angola to Masawa on the Red Sea coast of Ethiopia divides Africa into a more or less lowland northwestern zone and a highland area to the east. A rather unique depression—7,200 kilometers in length, 1 to 2,000 meters deep and 30 to 100 kilometers in width—forms a rift valley containing some of the largest lakes in the world. The 5,600-kilometer part of the Rift valley within Africa begins near Biera in Mozambique and runs northward through the valley of the Awash in Ethiopia into the Gulfs of Suez and Aqaba. Six major river basins dissect the plateau, with most of the rivers flowing into the Atlantic Ocean, Indian Ocean, and Mediterranean Sea.

Much of Africa lies within the tropics; with the exception of the highland areas, mean annual temperatures even during the coldest months exceed 18 °C. In general, mean annual temperatures range from 20 °C to 30 °C. They usually are above 30 °C on the desert margins of the Sahara and

Kalahara. Annual temperature ranges usually are lower than the diurnal ones. For example, at Samaru on latitude 11° 11′ North, in Nigeria, the diurnal range is 14° to 17°C. Mean temperatures and variations decrease with altitude. Solar radiation varies from about 330 calories per square centimeter per day in the rain forest areas near the coast to over 580 calories per square centimeter per day in savanna areas near the desert margins. Rainfall is determined by the relative positions of the Intertropical Convergence Zone (ITCZ) where winds moving northward or southward meet in relation to prevailing temperature, pressure, and wind changes. Annual rainfall ranges from up to 4,000 millimeters in areas close to the equator in Zaire and near the west coast, to below 100 millimeters in semidesert and desert areas.

The vegetation in different parts of Africa reflects the prevailing rainfall regimes. Areas with over 1,500 millimeters annual rainfall and over 270 days length of growing period (LGP) have a humid climate. Here the vegetation is tropical rainforest, consisting mainly of evergreen trees and shrubs in western and central Africa. People have greatly modified much of the vegetation and reduced it to secondary bush, except in parts of central Africa, especially in Zaire, Gabon, and Congo Brazaville, and western Africa in parts of Liberia and Ivory Coast. Areas with 1,200 to 1,500 millimeters annual rainfall and 180 to 270 days LGP are mainly forest/savanna mosaic or derived savanna woodland, consisting mainly of evergreen trees and shrubs with a few deciduous species. Areas having 600 to 1,200 millimeters annual rainfall and 120 to 170 days LGP have woodland vegetation of the *Parkia-Butyrospermum-Khaya* type, with trees, shrubs, and tall andropogonaceous grasses in the Guinea savanna areas of West Africa. In similar climatic zones of central, southern, and parts of eastern Africa, the dominant vegetation types consist of *Brachysteqia* and *Julbernadia* woodland of the "miombo" ecosystem of Acacia Central and southern Africa and the *Combretusm Aeaera* woodland of East Africa with perennial shorter grasses. Semiarid areas with annual rainfall between 400 and 600 millimeters and 75 to 119 days LGP have combretaceous trees and shrub grassland designated as the Sudan Savanna in West Africa and *Acacia-Commiphora* woodland in parts of East Africa. Arid areas with 100 to 400 millimeters annual rainfall and 1 to 76 days LGP have sparse scrub and some perennial grasses, with annual grasses dominant in moist areas; this consists of the Sahel Savanna of West Africa and part of the *Acacia-Commiphora* woodland of East Africa. Areas of less than 100 millimeters annual rainfall and less than one day LGP are desert with occasional dwarf desert thorny scrub and few perennial grasses. People have greatly modified the climax vegetation in all these climatic zones through cultivation, burning of vegetation, and overgrazing. Net annual primary productivity corresponding to the vegetation zones ranges from about 500 grams per square meter in desert margins with less than 100 millimeters annual rainfall to 3,000 to 4,000 grams per square meter in humid areas.

Much of Africa is made up of ancient crystalline rocks of Precambrian age, covered in places by sedimentary rocks and others of volcanic origin. These have been transformed into highly weathered soils of low inherent fertility, except those formed in valley bottoms on hydromorphic soils and on younger volcanic rocks. The major soil groups encountered in Africa consist of Aridisols (34.5 percent), Alfisols (22.4 percent), Oxisols (22.4 percent), Entisols (12.1 percent), Ultisols (4.1 percent), and Vertisols (1.6 percent).

These soils, under the prevailing climatic conditions with absence of frost except on very high mountains and beyond the tropics of Cancer and Capricorn, are capable of producing a wide range of tropical and subtropical crops in addition to farm animals. The humid tropical zone produces perennial tree crops, such as rubber, cocoa, oil palms, and food crops, including cassava, yams, maize, cocoyams, plantains, and bananas. The Guinea or moist savanna zone produces maize, cassava, sorghum, soybeans, sugarcane, some bananas, and livestock. The subhumid Sudan savanna zone produces sorghum, maize, groundnuts, cotton, millet, cowpeas, tobacco, cashew, and sheabutter, in addition to livestock, including cattle, goats, and sheep. The semiarid Sahelian zone produces mainly sorghum, millet, groundnuts, cowpeas, and gum arabic. In this zone, the many nomadic pastoralists, such as the Fulani in West Africa and the Masai in East Africa, raise large numbers of cattle, sheep, and goats. The arid zone produces some sorghum and millet in the wetter depressions, but the most important agricultural products are goats, sheep, and camels. Highland areas in eastern and southern Africa produce Arabica coffee, pyrethrum, beans, maize, sweet potatoes, Irish potatoes, wheat, and subtropical fruits and vegetables.

Tropical Africa has 42 countries and a population of about 400 million. In most countries, over 70 percent of the population is engaged in agriculture.

## Constraints to Increased Agricultural Production in Tropical Africa

Several physical, chemical, biological, and socioeconomic constraints affect sustainability of agricultural production in the various ecological zones of Africa. Although climate conditions are suitable for growing a wide range of crops throughout the year, some climatic constraints include (1) unreliability of rainfall in onset, duration, and intensity; (2) unpredictable periods of droughts, floods, and environmental stresses; (3) high soil temperatures for some crops, with adverse effect on growth, development, and some biological processes; and (4) cloudiness and reduced photosynthetic efficiency, especially in the humid tropical areas. Over 54 percent of the land area of Africa is deficient in rainfall, and 46 percent has only 0 to 74 days LGP (Table 1). In more than 50 percent of the land area in tropical Africa, rainfall reliability (expressed as the percentage departure from normal) ranges from 20 percent to over 40 percent (Food and Agriculture Organization, 1986a,b).

Soil-related constraints include (1) a high degree of weathering, sandiness, deficiency in clays, and high fragility and erodibility; (2) low values of CED and rapid rates of organic matter decomposition; (3) high levels of soil acidity and high tendencies for P fixation; (4) susceptibility to multiple nutrient deficiencies and toxicities under increasing intensities of cultivation; (5) proneness to leaching with high risk of erosion under prevailing rainstorms, especially at the beginning and end of the rainy season; (6) low inherent fertility; and (7) serious salinity problems under irrigation, especially with poor water management. Some 32.2 percent of the land resources of Africa exhibit specific management problems, with 30.5 percent and 8.6 percent having sandy texture and steep slopes, respectively (Table 2). Areas of land exhibiting various chemical constraints amount to 45.5 percent low in nutrient retention, 22.3 percent with aluminum toxicity, 13.5 percent with phosphorus fixation hazard, 22.4 percent low in potassium supply, 6.5 percent having excess calcium carbonate, 2.7 percent with excess soluble salts, and 1.1 percent with excess sodium (Food and Agriculture Organization, 1986b). Table 3 quantifies these soil prob-

**Table 1. Extent of climate zones length of growing period and annual rainfall in Africa (Food and Agriculture Organization, 1986a,b).**

| *Climate Zone* | *Area (10 bha) (%)** | *Length of Growing Period in Days* | *Rainfall (mm)* |
|---|---|---|---|
| Desert | 822.0 (29.1) | 0 | <100 |
| Arid | 488.0 (17.1) | 1-74 | 100-400 |
| Semiarid | 233.0 (8.1) | 75-119 | 400-600 |
| Dry subhumid | 314.0 (11.0) | 120-179 | 600-1,200 |
| Moist subhumid | 584.0 (20.4) | 180-269 | 1,200-1,500 |
| Humid | 409.0 | 7,270 | >1,500 |

*Figures in parentheses are percentages.

**Table 2. Areas of land and percentage of area exhibiting different kinds of physical constraints (Food and Agriculture Organization, 1986a).**

| *Region* | *Total Land Area (millions of ha)* | *Steep Slopes* | *Sandy Texture* | *Specific Management Problems* |
|---|---|---|---|---|
| | | | (%)* | |
| Sudano-Sahelian Africa | 828.2 | 50.9 (6.0) | 295.1 (34.9) | 370.2 (43.7) |
| Humid and sub-humid West Africa | 206.6 | 15.3 (7.3) | 57.5 (27.4) | 63.0 (30.0) |
| Humid central Africa | 398.8 | 11.9 (2.9) | 118.1 (29.6) | 35.7 (8.8) |
| Sub-humid and mountain East Africa | 251.0 | 54.6 (20.7) | 19.1 (7.2) | 52.2 (19.8) |
| Sub-humid and semiarid southern Africa | 559.2 | 49.1 (8.7) | 271.4 (48.2) | 51.1 (9.1) |

*Figures in parentheses are percentages.

**Table 3. Chemical constraints to agricultural production in Africa (Food and Agriculture Organization, 1986b).**

| *Chemical Constraint* | *Land Area Affected (millions of ha)* | *Percent of Total Land Area* |
|---|---|---|
| Low nutrient retention | 1,295.5 | 45.6 |
| Aluminum toxicity hazard | 635.0 | 22.3 |
| Phosphorous fixation | 382.5 | 13.5 |
| Low potassium supply | 637.0 | 22.4 |
| Soluble salts | 75.6 | 2.7 |
| Excess of sodium | 31.0 | 1.1 |
| Excess of calcium carbonate | 184.6 | 6.5 |
| Sulphate acidity | 3.8 | 0.1 |

lems in various regions in Africa. Desertification is rampant in Sahelian areas of West Africa and in semiarid areas. Many areas of the continent already are subject to various kinds of erosion, soil degradation, and crusting.

Biological constraints include unimproved crops and farm animals with low yields and low overall potential, susceptibility to diseases and pests, and high incidence of diseases, pests, and weeds. Parasitic diseases affect people and livestock, making it impossible for animals to be produced or used for work in certain areas or for human beings to live and practice productive agriculture in certain areas of the continent. Human activities have brought about drastic environmental changes that have adverse effects on ecological equilibrium and environmental quality. Rapid loss in genetic diversity is rife as a result of deforestation and overgrazing.

Socioeconomic constraints include (1) unfavorable land tenure systems, often resulting in fragmentation of holdings; (2) shortage of labor at peak periods of demand during planting, weeding, and harvesting; (3) low income and lack of credit; (4) poor marketing facilities and pricing structure; (5) high cost and unavailability of inputs; (6) poor rural infrastructure; (7) poor extension services; (3) high rates of illiteracy among farmers, which hampers adoption of technology; (9) lack of appropriate technologies for increasing production on a sustained basis and absence of a package approach to technology development, testing, and use; (10) extreme country or within-country regional specialization in agricultural production and reliance on only one agricultural commodity or mineral for foreign exchange; and (11) inappropriate agricultural development policies and various political constraints.

## African Agricultural Production Systems and Their Sustainability

African agriculture consists of a "mosaic of crops, traditions, and techniques that does not reveal a center, nuclear area or a single point of origin" (Harlan et al., 1976). The indigenous African crops and production systems that evolved in different parts of the continent were modified and reinforced with Asian crops during the first millennium AD and American crops after the discovery of America in 1492. Many production systems and their component technologies have been grafted into the traditional farming systems. Hence, some elements of traditional and transitional African agriculture pari—passu with some "modern" production systems and local adaptations—exist. In general, African agricultural production systems exhibit the following characteristics:

■ Objective of farming is mainly for subsistence, but increasingly commercial in whole or in part to satisfy farmers' needs for money for goods and services.

■ Farm size is usually small, with over 80 percent being 5 hectares or less.

■ Slash-and-burn clearance systems are widespread.

■ Labor is mainly manual, with most operations accomplished with simple tools, such as hoes and machetes. Use of livestock for work or draft is limited, due to prevalence of tryanosomiasis, which limits rearing of cattle in large areas of Africa. Mechanization in farming is very much limited to plowing or primary processing.

■ There is a marked division of labor between men and women with respect to operations performed and commodities produced.

■ Soil fertility maintenance is dependent on nutrient cycling and biological processes, such as nitrogen fixation and nutrient accumulation in biomass during the fallow period, for restoration of soil fertility. Manures and household refuse are used to maintain soil fertility, especially in home gardens. About 63 percent of the little fertilizer used is applied to nonfood cash crops.

■ Use of pesticides and chemicals inputs, such as growth regulators and fertilizers, is very limited or absent since poor farmers cannot afford them. Manual and cultural control of pests, diseases, and weeds is widespread.

■ Cropping systems usually are more complex than those in the agriculture of developed countries in terms of enterprise mix and range of commodities produced on each farm. Mixed cropping, or intercropping, is common among low-resource farmers as a strategy for reducing risk of crop failure and controlling or minimizing weed, pest, and disease damage. The number of commodities produced is highest in the humid tropics and lowest in the semiarid areas. In most traditional farming systems, crop production is associated with animal production.

■ Animal production systems involve mainly small livestock in the humid and subhumid areas, where they serve as sources of manure, meat, and cash in times of emergency. Large animals are kept in savanna areas by nomadic herdsmen, such as the Fulani and the Masai, who may also keep small livestock; nomads sometimes live in symbiotic relationship with cultivators. Among nomadic herders, animals are kept as much for food as for status and prestige.

■ Commodities produced, farming systems, and component technologies used in production usually are tailored to the prevailing rainfall regime. While limited irrigation is used in traditional African agriculture, there are traditional hydraulic systems of water management in dry areas.

■ The production systems under the management of one farmer or farm family range from extensive production systems, such as shifting cultivation in several fields at different distances from the homestead, to the permanent and continuous production systems in home gardens (Figure 1).

■ Farming is often associated with hunting, fishing, and gathering in varying degrees.

■ Traditional farming systems take advantage of microecological conditions. Various components of the field system ensure that commodities produced are located where they have obvious ecological advantages or are otherwise specially adapted.

■ Yields are usually low due to widespread use of unimproved crop varieties or breeds of animals and/or limited use of pesticides and other inputs. Production per unit of energy is usually higher than in "modern" agricultural production systems.

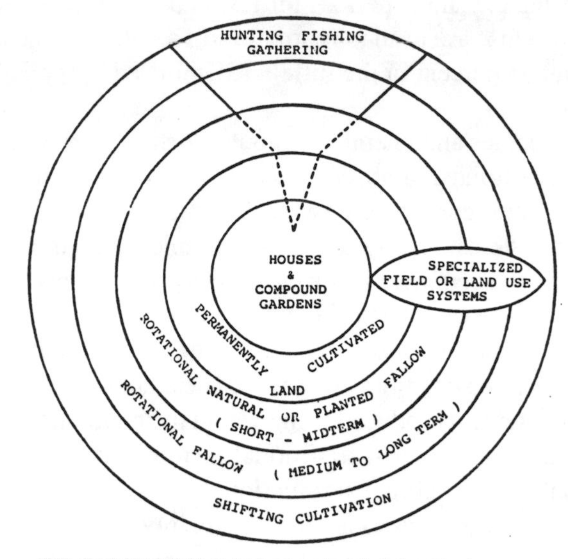


NOTE: Specialised field on land use systems include valley bottoms, terraces, termite mounds etc.

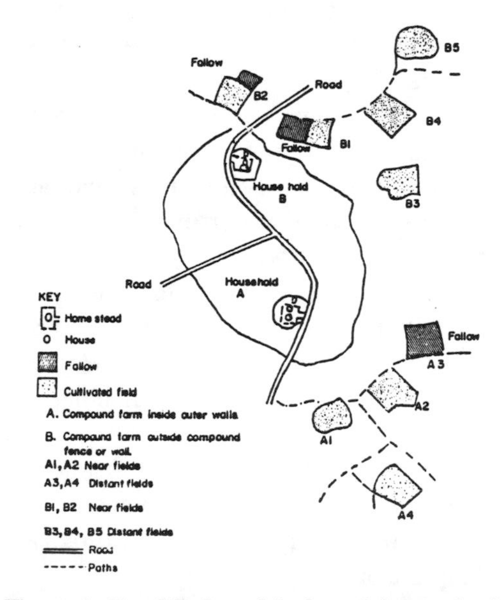


**Figure 1. Simplified model of spatial organization of fields and farming systems in tropical Africa. (top). Schematic diagram of compound farms in relation to associated field systems in traditional farming systems of the humid tropics of West Africa (bottom).**

■ Traditional farming systems, while ecologically sound and adapted to prevailing conditions and needs of the farmer when population density was low, are becoming increasingly outmoded and unable to meet demands of rapid population growth, high rates of urbanization, increased mobility, and rising incomes.

***Extent of Sustainability of African Farming Systems.*** There is currently no generally accepted topology of the farming systems of tropical Africa. The one used in this discussion is taken from Okigbo and Greenland (1976), based on a detailed review of classifications of farming systems of the tropics in Africa and elsewhere (Table 4). This classification takes into account the fact that African farming systems are not static. They are changing constantly. Developments such as the introduction of Asian and American crops, European colonization and commercialization of agriculture, communication improvements within and outside Africa, mechanization and other pressures of modernization have wrought great changes on the cultural landscape of the continent.

**Table 4. Classification of farming systems in tropical Africa.**

| Farming system | |
|---|---|
| Traditional and transitional systems | |
| 1. Nomadic herding: shifting cultivation (Phase 1) | L > 10[2] |
| 2. Bush fallowing or land rotation: shifting cultivation (Phase II) | L = 5-10 |
| 3. Rudimentary sedentary agriculture: shifting cultivation (Phase III) | L = 2-4 |
| 4. Compound or homestead farming and intensive subsistence agriculture: shifting cultivation (Phase IV) | L< 2 |
| 5. Terrace farming and floodland agriculture | |
| 6. Mediterranean agriculture (traditional) | |
| Modern farming systems and their local adaptations | |
| 1. Mixed farming | |
| 2. Livestock ranching | |
| 3. Intensive livestock production (poultry, pigs, dairying) | |
| 4. Large-scale farms and plantations: | |
| a. Large-scale food and arable crop farms based on natural rainfall | |
| b. Irrigation projects involving crop production | |
| c. Large-scale tree crop plantations | |
| 5. Specialized horticulture: | |
| a. Market gardening | |
| b. Truck gardening and fruit plantations | |
| c. Commercial fruit and vegetable production for processing | |

[2] L = C + F/C, where C = cropping period, F = fallow period, and L = land use factor.

***Nomadic Herding (Shifting Cultivation, Phase I).*** Shifting cultivation is ecologically sound and highly sustainable as long as periods of fallow are long enough for the land use factor (L) to exceed 10 and sufficient for soil fertility to be restored after each cropping phase. Sustainability of the system can be enhanced further by more controlled burning and use of improved cropping systems, improved varieties, and integrated pest management. Nomadic herding in savannas and drier areas is ecologically sound as long as the carrying capacity of the land is not exceeded. Sustainability breaks down when large numbers of animals are maintained above the carrying capacity, as is often the case when improved veterinary services reduce mortality of animals and when the movement of nomads is restricted. Sustainability can be enhanced by rotational grazing, controlled burning, improved range and water management, and integrated parasite/disease management.

***Bush Fallowing or Land Rotation (Shifting Cultivation, Phase II).*** This is not very sustainable when fallow periods are too short for rejuvenation of the soil before the subsequent cropping phase. It may be more sustainable at the upper ranges, especially on inherently fertile soils, where about five years is efficient for rejuvenation of the soil. Sustainability can be attained only with the intensification of production associated with better soil fertility management, of which fertilizers and organic residues are major components, as well as improved varieties and soil management.

***Rudimentary Sedentary Agriculture (Shifting Cultivation, Phase III).*** The situation here is the same as for bush fallowing, but the home-garden component is more sustainable.

***Compound or Homestead Farming and Intensive Subsistence Agriculture (Phase IV).*** This is a highly sustainable and reasonably productive agricultural production system in which fertility is maintained by optimum use of household refuse, animal manures, plant or crop residues, and nutrient cycling. Often it involves an efficient agroforestry system in which crop production is integrated in varying degrees with livestock production and should be more correctly termed an agrosilvo-pastural system. Permanent, intensive subsistence agricultural production systems occur in areas of high population density under various situations. The Kano close-settled area, Ukara Island in Tanzania, and the Igbo and Ibibio areas of southeastern Nigeria are good examples of this. Sustainability can be enhanced further by the use of improved crop varieties, more efficient cropping system design, and crop/soil management with judicious combinations of organic manure, biological processes, and fertilization where necessary. This system remains the most widespread traditional permanent system of cultivation in tropical Africa.

***Terrace Farming and Floodland Agriculture.*** Terrace farming systems are specialized, relatively intensive production systems in upland or hilly areas, which have been used since the time of the slave trade and intertribal warfare. They are as highly sustainable as home gardening and intensive subsistence production systems. Terraces are subject to occasional breakdown after heavy rainstorms; they must be maintained and continuously repaired. Moreover, because there is now no pressure for many communities to remain in such inaccessible locations, they are often abandoned, especially by the younger generation, which is unable to continue the system. In many locations where terrace farming communities descend to the plains or to more level land, farmers often fail to practice sustainable agricultural production because of their lack of experience in managing crops and soils in such situations. The loss of terrace farms often is followed by loss of sustainability in adjacent lowland areas, unless special orientation to the new situation takes place.

Floodland agriculture is practiced along the banks of major rivers and streams where rich alluvium is deposited annually during the rainy season flood. Some of the best crops of yams and rice are produced in this system. Special traditional hydraulic systems have been developed for growing different crops in the draw-down areas or banks of rivers, according to whether the flood is rising or receding, by use of deep-water rice and crops adapted to short periods of water availability. Specialized floodland agricultural production systems involve growing vegetables, fruits, rice, and other crops in the fadamas or valley bottom land and in savanna areas and using such structures as the "shaduf" for lifting water. These are highly sustainable production systems. Ironically, their sustainability can be as maintained only as long as water sediments and nutrients are brought down by runoff and erosion. Siltation in upstream areas above dams presents special problems that adversely affect sustainability.

***Modern Farming Systems and Their Local Adaptations.*** Modernization does not necessarily imply improvement. It mainly indicates the relative recentness of those systems in comparison with the traditional systems. The assessment that follows is summarized in table 5. The high degree of diversity that exists among ecological zones, countries, and cultural environments in Africa has to be taken into account in developing sustainable production systems.

*Mixed Farming*. This is an integrated crop and animal production system in which crops arc grown for subsistencc and/or salcs. Thc systcm is integrated, in that, the animals are reared for work, manures, draft purposes, meat, and other products. Crops are grown for forage or crop residues that are used as feed. Important by-products are farmyard manure and dung dropped in the field and used for maintainence of soil fertility. These systems are highly sustainable and can be enhanced by use of disease- and pest-tolerant crop species, improved pest and disease management programs, and improved crop and soil management, including good crop rotations

**Table 5. African agricultural systems and extent of sustainability.**

| System | L | Sustainability |
|---|---|---|
| Traditional and transitional systems | | |
| 1a. Shifting cultivation (Phase I) | L > 10 | HS |
| b. Nomadic herding | | SS |
| 2. Bush fallowing or land rotation | L = 5-10 | NS |
| 3. Rudimentary sedentary agriculture | L = 2-4 | NS |
| 4a. Compound farming (shifting cultivation, Phase IV) | L < 2-4 | S |
| b. Intensive subsistence agriculture | | SU |
| 5a. Terrace farming | | SU |
| b. Floodland agriculture | | HS |
| "Modern" farming systems and their local adaptations | | |
| 1. Livestock ranching | | SU |
| 2. Mixed farming | | S |
| 3. Intensive livestock production systems (poultry, pigs, and dairying) | | SU |
| 4a. Small-scale irrigated farms (lowland rice, vegetables, and arables) | | S |
| b. Small-scale fish farming | | S |
| 5. Large-scale farms and plantations | | |
| (a) Large-scale arable crop farms (unirrigated) | | NS |
| (b) Irrigated crop production projects/systems | | SU |
| (c) Tree crop plantations (oil palms, rubber) | | S |
| 6. Specialized horticulture | | |
| (a) Market gardening | | HS |
| (b) Truck gardening/fruit plantations | | SU |
| (c) Commercial fruit/vegetable production for processing | | SS |

L = (C+F)/C where C = cropping period, F = fallow period, L = land use factor
HS = Highly sustainable
S = Sustainable
NS = Not sustainable
SS = Sometimes sustainable
SU = Sustainable only under specificied circumstances

involving the use of legumes in the system.

*Livestock Ranching.* This is an extensive livestock production system that is fairly sustainable when the carrying capacity is not exceeded and the ranch is not located in humid areas. Sustainability can be enhanced by improved range management, rotational grazing, and care in provision of water and use of watering points.

*Intensive Livestock Production Systems.* These are sustainable only when operated as mixed farming systems with minimum use of purchased inputs. They are not very sustainable when they involve poultry and nonruminants that are raised on purchased inputs and feed, especially where the animals compete with people for grain that may not be grown on the farm. Poultry production is profitable, yet reliance on imported feed grains—especially maize—makes it a touch-and-go business in which the farmer often sells animals whenever feed is not available. Dairying, which is not based on the mixed farming system, is not as lucrative and sustainable as that based on alternative husbandry systems in tropical highland areas of eastern and southern Africa.

*Large-Scale Farms and Plantations.* Large-scale farms involving arable row crops and pastures and mechanized land development often are not sustainable in the humid tropics. Many of these enterprises (such as the famous groundnut scheme in Tanzania) were colossal failures. Those that are sustainable are based on good soil management, suitable rotations, and crop and input mixes that minimize cost of inputs and adverse environmental effects when sited in subhumid and savanna areas.

Irrigation projects involving small-scale irrigation in suitable locations in subhumid and savanna areas have been sustainable when good soil management and adequate drainage have prevailed. The most successful of this kind of irrigated production system in tropical Africa is the Asian lowland rice production system, which has almost completely displaced the traditional rice production. Large-scale systems often associated with hydroelectric dams frequently have not proved sustainable—because of inadequate drainage and salinization problems, siltation of dams upstream where watershed management has not been integrated and where no serious steps have been taken to control water-borne diseases of people and animals; and construction of dams without ecological measures to improve environmental quality and resettle people on suitable sites.

Large-scale tropical tree plantations are highly sustainable, especially when such crops as rubber, cocoa, oil palms, and tea are located in the humid tropics and suitable ecological zones and when care is taken in soil and vegetation management, especially during mechanized forest clearing and plantation establishment. Sustainability may be impaired in colonial and multinationally owned plantations with adverse political environment and exploitive farm, labor, and personnel management. A highly sustainable tree plantation system combines production of coconuts with pastures for rearing livestock.

*Specialized Horticulture.* Market gardening very much resembles home gardening, but is usually more intensive. It is highly sustainable when it is based on crop and animal wastes obtained from factories and urban centers, rather than mainly on inorganic fertilizers and pesticides with no efficient pest and disease management.

Truck gardening and fruit plantations are enterprises operated at a scale slightly larger than market gardens. They supply fruits and vegetables to urban centers and institutions. They usually are highly sustainable and successful when they are developed to take advantage of manure and organic waste or compost from urban areas, markets, and factories. Some use also is made of poultry manure from poultry farms and related enterprises. The most successful types are those located in highland areas, where subtropical vegetables are easier to grow.

Commercial fruit and vegetable production for processing has not often been successful and sustainable in tropical Africa. This is because adapted disease-resistant and high-yielding varieties of such crops as tomatoes are not always available to satisfy the processing capacity of plants installed. There is a lack of pesticides for numerous tropical pests and diseases. Breakdown of machinery and lack of parts are constant problems. Factories are only seasonally in operation. Highly commercial enterprises of the kind where a lot of mechanization is involved encounter failures when inappropriate machinery and techniques are used in land development. A great deal of care is required to minimize the adverse environmental hazards as well as the cost, number, and quantity of chemical inputs and energy used.

## Integrating Traditional and Modern/Emerging Technologies in Africa

Horizontal transfer of technologies from developed countries to Africa has been grossly inadequate and disappointing in meeting escalating demands for food. Modern agriculture of developed countries is now bedeviled by escalating costs of energy and inputs in relation to output and hazards to health and environment caused by pesticides. But the advent of farming systems research during the last two decades has made it possible for the demerits and merits of traditional and modern or conventional agricultural production systems to be identified. Based on this knowledge, the best strategy for developing new and improved sustainable agricultural production systems is to integrate what is good in both agricultural production systems in a compatible and economically and ecologically sound manner. This strategy is in line with the strategy for designing new and improved agroecosystems advocated by Hart (1979).

***Desirable Elements of African Agriculture.*** In determining desirable elements of traditional and transitional farming systems, due consideration should be given to the characteristics of the majority of farmers, their production systems, and technologies. These desirable aspects include:

■ Integration of crop and animal production systems in addition to development of farming systems that involve components of improved agroforestry and agrisilvopastoral systems as circumstances permit.

■ Utilization of nutrient cycling and biological nitrogen fixation potentials of plants wherever possible in order to reduce the use of costly fertilizers.

■ Cropping systems that make as much use as possible of indigenous and underutilized African crop plants.

■ Development of improved cropping patterns, grazing systems, and technologies ensuring that soil is adequately protected from erosion and degradation.

■ Integrated watershed development, including the development and utilization of relatively more fertile valley bottoms and hydromorphic soils for which solutions should be found to the various physical, biological, and socioeconomic constraints that limit their use.

■ Use of photoperiod sensitive and insensitive cultivars to achieve flexibility and special objectives in cropping systems.

"Modern" agricultural production systems and technologies that should be incorporated into improved farming systems for sustained yields include:

■ Mechanization and appropriate technology to minimize drudgery in farm work while significantly increasing productivity.

■ Integrated pest management to reduce losses in the field and in storage.

■ Techniques and methods of increasing the efficiency of those fertilizers that cannot yet be replaced by biological processes.

■ Intensification of production and increased productivity per unit area of land in order to curtail drastically the reliance on expansion of land under cultivation as the main strategy for increasing production.

■ Increased use of irrigation and water harvesting in semiarid and arid areas with measures to ensure adequate drainage and to minimize salinization.

■ Methods for minimizing or eliminating tillage.

- Greater use of techniques and potentials of conventional genetic improvement of crops and animals.
- Judicious use of agricultural chemicals.

***Relative Importance of Various Practices and Technologies.*** In various traditional and modern farming systems, different practices used during certain phases of the production process may interact with the crop, soil, or livestock in such a way as to contribute to sustainability. Examples are clearing the land of vegetation and developing land; planting; maintaining fertility and soil management, including fertilizer and chemical applications; and managing weeds, pests, and diseases. Or cultural operations, harvesting, post-harvesting, and the like might adversely influence the soil and crops growing on it so as to cause runoff, erosion, and soil degradation, thereby influencing the overall performance of subsequent crops.

Methods of removing vegetation, with respect to the extent of manual cleaning or mechanization and the kind of equipment used, in addition to the kind of preplanting cultivation technique used, can cause compaction and sealing of pore spaces, resulting in runoff and erosion. The possibility of runoff and erosion increases depending on the slope, soil type, and subsequent preplanting cultivation crop, cover, and so on. If the plant or crop residue is not burned, but applied as mulch, soil is protected from the beating action of rain, infiltration of water increases, and the organic matter content of the soil increases. The overall effect is reduced runoff and erosion and increased yield (Table 6). The effects of mulching vary with the kind of mulch applied and the cropping system and other practices used. Effects also vary from one place to another. The extent to which various operations may cause soil erosion is indicated in table 7.

In animal production, the method of pasture establishment and management, grazing system, harvesting methods, and species of livestock involved affect soils, soil fertility, and conservation. Uncontrolled burning of pastures could cause erosion and, depending on the time of burning and severity of the burn, soil may become exposed to erosion and the composition of the pasture may be changed. Delayed burning generates intense heat, which eliminates grass components of the pasture and causes dicots to become dominent. In developing sustainable agricultural production systems, not just the production system, but also the various component technologies associated with them determine the extent of environmental breakdown and hazards involved. Similarly, even where a production system is deemed highly sustainable, the use of an unadapted crop variety or breed of animals, or outbreak of pests, such as the desert locust, may result in at least temporary failure.

***Need for an Ecological and Systems Approach.*** No agricultural production system operates in a vacuum. Each system interacts with various factors in the environment. Many activities of man outside the farm may threaten the overall environment quality and the whole ecosystem in which the agricultural system occurs. For example, road construction, urban development, forest-logging operations, mining, and the construction of dams, industrial facilities, etc., may adversely affect agricultural production in adjacent or remote areas. In Nigeria, Grove (1952) noted that in the Anambra

**Table 6. Effects of methods of deforestation and tillage techniques on soil erosion.**

| *Method of Vegetation Removal* | *Sediment Density ($g\ l^{1}$)* | *Water Runoff ($mm\ y^{1}$)* | *Soil Erosion ($t\ ha^{1}\ y^{1}$)* |
|---|---|---|---|
| Traditional farming—incomplete clearing, no-till | 0.0 | 3 | 0.01 |
| Manual clearing—no-till | 3.4 | 16 | 0.4 |
| Manual clearing—conventional tillage | 8.6 | 54 | 4.6 |
| Shear blade—no-till | 5.7 | 86 | 3.8 |
| Tree pusher root rake—no-till | 5.6 | 153 | 15.4 |
| Tree pusher root rake—conventional tillage | 13.0 | 250 | 19.6 |

**Table 7. Different operations performed during different stages of crop production and utilization, and extent of likely erosion hazard involved.**

| *Operations at Different Stages of Crop Production and Utilization* | *Extent of Possible Erosion Hazard** |
|---|---|
| Clearing | Very high |
| Land development | High |
| Tillage and preplanting cultivations | High |
| Planting | Low |
| Subsequent soil management | Low |
| Water management | Low-high |
| Fertilization | Low |
| Weed, pest and disease management | Negligible to high |
| Harvesting | Medium to high |
| Primary processing (shelling, winnowing) | Negligible |
| Drying | None |
| Storage | None |
| Processing | None |
| Packaging | None |
| Preparation | None |
| Consumption | None |
| Waste disposal | Low-medium |

*Extent of erosion hazard depends on interaction of operations with environment and other factors.

State erosion hazard is usually high where footpaths run down the slope.

## Emerging Sustainable Agricultural Production Systems

Recent developments in farming systems research, at the international agriculture research centers and elsewhere; concerns about soil erosion and degradation; and the need for development of more sustainable agricultural production systems that address the needs and problems of low-resource farmers suggest the following emergency system of production and component techniques:

■ *Zero or reduced tillage.* A system of reduced tillage involving good residue management that effectively reduces soil erosion and gives yields as good or better than conventional tillage.

■ *Life mulching.* A system of growing field crops through living mulch of preferably leguminous cover crops, such as *Psophocareus pallustris* and *Centrosema*. This system promises to be good for maintaining soil fertility and for conserving the soil. It also gives good yields on steep slopes. The system may prove useful as well in humid areas with sufficient rainfall, but the potential in such areas has not been fully explored. The system may also make large-scale plantation production possible (Figure 2).

■ *Alley cropping and agroforestry.* These systems developed from traditional practices that combine the growing of ligneous species with field crops. Alley cropping has many advantages; rotations of food or arable crops are grown between hedge rows of preferably leguminous trees or shrubs that are periodically pruned to supply both mulch and fuelwood (Figure 3).

■ *Supplying mulch* for soil conservation and for maintenance of soil organic matter.

■ *Maintaining soil fertility* through nitrogen fixation and nutrient cycling.

■ *Supplying fuelwood* and staking material for viney crops.

■ *Supplying raw materials for crafts and industries,* such as paper manufacturing.

■ *Eliminating or minimizing fallowing,* thereby increasing area under cultivation without change in land tenure.

■ *Use of emerging biomass technology* (Figures 4 and 5).

■ *Integration of food crops, trees, and pastures/livetock.*

■ *Game ranching.* This is a method of rearing livestock and game animals in their natural environment by selecting species which compatibly feed on different strata and species of pasture plants, and then judiciously

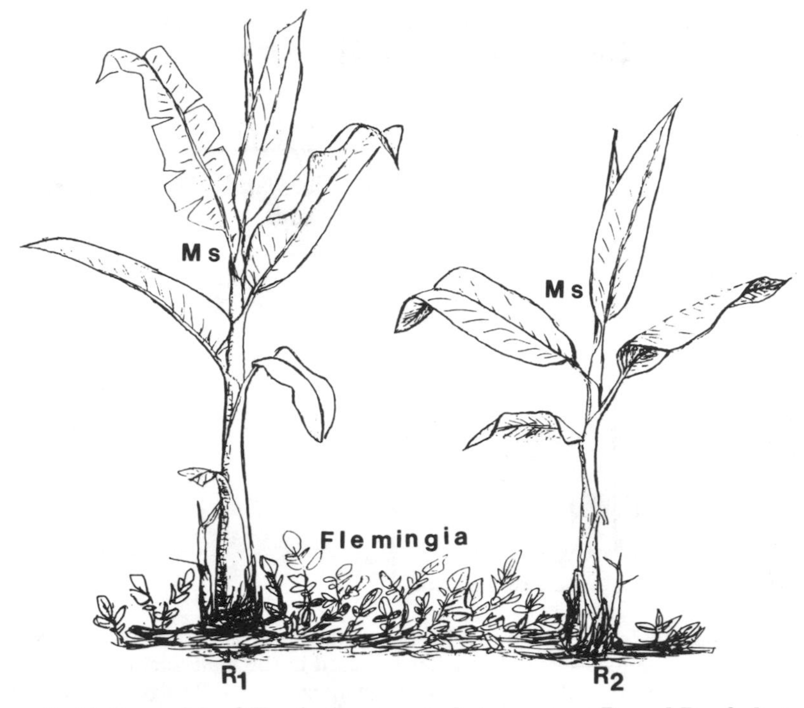


**Figure 2.** Living mulch of Flemingia congesta between rows $R_1$ and $R_2$ of plantain, *Musa paradisiaca* (MS). The Flemingia mulch supplies nitrogen to the plantain, minimizes stand decline of plantain, and ensures higher stable yields.

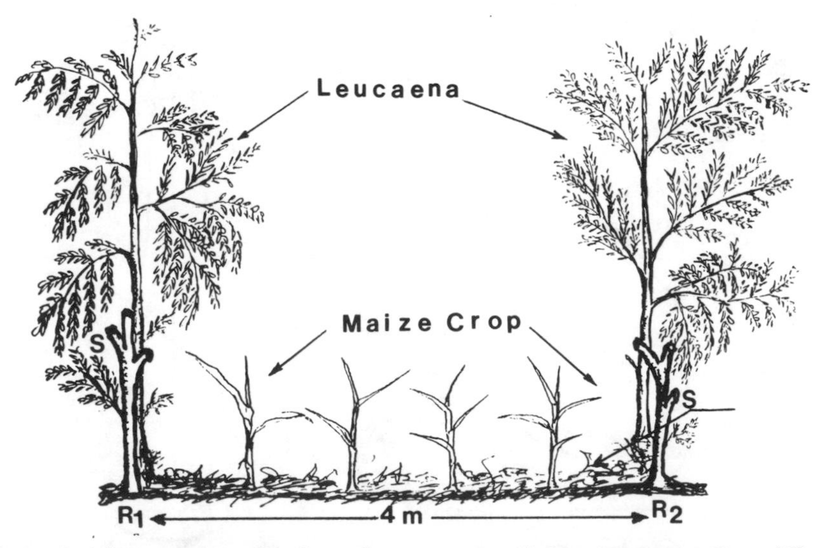


**Figure 3.** Alley cropping with the maize crop grown between two rows $R_1$ and $R_2$ of *Leucaena* planted 4 meters apart. The *Leucaena* is pruned periodically to mere stumps (s), and the pruning (PR) is applied as mulch to the maize crop.

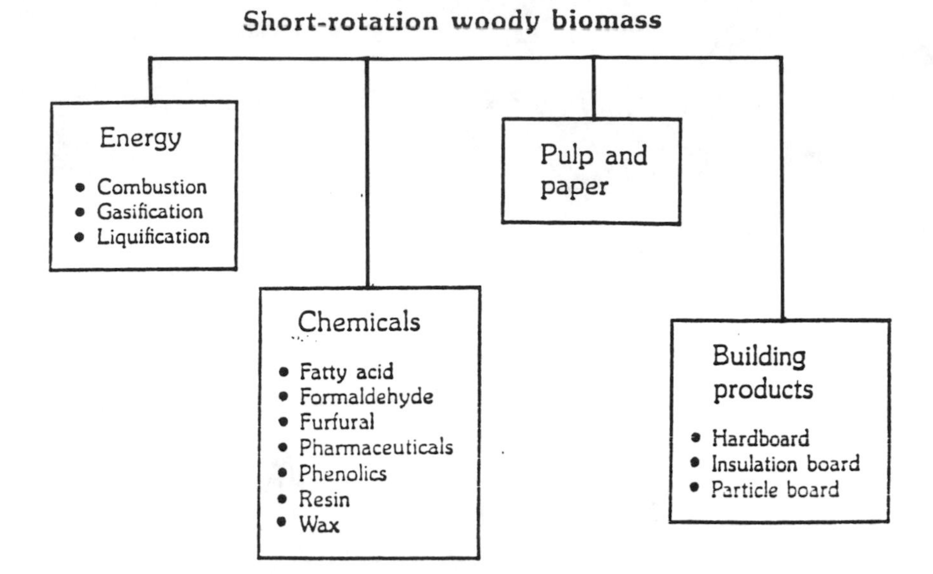


**Figure 4. Biomass from trees grown in short rotations is a source of many products, including chemical feedstocks. Phenolics are used in the manufacture of adhesives, fungicides, and plastics. Furfural is used in making industrial solvents (Chow et al., 1983).**

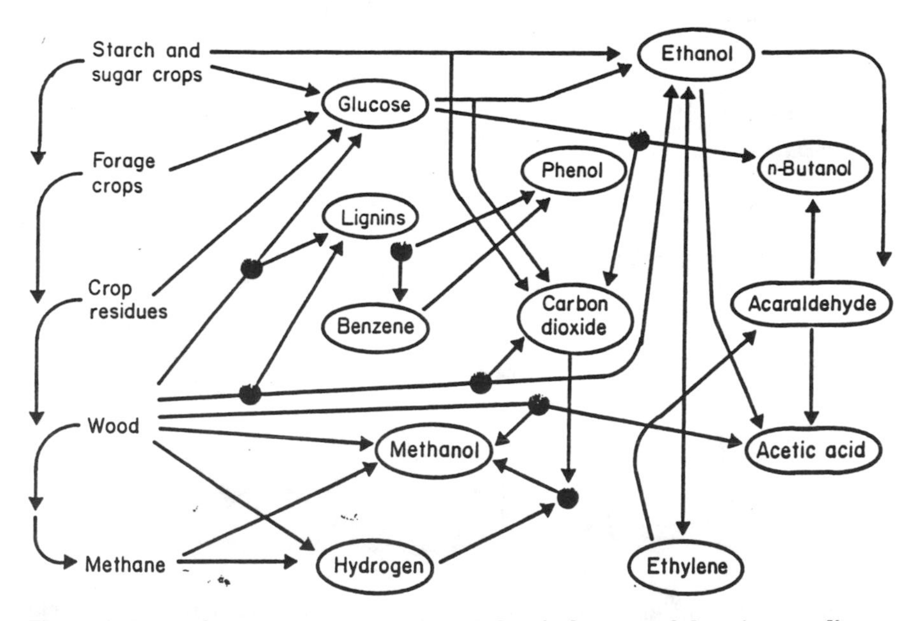


**Figure 5. Some of the many compounds and chemicals, most of them intermediates, produced from fields, crops, and trees; • = by-product (Rolfe and Moore, 1984).**

harvesting the animals (Figure 6). Benefits of these exceed many other pasture management systems (Table 8).

## Recommendations Toward Sustainable Agriculture in Africa

1. Develop policies, strategies, and measures aimed at integrated natural resource conservation and management, of which sustainable agricultural production is a component. This process might delineate specific areas for hunting and tourism and for germ plasm conservation; tree crop plantations; agricultural land, including fallow land; special agroforestry systems; grazing land; urban centers, airports, roads, and railways; landscape uses, including parks in urban areas; industrial uses; and mining.

2. Base all agricultural development projects on sound ecological principles to ensure that each program is carried out in areas that have the highest potential for producing the commodities (Figure 7). Large-scale development projects for arable crops, such as maize, are sometimes located in the humid tropics, where these crops have the least potential. Similarly, maize sometimes has replaced sorghum in areas subject to drought.

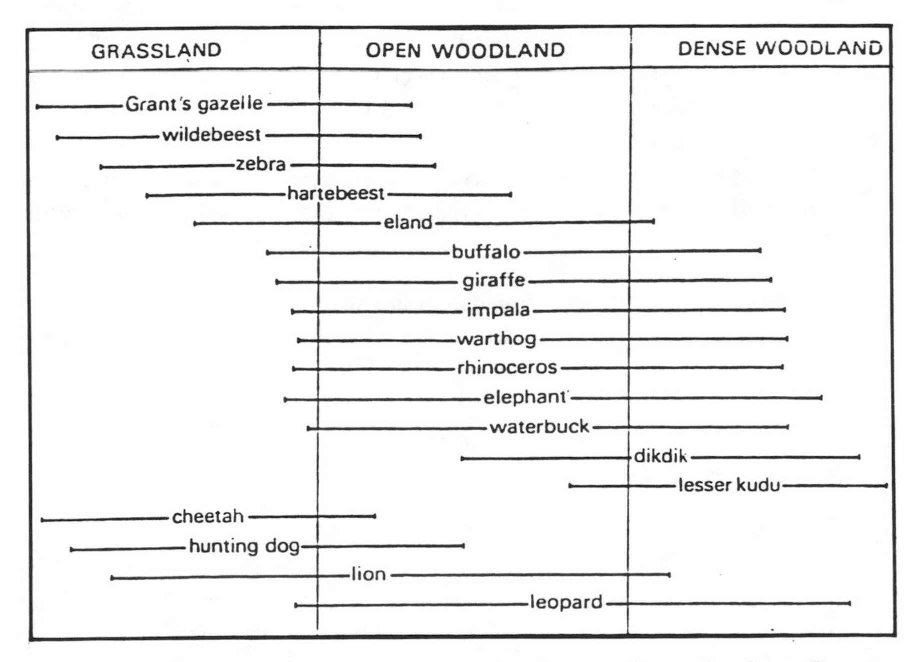


**Figure 6. Habitat and food preferences of animals in the Tarangire Game Reserve (Lind and Morrison, 1974).**

Land use capability classification and surveys are needed to provide information for use by foresters, agriculturalists, engineers, and so on. In agricultural development projects, sustainability can be attained in the most cost-effective manner where soils of the highest potential are developed first (Figure 8). This minimizes the cost of inputs needed to create a favorable environment for the crop or animal.

3. Use integrated watershed development principles to minimize soil degradation and damage to the landscape in agricultural production projects and to enable use of each toposequence for production of the commodities and activities most adapted to conditions. With this approach, flooded valleys can be used for aquaculture and rice production, followed, as one ascends the slope, by cocoyams, yams, cassava, and other upland crops, with the highest part of the slope used for pasture or tree crops (Figure 9). Integrated farming systems that must be given due consideration for suitable parts of the landscape include agroforestry systems, such as alley cropping, that combine food crops with the crops on sloped land; integrated crops and animal production systems; and integrated crops, livestock, and agricultural systems in which animal waste and crop residue are used to feed fish.

4. Tap the significant potential of highly productive valley bottoms of hydromorphic soils, about 20 million of which are underfertilized in tropical Africa, especially in the production of rice and off-season vegetables. Measures will be needed to combat river blindness and schistosomiasis. Lowland rice production is very sustainable and can reduce reliance on rice imports to satisfy demand.

5. Work to maintain soil fertility at reduced cost through the use of biological nitrogen fixation, mycorrhizal phosphate nutrition, and other

**Table 8. Benefits of alternative strategies of grassland management.**

| | *None* | *Livestock Ranching* | *Game Preserves* | *National Parks* | *Game Ranching* |
|---|---|---|---|---|---|
| Meat supply | (C) | N | N | N | C |
| Damate control | C | C | C | C | C |
| Animal export | (C) | I | C | C | C |
| Trophies and skins | (C) | I | N | N | C |
| Tourism | I | I | C | C | N |
| Economic development | I | N | C | C | N |
| Conservation research | I | N | C | N | N |
| Research | I | I | C | C | C |
| Disease control | C | C | I | I | C |
| Employment | I | N | C | C | N |

C = Compatible, (C) = Compatible but for limited duration in grassland area, N = Natural, I = Incompatible.

biological processes of reducing amount of fertilizers used. Tropical Africa relies currently on expansion of new areas to attain about 80 percent of annual increases in food production. In order to minimize deforestation

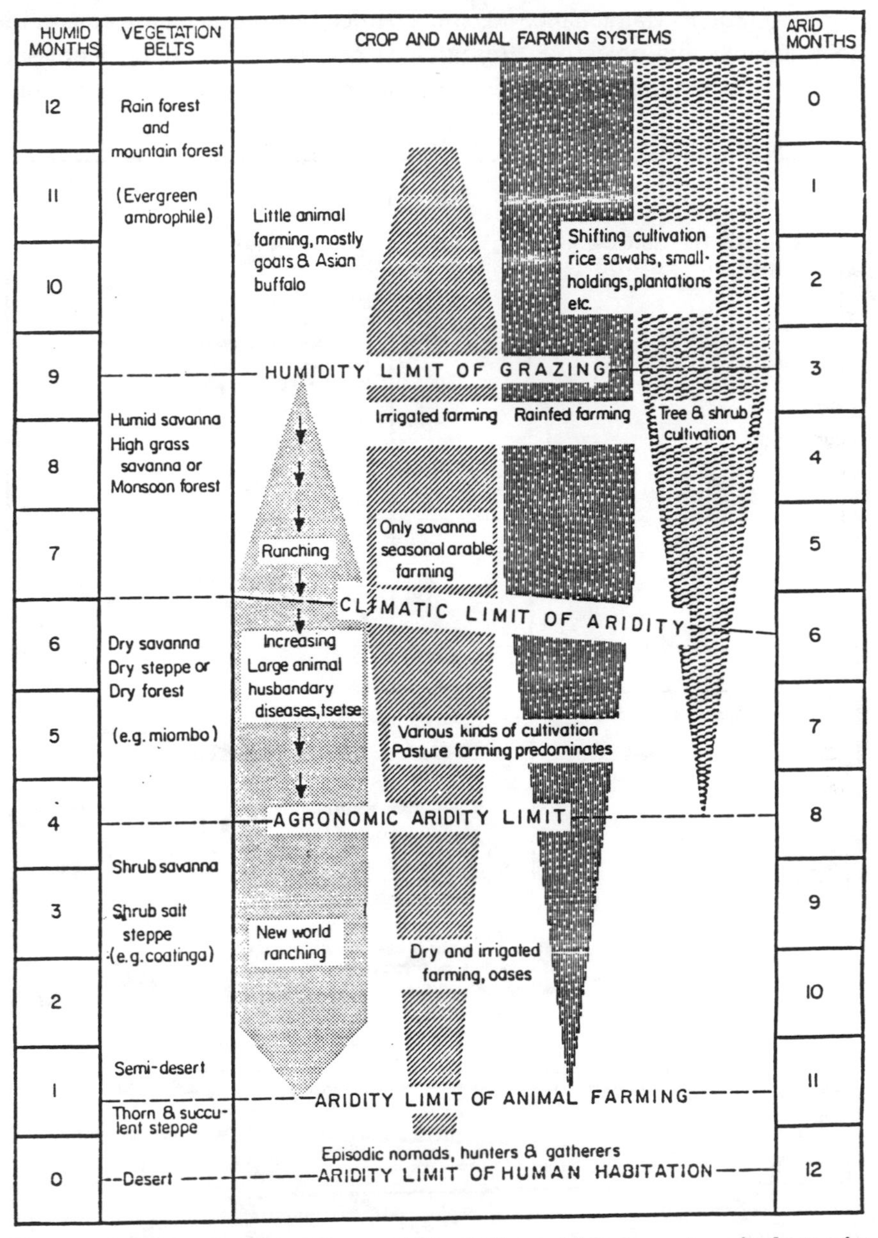


**Figure 7. Humid and arid months, vegetation belts, and farming systems in the tropics (Uhlig, 1965; Andreae, 1980; Okigbo, 1981).**

and expansion into marginal land, the Food and Agriculture Organization of the United Nations expects Africa to adopt a land-saving strategy aimed at increasing production 27 percent by expansion of arca, 22 percent by cropping intensity, and 51 percent by increased yield by the year 2000. Maintenance of soil fertility will be a major constraint to achieving this objective. Research is needed for attaining increased efficiency of fertilizer use and reducing the amount of fertilizers lost through

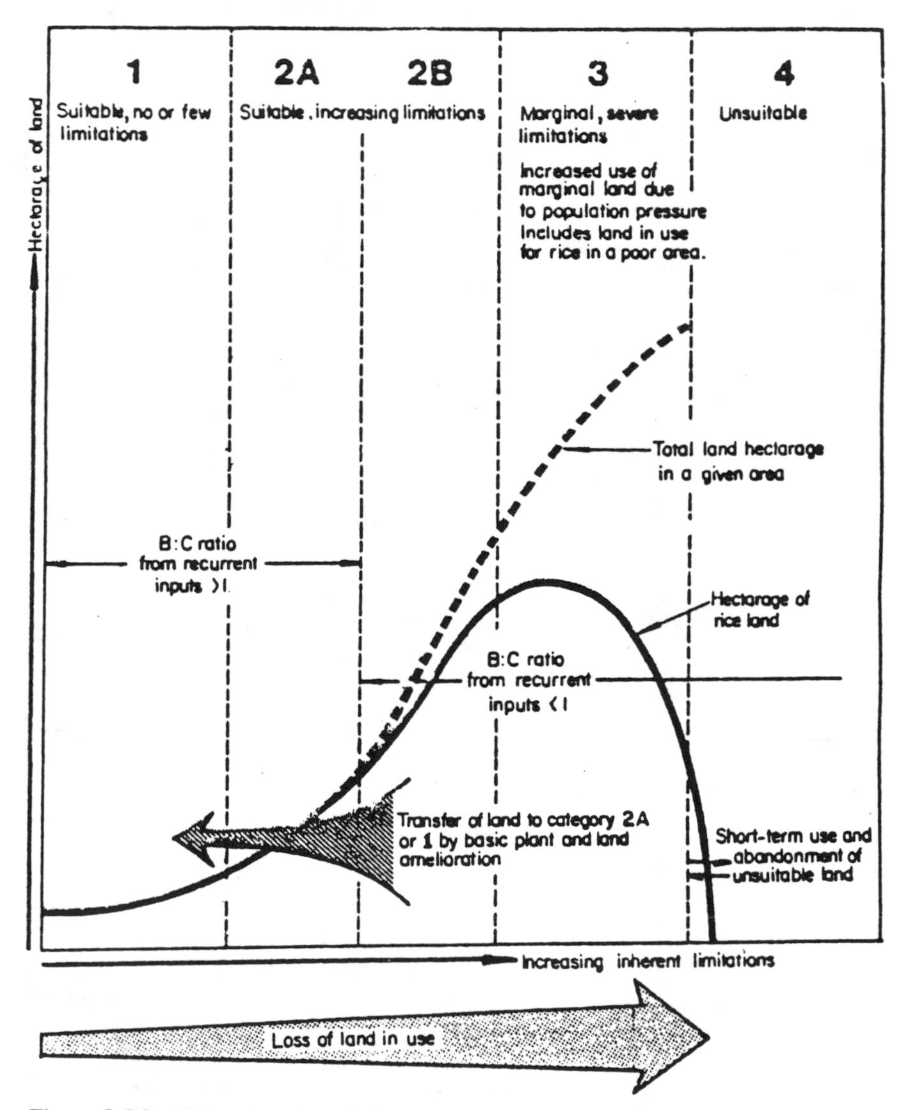


**Figure 8. Model showing the relationship of land quality to land use for rice growing (Moorman and Van Breeman, 1978).**

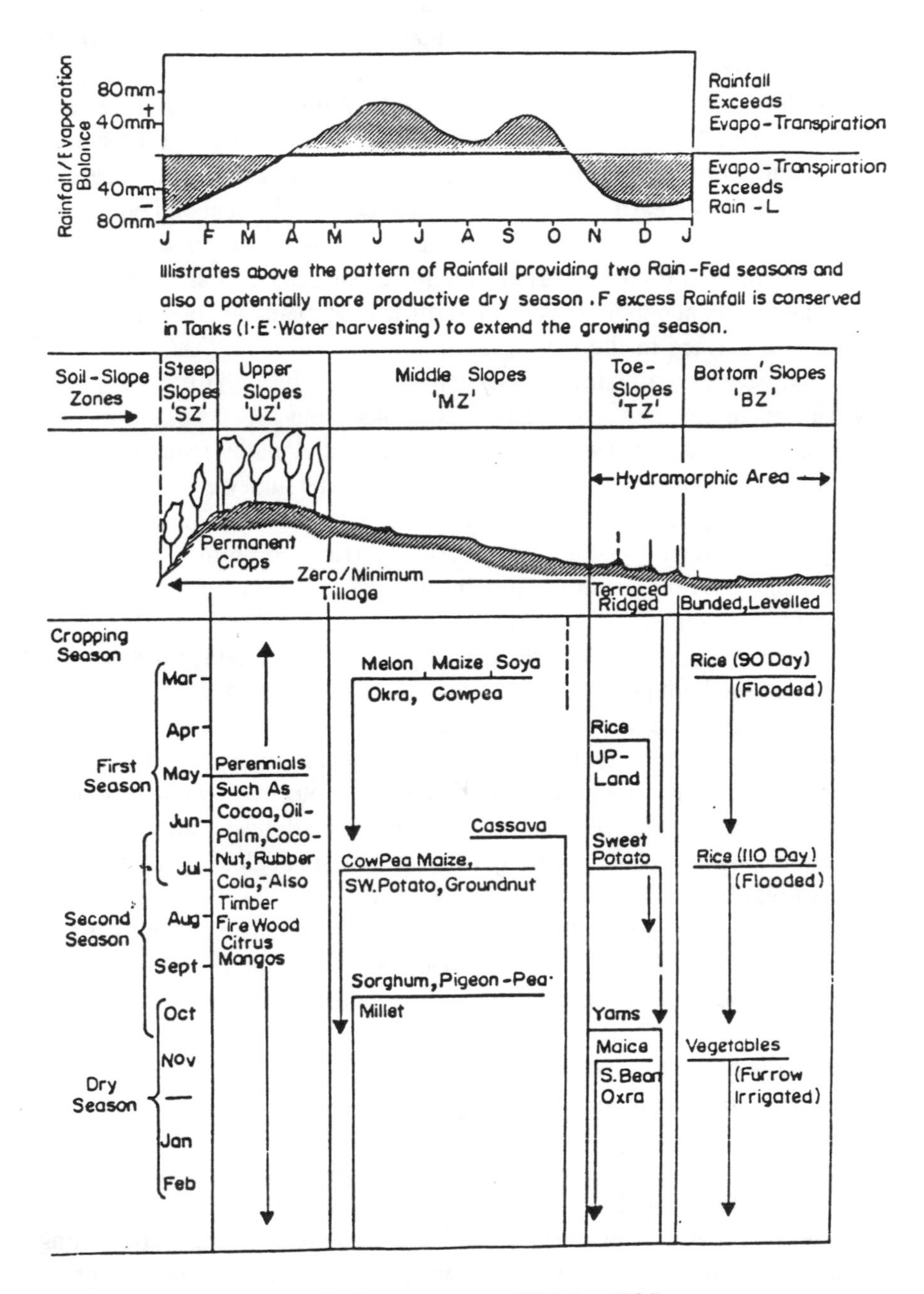


**Figure 9. Integrated watershed management (Okigbo, 1984).**

leaching, volatilization, and the like.

6. Develop integrated weed, pest, and disease management systems that combine physical, chemical, biological, and cultural methods, thereby minimizing the cost and amount of pesticides used and their adverse effects on the ecosystem.

7. Reduce soil erosion and degradation by improved soil management through reduced or zero tillage, maintenance of ground cover with adequate plant or crop residue management, various kinds of mulching, careful mechanization of clearing, and reduced reliance on physical and engineering methods of soil conservation. Special attention should be given to development of soil conservation systems in savanna and semiarid areas where large animal populations make it difficult to maintain good crop cover on the soil surface in the dry season.

8. Find ways of maintaining diversity of species through different planting patterns in time and space. Knowledge of various aspects of plant competition, expecially below the ground (such as allelopathy,) should be applied to manipulation of the rhizosphere of crops and weeds.

9. Devise a special plant-breeding agenda for developing sustainable agricultural production systems for tropical Africa, combining advantages of both photoperiod sensitive and insensitive cultivars for flexibility in polycultural systems.

10. Make special efforts to conserve, improve, and use indigenous African crops that have high potential as sources of food, drugs, new materials for industry, timber, and so on, but have remained underexploited and are increasingly in danger of extinction. Emerging technologies, such as tissue culture and the manipulations it enhances in biotechnological research, should be exploited for propagation, breeding, and conservation of these indigenous plants. About 200 indigenous species are neglected in the humid and sub-humid areas of Africa.

11. Develop technologies that address the needs and problems of women, who in certain operations are more involved in farming than men. Agricultural productivity and sustainability cannot be achieved readily in Africa as long as there is labor shortage due to rural-urban migration because of the drudgery and low incomes in agriculture. There is need for appropriate technologies for several operations ranging from land-clearing to post-harvest technologies.

12. Increase the use of irrigation. Some 56 percent of the area of Africa needs water for irrigation; only 5 percent now irrigated.

13. Seek institutional linkage and collaboration of sufficient scope among developing countries and with developed countries to take advantage of advances in emerging technologies, such as biotechnology; develop local

capabilities; and exchange information and materials. The development of sustainable agricultural production systems in tropical Africa, as a component of improved natural resource conservation, management, and utilization, is a complex problem requiring multidisciplinary research and development, institutional and manpower capabilities, and more political commitment on a long-term basis than many African countries can muster.

14. Provide more conferences on agriculture to ensure updating in research and development. A special project should be launched on the collation and assessment of indigenous knowledge systems and technologies in sustainable agriculture.

15. Provide priority research in biomass technologies as a means of obtaining chemicals from plants. The tropical environment, with its high primary productivity, has the potential for high yields of biomass and development of new industrial raw materials and sources of foreign exchange.

## REFERENCES

Andreae, B. 1980. The economics of tropical agriculture. English edition edited by Team Kestner. Commonwealth Agricultural Bureaux, Farnham Royal, England.

Chow, P., G. L. Rolfe, and L. E. Arnold. 1983. Fiber and energy from woody biomass. Illinois Research 25(1): 11-13.

Food and Agriculture Organization. 1986a. African Agriculture: The next 25 years. Atlas of African agriculture. United Nations, Rome, Italy.

Food and Agriculture Organization. 1986b. African agriculture: The next 25 years. Annex II: The land resource base. United Nations, Rome, Italy.

Food and Agriculture Organization. 1986c. African agriculture: The next 25 years. Main Report. United Nations, Rome, Italy.

Food and Agriculture Organization. 1986d. National resources and the human environment for food and agriculture in Africa. United Nations, Rome, Italy.

Harlan, J. R., J.M.S. DeWet, and A.B.L. Stemler. 1976. Plant domestication and indigenous African agriculture. pp. 3-19. *In* J. R. Harlan, J.M.S. DeWet, and A.B.L. Stemler [editors] Origins of African plant domestications. Mouton Publishers, The Hague, The Netherlands.

Hart, R. R. 1979. Methodologies to produce agroecosystem management plans for small farmers in tropical environments. Catie Turialba. pp. 15.

Johnston, B. F. 1958. The staple food economies of western tropical Africa. Stanford University Press, Palo Alto, California.

Kaufmann, Von R., B. N. Okigbo, and E.N.N. Oppong. 1983. Integrating crops and livestock in West Africa. FAO Animal Production and Health Paper No. 41. Rome, Italy.

Knight, C. G. 1976a. Food and Agriculture Organization of the United Nations, Rome. Prospects for peasant agriculture. pp. 195-211. *In* C. G. Knight and J. L. Newman [editors] Contemporary Africa. Prentice Hall Inc., Englewood Cliffs, New Jersey.

Knight, C. G. 1976b. Wild life. pp. 159-167. *In* C. G. Knight and J. L. Newman [editors] Contemporary Africa. Prentice Hall, Inc., Englewood Cliffs, New Jersey.

Lal, R. 1979. Soil tillage and crop production. International Institute of Tropical Agriculture, Ibadan, Nigeria.

Lal, R. 1981. Land clearing and hydrological problems. pp. 131-140. *In* R. Lal and E. W. Russell [editors] Tropical hydrology. J. Wiley and Sons, Chichester, United Kingdom.

Lind, E. M., and Mr. E. S. Morrison. 1974. East African vegetation. Longman Group,

London, England.
Manshard, W. 1974. Tropical agriculture, a geographical introduction and appraisal. Longman Group Limited, London.
Miracle, M. P. 1967. Agriculture in the Congo basin. University of Wisconsin Press, Madison.
Moorman, F. R., and N. Van Breeman. pp. 347. 1978. Rice, soil, water and land. International Rice Research Institute, Los Barnos, the Philippines.
Morgan, W. B. 1972. Peasant agriculture in tropical Africa. pp. *In* M. F. Thomas and G. W. Whittington [editors] Land use in Africa. Methuen & Co., Ltd. London, England.
Morgan, W. B., and J. C. Pugh. 1969. West Africa. Methuen & Co. Ltd., London, England.
Okigbo, B. N. 1981. Farming systems research and food policies in developing Africa. Paper presented at the IF PRI Board meeting, February 9-13, 1981. Ibadan, Nigeria.
Okigbo, B. N. 1982. Agriculture and food production in tropical Africa. pp. 11-67. *In* C. Christenson, E. B. Hogan, B. N. Okigbo, G. E. Schuh, E. J. Clay, and J. W. Thomas. The Development Council, New York, New York.
Okigbo, B. N. 1984. Improved permanent production systems as an alternative to shifting cultivation. pp. 9-100. *In* FAO Soils Bulletin No. 53, Improved production systems as an alternative to shifting cultivation. Food and Agriculture Organization of the United Nations, Rome, Italy.
Okigbo, B. N. 1986. Towards a new green revolution: From chemicals to new biological techniques in agriculture in the tropics. International meeting of the Italian Academy of Science, September 8-10, 1986, Rome.
Okigbo, B. N., and D. J. Greenland. 1976. Intercropping systems in tropical Africa. pp. 11-67. *In* R. I. Papendick, P. A. Sanchez, and G. B. Triplett [editors] multiple cropping. American Society of Agronomy, Madison, Wisconsin.
Oureshi, J. 1978. Sustained yields from tropical forest: A practical policy for resource and environmental management. East-West Environment and Policy Institute, East-West Center, Honolulu, Hawaii.
Pritchard, J. M. 1979. Africa: A study geography for advanced students. Longman Group, London, England.
Rolfe, G. L., and K. J. Moore. 1984. Renewable sources of chemicals. Illinois Research 26(2/3): 5-7.
Thomas, G. W. et al. 1988. Environment and natural resources: Strategies for sustainable agriculture. Task Force Report. Board for Multinational Food and Agricultural Development, Washington, D.C.
Uhlig H. 1965. Die geographischen Grundlagen der Weidewirtschaft in den Trockengebieten der Tropen und Subtropen, Shriften d. Tropeninstituts d. Justus-Liebig, Universitat Giessen, vol. 1 pp. 1-28.

# ECOLOGICAL CONSIDERATIONS FOR THE FUTURE OF FOOD SECURITY IN AFRICA

H.C.P. Brown and V. G. Thomas

Food security means ensuring that all members of a country have access to enough food throughout the year to lead an active, healthy life (World Bank, 1986). This assurance of a nutritionally balanced diet assumes that the biological bases for food production are sustainable over time—without massive fiscal, energy, and chemical subsidies. These bases can be sustained only with an agriculture that is integrated ecologically throughout the world.

A United Nations report (World Commission on Environment and Development, 1987) stated that short-sighted agricultural policies emphasizing increased production at the expense of environmental considerations have contributed greatly to the degradation of the agricultural resource base on almost every continent. The report emphasizes that application of the concept of sustainable development to ensure food security requires a holistic approach. The focus must be on ecosystems at national, regional, and global levels, with coordinated land use and careful planning of water use and forest exploitation. Furthermore, the report "For Whose Benefit?" (SCEAIT, 1987) recognized the importance of ecosystems in stressing that all Canadian government-funded development projects should be moderated by the ecological and environmental constraints of the regions for which they are intended.

The environment and development problem was best summarized by Tolba (1987): "Too often in the past ad hoc development plans and projects have taken place which have been destructive to the environment, and have thereby endangered the very basis on which continuity and sustainability of development depend."

The ability of people on the African continent to feed themselves has declined dramatically in recent years. Although it is still primarily an agrarian society, 140 million of the continent's 531 million people in 1984 were fed exclusively with grain from abroad. This increased to 170 million people in 1985 (Brown and Wolf, 1986). The actual declines in grain yields over a 30-year period indicate a dramatic decline in primary productivity, especially in the Sahelian-sub-Sahelian zone where, for example, yields in Sudan dropped 38 percent (Table 1). Even parts of Africa traditionally regarded as productive showed a decline. These lower yields are probably related to civil war and to environmental factors such as deforestation, overgrazing, soil erosion, prolonged drought, and generally inappropriate land use.

With a steadily increasing population in Africa, the need for sustainable development has become especially acute. The population growth rate of Africa is very high—2.8 percent per year—which results in an annual addition of 16.3 million people to Africa's present population of 583 million (Table 2). Kenya, with an annual rate of population growth at 4.2 percent, is experiencing the highest growth rate in the world. Nigeria's population, now just over 100 million, is projected to reach 532 million before the middle of the next century (Table 3). The population of Ethiopia is projected to quintuple before stabilizing in an area where poor land use and ill-conceived agricultural policies have already led to widespread starvation (Brown, 1987a).

The sheer numbers alone exert enormous pressures on the carrying capacity of the land. A detailed study by the World Bank of seven West African countries analyzed the carrying capacity of various ecological zones delineated by rainfall. This study indicated that these areas have almost reached or have exceeded their carrying capacity for food and fuel production (Table 4). The result is reduced nutritional self-sufficiency, which, in turn, leads to increased external debt and lower living standards. With such a large population and environmental deterioration undermining economic progress all across Africa, the only successful economic

**Table 1. Grain yields per hectare in four African countries with declining yields, from 1950 to 1952 and 1983 to 1985 (Brown and Wolf, 1986).**

| | *Average Yields* | | |
|---|---|---|---|
| *Country* | *1950-1952* | *1983-1985* | *Change* |
| | — *Kg* — | | — % — |
| Nigeria | 760 | 714 | −6 |
| Mozambique | 620 | 545 | −12 |
| Tanzania | 1,271 | 1,091 | −14 |
| Sudan | 780 | 479 | −38 |

**Table 2. World population growth by geographic region, 1986.**

| Region | Population | Population Growth Rate | Annual Increment |
|---|---|---|---|
| | —million— | —%— | — million — |
| Slow-growth regions | | | |
| Western Europe | 381 | 0.2 | 0.8 |
| North America | 267 | 0.7 | 1.9 |
| Eastern Europe and Soviet Union | 392 | 0.8 | 3.1 |
| Australia and New Zealand | 19 | 0.8 | 0.1 |
| East Asia* | 1,263 | 1.0 | 12.6 |
| Total | 2,322 | 0.8 | 18.6 |
| Rapid-growth regions | | | |
| Southeast Asia† | 414 | 2.2 | 9.1 |
| Latin America | 419 | 2.3 | 9.6 |
| Indian subcontinent | 1,027 | 2.4 | 24.6 |
| Middle East | 178 | 2.8 | 5.0 |
| Africa | 583 | 2.8 | 16.3 |
| Total‡ | 2,621 | 2.5 | 65.5 |

*Principally China and Japan.
†Principally Burma, Indonesia, the Philippines, Thailand, and Vietnam.
‡Numbers may not add up to totals due to rounding.

**Table 3. Projected population size at stabilization for selected countries.**

| Country | Population in 1986 | Annual Rate of Population Growth | Size of Population at Stabilization | Change from 1986 |
|---|---|---|---|---|
| | — million — | — % — | — million — | —— % —— |
| Slow-growth countries | | | | |
| China | 1,050 | 1.0 | 1,571 | +50 |
| Soviet Union | 280 | 0.9 | 377 | +35 |
| United States | 241 | 0.7 | 289 | +20 |
| Japan | 121 | 0.7 | 128 | + 6 |
| United Kingdom | 56 | 0.2 | 59 | + 5 |
| West Germany | 61 | −0.2 | 52 | −15 |
| Rapid-growth countries | | | | |
| Kenya | 20 | 4.2 | 111 | +455 |
| Nigeria | 105 | 3.0 | 532 | +406 |
| Ethiopia | 42 | 2.1 | 204 | +386 |
| Iran | 47 | 2.9 | 166 | +253 |
| Pakistan | 102 | 2.8 | 330 | +223 |
| Bangladesh | 104 | 2.7 | 310 | +198 |
| Egypt | 46 | 2.6 | 126 | +174 |
| Mexico | 82 | 2.6 | 199 | +143 |
| Turkey | 48 | 2.5 | 109 | +127 |
| Indonesia | 168 | 2.1 | 368 | +119 |
| India | 785 | 2.3 | 1,700 | +116 |
| Brazil | 143 | 2.3 | 298 | +108 |

development strategy will be one that promotes and sustains the natural ecological systems on which the economy depends (Brown and Wolf, 1986).

The two major challenges facing African agriculture are (1) how to manage land so that a continuous production is realized from areas characterized by erratic environmental constraints, and (2) how to effect a balance between input-intensive and the purely organic style of agriculture so that the practice remains productive and environmentally sympathetic.

## Traditional Food Production in African Ecosystems

The tropical ecological zone can be subdivided into six main tropical landscape belts (Figure 1). The variety of tropical environments is reflected in the different agricultural zones, which tend to conform closely with the natural landscape belts. This formerly resulted in a close adjustment of types of agricultural economy and enterprise to basic vegetation patterns. Altitude and soil type also contributed to the type of agriculture present within a generalized area.

In a continent of extreme diversity, Africans have developed different cultures in adjusting to their environment. The most basic is the hunter-gatherer, a system that is sustainable at a low population density. Shifting cultivation was the predominant mode of crop production before European colonization. It is well suited to nutrient-poor soils in areas of low human density. Pastoralists are people who obtain most of their sustenance from domestic animals and generally occupy areas too dry to sustain rain-

**Table 4. Measures of sustainability in seven African countries,* by ecological zone, 1980.**

| | *Food* | | | *Fuelwood* | | |
|---|---|---|---|---|---|---|
| *Zone* | *Agriculturally Sustainable Population* | *Actual Rural Population* | *Food Disparity* | *Fuelwood-Sustainable Population* | *Actual Total Population* | *Fuel Disparity* |
| | *million people* | | | | | |
| Sahelo-Saharan | 1.0 | 1.8 | −0.8 | 0.1 | 1.8 | −1.7 |
| Sahelian | 3.9 | 3.9 | 0.0 | 0.3 | 4.0 | −3.7 |
| Sahelo-Sudanian | 8.7 | 11.1 | −2.4 | 6.0 | 13.1 | −7.1 |
| Sudanian | 8.9 | 6.6 | 2.3 | 7.4 | 8.1 | −0.7 |
| Sudano-Guinean | 13.8 | 3.6 | 10.2 | 7.1 | 4.0 | 3.1 |
| Total | 36.3 | 27.0 | 9.3 | 20.9 | 31.0 | −10.1 |

*Burkina Faso, Chad, Gambia, Mali, Mauritania, Niger, and Senegal. The five ecological zones are delineated by amounts of rainfall.

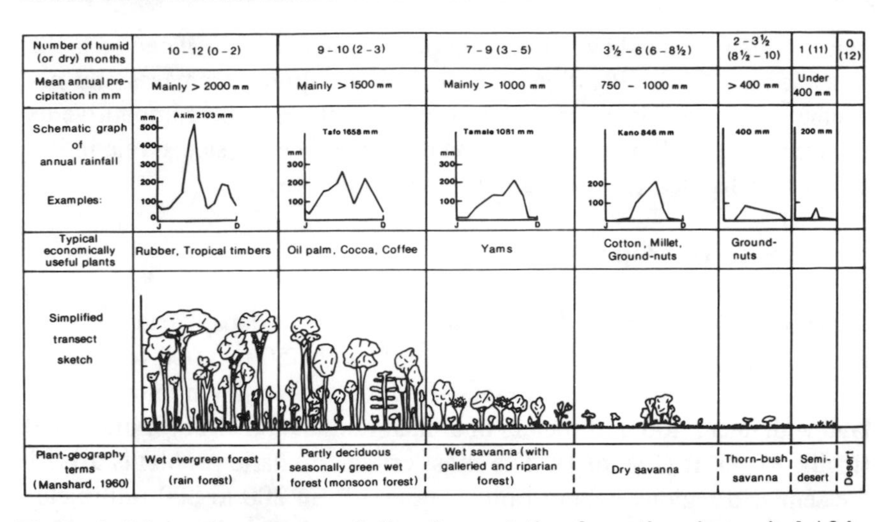


**Figure 1. Schematic summary of climatic vegetation formations in tropical Africa (Ruddle and Manshard, 1981).**

fed agriculture, although they move to wetter areas during dry periods. These groups are located in the savanna and desert zones, and their population densities are low because of low, unpredictable primary production. Pastoralism allows the conversion of low-quality plant food, which is unsuitable for human consumption, into high-quality meat and milk in areas that can otherwise support few people. Continuous subsistence cultivation in the tropics is possible only on unusually fertile soils where there is significant import of nutrients through periodic flooding, or where careful management minimizes losses. Mixed agriculture, with crops and livestock integrated into a single agroecosystem, has been widely practiced. Besides food, the animals provide fertilizer and domestic fuel in the form of dung and labor for plowing, operating irrigation systems, and transport (Deshmukh, 1986).

In all of these systems, the underlying principle was the selection of a sustainable culture so people could survive. The tribal traditions and practices maintained the ecosystem so that the environment, on which life depended, was not degraded at low population densities. Agriculture differs from the natural ecosystem in that it creates and maintains a highly productive, early successional stage of culture in contrast to the range of successional stages, lower production, and higher stability of natural ecosystems. Although traditional agriculture was integrated with the ecosystem, much of modern agriculture competes with ecosystems to maximize production. Modern agriculture is highly productive, but this high productivity is often obtained by application of external nutrients; it can be highly unstable as a result. Planting extensive monocultures increases and perpetuates crop-associated pest and disease problems. The competitive exclusiveness of modern agriculture and the denial of ecosystems' role in food production has led to many environmental problems.

## Problems in Development

***Arid and Semiarid Land.*** *Drought.* The drought-adapted arid and semiarid savanna ranges from the dry savanna, with 750 to 1,000 millimeters of rainfall per year and approximately three to five dry months, to arid land with under 400 millimeters of annual rainfall and 11 dry months and, finally, to desert (Figure 1). The most critical climatic problems arise in the semiarid regions, where rainfall ranges from 200 to 800 millimeters per year, and year-to-year variability in precipitation is relatively great (Rasmusson, 1987) (Figure 2). These dry areas, which cover much of Africa, are naturally drought-adapted because drought is a recurrent phenomenon in these regions. Climatic variability in Africa, when depicted in terms of the average departure from normal rainfall, indicates that the dry areas also have the greatest variation in the amount of rain and that recurring,

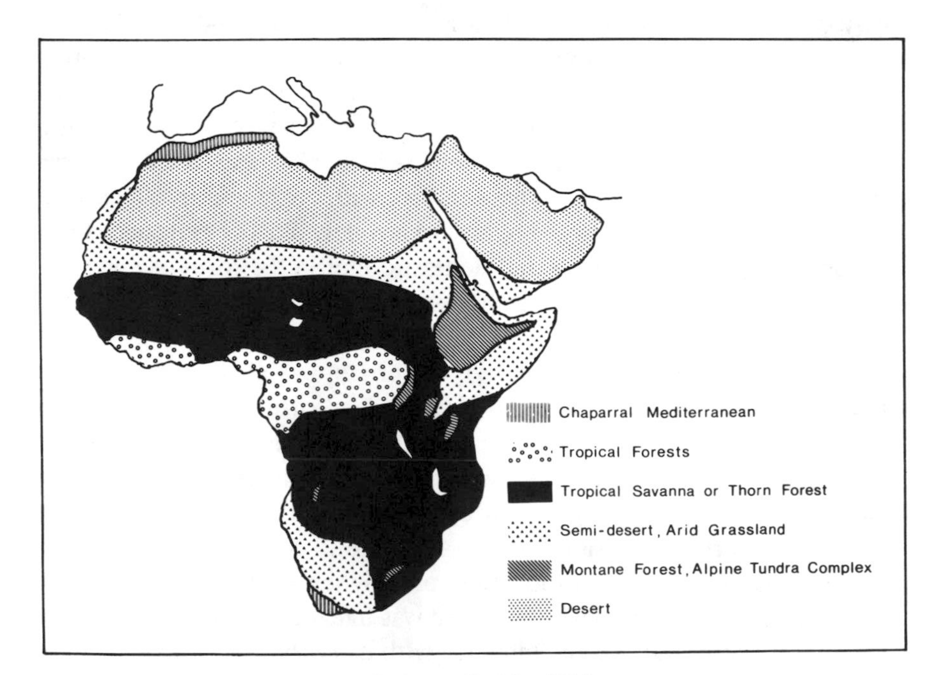


**Figure 2. Bioclimatic regions of Africa (Smith, 1974).**

unpredictable drought is a permanent feature of the climate (Glantz, 1987b). Although drought has been a common historical feature of much of Africa's climate, it did not result in widespread famine until recent years (Sinclair and Fryxell, 1985). The reality of repeated, unpredictable drought has been virtually ignored by development organizations, with consequent problems (Glantz, 1987a).

*Displacement of Pastoralism.* Pastoralism had been successfully practiced for centuries in arid and semiarid regions, but problems for this migratory-oriented system accelerated after World War II, when western countries increased their aid to African countries (Sinclair and Fryxell, 1985). Preference in development was given to crop production rather than to herding. This possibly was because there were more technological advances in crop production than in livestock production. Furthermore, most African politicians have ethnic roots in farming lifestyles that bias them toward sedentary rather than pastoral life (Horowitz and Little, 1987). This bias is further reinforced by the false idea that pastoralists always degrade the environment. A relatively humid climate during the 1950s and 1960s encouraged the northward expansion of settled agriculture into arid zones. The extent of pastoral rangeland was reduced because settled agriculture encroached onto marginal land (Ruddle and Manshard, 1981), and in some countries new crop varieties with short growing seasons allowed farmers to plant farther north on marginal land (Sinclair and Fryxell, 1985). The traditional symbiosis between pastoralist and cultivator was upset when pastoralists were no longer allowed to graze cattle on stubble in the farmers' fields, due to planting of the normally fallow area and dry-season planting of crops such as cotton (Ruddle and Manshard, 1981; Sinclair and Fryxell, 1985). This also deprived the soil of organic fertilizer in cattle manure.

Aid, reflected in medical, veterinary, and irrigational facilities, compounded by definition of national frontiers, caused an aglomeration of expanding human and animal populations. Persistent, excessive use of the land, beyond its carrying capacity, ultimately led to land degradation, diminished production, and soil erosion (Sinclair and Fryxell, 1985; Wade, 1984; Wijkman and Timberlake, 1984).

*Deforestation.* The arid and semiarid savanna has been further degraded by land-clearing for agriculture and wood-gathering for fuel. The degree of imbalance between demand and sustainable yields of fuelwood varies. In semiarid Mauritania firewood demand is 10 times the sustainable yield of remaining forests. In Kenya the ratio is 5:1; in Ethiopia, Tanzania, and Nigeria demand is 2.5 times the sustainable yield; and in Sudan it is roughly double (Brown and Wolf, 1986). This deforestation is a response to the increasing energy needs of an increasing human population, coupled with the need for productive land to replace degraded land. The result is a spiraling decline in forest resources.

*Changes in the Ecosystem.* As the intensity of grazing and wood-gathering increases in semiarid regions, perennial grasses and woody perennial shrubs are replaced by rapidly reproducing annuals comprising shallow-rooted, unpalatable legumes and other species (Brown and Wolf, 1986; Sinclair and Fryxell, 1985). The loss of trees, such as the acacia (*Acacia albida*), in the Sahel means less forage in the dry season, when the protein-rich acacia pods formerly fed livestock on otherwise barren rangeland (Brown and Wolf, 1986). Another serious effect occurs when all plant growth is completely eaten during the wet season, leaving nothing to eat during the dry season. In a very dry year starvation may occur.

Soils completely denuded from overgrazing, deforestation, and the trampling by animals are prone to rapid erosion, especially by hot, persistent winds. When rain does fall, puddling occurs on the soil surface because of reduced infiltration, and the accelerated runoff results in further soil loss. Surface evaporation losses increase on bare soils, leading to local salinization where deeper rooted plants have been removed and water tables have risen (Ruddle and Manshard, 1981). In other areas reduced infiltration contributes to falling water tables. Collectively, these conditions produce soil conditions unfavorable for any plant reestablishment (de Vos, 1975).

*Crop Production.* The cultivation of marginal land during the wetter conditions in the 1950s and 1960s increased the vulnerability of farmers' food security, aside from depriving pastoralists of their dry and wet season pastures. The multiyear drought in the late 1960s brought an inevitable return to drier conditions. Unsustainable crop production in those regions accelerated environmental degradation. The cultivation of land subject to a high degree of rainfall variability makes the land extremely susceptible to wind erosion during prolonged drought because the cultivated soils lack the vegetative cover necessary to minimize these erosive processes (Glantz, 1987a). The most dramatic effect of overgrazing, overcultivation, and settlement in the Sahel seems to be the famines of 1973 and 1984. Although drought was a proximate trigger for the famine, the prior overgrazing and overcultivation were the ultimate causes (Sinclair and Fryxell, 1985; Wijkman and Timberlake, 1985).

Marginal land is often put into agricultural production because governments tend to appropriate the relatively better agricultural areas for cash crops to export. These crops were considered so important that the export of cash crops continued or increased during the droughts and famines in the Sahel and Ethiopia in the early 1970s and 1980s (Glantz, 1987a). Little of the foreign exchange so earned goes back to agricultural development

Cash crops receive costly inputs, such as irrigation and well-drilling technology, fertilizers, herbicides, and pesticides. Traditional food crops are usually not able to bear the costs of high-input systems because food prices and the prices farmers get for their crops are kept artificially low by government policy in most African countries (Glantz, 1987a).

Irrigation is expensive. It also has negative effects on the environment, including reduced river flow and, therefore, increased soil erosion. Use of groundwater can lead to a severe decline in regional water reserves that cannot be recharged by rainfall. Other problems concern the impact on human health of irrigation runoff contaminated by pesticides and fertilizers and infectious diseases transmitted by organisms in irrigation water (Ruddle and Manshard, 1981). About 7 percent of the world's land area is affected detrimentally by salt; an estimated 50 percent of all irrigated land has been damaged by secondary salinization or sodification and waterlogging. Salt-affected soils are found on all continents (Ruddle and Manshard, 1981). A report on irrigation in the Sahel indicated that development of new irrigation areas had barely surpassed the area of older ones that had to be abandoned (Brown, 1987b).

*Desertification and Rainfall Trends.* The long-term result of inappropriate land use in dry areas is desertification—the process whereby the intensity or extent of desert conditions is magnified as a consequence of reduced biological productivity. This chain of events results in a reduction of plant biomass and an area's capacity to support livestock, crops, and, hence, people (Ruddle and Manshard, 1981). The trend toward increased sub-Saharan drought and the prospect of climatic change apparently is arising from this inappropriate land use and associated desertification (Glantz, 1987a). The trend toward marked declines in rainfall in the Sahel from 1950 to the present indicates that such climatic change may have begun (Figure 3). Even in normally wetter areas of Africa, the trend in the last 20 years has been toward reduced rainfall (Figure 4). Sinclair and Fryxell (1985) suggest that prolonged drought has become a reality, particularly in the Sahel, and it is against this backdrop that new agricultural practices must be evaluated. Aside from the land surface processes, larger scale anomalies in sea surface temperature may be responsible for a significant component of the vacillation of sub-Saharan rainfall (Semazzi et al., 1988).

*Livestock Production.* Beef production per head of cattle per annum averages 93 kilograms in the industrialized world and 25 kilograms in the Third World (Preston, 1986). This disparity has led many development workers to suggest that a similar increase in livestock production is needed to help solve food security problems in developing countries. However, the high rate of animal productivity is achieved in the industrialized coun-

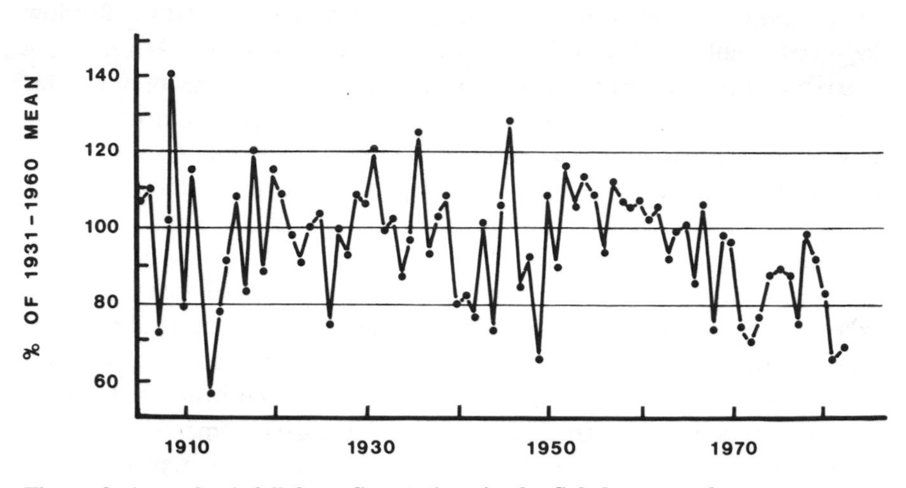


Figure 3. Annual rainfall from five stations in the Sahel expressed as a percentage of the 1931-1960 mean (Sinclair and Fryxell, 1985).

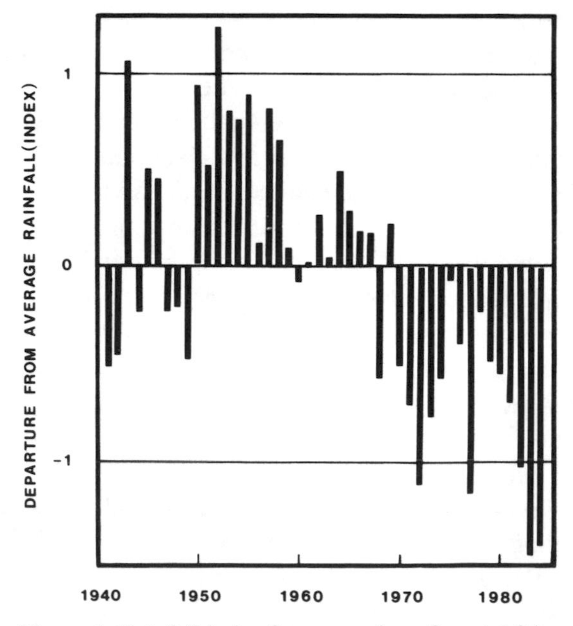


Figure 4. Rainfall index from a region of west Africa including the Sahel calculated on the basis of data from 20 stations (Glantz, 1987b).

tries through a disproportionate use of fossil fuels and the protein-rich cakes and meals (Preston, 1986). Developing tropical countries do not have the economies to generate the necessary foreign exchange to import the quality feeds used in intensive animal production. The import of western livestock breeds and production technologies necessitates long-term dependency on imported feeds and the "superior" animals to take maximum advantage of the transferred systems. This has led to serious neglect of indigenous breeds and feed resources (Preston, 1986).

African livestock is genetically diverse, tolerant to tropical pests, and produces milk and meat under poor range conditions. To replace them with western breeds that produce either milk or meat separately is not energy-efficient. The conversion efficiency for milk production with North American livestock under ideal conditions is approximately 25 percent. Beef production is much lower under similar conditions, at 4 percent on a protein-to-protein conversion efficiency basis (Table 5). Because many Africans (e.g., Masai) now eat milk and blood produced at an efficiency level of 20 to 25 percent, killing cattle for meat would lower their trophic efficiency at least five times. Furthermore, the less-than-ideal conditions in Africa would make these conversion efficiencies of western breeds much lower (Preston, 1986). Grasses on African rangeland, especially in the dry season, are much lower in protein content than are supplemented forages in the West, which would lead to even lower conversion efficiencies. Furthermore, western breeds are not resistant to the various African pests, such as ticks.

The lower conversion efficiencies of beef production and the instability of beef prices on the world market makes it unwise to begin a major, international export beef industry in Africa. For Africans the total efficiency

**Table 5. Efficiency of various classes of domestic livestock in converting feed nutrients into edible products.**

| *Class of Livestock* | *Efficiency of Conversion (%) (Plant protein to animal protein)** |
|---|---|
| Broilers (meat) | 23 |
| Turkeys (meat) | 22 |
| Hens (eggs) | 26 |
| Swine (meat) | 14 |
| Dairy cattle (milk) | 25 |
| Beef cattle (meat) | 4 |

*Assumes that nonessential and essential amino acid balance has been maintained.

of an animal is most important. Therefore, African livestock that produces milk, blood, and meat under poor habitat conditions and that can be used as draft animals, presents the best option for the African farmer.

Many livestock development schemes in Africa have been based on range management initiatives with the introduction of ranching systems or the provision of new watering points designed to regulate grazing. These approaches have not been successful because they failed to recognize, or ignored the fact, that an African livestock owner is, in many instances, a herdsman or a shepherd rather than a rancher. The livestock owner tends to live on the fringe of the monetary economy, and livestock is the major source of capital. Therefore, more importance is usually attached to numbers of livestock than to their productivity (Nestel, 1986). Furthermore, most imported technology has been developed on large farms that do not resemble most African farms, while the subsistence farmers, who occupy the largest group of workers, are largely ignored (Leng and Brumby, 1986).

*Humid and Subhumid Region.* Subhumid and humid areas encompass the wet savanna, which receives more than 1,000 millimeters of rain per year; the monsoon forest; and the rain forest, with more than 2,000 millimeters of rain (Ruddle and Manshard, 1981) (Figure 1). Traditional tropical agriculture exhibits one of two patterns. The first of these takes advantage of fertile soils that are enriched periodically by river-borne or wind-borne materials. The second, found on poorer soils, takes advantage of the great supply of nutrients available in tropical vegetation and the resilience of tropical biota (Dasmann et al., 1973). On poorer soils, the two major patterns of agriculture are shifting cultivation and the intensive village garden method of agriculture (Dasmann et al., 1973). The stability and ecological compatibility of these systems depend upon a low human population density, improvement in soil structure by plant roots, and erosion control through leaf litter, mulch, and a continuous canopy cover. Nutrient contributions are made to the soil through ash and recycling by deep-rooted perennials, and pests are controlled by growing a wide range of crop species simultaneously (Lal, 1987).

Given Africa's increasing population, these traditional systems have begun to break down because of a reduced forest fallow period for the land or because the soils are laid bare of vegetation. The structurally weak soils of the tropics suffer the greatest damage by erosion, and the problems of soil compaction are exacerbated by the use of mechanized farm vehicles. The absence of a forest fallow period and reduced additions of organic matter eventually result in reduced soil fertility because these highly leached soils are not naturally rich in nutrients. Tropical Africa has one of the worst erosion problems on arable lands in the world, often attended by irreversible reductions in crop yields (Aneke, 1986; Lal, 1987).

The need for foreign exchange encourages many developing countries to cut timber faster than forests can be regenerated (WCED, 1987). Shifting cultivation accounts for 70 percent of the clearing of closed-canopy forests and 60 percent of the cutting of savanna forests. Cutting for fuelwood also accounts for much deforestation (Brown and Wolf, 1986). More than 3.6 million hectares of Africa's forests are cleared each year—half of one percent of remaining forests (WECD, 1987). More than 5 percent of the moist forests of coastal west Africa were being cleared each year in the early 1980s (Brown and Wolf, 1986). A United Nation's Food and Agriculture Organization (FAO) survey of African forest resources showed that the area cleared each year exceeds that on which trees are deliberately planted by a ratio of 29 to 1, far higher than any other region in the world (Brown and Wolf, 1986).

Some humid forest land is also being cleared for cattle pasture. Much of this land is not suitable for pasture beyond a few years because of soil erosion and low soil fertility. Then, as degraded pasture land is abandoned, more forest is cleared, and the deforestation continues. Cattle production is also hindered by the animal's susceptibility to trypanisomiasis. Tsetse fly eradication projects are extremely expensive, and the massive use of chemicals has a detrimental effect on the natural environment (Goodland et al., 1984; Linear, 1985).

Evidence now shows that degradation of the world's tropical rain forests has implications for global climate. The earth's hydrologic cycle, the transfer of heat from the tropics toward the poles, the chemistry of the atmosphere, and global climate are all influenced by the tropical biota. Changes resulting from the vast destruction by humans are as yet unknown, but are likely widespread (Bunyard, 1985).

The indiscriminate transfer of technology from temperate zones to the tropics presents other problems. Inappropriate mechanization and reliance on cash crops, [cotton (*Gossypium barbadense*) and tobacco (*Nicotiana tabacum*)], without rotations, increase soil erosion and reduce soil productivity (Goodland et al., 1984). Persistent monoculture increases vulnerability to pest problems (Goodland et al., 1984), and high use of insecticides, fungicides, and herbicides, to suppress plant and animal pests (Dasmann et al., 1973), often has the unwanted effect of increasing pest problems (Charboussou, 1986). Inappropriate use of pesticides causes many environmental and health problems. The World Health Organization estimated in 1974 that about 500,000 people suffer acute pesticide poisoning worldwide annually (Wasilewski, 1987). Often, chemicals that are banned in western countries enjoy widespread use in the Third World. Illiteracy and lack of proper application equipment contribute to the contamination of water supplies and the environment generally and damage the health of farmers (Goodland et al., 1984; Linear, 1985; Wasilewski, 1987).

## Prospects for Sustainable Development

***The Green Revolution.*** When approaching agricultural development, it is vital to recognize that one cannot transfer en masse the Green Revolution from Asia to Africa. In the wet tropics of southeastern Asia, ecosystems are limited by low levels of soil nitrogen and phosphate, but rainfall is regular and adequate. Therefore, the addition of fertilizers and new plant hybrids has made possible the production wonders of the Green Revolution. Africa is different from southeastern Asia in that water is the major limiting factor. Periodic drought is common, and rain, when it comes, often is torrential. Africa also has more wind, which combines with the hot sun to result in very high evapotranspiration rates (Shipley, 1987). The biomass production of a plant relates directly to the evapotranspiration index (Pianka, 1974). Growth will not occur if there is insufficient water movement through a plant. Development of drought-resistant varieties of Green Revolution crops would allow these crops to survive under African conditions. However, they could not be expected to yield the same as in Asia because there is a significant trade-off between yield ability under ideal conditions and yield stability under drought conditions. With a lack of available water in much of Africa, it is inadvisable to expect an Asian Green Revolution to happen. Although some of Africa's soils are potentially fertile, most of the fertile areas lack the necessary water for high crop yields (Chesworth, 1982). Research and support for indigenous knowledge rather than more transferred western technology is needed so that Africa's huge rural work force can be supported in creating its own characteristic agricultural revolution (Shipley, 1987).

***Drought Relief.*** In arid and semiarid regions, the reality of drought must be incorporated into the agricultural development process. Drought relates closely to the problem of achieving sustained and adequate agricultural production in most countries of sub-Saharan Africa, so it can no longer be ignored in planning for development. Leaders in vulnerable countries must be educated about the nature of drought as a constraint on development (Glantz, 1987a) and alternative ways of dealing with it.

It is important that governments adopt policies that restrain the tendency to extend farming and grazing to marginal rainfall areas or unsuitable land. This may help to avoid continued desertification and the results of agricultural drought. Good rain-fed agricultural land should be switched from cash crops to food crops to enhance local self-sufficiency and food security. During the current (1987-1988) famine in Sudan, the government imposed a ban on exports of sorghum, which is the country's main export, so as to ensure sufficient domestic food supplies (Stover, 1987).

***Rangelands and Livestock Production.*** Because many grasslands have been damaged by overgrazing, effort must be devoted to the rehabilitation and reestablishment of this ecosystem. Destocking of animals should take place on a rotational basis to control numbers and to allow for marketing of animals (Dasmann et al., 1973). Measures to assist the regeneration of grasslands, such as rotational grazing, temporary protection of grazing land, seeding, or replanting to help re-establish plants in denuded areas, must be implemented (Ruddle and Manshard, 1981). Vegetational succession on degraded land must be monitored and appropriate grazing densities maintained. These ecological measures will affect the human population; therefore, any solution to these ecological problems must be resolved in the sociological context.

People should be encouraged to continue the traditional practice of mixed herds, instead of keeping just one kind of livestock (Coppock et al., 1986). Herded animals have distinct food and water requirements, and use of several livestock species enables pastoralists in marginal environments to exploit ecological niches that would otherwise remain underutilized. Camels and goats are the domestic stock best adapted to arid and semiarid ecosystems. The water and food requirements of sheep and cattle are more critical, so they are more restricted in their range (Ruddle and Manshard, 1981). Keeping a mixed herd would enable a buffer against periods of environmental hardship.

Indigenous breeds of livestock, such as the zebu, are well-suited to their native environment; they should not be replaced by western breeds (Seifert, 1975). Instead of having separate milk and beef production systems as in North America, the most economical way of meeting demand is through improvement of existing livestock production systems based on multipurpose animals (Preston, 1986). These can be bred for milk, meat, and use as draft animals to produce an improved multipurpose breed.

Although some have suggested the use of wildlife as protein for people, this is probably not feasible on a large scale. Wildlife parks hold a small remnant of the wildlife populations, so these are not really suitable for cropping. Cropping can be done in areas of diffuse animal populations, but only on a localized scale because of the logistical costs and problems with refrigeration and transportation to human centers. Cheap wild meat can also have an adverse effect on the local domestic meat economy. However, wildlife husbandry can be successful in a localized area, as exemplified by the Nazinga Project in Burkina Faso (Lungren, 1987). In most other situations it is probably more profitable to promote tourism relative to wildlife to generate foreign exchange (Dasmann et al., 1973).

***Crop Production.*** Problems in cultivation of land are broadly similar in dry and humid environments. Dry, marginal land should not be cultivated, especially with cash crops. It should instead be devoted to livestock production or production of appropriate drought-resistant forage crops. Rather than growing crops in arid and semiarid regions, which have adequate rainfall requirements only in unusually good years, one should plant drought-resistant crops. Such crops include sorghum (*Sorghum sp.*), cowpeas (*Vigna sinensis*), and millet (*Sorghum vulgare*), for example (Hunt, 1977). Native African species of these crops are genetically diverse and suited to the unpredictable environment (Deshmukh, 1986). Although they may not have a level of production as high as western varieties under ideal conditions, they are better able to produce consistently within a highly variable environment. Moreover, planting a diversity of crop species ensures a relatively stable production level over time, particularly in areas of high climatic variability. The crops may have varying environmental requirements, assuring in any particular year that at least one or more would produce a good yield relative to the environmental conditions. Although this has become standard practice in some African countries, more widespread adoption of this practice is advised.

In humid areas it is probably best to produce crops rather than to keep livestock because of the ecological problems resulting from the tsetse fly eradication program. Although research is being done on trypano-tolerant cattle, the introduction of cattle in humid areas would probably result in severe environmental degradation, unless closely monitored (Nestel, 1986). Furthermore, with the instability of beef prices on the international market, it is inadvisable to begin an African beef export industry even with trypano-tolerant cattle. For Africans, a secure food supply does not rely upon production of beef cattle. In these humid areas adequate nutrition can be gained from suitable crop production. These crops could be grown easily and would be a much more ecologically efficient way of attaining nutritional self-sufficiency. And such a production system would not produce the widespread environmental degradation that cattle in humid areas would produce.

Much soil erosion and nutrient loss on farmland can be controlled through tillage practices and mulch farming. In both dry and humid areas it is best to clear new land by hand. Although slow and labor-intensive, hand-stumping is the most ecologically sound method. If mechanical land-clearing must be done, then methods that minimize damage should be employed. However, mechanically cleared land must be planted immediately with a suitable cover crop for an appropriate duration to improve soil structure, prevent compaction, and provide the protective mulch cover (Jones, 1986; Lal, 1987).

Tillage methods can also be useful in preventing soil erosion and improving productivity. Semiarid soils can be plowed every other year or every other season in the row zone, depending upon soil conditions. This can improve crop yield. Ridge-cropping, which evolved in subsistence farming, can be improved by putting cross ties in the furrows. This practice is designed to allow more time for water to infiltrate into the soil. In humid areas no-till farming, which involves seeding through a crop residue mulch or sod without plowing, has definite advantages in conserving soil and water. Other advantages are its lowering the maximum soil temperature and maintaining high levels of organic matter. A comparison of grain yields for 24 consecutive no-till crops of maize showed definite yield advantages for no-till over the plow-based system (Jones, 1986; Lal, 1987). Such advances in these types of technology can benefit the small, family farmer. Mulch farming not only improves soil structure but also benefits crop yields by enhancing the activity of the soil fauna and adding plant nutrients to the soil.

An increase in soil fertility can also be gained with proper crop rotations and intercropping with other species. In arid and semiarid areas, growing cereals, such as millet (*S. vulgare*) and maize (*Zea mays*), in mixed or relay cropping patterns with groundnuts (*Arachis hypogaea*) and cowpeas (*V. sinensis*) may increase returns 50 to 80 percent and involves less risk of failure in a bad year (Lal, 1987). Some tree species, such as *A. albida*, can be grown with the crops to reduce soil erosion, enhance soil characteristics, and provide fodder for livestock. In the Sahel, it was found that *A. albida* trees return organic matter to the soil and thus increase the yield and protein composition of crops. Each tree improves 100 to 300 square meters of soil; 10 to 15 percent of the land is thus fertilized.

Prunings can also be used as sources of fodder for livestock and of mulch. A symbiosis between cultivators and herdsmen has existed for centuries. Use of this fodder source will help keep or reestablish the relationship. Traditionally, some subsistence farmers kept livestock and allowed grazing in their fields. The livestock, in turn, provided organic fertilizer for the garden. This use of compost and manure is another viable alternative to supplement chemical fertilizers.

Integrating livestock with tree crops and food crops is important in providing the needed diversity for an ecologically sustainable system in dry areas. In subhumid regions intercropping stabilizes yields even at low inputs. Mixed cropping of legumes and cereal is beneficial, as is maize (*Z. mays*) monoculture when rotated with a leguminous cover crop (Lal, 1987).

Humid areas are most suitable for crops such as plantain (*Musa paradisiaca*), bananas (*Musa sp.*), cassava (*Manihot sp.*), and yam (*Dioscorea sp.*). Some plantation crops, such as cocoa (*Theobroma sp.*), might be grown as well. Perennial tree crops adapted to tropical conditions can provide a valuable source of nutritious food; soil, water, and fertility can be maintained by growing a cover crop between rows. In some agroforestry systems food crops are grown in alleys formed by hedgerows of trees or shrubs, cut back at planting and kept pruned during cropping to prevent shading and to reduce competition with food crops. This practice provides mulch for the food crops, suppresses weed growth, creates a favorable microclimate, recycles nutrients from deeper soil layers, provides biologically fixed nitrogen to the companion crop, and produces a source of firewood. This type of planting decreases runoff amount and velocity, especially on sloping land (Cashman, 1988; Lal, 1987).

All of the above methods are important to increase soil fertility without the expense of imported fertilizers. Fertilizers will still be required in some situations, especially to balance the nutrients exported from a locality in food and fodder, but the emphasis must be on methods to enhance soil fertility in other inexpensive ways. This improvement will be particularly beneficial to subsistence farmers, who will be able to increase their yields without costly inputs. The requirements for pesticides will be reduced substantially by adopting ecologically compatible agricultural practices, such as mixed cropping, crop rotation, cover crops, and biological control methods. Breeding of plants to improve pest and disease resistance will also reduce input costs for small farmers (Lal, 1987). Although pesticides may be needed in intensive crop systems, their role in more extensive systems must be regarded as minor, especially when considering the total economic cost associated with their use and the growing awareness of alternatives to their use that ultimately result in comparable production. Detailed knowledge of pest ecology can facilitate cultural control. The avian species *Quelea quelea* forms large migratory flocks that become a serious pest in much of Africa. In Chad and Cameroon, timing the rice harvest to coincide with the seasonal absence of the birds, from mid-May to mid-June, reduced damage to less than one percent of the crop (Elliott, 1979).

***Land Resources.*** The tendency in western agriculture has been to cultivate large tracts of land to the extent that little unfarmed land remains in any one area. Additionally, this unfarmed land tends to exist in very small islands that are widely separated from each other. The result is that these wildlands contain fewer species of flora and fauna then would otherwise be expected (Diamond, 1975). If more extensive methods of agriculture are adopted, wildlands should exist in the form of networks interspersed with farmland to allow for the necessary species contact and movement. Such shelterbelts and wildland will then act as dispersal corridors for various species (Arnold, 1983; Wegner and Merriam, 1979). Networks like this would increase the species richness and diversity in wildlands, which would benefit a wildlife tourist industry. Furthermore, the different components of unfarmed land within agroecosystems contribute directly and indirectly in a variety of ways to benefit the farmland. Trees allow the cycling of minerals from deep within the soil. Trees and other forms of vegetation enhance the capture and retention of rainwater in the soil and participate in the hydrological cycle. They provide leaf fodder for livestock and wild animals and, ultimately, wood. They also offer the possibility for reducing pest populations in farmed areas because of the increased presence of insectivorous animals (Arnold, 1983). Agroforestry systems, as well as wildland networks, can result in soil erosion control, soil enhancement, and fuelwood supply (Lal, 1987). The Green Belt movement, founded and continued by women in Kenya, has been successful in reforesting areas around local farms and producing better soil, livestock fodder, and a local supply of firewood (Vollers, 1988). This type of approach to agricultural land use in Africa would benefit the agroecosystem and, most important, provide the bases for sustainability.

***Water Sources and Uses.*** The lack of surface water in arid and semiarid regions means finding new approaches to managing water that preclude most of the drawbacks of traditional irrigation. Much of Africa is still not irrigated, and large-scale irrigation schemes are not only capital-intensive and relatively short-lived (several decades) but many times are logistically infeasible (Lal, 1987). The efficiency of water use in large irrigation projects can be increased by replacing open irrigation canals with closed conduits to reduce evaporation, or by lining the canals to reduce seepage loss. At the farm level, water can be managed better by designing more efficient water distribution (e.g., trickle irrigation) and drainage systems between fields and by training farmers to use such methods. In areas where the soil has become saline because of misuse of irrigation water, salt-tolerant plants should be cultivated if possible (Ruddle and Manshard, 1981). The appropriate concepts for irrigation in Africa are small-scale, labor-intensive methods.

Although there are few rivers, lakes, and reservoirs in the dry areas, there is considerable potential for exploiting the rain that falls. The average local runoff coefficient in Africa is 20 percent, so rain can be retained in catchment areas (Stern, 1977). Rain-harvesting schemes are small-scale projects with low capital expenditures, often involving little more than small retaining walls or ridges along contours, constructed from locally available materials (Piper, 1986).

Other small-scale methods of providing water for crops and livestock are runoff agriculture, terrace-farming on steep hillsides, seepage control, and evaporation reduction from the soil surface, using gravel or rock mulches. Some of these techniques have been used for several millenia in certain arid zones (Ruddle and Manshard, 1981). Groundwater use can be beneficial, particularly on a small scale, as long as it is recharged by freshwater to prevent increasing salinization (Ruddle and Manshard, 1981). Although it is important to control the amount of water removed, so the water table does not fall, this is difficult to accomplish in practice, especially in densely populated areas.

***Policies Affecting Agriculture and Ecosystems.*** When considering the prospects for sustainable development in African countries, one must consider the relevance of their agricultural policy goals. In Europe and North America, agricultural policy has been dominated by an emphasis on increasing productivity. The great increases in agricultural output can be attributed to cheap, readily available energy; intense mechanization; and the widespread use of chemical fertilizers, pesticides, and herbicides (Thomas, 1988). Under western advice, Africa adopted many of the same agricultural policies—policies that ignored and in many cases damaged ecosystems irreparably. However, agriculture in the developed world is now in crisis because revenues have not kept pace with rising input costs; farm profits have, therefore, fallen dramatically. Increased production does not necessarily equal increased profit. As cash costs of production make up a larger and larger proportion of total costs, farmers in the developed world are more vulnerable to gambler's ruin (Ehrenfeld, 1987). Given the current economic situation, most farmers believe they are currently unable to afford to practice wise, ecologically sympathetic agriculture. Furthermore, the natural resource base for agriculture in the West is being severely degraded. In North America and Europe, as well as in Africa, it is time for agricultural policy to de-emphasize increased production and emphasize sustainable production at levels profitable to producers (Thomas, 1988).

Although the ecosystems are very different, agricultural policies focusing on increased production without concern for the ecosystem have produced similar problems in Africa and the western nations. The problem of low revenues, high costs, and declining profits has been exacerbated by the low rainfall and drought in Africa. This is due mainly to the fact that the transplanted farming practices are not adjusted to African ecosystems. The poor economic conditions in Africa, and the changing environment, mean that agricultural policy should focus on low-input, intrinsically sustainable production systems rather than high-input systems aimed at increased productivity. This would involve a change from intensive (chemical-energy-dependent) production of a limited crop to an extensive (low dependency on high-cost chemical inputs), small-scale production of a variety of types, particularly mixed crop-livestock systems. Paradoxically, such traditional farming systems were once common in Africa.

Africans have undergone their cultural evolution within distinct biomes, and the components of their culture have been shaped by the principal ecological variables of their region. Therefore, development and implementation, instead of stressing adoption of new, exotic technology, should emphasize adaptation of existing traditional practices (Nestel, 1986), from pastoralists and farmers alike.

In North America and Europe, the number of people involved in farming represents a small percentage of the population. In spite of this, there are still huge surpluses of agricultural products, due mainly to intensive, high-input agricultural practices and government subsidies. The result has been an exodus from rural areas to urban centers. Similarly, the advent of intensive, mechanized farming in Africa means that fewer people are required for agricultural labor, and many are moving to the cities. An extensive approach to agriculture requires more manpower in the rural areas and involves more people in the production of their own food. This would help to ensure food security because greater numbers of Africans would participate directly in their own food production.

Another problem in the transfer of western agriculture to Africa has been the emphasis on growing cash crops for export in world markets. This is one option available for countries; the return from sale of cash crops can be used to buy food for the nation. However, this approach has its limitations. Often, there is an excess of these or similar goods in international markets so that prices are low and yield little foreign exchange to the African country, and even less revenue to the producer. In 1983-1984, at the beginning of the drought and famine in the Sahel, five Sahelian nations produced record amounts of cotton. In those years, record amounts of cereals were imported to the Sahel. Over the period that Sahelian cotton harvests were steadily rising, world cotton prices were steadily falling in real terms. If farmers who produce cotton cannot feed themselves, this shows that cash crops are getting too much attention and food crops too little (WCED, 1987). Often, the foreign exchange derived from the export of such goods benefits the urban dwellers with western tastes; little goes back to the farmer. Therefore, most local inhabitants are not assured of immediate food security, and the ecosystem is also damaged from the monocultures of cash crops. This conflict between the values of political leaders and the ecological realities of their countries must be resolved.

Biotechnology has been put forward in recent years as a means to help farmers raise their incomes. It is unlikely that biotechnology will have much effect on agricultural output or farm profits, however. The reasons are, first, biological. There are limits to what can be achieved within the context of gene manipulation, and the evolutionary constraints on such technology because of ecological interaction and adaptation. Second, biotechnology is unlikely to alleviate the general problems in agriculture because of its exceedingly high initial cost and the nature of the agricultural treadmill. The added expense of biotechnological products will mean more high-input farming and a greater likelihood of gambler's ruin for farmers (Ehrenfeld, 1987). The high cost of biotechnology and its underlying assumptions will limit its usefulness, particularly for African countries with unstable economies.

As agricultural policies are revised, it is essential for planners to recognize that short- and long-term economic returns and environmental welfare are inseparable. Investment in the environment, especially its rehabilitation, makes sound business sense because, in the long run, it assures the continuous, sustained production of the ecosystem, and the total revenues will far exceed the costs of protection and rehabilitation (Thomas, 1988). Realization of the synonomy of environment and economy requires that government policies directing agriculture in developing nations and criteria for foreign aid to such nations give priority to extensification of agriculture and its role as the basis of food security.

## Conclusion

The problem of food security in Africa is linked to the continent's diverse ecosystems and the climatically variable environments. Any attempts to realize assured, continuous agricultural production from environments that are often constrained climatically must recognize the limitations of widespread monoculture and favor systems of diversified production. Transferring an intensive, production-oriented approach to agricultural development from the West to Africa has led to environmental problems and unsustainable development. The focus of African development in the future must not be on increased agricultural productivity solely, but rather on sustainable development that is appropriate to the environment, and profitable for the small farmer. Ecologically efficient food production should be emphasized. Furthermore, sociological solutions, in addition to ecological ones, must be sought.

Of course, there will always be a need for some intensive food production, particularly around urban centers, because of the high population growth rate. The intensive and extensive approaches to agriculture represent the two extremes of a continuum. The question of where the optimal point is along the continuum has not been answered. Moreover, that point will differ according to biome. Research should concentrate on answering this question because environmental integrity is inextricably linked to food security.

## REFERENCES


Aneke, D. O. 1986. Coping with accelerated soil erosion in Nigeria. Journal of Soil and Water Conservation 41: 161-163.

Arnold, G. W. 1983. The influence of ditch and hedgerow structure, length of hedgerows, and area of woodland and garden on bird numbers on farmland. Journal of Applied Ecology 20: 731-750.

Brown, L. R. 1987a. Analyzing the demographic trap. pp. 20-37. *In* L. Starke [editor] State of the world 1987. W. W. Norton and Company, New York, New York.

Brown. L. R. 1987b. Sustaining world agriculture. pp. 122-138. *In* L. Starke [editor] State of the world 1987. W. W. Norton and Company, New York, New York.

Brown, L. R., and E. C. Wolf. 1986. Assessing ecological decline. pp. 22-39. *In* L. Starke [editor] State of the world 1986. W. W. Norton and Company, New York, New York.

Brown, L. R., and S. Postel. 1987. Thresholds of change. pp. 3-19. *In* L. Starke [editor] State of the world 1987. W. W. Norton and Company, New York, New York.

Bunyard, P. 1985. World climate and tropical forest destruction. The Ecologist 15: 125-136.

Cashman, K. 1988. The benefits of alley farming for the African farmer and her household. pp. 231-237. *In* P. Allen and D. Van Dusen [editors] Global perspectives on agroecology and sustainable agricultural systems. Proceedings. Sixth International Scientific Conference of the International Federation of Organic Agriculture Movements, Volume 1. Regents of the University of California, The Agroecology Program, Santa Cruz.

Charboussou, F. 1986. How pesticides increase pests. The Ecologist 15: 125-136.

Chesworth, W. 1982. Late cenozoic geology and the second oldest profession. Geoscience 9: 56-61.

Coppock, D. L., J. E. Ellis, and D. M. Swift. 1986. Livestock feeding ecology and resource utilization in a nomadic pastoral ecosystem. Journal of Applied Ecology 23: 573-583.

Dasmann, R. F., J. P. Milton, and P. H. Freeman. 1973. Ecological principles for economic development. John Wiley and Sons, London, England.

Deshmukh, L. 1986. Ecology and tropical biology. Blackwell Scientific Publications, Palo Alto, California.

de Vos, A. 1975. Africa, the devastated continent? Dr. W. Junk Publishers, The Hague, The Netherlands.

Diamond, J. M. 1975. The island dilemma: Lessons of modern biogeographic studies for the design of nature reserves. Biological Conservation 7: 129-146.

Ehrenfeld, D. 1987. Implementing the transition to a sustainable agriculture: An opportunity for ecology. Bulletin of the Ecological Society of America 68: 5-8.

Elliott, C.C.H. 1979. The harvest time method as a means of avoiding Quelea damage to irrigated rice in Chad/Cameroon. Journal of Applied Ecology 16: 23-35.

Glantz, M. H. 1987a. Drought and economic development in sub-Saharan Africa. pp. 37-58. *In* M. H. Glantz [editor] Drought and hunger in Africa. Cambridge University Press, Cambridge, England.

Glantz, M. H. 1987b. Drought in Africa. Scientific American 256: 34-40.

Goodland, R.J.A., C. Watson, and G. Ledec. 1984. Environmental management in tropical agriculture. Westview Press, Boulder, Colorado.

Horowitz, M. M., and P. D. Little. 1987. African pastoralism and poverty: Some implications for drought and famine. pp. 59-82. *In* M. H. Glantz [editor] Drought and hunger in Africa. Cambridge University Press, Cambridge, England.

Hunt, D. 1977. Poverty and agricultural development policy in a semi-arid area of eastern Kenya. pp. 74-91. *In* P. O'Keefe and B. Wisner [editors] Landuse and development. African Environment Special Report 5. International African Institute, London, England.

Jones, M. J. 1986. Conservation systems and crop production. pp. 47-59. *In* L. J. Foster [editor] Agricultural development in drought-prone Africa. Overseas Development Institute, London, England.

Lal, R. 1987. Managing the soils of sub-Saharan Africa. Science 236: 1,069-1,076.

Leng, R. A., and P. Brumby. 1986. Cattle production in the tropics. pp. 884-887. *In* T. G. Taylor and N. K. Jenkins [editors] Proceedings of the XIII International Congress of Nutrition. John Libbey & Company, Ltd., London, England.

Linear, M. 1985. The tsetse war. The Ecologist 15: 27-35.

Lungren, C. 1987. The Nazinga report. Quarterly Report No. 31, April-June. African Wildlife Husbandry Development Association, Burkina Faso. (unpublished).

Nestel, B. 1986. Livestock and range resources. pp. 79-86. *In* L. J. Foster [editor] Agricultural development in drought-prone Africa. Overseas Development Institute, London, England.

Pianka, E. R. 1974. Evolutionary ecology. Harper & Row Publishers, New York, New York.

Piper, B. 1986. Note on surface water resources. pp. 43-45. *In* L. J. Foster [editor] Agricultural development in drought-prone Africa. Overseas Development Institute, London, England.

Preston, T. R. 1986. Matching livestock systems with available feed resources in tropical countries. pp. 873-880. *In* T. G. Taylor and N. K. Jenkins [editors] Proceedings of the 13th International Congress of Nutrition. John Libbey & Company Ltd., London, England.

Rasmusson, E. M. 1987. Global climate change and variability: Effects on drought and desertification in Africa. pp. 3-22. *In* M. H. Glantz [editor] Drought and hunger in Africa. Cambridge University Press, Cambridge, England.

Ruddle, K., and W. Manshard. 1981. Renewable natural resources and the environment. Natural Resources and Environment Series, Volume 2. Tycooly International Publishing Limited, Dublin, Ireland.

Seifert, H.S.H. 1975. Animal production and health in the Sahelian zone. pp. 55-60. *In* The Sahel: Ecological approaches to land use. M.A.B. Technical Notes. The Unesco Press, Paris, France.

Semazzi, F.H.M., V. Mehta, and Y. C. Sud. 1988. An investigation of the relationship between sub-Saharan rainfall and global sea surface temperatures. Atmosphere-Ocean 26: 118-138.

Shipley, K. G. 1987. Agriculture in Africa. Agrologist 16: 5-6.

Sinclair, A.R.E., and J. M. Fryxell. 1985. The Sahel of Africa: Ecology of a disaster. Canadian Journal of Zoology 63: 987-994.

Smith, R. L. 1974. Ecology and field biology (2nd edition). Harper & Row Publishers, New York, New York.

Standing Committee on External Affairs and International Trade (SCEAIT). 1987. Trade on Canada's official development policies and programs, for whose benefit? Report of the Standing Committee on External Affairs and International Queen's Printer, Ottawa, Canada.

Stern, P. H. 1977. Desertification and man in the Sahel. pp. 171-178. *In* P. O'Keefe and

B. Wisner [editors] Landuse and development. African Environment Special Report 5. International African Institute, London, England.

Stover, E. 1987. A famine of wandering. New Scientist 1589: 30.

Summers, J. D., and S. Leeson. 1985. Poultry nutrition handbook. Ontario Ministry of Agriculture and Food, Toronto.

Thomas, V. G. 1988. Agro-ecology: Past and future trends, chapter 8. pp. 67-73. *In* M. Moss [editor] Landscape ecology and management. Polyscience Publications Inc., Montreal, Canada.

Tolba, M. K. 1987. Forward. pp. IX-X. *In* A. K. Biswas and Q. Geping [editors] Environmental impact assessment for developing countries. Natural Resources and the Environment Series, Volume 19. Tycooly International Publishing Limited, Dublin, Ireland.

Vollers, M. 1988. Healing the ravaged land. International Wildlife 18: 4-11.

Wade, N. 1974. Sahelian drought: No victory for western aid. Science 185: 234-237.

Warren, A., and J. K. Maizels. 1977. Ecological change and desertification. pp. 169-260. *In* Desertification: Its causes and consequences. The Secretariat, United Nations Conference on Desertification. Pergamon Press, Oxford, England.

Wasilewski, A. 1987. The quiet epidemic. Agrologist 16: 16-17.

Wegner, J. F., and G. Merriam. 1979. Movements by birds and small mammals between a wood and adjoining farmland habitats. Journal of Applied Ecology 16: 349-357.

Wijkman, A., and L. Timberlake. 1985. Is the African drought an act of God or of man? The Ecologist 15: 9-18.

World Bank. 1986. World development report 1986. Oxford University Press, Oxford, England.

World Commission on Environment and Development (WCED). 1987. Our common future. Oxford University Press, Oxford, England.

# UNDERSTANDING THE BASIS OF SUSTAINABILITY FOR AGRICULTURE IN THE TROPICS: EXPERIENCES IN LATIN AMERICA

Stephen R. Gliessman

In attempting to establish sustainable agricultural systems in the tropics, it has become necessary to include parameters other than yields and economic profitability in determining sustainability (Altieri and Anderson, 1986; Gliessman, 1985b). This has become especially true during the past decade or so because the debt load of most tropical countries has risen dramatically. Correspondingly, these countries have become increasingly dependent on imported food to meet the basic needs of rising populations (Brown, 1988).

Agricultural research of the last several decades, in its struggle to solve hunger and production problems in the developing world, has focused most of its resources on the development and transfer of technologies that are often not appropriate to the cultural needs or resource base of the receiving countries. Despite the achievement of dramatic yield increases in some sectors of the agricultural communities of many tropical countries, these yields increasingly are dependent on costly imported inputs. Often, this has brought about a massive shift to the production of crops destined for export to generate the cash flow necessary to purchase the inputs. As a result, less land is used for growing basic food crops, especially land with the best agricultural capabilities. Food imports by these countries have risen dramatically, hunger in rural areas has increased, and the movement of farmers to urban centers is well-known. As the better land is transferred to export crops, local farmers who do not leave the land are forced to move onto less productive marginal land or areas of uncut tropical forest, thereby promoting further deforestation and ecological degradation of the land base. These farmers often receive the primary blame for such environmental disasters, when most often they are victims of shifting economic and political priorities over which they have little control (Barkin, 1978; Wright, 1984). This places them in the position of being even less able to afford new technologies and inputs on their farms.

During the past several years, there has been a growing awareness of the need to reorient agricultural development programs to address more directly the needs of the resource-poor small farmers of the developing world, especially in the tropics (Altieri, 1984; Altieri and Anderson, 1986; Dover and Talbot, 1987; Marten, 1986). At the same time, there has been a growing recognition of the value of local, traditional agroecosystems that have enabled peasant farmers to meet their subsistence needs for centuries under adverse environmental and economic conditions (Altieri, 1987).

Therefore, throughout much of the tropics, and especially in Latin America, traditional knowledge continues to form the basic foundation of agriculture and small farm management. Such knowledge reflects experience gained from past generations, yet continues to develop in the present as the ecological and cultural environment of the populations involved go through a continual process of adaptation and change. Studies of traditional agriculture are beginning to show the great value such systems have for contributing to the development of ecologically sound management practices that are understandable and acceptable to rural people (Altieri, 1984; Gliessman, 1984; Klee, 1980; Wilken, 1988). These systems often make use of locally available resources rather than relying on costly inputs imported from distant sources. They allow for the simultaneous satisfaction of local needs and a significant contribution to demands on a larger scale. Most important, protection takes place in ways that focus more on the long-term sustainability of the system rather than an overemphasis on the maximization of yields. The ability of these systems to keep the land productive on a permanent basis, reduces the need for the development of new land. An agroecological approach to understanding how such systems function can provide information that can contribute significantly to establishing sustainable small farm systems for the tropics.

## Sustainability of Traditional Tropical Agroecosystems

Sustainability refers to the ability of an agroecosystem to maintain production through time, in the face of long-term ecological constraints and disturbances as well as an array of socioeconomic pressures (Altieri, 1987; Conway, 1985). The emphasis of modern agriculture, under the criteria of sustainability, is undergoing a gradual shift from a primary goal of maximizing production and profit for the short term to a perspective that also considers the ability to maintain production in the long run (Allen and Van Dusen, 1988). This ability is beginning to be evaluated on an expanding set of criteria, including aspects such as soil and water conservation, genetic diversity, and appropriate management, to ensure a stable food supply, a reasonable quality of rural life, and a safe and healthy environment (Allen and Van Dusen, 1988; Altieri, 1987; Douglass, 1984; Edens et al., 1985; Jackson et al., 1984; Lowrance et al., 1984). At the same time, there is considerable concern for the possible trade-offs between the goals of maximizing production and maximizing sustainability (Conway, 1985). This is what makes many traditional agroecosystems so valuable, especially those that have been studied in the tropics of Mexico and Central America (Altieri, 1987; Gliessman et al., 1981; Wilken, 1988). They have been in use for a long time, and during that time have gone through many changes and adaptations. The fact that they still are in use is strong evidence for a social and ecological stability that modern, mechanized systems could well envy. We may have much to learn from them.

The context of sustainability includes the following criteria:

- A low dependence on external, purchased inputs.
- The use of locally available and renewable resources.
- Benign or beneficial impacts on both the on- and off-farm environments.
- Adapted to or tolerant of local conditions rather than dependent on massive alteration or control of the environment.
- The long-term maintenance of productive capacity.
- Biological and cultural diversity.
- The knowledge and culture of local inhabitants.
- Adequate domestic and exportable goods.

Sustainable agriculture depends upon the integration of all of these components. This involves understanding the agroecosystem at all levels of organization, from the crop plant or animal in the field, to the entire farm, to the region or beyond (Hart, 1984). Knowledge of this integration has been generated by the work of a diverse group of researchers, organizations, institutions, and, most important, farmers in Latin America. Examination of a few traditional agroecosystems in this region can begin to provide a valuable means of understanding this integration.

## Ecology and Management of Traditional Agroecosystems

***Multiple-Crop Agroecosystems.*** The use of crop associations or multiple cropping can contribute greatly to increasing the sustainability of a cropping system (Amador, 1980; Gliessman, 1985a). Multiple cropping means that more than one crop occupies the same piece of land either simultaneously or in some type of rotational sequence during the season. Production can be increased, more efficient use of resources takes place, and the land can be occupied productively more continuously. The importance of multiple cropping is being recognized (Francis, 1986), and the need for intensive agroecological studies of such mixed cropping has become more evident.

One traditional tropical multiple-cropping system that has been studied in some detail is a polyculture of maize, beans, and squash. There is evidence that intercropping maize (*Zea mays*) and beans (*Phaseolus vulgaris*) has been practiced in Central America since prehispanic times, and it continues to form an important part of the patterns of food production in this region today (Pinchinat et al., 1976). In a series of studies done in Tabasco, Mexico, researchers found that maize yields could be stimulated as much as 50 percent above monoculture yields when planted with beans and squash (Table 1) (Amador, 1980). There was some yield reduction for the two associated crop species, but the summed yields for the crops planted together were higher than for an equivalent amount of land planted to the crops in monoculture (overyielding).

Studies that provide an understanding of the ecological mechanisms of this yield increase are important for establishing a strong basis for recommending widespread use of the cropping system. On the one hand, it appears that beans in polyculture with maize nodulate more and potentially are more active in fixing nitrogen biologically, which could be made directly available to the maize (Boucher, 1979). Net gains of nitrogen have been observed when the crops are associated, despite the removal of this element with the harvest (Gliessman, 1982). This contributes to both the long-term reduction in dependence on external purchased inputs of fertilizer and an over-

**Table 1. Yields of a polyculture of maize, beans, and squash as compared to monocultures planted at four different densities, Cardenas, Tabasco, Mexico (Amador, 1980).**

| | *Monoculture* | | | | *Polyculture* |
|---|---|---|---|---|---|
| Maize | | | | | |
| Densities (plants/ha) | 33,000 | 40,000 | 66,000 | 100,000 | 50,000 |
| Yield (kg/ha)* | 990 | 1,150 | 1,230 | 1,170 | 1,720 |
| Beans | | | | | |
| Densities (plants/ha) | 56,800 | 64,000 | 100,000 | 133,200 | 40,000 |
| Yield (kg/ha)* | 425 | 740 | 610 | 695 | 110 |
| Squash | | | | | |
| Densities (plants/ha) | 1,200 | 1,875 | 7,500 | 30,000 | 3,330 |
| Yield (kg/ha)* | 15 | 250 | 430 | 225 | 80 |

*Yields of maize and beans expressed as dried grain, squash as fresh fruits.

all, more stable basis for managing resources within the system.

At the same time, studies of the management practices employed in this polyculture have demonstrated the ecological basis upon which the practices can function. For example, despite the lower squash yields in the mixed planting, farmers insist that the crop system benefits from the presence of squash through the control of weeds (Gliessman, 1983). The thick, broad, horizontal leaves of squash cast a dense shade that blocks sunlight, while leachates in rains washing off the leaves contain allelopathic compounds that potentially inhibit weed growth. Herbivorous insects are at a disadvantage in the intercrop system (Risch, 1980), and the presence of beneficial insects is promoted (Letourneau, 1983). Leaving weeds in the intercropped system can be advantageous as well (Chacon and Gliessman, 1982). *Chenopodium ambrosioides*, for example, has the potential for inhibiting plant pathogenic nematodes through the release of toxic root exudates (Garcia, 1980). *Lagascea mollis* can control the invasion of weeds detrimental to the crop's allelopathically if it is allowed to form a dense cover after the critical establishment stage in crop development (the first three to four weeks), thus avoiding inhibition of the crop as well (Gliessman, 1983). Many other factors are coming into play in such an interactive system, and detailed and long-term research is necessary to determine their relative importance and contribution to its long-term sustainability.

***Diverse Home Gardens.*** One agroecosystem that seems to incorporate most of the criteria for sustainabillity is the tropical home garden or kitchen garden system (Allison, 1983; Gliessman, 1988). Such gardens are structurally diverse, with an overstory of trees and an understory mixture of herbs, shrubs, small trees, and vines. This diversity permits year-round harvesting of food products as well as a wide range of other products used on small farms in developing countries, such as firewood, medicinal plants, spices, and ornamentals. In an ecological analysis of home gardens on both lowland and upland sites in Mexico, researchers found that in quite a small area (0.3 to 0.7 hectare) high diversity permitted a high degree of similarity between the managed agroecosystem and local natural systems. Relatively high diversity of species for a cropping system also was achieved (Table 2).

A home garden on the outskirts of Canas, Guanacaste Province, Costa Rica, included 71 plant species in an area of 1,240 square meters (Table 3). The garden served as a source of food, firewood, medicine, and color and enjoyment for the household. Some of the plant species served more than one function. The Shannon-Weaver species diversity index for the garden was 3.55, a relatively high value for an agricultural system. To a certain extent, plants also were distributed in the garden depending on their uses. Trees were concentrated toward the back of the plot, providing shade for the work area at the back of the house, as well as providing a stabilizing border along a riverbank that parallels the back of the property. Annual food crops were concentrated toward the front of the garden in full sunlight. The large number of ornamental species was clustered in beds or containers around the walls of the house or along the pathway leading from the front of the property to the house. Animal pens behind the house in the shade of the trees contained two pigs, a goat, and a guinea pig. An undetermined number of chickens roamed freely throughout the plot, as did several small dogs and two cats. Mango was the principal tree species, with maize, squash, beans, papaya, bananas, and yucca (cassava) playing the most important roles in food production. The man of the household had a full-time job in the nearby town, so the garden played more of a supplemental role in the family economics.

Ecological as well as sociological studies are needed to understand further the structure and diversity of the garden. Home gardens are extremely

**Table 2. Species types and characteristics of home garden agroecosystems on upland (Tepeyanco, Tlaxcala) and lowland (Cupilco, Tabasco) sites in Mexico.***

| *Characteristics* | *Cupilco* | *Tepeyanco* |
|---|---|---|
| Average garden size (ha) | 0.70 | 0.34 |
| Number of useful species per garden | 55 | 33 |
| Diversity (bits) | 3.84 | 2.43 |
| Leaf area index | 4.5 | 3.2 |
| Cover (%) | 96.7 | 85.3 |
| Light transmission (%) | 21.5 | 30.5 |
| Perennial species (%) | 52.3 | 24.5 |
| Tree species (%) | 30.7 | 12.3 |
| Ornamental plants (%) | 7.0 | 9.0 |
| Medicinal plants (%) | 2.0 | 2.8 |

*Data from four gardens in Tepeyanco and three in Cupilco (Allison, 1983).

**Table 3. Number of plant species and individuals of each species in a home garden agroecosystem on the outskirts of Canas, Guanacaste Province, Costa Rica, according to common uses.***

| *Plant Uses* | *Number of Species* | *Number of Individuals* | *Percent of Species* | *Percent of Individuals* |
|---|---|---|---|---|
| Ornamental | 36 | 517 | 48 | 21.6 |
| Food | 26 | 164 | 36 | 68.2 |
| Medicinal | 6 | 1 | 8 | 1.6 |
| Firewood | 5 | 17 | 7 | 1.7 |
| Animal feed | 1 | 51 | 1 | 6.7 |
| Total | 71 | 758 | | |

*Total species and individual numbers are less than the sum of the columns due to the multiple function of some species (Gliessman, 1988).

variable in size and design. They respond to local variations in soil type, drainage patterns, cultural preferences, economic standing of the family, family size and age patterns, and other factors, reflecting a multiplicity of both ecological and cultural components. At the same time, they are flexible, dynamic, and changing, depending on the needs of the family (Gonzalez Jacome, 1985).

In a home garden in the Atlantic lowlands of Costa Rica, near Puerto Viejo, Sarapiqui, mapping revealed considerable diversity and complexity in an area of about 3,250 square meters. There were 26 species of trees, 16 perennial ornamentals, 8 annual/biennial crops, and 6 herbaceous species in the garden at the time of the study. The plants were distributed into what could be characterized as five functional areas, as follows:

- A low-diversity, regularly patterned planting of crops of potential cash value, including tuber crops, pineapple, and young coconuts.
- A high-diversity, irregularly patterned planting of trees, shrubs, herbs, and vines of many uses designed to satisfy domestic needs.
- A low-diversity, widely spaced planting of trees, most often with low grass or bare soil below, often used for social or recreational purposes.
- A very high-diversity, intercropped planting of ornamental herbs and shrubs planted close to the house and cared for by the women in the household.
- A moderate-diversity, alternately-planted fencerow surrounding the property, primarily composed of fruit and firewood tree species.

The garden reflected interactions between the need for domestic food or use items, the desire or need for cash income, personal preference and enjoyment, and the constraints of time and space. A move into cash cropping, relatively new to this particular garden, has changed its structure dramatically. This trend will continue as trees mature, markets change, and the socioeconomic status of the family changes. Home gardens seem to incorporate this flexibility and dynamism, while at the same time maintaining a sustainable basis to their design and management.

***A Traditional Wetland Agroecosystem.*** Environmental factors that limit agricultural production, be they physical or biological, require special adaptations for agriculture to be sustainable. Agricultural development projects in much of the tropics normally have approached such limitations intent upon eliminating or altering them to fit the needs of the cropping systems being introduced. This usually involves high levels of external inputs of energy or materials. There are many well-known examples of massive irrigation, drainage, and desalination projects that have attempted to alter existing ecological conditions but have achieved only limited success when evaluated in terms of crop productivity and economic viability; at the same time they have little applicability for meeting the needs of small farmers (Barkin, 1978). In many areas of Mesoamerica, where heavy rainfall and low-lying topography combine to generate conditions of excess soil moisture, local farmers have found ways to accommodate this factor into the design and management of sustainable traditional agroecosystems.

A very interesting and productive use of wetland areas has been observed in the state of Tlaxcala, Mexico, for the production of maize and other crops (Anaya et al., 1987; Crews, 1985; Gonzalez Jacome, 1986; Wilken, 1969; Wilken, 1988). In an area known as the Puebla Basin, a triangular floodplain of about 290 square kilometers is formed where the Atoyac and Zahuapan Rivers meet in the southern part of the state. A large part of the basin floor has a water table less than one meter below the surface during much of the year (Gonzalez Jacome, 1986). Soils are poorly drained and swampy (Wilken, 1969). To make such land agriculturally productive, most present-day agronomists probably would recommend draining the region so that large-scale mechanized cropping practices could be introduced. But an examination of local, traditional cropping systems provides an alternative that makes use of the high water table and hydrological characteristics of the basin (Figure 1).

Using a system that is prehispanic in origin (Wilken, 1969, 1971), raised platforms, locally called "camellones," have been constructed from soil ex-

**Figure 1. Drawing of a cross-section of platform-canal agroecosystem in the Puebla Basin of Tlaxcala, Mexico. Drawing by Peg Mathewson (Crews, 1985).**

cavated from their borders, creating a system of platforms and canals, called *zanjas*. Individual platforms vary from 15 to 30 meters wide and 150 to 300 meters long. A diverse mixture of crops is grown on the platforms, including intercropped maize, beans, squash, vegetables, alfalfa, and other annuals (Crews, 1985; Gonzalez Jacome, 1986). Crop rotations with legumes, such as alfalfa or fava beans, help maintain soil fertility, and the crop mixtures themselves help control weeds (Anaya et al., 1987). Soil fertility is maintained with frequent applications of composted animal manures and crop residues. Much of the feed for the animals comes from alfalfa grown on the platforms, or from residues that cannot be directly consumed by humans from other crops, for example, corn stalks. Supplemental feed for animals is derived from the noncrop vegetation, such as weeds, that is removed selectively from the crop area or periodic harvests of the ruderals and native plants that grow, either along the canals or directly in them, as aquatic species. This latter source of feed can constitute a significant component of livestock diets during the dry season or between cropping periods (Crews, 1985).

A final and important aspect of this traditional agroecosystem is the management of the complex set of canals themselves. Besides serving as a major source of soil for raising the platform surfaces, canals also serve as a major reservoir of both water and nutrients. Organic matter accumulates in the canals as aquatic plants die, leaves from trees along the canal borders fall into the water, and weeds from the crop field are thrown into the canals. Undoubtedly, some soil is washed into the canals periodically or even transported with surface waters that enter the basin from surrounding hillsides (Gonzalez Jacome, 1986). Every two to three years, the canal is cleaned of the accumulated soil and muck, and the excavated materials are applied as a top dressing to the adjacent field. Farmers perceive this material to be the most important input to their fields and value it greatly (Crews, 1985). Interestingly, one of the primary trees planted along the borders is a nitrogen-fixing alder, *Alnus firmifolia*, providing a nitrogen-rich litter to the organic matter-trapping system of the canals (Gliessman, unpublished data).

The canals play an important role in the sustainability of this agroecosystem. They function as a nutrient "sink" for the farmer and are managed in ways that permit the capture of as much material as possible (Crews, 1985; Gonzaelz Jacome, 1986). The canals provide supplemental irrigation water in the dry season, and plants rely greatly on moisture that moves upward through the soil from the water table by capillary action (Wilken, 1988). Water levels in the canals are controlled by an intricate system of interconnected canals that lead eventually to the rivers of the basin, but flow in the canals is very limited. Farmers often block the flow of canals along their fields during the dry season to maintain a higher water table. Even in the wet season, water flow out of the system is minimal. Only at times of excessive rainfall do appreciable quantities of water drain from the area (Gonzalez Jacome, personal communication). Water is both an input and a tool in management of the system. Inputs from the canals, along with those gained from certain crop rotations and associations and those returned to the system with manures and compost, form the basis for the system's sustainability.

At the same time, it is important to remember that sustainability of any agroecosystem must include cultural components as well. Significant alterations to the basin have come about because of industrialization of the region, with groundwater supplies and runoff patterns subject to serious modification by water use for industries (Gonzalez Jacome, 1986). Water quality can be affected adversely by effluents from these same industries (Gonzalez Jacome, 1986). At the same time, maintenance of the canal platform complex requires constant inputs of human energy. The allure of salaried jobs in factories or in the cities attracts labor away from agriculture, and management aspects of the agroecosystem begin to suffer. Details, such as less frequent pruning of trees along the borders, a shift to the use of chemical fertilizers to save time, or less coordinated maintenance of the canal network, all begin to offset the stability of the system. But worst of all is the threat of losing the information and knowledge of how to design and manage these systems. Such information is the result of a long period of coevolution between a culture and its surrounding environment and should serve as a beginning point for any future agricultural development in the area. Many of the components of sustainability are already in place, but they should be added to, not replaced. Otherwise, there is the danger of greatly increasing the dependence on purchased inputs, as well as increasing the need for imported food if yields in these systems fall. The agroecosystem then loses its applicability to small farm needs as well as the locally sustainable base it once had.

## Future Directions

An agroecological focus on small farm development goes beyond crop yields, delving deeply into the complex set of factors that may contribute to agroecosystem sustainability. Local, traditional agroecosystems that have evolved under the diverse and often limiting conditions facing small farmers are adapted to this set of factors. They have evolved through time as reduced-external-input systems, with a greater reliance on renewable resources and an ecologically based management strategy. A research focus in agriculture that can take advantage of this knowledge and experience can permit exploration of the multiple bases upon which sustainability rests. It represents the blending of knowledge gained by ecologists studying the dynamics and stability of natural ecosystems with the knowledge of farmers and agronomists on how to manage the complexities of food-producing agroecosystems. From this can come the sustainability in the production base, so critical for giving small farmers the stability and viability they need to provide their own needs and contribute to meeting the needs of the greater society.

## REFERENCES

Allen, P. A., and D. Van Dusen, editors. 1988. Global perspectives in agroecology and sustainable agriculture. University of California, Santa Cruz.

Allison, J. 1983. An ecological analysis of home gardens (huertos familiares) in two Mexican villages. M.A. thesis. Biology Department, University of California, Santa Cruz.

Altieri, M. A. 1984. Towards a grassroots approach to rural development in the Third World. Agriculture and Human Values 1: 45-48.

Altieri, M. A. 1987. Agroecology: The scientific basis of alternative agriculture. Westview Press, Boulder, Colorado.

Altieri, M. A., and M. K. Anderson. 1986. An ecological basis for the development of alternative agricultural systems for small farmers in the Third World. American Journal of Alternative Agriculture 1: 30-38.

Amador, M. F. 1980. Comportamiento de tres especies (Maiz, Frijol, Calabaza) en-policultivos en la Chontalpa, Tabasco, Mexico. Tesis Professional. Colegio Superior de Agricultural Tropical, Cardenas, Tabasco, Mexico.

Anaya, A. L., L. Ramos, R. Cruz, J. G. Hernandez, and V. Nava. 1987. Perspectives on allelopathy in Mexican traditional agroecosystems: A case study in Tlaxcala. Journal of Chemical Ecology 13: 2,083-2,102.

Barkin, D. 1978. Desarrollo regional y reorganizacion campesina: La chontalpa como reflejo del problema agropecuario Mexicano. Editorial Nueva Imagen, Mexico City, Mexico.

Boucher, D. H. 1979. La nodulacion del frijol en policultivo: El efecto de la distancia entre las plantas de frijol y maiz. Agricultura Tropical Colegio Superior de Agricultura Tropical 1: 276-283.

Brown, L. R. 1988. The changing world food prospect: The nineties and beyond. Paper No. 85. Worldwatch Institute, Washington, D.C.

Chacon, J. C., and S. R. Gliessman. 1982. Use of the "nonweed" concept in traditional tropical agroecosystems of southeastern Mexico. Agro-Ecosystems 8: 1-11.

Conway, G. 1985. Agroecosystem analysis. Agricultural Administration 20: 31-55.

Crews, T. E. 1985. Raised field agriculture, Tlaxcala, Mexico, with a focus on nutrient cycling. Senior thesis. Environmental Studies, University of California, Santa Cruz.

Dover, M. J., and L. M. Talbot. 1987. To feed the earth: Agro-ecology for sustainable development. World Resources Institute, Washington, D.C.

Douglass, G. K., editor. 1984. Agricultural sustainability in a changing world order. Westview Press, Boulder, Colorado.

Edens, T. C., C. Fridgen, and S. L. Battenfield, editors. 1985. Sustainable agriculture and integrated farming systems. Michigan State University, East Lansing.

Francis, C. A. 1986. Multiple cropping systems. Macmillan, New York, New York.

Garcia, R. 1980. Chenopodium ambrosioides L., planta con uso potencial en el combate de nematodos fitoparasitos. Agricultura Tropical Colegio Superior de Agricultura Tropical 2: 92-98.

Gliessman, S. R. 1982. Nitrogen distribution in several traditional agroecosystems in the humid tropical lowlands of southeastern Mexico. Plant and Soil 67: 105-117.

Gliessman, S. R. 1983. Allelopathic interactions in crop-weed mixtures: Applications for weed management. Journal of Chemical Ecology 9: 991-999.

Gliessman, S. R. 1984. Resource management in traditional tropical agroecosystems in southeast Mexico. pp. 191-198. *In* G. Douglass [editor] Agricultural sustainability in a changing world order. Westview Press, Boulder, Colorado.

Gliessman, S. R. 1985a. Multiple cropping systems: A basis for developing an alternative agriculture. pp. 69-83. *In* Innovative biological technologies for lesser developed countries. Workshop Proceedings. Office of Technology Assessment, U.S. Congress, Washington, D.C.

Gliessman, S. R. 1985b. Economic and ecological factors in designing and managing sustainable agroecosystems. pp. 56-63. *In* T. C. Edens, C. Fridgen, and S. L. Battlefield [editors] Sustainable agriculture and integrated farming systems. Michigan State University Press, East Lansing.

Gliessman, S. R. 1988. The home garden agroecosystem: A model for developing more sustainable tropical agricultural systems. pp. 445-453. *In* P. A. Allen and D. Van Dusen [editors] Global perspectives in agroecology and sustainable agriculture. University of California, Santa Cruz.

Gliessman, S. R., R. Espinosa, and M. F. Amador. 1981. The ecological basis for the application of traditional agricultural technology in the management of tropical agroecosystems. Agro-Ecosystems 7: 173-185.

Gonzalez Jacome, A. 1985. Home gardens in central Mexico. *In* I. S. Farrington [editor] Prehistoric intensive agriculture in the tropics. BAR International Series 232, Oxford, England.

Gonzalez Jacome, A. 1986. Agroecologia del suroeste de Tlaxcala. pp. 201-220. *In* Historia y Sociedad en Tlaxcala. Gobierno del Estado de Tlaxcala, Tlaxcala, Mexico.

Hart, R. D. 1984. Agroecosystem determinants. pp. 105-120. *In* R. Lowrance, B. R. Stinner, and G. J. House [editors] Agricultural ecosystems. John Wiley and Sons, New York, New York.

Jackson, W., W. Berry, and B. Colman, editors. 1984. Meeting the expectations of the land. Northpoint Press, Berkeley, California.

Klee, G. 1980. World systems of traditional resource management. Halstead, New York, New York.

Letourneau, D. 1983. Population dynamics of insect pests and natural control in traditional agroecosystems in tropical Mexico. Ph.D. dissertation. University of California, Berkeley.

Lowrance, R., B. J. Stinner, and G. J. House, editors. 1984. Agricultural Ecosystems. Wiley and Sons, New York, New York.

Marten, G. G. 1986. Traditional agriculture in southeast Asia: A human ecology perspective. Westview Press, Boulder, Colorado.

Pinchinat, A. M., J. Soria, and R. Bazan. 1976. Multiple cropping in tropical America. pp. 51-64. *In* R. I. Papendick, P. A. Sanchez, and G. B. Triplett [editors] Multiple cropping. Special Publication 27. American Society of Agronomy, Madison, Wisconsin.

Risch, S. 1980. The population dynamics of several herbivorous beetles in a tropical agroecosystem: The effect of intercropping corn, beans, and squash in Costa Rica. Journal of Applied Ecology 17: 593-612.

Wilken, G. C. 1969. Drained-field agriculture: An intensive farming system in Tlaxcala, Mexico. The Geographical Review 59: 215-241.

Wilken, G. C. 1971. Food producing systems available to the ancient Maya. American Antiquity 36: 432-448.

Wilken, G. C. 1988. Good farmers: Traditional agricultural resource management in Mexico and Central America. University of California Press, Berkeley.

Wright, A. 1984. Innocents abroad: American agricultural research in Mexico. pp. 134-151. *In* W. Jackson, W. Berry, and B. Colman [editors] Meeting the expectations of the land. Northpoint Press, Berkeley, California.

# SUSTAINABLE AGRICULTURAL SYSTEMS IN THE HUMID TROPICS OF SOUTH AMERICA

Hugo Villachica, Jose E. Silva,
Jose Roberto Peres, and Carlos Magno C. da Rocha

The total land area on this planet that is potentially available for food and fiber production is estimated at 3.2 billion hectares (20 percent of the world's continental surface), of which 50 percent is already in use by farming enterprises (Meadows et al., 1972). Given a world population of about 5.2 billion inhabitants, this equates to 0.3 hectare for food and fiber production per inhabitant. These values vary among countries, depending on their degree of development, type of farming enterprise, available land, population, and agricultural development policy (USDA, 1976).

The world population might double in the next 30 years if it keeps increasing at the average growth rate registered in past years. This population will require more food and fiber and the development of additional urban areas, usually at the expense of surrounding rural areas. About 250,000 hectares of agricultural land are replaced annually by urban constructions in the United States (Brown, 1975).

The increase in productivity of currently cultivated land and the opening of new areas for agricultural purposes are, therefore, essential to maintain or improve food and fiber supplies. The priority that every country will give to each alternative depends upon their particular conditions.

Evaluation of problems and answers for food and fiber production has to consider worldwide trends in agricultural production. In fact, most of the social and economic pressures that arise from the accelerated world population growth are converging on countries that have the potential for agricultural expansion.

South America is the continent with largest potential to advance its agricultural frontiers to new areas of savannas and forests (Tergas et al., 1979). These ecosystems cover more than 850 million hectares of acidic, low-fertility soils (Centro Interamcricano de Agricultura Tropical, 1978), mixed with small proportions of medium- to high-fertility soils (Villachica, 1986). The largest portion of this area is in the Amazon watershed and the Cerrado regions. In both of these areas, climatic conditions, technology, and native species offer a high potential for food and fiber production. In fact, research carried out during the last two decades has led to the development of technology that makes these areas ready for the implantation and maintenance of sustainable agricultural systems.

A program to increase food and fiber production in a short period requires the establishment of priorities that can be used in a given region and for a given time. These priorities have to be implemented in such a way that the occupation of new areas makes use of the adequate technology, considering the ecological balance of the system. Sustainable agriculture can be defined as a production system that meets the needs and aspirations of the present without compromising the ability to meet those needs in the future (World Commission on Environment and Development, 1987). This definition does not apply to traditional farmers in tropical South America because they are more concerned with present needs. It also does not apply to agricultural systems that preserve the future while limiting availability for present needs.

An adequate balance between present and future needs is a requirement for sustainable agriculture in a region, with the balance being more fragile in areas having acidic, low-fertility soils and low-input systems for annual crops and areas establishing perennial crops. Accordingly, an increase in present needs and aspirations or an increase in the need to maintain the ability of the system for the future can break the balance and the sustainability of a given agricultural system. Therefore, sustainable agriculture, as defined herein, refers to systems that satisfy short-term requirements and concurrently maintain or improve the ability to satisfy long-term production needs.

The two ecosystems that predominate in tropical South America are rain forests and savannas. This chapter presents results and advances of research on sustainable agricultural systems for the Amazon rain forest and the Brazilian Cerrados. There is a clear difference between these two geographical areas, mainly because of the differences in agricultural development; the Amazon is almost undeveloped, and the Cerrado is developed almost as much as agriculture is in temperate regions. Thus, the results are different. The Amazon rain forest is envisioned more in terms of expansion, with several technological options and a large potential. The Cerrado, on the other hand, is already contributing largely to Brazilian agricultural production.

## Sustainable Systems in the Amazon Rain forest

***Characteristics of the Region.*** *Geography and Population.* The Amazon region comprises the heart of the humid tropics of South America. It is bounded by Bolivia, Brazil, Colombia, Ecuador, French Guyana, Guyana, Peru, Surinam, and Venezuela. The Amazon watershed has an area of 6,831,314 square kilometers, about 47 percent of the total areas of the indicated countries. The Amazon basin occupies 60 percent of Peru, 55 percent of Brazil, 48 percent of Ecuador, 35 percent of Colombia, 33 percent of Bolivia, 19 percent of Guyana, 6 percent of Venezuela, and 5 percent of Surinam.

The need to provide food for local people; to absorb surplus population and relieve political pressure from other areas; to provide the economic base for regional development; to obtain additional production to reduce the national balance of payments deficits; and to integrate and occupy national territory led Brazil, Colombia, Ecuador, Peru, and Bolivia to encourage migration into the Amazon region. The process is not new, but it has intensified during the last 25 years. Spontaneous migration also has occurred in areas not under the government colonization programs. However, the Amazon's tropical forest is still essentially intact because colonization and migration are happening only along main rivers and the few roads, around major cities, and on country borders.

Population data are not very precise, but they demonstrate a very low density for the region. Populations are concentrated in small towns in the Andes foothills, also known as the high jungle, and small to medium-size cities in the Amazon plain or low jungle (Iquitos, Peru; Manaos and Belem, Brazil). The population of the Amazon plains is around 15 million people. The population higher in the Andes foothills is not known precisely, due to the high rate of migration to and from the Andes. People are concentrated along the coast and in the highlands of the Andean countries (Table 1) and in southern Brazil. Population density on the coast and the highlands of the Andean countries is 10 to 40 times that observed in their respective Amazon regions (Table 2).

*Climate.* The Amazon rain forest is usually thought of as having a uniform climate: hot and humid all year. This is not necessarily true, because variations in temperature and rainfall occur. Night temperatures in the Andean foothills, or high jungle, decrease to 15 to 16°C, and daytime temperatures may reach 28 to 30°C, making the weather very comfortable for people and adequate for most crops. On the other hand, some places in the Amazon plains with daily mean monthly averages of 30°C experience very small differences between day and night temperatures, limiting development of certain crops and favoring others. In general, mean annual temperatures are in the order of 20 to 22°C for the high jungle, increasing progressively to 28°C in the low jungle (Villachica, 1986).

Rainfall is not constant in the rain forest areas. There are some areas where rainfall can be as low as 800 millimeters per year, and others where it can be as high as 5,000 millimeters per year, with most of the Amazon in the range of 1,500 to 3,500 millimeters per year. In general, the region does not suffer from moisture limitations because rainfall usually exceeds potential evapotranspiration. Most of the rain falls in nine months; the remaining three months have less than 100 millimeters per month, producing the "dry" season. This distribution is not limiting for perennial crops, but it does limit the growth of annual crops growing all year around.

*Soils and Soil Dynamic with Cropping.* Soils of the Amazon are acidic, having high aluminum saturation, low exchangeable bases, and low nutrient content—hence, low fertility. This widespread generalization masks considerable regional and local variations in soil fertility. Medium- to high-fertility soils are found in the Amazon, although they do not represent a high percentage of the area. The best soils, from a nutrient standpoint, are located in the Andes foothills and along the river networks, where alluvial soils are formed and renewed yearly.

A study covering 484 million hectares, 71 percent of the Amazon basin, has described the soils of the region (Cochrane and Sanchez, 1980). The study covers areas located between 4° North and 12° South, 48° East, and a variable line along the Andes piedmont. Oxisols and Ultisols are dominant, representing 45 and 29 percent of the area, respectively (Table 3).

**Table 1. Populations in some Amazonian countries, 1985.**

| *Country* | *Total* | *Coast* | *Highland* | *Amazon* |
|---|---|---|---|---|
| | | | *percent* | |
| Colombia | 27,456,026 | 38.6 | 59.0 | 2.4 |
| Ecuador | 8,060,712 | 49.8 | 48.0 | 2.2 |
| Peru | 18,734,543 | 51.5 | 36.6 | 11.9 |

**Table 2. Population density in some Amazonian countries, 1985.**

| *Country* | *Coast* | *Highland* | *Amazon* | *Total* |
|---|---|---|---|---|
| | | *persons/km²* | | |
| Colombia | 56.6 | 46.2 | 2.0 | 24.5 |
| Ecuador | 56.4 | 54.2 | 1.3 | 29.9 |
| Peru | 75.4 | 17.9 | 2.9 | 14.6 |

**Table 3. Distribution of major soils in the Amazon by physiography.**

| Soil Grouping | Poorly Drained Flat | Well-Drained Slope 0-8% | Well-Drained Slope 8-30% | Well-Drained Slope 30% | Total |
|---|---|---|---|---|---|
| | million ha | | | | |
| Acid low fertility (Oxisols) | 20 | 116 | 64 | 19 | 219 |
| Acid low fertility (Ultisols) | 23 | 91 | 24 | 4 | 142 |
| Poorly drained alluvial soils | 56 | 14 | - | - | 70 |
| Moderately fertile, well-drained | - | 17 | 13 | 7 | 37 |
| Very low fertility, sandy soils | 10 | 5 | 1 | - | 16 |
| Total | 109 | 243 | 102 | 30 | 484 |

Oxisols are usually deep and well-drained, having uniform properties with depth, firm granular structure, acidic with some high base status areas, low fertility, and a red or yellow color. They are predominant in the savannas of Colombia and Venezuela, the area under the influence of the Guyana and Brazilian shields, and the area east of Manaos, Brazil, where they are mixed with ultisols. Ultisols are deep, are variable in drainage, do not have uniform properties with depth, usually have a sandier topsoil over a more clayey subsoil, are low in weatherable minerals, are acidic, have low-fertility, and are red or yellow in color. Constraints for developing these soils are more chemical than physical. Sustainable agriculture systems of perennial crops are found on Oxisols of Brazil and to a lesser degree on Ultisols of Peru and Brazil.

Poorly-drained alluvial soils represent 14 percent of the area (Cochrane and Sanchez, 1980). These soils are located mainly in floodplains or in inland palm swamps. The soils located in floodplains are important for food production. They are rejuvenated yearly, and good rice yields can be obtained with minimum inputs, but with a high risk by flooding. Well-drained soils of moderate fertility represent about 8 percent of the area; they are classified as Alfisols, Eutropepts, Tropofluvents, Argiudolls, Eutrophic Oxisols, and Chromuderts. They are deep (except Tropofluvent) and well-drained. Alfisols and Inceptisols change properties with depth, are slightly acidic to alkaline, have medium to high fertility and are red, brown, or dark brown in color. Sustained agriculture systems for cocoa coffee, oil palm, and other perennial crops have been developed on these soils and for flooded rice in Eutropepts and Tropofluvents.

Very low-fertility sandy soils, represented by Spodosols and Entisols (Psamments), occupy 3 percent of the surveyed area. These soils are derived from coarse sand materials and are considered of no use for agriculture or forestry purposes. Given their special features, they have been extensively studied with regard to the nutrient cycling. However, extrapolation of these studies has produced misleading opinions about Amazonian soil fertility.

The study of Cochrane and Sanchez (1980) does not include areas in the high jungle, where medium- to high-fertility soils are found, and does not include the Peruvian, Bolivian, and Brazilian Amazon plain more than 12 C⁰ South, where there are higher ratios of fertile soils. A study made in the Madre de Dios department of Peru (1982) showed that of 120,641 hectares mapped at the detailed level, 56.1 percent were Inceptisols, 32.6 percent were Alfisols, 6.1 percent were Entisols, and 4.6 percent were Ultisols. This area is mapped at the general level as an Ultisol area. Similar results were observed in the adjacent areas of Bolivia and Brazil, where higher proportions of Alfisols are found than those predicted by general maps.

Though Oxisols and Ultisols are dominant in the Amazon plain, Inceptisols, Entisols, and Alfisols predominate in the high jungle. The latter is the area of greater agricultural development for the Andean-Amazon countries. Some sustainable agricultural systems are used there. However, the region has a high risk of soil erosion, given slopes of up to 50 percent on which agriculture is practiced. Farm development and local development have to consider the combinations of soils existing in the area, which will result in a mosaic type of agriculture, with areas suitable for annual crops, perennials, pastures, forests, and protection.

Soils are classified by criteria based on subsoil characteristics and may or may not reflect the surface or plow layer properties. Surface characteristics of the plow layer usually are changed by slash-and-burning of the vegetation prior to first cropping. The amount of ash added, and its composition and degree of burning, varies depending upon the degree of burning. Differences in ash composition in the same forest growing on Ultisols were observed for nitrogen, phosphorus, potassium, and magnesium, but only for nitrogen, potassium, and iron in Oxisols (Table 4). Nutrient addition to the soil also occurs in other forms that are not measured through surface ash collection (root burning, nutrient release, and pH increase by heat) (Fassbender, 1975). The overall consequence is a decrease in exchangeable aluminum, and an increase in pH and available nitrogen, phosphorus, potassium, calcium, magnesium, zinc, and copper (Villachica, 1978).

Soil fertility gained by burning the forest declines with time due to organic matter decomposition, nutrient uptake by plants, and nutrient leaching at a rate that depends upon soil characteristics and on soil and crop management. In sandy Ultisols this takes about 14 to 24 months under the conditions of mechanized land preparation, continuous cropping systems, and maximum weed and residue removal (Table 5). It occurs before 35 months in no-till and low-input cropping systems where residues are not removed (Table 6). The fertility level of the soil at the end of the decline will be different—higher in the high-input system. Crop yields depend upon their adaptation to these high-aluminum, low-fertility conditions.

Understanding soil dynamics is the agronomic key for successful sustained agriculture in the Amazon. Once soil dynamics are understood, fertilizer treatments can be applied to maintain or raise the fertility level of the soil as a consequence of burning. This fertility treatment is based upon a maximum application of 3 tons of lime per hectare per three years, 80 to 100 kilograms of nitrogen per hectare per crop (except for legumes), 25 kilograms of potassium per hectare per crop, 80 to 100 kilograms of potassium per hectare per crop, 25 kilograms of magnesium per hectare per crop (unless dolomitic lime is used), 1 kilogram of copper per hectare per every one to two years, 1 kilogram of zinc per hectare per one to two years,

**Table 4. Nutrient contribution of the ash after burning forests on an Ultisol in Yurimaguas, Peru, and an Oxisol in Manaos, Brazil.**

| | *Ultisol* | | *Oxisol* | |
|---|---|---|---|---|
| | *Forest Fallow* | | *Forest Fallow* | *Virgin Forest* |
| *Element* | *17 Years* | *20 Years* | *12 Years* | |
| | *kg/ha* | | | |
| Nitrogen | 67 | - | 41 | 80 |
| Phosphorus | 6 | 24 | 8 | 6 |
| Potassium | 38 | 92 | 83 | 19 |
| Calcium | 75 | 69 | 76 | 82 |
| Magnesium | 16 | 51 | 26 | 22 |
| Iron | 8 | 7 | 22 | 58 |
| Manganese | 7.3 | 8.1 | 1.3 | 2.3 |
| Zinc | 0.5 | 0.6 | 0.3 | 0.2 |
| Copper | 0.3 | 0.4 | 0.1 | 0.2 |

**Table 5. Topsoil (0-15 cm) fertility dynamics within the first 24 months in a rice-corn-soybean tilled cropping system (unfertilized) on an Ultisol at Yurimaguas, Peru.**

| *Months After Burning* | *pH* | *Exchangeable Cations* | | | | *Available P* |
|---|---|---|---|---|---|---|
| | | *Al* | *Ca* | *Mg* | *K* | *Mg dm$^{-3}$* |
| | | *cmol / dm$^{-3}$* | | | | |
| 0* | 4.1 | 1.8 | 1.2 | 0.20 | 0.25 | 15 |
| 1 | 4.8 | 1.3 | 2.8 | 0.66 | 0.57 | 31 |
| 6 | 4.6 | 1.3 | 1.3 | 0.47 | 0.26 | 20 |
| 14 | 4.7 | 1.5 | 1.3 | 0.39 | 0.17 | 11 |
| 24 | 4.6 | 2.0 | 1.5 | 0.27 | 0.13 | 11 |

*Sampled in an adjacent forest, unburned.

and 1 kilogram of boron per hectare per year (Nicholaides et al., 1984; Sanchez et al., 1982; Sanchez et al., 1983; Villachica, 1978).

Results of soil analyses in plots that have been cultivated for 15 years and 37 consecutive crops (Table 7) showed that prediction of phosphorus, magnesium, and micronutrient deficiencies, made in the early years of research, could have been overemphasized and produced fertilizer recommendations in excess of those required. This resulted in a buildup of phosphorus, magnesium, and copper in the soil. Lime and fertilizer application also result in an increase in subsoil exchangeable calcium, magnesium, potassium, pH, and available phosphorus, and decreases in exchangeable aluminum and aluminum saturation (Table 8). Thus, continuous cultivation with lime and fertilizer application improves topsoil and subsoil and

**Table 6. Topsoil (0-15 cm) fertility dynamics within the first 35 months in a low-input cropping system on an Ultisol at Yurimaguas, Peru.**

| *Months After Burning* | *Fertility* | *pH ($H_2O$)* | *Exchangeable ($cmol\ dm^{-3}$)* | | | | *Al Saturation (%)* | *Available P ($mg\ dm^{-3}$)* |
|---|---|---|---|---|---|---|---|---|
| | | | *Al* | *Ca* | *Mg* | *K* | | |
| 3 | No | 4.4 | 1.1 | 0.30 | 0.09 | 0.13 | 68 | 20 |
| 11 | No | 4.6 | 1.5 | 0.92 | 0.28 | 0.19 | 51 | 13 |
| | Yes | 4.7 | 1.1 | 0.97 | 0.27 | 0.19 | 45 | 18 |
| 35 | No | 4.6 | 1.7 | 1.00 | 0.23 | 0.10 | 53 | 5 |
| | Yes | 4.6 | 1.2 | 1.16 | 0.20 | 0.16 | 44 | 16 |

**Table 7. Topsoil (0-15 cm) fertility dynamics after 15 years of continuous cultivation and 37 crops of upland rice, corn, and soybean on an Ultisol at Yurimaguas, Peru.**

| *Months After Burning* | *Fertility Treatment* | *pH ($H_2O$)* | *Exchangeable ($cmol\ dm^{-3}$)* | | | | *Al Saturation* | *Available ($mg\ dm^{-3}$)* | | |
|---|---|---|---|---|---|---|---|---|---|---|
| | | | *Al* | *Ca* | *Mg* | *K* | | *P* | *Zn* | *Cu* |
| −1 | None | 4.0 | 2.3 | 0.26 | 0.15 | 0.10 | 82 | 5 | - | - |
| 1 | Burning | 4.5 | 1.7 | 0.70 | 0.29 | 0.32 | 56 | 17 | (1.5)* | (0.9) |
| 185 | None | 4.5 | 3.0 | 0.65 | 0.26 | 0.09 | 75 | 4 | 2.6 | 0.6 |
| 185 | Lime, fertilizers | 6.0 | 0.0 | 2.65 | 1.65 | 0.13 | 0 | 25 | 3.9 | 3.4 |

*Data in parenthesis are for 31 months after burning.

**Table 8. Topsoil and subsoil fertility dynamics after 15 years of continuous cultivation and 37 crops of upland rice, corn, and soybean on an Ultisol at Yurimaguas, Peru.**

| *Treatment* | *Depth (cm)* | *pH ($H_2O$)* | *Exchangeable ($cmol\ dm^{-3}$)* | | | | *Al Saturation (%)* | *Available P ($mg\ dm^{-3}$)* |
|---|---|---|---|---|---|---|---|---|
| | | | *Al* | *Ca* | *Mg* | *K* | | |
| Check | 0-20 | 4.5 | 3.0 | 0.65 | 0.26 | 0.09 | 75 | 4 |
| | 20-40 | 4.5 | 3.8 | 0.21 | 0.05 | 0.05 | 92 | 2 |
| | 40-60 | 4.4 | 4.1 | 0.15 | 0.04 | 0.03 | 95 | 1 |
| Lime and fertilizer | 0-20 | 6.0 | 0 | 2.65 | 1.65 | 0.13 | 0 | 25 |
| | 20-40 | 4.9 | 2.5 | 1.75 | 0.39 | 0.08 | 54 | 16 |
| | 40-60 | 4.6 | 2.9 | 0.87 | 0.12 | 0.04 | 74 | 5 |

increases crop growth.

*Native Agricultural Systems.* The native agricultural system in the Amazonian rain forest is shifting cultivation. Usually the crops grown involve the association of rice or corn with cassava and plantains, but they can include other native tuber or root crops or some grain legumes.

When shifting cultivation is practiced close to huge towns and cities, socioeconomic aspects are considered and the system includes more market-oriented crops, such as rice, plantain, and papaya. One example is the cyclic agroforestry system developed by the residents of the Peruvian village of Tamshiyacu, near Iquitos. The Tamshiyacu system (Padoch et al., 1985) begins when the standing vegetation in an area is cut. Then, rather than burning all the slash in the manner typical of shifting cultivation, the larger woody vegetation is converted to charcoal and sold in the Iquitos market. Following clearing, the field is planted to a variety of annual and semiperennial crops, of which the most important commercially are rice, cassava, pineapple, peaches, tomatoes (*Solanum sesiliflorum*) as annuals, and plantain, cashew, and uvilla as semiperennials. In the second year, some of these crops are replanted and a number of penile tree crops, such as peach palm (*Bactris gassipaes*), umari (*Poraqueiba paraensis*), and Brazil nut, are planted.

After the initial period of two to five years, annual crop production is gradually phased out and perennial tree crops become the most important producers on the farm. Such production often can continue for 25 to 50 years if care is taken to maintain the fields. Clearing of the plot, done several times a year while annuals and semiperennials predominate, gradually is reduced in frequency to once or twice a year, just before umari, the most important tree crop, is to be harvested. As soon as the yields began to fall significantly, the larger vegetation (mostly umari and Brazil nut) is cut and converted to charcoal. Then the field is left in fallow for six years or so, after which a new cycle is initiated.

This system bears a close resemblance to, and probably is derived from, the swiden fallow techniques used by native groups throughout the Amazon region. However, the system differs from native patterns because (a) the market-orientation is toward the city of Iquiitos (Padoch et al., 1985), (b) the charcoal production removes some of the big logs and establishes an almost pure stand of umari and Brazil nut trees, and (c) weeding of the plots is continued for a longer period.

Another example of a traditional agriculture system is annual cropping of seasonally flooded alluvial soils during the dry season. Farmers plant longer-growing species (rice, peanut) first as the rivers decrease their volume. Shorter period species (beans, horticultural crops) are planted later.

The growing season extends from May to September in rivers originating south of the equator and from October to February in the north. Rice is by far the dominant species, but if farmers have a choice, they prefer silty soils for rice and sandy soils for beans and peanuts. Advantages of this alluvial system are the higher soil fertility renewed yearly, good water availability, limited hand labor required to clear the land because no forest has developed in these soils, annual control of pests and diseases by flooding, limited presence of weed seeds, and a transportation system formed by the rivers. The system can be improved by establishing perennial tree crops that grow well under these seasonally flooded conditions, such as camu camu (*Myrciaria dubia*). A major limitation of the system is the high risk involved because flooding is not predictable and much less controllable. A second limitation is the decrease in rivers' navigability during the dry season. This can be managed through a good storage infrastructure.

***Sustainable Systems for Annual Crops.*** *Irrigated Rice.* Rice, both irrigated and upland, is the most important annual crop in the Amazon basin. Irrigated rice produces higher yields than upland rice—3.5 to 5.0 tons per hectare versus 1.5 to 2.5 tons per hectare. Advantages for irrigated rice in the Amazon include water availability and high temperatures all year, which, together with a greater availability of land, present an enormous potential for increasing rice production to cover the internal demand of each country. For example, in 1982, about 72 percent of Peru's national rice production (655,000 tons) was harvested on the coast, with the remaining 38 percent in the Amazon (INIPA, 1982). In 1987, equal amounts of rice were harvested along the coast and in the Amazon.

Irrigated rice fields are usually installed on Entisols, Inceptisols, and Alfisols that are close to a water source. Leveling and diking are done on soils of different depths. Sometimes, puddling is practiced. Transplanting is used most often, but labor costs are forcing farmers to direct seeding. Moderate rates of fertilizers are applied. Fields are dried 15 days before harvesting, but because of lack of a drainage system, they might not dry adequately. Harvesting is usually by hand, and the grain is threshed with a stationary thresher.

Irrigated rice yields can be increased with the use of improved varieties, water management, land leveling, fertilizer, machinery usage, and improved drainage. Traditional irrigated rice farmers in other areas of tropical South America control most of these factors adequately, obtaining higher yields than in the new areas of the Amazon (Table 9). Rice varieties for the Amazon are site-specific and usually do not yield as well in other areas of the region, due to differences in soil, climate, management, or disease incidence.

Research into developing new rice varieties in the region is relatively recent and frequently lacks resources.

Water management and land leveling usually work together, but each can limit rice yields independently. Yields in fields irrigated with only rainfall water can be increased by 30 percent to 50 percent when supplementary irrigation is provided every two weeks from a nearby river (Table 10). Land leveling is important for adequate water control and puddling, for weed control, and for reducing water percolation. Puddling is not practiced extensively in the Amazon. Farmers practice land leveling, but the existing technology, scarcity of machinery, and lack of experience do not result in well-leveled fields. This imperfect leveling is compensated for by transplanting seedlings, which yield better than direct seeding of pregerminated seeds (Table 11). Continuous use of the same field produces better leveling, more experience for the farmer, and better water management. In places with shallow topsoils, land leveling removes topsoil and occasionally exposes the subsoil, resulting in lower rice yields. The effect of fertilizer applications on increased yields in these truncated soils has to be investigated.

Swampland and areas with poor drainage are being used for irrigated rice. Yields are in the range of 3.0 to 4.0 tons per hectare, with no fertilizer use, but they have the potential to be increased to 5.0 to 6.0 tons per hectare.

**Table 9. Irrigated rice grain yields in selected ecosystems of Peru, regional averages.**

| *Ecosystem* | *Place* | *Main Soils* | *Yield (t/ha)* |
|---|---|---|---|
| Coastal valley | Chiclayo | Entisol | 5.0-6.0 |
| High jungle | Jaen | Entisol, Inceptisol | 4.5-5.2 |
| High jungle | Moyobamba | Entisol, Inceptisol | 3.5-4.5 |
| Low jungle | Yurimaguas | Entisol, Inceptisol | 3.5-4.0 |

**Table 10. Rice response to supplementary irrigation on a leveled Tropaquept at Yurimaguas, Peru.**

| | *Harvest (t/ha)* | | | |
|---|---|---|---|---|
| *Water Management* | *First* | *Second* | *Third* | *Mean* |
| Rainfall alone | 4.1 | 5.1 | 4.0 | 4.4 |
| Rainfall plus irrigation | 5.8 | 6.7 | 6.0 | 6.2 |

**Table 11. Performance of irrigated rice with two planting systems and dry land preparation on a Haplaquept at Yurimaguas, Peru.**

| | *Harvest (t/ha)* | | | | | |
|---|---|---|---|---|---|---|
| *Planting System* | *First* | *Second* | *Third* | *Fourth* | *Fifth* | *Mean* |
| Transplanted | 8.3 | 6.7 | 6.2 | 5.6 | 6.3 | 6.6 |
| Broadcast/ direct seeding | 6.3 | 5.6 | 4.9 | 4.6 | 6.0 | 5.5 |

Poor drainage is a problem for harvesting in these areas. Iron toxicity is observed in soils with high organic matter, poor drainage, or exposed clayey, low-base saturation subsoils.

Rice yields usually are greater for the first crops and decrease with time, even though moderate amounts of nitrogen, phosphorus, and potassium are applied (Table 11). Scarcity of hand labor for weeding and of harvesting machinery also limits yields. The main crop diseases are *Helminthosphorium* and *Piricularia*. Plant breeders must find varieties resistant to these fungal diseases. Fungicides are used sometimes to control or prevent the diseases, but their use is not widespread in the Amazon. Price and marketing constraints of inputs and rice grain limit the use of the technology available.

Future research needs for irrigated rice in the Amazon region should focus on high-yielding varieties tolerant to diseases and suited for every ecological condition existing within the region, better soil and water management, and the elimination of harvesting constraints. However, these agronomic constraints are secondary to the socioeconomic limitations given by the deficient marketing of both inputs and rice grain.

*Annual Grain Crops Rotations in Upland Positions.* Annual grain crop production is technically feasible on the acidic, low-fertility soils of the Amazon basin. Its economic feasibility varies within the region. Production systems have been described for Ultisols (Nicholaides et al., 1984; Sanchez et al., 1982; Sanchez et al., 1983; Villachica, 1978) and are applicable to Oxisols in the region (North Carolina State University, 1987). Each system begins when the field is cleared by slash-and-burn because mechanical clearing has a detrimental effect on soil physical properties and does not allow for the beneficial effects of residual ashes (Seubert et al., 1977). Sustainability can be enhanced by making a few refinements on clearing and land preparation over the traditional slash-and-burn technique: (a) using a chain saw for clearing; (b) allowing the larger, valuable timber trees to stand while burning the rest of the slash, cutting it after burning, and rolling it to a side of the field for sale; and (c) cutting small trees 15 centimeters below the surface and returning the disturbed topsoil to its place before plowing and disking. Usually one upland rice crop can be obtained in unfertilized fields that are cleared from mature secondary forests. This differs little with respect to limed or fertilized fields (Villachica, 1978).

Lime and fertilizers have to be applied after the first rice crop is harvested; the rates must be based on soil analyses. These analyses have to be made once a year during the first two to three years, after which the soil will reach a steady state and analysis can be made every 18 to 24 months. Recommended rates of fertilizer application for rotations of rice, corn, and soybeans or rice, soybeans, and peanuts on Ultisols of Peru (Sanchez et al.,

1983; Villachica and Raven) are proposed in table 12. Grain yields obtained with the fertility levels described in table 12 are in the order of 3.0 tons of rice per hectare per 135-day growing season, 3.5 tons of corn per hectare per 110 days, 2.5 tons of soybeans per hectare per 105 days, and 3.5 tons of unshelled peanuts per hectare per 110 days.

Major constraints for expanding the system are socioeconomic and are related to the price of inputs and the market for farm outputs. Availability of existing roads and marketing systems results in high production costs and low profits. Technology and infrastructure development must go hand-in-hand, but, in practice, technology is ahead in some areas while it is behind in others. The large number of crops and options that can be developed in the region enhance this difference, augmenting the need for research investment in countries with limited economic resources.

Soil management constraints have been largely overcome (Nicholaides et al., 1984; Sanchez et al., 1982; Sanchez et al., 1983) but might have been due to excessive fertilization. Improved efficiency of soil amendments and fertilizers is needed, with special emphasis on the use of local lime and rock phosphate sources in the timing and rates of application of nitrogen and potassium, in the selection of species and varieties for tolerance to aluminum, and in the ability to grow crops under low nutrient levels in the soil. Lime deposits, calcitic and dolomitic, are abundant in the Andes piedmont and are found frequently in the Amazon plain. Rock phosphate deposits have been identified in the region (Sanchez and Salinas, 1981). Times and rates of nitrogen and potassium application should be studied to increase their efficiency of use, because heavy rainfall is common after applying nitrogen fertilizers. Potassium also moves down through the profile in Ultisols and Oxisols. Thus, split applications produce better responses than broadcast applications. Slow-release fertilizers are not recommended due to their higher costs.

The importance of plant species and varieties that are tolerant to aluminum

**Table 12. Lime and fertilizer recommendations for continuous cropping of rice and corn, and soybeans on Ultisols in Peru.**

| *Input** | *First 18 Months* | *After 18 Months* | *Crop* |
|---|---|---|---|
| Lime | 3 t/ha as dolomite | 1 t/ha/2 crops | - |
| N | 80-100 Kg N/ha/crop | 100 Kg N/ha/crop | Corn and rice |
| P | 100 Kg P/ha/crop as rock phosphate | 25 Kg P/ha/crop | Every crop |
| K | 80 Kg K/ha/crop | 80-100 Kg/ha/crop | Every crop |

*Three crops in 18 months. Use dolomitic lime. Use of 1 Kg Zn/ha or 1 Kg B/ha might be needed once every two to three years for corn and grain legumes. Legume seeds require inoculation.

and capable of growing at low nutrient levels cannot be overemphasized. National and international research institutions are coordinating activities to develop improved varieties of the main crops, such as rice, cassava, cow peas, and pastures, that are tolerant to acid soil conditions. However, little is being done with corn, beans, soybeans, peanuts, plantains, and native species. In other crops, such as rice, cassava, and plantains, the acid-soil-tolerant germplasm should be improved to enhance tolerance to insects and diseases.

The management system, as described above, is proposed for flat to minimum-slope lands and can be based on the use of hand labor. When plenty of land is available, hand labor is scarce because everybody owns land. Under this situation, machinery will have to be used to increase operational efficiency and decrease the costs of activities, such as land preparation, sowing, weeding, insect and pest control, and especially harvesting.

***Sustainable Systems for Perennial Crops.*** Perennial crops are the most appropriate form of agricultural use for Amazon soils, especially if they are associated with agroforestry systems. Perennial crops offer several advantages over annual crops. They protect the soil against leaching, compaction, and erosion. They have lower nutrient requirements. They are better adapted to acidic, low-fertility soils. And they are more efficient nutrient recyclers. Additionally, products harvested from perennial crops remove fewer nutrients. Other advantages of perennial crops are their lower hand labor requirement once the plantation is established, the better use of available hand labor, and the possibility to settle the farmer in a particular place.

Perennial crops are important to the economy of South American countries—coffee being the most evident example. Nevertheless, the Amazon region supplies less than 10 percent of the world production of commodities from its best-known native trees, cocoa and rubber. Inadequate agronomic research and lack of technical assistance to farmers were the reasons for many unsuccessful attempts at growing traditional perennial crops in the Amazon (Alvim, 1980).

These problems now are receiving more attention. Oil palm, cocoa, rubber, coffee, black pepper, and guarana are being grown successfully in the Amazon. Frequently, where perennial crops are cultivated successfully, agronomic practices that farmers use are highly technical, and farmers often are ahead of research at experimental stations. In fact, Amazonian countries are making less investment in tropical perennial crops research than on annual crops (Evenson, 1984). Much research is still needed with tropical perennials, not only to improve yields for the existing plantations, but also to decide which known or potential species should be recommended for given conditions, what production systems and cultural practices should be used, and ways to sustain the farm budget during the establishment period of the perennial.

*Coffee.* Arabiga coffee (*Coffea arabiga*) is grown in the high jungle of Andean countries and in the southern part of the Brazilian Amazon plain, particularly in the state of Rondomia. Most of the traditional growing areas in Brazil are in the southern part of the country, out of the Amazon basin, with 40 percent of Brazil's coffee produced in the Cerrado region. Robusta coffee (*Coffea canephora*), which is more tolerant to higher temperatures, could be developed as a commercial crop in areas of the Amazon plain close to the Andes and to the Tropic of Capricorn. The data presented here refer only to arabiga coffee that is grown in the high jungle.

Coffee yields and quality are excellent in areas with mean monthly temperatures of 23 °C daily and 17 °C nightly; rainfall on the order of 1,500 to 2,500 millimeters per year; and 1,500 to 2,500 sunlight hours per year. Temperatures over 26 °C produce intermittent flowering, so that a tree branch may have ripe fruits, green fruits, and flowers simultaneously, thereby increasing harvesting costs. Rainfall below 1,000 millimeters per year with uneven distribution reduces opening of flowers. Rainfall greater than 3,000 millimeters produces premature flowering and out-of-season intermittent flowering. Coffee is shade-tolerant; therefore, solar radiation in the Amazon would not limit production but excess shade might limit yields. Coffee grows well in soils of pH 5.0 to 7.5; however, yields are better in soils of pH 5.5 to 6.5. The high clay content, usually found in soils of the Amazon, is not a problem for coffee, if drainage is good.

The many coffee varieties have different names in every country, but most of them are arabiga-derived. Production starts two to four years after seedlings are transplanted to the field, with a productive phase of more than 20 years, depending upon management. Yields in commercial areas are on the order of 550 to 4,400 kilograms per hectare.

Among the several insects affecting coffee growth are the cherry borer (*Hypothenemus hampei* Ferr) and the leaf miner (*Leucoptera coffeella*); and fungal diseases include yellow rust (*Hemileia vastratrix*), cercospora (*Cercospora*), ruster eye (*Omphalia flavidacoffeicola*), and dye back (*Pellicularia filamentosa*). There are also some nematode pests. These insects and fungi are affected by climate; for example, low temperatures and high rainfall limit yellow rust. The presence of the cherry borer and yellow rust have fostered research and technology development for coffee.

Some sustained cropping systems for establishing coffee in the Amazon have been developed. Farmers have established 7- to 12-month-old coffee seedlings on soils after burning when soil fertility is high. Some of the original forest trees are allowed to remain for shade until legume trees with an open canopy planted at the same time reach maturity. Frequently, some annual or semiperennial crop will be grown during a two- to three-year establishment period to provide a source of income. A sustained production system involving coffee may include corn as an annual crop, bananas as a semiperennial crop, and inga (*Inga* sp.) as the shade species.

The system would be managed in the following manner: In April-May of year one, the farmer collects the seeds of the species and varieties to be used, germinates them in boxes, and transplants them to nursery beds or plastic bags filled with earth. In July-August of the same year, the land is cleared, and the forest is burned in late September. Corn is sown in October-November, and bananas in November-December. Coffee seedlings are transplanted in January-February under corn shade; corn is harvested in March; and the inga seedlings are transplanted. At this time, the bananas have produced enough growth to provide shade for coffee, and sometimes cassava is planted with the legume trees, to be harvested after eight to 10 months, usually for on-farm consumption. Banana trees begin to produce 14 months after transplanting and can be harvested commercially for three to four years. Plant density is reduced 30 percent to 40 percent annually, arriving at a final plant density of 10 percent or lower of the original planting. At this time, year four or five, coffee will be in full production and the legume-shade trees will be large enough to fulfill their function. Table 13 presents the sustainability of the system for one hectare.

Figure 1 shows the distribution of coffee, bananas, and inga in the system equilibrium. Bananas could be replaced in the first years by cassava or

**Table 13. Estimated income for a coffee-banana-inga production system in Chanchamayo, Peru.**

| *Year After Planting* | *Crop* | *Yield (Kg/ha)* | *Gross Income (US$/ha)* |
|---|---|---|---|
| 1 | Corn | 1,500 | 180 |
| 2 | Banana | 5,000 | 500 |
| 3 | Banana | 2,000 | 200 |
| 3 | Coffee | 275 | 210 |
| 4 | Banana | 1,500 | 150 |
| 4 | Coffee | 440 | 335 |
| 5 | Banana | 500 | 50 |
| 5 | Coffee | 660 | 503 |
| 6 | Coffee | 990 | 754 |
| 7 | Coffee | 1,155 | 880 |
| 8 | Coffee | 1,375 | 1,048 |

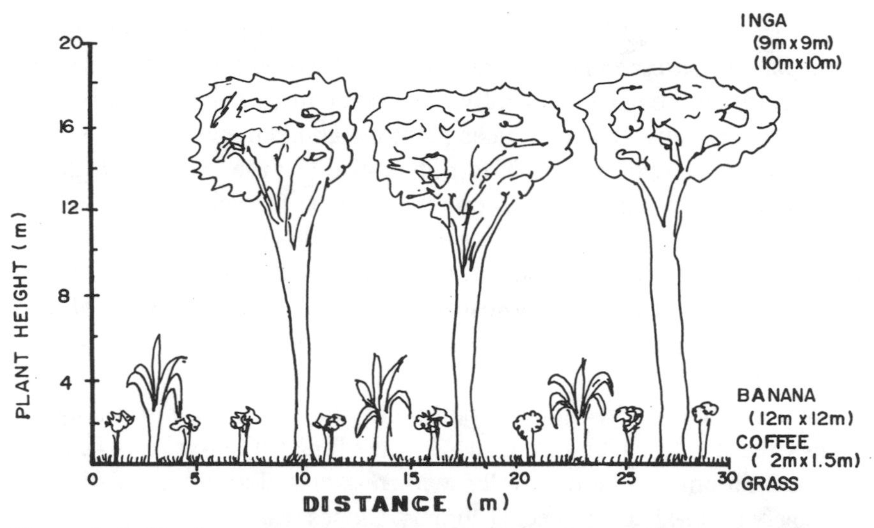


**Figure 1. Coffee, banana, and inga canopies in a mature system.**

papaya and not be present in the system after three years. Inga also could be replaced by other legume trees, such as albizzia (*Albizzia spp.*). The system also has a weed grass cover, which on sloping land is two to three centimeters high, thereby protecting against erosion. Table 14 presents the composition and height of the species in the system. The system has proven to be sustainable in the Andean piedmonts for more than 25 years, with some technical improvements made during this time. The changes include high-yielding varieties with resistance to yellow rust, higher plant density, inclusion of cover crops, and better fertility rates.

*Cocoa.* Cocoa is believed to have originated in the Amazon subregion that borders the Andes. It is well-adapted to hot climates with mean annual temperatures of 22 °C to 28 °C. However, better grain quality is observed in those areas that have a lower night temperature (18 °C to 20 °C). Most cocoa-growing areas in tropical South America have rainfall from 1,400 to 2,000 millimeters per year, but cocoa also can be grown adequately in locations with as much as 3,000 millimeters of rain per year.

Two factors limit cocoa development in the Amazon region, as well as in other tropical areas: soil fertility and diseases. Native cocoa is found mostly on high-fertility soils. Consequently, almost all commercial clones and hybrids do not tolerate low-fertility soils. However, some native clones are tolerant of acidic, low-fertility soils. During the last 10 years, CEPLAC, the Brazilian Cocoa Institute, has developed about 150,000 hectares of cocoa on Alfisols in the Amazon, using moderate amounts of fertilizer. The need for additional fertilizer applications after 10 or more years of harvesting has to be determined for these soils. Cocoa nutrient requirements depend upon plant genetics, soils, climate, and shade. Plantations without shading trees require more nutrients. Native Amazon cultivars require more shade than others because shade reduces plant metabolism and nutrient uptake, which are more relevant in low-fertility soils. The use of shade trees, fertilizer, and some limestone has produced cocoa plantations with average yields of up to 1,000 kilograms per hectare in Ultisols and acid Inceptisols in Peru. When the use of lime and fertilizers is not profitable, farmers are planting cocoa, clearing the forest partially, and obtaining yields of up to 500 to 600 kilograms per hectare, using improved native clones at low plant densities.

Several diseases affect cocoa worldwide, but in the Amazon only one is important: witches' broom. Caused by the fungus *Crinipellis perniciosa,* witches' broom can be controlled with resistant clones and hybrids and the timely removal of the infected sprouts. Clones resistant to witches' broom are native to the Amazon (SCA-6, SCA-12, IMC-67, Pound 7, and others). Sanitary pruning at the right time, twice a year, is recommended for efficient witches' broom control and can be complemented with chemical sprays to prevent pod infection. Moniliasis, caused by the fungus *Monilia roeri,* has reduced yields drastically in many traditional cocoa areas of Costa Rica, Ecuador, and Colombia, but its presence in the Amazon region has not been classified.

Sustained systems are known for the establishment of a cocoa plantation—most commonly intercropping with corn and bananas or plantains. Soil selection is the first step (usually in May-June), followed by slash-and-burn in June-August of year 1. Temporary shade (bananas or plantains) is planted in October-November of year 1, at 7 meters by 3 meters. Farmers also plant corn at 2 meters by 2 meters or cajanus at 2 meters by 2.5 meters at the same time. Cocoa seedlings are transplanted in October-February, depending on rainfall, at 4 meters by 4 meters, and more recently at 3.5 meters by 3.5 meters. Permanent shade and windbreak trees are also planted at this time. Permanent shade trees are *Inga sp.* or *Pithecellobium edwalii* and, as windbreaker trees, *Grevilea robusta.* Permanent shade trees are

**Table 14. Composition and height of the especies found in a 15-year coffee plantation on an Inceptisol at Chanchamayo, Peru.**

| *Crop* | *Spacing (m)* | *Height (m)* | *Plants/ha* | *Yield (kg/ha)* |
|---|---|---|---|---|
| Inga | 9 × 9 or 10 × 10 | 15-20 | 100-120 | — |
| Banana | Random | 5-6 | 60-100 | 600 |
| Coffee | 2 × 1.5 | 2 | 3,300 | 2,200 |

Fertilizer rate (kg/ha): 90-26-75 of N-P-K for coffee.

**Table 15. Estimated income for a cocoa-banana production system in Alfisols at Rondonia, Brazil.**

| *Year* | *Crop* | *Yield (kg/ha)* | *Gross Income (US$/ha)* |
|---|---|---|---|
| 1 | Corn | 1,500 | 165 |
| 2 | Banana | 2,250 | 135 |
| 3 | Banana | 1,750 | 35 |
| 3 | Cocoa | 150 | 200 |
| 4 | Cocoa | 400 | 534 |
| 5 | Cocoa | 600 | 800 |
| 6 | Cocoa | 700 | 934 |
| 7 | Cocoa | 800 | 1,068 |
| 8 | Cocoa | 900 | 1,200 |

planted at 18 meters by 18 meters spacing, and windbreaks in twin rows at 5 meters between them, with 3.5 meters between plants and 120 to 150 meters between windbreak trees.

Corn is harvested after five months and bananas or plantains after 12 to 15 months. Some cocoa plants will flower during the second year after transplanting but will be of commercial importance only after the third year. If the plantation is established on soils with a pH between 5.0 and 5.8, applying 500 grams of dolomitic limestone per plant is recommended at planting time. Further liming once a year in October-November improves growth and yields. Even soils that are more acidic or soils with a high aluminum saturation and low calcium content require a basal application of lime before planting. Bananas are eliminated gradually, from the third year up to the sixth year. Then the inga will provide sufficient shade. Average yields with adequate management, especially fertilizing, weeding, pruning, and systematic harvesting, are on the order of 150, 350, 500, 600, 700, and 800 kilograms-per-hectare for years 3 to 8, respectively, and can be maintained for 30 years or more at the 800 kilograms-per-hectare level. Table 15 shows the estimated gross income for a cocoa-banana system in Rondonia, Brazil.

An established cocoa plantation is considered to be a closed system; nutrients are lost only by bean exportation and through leaching. Superficial accumulation of organic matter by cocoa and *Erythrina fusca,* a shade tree, equals 8 tons per hectare per year, containing 143 kilograms of nitrogen per hectare per year (Alvim, 1980). This is six times the amount of nitrogen exported from the field in a bean yield of 1,000 kilograms per hectare.

*Commercial Fruit Trees.* Commercial fruit trees grown extensively in the Amazon include citrus, bananas, and pineapple, mainly in the Andes piedmonts and around Belem, Brazil. All are grown to satisfy regional or national markets. Citrus growing by small farmers who were formerly shifting cultivators of the Peruvian Andes foothills is an example of private initiative to develop sustained production systems (Villachica, 1986). These plantations resemble the coffee or cocoa production systems. To establish an orange plantation, farmers slash-and-burn the forest as in shifting cultivation, with the second burning done more thoroughly to eliminate stumps and logs as much as possible. In November, the farmer plants corn at a low density, 2 meters by 1 meter, and in December-January, transplants 30-day-old papaya seedlings at 3 meters by 3 meters. Using stock bought in commercial nurseries, 6- to 12-month-old orange seedlings are transplanted in January-March at 7 meters by 7 meters, eliminating any corn that interferes. The corn is harvested in March-April at the end of the rainy season.

During the dry season, the field has two crop species, papaya and oranges. The rate of plant growth usually declines because of lower rainfall, but the plants continue growing. Eleven to 12 months after the papaya was transplanted, fruits begin to ripen and are harvested for about 18 to 24 months. Fruit yields are 20 tons per hectare per 12 months. The sale of the papaya maintains the system for the second and third year after transplanting the papaya. Low-yielding papaya plants are removed during the fourth year to form an orange tree plantation. Oranges start significant production after year 4 (6 tons per hectare), increasing in years 5 through 8 to about 10, 14, 18, and 23 tons per hectare, respectively. Maximum yields of 25 to 30 tons per hectare occur in year 10; these are sustained for more than 40 years. Figure 2 shows the crop and gross income data for annual and perennial species in the system.

Several virus diseases of papaya have decreased yields of this crop. Research is very limited, and the system might break down if alternatives are not found. Farmers are replacing papaya with other crops, such as cassava. But the income is not as high as with papaya. Bananas cannot be used because they provide excessive shade for citrus. Virus-tolerant papaya coupled with good fertilization programs would allow the system to continue (Villachica and Raven).

*Native Fruits.* Many native fruits of the Amazon can be incorporated in sustainable agricultural systems. Native trees have the advantage of being adapted to the acidic, low-fertility soils and the hot and humid climates of the region. Large genetic variability exists, from which the best clones can be used for commercial agriculture. Marketing, agricultural technology, and processing techniques currently limit the establishment of these crops over large areas. Marketing studies (INIPA, 1987) indicate that Brazil nuts (*Bertholletia excelsa*), guarana (*Paullinia cupana*), camu camu (*Myrciaria dubia*), and peach palm for heart palms (*Bactris gassipaes*) have limited world markets. Araza (*Eugenia atipitata*), cocona or peach tomato (*Solanum sessiliflorum*), and others have only a local market. Technology for growing guarana and Brazil nuts has been developed by EMBRAPA, Brazil, and for peach palm by INIAA, Peru. INIAA is also developing the technology for growing camu camu, a species native to areas with seasonally flooded soils.

Guarana is a native plant from the border region of Brazil and Venezuela, Colombia, and Peru. It is cropped commercially only in Brazil. In 1985, 12,000 hectares were cultivated in Brazil, 10,000 of which were in the state of Amazonas, and 65 percent of that in Maues County (EMBRAPA, 1986). Total yield was about 939 tons, 87 percent in the state of Amazonas. Guarana grows adequately in areas with maximum and minimum mean monthly temperatures of 33 °C and 17 °C, respectively. Rainfall needs are between 2,000 and 3,300 millimeters per year, with less than four months of rainfall at less than 60 millimeters. The species is native to the Ultisols and Oxisols of the Amazon. It is not limited by acid soils with high aluminum saturation and low nutrient content. Yields on commercial plantations vary, with plants yielding between 0.1 and 3 kilograms per plant. This is the

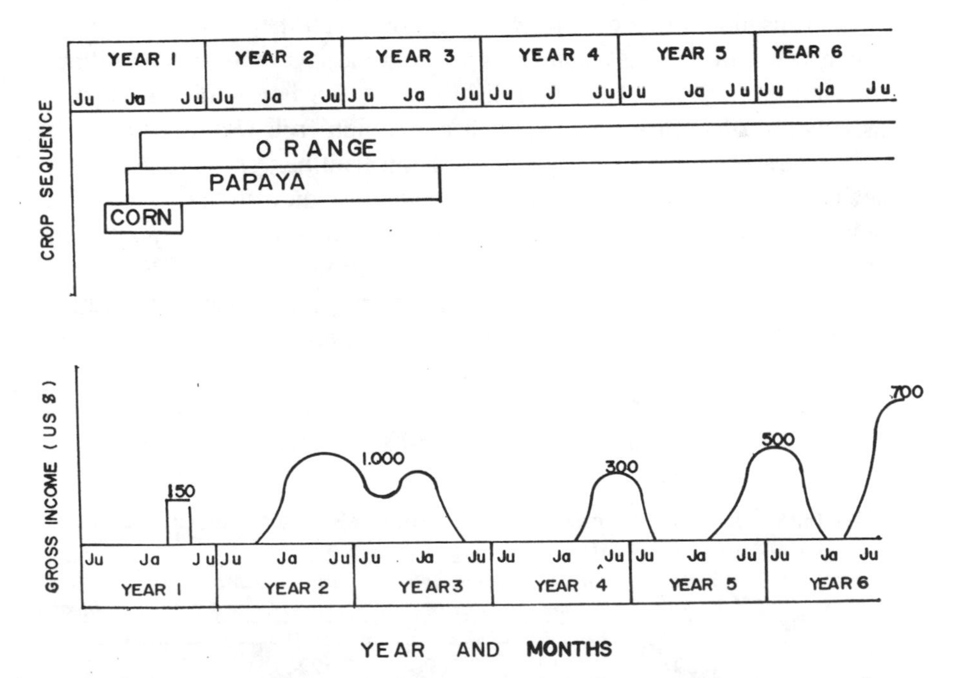


**Figure 2. Crop planting and gross income sequence in the corn-papaya-orange system in Chanchamayo, Peru.**

consequence of open pollonization. The average yield in traditional areas is 130 kilograms of dry guarana beans per hectare, but yields can be as high as 400 kilograms per hectare with better technology on more fertile soils. New clones have been developed that yield up to 2,000 kilograms per hectare. These clones, coupled with new propagation techniques and adequate fertilization, will improve yields significantly. The main disease affecting guarana is anthrocnose (*Colletrotrichum guaranicola*), which affects plants in nurseries and orchards and is favored by high rainfall. Superbrotamiento or supersprouting (*Fusarium decemcellulare*) is another disease of importance; it slows growth. Both fungi can be controlled with chemical sprays.

Three phases can be defined for development of guarana plantation systems: seedling preparation in a greenhouse (one year), growth after transplanting (three to four years), and the productive phase (fourth year and beyond). Guarana seedlings need shade during the first year after transplanting. This is provided by palm leaves or by a living plant, usually cassava. Several systems have been developed to decrease guarana establishment costs, increase the efficiency of land use during the first few years, and provide shade for the transplanted seedling (Carrie, 1983; EMBRAPA, 1986). One of these systems includes a rotation of four cowpea and three corn crops, intercropped with guarana rows at 5 meters by 5 meters. A second system is to associate cowpea and cassava with guarana during the first three years of the establishment. A third system consists of an association of semiperennials, such as passion fruit (*Passiflora edulis*), with guarana. In the latter system, passion fruit harvest starts during the second year and lasts 18 months. Yields of passion fruit are 12 to 16 tons per hectare (Table 16), which compare favorably with average unfertilized monoculture yields of 10 tons per hectare (Correa, 1983). The advantage of these systems is that the temporary crops pay for establishment costs, leaving a three-year guarana plantation almost ready to start production.

Peach palm is another fruit tree native to the Ultisols of the Amazon. It is found in areas having 1,500 to 4,500 millimeters of annual rainfall

**Table 16. Yield, production costs, and income of passion fruit (maracuya) associated with guarana (Manaos, 1981).**

| *Maracuya* (m)* | *Yield (kg/ha)* | *Gross Income* | *Guarana Cost* | *Partial Income* | *Maracuya Cost* | *Net Income* |
|---|---|---|---|---|---|---|
| | | | | *US $* | | |
| 3 × 3 | 16,303 | 4,891 | 2,769 | 2,122 | 921 | 1,201 |
| 6 × 3 | 12,376 | 3,713 | 2,769 | 944 | 558 | 386 |

*70-66-75 kg N-P-K/ha applied for maracuya at 3 × 3, and half the rate for 6 × 3.

**Table 17. Yield, production cost, and income of rice and cowpea associated with peach palm at Yurimaguas and Iquitos, Peru.**

| *Crop* | *Yield* | *Gross Income* | *Production Cost* | *Cumulative Net Income* | *Months After Palm Planting* |
|---|---|---|---|---|---|
| | | US $ | | | |
| Rice | 1,600 | 175 | 116 | 59 | 5 |
| Rice | 1,300 | 142 | 86 | 115 | 11 |
| Cowpeas | 1,000 | 143 | 81 | 177 | 15 |
| Rice | 1,300 | 142 | 83 | 236 | 21 |
| Desmodium* | - | - | 100 | 136 | 48 |

*Cost of maintenance for two years.

per year and a temperature range of 22 °C to 30 °C mean annual temperature. The palm grows well in acid soils of pH 4.6 and 50 percent to 60 percent aluminum saturation, although better yields are obtained on more fertile soils. Fruits are used for both human and animal feeding, and the young stem is used for hearts ("palmito"). Peach palm has the advantage that many suckers are formed from the same tree, which does not occur with other trees used for heart-of-palm production.

Few commercial plantations are found in the Amazon, but they are present in other tropical areas of Colombia, Venezuela, and Costa Rica. A comprehensive germplasm collection has been made in the Amazon, and ecotypes are being evaluated for fruit and heart-of-palm production. No diseases and insects have been found that would severely limit production. Fruit production starts during the fourth year after transplanting and can last for 10 to 20 or more years, depending upon management. In contrast, the first harvest of hearts of palm is made 18 months after transplanting, with subsequent annual harvests.

A productive system based on two crops each of rice and cowpeas in rotation, followed by planting of legume cover crops, has been tested successfully for peach palm establishment in Yurimaguas and Iquitos, Peru (Table 17). This system, with 5-meter by 5-meter spacing of palms, allows the annual crops to pay for the cost of investment, while the legume cover crop is used to improve soil fertility and minimize maintenance costs during the third and fourth years. The sustainability of the system could be improved further if another harvest income could be obtained during years three and four. Thus, peach palm, with the rice-cowpea rotation, is being tested at a close spacing (2.5 meters by 2.5 meters). This allows plants to be thinned out for heart palm production after 24 and 36 months of transplanting to establish a final spacing of 5 meters by 5 meters. The thinned-out peach palm will give an additional gross income of $150 US from each harvest, with minimum additional costs. Fruit production will start in year four and by the fifth year will be large enough to sustain the plantation, with yields increasing with time until full development is reached.

The system described here does not use fertilizers if the palm and the annual crops are planted immediately after clearing and burning of the forest—taking advantage of the initial higher soil fertility. It is also a good alternative for shifting cultivation in the region because farmers are familiar with the palm, the fruit is consumed locally, and the palm hearts are canned for exportation. Farmers' interest in the region is evident, with increasing enthusiasm to start growing this crop.

***Other Perennial Crops and Pastures.*** *Rubber.* Another tree native to the Amazon, rubber has a history associated with the growth and decline of several towns in the region during the so-called rubber boom between 1890 and 1912. Attempts made early in the century to grow rubber trees in the Amazon met with little success because of the very serious leaf disease "South American leaf blight," caused by the fungus *Microcyclus ulei*. This disease, endemic to the Amazon, is not present on other continents where rubber is grown successfully. Rubber growing in the Amazon also has problems from the leaf-eating caterpillar (*Erynnis ello*) and the panel disease caused by *Phytophthora palmivora.* But *Microcyclus* has been the main limiting factor in the region.

Methods of controlling *Microcyclus ulei* have recently been developed and have become available to farmers, opening up new possibilities for expanding the crop. The development of mechanized sprayers allowed fungicides to be applied to the leaves of the trees to control the fungus in old plantations with traditional susceptible clones such as Fx 25. The fungicide has to be applied in the dry season, July and August, when new leaves, which are susceptible to the fungus, begin to develop. A second approach being used in Brazil to control the fungus consists of the selection of resistant clones (Alvim, 1980). Research in Brazil also has shown that damage caused by the leaf blight can be minimized or avoided in areas having three consecutive months of rainfall below 50 to 60 millimeters per month (Moraes, 1974). In areas of the Amazon where the dry season is not well-defined or no resistant clones are available, *Mycrocyclus* can be controlled by the use of rootstocks of *H. brasiliensis* with canopies of *H. panciflora,* the first for latex production and the latter for photosynthesis because of its resistance to the fungus.

*Black Pepper.* The most economically important crop for the Brazilian Amazon is black pepper, with fields concentrated close to Belem, Para, and, to a lesser extent, in Manaos, Amazonas. The total area in the Brazilian Amazon is around 20,000 hectares, with a production of 50,000 tons per year, making Brazil the third largest pepper-producing country, after Malaysia and India (Alvim, 1980). Almost all pepper production in the state of Para occurs on well-drained, low-fertility Oxisols with high rates of fertilization. About 60 percent to 70 percent of the production cost is for chemical inputs and labor. Nevertheless, because of the high prices in the international market, pepper cultivation is one of the most profitable activities in the Brazilian Amazon. It is also one of the most profitable crops in the Peruvian Amazon, with a net income around $2,000 US per hectare per year in year four.

A root disease caused by *Fusarium solani f. piperi* is the main factor limiting pepper culture. The fungus reduces the average productive life of pepper to only 8 to 10 years. However, when a plant dies, farmers apply fungicides and organic matter to the soil and plant new pepper seedlings, thereby obtaining continuous production in the rest of the field. If the attack is too severe, farmers abandon the land or, taking advantage of the residual effect of several years of fertilization, change to other crops, such as cocoa, rubber, or papaya. Another limiting factor for pepper establishment is the high cost of support for pepper vines. This is being solved by planting native forest trees in a row at an adequate spacing and using them as living supports.

*Oil Palm*. The African oil palm (*Elaeis guianensis*) is recognized as the most efficient oil-producing cultivated plant. Predominant soils and the climate in the Amazon are suitable for oil palm, which is known for its tolerance to soil acidity, apparently with no need for liming. It has been grown successfully in soils of pH 4.0 to 4.5, with fertilizer applied. It is one of the most promising crops for the acidic, low-fertility soils of the Amazon.

The technology for growing oil palm is well known and is being used in the Amazon. In Peru, the government has established 5,000 hectares near Tingo Maria, with an average yield of 5.0 tons of oil per hectare. An adjacent private enterprise covers another 5,000 hectares, and there are new developments of 2,000 hectares near Iquitos and also near Pucallpa, both on Ultisols. The main limiting factors for expanding the crop include the cost of establishment, the large area required to be under production for a single processing facility, and the lack of locally produced seeds of improved hybrids. Seeds from the high-yielding hybrids have to be imported from outside the region, at high costs. The use of tissue culture could permit propagation of superior germplasm adapted to Amazon conditions.

*Annato.* Annato or achiote *(Bixa orellana)* is a small tree native to the Amazon that is grown as a source of edible red dye. The demand for natural dyes has produced a demand for achiote. Farmers are planting the crop using technologies suitable for other crops. Around 1980, annato was not grown extensively in the Amazon but was collected only from backyards and from wild plants.

Annato grows well in acidic, low-fertility soils where other crops will not. Yields in commercial plantations of 625 plants per hectare are about 1,000 kilograms per hectare. Technology for the crop requires improvement, but the potential is certainly there. Results of evaluation trials in the acidic soils of Peru show that promising selections yield 2.0 tons of seed per hectare in the fourth year, with bixine contents of up to 6.5 percent (the average bixine content is only 2.6 percent). Usually, the plant starts commercial production during the second or third year after seedlings are transplanted. Some ecotypes start production eight months after transplanting, even before traditional cassava varieties, with seed yields of 110, 495, and 650 kilograms per hectare for the first, second, and third years, respectively. Yields continue to increase, reaching maximum yields in years six to eight.

*Pastures*. Research with pastures during the last few years has focused on developing a strategy for reclaiming and improving areas with cultivated pastures. The basic components of pasture reclamation systems include the method of land-clearing, a first crop of a grain legume, establishment of the grass or the grass-legume association, and minimum soil fertilizer and pesticide application. Land-clearing is done with the help of chemical herbicides (in pure grasses or degraded pastures) or by cutting and burning of bushes. This is followed by the application of 200 kilograms per hectare of both dolomitic limestone and rock phosphate, and disking with conventional machinery. This results in a harvest of 1.0 to 1.5 tons of cowpeas per hectare.

Once the degraded pasture is cleared, several species can be used for the new pasture. *Brochiana decumbens* and *B. humidicola* are the species

**Table 18. Average annual productivity of several pasture systems on an Ultisol at Yurimaguas, Peru.**

| *Pasture System** | *Years Grazing* | *Stocking Rate (an/ha)* | *Live-weight Gain (kg/ha/yr)* | *Legume Content (%)* |
|---|---|---|---|---|
| Torourco (Native grass) | 6 | 1.0 | <100 | 0 |
| *Centrosema pubescens* 438 | 5 | 3.8 | 573 | 96 |
| *B. decumbens/D. ovalifolium* | 6 | 4.7 | 532 | 35 |
| *B. humidicola/D. ovalifolium* | 4 | 4.6 | 671 | 38 |
| *A. gayanus/S. guianensis* | 6 | 3.2 | 477 | 31 |

*Improved pastures fertilized with 22-42-12 kg N-P-K/ha/year and 500 Kg lime/ha.

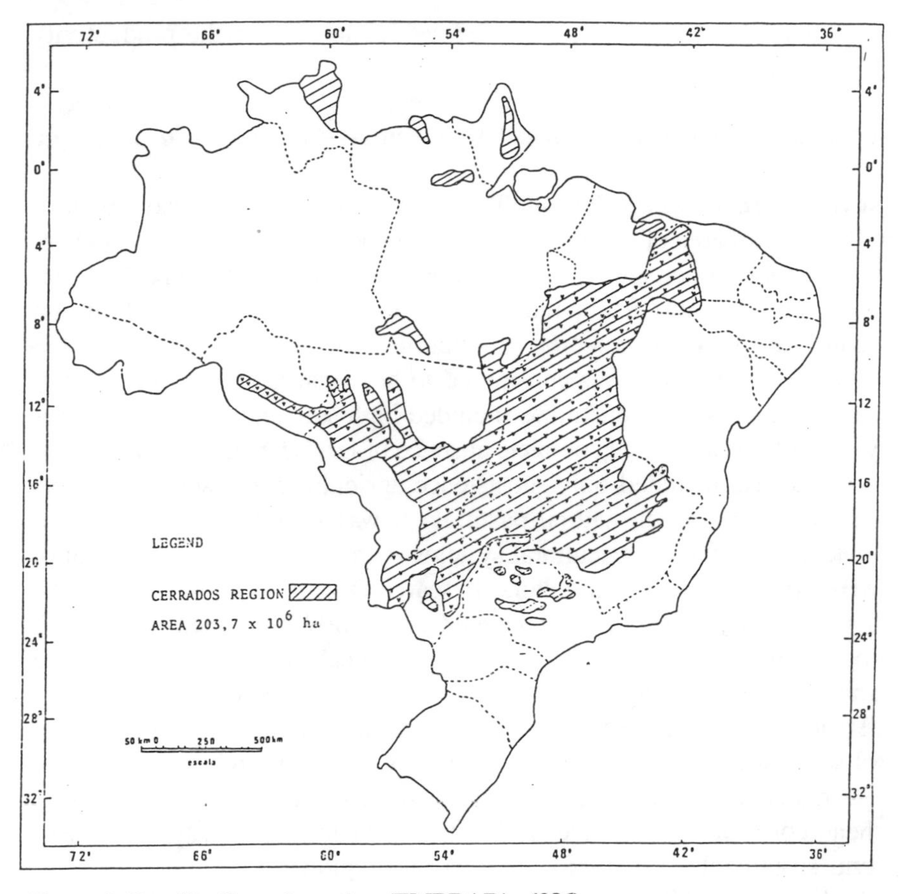


**Figure 3. Brazil's Cerrado region (EMBRAPA, 1986).**

most commonly used during the last 15 years (Toledo and Serrao, 1980). New species and varieties were found to solve several problems, especially spittle bug infestations. Research conducted during the last nine years in Yurimaguas, Peru (North Carolina State University, 1987), has shown the feasibility of using grass-legume mixtures or pure legume pasture for livestock-raising on Ultisols of the Amazon. Live-weight gains are about 477 to 671 kilograms per hectare per year (Table 18), which compares favorably with those of pure grasses. The amounts of fertilizer that have to be added yearly and the lime applied at establishment are considered to be low. These costs can be compensated for by the increase in live-weight gain. Fertilizer application and proper grazing management can sustain the pasture for about 10 to 12 years. After that period, another lime application, tillage, and reseeding might be needed to recover the productivity of the pasture.

## Sustainable Systems in the Brazilian Savannas—The Cerrado Challenge

***Characteristics of the Region.*** The Cerrado region in the central plateau of Brazil is an area of about 2 million square kilometers (200 million hectares), about 25 percent of the total area of the country (Figure 3). It extends from 4° to 24° South and from 42° to 66° West. Nearly 50 million hectares are available for crop production (Goedert et al., 1982).

The Cerrado region was considered to be a marginal area for agriculture until the 1960s. Constraints included high soil acidity and low fertility, lack of research, and complete absence of fertilizers and farm machinery. Land use included only subsistence agriculture and some extensive beef cattle grazing on natural pastures. It was impossible to plan or to establish sustainable agricultural systems at that time. With the advent of EMBRAPA (Brazilian Agricultural Research Organization) in 1973 and the CPAC (Cerrados Agricultural Research Center) in 1975, an intensive research program was initiated to look for technological alternatives that would enhance the production potential of this savanna-type ecosystem.

Positive results began early, with the development of technologies to control soil acidity and to increase soil fertility. This research made possible a boom in land use and increase in yields of many crops, such as soybeans, wheat, corn, beans, and rice. Today, farmers can choose from many efficient, sustainable systems for the Cerrado region.

*Soils.* Latosols (Oxisols and Ultisols in the U.S. taxonomy) occupy about 46 percent of the area, followed by quartz sands (15.2 percent), Spedosols (15.1 percent), and less common soil orders (Adamoli et al., 1986). Within the Latosols, the Red Yellow Latosol (RYL) and the Dark Red Latosol (DRL) predominate where most agriculture has developed since the beginning of the last decade. Latosols are very deep soils, with little differentiation between horizons and a variable texture, but most are high in clay (EMBRAPA/SNLCS, 1978). The clay fraction presents a mineralogy dominated by kaolinite, gibbsite, and iron oxides (Table 19). Detectable amounts of hydrosy-interlayered clays, probably vermiculites, have also been found in some Oxisols (Bigham, 1977; Weaver, 1975). Such a mineralogy, with a predominance of low-activity clays, reflects the physiochemical behavior of the soil, especially with respect to ion exchange, fertility, water-holding capacity, and phosphorus fixation (Lopez, 1974; Wolf, 1975).

*Soil Chemistry, Fertility, and Conditions for Agriculture.* The effective cation exchange capacity is very low, mostly dominated by aluminum

High values of aluminum saturation (Table 20) restrict plant growth severely unless corrected.

The distribution of aluminum and calcium in an Oxisol profile reveals two major problems that restrict root growth in the subsoil: if aluminum is the dominant cation, aluminum toxicity is the primary restriction; and if aluminum and calcium are very low, calcium deficiency becomes the restriction (Ritchey et al., 1982). Nevertheless, despite the very low fertility and the problems associated with aluminum toxicity and calcium deficiency, these soils can be put into production with a satisfactory economic return.

**Table 19. Selected chemical and mineralogical properties of the Ap horizon of the Dark Red Latosol (Haplustox) and Red Yellow Latosol (Acrustox).**

| *Properties* | *Haplustox* | *Acrustox* |
|---|---|---|
| pH ($H_2O$, 1:1) | 4.55 | 5.05 |
| Al ($cmol_c$* $kg^{-1}$) | 2.50 | 0.38 |
| Ca ($cmol_c$ $kg^{-1}$) | 0.31 | 0.08 |
| Mg ($cmol_c$ $kg^{-1}$) | 0.15 | 0.11 |
| K ($cmol_c$ $kg^{-1}$) | 0.10 | 0.14 |
| ECEC ($cmol_c$ $kg^{-1}$) | 3.06 | 0.71 |
| P (mg $kg^{-1}$) | 1.50 | 0.90 |
| C (g $kg^{-1}$) | 22.00 | 20.10 |
| CEC (pH 4.7) ($cmol_c$ $kg^{-1}$) | 3.22 | 1.05 |
| CEC (-H 6.0), ($cmol_{c-}$ $kg^{-1}$) | 5.60 | 2.17 |
| Total K (g $kg^{-1}$) | 3.79 | 3.77 |
| Clay content (g $kg^{-1}$) | 500.00 | 750.00 |
| Clay mineralogy† | $K_{63}G_8Go_{14}HIV_{15}$ | $K_{31}G_{66}Go_8$ |

*$cmol_c$ = cmol of charge.
†K = kaolinite, G = gibbsite, Go = goethite, HIV = hydroxy-interlayered vermiculite. Subscripts denote percentage in clay fraction.

**Table 20. Some physical and chemical properties of a Dark Red Latosol (Haplustox) and of a Red Yellow Latosol (Acrustox) at the Cerrado Research Center, Brasilia, Brazil.**

| *Profile* | *Depth (cm)* | *Particle Size* Sand | Silt | Clay | *pH ($H_2O$)* | *Exchangeable Cations* Al | Ca+Mg | K | *Al Saturation (%)* |
|---|---|---|---|---|---|---|---|---|---|
| | | % | | | | cmol $kg^{-1}$ | | | |
| DRL | 0-10 | 36 | 19 | 45 | 4.9 | 1.9 | 0.4 | 0.10 | 79 |
| (Haplustox) | 10-35 | 33 | 19 | 48 | 4.8 | 2.0 | 0.2 | 0.5 | 89 |
| | 35-70 | 35 | 18 | 47 | 4.9 | 1.6 | 0.2 | 0.03 | 88 |
| | 70-150 | 35 | 18 | 47 | 5.0 | 1.5 | 0.2 | 0.01 | 88 |
| RYL | 0-12 | 28 | 27 | 45 | 5.1 | 1.8 | 0.2 | 0.08 | 86 |
| (Acrustox) | 12-30 | 26 | 30 | 44 | 5.0 | 1.4 | 0.2 | 0.05 | 82 |
| | 30-50 | 25 | 27 | 48 | 5.2 | 0.6 | 0.2 | 0.03 | 67 |
| | 50-85 | 22 | 28 | 50 | 4.9 | 0.0 | 0.2 | 0.02 | 0 |
| | 85-125 | 22 | 28 | 50 | 5.3 | 0.0 | 0.2 | 0.01 | 0 |

Aluminum toxicity and fertility problems (Table 20) in the plow layer (20 to 30 centimeters) have been solved by liming and fertilizer applications— adequate amounts of phosphate, nitrogen, potassium, and sulfur, as well as micronutrients (mainly zinc). Liming is essential for crop production, and its effects on soil properties and crop performance are impressive (Lathwell, 1979b); Miranda et al., 1980). Phosphorus fixation occurs, but all results show a strong crop response to phosphate application (Goedert et al., 1986; Lathwell, 1979a; Miranda et al., 1980; Sanchez and Salinas, 1981; Smyth and Sanchez, 1980).

Liming materials, abundant in the Cerrado region (Goedert et al., 1982), can be supplied by lime-processing plants distributed strategically in the region. Despite these facilities, the price of lime plus transportation is increasing, thereby increasing production costs.

Supplies of phosphate fertilizer improved after construction of two phosphate processing plants in the region, one located in Araza and the other in Uberaba, both in the State of Minas Gerais. The installation of these two plants has had a remarkable effect on reducing the price of phosphorus and speeding the distribution of phosphates throughout the region. In addition, the Brazilian government has created several mechanisms and conditions to subsidize these products and to guarantee their supply at a reasonable cost.

Potassium, another important nutrient, is added to the soil mainly as potassium chloride, supplied by a recently built processing plant in northeastern Brazil. Presently, domestic production supplies only 25 percent of Brazil's requirements. Potassium leaching in soils with low cation exchange capacity is important, especially during the rainy season.

Use of nitrogen fertilizers depends on the crop species. For legumes (soybeans, beans, peas), the use of efficient and specific *Rhizobium* has been a major contribution of research to the Cerrados agriculture. By using such technology, which is simple and inexpensive, the farmer and ultimately the country saves about $1 billion US annually. For other crops, which do not fix nitrogen, this nutrient is provided by conventional sources, such as urea, which is produced domestically. Other sources are also available, but their use is related to their cost.

***Research and Advance of the Agricultural Frontier.*** Prior to the 1960s, existing farming enterprises in the Cerrado included subsistence agriculture and extensive beef cattle production on native pastures. The productivity of these systems, in general, was low; they supplied only the minimum needs of the farmer and his family. Furthermore, they did not contribute to the economic development of the region.

Land management in new areas has to contribute to the welfare of rural populations, as well as that of the country. Under these circumstances, the economic development of a region has to drive agriculture to produce commercial commodities, with good market acceptance (Mosher, 1970). Land use for commercial purposes in the Cerrado region began in the 1960s, but it was problematic because of technological gaps for profitable production. For a long time, agricultural research in Brazil had been directed to production in new and fertile soils in temperate regions. Old and weathered soils received little attention from researchers. Despite the lack of technology, the Cerrado region was considered to be the best option to expand Brazil's agricultural frontier (Azevedo and Caser, 1980). The greatest challenge was to make the acidic and low-fertility soils ready for planting annual crops that were susceptible to aluminum toxicity and that had high nutrient requirement.

Research programs carried out in the last 15 years by EMBRAPA/CPAC, have advanced the knowledge of natural resources and socioeconomics, soil-waterplant relationships, and production systems. This research has provided the basis for developing a technology to implement and maintain sustainable agricultural systems. In 1980, five years after the inauguration of CPAC, the Cerrado region already was contributing a reasonable proportion of the domestic agricultural production (Table 21). Many crops had higher yields than the national average, but those averages were still considered low. Research data (Figure 4) indicate that it is possible to double the productivity of soybeans, corn, and wheat by managing phosphorus fertilization. In addition to the efficient use of phosphorus, other factors can triple productivity.

**Table 21. Contribution of the Cerrado region to Brazilian agricultural production in 1980.**

| *Crop* | *Productivity (kg/ha)* | | *Participation of Cerrado in Brazilian Production (%)* |
|---|---|---|---|
| | *Cerrado* | *Brazil* | |
| Rice | 1,200 | 1,600 | 45.5 |
| Corn | 1,900 | 1,800 | 17.0 |
| Beans | 400 | 700 | 12.0 |
| Cotton | 1,400 | 1,100 | 11.5 |
| Soybeans | 1,700 | 1,700 | 10.5 |
| Cassava | 15,100 | 12,600 | 6.5 |
| Sugar cane | 78,400 | 56,100 | 6.5 |
| Coffee | 1,700 | 1,100 | 27.0 |
| Dryland wheat | 1,200 | 900 | — |

One important characteristic of the Cerrado region is its agrarian structure, with the predominance of large, private landholdings (Table 22). About 70 percent of the area has farms of more than 500 hectares; more than 90 percent of these are privately owned. On the other hand, farming activities in the Cerrado region are very heterogeneous, and systems of production are not well-defined. Presently, pastures (native and cutivated) occupy about 55 percent of the area, but these figures can change with government agricultural policy (Table 23).

The Cerrado region, despite the high acidity and low fertility of its soils, presents many favorable conditions for intensive agricultural development. The mean annual temperature is about 21°C, with little variation from one season to another. This allows cultivation of almost any crop in the world. The accumulative annual precipitation averages 1,500 millimeters. This amount of water, if well-managed, is sufficient for up to two harvests per

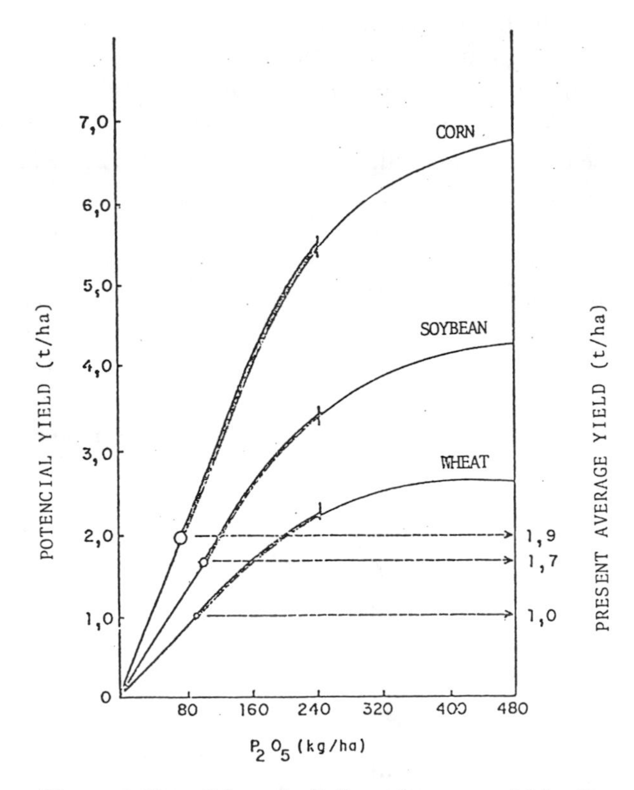


**Figure 4. Potential productivity and average yields of corn, soybeans, and wheat in nonirrigated systems (EMBRAPA, 1986).**

**Table 22. Characteristics of the agrarian structure in the Cerrado region.**

| | *Number of Properties* | *Percent of Total* | *Area of Properties (ha)* | *Percent of Total* |
|---|---|---|---|---|
| Stratification | | | | |
| Less than 100 ha | 792,126 | 79.3 | 11,860,399 | 7.5 |
| 100 to less than 500 ha | 150,991 | 15.1 | 33,049,486 | 21.0 |
| 500 to less than 1,000 ha | 27,010 | 2.7 | 18,784,898 | 11.8 |
| 1,000 to less than 5,00 ha | 22,112 | 2.2 | 43,574,852 | 27.6 |
| More than 5,000 ha | 3,623 | 0.4 | 50,734,479 | 32.1 |
| Unknown | 2,506 | 0.3 | | |
| Total | 998,368 | 100.0 | 158,004,114 | 100.0 |
| Farmer status | | | | |
| Owner | 602,828 | 60.4 | 149,269,927 | 92.9 |
| Tenant farmer | 141,657 | 14.2 | 3,156,628 | 2.0 |
| Associate farmer | 71,862 | 7.2 | 689,770 | 0.4 |
| Squatter | 182,021 | 18.2 | 7,565,263 | 4.7 |
| Total | 998,368 | 100.0 | 160,681,588 | 100.0 |

**Table 23. Land occupation according to farming activity (adapted from Kornelius et al., 1988).**

| *Activity* | *Area (ha)* | *Participation (%)* |
|---|---|---|
| Annual crops (grain production) | 18 | 9.0 |
| Perennial crops (including forestry) | 2 | 1.0 |
| Cultivated pastures | 30 | 15.0 |
| Native pastures | 80 | 40.0 |
| Native forests | 70 | 35.0 |
| Total | 200 | 100.0 |

year. The level topography, together with the good drainage properties of the soils, allows for the mechanization of large areas in the region. The basic infrastructure (roads, energy supplies, grain storage capacity) is developing at a high rate, under programs sponsored by the government and private business.

Recognizing that the technology for establishing agriculture in these soils is known, the availability of land at a low price is another factor that makes farming enterprises very attractive in the region (Goedert et al., 1986). Even considering the high costs of land clearing and fertilization, the final land price is still competitive (Table 24). It is clear that the initial investment cannot be returned in the first year regardless of crop. But if an agricultural development policy is planned adequately and applied, farmers can take advantage of the opportunity to start some farms on a reasonably structured soil. A sustainable agricultural system in the Cerrados includes annual crops (soybeans, corn, wheat, rice), perennial crops (fruit crops, coffee, forestry), beef cattle production, and irrigated systems.

***Sustainable Systems for Annual Crops.*** Annual crop agriculture is perhaps the best kind of land use of the farm activities in the Cerrado region. The conditions that stimulated annual cropping were the availability of technology, inputs (lime and fertilizers), farm machinery, and equipment. In addition, reasonably well-paved roads link the region to the consumer centers (Villas, 1980). Very recently, two soybean oil plants were constructed

**Table 24. Average cost of transformation of one hectare of Cerrado highly productive soil, with and without irrigation, June 1984.**

| *Items* | *Cost (US $)* | *Participation (%) System* | |
|---|---|---|---|
| | | *Nonirrigated* | *Irrigated* |
| Inputs | | | |
| Lime | 69.60 | 13 | 5 |
| Phosphorus | 145.00 | 26 | 10 |
| Potassium | 58.00 | 10 | 4 |
| Micronutrients | 40.60 | 8 | 4 |
| Subtotal | 313.20 | 57 | 23 |
| Services | | | |
| Opening | 58.20 | 10 | 4 |
| Cleaning | 20.26 | 4 | 1 |
| Lime incorporation | 17.38 | 3 | 2 |
| Soil conservation (Terraces) | 17.38 | 3 | 1 |
| Other | 8.73 | 1 | 1 |
| Subtotal | 121.95 | 22 | 8 |
| Land price | 115.97 | 21 | 8 |
| Total without irrigation | 551.12 | 100 | — |
| Investment in irrigation | 869.62 | — | 61 |
| Total with irrigation | 1,420.74 | — | 100 |

**Table 25. Area and production of soybeans, corn, wheat, and rice for 1987 in the Cerrado region and in Brazil.**

| *Crop* | *Area (millions ha)* | | | *Production (1,000 tons)* | | |
|---|---|---|---|---|---|---|
| | *Cerrado* | *Brazil* | *Percent* | *Cerrado* | *Brazil* | *Percent* |
| Soybeans | 3,471 | 9,161 | 37.9 | 6,838 | 16,581 | 40.6 |
| Corn | 3,310 | 13,649 | 24.3 | 6,501 | 26,978 | 24.1 |
| Wheat | 370 | 3,295 | 11.2 | 464 | 4,652 | 10.0 |
| Rice | 3,356 | 6,046 | 55.5 | 2,845 | 10,562 | 36.4 |
| Total | 10,507 | 32,151 | 32.7 | 17,648 | 59,043 | 29.9 |

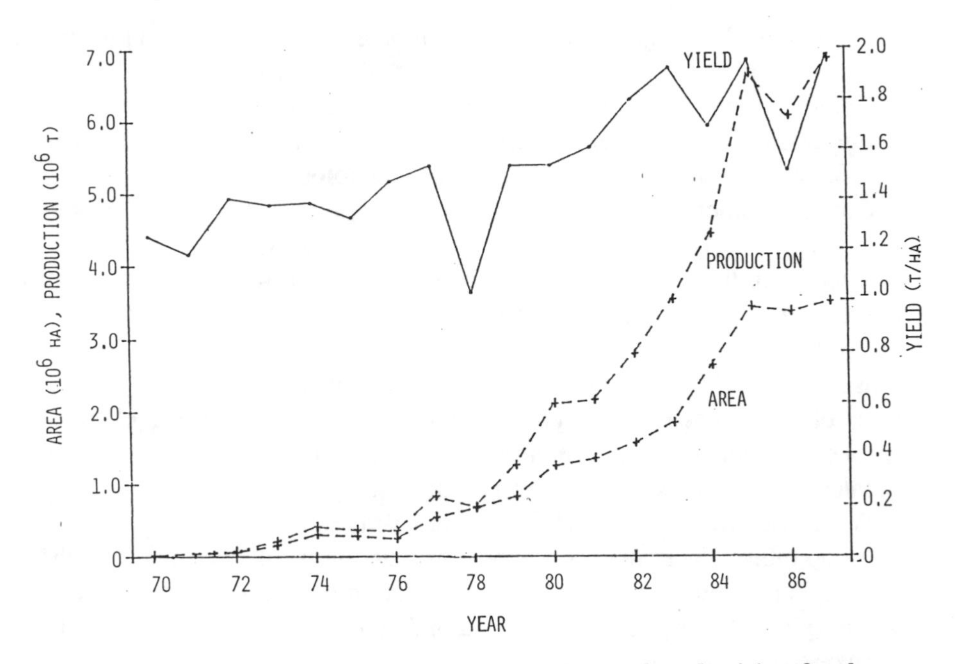


**Figure 5. Evolution of the cultivated area, production, and productivity of soybeans in the Cerrado region.**

near Brasilia, opening another opportunity for soybean farmers.

Cultivated annual crops have specific nutrient requirements, and most cultivars are not tolerant to aluminum toxicity. Thus, the acidic and infertile soils of the Cerrados can be used for annual crops only if a minimum level of technology is applied. In a broad sense, such technology consists of aluminum neutralization, by liming; improving soil fertility by fertilizer application; and soil tillage and management to allow for adequate root development without compacting the soil and causing soil erosion (Goedert et al., 1986).

Soybeans, corn, wheat, and rice occupy about 10 million hectares of the available land in the Cerrado region; soybeans occupy the largest portion (Table 25). Total grain production of these four crops in the Cerrado region represents about 30 percent of the national gross production. The area of cultivated soybeans increased from 10,000 hectares in 1970 to 3.5 million hectares in 1987 (Figure 5). This increase in production in 1987 represented 41 percent of the total soybean production in Brazil and about 26 percent of the total grain production in the country (65.5 million tons). One of the factors that has contributed to the expansion of soybeans is their price relative to that obtained for other crops. The development of technology to improve soil management and nitrogen fixation has made it possible to obtain yields of 4,000 kilograms per hectare (Figure 4). However, the average yield is about 2,000 kilograms per hectare, which means that there is still room for the development and transference of technology.

Corn is another important cash crop in the region. In 1987, the Cerrados accounted for 24 percent of Brazil's total corn production. Corn was introduced into the region prior to soybeans. The area in corn increased 50 percent, from 1.67 million hectares to 3.31 million hectares, between 1970 and 1987 (Figure 6). Total grain production increased from 2.32 million tons in 1970 to 6.5 million tons in 1987, an appreciable increase in productivity of 42 percent in terms of the national average. However, data obtained from experimental research in nonirrigated systems show a yield potential greater than 6.0 tons per hectare (Figure 4).

Wheat was first introduced in southern Brazil and later brought to the Cerrado region. The success of wheat production is related directly to the production system used. The best combination has been irrigated winter wheat (planted in April and harvested in August-September), followed by corn or soybeans during the rainy season. The advantage of wheat produc-

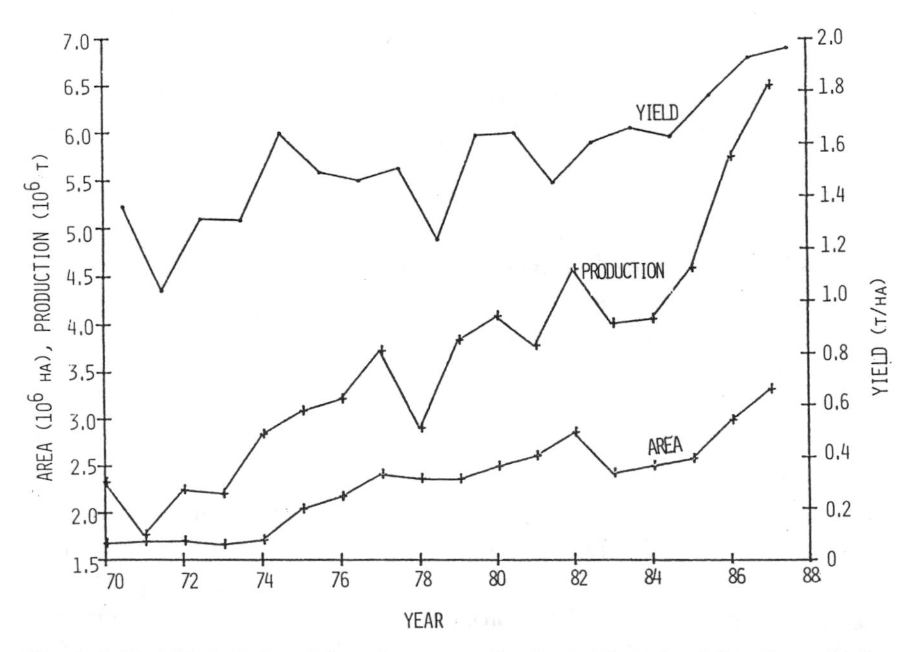


**Figure 6. Evolution of the cultivated area, production, and productivity of corn in the Cerrado region.**

**Table 26. Maximum yield and average daily gain of irrigated winter wheat obtained in different situations on Cerrado soils and in the United States.**

| Activity | Place | *Average Cycle (days)* | *Average Yield (kg/ha)* | *Average Daily Gain (kg/ha)* |
|---|---|---|---|---|
| Experimental results* | CPAC/DF | 115 | 6,372 | 55.4 |
| Validation field* | Minas Gerais | 115 | 5,535 | 48.1 |
| Commercial field* | Minas Gerais | 115 | 5,406 | 47.0 |
| Record wheat yield† | U.S.A. | 330 | 14,100 | 42.7 |

*Source: EMBRAPA, 1986.
†Source: Hanson et al., 1982.

tion in the Cerrado, compared to other regions in the world, is clear (Table 26). The average daily gain of grain production in a commercial field managed with high technology is higher than the record yield of wheat (Hanson et al., 1982). This is a clear demonstration of the potential of the region for grain production. In terms of relative expansion, wheat is the second leading crop in the Cerrado region (1,800 percent increase from 1970 to 1987), but the total cultivated area in 1987 was only 37,000 hectares. The average yield increase jumped from 264 kilograms per hectare in 1972 to 1,254 kilograms per hectare in 1987 (Figure 7). Yields up to 5,400 kilograms per hectare were obtained on irrigated fields.

Rice is the second leading crop after soybeans in area planted in the Cerrados. It is produced with a low level of technology, as a starter crop after the opening up of new areas. This approach is used less frequently because of the alternative technological packages available. For this reason, and because of its substitution by other crops, the area occupied by rice increased only 50 percent (2.29 million hectares to 3.36 million hectares) from 1970 to 1987 (Figure 8). The average productivity of rice is still low, even though recently developed cultivars have yield potentials of 4,500 kilograms per hectare. Other annual crops, such as sorghum, cotton, beans, and peanuts, are cultivated with reasonable success, but on a smaller scale.

***Sustainable Systems for Perennial Crops.*** The ecological conditions of the Cerrados are also adequate for many perennial crops, such as coffee, tropical fruit, and forestry. Coffee, a crop cultivated traditionally in the southern states of Brazil, is now taking over large expanses of land in the Cerrado region. The reasons for the changes are the low probability of frost, low land costs, a level topography, and the availability of labor. These factors compensate for the reduced soil fertility and irregular rain distribution in the region. At present, coffee occupies about 900,000 hectares, contributing 40 percent of the total production of coffee in Brazil

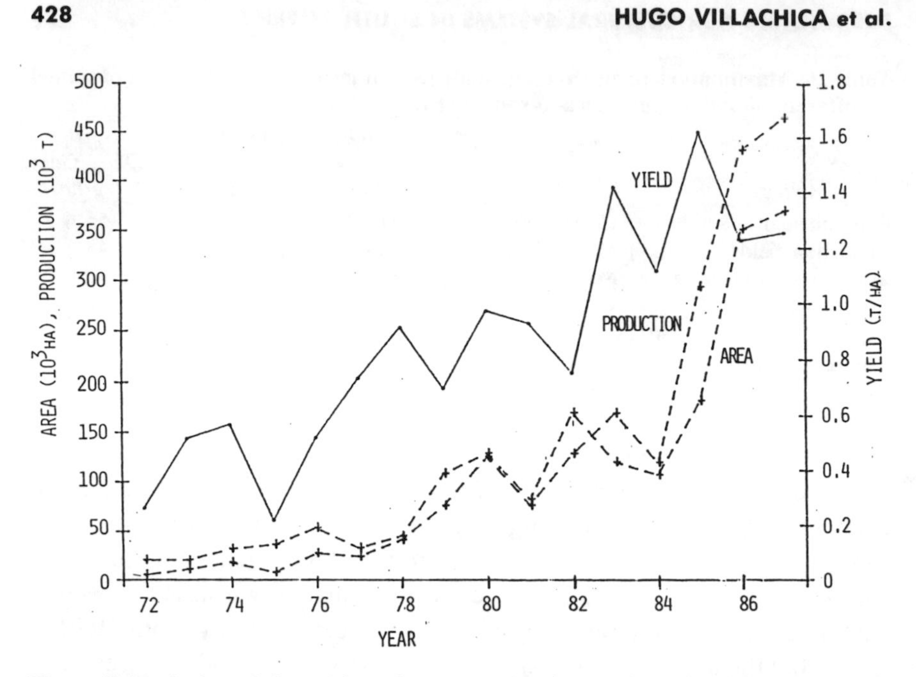


**Figure 7. Evolution of the cultivated area, production, and productivity of wheat in the Cerrado region.**

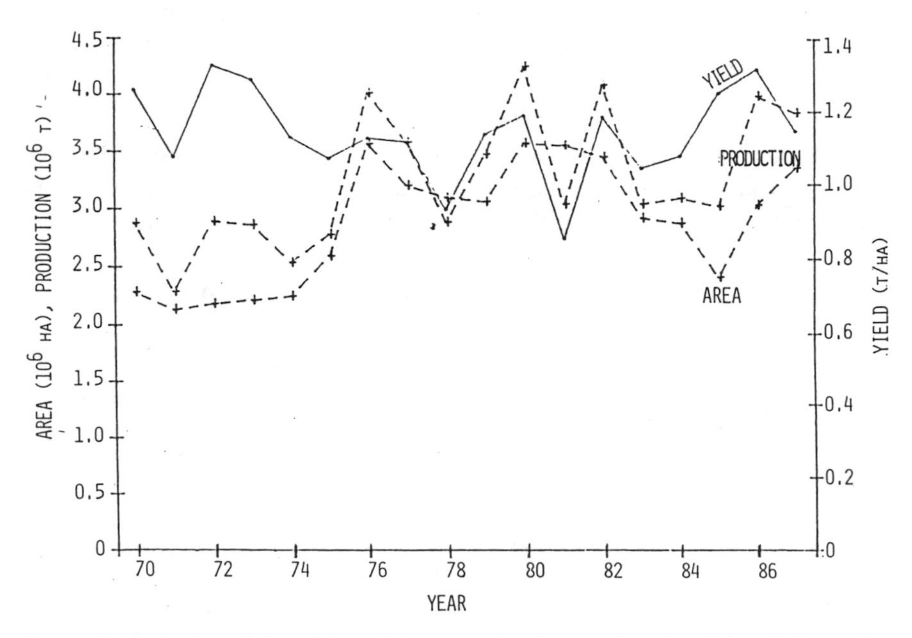


**Figure 8. Evolution of the cultivated area, production, and productivity of rice in the Cerrado region.**

(Brasil Ministerio da Industria e Comercio/IBC, 1987). Coffee productivity in the Cerrado region has been higher than the national average due to the use of improved technology, such as adequate spacing, productive cultivars, liming, and fertilizers. Coffee in the Cerrado generates a gross income of $1.5 billion US and a production of 12 million bags per year. Irrigation can eliminate the problem of irregular rain distribution, but most of the area cultivated with coffee is established with nonirrigated systems.

Tropical fruit crops, such as citrus, mango, and avocado, have increased in cultivated areas. At present, the area occupied by these crops is about 207,000 hectares. The economic possibility of these crops represents another alternative to be included in the production systems of the region.

Silviculture is another activity that is expanding in Brazil and in the Cerrado region as well. Until the late 1960s, forestry was not important in the country's economy, occupying an area of about 30,000 hectares. Now, tree plantations cover about 5.5 million hectares, of which 2.25 million hectares were established between 1979 and 1984, an average increase of 400,000 hectares per year. In this period, 42.5 percent of that area (2.25 million hectares) was established in the Cerrado region. That expansion is attributed to the increasing demand for charcoal, which substitutes for coal used in steel production, as well as for pulp and cellulose. Recently, the possibility of substituting wood for other energy sources has been considered. *Eucalyptus* and *Pinus* are the two most important trees introduced into the region (Brasil Ministerio da Agricultura/IBDF, 1986).

***Irrigated Systems.*** Irrigation is becoming an important technology for agriculture in the Cerrado region. It can be used to expand cropping systems in the dry season (April to September) or to provide supplemental water to plants during short, dry spells known as "veranico." About 5 percent of the Cerrados (10 million hectares) can be irrigated using available surface water (rivers, creeks) and water stored from rainfall (Pruntel, 1975). The area of irrigated land in the Cerrado region presently is not extensive, but the area is increasingly annually (Table 27). This increase has been made possible by subsidized programs oriented to stimulate the expansion of irrigated agriculture in high lands. Systems include the use of pressurized processes (conventional self-propelled and center pivot systems) and drip irrigation.

One problem that has hindered the development of irrigation capabilities is the cost of equipment, which amounts to 61 percent of the total expense per hectare (Table 24). This cost is increasing annually (Table 27), which makes it more difficult for farmers to use this technology. In irrigated systems, it is important to maintain adequate soil moisture to optimize the conditions for plant growth. However, irrigation alone does not guarantee high yields. The efficiency of irrigation is conditional upon the adequate use of other inputs. The costs of irrigation, financial conditions, and market prices of some commodities limit the use of this technology to a few crops. Once an irrigation system is installed, farmers have to plan their activities precisely to obtain the highest land use efficiency. One example is combining irrigated winter wheat and soybeans, whereby wheat yields can reach 5,400 kilograms per hectare and soybeans can yield 4,000 kilograms per hectare.

***Beef Cattle Production.*** *Livestock Contribution.* Beef cattle production in the Cerrado region has been a traditional activity, as a pure system or as a component in a production system (bovine, annual and perennial crops, etc.). There are about 46 million head of cattle in the region, 42 percent of the Brazilian total (Santos and Aguilar, 1984). Beef cattle production in the Cerrados contributes 35 to 40 percent of the gross farm income and 70 percent to 80 percent of the total gross value of the agricultural sector. Nevertheless, the significant socioeconomic importance of beef cattle productivity in the Cerrados is still low compared to its potential.

*Types of Beef Production.* Beef cattle production in the Cerrado region is characterized by cow-calf systems, involving 40 percent of the beef cattle population. This phase requires 17 percent of the total time of the production cycle and 60 percent of producers. Recent studies in the Brasilia region showed that 62 percent of the surveyed farms work with a dual-purpose system. In this system, producers milk their cows, particularly in the rainy season, to have a monthly cash flow. The animals on pasture represent 48 percent of the bovine population and require 58 percent of the total time of the production cycle. The finishing phase is not yet as

**Table 27. Incorporation of irrigated areas and average cost of implantation from 1982 to 1987.**

| | Area (ha) | | | Average Cost |
|---|---|---|---|---|
| Year | Cerrado | Brazil | Participation | (US $/ha) |
| 1982 | 14,413 | 26,109 | 55.2 | 1,100 |
| 1983 | 10,218 | 26,462 | 38.6 | 1,050 |
| 1984 | 5,033 | 13,450 | 37.4 | 1,080 |
| 1985 | 3,817 | 19,634 | 19.5 | 1,100 |
| 1986 | 29,597 | 49,342 | 60.0 | 1,300 |
| 1987 | 10,734 | 19,787 | 54.2 | 1,800 |
| Total | 73,812 | 154,782 | 47.7 | |

significant in the Cerrados. Presently, it represents only 12 percent of the animal population and consumes daily 25 percent of the total time of the production cycle.

Beef cattle production is a very extensive system, comprising an important economic activity on about 95 percent of Cerrados properties. Both the cow-calf and the growing phases show extremely low productivity levels (Table 28). Normally, beef cattle use native pastures almost exclusively, without any improved pasture or animal management and without animal health care.

*Constraints for Beef Cattle Production.* Beef cattle nutrition, management, and health care are important constraints for beef cattle production in the region, but the lack of available forage (quantity and quality) is the main constraint, especially during the dry season. During the dry period, the native pasture has a crude protein content below 7 percent, which drastically reduces forage dry matter consumption (Milford and Minson, 1966) and results in loss of body weight.

***Perspectives on Utilization of the Cerrado.*** *Sustainable Agricultural Systems and Levels of Technology.* Sustainable agricultural systems depend on the level of technology applied to farms. Presently, there are four levels of farm activities in the region; extensive, intermediate, intensive, and advanced (Goedert et al., 1986).

Extensive activities, characterized by the exploitation of natural resources with a low technology, absence of investments, and hand labor, predominated in the region until recently and are still common in the less developed areas. Wood production and beef cattle production on extensive grazing systems are the most typical activities. The economic stability of these systems is related directly to the price of land and the farmer's level of knowledge. Because they allow only for subsistence of the family, the systems are subjected to pressures for technological changes or for sale as land price increases.

Intermediate farming activities include some kind of diversification and a slightly higher level of technology, investment, and hand labor. In these systems, after land clearing, upland rice is cultivated for two or three years, and then replaced by a grass pasture. Although there is some concern for soil management and conservation, lime is usually applied in insufficient amounts to neutralize soil acidity, and minimal fertilizer applications do not correct soil nutrient deficiencies.

Extensive and intermediate systems predominate in the region in terms of total land area. The inefficient use of inputs is responsible for their low productivity.

Intensive systems require a high level of investment and technology, with less attention given to land. Soil management is very important and is maximized to meet its objectives. Many details have to be taken into consideration, including maintenance of the surface horizon during land clearance and adequate soil preparation to avoid the risk of soil erosion. The common sequence of activities is clearing and cleaning the area, liming, soil preparing, fertilizing, seeding, cropping, and harvesting. All of these activities are done with farm machinery. Adequate liming, to neutralize aluminum (water pH 5.5), and adequate fertilization, especially phosphate, are essential to the success of the farming enterprise. After clearing, the most common crop sequence is rice-pasture, rice-soybeans, or directly to soybeans. Corn and wheat crops can be introduced later in the rotation. At this level, water supply for plant growth is provided by rainfall during the rainy season. Improving subsoil chemical properties allows for root penetration and enhanced water uptake from subsurface horizons, which, in turn, attenuates the effects of the "veranico."

Advanced technology systems can be characterized by the intensive and continuous use of the land throughout the year. Irrigation is often necessary for systems that use annual or seasonal crops, but is not essential for farming enterprises that combine annual crops and perennial crops, such as coffee and fruit crops. In irrigated systems, all soil and plant factors have to be optimized to maximize yields and to compensate for the high cost of the

**Table 28. Beef cattle production levels in the Cerrado region.**

| | |
|---|---|
| Birth rate (%) | 40-45 |
| Calf mortality rate (%) | 7-8 |
| Weaning age (months) | 8-10 |
| Heifer age at first calving (years) | 3.5-4.5 |
| Calving interval (months) | 25-30 |
| Steer slaughtering age (years) | 4.5-5.5 |
| Slaughtering rate (%) | 12 |
| Carcass weight (kg) | 192 |
| Carcass performance (%) | 43-52 |

**Table 29. Perspectives of land occupation in the Cerrado region in the future.**

| | *Area* | |
|---|---|---|
| *Activity* | *Percent* | *Millions of ha* |
| Environmental protection | 20 | 40 |
| Animal production | 40 | 80 |
| Crop production | 40 | 80 |

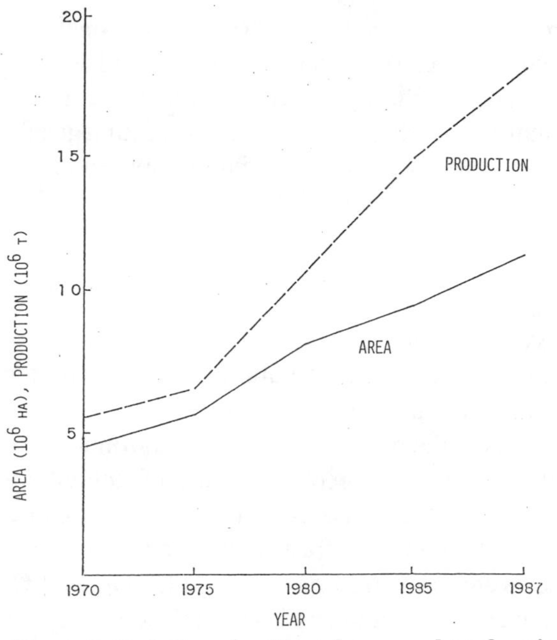


**Figure 9. Evolution of cultivated area and total grain production in the Cerrados region.**

advanced technology. The cost of irrigation decreases the relative financial importance of other practices in the total cost of the technology, with respect to nonirrigated systems (Table 24). The high productivity per unit of cultivated area and per unit of capital invested can be obtained only by continuous cropping of the land throughout one year.

***Perspectives on Production in the Cerrados.*** Currently, the potential for use of the Cerrado region for food and fiber production cannot be ignored. Consideration must be given to products used as renewable energy sources, that is, cassava and sugar-cane as raw material for alcohol. Because technology for development is available, the government must be involved in planning land use in the Cerrado region to guarantee the preservation and maintenance of this agroecological system. The potential for expansion of cultivated areas and pastures is limited, and in the future an equilibrium must be established (Table 29). If such an equilibrium is reached, using the available technology, it would be possible to produce 1.2 tons per hectare of rice, 4.0 tons per hectare of corn, 4.0 tons per hectare of sorghum, 1.0 ton per hectare of beans, 2.5 tons per hectare of soybeans, and 1.5 tons per hectare of wheat. This would increase the overall grain production of the region up to 200 million tons (Table 30), or three times the record Brazilian grain production obtained in 1987 (67 million tons). Similar response can be expected with respect to animal (meat) and wood production.

Fortunately, trends in the cultivated areas and overall crop production are upward (Figure 9). Farmers are thus taking advantage of the technology developed for the region, which is providing an opportunity for establishing sustainable agricultural systems.

## Conclusion

Sustainable systems that satisfy short-term requirements for production, while maintaining or improving the ability to meet long-term production needs, are available in tropical South America. Understanding soil dynamics and adapting plants to the environment are the agronomic keys to success.

Socioeconomic constraints preclude the use of technology for sustained agricultural systems in the Amazon rain forest. The main constraints are lack of markets, inadequate infrastructure and credit, and unfavorable economic policies and some uncertainty about land tenure. Once these constraints are resolved, the technology is available to farmers to make significant increases in food and fiber production in the Cerrado region. Advances in knowledge about the Cerrado region and its ecosystem, development of the basic infrastructure, and availability of inputs have permitted the evolution from subsistence to sustainable agriculture in a short time.

Strategies for sustainable agriculture can be developed through soil modifications to suit the needs of the crops (a strategy usually applied with annual grain crops), or through crop adaptation to adverse conditions and low input (a strategy applied more easily with perennial crops). Tropical South America has the natural resources to implement either strategy, depending upon national policies and priorities.

Research and extension activities appropriate to annual crops receive more support from national and international institutions than perennial crops

**Table 30. Perspectives of Cerrado agricultural production after plentiful occupation of the total land, using improved technology.**

| *Activity* | *Area (million ha)* | *Productivity* | *Annual Production (millions)* |
|---|---|---|---|
| Grain* | 80 | 2.5 t/ha/year | 200 t |
| Meat† | 80 | 100 kg/ha/year | 8 t |
| Wood | 10 | 30 $m^3$/ha/year | 300 $m^3$ |

*Rice (1.2 t/ha, 15%), beans (1.0 t/ha, 4%), corn (4.0 t/ha, 20%), soybeans (2.5 t/ha 35%), soybeans (4.0 t/ha, 4%), wheat (1.5 t/ha, 4%), other crops (18%).

†60% of the area with improved pasture, charge of 1 AU/ha and slaughtering at 2.5 years.

do. The contribution of annual crops is mainly as basic food; however, perennial crops are likely to play an increasingly important role in raising agricultural production in the region. To assure agricultural sustainability in the region, more support will be needed from aid sources, and a much greater effort will be required from national and international research institutions to solve the problems that remain.

REFERENCES

Adamoli, J., J. Macedo, L. G. Azevedo, and J. M. Neto. 1986. Caracterizacao da regiao dos Cerrados. pp. 33-74. *In* W. J. Goedert [editor] Solos do Cerrado: Tecnologias e estrategias de Manejo. Nobel, S.A., Sao Paulo, Brazil.

Alvim, P. de T. 1980. An appraisal of perennial crops in the Amazon Basin. pp. 311-328. *In* S. B. Hetch et al. [editors] Amazonia: Agriculture and land use research. Center Interamericano de Agricultura Tropical, Cali, Colombia.

Azevedo, L. G., and R. L. Caser. 1980. Regionalizacao do Cerrado. *In* IV Simposio sobre Cerrado: Uso e Manejo. Editerra, Brasilia, Brazil.

Barcelos, J. M., L.C.R. Echevarria, D. M. Pimentel, W. V. Soares, and L. S. Valle. 1982. Producao de carne bovina em solos de baixa fertilidade no Brasil: Estudo de dois sistemas de producao simulados no Mato Grosso do Sul. pp. 327-335. *In* P. A. Sanchez [editor] Producao de pastagens em solos acidos dos tropicos. Ed. Editerra-CIAT-EMBRAPA. Brasilia, Brazil.

Bigham, J. M. 1977. Iron mineralogy of red-yellowed hued Ultisols and Oxisols as determined by Mossbauer spectroscopy, X-ray diffractometry and supplemental laboratory techniques. Ph.D. thesis. North Carolina State University, Raleigh.

Brasil Ministerio da Agricultura/IBDF. 1986. O setor florestal Brasileiro 1979/1985. Brasilia, Brazil. pp. 65.

Brasil Ministerio da Industria e Comercio/IBC. 1987. A moderna cafeicultura nos Cerrados: Instrucoes tecnicas sobre a cultura do cafe no Brasil. Brasilia, Brazil. pp. 147.

Brown, L. R. 1975. Bread alone. Pergamon Press, Oxford, England.

Centro Interamericano de Agricultural Tropical. 1978. Annual Report, Cali, Colombia. pp. A-1-114.

Cochrane, T. T., and P. A. Sanchez. 1980. Land resources, soils and their management in the Amazon region: A state of knowledge report. pp. 137-210. *In* S. B. Hetch et al. [editors] Amazonia: Agriculture and land use research. Cali, Colombia.

Correa, M. P. 1983. A pesquisa do guarana. pp. 42-47. *In* ler Simposio Brasileiro do Guarana. EMBRAPA. Manaus, Brazil.

EMBRAPA. 1986. Curso de tecnologia do guarana. UEPAE-Manaus and Asociacao dos Engenheiros Agronomos do Rondonia, Brazil.

EMBRAPA/SNLCS. 1978. Levantamento de reconhecimento de solos do Distrito Federal. Bolivia Technical Number 53. Rio de Janeiro, Brazil. pp. 445.

Evenson, R. E. 1984. Observations on Brazilian agricultural research and productivity. pp. 247-276. *In* L. Yeganiantz [editor] Brazilian agriculture and agricultural research. Brasilia, Brazil.

Fassbender, H. W. 1975. Experimento de laboratorio para el estudio del efecto del fuego de la quema de restos vegetales sobre las propiedades del suelo. Turrialba 25: 249-254.

Goedert, W. J., D.D.G. Scolari, and E. Lobato. 1986. Estrategias de uso e manejo do solo. pp. 409-422. *In* W. J. Goedert [editor] Solos dos Cerrado: Tecnologias e estrategias de Manejo. Nobel S.A., Sao Paulo, Brazil.

Goedert, W. J., D. M. de Sousa, and E. Lobato. 1986. Fosforo. pp. 129-166. *In* W. J. Goedert [editor] Solos dos Cerrados: Tecnologias e estrategias de Manejo. Nobel S.A., Sao Paulo, Brazil.

Goedert, W.J., E. Lobato, and M. Resende. 1982. Management of tropical soils and world food prospects. pp. 338-364. *In* Proceedings of the 12th International Congress of Soil Science.

Hanson, H., N. E. Borlang, and R. G. Anderson. 1982. Trigo en el Tercer Mundo. CIMMYT, Mexico.

INIPA. 1982. Programa nacional de arroz. Documento Base. Lima, Peru. pp. 67.

INIPA. 1987. Estudio del mercado de frutales nativos de la selva Peruana. I. Resumen ejecutivo. Informe Tecnico No. 1. Programa Nacional de Cultivos Tropicales, Lima, Peru. pp. 18.

Kornelius, E., W. J. Goedert, J.L.F. Zoby, and C.M.C. Rocha, 1988. Diagnostico e prioridades de pesquisa en Cerrados nativos. Trabalho Apresentado na XI Reuniao Latinoamericana de Producao Animal. Havana, Cuba.

Lathwell, D. J. 1979a. Crop response to liming in Oxisols and Ultisols. Cornell International Agriculture Bulletin 35. Cornell University, Ithaca, New York.

Lathwell, D. J. 1979b. Phosphorus response in Oxisols and Ultisols. Cornell International Agriculture Bulletin 33. Cornell University, Ithaca, New York.

Lopez, A. S. 1974. A survey of the fertility status of soils under Cerrado vegetation in Brazil. M.S. thesis. North Carolina State University, Raleigh. pp. 138.

Luciari, A. L., Jr., M. Resende, K. D. Ritchey, E. Freitas, Jr., and P.I.M. Souza. 1986. Manejo do solo e aproveitamento de agua. pp. 285-322. *In* W. J. Goedert [editor] Solos dos Cerrados: Tecnologias e estrategias de Manejo. Nobel S.A., Sao Paulo, Brazil.

Meadows, D. H., D. L. Meadows, J. Randers, and W. W. Behrens, III. 1972. The limits to growth. Universe Books, New York, New York.

Milford, R., and D. J. Minson. 1966. The feeding value of tropical Pastures. pp. 106-114. *In* W. Davies and C. L. Skidmore [editors] Tropical pastures. Faber and Faber Limited, London, England.

Miranda, L., J. Mielniczuk, and E. Lobato. 1980. Calagem e adubacao corretiva. pp. 521-578. *In* V. Simposio sobre o Cerrado. Brasilia, Brazil.

Moraes, V.H.F. 1974. Fatores condicionantes e perspectivas atuais de desenvolvimento de cultivos perennes na Amazonia brasileira. pp. 37. *In* Reuniao do Grupo Intedisciplinar Sobre Diretrizes de Pesquisa Agricola Para a Amazonia. EMBRAPA, Brasilia, Brazil.

Mosher, A. T. 1970. The development problems of subsistence farmers: A preliminary review. pp. 6-10. *In* C. R. Wharton, Jr. [editor] Subsistence agriculture and economic development. Frank Cass and Co., Ltd.

Nicholaides, J. J., D. E. Bandy, P. A. Sanchez, J. H. Villachica, A. J. Coutu, and C. S. Valverde. 1984. From migratory to continuous agriculture in the Amazon basin. pp. 141-168. *In* Improved production systems as an alternative to shifting cultivation. FAO Soils Bulletin 53. Rome, Italy.

North Carolina State University. 1987. Technical report for 1985-86. Tropical Soils Research Program, Raleigh, North Carolina. pp. 1-140.

Padoch, C., J. Chota, W. de Jong, and J. Unruh. 1985. Amazonian agroforestry: A market oriented system in Peru. Agroforestry Systems 3: 47-58.

Peru, Proyecto Especial Madre de Dios. 1982. Estudio de suelos y de clasificacion de tierras de la microregion Iberia-Inapari del Dpto. de Madre de Dios. Presidencia del Consejo de Ministros, Lima, Peru. pp. 206.

Pruntel, J. 1975. Water availability and soil suitability for irrigation water impoundments in the Federal District of Brazil. M.S. thesis. Cornell University, Cornell, New York. pp. 113.

Reis, E. G., and J. B. Rassini. 1986. Aproveitamento de Varzeas. pp. 353-384. *In* W. J. Goedert [editor] Solos dos Cerrado: Tecnologias e estrategias de Manejo. Nobel S.A., Sao Paulo, Brazil.

Ritchey, K. D., J. E. Silva, and U. F. Costa. 1982. Calcium deficiency in clayey B horizons of savanna Oxisols. Soil Science 133: 378-382.

Sanchez, P. A. 1984. Nutrient dynamics following rainforest clearing and cultivation. pp. 52-56. *In* Australian Center for International Agricultural Research: Proceedings on the International Workshop on Soils. ACIAR. Townsville, Australia.

Sanchez, P. A., D. E. Bandy, J. H. Villachica, and J. J. Nicholaides. 1982. Amazon basin soils: Management for continuous crop production. Science 216: 821-827.

Sanchez, P. A., and J. G. Salinas. 1981. Low-input technology for managing Oxisols and Ultisols in tropical America. Advances in Agronomy 34: 248-398.

Sanchez, P. A., J. H. Villachica, and D. E. Bandy. 1983. Soil fertility dynamics after clearing a tropical rainforest in Peru. Soil Science Society of America Journal 47: 1,171-1,178.

Sanchez, P. A., and J. R. Benites. 1987. Low-input cropping for acid soils of the humid tropics. Science 238: 1,521-1,527.

Santos, C. A., and J. P. Aguilar. 1984. Evolucao agropecuaria da regiao nuclear dos Cerrados. 1978-1980. Documentos 16. EMBRAPA-CPAC, Planaltina, Brazil.

Santos, C. A., S. Esterman, P. Esterman and A. Esterman. 1980. Aproveitamento da pastagem nativa no cerrado. pp. 419-435. *In* V Simposio Sobre o Cerrado: Uso e Manejo. Ed. Editerra. Brasilia, Brazil.

Seubert, C. E., P. A. Sanchez, and C. Valverde. 1977. Effects of land clearing methods on soil properties of an Ultisol and crop performance in the Amazon jungle of Peru. Tropical Agriculture (Trinidad) 54: 307-321.

Smyth, T. J., and P. A. Sanchez. 1980. Effects of lime, silicate, and phosphorus applications to an Oxisol on phosphorus and ion retention. Soil Science Society of America Journal 44: 500-505.

Tergas, L. E., P. A. Sanchez, F. Kramer, and D. Evans. 1979. Prefacio. *In* L. E. Tergas, P. A. Sanchez, and S. S. Salcedo [editors] Produccion de pastos en suelos acidos de los tropicos. CIAT, Cali, Colombia.

Toledo, J. M., and E. S. Serrao. 1980. Pasture and animal production in Amazonia. pp. 281-310. *In* S. B. Hetch et al. [editors] Amazonia: Agriculture and land use research. CIAT, Cali, Colombia.

United States Department of Agriculture. 1976. Working paper, Agriculture in the Americas. Statistical data. Washington, D.C.

Villachica, J. H. 1978. Maintenance of soil fertility under continuous cropping in an Ultisol of the Amazon jungle of Peru. Ph.D. thesis. North Carolina State University, Raleigh. pp. 268.

Villachica, J. H. 1986. La agricultura en la selva peruana. pp. 51-95. *In* La agricultura en el Peru (Vol. 5). Gran geografia del Peru: Naturaleza y hombre. Manfer J. Mejia Baca, Barcelona, Spain.

Villachica, J. H., and K. Raven. Deficiencias nutricionales del papayo en la selva central del Peru. Turrialba 36:523-531.

Villas, A. T. 1980. Utilizacao de insumos para a agricultura na regiao dos Cerrados. pp. 161-181. *In* V. Simposio Sobre o Cerrado: Uso e Manejo. Ed. Editerra. Brasilia, Brazil.

Wagner, E. 1986. Desenvolvimento da regiao dos Cerrados. pp. 19-31. *In* W. J. Goedert [editor] Solos dos Cerrado: Tecnologias e estrategias de Manejo. Nobel S.A., Sao Paul, Brazil.

Weaver, R. M. 1975. Quartz presence in relationship to gibbsite stability in some highly weathered soils of Brazil. Clays/Clay Miner 23:431-436.

Wolf, J. M. 1975. Soil-water relations in Oxisols of Puerto Rico and Brazil. pp. 145-154. *In* E. Bornermiza and A. Alvarado [editors] Soil management in tropical America. North Carolina State University, Raleigh.

World Commission on Environment and Development. 1987. Our common future. Oxford University Press.

# SUSTAINING AGRICULTURE: THE INDIAN SCENE

N. S. Randhawa and I. P. Abrol

Sustainable agriculture implies the ability to produce food and provide nutritional security for the increasing population without reducing or putting at risk the ability to feed future generations. Although past efforts in India have brought that country largely to a stage of self-sufficiency in production of food grains, these achievements have been a mixed blessing. It is time to assess the implications of these past efforts on India's ability to continue meeting the future challenges.

## Food Production—The Past Experience

Over the past three decades, India has moved successfully from a food-deficit state to one that is, by-and-large, self-sufficient in production of food grains, although there are still deficits of pulses and oilseeds. Total food grain production was about 50 million tons in 1950-1951 and increased to 82 million tons, 108.4 million tons, and 129.6 million tons in 1960-1961, 1970-1971, and 1980-1981, respectively. Production reached an all-time high of 152.4 million tons in 1983-1984. Total cropped area increased from 118.7 million hectares in 1950-1951 to 140.8 million hectares in 1970-1971 and has remained almost constant at that level.

Cropland expansion has played a relatively minor role in increased food grain production. Most increases in cropped area were due to conversion of traditional grazing lands of low productivity to cropland. The reduction of grazing land and resultant overstocking of that land has resulted in increasingly severe resource degradation. This has had adverse effects on livestock nutrition, health, and productivity, as well as the economic conditions of the landless and poor livestock holders.

Increased production since the mid-1960s has come about, largely, through increased cropping intensity and increased crop productivity in areas favored by agroclimatic resource conditions and where irrigation facilities already existed or could be developed relatively rapidly. The major elements of strategy for increased production included (a) expansion of irrigated areas; (b) improved availability and use of key inputs, including high-yielding varieties of wheat and rice crops and increased use of fertilizers; and (c) improvement and expansion of institutional support services (Table 1). The availability and adoption of improved crop production technologies can be credited to research that became available after establishment of several agricultural universities in the country in the early 1960s.

The high-yielding varieties performed particularly well under irrigated conditions and responded well to fertilizer. Availability of short-duration varieties further enabled double-cropping, particularly under irrigation.

Irrigation was the key factor in this strategy. India's net irrigated area totals about 40 million hectares, compared with less than 25 million hectares in 1960-1961. Over the same period, the gross irrigated area has almost doubled, to nearly 55 million hectares. The percentage of the country's gross cropped area under irrigation increased from 18.3 percent in 1960-1961, to 23 percent in 1970-1971, and to over 30 percent at present.

Fertilizer use increased from less than 0.3 million tons in 1960-1961 to 2.18 million tons in 1970-1971, 5.51 million tons in 1980-1981, and 9.7 million tons in 1987-1988. Increases in fertilizer use were confined largely to irrigated areas.

The combination of expanding irrigation coverage and widespread adoption of short-duration crop varieties led to significant increases in cropping intensity. From 1950-1951 to 1965-1966, the area cropped more than once a year increased by about 5 million hectares. Since 1965-1966, the area cropped more than once annually increased from about 19 million hectares to more than 35 million hectare by the mid-1980s. This was possible because of the spread of improved crop varieties, including wheat, rice, cotton, and sorghum. Double-cropping also became possible in some rainfed areas. In the country as a whole, the average cropping intensity increased from 115 percent in 1960-1961 to 125 percent in 1983-1984. In states with a large proportion of irrigated area—Punjab, Haryana, and Uttar Pradesh—the cropping intensity increased to 165 percent, 148 percent, and 143 percent, respectively, in 1982-1983.

Analyzing the contribution of various inputs toward additional food grain production between 1960-1962 and 1975-1977, Sarma and Roy (1979) concluded that the coefficient of increase due to irrigation was rather low, and the contribution of fertilizers was the highest (Table 1).

The greatest impact of irrigation, high-yielding varieties, and fertilizer technology has been to increase production of two main crops: rice and wheat. Over the past two decades, average wheat yields have risen about 125 percent and rice yields about 30 percent. By 1980-1981, rice and wheat accounted for 31.8 million hectares of the gross irrigated area, 85 percent of the total irrigated food grains, and 64 percent of the country's gross irrigated area.

Due to the spread of high-yielding varieties of maize and sorghum in nonirrigated areas with reliable rainfall, grain yields ranging from 30 percent to 35 percent also have been achieved for these crops. Progress in achieving increased production has been particularly slow in rain-fed agriculture, which accounts for 70 percent of the gross cropped area.

## The Challenge of the Future

According to current assessments, India's present population of an estimated 800 million will reach one billion by the year 2000. An assumed food grain production of 225 million tons will be required to feed this population (production at present is 150 million tons). The per capita availability of land fell from 1.37 hectares in 1901 to 0.50 hectare in 1981. Per capita availability of arable land (net cultivated area) decreased from 0.48 hectare in 1950-1951 to 0.20 hectare in 1980-1981, even though the net cultivated area increased from 118.7 million hectares to 140.27 million hectares during that period. The availability of cultivated land will fall further to 0.15 hectare toward the end of this present century. This fast-declining people-to-land ratio trend cannot be viewed with complacency and calls for all-out efforts to produce more and more from a limited area.

Although food production needs have been met, the past few decades have witnessed a sharp degradation, both in qualitative and quantitative terms, of natural resources. Forests and grasslands play a crucial role in

**Table 1. Contribution of various inputs to increased food production between 1960-1962 and 1975-1977 (Sarma and Roy, 1979).**

| *Input* | *Additional Food Production (million tons)* |
|---|---|
| Irrigation | 6.24 |
| Fertilizers | 19.90 |
| Area | 3.70 |
| Shift in cropping system | 3.60 |

**Table 2. Soil erosion in India (Narayana et al., 1983).**

| | |
|---|---|
| Average soil loss due to agricultural activities (t/ha) | 16.35 |
| Total soil detached annually (million tons) | 5,333 |
| Soil carried to the sea (%) | 29 |
| Soil deposited in surface reservoirs (%) | 10 |
| Soil deposited in lower reaches (%) | 61 |

providing and maintaining ecosystems so vital for crops and livestock. India has 15 percent of the world's population and only 2 percent of the world's total forest area. The per capita growing stock is 5.2 cubic meters compared to the world average of 46.7 cubic meters. Interpretation of satellite imagery obtained during 1972-1975 and again during 1980-1982 showed that the country was losing closed forests (crown cover greater than 40 percent) at a rate of nearly 1.5 million hectares per year. At present, India may be left with only about 8 to 10 percent of its geographical area under closed forests, in contrast to the figure of 33 percent recommended as minimum for ecological security by the National Commission on Agriculture (Ministry of Agriculture and Irrigation, 1976). The per capita availability of forest areas is only 0.08 hectare compared to the world average of 1.0 hectare. Although the need and urgency for reversing the above trends are obvious, any action plan has to recognize that in rural areas noncommercial energy accounts for nearly 90 percent of energy consumption and that firewood constitutes the single most important source (48%), followed by dung cake (30%) and crop wastes (15%). The country's forestry requirements, including fuel and industrial wood, were estimated to be about 200 million cubic meters in 1980. This is about five times the reported 40-million-cubic-meter production from national forests. The National Commission on Agriculture estimated that by the year 2000, the forestry requirement will rise to 300 million cubic meters. Unless forest yields can be increased considerably, overexploitation, further resource depletion, and environmental degradation are inevitable.

According to Narayana et al. (1983), soil erosion rates in India were 16.35 tons per hectare per year, which is more than the permissible value of 4.5 to 11.2 tons per hectare (Table 2). About 29 percent of the total eroded soil is lost permanently to the sea. Ten percent of this is deposited in reservoirs, reducing their capacity, which was created at huge costs (Table 3). The remaining 61 percent of the eroded soil is dislocated from one place to another. Soil erosion problems are particularly acute in the rain-fed regions and are continually reducing the soil's productive capacity.

Future strategies will, therefore, have to address not only the problems of aggregate food production, but also the regional imbalances and improvement of socioeconomic conditions of the majority of the rural population, whose existence is based on rain-fed agriculture. Future increases in production must come solely through increased crop productivity in both irrigated and rain-fed areas. There is no scope for expanding the net cropped area without substantially further impoverishing the already deteriorated ecological balance. Concerted efforts will be required to bring marginal land that has site or environmental limitations under permanent vegetative cover of grassland and forests.

## Irrigated Farming

Although expansion of irrigation has been one of the major factors in achieving self-sufficiency in food grain production in the past decades, the gains in productivity have not been commensurate with investments in major and medium-size irrigation projects. Some considerations that have relevance to long-term productivity in the irrigated regions are briefly discussed here.

***Expansion of Irrigation in the Humid Regions.*** Of the estimated ultimate gross irrigation potential of 113 million hectares, nearly 33 percent of that will be in the states of Assam, Bihar, Madhya Pradesh, Orissa, and West Bengal. Mean annual rainfall in these states is more than 1,150 millimeters. Of the gross potential created through major, medium, and minor irrigation projects, nearly one-fourth reportedly are used in these five states (Table 4). Most of these irrigation projects have been developed in the past two decades. Rice accounts for nearly 90 percent of the irrigated area

**Table 3. Assumed and observed rates of sedimentation of some reservoirs (Narayana et al., 1983).**

| *Name of Reservoir* | *Catchment Area (square km)* | *Sedimentation Rate ha m/100 square km* | |
|---|---|---|---|
| | | *Assumed* | *Observed* |
| Hirakud (Mahanadi) | 82,650 | 2.54 | 3.60 |
| Tunghabhadra | 25,830 | 4.32 | 6.57 |
| Mahi | 25,400 | 1.29 | 8.99 |
| Rana Pratap (Chambal) | 22,700 | 3.61 | 5.29 |
| Pong (Beas) | 12,500 | 4.29 | 17.30 |
| Nizamnagar | 18,470 | 0.28 | 6.38 |
| Pamchet (Damodar) | 9,620 | 2.49 | 10.08 |
| Tawa | 5,980 | 3.61 | 8.10 |
| Kaulagarh (Ramganga) | 2,000 | 4.32 | 18.33 |
| Mayurakshi | 1,860 | 3.61 | 20.89 |

**Table 4. Use of developed irrigation potential and share of principal states toward ultimate potential.**

| Rainfall Zone | State | Estimated Ultimate Potential Share of the Total Percentage | Potential Use | | |
|---|---|---|---|---|---|
| | | | Major and Medium | Minor | Total |
| <750 m | Punjab | 5.8 | 2.444 | 3.139 | 5.583 |
| | Haryana | 4.0 | 1.745 | 1.361 | 3.106 |
| | Rajasthan | 4.5 | 1.476 | 1.937 | 3.413 |
| | Karnataka | 4.1 | 1.076 | 1.105 | 2.181 |
| | Gujarat | 4.2 | 0.748 | 1.609 | 2.357 |
| | Subtotal | 22.6 | 7.489 | 9.151 | 16.640 |
| 750-1,150 m | Andhra Pradesh | 8.1 | 3.017 | 2.196 | 5.213 |
| | Maharashtra | 6.4 | 0.832 | 1.832 | 2.664 |
| | Uttar Pradesh | 22.7 | 5.513 | 10.977 | 16.490 |
| | Tamil Nadu | 3.4 | 1.225 | 1.943 | 3.168 |
| | Subtotal | 40.6 | 10.587 | 16.948 | 27.535 |
| >1,150 m | Assam | 2.4 | 0.078 | 0.349 | 0.427 |
| | Bihar | 10.9 | 2.176 | 3.150 | 5.326 |
| | Madhya Pradesh | 9.0 | 1.305 | 1.870 | 3.175 |
| | Orissa | 5.2 | 1.508 | 0.980 | 2.488 |
| | West Bengal | 5.4 | 1.475 | 1.600 | 3.075 |
| | Subtotal | 32.9 | 6.542 | 7.949 | 14.491 |
| | Total | | 24.618 | 34.048 | 58.666 |

Source: Directorate of Economics and Statistics, Ministry of Agriculture, New Delhi.

during the rainy season in these states. Rice yields have not shown significant improvements during the past 20 years. Availability of irrigation in humid regions likely will result in loss of crop productivity, including that of rice, because of the difficulties of managing water in years of more than normal rainfall and lack of water in years of low rainfall. For these reasons, Bhumbla (1982) concluded that any emphasis on major and medium irrigation works in high rainfall areas would not likely be beneficial and could have negative effects. High investments needed for creating irrigation potential through major and medium projects demand a rethinking of existing plans for expanding irrigation in different parts of the country.

***Improving Irrigation Efficiency.*** Irrigation efficiency, particularly in canal-irrigated regions, is low (Table 5). Dhawan (1988) reported that productivity per hectare of net irrigated land in two northern states was between 5.5 and 5.7 tons per hectare in well-irrigated areas and 2.4 tons to 3.2 tons per hectare in canal-irrigated areas. Low productivity in canal-irrigated areas is ascribed to large losses of water due to transportation and on-farm use, unscheduled cropping patterns, inequitable distribution between the upper and lower reaches of the canal, excessive water use during each irrigation because of uncertainties in availability of canal supplies, absence of land consolidation, and other factors.

A study by the International Land Reclamation Institute showed that for 90 projects in India, irrigation efficiency was between 20 percent and 40 percent from the reservoir to the field (Table 5). Apart from the low efficiency of water use, extension of canal irrigation has resulted invariably in the rise of groundwater, causing problems of water logging. Because groundwater in most arid and semiarid regions contains appreciable quantities of soluble salts, rises in the water table are accompanied by serious soil salinization problems, reducing the productivity of prime, highly productive agricultural land.

Water from the Bhakhra canal system was introduced into parts of the Hissar district in north India in 1965. During the past 25 years, water table levels have risen from more than 15 meters to less than 2 meters, causing widespread salinity problems.

The left bank canal of the Tunghabhadra irrigation project in Karnataka was commissioned in 1953. A study 30 years later showed that 33,000 hectares had been severely affected by waterlogging and salinity. Furthermore, this area was expanding at an estimated rate of 6,000 hectares annually and production from about 20,000 hectares already had fallen to zero. In the Nagarjunasagar project command area, nearly 25,000 hectares of the 140,000 hectares in irrigation have been affected by salinity and waterlogging in a period of about 14 years.

***Drainage.*** Both preventive and corrective measures are called for to maintain the productivity of canal-irrigated areas. A top priority must be to prove the efficiency of the system through improved on-farm water management. Although it has been recognized that lasting success in irrigated agriculture requires adequate drainage of excess water, in practice very little has been accomplished. Drainage provisions, where they exist, are intended, for the most part, to remove excess rainwater of excellent quality. Drainage to control profile salinity and groundwater tables should be integrated into the planning stages of irrigation development. Due consideration will have to be given to the on- and off-site environmental aspects of recycling and/or disposing of saline groundwater. Efforts toward sustained use of available water resources further demand coordinated and harmonious development of surface water and groundwater resources, such that use of the total

**Table 5. Approximate order of losses and use of water in some northern India canals.**

| Particulars | *Approximate Value Percentage* |
|---|---|
| Losses in main canal and branches | 15 to 20 |
| Losses in major and minor distributaries | 6 to 8 |
| Losses in field channels | 20 to 22 |
| Losses during application including deep percolation evaporation, etc. | 25 to 27 |
| Use via evapotranspiration | 28 to 30 |

resource is maximized at minimal cost and with least environmental degradation.

***Maintaining Groundwater Balance.*** Yet another serious problem that requires careful scientific examination is overexploitation of groundwater in many parts of the country, where harnessing of groundwater is not commensurate with the recharge of aquifers. In Tamil Nadu, overpumping of groundwater has resulted in drops in the water table of more than 25 to 30 meters in the 1970s, causing serious water availability problems. In parts of the Indo-Gangetic plains, where widespread use of groundwater for irrigation has been considered a key to the spread of the green revolution, withdrawals of groundwater exceed recharges. Problems arising from excessive exploitation of groundwater in the coastal regions are more serious. Large-scale pumping of groundwater in Minjur, about 30 kilometers north of Madras, has caused an alarming intrusion of sea water into the coastal-area aquifer, 350 square kilometers of which has been rendered saline and unusable. Reports of increasing sea water intrusion in other coastal areas are of great concern. Maintaining a favorable hydrological balance for optimal, sustained use of groundwater under different agroclimatic regions will be a major concern in achieving sustained high productivity in areas where groundwater use is a major component of irrigation.

## Rain-fed Farming

Continued reliance on irrigation, high-yielding varieties, and fertilizers alone will not bring about the required increases in food grain production and simultaneously expand the production of other food and industrial crops. Improving productivity and the stability of that production in rain-fed areas will, therefore, be crucial in meeting the needs of the increasing population and in reducing the regional imbalances in agricultural development, food availability, and the socioeconomic conditions of most of the rural population, whose existence is based on rain-fed agriculture.

Presently, about 125 million hectares, or 70 percent of the country's gross cropped area, is farmed under rain-fed conditions. Rain-fed agriculture accounts for more than 40 percent of total food grain production, 75 percent of all oilseeds, 90 percent of pulses, and 70 percent of cotton. Even if 20 million hectares were to be added to irrigated areas by the end of the century, increasing the gross irrigated area to 75 million hectares, more than 55 percent of the country's gross cropped area still would depend exclusively upon rainfall. This would call for major research and development efforts to increase and stabilize production from rain-fed areas.

Rain-fed farming is practiced under a wide range of soil and climatic conditions. Nearly 80 percent of total rainfall occurs from June to September, the rainy season. Late onset of, long dry spell during, and early withdrawal of the monsoon are common aberrations that reduce production and result in crop failures. The high variability in the total and seasonal distribution of rainfall and uncertainties associated with rainfall events render farming highly risky. Soils in many rain-fed areas are characterized by poor physical conditions, low water-retaining capacity, and low fertility. These soils are often affected by serious drainage problems.

In most rain-fed areas, farmers have expanded cultivation into areas of poor potential, especially in shallow red soils. This has aggravated water erosion problems, further reducing the capability of agroecological resource systems to sustain the growing human and livestock population. Extensive tracts of such marginal sloping lands have been abandoned after a few years of unproductive agriculture. In the Deccan plateau, nearly 40 percent of land falls into this class.

A key requirement for raising farm productivity in the rain-fed areas is improving the nutritional status of livestock. This will require major changes in the management of grazing land. Any effort to rehabilitate and stabilize grazing land must ensure close involvement of local communities in planning and implementation.

Research efforts to improve the productivity of rain-fed areas have centered on improved characterization of soil and water (particularly rainfall) resources, developing crop varieties to suit growing seasons with limited water availability periods, defining production technologies for cropping systems, and developing techniques for optimum rainfall management.

In recent years, there has been an increased emphasis on improving farming in rainfall areas through management on a watershed basis. The essential objectives are stabilization of the natural resource systems of watersheds, so as to facilitate more productive and sustainable use. Although the main thrust is on improving conditions for production of field crops, in many arid, semiarid, and hilly regions sustainable alternatives include land use systems and practices that would promote horticulture, agroforestry, silvipasture, and animal husbandry programs. A watershed-based development approach is an improvement over the piecemeal approach to planning and implementing within the framework of the natural unit of the watershed. Effective watershed development requires treatment of all land—government, community, and private—that requires treatment. Certain areas should not be left out because they are privately owned or because a particular owner is not qualified for a loan.

Although there is an urgent need to extend the above approach to management of resources in the rain-fed areas, there is a concurrent need to develop models of development that will suit resource-poor farmers. Particular attention must be given to improving the soil resource base, which will have long-term impacts in making agriculture in rain-fed areas more sustainable.

## Efficient Cropping Zones

Each crop requires specific soil and climatic conditions for best expression of its potential. Although moisture deficits during the growing season and the lack of soil nutrients could be moderated to a certain extent by irrigation and fertilization, the photo period, temperature, and other climatic variables determine, to a large extent, the choice of crops. Sugarcane yields best under tropical conditions, while wheat yields best in situations of a prolonged but not intense winter season. Cotton will not be a successful crop where rainfall coincides with flowering stage, and maize where water stagnation is a perpetual problem. At present, crops are grown under a wide range of soil and climatic conditions that are not necessarily ideal for the particular crop. Reasons for growing crops in environments other than best-suited ones often center on the necessity of meeting household needs; nonavailability of better alternative crops, particularly from net income considerations; availability of marketing infrastructure; and constraints of inputs, including labor and other factors. In the coming years, there will be a need to identify and adopt cropping systems based on efficient zones, long-term effects on productivity, and availability of resources.

## Managing Soil Fertility

Fertilizer use has been recognized as the key input for increased agricultural production. The annual use of inorganic fertilizers increased from

69,000 tons in 1950-1951 to 9.7 million tons in 1987-1988. During the same period, per-hectare use of fertilizers increased from 0.5 kilograms to 50 kilograms. Progress has been particularly significant during the last 10 years; fertilizer use rose from 2.9 million tons in 1975-1976 to 8.7 million tons in 1985-1986. Fertilizer use will increase to an estimated 13.5 to 14 million tons by 1989-1990 and to 20 million tons by 2000. Although India now ranks as the fourth largest consumer of fertilizers in aggregate terms, per-hectare use of fertilizers is still among the lowest in the world. For example, fertilizer use in China in 1983-1984 was 180.6 kilograms per hectare compared to 39.4 kilograms per hectare in India, 74.0 kilograms per hectare in Sri Lanka, and 331 kilograms per hectare in the Republic of Korea. In the same year, per-capita use was 17.5 kilograms in China, 9.0 kilograms in India, and 18.0 kilograms in the Republic of Korea.

Fertilizer use in India is highly skewed. Although fertilizer use increased from 0.5 kilograms per hectare to 50 kilograms per hectare in the last four decades, there are still 24 districts where fertilizer use is less than 5 kilograms per hectare; 25 districts use 5-10 kilograms per hectare, 68 districts use 10-25 kilograms per hectare, 86 districts use 25-50 kilograms per hectare, 48 districts use 50-75 kilograms per hectare, 28 districts use 75-100 kilograms per hectare, and 48 districts use more than 100 kilograms per hectare. Wide variation similarly exists between different states—158.5 kilograms per hectare in Punjab on the one end to less than 5 kilograms per hectare in Assam on the other.

Fertilizer use is also restricted to a few crops. For example, rice and wheat crops used nearly 60 percent of the total fertilizer. Sugarcane accounted for nearly 8 percent of fertilizer use, followed by cotton at 5.5 percent. Very little fertilizer was being used for coarse grains, oilseeds, and pulses, the major crops of the rain-fed agriculture. Another study showed that for the country as a whole, only about 33 percent of the cultivated area, accounting for 45 percent of cultivators' holdings, received any fertilizer. Thus, a large portion of the cultivated land is continually being depleted of nutrients by crop production because of continued removal by crops (Table 6). In 1983-1984, against an estimated nutrient removal of 18.9 million tons, application in the form of chemical fertilizers was only 7.7 million tons. This points to the need for intensifying fertilizer use in regions where use currently is low.

In intensively cultivated areas with relatively high consumption of fertilizers, nutrients other than nitrogen, phosphorus, and potassium have been increasingly limiting production. Sulfur deficiencies have been reported in about 90 districts in the country, affecting about 25 million hectares of cultivated land. Among micronutrients, zinc deficiency was most wide-

**Table 6. Nutrient removal with crop production.**

| Crop | *Production (1983-1984) (million tons)* | *Nutrient Removal (mt)* | | | *Total* |
|---|---|---|---|---|---|
| | | *N* | *P* | *K* | |
| Rice | 60.1 | 1.26 | 0.52 | 2.69 | 4.47 |
| Wheat | 45.5 | 1.24 | 0.68 | 1.64 | 3.56 |
| Sorghum | 11.9 | 0.55 | 0.19 | 0.92 | 1.66 |
| Pearl millet | 7.7 | 0.26 | 0.09 | 0.87 | 1.22 |
| Maize | 7.9 | 0.33 | 0.14 | 0.31 | 0.78 |
| Other cereals | 6.4 | 0.14 | 0.07 | 0.32 | 0.53 |
| Pulses | 12.9 | 0.75 | 0.09 | 6.12 | 0.96 |
| Total | 152.4 | 4.53 | 1.78 | 6.87 | 13.18 |
| Commercial crops | — | — | — | — | 5.72 |
| Total removal (million tons) | | | | | 18.90 |
| Nutrients applied as fertilizers (million tons) | | | | | 7.71 |

spread. Iron, as well as manganese and copper, deficiencies were being reported more frequently. These deficiencies have reflected on crop productivity in the intensively cultivated areas. The trends call for intensified research to monitor and quantify the emerging deficiencies and to develop practices for maintaining and improving soil fertility through appropriate integrated management systems, involving optimum combinations of cropping systems and use of organic nutrient sources, including practices such as green manuring, recycling farm wastes, and using biofertilizers in combination with the nutrients from fertilizers for sustained productivity.

High fertilizer use in specialized farming tracts (for example, sugarcane and grape-growing areas in Maharashtra) have shown signs of overuse, reflected in the quality of groundwater—a trend that has to be reversed to enable sustained use of resources.

Special efforts will be required to maintain and improve the nutrient-supplying capacity of soils under rain-fed agriculture. These areas presently receive only 20 percent of the total fertilizer but account for more than 80 percent of the production of sorghum, pearl millet, pulses, and oilseed, and 30 to 40 percent of the production of rice and wheat.

## REFERENCES

Bhumbla, D R. 1982. Small reservoirs—A programme for improved rainfed agriculture. Society of Promotion of Wastelands Development, New Delhi, India.

Dhawan, B. D. 1988. Output impact according to main irrigation sources—Empirical evidence from four selected states. *In* J. S. Kanwar [editor] Water management—The key to developing agriculture. Agriculture Publishing Academy, New Delhi, India.

Ministry of Agriculture and Irrigation, Government of India. 1976. Report of the National Commission on Agriculture. New Delhi, India.

Narayana, V., V. Dhruva, and R. Babu. 1983. Estimation of soil erosion in India. Journal of Irrigation and Drainage Engineering 109: 419-434.
Sarma, J. S., and S. Roy. 1979. Behavior of foodgrain production in India 1966-1970. Staff Working Paper No. 339. World Bank, Washington, D.C.

# POLICY DEVELOPMENT FOR THE LOW-INPUT SUSTAINABLE AGRICULTURE PROGRAM

# POLICY DEVELOPMENT FOR THE LOW-INPUT SUSTAINABLE AGRICULTURE PROGRAM

Paul F. O'Connell

In 1986, policymakers in the U.S. Department of Agriculture (USDA) began discussions with proponents of reduced-input farming, particularly for those inputs purchased off the farm. Farmers were looking for ways to increase net returns and to achieve greater compatibility between environmental and production goals. USDA was responding to this concern.

Orville Bentley, assistant secretary for science and education, and John Patrick Jordan, administrator of the Cooperative State Research Service (CSRS), visited the Rodale Institute at Emmaus, Pennsylvania, and had extensive discussions with Robert Rodale and others about the merits of reduced-input farming. Peter Meyers, USDA deputy secretary, also visited the Rodale Institute at a different time in 1986 and made an on-farm visit to the Dick Thompson farm in Boone, Iowa. These and other interactions with high-ranking USDA officials began to build a mutual understanding between and among individuals and interest groups having different philosophies regarding farm management.

This improved atmosphere led to a conference on low-input agriculture systems in Racine, Wisconsin, on January 18-20, 1987. At that meeting were representatives from foundations, federal and state agencies, private research institutions, and academia. When current research and education programs were highlighted for about eight states, it became obvious to everyone present that this topic was receiving little attention. The primary goal of the conference then focused on ways to get more resources into this activity. Options examined included foundation support, congressional action, and redirection of current programs. Each received considerable discussion, but attendees all agreed that funding of Subtitle C (The Agriculture Productivity Act) in the Food Security Act of 1985 should get highest priority. The consensus of conference participants was that a federal appropriation to implement that approved legislation would stimulate action in other programs sponsored by the federal government, state agencies, universities, and the private sector.

Throughout 1987, congressional contacts and hearings ensued. Fortunately, the initiative had a receptive hearing because it could address major concerns facing U.S. agriculture. Low-input technologies provide opportunities to reduce a farmer's dependence on certain kinds of purchased inputs in ways that increase profits, reduce environmental hazards, and ensure a more sustainable agriculture in the future.

## Why is Low-Input, Sustainable Agriculture Acceptable Now?

Conventional agriculture involves highly specialized systems that emphasize high yields achieved by inputs of fertilizers, pesticides, and other off-farm purchases. Alternative farming systems, on the other hand, range from systems with only slightly reduced use of these inputs (through soil tests, integrated pest management, and capital inputs) to systems that seek to minimize their use (through appropriate rotations, ridge tillage, integration of livestock with crops, mechanical/biological weed control, and less costly buildings and equipment).

Low-input, sustainable agriculture addresses multiple objectives, from increasing profits to maintaining the environment, and may incorporate and build on multiple systems and practices, such as integrated pest management and crop rotations. In contrast, the conventional approach to farming features a capital-intensive system, continuous cropping, and a substantial reliance on manufactured inputs and extensive use of credit. Conventional agriculture also stresses high levels of production—"more is better." Yet, agricultural economists point out that the most profitable output on a farm is usually something less than maximum physical output, and at some point dollar returns from higher increments of output may not cover additional costs.

Factors fostering the development and expansion of conventional agriculture have included the following:

- Ample credit.
- Suitable infrastructure.
- Availability of research-based information and education assistance from land grant colleges and USDA.
- Farm price and income support policies.

■ Other public programs.

While not ignoring resource conservation or environmental quality, the conventional approach tends to view these factors as constraints on profit maximization. For example, soil conservation has traditionally played a secondary role to production. However, when soil conservation and protection of the environment are socially desirable but not profitable for the farmer, the government has provided financial and technical assistance.

In low-input or sustainable agriculture, farmers seek to complement conservation and production goals; rotations can serve both goals. Average annual soil erosion from land planted in one year to corn, but in the previous year to hay or a legume crop, is less than erosion from the same land used to grow corn continuously. Rotations break cycles of crop-specific diseases and pests, thereby reducing the need for pesticides. When legumes are included in the rotations, atmospheric nitrogen is synthesized into a form used by crops.

## Highlights of the 1988 Appropriation Language

Growing numbers of farmers are looking for reliable information on reduced-input systems that reduce cost, control erosion, abate pollution from heavy fertilization and pesticide use, and alter monocultural cropping systems.

Priority is given to providing information to farmers in a readily usable form so that past and ongoing research can be applied immediately. CSRS is directed to coordinate activities by assuring participation of private foundations, land grant institutions, nonprofit organizations, the Extension Service, the Agricultural Research Service, local farmer groups, and the Soil Conservation Service. Congress appropriated $3.9 million for this joint program in fiscal year 1988, followed by a 14 percent increase to $4.45 million in fiscal year 1989.

## Activities in USDA

On March 8, 1987, Assistant Secretary Bentley established a task force on alternative farming systems to explore the implications for USDA of the growing interest in this topic and to recommend actions for dealing with Subtitle C of the Food and Security Act.

On November 5, 1987, Bentley formed the Research and Education Subcommittee on Alternative Farming Systems, which I chaired. Membership includes representatives from the following groups:

■ Agricultural Research Service (ARS).

- Agricultural Stabilization and Conservation Service (ASCS).
- Cooperative Extension Service (CES).
- Cooperative State Research Service (CSRS).
- Economic Research Service (ERS).
- National Agricultural Library (NAL).
- Soil Conservation Service (SCS).
- Other USDA agencies that have an interest.
- Extension Committee on Organization and Policy (ECOP).
- Experiment Station Committee on Organization and Policy (ESCOP).
- Private institutions.

Reponsibilities of the subcommittee include development of policy recommendations, establishment of procedures for awarding funds, coordination of research and extension activities, and preparation of a secretary's memorandum. In January 1988, Secretary of Agriculture Richard Lyng issued a memorandum of major historical importance to USDA policy on low-input agriculture. The text of that memorandum is as follows:

"The purpose of this memorandum is to state the Department's support for research and education programs and activities concerning 'alternative farming systems,' which is sometimes referred to as 'sustainable farming systems.'

"Many of the nation's farmers have experienced financial stress in the 1980s due to the downturn in exports of farm products, commodity prices, and land values. The traditional solution of increased production will only depress commodity prices further. Also, farmers are under increased pressure to reduce nonpoint pollution from fertilizers and pesticides and reduce erosion. Alternative farming systems that decrease or optimize the use of purchased inputs and that can increase net cash returns to the farmer through decreased costs of production may effectively improve the competitive position of the farmer and decrease the potential for adverse environmental impacts.

"Alternative farming systems are defined here as alternatives to current farming systems that tend to have a high degree of specialization. The current systems emphasize high yields which are achieved by the use of major inputs of fertilizers, pesticides, and other off-farm purchases. Alternative farming systems range from systems with only slightly reduced use of these inputs through the better use of soil tests, integrated pest management, and capital inputs to systems that seek to minimize their use through appropriate rotations, integration of livestock with crops, mechanical/biological weed control, and with less costly buildings and equipment.

"The Department encourages research and education programs and activities that provide farmers with a wide choice of cost-effective farming systems, including systems that minimize or optimize the use of purchased inputs and that minimize environmental hazards. The Department also encourages efforts to expand the use of such systems.

"The Assistant Secretary for Science and Education is responsible for encouraging and guiding the development of research and extension programs that best meet farmers' needs for facts, information, and guidance concerning alternative farming systems.

"Each agency head shall implement the programs for which the agency head is responsible in ways that are consistent with this policy on alternative farming systems. Activities involving more than one agency will be coordinated through the Department's Research and Education Committee."

## Management of Program

Pursuant to the secretary's memorandum and the legislation, the Low-Input/Sustainable Agriculture (LISA) program was established. Each of four regions—Northeast, South, North, Central, and West—has an administrative council and a technical committee. Each regional administrative council includes representatives from ARS, CES, state agricultural experiment stations, private research and education organizations, SCS, and producers. Responsibilities of the administrative council include:

- Overall policy formulation at the regional level, including program goals, priorities, and project evaluation criteria.
- Appointment of a technical committee.
- Involvement of all eligible institutions.
- Review and approval of actions of the technical committee.
- Submission of a plan of work to USDA.

Each of the four regional technical commmittees, appointed by administrative councils, includes researchers, extension specialists, producers, and farm management experts. These committees serve as the key action level for regional programs. They integrate activities of all participating institutions and evaluate project proposals submitted for funding.

## Progress to Date

The projects accommodated by 1988 funds have been selected; 53 were funded. In general, the projects are interdisciplinary team efforts involving public and private organizations, with the meaningful involvement of farmers. The emphasis is on providing readily usable information to farmers. With fiscal year 1989 funding of $4.45 million, the regional administrative councils have issued calls for proposals for LISA's second year of operation.

The potential impact of this program can best be stated by excerpting a few quotes from an April 8, 1988, *Kiplinger Newsletter:*

"Makes no difference what you call it, either alternative, regenerative, renewable, low-input, organic, sustainable, or some other terminology. There's a new kind of agriculture starting to emerge in the U.S. which has implications for producers, processors, all agribusiness, and ties to fewer chemicals, better soils stewardship, a cleaner environment. What many considered to be an embryonic movement that would die a few years ago is now starting to bloom and fast approaching full flower.

"It is driven by consumers who are worried about chemicals in their food and environmentalists concerned with groundwater, soil erosion, etc. Safer food and cleaner environment advocates are on the offensive. They are gearing for a major effort to reshape legislation to their liking. Farm price supports, animal health, land use, food safety, you name it. Make no mistake, they plan on having a major say in the next Farm Bill debate.

"Food processors and retailers are getting involved and are taking the initial steps to reduce or eliminate chemicals from the raw products they process or sell. Leaning on producers to move in this direction provides them with raw materials that have had less exposure to chemicals. Increased talk of checking raw products for chemicals at the loading dock.

"Congress is feeling the heat, reacting to the rising pressures. It directed USDA to conduct a $4-million study of low-input agriculture. And even pro-farmer, pro-agribusiness lawmakers take notice, pay attention to demands from groups pushing to clean up the water, air, and food supply."

# THE ROLE OF ECONOMICS IN ACHIEVING LOW-INPUT FARMING SYSTEMS

J. Patrick Madden and Thomas L. Dobbs

Economic analysis can and should play an important role in the development and adoption of low-input farming methods and systems. A low-input farming system is a combination and sequence of low-input farming methods or technologies integrated into the whole-farm management plan. Called by many different names (reduced-input, sustainable, regenerative, alternative, etc.), low-input farming encompasses a wide array of approaches that depart in important ways from the conventional norms. It includes a diverse array of farming methods such as integrated pest management, biological control, and legume-based crop rotations—methods that have been widely and profitably adopted by conventional farmers. But it also includes innovative approaches not yet fully understood by scientists or widely adopted by farmers. The fundamental goal of low-input agriculture is an abundance of food and fiber produced in ways that are harmless to humans and the environment, as well as sustainable for generations to come. Farmers using low-input methods do not necessarily totally eliminate the use of all synthetic chemical pesticides[1] and fertilizers, especially during a transitional phase, but they attempt to replace synthetic chemical inputs with more harmless and sustainable methods, and with on-farm inputs, to the extent that is technically and economically feasible.[2]

[1]The term "synthetic chemical" is used to differentiate between natural substances, such as manure and sulphur (which is used both as a pesticide dust and as a soil amendment), versus chemically manufactured compounds, such as aldicarb or anhydrous ammonia.

[2]The concept of regeneration was first offered as a norm for modern agriculture by Bob Rodale, "Breaking New Ground: The Search for a Sustainable Agriculture," *The Futurist,*

The potential profitability of a specific farming method often cannot be accurately anticipated except in the context of a whole-farm plan—taking into account managerial, labor, and capital requirements, including the complex interactions among crops, livestock enterprises, soils, and populations of pests and their natural enemies. Developing and evaluating low-input farming systems require a holistic perspective, incorporating the insights of operating farmers and various agricultural disciplines alike. We view economics as an essential integrating framework for whole-farm analyses of low-input systems (Dobbs, 1987; Madden, 1987b). Properly used, economics can provide the broad farm management perspective needed to incorporate both hard and soft data into a framework for assessing whether farmers are likely to be better off by adopting particular low-input systems. Defining the whole-farm perspective and conducting research to provide necessary data require not just economists, however, but a multidisciplinary team. Economists must be actively involved from the outset, but whether they are team leaders in a formal sense will depend on a variety of circumstances (Dobbs, 1987). Likewise, farmers must be actively involved in the technology development and evaluation process. How farmers become involved will depend upon a variety of needs and circumstances.

Farmer incentives constitute the driving force for individual decisions about farming systems or technology (Dobbs and Foster, 1972). Incentives are influenced by a variety of factors, both internal and external, the individual farm operation. Thus, in assessing incentives to adopt low-input systems, these various factors must be accounted for. The whole-farm analysis approach, for example, has to incorporate internal factors such as how the interplay of crops in a system affects weed control and soil fertility. On a broader, external level, the impact of factors such as federal farm policy on the profitability of particular farming systems must be incorporated into the analyses.

A major role of economics in low-input farming systems is to determine the present and necessary incentives for farmers to adopt low-input systems. Degree of profitability is a major aspect of incentives, but other aspects, such as the farmer's risk of financial loss, environmental hazards, and human health risks, are also relevant and sometimes are more important than

Volume 1, January 1983, pp. 15-20. This concept is further developed by Patrick Madden, "Regenerative Agriculture: Beyond Organic and Sustainable Food Production," The Farm and Food System in Transition (East Lansing, Michigan: Michigan State University Press), Extension Publication No. FS 33, 1984. The concept of "reduced-input" agriculture has been proposed as a less value-laden term than organic or regenerative, by Frederick H. Buttel, Gilbert W. Gillespie, Jr., Rhonda Janke, Brian Caldwell, and Marianne Sarrantonio, "Reduced-Input Agricultural Systems: Rationale and Prospects," American Journal of Alternative Agriculture, Volume 1, pp. 58-64. Spring 1986.

profit considerations.

Another role of economics is to help assess the impacts of various farming systems on society at large, including consumers of farm products and users of the environment. Various public policies that affect farmer incentives are increasingly being introduced in the United States because of concerns about impacts on society. One example is recent legislation in the state of Iowa that places special taxes on nitrogen fertilizer and on pesticides because of concerns about groundwater contamination. Whole-farm analysis methods must be capable of assessing the effects of such policies on farmers' decisions about systems or technology.

## Methods of Assessing Profitability

Understanding the economic implications of alternative farming practices requires research at several levels of aggregation, including the individual component of a crop or livestock enterprise, the entire enterprise, a whole farm, commodity markets, and national and international agricultural economies. Some of these methods are discussed briefly here.

***Enterprise or Component Analyses.*** Economic analyses of single enterprises or their components focus only on the costs and returns associated with particular activities of a farm. Such analyses may suffice for certain kinds of alternative agriculture decisions. For many farmer decisions in the area of alternative or low-input agriculture, however, the whole-farm perspective is needed. In those cases, enterprise budgets are essential building blocks for the whole-farm analyses.

Enterprise budgets for crop and livestock operations may be "complete," encompassing all of the fixed costs, variable costs, and returns associated with a particular enterprise. Alternatively, "partial" budgets are sometimes used, dealing only with those variable production costs and returns that change as a result of a specific farming practice change (Kay, 1986). Both complete and partial enterprise budgeting require somewhat restrictive assumptions about what happens (or does not happen) in the rest of the farm operation when a farming practice changes.

A special type of enterprise budgeting deals with a sequence of crops constituting a specific rotation. Whereas an individual crop enterprise ordinarily presents data for one acre (or hectare) of that crop, an enterprise rotation budget in effect splits the acre into fractions representing each crop's time share in the rotation.

A primary appeal of enterprise budgeting analysis for purposes of assessing the economic farm-level implications of alternative farming practices is that it is practical and readily understood. Consequently, the research is open to constructive criticism by non-economist experts having specialized knowledge of the biological or physical science subject matter at hand. As a result, this approach is more likely to yield correct conclusions than is a more abstract and methodologically opaque analytical method. Enterprise budgeting is reported to be the most widely used method of estimating changes in income of individual farms as a result of adopting integrated pest management practices (Allen et al., 1987; Osteen et al., 1981). The landmark research on the economics of crop rotations by Heady (1948) and Heady and Jensen (1951) was essentially based upon the enterprise budgeting approach, in that the only aspect of the farm operation assumed to vary was the crop rotation. Ali and Johnson (1981) effectively used enterprise budgeting to assess the short-term economic benefits to North Dakota farmers of summer-fallow wheat ground. Alternative cropping systems in Nebraska, including organic system rotations, were compared with a budgeting approach by Helmers et al. (1986). The Nebraska study also included a risk analysis of budgeted annual net returns by examining the variability among rotations over time.

Whether budgets are used for individual enterprise or component analyses or as the building blocks of whole-farm analysis, sources of data constitute a critical focus of concern. To what extent can and should researchers draw on experiment station trials, cooperating farmers, farm surveys, case studies, or other sources of data? Historically, agricultural economists relied heavily upon farmer surveys (usually conducted by personal interviews) for enterprise budget data. Personal interviews constitute a very expensive approach, however, and because the methods involved are neither new nor sophisticated, the returns to academic researchers facing peer review tend to be limited or negative at many institutions. Given the costs and the lack of professional rewards for such survey work, this approach has been used less within the U.S. in the last two decades than previously.

Collaborative research between agronomists and agricultural economists with experiment station trials is another approach that has been used over the years. But the ever-increasing specialization of agricultural disciplines, especially since the 1960s, has tended to diminish the attractiveness of this work in the U.S., although it is prominent in a number of the international agricultural research centers supported by the Consultative Group on International Agricultural Research (CGIAR). Where such collaborative research still exists in the U.S., agricultural economists are often brought in too late in the effort, after the trials have been designed and underway for a number of years. Thus, data important to enterprise budgeting have sometimes not been incorporated into the trials and measurements. Research on low-input agriculture will require greater emphasis on truly collaborative efforts when experiment station trials are part of the research. In early stages of the research, agricultural economists should be involved with agronomists, as well as animal scientists and agricultural engineers when appropriate.

Likewise, on-farm research, in which university and other professional researchers collaborate with farmers in trials and other research on operating farms, can play an important role in low-input agriculture (Lockeretz, 1987). On-farm work has had a prominent place in the "farming systems research" that gained widespread recognition in developing countries during the 1970s. Good on-farm research complements, but does not obviate the need for, good experiment station trial research.

A combination of data collection methods will be needed in developing enterprise budgets for research on low-input agriculture. Because it is important to do preliminary economic analyses of low-input farming system alternatives, somewhat eclectic approaches will be necessary in the early stages of most research efforts. For example, at South Dakota State University (SDSU), a combination of experiment station crop trial data, various research and extension sources, and "expert judgment" have been used to estimate "normalized" input-output relationships and costs for initial crop rotation budgets in the university's low-input agriculture research program (Dobbs et al., 1987). Budgets will be refined as research trial data continue to be collected over time. Year-to-year variations in yields and net returns will be analyzed to estimate the risks associated with various alternative farming systems. Additional budgets will also be estimated for rotations of regenerative farmers who have collaborated with SDSU researchers in on-farm work for several years. A mail survey of all known regenerative farmers in South Dakota was also recently conducted, in part to ascertain the range of most common practices employed by these farmers. Research plans include personally interviewing a sample of the farmers who responded to a mail survey regarding common farming practices, and developing enterprise budgets and whole-farm analytic models for a representative selection. Thus, over time the SDSU research will draw upon a variety of data sources for the enterprise budgets used in whole-farm analyses. Other universities are engaging in similar research.

***Whole-Farm Analyses.*** Frequently, a farming method that appears to be very profitable or otherwise advantageous per acre, per cow, or at the individual enterprise level may prove to be much less attractive from the perspective of the whole farm or the household. Analysis at the whole-farm level recognizes that the farmer's decision to adopt one or more alternative farming practices is not made in isolation from the rest of the farm business and the farm household. The successful, commercial-scale farmer must assess the compatibility of proposed alternatives with the various practices already in place, taking into account the farm's physical and biological resources and anticipated changes in crop yields, livestock enterprise productivity, and production costs, all of which strongly affect the farm operator's cash flow and equity position. The experienced farmer would also evaluate the proposed alternative methods in terms of their impact on the farm's critical-period labor and machinery requirements, particularly at harvest and other times of peak labor loads. The farmer would further consider the size of farming operation that could profitably be managed using the alternative methods, and the compatibility with off-farm employment opportunities and other interests of the farmer and members of the farm family. Clearly, analysis at the whole-farm level is essential in determining the economic suitability of alternative farming practices.

A key decision at the outset of any whole-farm analysis involves the type and degree of sophistication of the economic models to employ. Options range from relatively simple microcomputer spreadsheet models to the most complex computer optimization and simulation models. Each has its place. Disciplinary pressures tend to encourage model "sophistication," regardless of whether available data and research resources warrant that.

Economists and multidisciplinary teams involved in research on low-input agriculture are urged to start with relatively simple models and to add greater complexity over time as circumstances require. It does little good to have a cutting-edge model that has taken so long to develop and acquire data for that the questions it was originally designed to address are no longer pertinent. Moreover, many of the extremely complex models require data for which no reliable estimates exist. The "Rube Goldberg" nature of the data generated to feed such models is a major source of potential error in the findings. Many of these models also require so much effort to complete and produce consistent and sensible results that little time and research resources are left to examine important but less easily quantifiable issues.

Low-input agriculture research in South Dakota has started with relatively simple whole-farm models on microcomputer spreadsheet formats (Dobbs et al., 1988; Leddy, 1987). The initial models explicitly examined only the crop component of low-input and conventional farms. Care was taken to assure that the interrelationship of cropping decisions and federal farm program requirements was accounted for in the models. Sensitivity analyses with these relatively simple models has shown that many insights can be gained early in a research project about the impacts of a variety of internal and external factors on farm profitability.

A somewhat greater degree of complexity has been added to the South Dakota research by incorporating both crop and livestock enterprises in the microcomputer farm financial planning and analysis package called FINPACK (Hawkins et al., 1986). FINPACK versions of the whole-farm models facilitated examination of the implications of low-input crop farming systems for features such as livestock enterprises of the farm; labor utilization; and farm profitability, liquidity, and solvency (Leddy, 1987). The FINPACK models do not show optimum allocations of a farm's resources as do linear programming models. They are basically computerized accounting models of the whole-farm operation. For some questions and types of analyses, the FINPACK models proved useful; for others, the more simple spreadsheet models involving only the crops component were sufficient and easier to interpret (Dobbs et al., 1988; Leddy, 1987).

Since the 1950s, agricultural economists have used numerous versions of mathematical programming or optimization to model whole-farms. These models typically are based on either profit maximization or cost minimization concepts. The most widely used optimization method is linear programming. To a greater extent than models described in the immediately preceding paragraphs, linear programming models can identify enterprise interconnections and constraints (e.g., land of particular quality, family labor, credit) that must be accounted for in farming system choices.

Walker and Swanson (1974) used linear programming several years ago to examine the likely income effects of policies to reduce groundwater and surface water contamination by restricting nitrogen fertilizer use on a typical Illinois farm. Hunter and Keller (1983) used linear programming to analyze alternative crop and soil management systems for reducing soil erosion losses on Tennessee farms. Domanico et al. (1986) made a similar application of linear programming to determine income effects of using various management practices to limit soil erosion on a Pennsylvania farm; analysis included an organic farming system as one of the alternative sets of practices. Dabbert and Madden (1986) used another linear programming model to simulate income trends during the transition from conventional to organic practices on a Pennsylvania farm. Examples of linear programming applications to farming system decisions in other countries include analyses by Nadar and Rodewald (1980) in Kenya and by Jones (1986) in the United Kingdom. The United Kingdom application focused on the probable impacts of increased energy prices on the economic viability of organic farming systems.

A variety of other economic models are available that could be adapted for research on low-input agriculture. The FLIPSIM model, developed at Texas A&M University (Richardson and Nixon, 1986), is an example of a relatively sophisticated model that does not employ optimization. The model is capable of simulating a variety of economic features of case farms over a multiple-year planning horizon.

Researchers involved in low-input agriculture must be open to consideration of a variety of modeling approaches, including simulation models such as FLIPSIM. However, considerations such as problem complexity, data availability, time, and available research resources should guide the choice of whole-farm model. Researchers should take care to not fall into the "have model, looking for research problem" syndrome.

***Macroeconomic Analyses.*** Recognizing the limitations of studies at the component, enterprise, or whole-farm level, a few economists have attempted to estimate or predict the likely impact of widespread adoption of alternative farming methods and systems. If widespread adoption of alternative farming methods is possible, comprehensive assessment of the potential impact requires examination of effects on market prices, various farming regions, international trade competitiveness, employment, incomes of various categories of farmers and consumers, and other macroeconomic variables, as well as on human health risks, environmental hazards, and impending shortages of phosphates, fossil sources of energy, and water.

The one study that has attempted to estimate quantitatively, in a comprehensive way, the macroeconomic or market-level impact of widespread adoption of organic farming is seriously flawed. However, because the study is widely quoted as a macroeconomic analysis of low-input farming, it should be discussed briefly here. An interregional competition linear programming model was used to predict the potential effect of what the authors defined as "organic farming practices." The study concluded that total production of many commodities would decrease substantially (Langley et al., 1983). However, the crop yields assumed in the analysis were based on historical 1944 yields, with some adjustment for improvements in cultivars. The study assumed no fertilizer would be applied. Recent experimental results (Helmers et al., 1986) suggest this procedure seriously understates the productivity of organic agriculture. Furthermore, organic or other alternative farming practices are likely to be adopted very slowly (if at all), causing gradual shifts in prices and resource use. If farm commodity prices would begin to increase significantly, the resulting induced change in investments in research and technology would facilitate innovations that would tend to ameliorate the study's predicted long-term impact on production, prices, incomes, and exports (Ruttan, 1982). Among other deficiencies, the study also seems to have overstated the dependence of organic farms on livestock manure, erroneously assumed that organic farmers apply no fertilizers, and underestimated the contribution of legume-based crop rotations to soil fertility. Because of these procedural flaws, the substantive findings of this study are of no value.

After reviewing the literature on organic farming, Cacek and Langner (1986) predicted that widespread adoption of organic farming methods would yield several benefits to society, including a reduction in the taxpayer cost of federal price support programs, reduced depletion of fossil fuels, reduced environmental damage from agricultural chemical and soil erosion, and enhanced sustainability of agriculture for future generations. While these benefits seem plausible, widespread adoption of organic farming methods would undoubtedly carry some adverse side-effects, possibly including higher prices of some foods, a further reduction in the nation's balance of international trade, and a decline in the incomes of agribusiness firms supplying synthetic agricultural chemical pesticides and fertilizers. Furthermore, regional shifts in production toward areas where production is less dependent on these inputs would tend to enrich some regions at the expense of others.

Additional studies will be needed to further clearly identify and quantify the macroeconomic impacts of widespread adoption of sustainable agriculture practices. Of particular interest are the positive and negative effects sustainable agriculture might have on the rural economies of farming-dependent regions. Does low-input or sustainable agriculture have the potential to be a vital force in the rural revitalization of such regions? From the standpoint of added on-farm employment and enterprise diversification, possibly yes. From the standpoint of reduced demand for purchased farm inputs, possibly no. The net, overall impact is not known at this time.

## Empirical Findings on Profitability of Adoption

Extensive literature exists on the profitability of some aspects of low-input agriculture, such as integrated pest management (IPM), while only a small body as yet exists on other aspects, such as cropping systems in which management strategies are substituted for all or a significant share of the synthetic chemical fertilizers and herbicides. The jury is still out, however, on the relative profitability of many low-input practices and systems in different agroclimatic regions. A new federal program is providing grants to support research and education that will ultimately improve the profitability and reduce the risk inherent in adopting many low-input methods of farming (Madden, 1988b).

Several studies have estimated the farm-level and aggregate monetary benefits and costs associated with development and adoption of IPM programs (Osteen et al., 1981). Economic analyses at the farm level take into account the increase in sales value, cost of pest scouting, and changes in pesticide application costs.

A recent national study of extension IPM programs found that the various IPM programs typically resulted in significant increases in farm profits (Allen et al., 1987). The findings are somewhat problematical, however, in that the bases for comparison were not always apparent. Nonetheless, the evidence seems to indicate that IPM increases the profits of farmers who use it and may also decrease the environmental loadings of certain pesticides, primarily insecticides. In some instances, pest scouting results in elimination of unnecessary sprayings. In other cases, detection of potential insect damage results in an increase in insecticide use as compared with farmers who do not use IPM monitoring of pest populations (Allen et al., 1987). IPM generally does not result in decreased use of fungicides or herbicides.

A recent report (Allen et al., 1987) included a review of 42 IPM evaluation studies. In the vast majority of cases, crop yields were reported to have increased as a result of adopting IPM, and in all instances that reported pesticide use and/or production costs, a lower cost per acre was noted. The difference in production costs between IPM users and other growers varied greatly from state to state and by crop grown. In this national study pesticides were estimated to account for 2 to 22 percent of individual farmer total production costs.

Whereas conventional agriculture has generally gravitated toward higher degrees of specialization, alternative farming methods encompass a number of diversification strategies. Perhaps most prevelant among these is a multi-year crop rotation that alternates forage legumes with row crops and small grains. Early evidence of the economic advantages of legume-based crop rotations came from results of a series of rotation experiments conducted in various midwestern states during the 1930s and 1940s (Heady, 1948; Heady and Jensen, 1951). Analysis of data from experiments in Illinois, Iowa, and Ohio found that a greater total volume of grain was produced per acre (over and above the legumes produced) using certain rotations including clover or alfalfa as compared with continuous corn. The net return over variable cost was calculated for each of the rotations under a variety of pricing assumptions. In most instances, continuous corn was found to be less profitable than legume-based rotations even when the forage was assumed to have no monetary value. It is important to realize, however, that these findings were based on the prices and technology prevailing in the 1930s and 1940s when pesticides and modern cultivars were not available.

A more recent study in southeastern Minnesota examined both the nitrogen contribution and other benefits of legumes in a crop rotation with corn and/or soybeans (Kilkenny, 1984). Using a linear programming model of a 400-acre farm, a corn-soybean rotation was found to be more profitable than continuous corn when nitrogen fertilizer prices were at 1980-82 levels ($0.115 per pound). If the price of nitrogen were to increase dramatically to $0.69 per pound (three to five times 1988 levels, depending upon the nitrogen fertilizer form), the most profitable rotation would shift toward continous soybeans on more of the acreage in combination with a three-year rotation of corn-oats/alfalfa-alfalfa.

An eight-year experiment conducted recently by University of Nebraska scientists compares 13 cropping systems, including an essentially "organic" rotation that used manure for fertilizer and no herbicides or synthetic chemical fertilizers (Helmers et al., 1986). The crops grown included corn, soybeans, grain sorghum, and oats with sweet clover in various rotations and in continuous cropping systems. The results confirmed the findings of studies done in the first half of this century using more primitive cultivars and no synthetic chemical pesticides: rotations have higher yields and higher net returns per acre than continuous mono-cropping systems involving crops such as corn, soybeans, or sorghum. Different fertilization regimes were found to have little impact on profitability. The continuous cropping systems were found to require a higher expenditure for pesticides and to be subject to greater year-to-year variation in yields and profits per acre compared to the various rotations.

A study of a 305-acre mixed crop and livestock farm in Pennsylvania simulated operation of the farm over several years in transition from conventional to organic management. The study concluded that the farm's income would be reduced 43 percent in the first year of the transition but that income would increase over a five-year period, reaching an equilibrium at about 7 percent below the income expected under conventional management (Dabbert and Madden, 1986). This study is flawed, however, in that transitional yields were assumed, rather than based on empirical findings, which were not available.

Two of the major financial disincentives to using legumes are the high cost of establishing a stand and the opportunity cost (profit foregone) in delaying production of higher value crops. Both of these disadvantages seem to have been overcome at least partially by an alternative rotation studied in the Palouse area of eastern Washington (Goldstein and Young, 1987; Young and Goldstein, 1987). This rotation, called the perpetual-alternative-legume-system, or PALS, features a biennial legume (black medic) that has been observed to reseed itself for as long as 30 years following establishment. The PALS rotation is three years, consisting of spring peas plus medic-

medic-winter wheat (P/M-M-W). The only synthetic chemical used during this rotation is an insecticide applied to the peas. The rotation controls almost all the weeds in wheat; harrowing is adequate to control the rest. The conventional comparison rotation, four years, consists of wheat-barley-wheat-peas (W-B-W-P). Chemical pesticides and fertilizers are used on the conventional rotation.

Crop yields under the two rotations were similar during two trial years at three sites. PALS wheat yields averaged 62.6 bushels per acre compared to 60.3 bushels on the conventional plots. The largest differences occurred during the drought of 1985, when yields for the PALS experimental plots averaged 83 percent more than those of the conventional plots. In 1984, when rainfall was close to normal, the PALS wheat yields were 3 percent less than the conventional yields.

Relative profitability of the PALS and conventional systems depended upon how the crops were valued. The conventional system was more profitable with 1986 federal farm programs in place. However, the PALS system was more profitable when production was valued only at 1986 market prices with no government farm program payments.

An interdisciplinary team of scientists at South Dakota State University started a crop rotation study in 1985 to compare various conventional, reduced tillage, and low-input farming systems. The conventional systems use the moldboard plow and rely on various synthetic chemical pesticides and fertilizers to produce three-year crop rotations. The reduced tillage systems also use chemicals and three-year rotations, but ridge tillage or minimum tillage systems are substituted for the moldboard plow. The low-input systems are essentially "organic," using no synthetic chemical pesticides or fertilizers (Dobbs and Mends, 1989). Economists on the project team have used the findings from the first four years of experimentation to simulate the operation of a typical 640-acre family farm (with 540 tillable acres), assuming various management systems. The experimental plots are testing crop rotations typical at two locations, Watertown and Madison, South Dakota. Preliminary (1985-86) simulation findings, based in part on experimental findings and adjusted or "normalized" with data from various sources, suggested the low-input methods would be profitable. Simulations representing the Madison area indicated the low-input system would earn substantially lower net returns than conventional or reduced tillage systems in that area. Low-input simulations representing the Watertown area found that the low-input systems would earn about the same profits as the conventional and reduced tillage systems (Dobbs et al., 1987; Dobbs et al., 1988).

However, subsequent findings based on actual (rather than "normalized") yields for 1988 indicate that the low-input cropping systems were much more profitable than the conventional and reduced tillage systems during that drought year (Dobbs and Mends, 1989). In fact, the only systems tested in this South Dakota study that were estimated to earn a profit in 1988 were the low-input systems. The low-input farming system for the Madison area was estimated to earn a profit of about $4,900, using a crop rotation of oats, alfalfa, soybeans, and spring wheat. The conventional and ridge till rotations for this area (corn, soybeans, and spring wheat), using chemical pesticides and conventional tillage, were estimated to incur net losses of about $24,000—a difference of about $29,000 compared to the low-input system. The differences in earnings from the 1988 simulations were much smaller for the Watertown area. The low-input system for this area, consisting of oats, clover, soybeans, and spring wheat, was estimated to about break even, with a nominal profit. But the minimum tillage and conventional systems—producing soybeans, spring wheat, and barley—each lost about $15,000. These findings are based on the assumption that current farm price support programs are in effect. Under different policy assumptions, the profitability comparisons can change substantially.

Overall, the emerging literature on U.S. farming systems that emphasize legumes in the rotation and minimize or eliminate the use of synthetic chemicals for fertility and pest control tends to offer encouraging farm-level profitability prospects.

## Challenges in Assessing the Economics of Low-input Farming Systems

Research on the economics of low-input farming systems should be significantly expanded over the next decade. Results presently available indicate that many low-input systems have promise. To what extent they will be economically competitive with more conventional systems when additional comparative yield and cost data become available remains to be seen. Also, systems have to be analyzed in different agroclimatic settings and with various assumptions about the external economic environment (e.g., federal farm program provisions). Several challenges remain to be fully confronted in embarking on this research.

***Strengthening Multidisciplinary Teamwork.*** Research and extension efforts designed to better understand and disseminate information about alternative farming systems require multidisciplinary teamwork. Understanding the interactions of components and enterprises comprising the whole farm requires contributions from agricultural economists, plant and soil scientists, animal scientists, agricultural engineers, and sometimes individuals from still other disciplines. Although the precise discipline mix will vary from one project to another, it is essential that economists (and, in some cases, rural sociologists) and appropriate representatives of the biological and physical sciences work together, that research and extension be functionally integrated, and that farmers be meaningfully involved in testing and evaluating the alternative methods.

However, the reward systems of academic institutions tend to militate against such multidisciplinary teamwork (Dobbs, 1987; Johnson, 1983; Johnson, 1984; Madden, 1988a; Schuh, 1986). Disciplinary work generally receives greater recognition and acceptance than does multidisciplinary work in peer-oriented professional journals, university tenure and promotion processes, and university salary policies. Add to that the fact that multidisciplinary work has several inherent tensions (Dobbs, 1987) and is sometimes slower to bear visible fruit than is disciplinary work. All of these factors cause multidisciplinary farming systems work to be avoided or to be given only lip service by many agricultural researchers.

Perhaps what is needed—to borrow a phrase from a U.S. military recruiting advertisement—is "a few good men" (and women) for expanded multidisciplinary research and extension work on low-input farming. There is no need to draw the entire agricultural research establishment into multidisciplinary farming systems studies. But there is a need for an administrative and peer-recognition environment in academic institutions that supports and encourages such studies by an expanded core of professionals. Currently, that environment more often appears to be present at some of the smaller land-grant universities than at larger institutions. One challenge is to maintain and strengthen such a multidisciplinary environment where it exists presently, and to create and foster that environment in institutions where it is lacking. This is a challenge not only to individual academic institutions but to respective agricultural disciplines and professional societies as well.

***Improving the Data Base for Whole-Farm Analyses.*** Various data sources for enterprise budgets constitute essential building blocks for whole-farm analyses. The importance of good, empirical information for these budgets and whole-farm analyses cannot be overemphasized. However, development of farm management budgets is precisely the kind of work that is generally considered "unglamorous" and lacking in professional rewards within most academic institutions. At many institutions it is considered "beneath" research faculty to spend much time developing budgets, although many agricultural economics researchers expect to have budgets available to use in their computer models, or they make simplifying assumptions that sometimes bear little resemblance to reality. Even within Cooperative Extension, where enterprise budgets are the bread-and-butter components of many farm and financial management programs, few faculty-level economists are willing or able to spend much time developing and maintaining budgets. Because there is usually an acute lack of funds and little or no personal or professional reward for spending long hours compiling and reconciling data for enterprise budgets, the unwillingness of most faculty to engage in such endeavors is quite rational.

A major challenge to those concerned about meaningful economic research on low-input agriculture is to enhance the rewards and acquire the necessary resources for developing an improved data base for whole-farm analyses. Administrators and discipline peers need to realize and recognize the effort and professional competence required in collecting and reconciling data from a variety of experiment stations and farmer sources for enterprise budgets. Adequate financial resources must be provided for research assistants and technicians to be employed in ongoing capacities for such work. Much of the money for this assistance has to come from core budgets of experiment stations and Cooperative Extension service units because grant funds are difficult to attract for enterprise budgeting work and, even when they can be obtained, generally do not provide the necessary continuity.

Improving the data base for whole-farm analyses also requires expanded horizons on possible information sources. Much more thought should be given to when and how to efficiently and effectively use farmers as sources of information. Traditional one-shot mail surveys and personal interviews of farmers continue to have a place. However, other ways of involving farmers in the data generation process require greater exploration. For example, panels of farmers (including paired comparisons) and key informants should be encouraged to cooperate over a period of years in order to monitor and obtain feedback on both the successes and the difficulties experienced with particular low-input systems. Attributes of panel members' farms that must be monitored include profitability, yields, amounts required of certain inputs, crop cover, and other behavioral characteristics of the farm decision-maker and his or her operation (Madden, 1988b). Cooperation of a longitudinal panel of farmers over several years could provide extremely important information for calibration and self-correction of low-input farming systems.

***Incorporating Price Effects and Changes in Government Programs.*** A third challenge in examining the economics of low-input farming systems is to adequately account for the "macro" perspective in farm-level analyses. For example, specified input and product prices as well as a given structure and level for government farm programs are often assumed in whole-farm analyses. Such techniques are sometimes appropriate. If particular low-input systems were to become widely adopted, even within a particular farming region, the prices of certain inputs used, for example, labor, livestock, manure, and chemical fertilizers and pesticides, and farm products sold, for example, alfalfa hay and organically grown grain, could be significantly altered. Projections a few years into the future should allow for at least the possibility of much different federal farm program provisions than currently exist. Even minor changes in crop price-support programs would have major impacts on the profitability of farms using low-input systems. In addition, the possibility of expanded interventions by state governments in areas such as soil erosion control and groundwater contamination could affect the prices and constraints incorporated into whole-farm analyses.

In light of the current dearth of macroeconomic studies on low-input agricultural systems, as well as the methodological difficulties sometimes associated with such studies, researchers involved in whole-farm analyses probably must continue to operate for at least the next several years with rather limited quantitative information on the macroeconomic impacts. However, this is no excuse for not including the possible or likely external factors emanating from the macro environment in whole-farm analyses. A great deal can be done with sensitivity analysis. For instance, research studies by Goldstein and Young (1987) and Dobbs et al. (1988) employed sensitivity analysis to ascertain the implications of alternative provisions in federal farm programs for the relative profitability of low-input agricultural systems.

The work in South Dakota was quite detailed in its treatment of alternative farm program scenarios (Dobbs et al., 1988; Leddy, 1987). In general, the South Dakota sensitivity analysis indicated that reductions in farm program benefits tend to increase the competitiveness of low-input systems, relative to more conventional systems. However, there are important exceptions to this outcome. Results indicated that the level of farm program benefits and the form of program provisions and compliance requirements affect the relative competitiveness of low-input systems. Similar sensitivity analyses were conducted in the South Dakota research to determine the effects on farm profitability of the prices for various chemical fertilizers and herbicides.

Strengthening multidisciplinary teamwork, improving the data base for whole-farm analyses, and adequately accounting for macroeconomic or external factors in farm profitability analyses are some of the challenges to be faced in assessing the economics of low-input agricultural systems. However, these challenges must be met to generate the information necessary for sound decision-making, not only by farmers, but by public policymakers as well.

## REFERENCES


Ali, M. B., and R. G. Johnson. 1981. Summer fallow in North Dakota—An economic view. North Dakota Farm Research 38(6): 9-14.

Allen, W. A., R. F. Kazmeirczak, M. T. Lambur, G. W. Norton, and E. G. Rajotte. 1987. The national evaluation of Extension's Integrated Pest Management (IPM) programs. VCES Publication 491-010. Virginia Cooperative Extension Services, Blacksburg.

Buttel, F. H., G. W. Gillespie, Jr., R. Janke, B. Caldwell, and M. Sarrantonio. 1986. Reduced-input agricultural systems: Rationale and prospects. American Journal of Alternative Agriculture 1(2): 58-64.

Cacek, T., and L. Langner. 1986. The economic implications of organic farming. American Journal of Alternative Agriculture 1(1): 25-29.

Dabbert, S., and P. Madden. 1986. The transition to organic agriculture: A multi-year model of a Pennsylvania farm. American Journal of Alternative Agriculture 1(3): 99-107.

Dobbs, T. L. 1987. Toward more effective involvement of agricultural economists in multidisciplinary research and extension programs. Western Journal of Agricultural Economics 12(1): 8-16.

Dobbs, T. L., and C. Mends. 1989. Economic results of SDSU alternative farming systems trials: 1988 compared to 1987. Economics Commentator No. 270. Economics Department, South Dakota State University, Brookings.

Dobbs, T. L., L. A. Weiss, and M. G. Leddy. 1987. Costs of production and net returns for alternative farming systems in northeastern South Dakota: 1986 and "normalized" situations. Research Report 87-5. Economics Department, South Dakota State University, Brookings.

Dobbs, T. L., M. G. Leddy, and J. D. Smolik. 1988. Factors influencing the economic potential for alternative farming systems: Case analyses in South Dakota. American Journal of Alternative Agriculture 3(1).

Dobbs, T., and P. Foster. 1972. Incentives to invest in new agricultural inputs in north India. Economic Development and Cultural Change 21(1): 101-117.

Domanico, J. L., P. Madden, and E. J. Partenheimer. 1986. Income effects of limiting soil erosion under organic, conventional and no-till systems in eastern Pennsylvania. American Journal of Alternative Agriculture 1(2): 75-82.

Goldstein, W. A., and D. L. Young. 1987. An economic comparison of a conventional and a low-input cropping system in the Palouse. American Journal of Alternative Agriculture 2(2): 51-56.

Hawkins, R. O., D. W. Nordquist, R. H. Craven, and D. A. Judd. 1986. FINPACK: A computerized farm financial planning and analysis package. Center for Farm Financial Management, Department of Agriculture and Applied Economics, Extension Service, University of Minnesota, St. Paul.

Heady, E. O. 1948. The economics of rotations with farm and production policy applications. Journal of Farm Economics 30(4): 645-664.

Heady, E. O., and H. R. Jensen. 1951. The economics of crop rotations and land use. Agricultural Experiment Station Bulletin 383. Iowa State University, Ames. pp. 421-459.

Helmers, G. A., M. R. Langemeier, and J. Atwood. 1986. An economic analysis of alternative cropping systems for east-central Nebraska. American Journal of Alternative Agriculture 1(4): 153-158.

Hunter, D. L., and L. H. Keller. 1983. Economic evaluation of alternative crop and soil management systems for reducing soil erosion losses on west Tennessee farms. Agricultural Experiment Station Bulletin 627. University of Tennessee, Knoxville.

Johnson, G. L. 1983. Ethical dilemmas posed by recent and prospective developments with respect to agricultural research. Paper delivered at American Association for Advancement of Science Annual Meeting, Detroit, Michigan.

Johnson, G. L. 1984. Academia needs a new covenant for serving agriculture. Agricultural and Forestry Experiment Station Special Publication. Mississippi State University, Mississippi State.

Jones, M. R. 1986. The effect of increased energy prices on the viability of organic agricultural systems on arable farms in the U.K. Agriculture, Ecosystems and the Environment 15: 221-229.

Kay, R. D. 1986. Farm management: Planning, control, and implementation. McGraw-Hill Book Co., New York, New York.

Kilkenny, M. R. 1984. An economic assessment of biological nitrogen fixation in a farming system of southeast Minnesota. M.S. thesis. University of Minnesota, St. Paul.

Langley, J. A., E. O. Heady, and K. D. Olson. 1983. The macro implications of a complete transformation of U.S. agricultural production to organic farming practices. Agriculture, Ecosystems, and the Environment 10(4): 323-333.

Leddy, M. G. 1987. An economic analysis of alternative farming systems in northeastern South Dakota. M.S. thesis. Economics Department, South Dakota State University, Brookings.

Lockeretz, W. 1987. Establishing the proper role for on-farm research. American Journal of Alternative Agriculture 2(3): 132-136.

Madden, P. 1984. Regenerative agriculture: Beyond organic and sustainable food production. *In* The farm and food system in transition. Extension Publication No. FS 33. Michigan State University Press, East Lansing.

Madden, P. 1987. Can sustainable agriculture be profitable? Environment 41(4): 18-34.

Madden, P. 1988a. Low-input sustainable agriculture research and education: Challenges to the agricultural economics profession. American Journal of Agricultural Economics 70(5): 1,167-1,172.

Madden, P. 1988b. Policy options for a more sustainable agriculture. pp. 134-142. *In* Increasing understanding of public problems and policies—1988. Farm Foundation, Oak Brook, Illinois.

Nadar, H. M., and G. E. Rodewald. 1980. Interaction between agronomic research and agricultural economic analysis to develop successful dryland cropping systems in Kenya. pp. 146-154. *In* C. L. Keswani and B. J. Ndunguru [editors] Intercropping: Proceedings of the Second Symposium on Intercropping in Semi-Arid Areas, Morogora, Tanzania.

Osteen, C. D., E. B. Bradley, and L. J. Moffitt. 1981. The economics of agricultural pest control. Bibliographies and Literature of Agriculture No. 14. Economic and Statistics Service, U.S. Department of Agriculture, Washington, D.C.

Richardson, J. W., and C. J. Nixon. 1986. Description of FLIPSIM V: A general firm level policy simulation model. Agricultural and Food Policy Center Bulletin 1528. Texas Agricultural Experiment Station, Texas A&M, College Station.

Rodale, R. 1983. Breaking new ground: The search for a sustainable agriculture. Futurist. 1: 15-20.

Ruttan, V. W. 1982. Agricultural research policy. University of Minnesota Press, Minneapolis.

Schuh, G. E. 1986. Revitalizing land grant universities: It's time to regain relevance. Choices 2nd Quarter: 6-10.

Walker, M. E., and E. R. Swanson. 1974. Economic effects of a total farm nitrogen balance approach to reduction of potential nitrate pollution. Illinois Agricultural Economics 14(2): 21-27.

Young, D. L., and W. G. Goldstein. 1987. How government farm programs discourage sustainable cropping systems: A U.S. case study. Paper delivered at University of Arkansas Farming Systems Research Symposium, Fayetteville.

# THE ECONOMICS OF SUSTAINABLE AGRICULTURAL SYSTEMS IN DEVELOPING COUNTRIES

Randolph Barker and Duane Chapman

History records the decline and disappearance of earlier civilizations that were unable to sustain agricultural production (Douglas, 1984). Since the time of Malthus there have been repeated warnings about the "limits to growth," first by economists (economics gained an early reputation as the dismal science) and more recently by environmentalists. However, scientific progress and new technology have always postponed the day of reckoning (Barnett and Morse, 1962).[1] There is now a faith among many that science and new technology will continue to remove the environmental constraints to growth. Like the story of the boy who cried wolf, there is a danger that society will discover too late that the wolf has already arrived.

What constitutes a sustainable agricultural system? From a macroeconomic perspective, sustainable agricultural systems must be viewed in the context of national, regional, and world agricultural development.

Douglas, writing on the "meaning of sustainability," distinguishes between sustainability as "food self-sufficiency" (an economic perspective),

[1]One of the most comprehensive investigations of this issue was conducted by Barnett and Morse (1962). The authors state that: "Malthusian scarcity no doubt has characterized many relatively primitive societies which possessed limited knowledge and skill. They not only failed to develop cultural taboos which stabilized population but also were able to extract only a small portion of services available in their natural resources. Thus, the limits of their resources were quickly reached.... Under primitive conditions of isolation, a relevant question is whether it is the limited availability of natural resources or the limited stock of knowledge which produces diminishing returns and inhibits economic growth.... Recognition of the possibility of technological progress clearly cuts the ground from under the concept of Malthusian scarcity." pp. 6-7.

"stewardship" (an ecological perspective), and "community" (a sociological perspective) (Douglas, 1984). Achieving food self-sufficiency is itself, by definition, an illusive concept. Meeting food demand in most economies in the developing world is a necessary but not a sufficient condition for take-off to sustained economic development. For example, in the mid-1980s, India actually exported food grains while millions of its citizens lacked the purchasing power to meet minimum basic needs.

## Sustainable Agricultural Systems: A Perspective

Sustainable development in a world economy implies a stable and satisfactory relationship between agricultural production and consumption. It implies a world population level or growth rate that is supportable on a long-term basis. It implies that negative products, such as hazards from pesticides, are controlled. Sustainability requires sufficient equity in access to production capacity and distribution to ensure political stability.

Agricultural sustainability has extraordinarily different implications in today's world than in even the recent past. Some traditional systems were able to sustain growth rates in agricultural production of 1 percent per year (Hayami and Ruttan, 1985). Prior to World War II, Japan was among the first countries to make use of chemical fertilizers. However, between 1918 and 1940 rice output in Japan grew at less than 1 percent, although the Japanese colonies grew at rates of about 2 percent. Agricultural production, slowed by the great depression and the drought in the 1930s, grew at 1 percent per annum in the U.S. between the two World Wars.

Following World War II, advances in medicine, health care, and nutrition greatly reduced mortality, leading to a population explosion in the developing world. Population growth rates of 2 to 3 percent became normal. The annual increase in food demand now ranges from 2 to more than 4 percent. This unprecedented expansion in food demand was met by expanding cultivated land and land under irrigation, and by increasing yield per hectare. Increases in yield per hectare in the order of 2 to 3 percent were made possible only through the rapid increase in use of purchased inputs, particularly chemical fertilizers. Not all of the developing world could achieve these growth rates. Some countries, particularly in Africa, have become increasingly dependent upon food imports as production per capita has declined.

The situation that the world now faces with respect to sustaining growth in agricultural production is described by Ruttan (1987) as follows:

"We are during the closing decades of the 20th century, approaching the end of the most remarkable transition in the history of agriculture. Before the beginning of this century, almost all increases in agricultural production occurred as a result of increases in the area cultivated. By the end of this century, there will be few significant areas where agricultural production can be expanded by simply adding more land to production. Agricultural output will have to be expanded almost entirely from more intensive cultivation in areas already being used for agricultural production."

Sustainable agricultural systems are those systems that support sustainable development in a world economy. Today's hope is to develop flexible agricultural systems with the capacity to incorporate new knowledge and increase yield per hectare by 2 percent or more with a system that maintains both the quality and production potential of the physical environment. At present, these growth rates can be achieved only by exploiting nonrenewable resources. During the time it will take to find a more sustainable long-run alternative, the production potential of the environment must not be destroyed. At the same time, enough must be invested in research to ensure that the technology is on hand to provide continuous gains in yield.

As with all processes of development, the question is one of balance in investments. There is always the danger that the effort to satisfy short-run demand will do irreparable damage to long-run agricultural production potential. This threat to the environment comes in the developing world from farmers in less favorable areas struggling to meet basic needs and from farmers in more favorable areas using modern inputs to sustain high growth in yields. The same threat comes in the developed world from consumers whose energy demands are depleting resources and polluting the environment at an alarming rate.

## The Asian Experience

The predominant food grain in Asia is rice. The wetbed-paddy rice culture practiced throughout Asia today had its origins in China and South Asia centuries ago (Barker et al., 1985). Irrigation was practiced widely in China in the second and third centuries B.C. Transplanting and fertilizing with manures were practiced in the early Christian era. Many of the tools used today, such as the combtooth harrow, were developed in the eighth to the 12th century. With the expansion of irrigation during the colonial period, the paddy rice culture spread widely. The development of irrigation, together with the newer technologies, permitted Chinese farmers between the 14th and mid-20th centuries to raise grain output in more or less equal measure by expanding the cultivated acreage and by raising yield per acre (Perkins, 1969). But population grew at less than one-half of 1 percent per annum.

The rice culture practiced throughout Asia today is remarkable in both its longevity and degree of homogeneity. With the exception of East Asia (Japan, South Korea, and Taiwan), which is largely mechanized, land is puddled with animal power and rice is transplanted, cultivated, and harvested by hand in much the same fashion throughout the region.

Adding new seed varieties and chemical fertilizers to this ancient system has made it possible to raise growth in yield per hectare from less than 1 percent to 2 to 3 percent in many countries. Although similar success has been achieved with wheat and a limited number of other crops in terms of geographic coverage, the so-called modern rice technology (making relatively minor management changes to a traditional system) has had the biggest impact in Asia.

However, much more was needed for this seemingly simple technological achievement in rice to succeed. It was necessary to distribute seeds and fertilizer and to provide credit, irrigation facilities, transportation, storage, and stable grain markets for the large increases in marketable surplus. All of these inputs are part and parcel of Asia's sustainable rice system.

Indonesia illustrates the problems that can occur with the growing dependency on agricultural chemicals in a developing country. Indonesia is regarded as one of the recent success stories in agriculture. In the late 1970s, Indonesia imported 1.5 million metric tons of rice per year but in 1985 exported one-half million metric tons. Rice production grew at 5 percent per year between 1968 and 1984, but roughly half of that growth is attributable to improved financial incentives generated by a massive fertilizer subsidy (Timmer, 1988). Farm-gate fertilizer prices were less than half of world prices, and fertilizer consumption grew at 25 percent per year.

Farmers in Indonesia also paid only 10 to 20 percent of the cost for pesticides, and the extremely low price led to widespread and heavy application (Repetto, 1985). The high rates of application in Java caused serious damage to the 1986 rice crop and created serious ecological problems, poisoning the breeding grounds for fish and shrimp in the coastal waters.[2] The heavy application of chemicals promoted the build-up of brown planthopper by destroying the predators of the planthopper and by encouraging the development of new planthopper biotypes.

The problems in Indonesia came as no surprise to rice scientists in Asia who have been breeding for varieties resistant to insects and diseases for more than two decades, and who have been researching ways to make more efficient use of chemical fertilizers and to find alternative sources for nitrogen

[2]International Rice Research Institute, News Release, "Indonesia Backs Beneficial Insects," Los Banos, Philippines: International Rice Research Institute, January 1987. The brown planthopper was said to have reached epidemic proportions in Central Java, resulting in a rice harvest shortfall of an estimated 100,000 tons.

since the energy crisis in the mid-1970s. In fact, much of the research at the International Rice Research Institute and elsewhere is described as "maintenance research" designed to sustain the recent gains in productivity.

Despite achievements of the past, the future of the Asian rice economy remains more uncertain than is commonly recognized. First, sustaining the current yield levels is not enough. Figure 1 shows the long-term yields of rice under experimental conditions in four locations in the Philippines. In all four locations, experimental yields over two decades have been either constant or declining. With the exception of hybrid rices that are now grown widely in China, there has been no significant breakthrough in the yield ceiling in Asia since IR8 was first released in 1966. There are already signs in parts of China and other intensively cultivated areas, such as Central Java, that we may be approaching these yield ceilings.

Another mainstay of output growth in Asia has been the expansion of irrigation. The downward trend in world grain prices, due in no small measure to surpluses generated through subsidized production in developed countries, has been a major factor discouraging investments in irrigation. Table 1 and figure 2 show the sharp decline in the 1980s in growth in new area irrigated and in irrigation investments by the World Bank and other major international lending institutions.

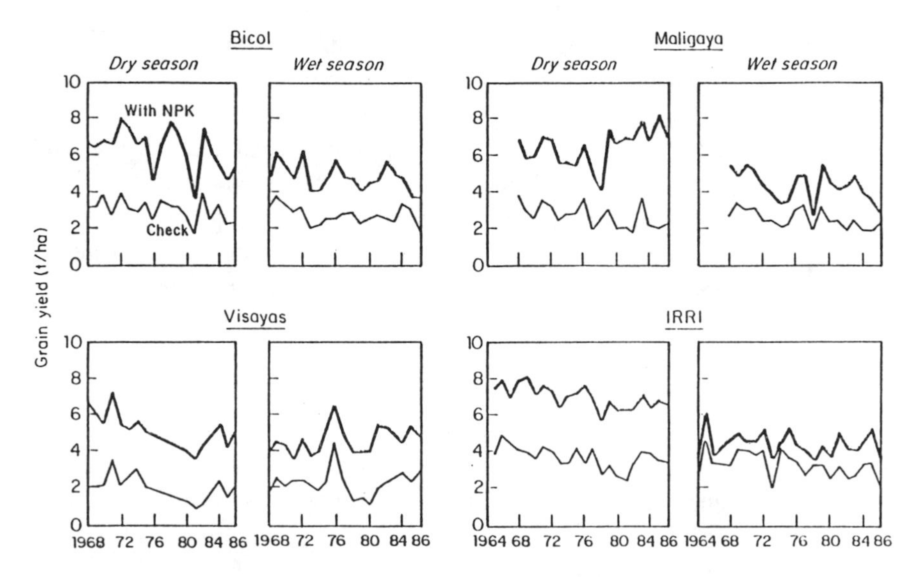


**Figure 1. Changes in yield response to NPK with successive croppings of improved rice varieties grown for 23 years (1964-1986) at IRRI and 19 years (1968-1986) at Bicol, Maligaya, and Visayas Experiment Stations (DeDatta et al., 1988).**

**Table 1. Compound growth rates of net irrigated area, world.**

| | *Rate* (percent)* | |
|---|---|---|
| *Period* | *World* | *Asia* |
| 1965-1984 | 2.0 | 1.6 |
| 1965-1969 | 2.2 | 2.5 |
| 1970-1974 | 2.3 | 2.1 |
| 1975-1979 | 2.5 | 1.9 |
| 1980-1984 | 0.9 | 0.7 |

Source: Food and Agriculture Organization.
*Computed from table 1 using semi-log regression techniques.

The sustainability of Asia's rice production systems depends upon a healthy irrigation system. The past emphasis on investment in physical structures (hardware) as opposed to management (software) means that many systems are not operating efficiently. Increased siltation because of loss of tree cover in the catchment areas reduces production potential in many irrigation systems. Waterlogging and salinity threaten the sustainability of agriculture in the Indus Basin.

Some major regions of Asia have benefitted little, if at all, from the Green Revolution. Of particular concern are the heavily populated, rice-dependent areas of Eastern India and Bangladesh. In most of these areas

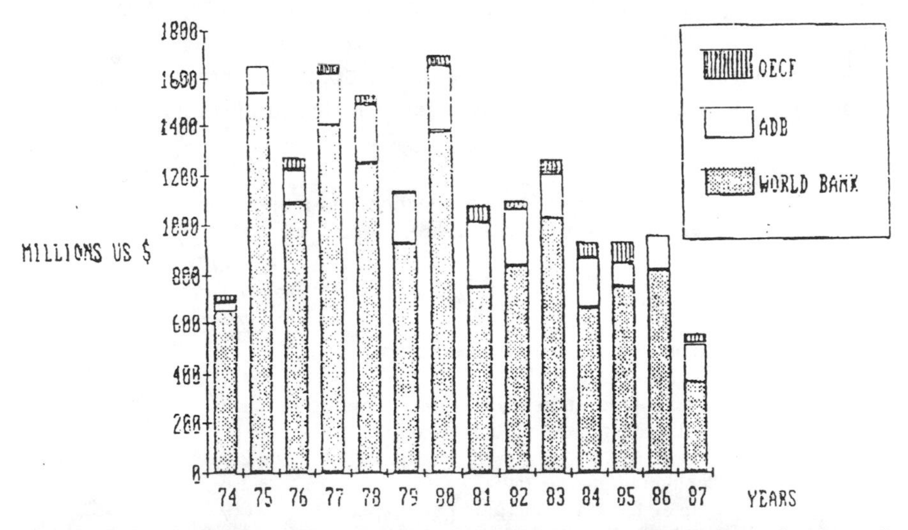


**Figure 2. World Bank, ADB, and OECF Irrigation Loans—Middle East, South, and Southeast Asia (1980 prices). Sources: World Bank, Asian Development Bank (ADB) and Japanese Overseas Economic Development Fund (OECF).**

of uncontrolled water, constant flooding, poor drainage, and drought, rice yields have remained at about 1.5 tons per hectare for almost three decades. Etienne (1985) writes about this region as follows:

"The advanced districts which are already highly productive cannot feed India forever. The future battles will increasingly be fought in the eastern plains where the untapped potential is enormous" (p. 147).

Yet the battle has scarcely been joined. The rivers that flow from the Himalayas are short, frequently change course, and are extremely difficult to harness. Massive investments are needed for the development of groundwater and the control of flooding. There is a chronic shortage of power for operating pumps. Any comprehensive, long-term plan for development of the Ganges Brahmaputra basin will require cooperation of the governments of India, Bangladesh, and Nepal.

## The Sustainability of Sub-Saharan African Agricultural Systems

Population growth in Africa has been more rapid than growth in food crop production (Table 2). The result has been a decline in staple food production per capita and a sharp rise in food imports to hold food consumption per capita almost constant. It would be necessary to double Africa's annual rate of growth in food production to attain regional self-sufficiency (Eicher, 1988).

It would be presumptuous to pinpoint the reasons for this problem or to indicate precisely how this rate can be doubled. A complex of factors appears to be involved.

In comparing regions of Africa with the upland or nonirrigated areas of much of Asia, the problems in increasing agricultural production appear

**Table 2. Growth of food production, consumption, and trade in Africa (Staatz, 1988).**

| | *Annual Growth Rates in Percent* |
|---|---|
| Production of all major food crops (1961-1980) | +1.7 |
| Cereals production (1961-1980) | +1.8 |
| Total human consumption of staple food | +2.5 |
| Population growth rate | +2.6 |
| Staple food production per capita | −0.9 |
| Staple food consumption per capita | −0.1 |
| Food exports (1966/70-1976/80) | −7.1 |
| Food imports (1966/70-1976/80) | +9.2 |

to be similar. For example, prior to the 15th century and the spread of paddy rice, a tree-crop, root-crop culture predominated in Southeast Asia. Today, the vestiges of that culture can still be found in the marginal and upland areas and on the islands of the South Pacific. The shifting cultivation crop rotations are remarkably similar to those of West and Central Africa. A crop or two of upland rice or maize is followed by a cassava or sweet potato crop and then several years of fallow. The problems faced by farmers—controlling emparada weed, shortening the fallow, and maintaining soil fertility without access to chemical fertilizers—are virtually the same in both regions. Food production has not increased rapidly in these areas of Asia and Africa.

One of the most convincing arguments for the failure of farmers in Sub-Saharan Africa to adopt modern varieties and purchased inputs is advanced by Binswanger and Pingali (1988). The person/land ratios in Africa have usually been lower than in Asia, invalidating the focus on high yields per acre as these farmers have sought high production per farmer.

However, this situation is changing rapidly. Some farming systems in Africa have been able to achieve sustainability under intense population pressure, as described by Okigbo (1989) and Hahn (1989). Goldman and colleagues at the International Institute of Tropical Agriculture are conducting research to understand how the process of adjustment to population pressure takes place.

Goldman (1988) lists ways in which farming systems on the high-density acid soil regions of Imo State in Southeastern Nigeria have adapted to the problems of generating increased food and income:

- Extending the margins of cultivation to previously unused land.
- Shortening the fallow period, increasing the intensity of land use overtime.
- Using soil amendments to increase productivity by enhancing the recycling the of soil nutrients in vegetation, crops, etc., and importing soil nutrients.
- Differentiating fields in relation to soil and crop management.
- Creating high-intensity production niches: compound farms and wetland development.
- Adopting crops less sensitive to existing soil conditions.
- Managing fallow vegetation and enhancing regeneration of soil productivity during the fallow period.
- Transferring land use entitlements that diffuse intensified land use.
- Importing food from other areas.
- Increasing non-farm activities for income generation.
- Out-migration from the area, seasonal or permanent.

This list is extremely useful because it is readily generalizable to other areas and because it allows for systematic examination of the opportunities and constraints for expanding food production in a given farming system, region, or country.

In much of the humid tropics where soils are notoriously infertile, migration has been a major form of adjustment to growing population pressure. Migration to urban areas has been encouraged by the lack of investment in rural infrastructure and the subsidization of cheap food imports for urban consumers. The unfavorable climate, matched by unfavorable policies for rural development, makes it impossible to generate the marketable surplus to feed the rapidly growing (6 percent per annum) urban population. It is estimated that by the year 2010 approximately half of Africa's population will reside in the cities (IITA, 1988, p. 11).

Recognizing that rapid growth in food production has been achieved by using chemical inputs, scientists in both Africa and Asia have attempted this approach on nonirrigated land in the humid tropics, hoping to find an alternative for shifting cultivation or at least to shorten the fallow period. At the experiment station at the International Institute of Tropical Agriculture, where neither capital nor access to purchased inputs is a constraint, researchers began clearing the land in 1967 and practiced conventional plow/harrow seedbed preparation with contour farming, hoping to maintain fertility with chemical fertilizers.[3] Since 1974, increasing acreages have been planted in a no-till system until, by 1985, only the root crops received conventional seed-bed preparation. The replacement of mechanical weed control with chemical weed control resulted in a dramatic decrease in soil erosion. However, despite improved soil conservation, crop yields continued to decline because of inadequate organic matter levels and soil compaction, and possibly phyrotoxicity (yield reduction linked to pesticide residues in soil and water). Beginning in 1980, the legume *Mucuna pruriens* was used in a rotation every second or third year. From 1980 onward, experimental yields of both cassava and maize have remained fairly constant but have shown no tendency to increase (Table 3). Furthermore, yields reported have not been adjusted for the fallow period.

Much has been learned over a period of two decades. However, whether the practices currently being followed at IITA constitute a viable alternative to the shifting cultivation widely practiced in this area is still open to question.

The mixture of trees with annual crops has been a traditional feature

[3]The remainder of the paragraph is based on "The Ibadan Farm as a Research Resource," a note prepared for management by the Farm Management Unit, International Institute of Tropical Agriculture, in 1987.

**Table 3. Yields of cassava and maize varieties from experiment conducted at IITA, 1972-1987.***

| Year | *Cassava Variety TMS30572* | Year | *Maize Varieties* | |
|---|---|---|---|---|
| | | | *TZB* | *TZ5RW* |
| 1973-1974 | 66.2 | 1972 | 4.3 | |
| 1974-1975 | 54.1 | | | |
| 1975-1976 | 35.2 | | | |
| 1976-1977 | 22.5 | | | |
| 1977-1978 | 40.5 | | | |
| 1978-1979 | 24.6 | 1978 | 3.1 | |
| 1979-1980 | 30.7 | 1979 | 6.7 | 4.4 |
| 1980-1981 | 15.6 | 1980 | 3.6 | 3.7 |
| 1981-1982 | 19.0 | 1981 | | 3.8 |
| 1982-1983 | 18.4 | | | |
| 1983-1984 | 21.8 | | | |
| 1984-1985 | 20.1 | | | |
| 1985-1986 | 17.1 | 1985 | 4.9 | 5.0 |
| 1986-1987 | 20.4 | 1986 | 3.7 | 4.8 |

Source: Cassava data: Root and Tuber Improvement Program, IITA, Ibadan, Nigeria. Maize data: Farm Management Unit, IITA, Ibadan, Nigeria.

*Data should be interpreted with caution. Experiments are not conducted at the same location on the farm each year. Beginning around 1980, mucuna is planted every second or third year to nurture soil organic matter, and yields have not been adjusted for this.

of farming systems in the humid tropics.[4] Alley farming with the use of *Leucaena* has been widely publicized in both Africa and Asia as a "new" technology with great potential for raising crop productivity. It has been tested in the experiment station and in farmers' fields with mixed results. The fast-growing *Leucaena*, with its high demand on labor and management, appears to be an appropriate technology for only a few farmers. *Leucaena* is very sensitive to acid soils, and in many areas where labor is in short supply, it has the potential of becoming an unwanted weed.

The wide range of differences in climate, soil, and labor in Sub-Saharan Africa and in upland Asia leads to a variety of farming systems. Technology strategies must take into account local differences. For example, the moist savannahs offer more potential than the humid tropics for creating the needed food surpluses. However, in these more favorable ecologies, inadequate transportation, markets, and facilities for supplying inputs and credit constrain the growth of agriculture. Eicher (1988) states the issue as follows:

"Food policy analysts must, by necessity, include both food demand and

[4]For an excellent review of alley farming, see B. T. Kang, L. Reynolds, and A. N. Atta-Krah, "Alley Farming." This publication, prepared by scientists from the International Institute of Tropical Agriculture and the International Livestock Center for Africa, was in draft form as of 1988.

supply issues in their analyses instead of assuming that Africa's food gap can be closed by action on the supply side, that is, stepping up food production. More research is urgently needed on food consumption, marketing, and food systems."

The growth of food production in the nonirrigated areas of Africa and Asia has been extremely slow—too slow to keep pace with the growth in food demand of 2 to 4 percent. Asia, however, has extensive areas of irrigated land, representing an investment in land infrastructure that has occurred over centuries. Africa is poorer, not only in terms of natural resources but also in terms of the institutions and infrastructure needed for rapid growth in agricultural production. It is likely to take at least a half century of major investments at extremely low rates of return (internal rates of return that the World Bank would find unacceptable), under a relatively stable political situation with favorable policies toward the rural areas, to lay the foundation for a truly productive African agriculture.[5] Countries such as Ivory Coast, Cameroon, and Kenya already show signs that programs of this nature can have a high, long-term payoff. Is such a long-term plan for Africa feasible? How will Africa feed itself in the meantime? Are these not the issues that must be addressed when considering the sustainability of African agriculture?

## Impact of Developed Country Policies

The policies and practices of developed countries with respect to resource use represent probably the greatest long-term threat to the sustainability of developing countries' agricultural systems. One such policy is the tendency

[5]The fact that many essential investments are unlikely to pay off in the short run or even in 10 to 20 years raises questions about the relevance of widely practiced benefit-cost studies for judging the feasibility of projects. Even within the World Bank, those who develop African projects recognize that some form of "creative" project development is required to get project approval. But the results are often projects with faulty design. For example, in the early 1980s a comprehensive study of the Gambia River Basin was accomplished by the University of Michigan for the U.S. Agency for International Development and the Gambia River Basin Development Organization. (See the University of Michigan Gambia River Basin Studies). For the irrigation feasibility study, the internal rate of return was projected to be 4 percent. The Gambia River Basin Development Organization rejected the report and hired a new consultant, who raised the estimated internal rate of return primarily by increasing the assumed potential irrigable area and the yield of rice to unrealistically high levels. Whether the project should be undertaken is an open question that probably should not be judged on the basis of the internal rate of return. But if the project is designed on the basis of the assumptions used by the second consultant, it is almost certain to fail.

of developed countries to subsidize agricultural production, treating agriculture much like a utility or, in other words, regarding a high level of national self-sufficiency as necessary and desirable. The effect on developing countries is to destroy potential export markets and encourage greater dependency upon developed countries' surplus food-grain exports. This slows the growth of agriculture and threatens the viability of developing countries' agricultural economies.

Of more widespread concern is the potential impact of the so-called greenhouse effect, due largely to the emission of gases into the atmosphere through vehicular and utility fuel consumption in developed countries. A report released by the United Nations (Jaeger, 1988), prior to any awareness of the magnitude of the drought of 1988, states that the impact of "greenhouse gases" is expected to be greatest in three general areas: (1) semi-arid regions of Africa where the hotter days would aggravate famine and drought; (2) humid, tropical parts of Asia where higher sea levels would increase risk of flooding; and (3) high latitudes of Alaska, Canada, and Scandinavia where more extensive ice thaws would complicate everything from marine transportation to construction practices.

As noted earlier, both developed and developing countries are moving rapidly toward dependence upon increases in yield per hectare as the major source of growth in food production. This means a growing dependency, at least in the immediate future, upon nitrogen fertilizers as the single largest category of energy use. Developed countries rely almost exclusively upon chemical fertilizers, and chemical fertilizers are rapidly replacing organic fertilizers in developing countries. In fact, agricultural systems that have been able to sustain growth rates of 2 to 4 percent per annum over a period of 2 to 3 decades have all relied upon purchased chemical fertilizers as a major source of output growth.

Agricultural production itself uses a modest 1 percent of the U.S. total annual energy consumption of 75 Q.[6] With world energy consumption at 300 Q, global agricultural energy consumption is probably much smaller than 3 Q. Why, then, should we be concerned?

The supply relationship for energy use in agriculture depends upon nonagricultural consumption of energy. The question is particularly important for petroleum. In 1987, proven world reserves were 700 billion barrels. In addition to proven reserves, recent geological estimates of un-

[6] Q is a quadrillion Btu or 252 trillion kcal. The 1 percent estimate is by Heady and Christiansen, 1984. International data are published in U.S. Energy Information Administration, International Energy Annual. Energy used to process, transport, refrigerate, and cook food is much greater than the conventional energy used in on-farm production. To raise corn, farm production is only one-seventh of the total energy requirement.

discovered oil are about 450 billion barrels.[7] At a world consumption of 20-25 billion barrels per annum, the supply of 1.2 trillion barrels will last about 50 years. Figure 3 shows one solution to the problem of pricing remaining oil resources and related depletion of reserves (Chapman, 1987). With a stable world population and no effect of rising Third World income level on demand, prices will stay near present levels (in real dollars) for many years and begin rising sharply in the next century (Figure 3a).

Most countries strive to emulate the high living standard in the U.S., which is sustained with an annual per-capita oil consumption of 24 barrels, most of it burned in transportation. If the current population of 5 billion were to obtain only half the U.S. consumption level, the global use would be 60 billion barrels annually and the supply would last only 18 years. Depending upon the actual future rate of population and income growth, oil reserves are likely to become scarce resources in less than the 50 years suggested in figure 3b.

An important generalization follows: Energy-intensive, high technology agriculture can continue to expand, perhaps for several decades. At some point in the next century, accelerating energy prices will require a new direction in production technology. What are the alternatives? Can today's alternative energy technologies provide substitutes for conventional oil and gas? No, because they are too costly. Synthetic gas from coal will cost $16 to $20 per 1,000 cubic feet. Synthetic gasoline from coal is equally costly, in the range of $2 plus per gallon production cost (Chapman, 1983).

Can biomass energy substitute for oil and gas in agriculture? Again, the cost seems prohibitive. Brazil has demonstrated that sugar-based ethanol can be the fuel basis for automotive transportation. But one cannot visualize technically an economy wherein tractors and trucks are manufactured with biomass energy and used on biomass farms with biomass fuel to produce liquid fuels for general nonagricultural use. The most widely used U.S. process today requires one gallon of conventional petroleum to produce one gallon of biomass ethanol (Chapman, 1983). Brazil's debt problem is caused in part by the massive subsidies necessary to support its sugar-based ethanol program.

Can coal or nuclear power replace oil and natural gas as an energy source for agricultural inputs? The answer is not clear. Increased use of global coal may create global environmental problems with respect to climate change, upper atmosphere ozone depletion, lower atmosphere ozone pollution, and acid deposition. Nuclear power in the 1980s is much more

[7]Proven reserve data are published in year-end issues of the Oil and Gas Journal. Geological estimates of undiscovered oil and gas are from Masters (1985).

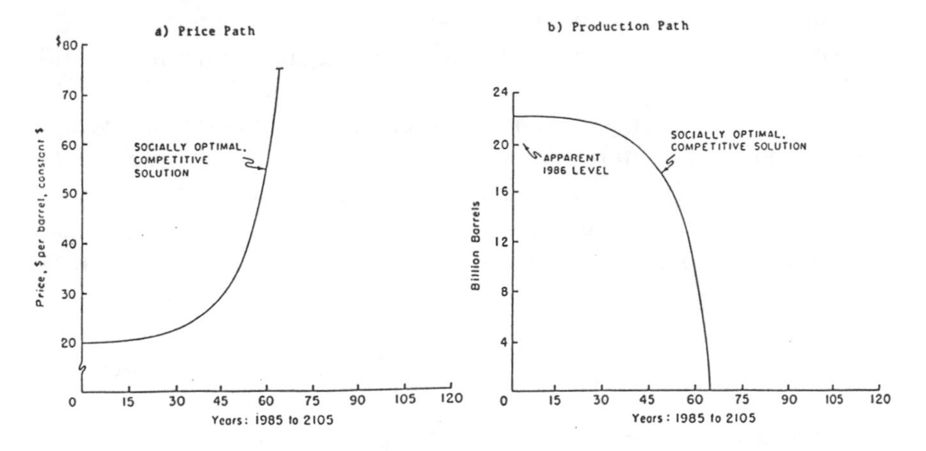


**Figure 3. Competitive world oil market.**

costly than conventional electricity sources or conservation investments.

Global climatic problems caused by the use of fossil energy and chemicals and the limited nature of world petroleum resources both imply the need for reduced dependency upon fossil-fuel based energy and agricultural chemical use. However, none of these alternatives is presently cost-effective.

One conclusion is that agricultural research will have to focus on obtaining high yields with less dependency upon conventional oil and natural gas sources. Recent advances in the biological sciences offer the greatest hope for developing the needed technologies. This kind of research, both basic and applied, is typically termed "biotechnology." In the broad generic sense, it includes traditional areas such as plant breeding, crop physiology, tissue and anther culture, biocontrol, and wide crossing with related species, as well as recombinant DNA and other new biotechnological approaches. Advances in biotechnology might make it possible to reduce the use of insecticides, enhance the nitrogen-fixing capacity of plants as a substitute for chemical fertilizers, improve the tolerance of certain plant species to stresses such as drought or cold temperature, and increase the biomass or yield potential of plants. However, major gains from biotechnology are still decades away and will require substantial investment in research.

## Conclusions

The sustainability of agricultural systems in developing countries must be viewed from a macro- or world-economy perspective. Population growth rates are projected to decline only gradually in the decades ahead. The world's population is becoming increasingly dependent upon yield per hectare as the major source of added food-output growth.

Asian agriculture has grown at an extraordinarily rapid rate since World War II. The capacity of Asian farmers to effectively utilize modern technology (seeds, fertilizer, and chemicals) is based upon centuries of development of land infrastructure and institutions. Nevertheless, the capacity to sustain these growth rates in the future is less certain than generally recognized. Furthermore, agricultural productivity has grown slowly, if at all, in major regions of Asia. Continued growth of Asian agriculture will require greater investment in research and infrastructure than has occurred in the past.

The situation in Africa is even more problematic. Evidence shows that the growth rate of African agriculture would have to double to achieve sustainability as defined here. Achieving such growth is not simply a matter of taking steps to increase food production. Markets and transportation networks must also be improved. Achieving a sustainable African agriculture will require major investments in research, infrastructure, and institutions over a period of at least half a century.

A major long-term threat to the sustainability of developing countries' agricultural systems comes from consumers in developed countries, whose energy demands are depleting resources and polluting the environment at an alarming rate. Because fossil fuel supplies are limited and because these energy sources perhaps do irreparable damage to the environment, alternative energy sources will have to be found. At present, there are no cost-effective alternatives.

The transition to a heavy reliance upon purchased chemical inputs as a major source of output growth, already occurring in much of Asia, threatens the environment in terms of pollution and health hazards. A clear understanding of the nature and magnitude of problems associated with the increased use of purchased inputs is needed in seeking ways to maintain productivity growth in these more favorable environments.

In the less favorable agroclimatic regions, where modern inputs cannot be used effectively, farmers are reducing the production potential of the environment in their efforts to meet food demands in the short run. More research and education are needed to assist farmers in finding ways to sustain or enhance the production potential of these environments in the face of increasing population pressure. Opportunities for enhancing environmental protection also may arise by changing ownership of resources—transferring public lands to private ownership.

Consumers in developed countries are becoming increasingly aware that the growing demand for energy poses a serious long-term threat to the global environment. Yet, given the potential magnitude of the problem, the research effort in this area is very modest.

Sustaining agricultural systems in developing countries requires a greater commitment to long-term investment in research, infrastructure, and institution building on the part of developed and developing countries. Unfortunately, declining world grain and oil prices in the short run have discouraged such investments. The energy crisis in the 1970s and the 1988 drought are reminders of the tenuousness of the situation. But memories are short, and thinking about agricultural development is dominated by short-run project analyses and the search for quick solutions to development problems. As Schultz (1987) states:

"The adverse consequences of the short view in economic policy carry a high price. Though theoretical elaboration of the short view is being made by economists with increasing subtlety, refinement, and elegance, it is nevertheless a structure built upon shifting sand."

## REFERENCES

Barker, R., R. W. Herdt, and B. Rose. 1985. The rice economy of Asia. Resources for the Future, Washington, D.C.

Barnett, H. J., and C. Morse. 1962. Scarcity and growth: The economics of national resource availability. Johns Hopkins University Press, Baltimore, Maryland.

Binswanger, H., and P. Pingali. 1988. Technological priorities for farming in sub-Saharan Africa. Research Observer 3(1): 81-98.

Chapman, D. 1983. Energy resources and energy corporations. Cornell University Press, Ithaca, New York.

Chapman, D. 1987. Computational techniques for the intertemporal allocation of natural resources. American Journal of Agricultural Economics 69(1): 134-142.

DeDatta, S. K., K. A. Gomez, and J. P. Descalsota. 1988. Changes in yield response to major nutrients and in soil fertility under intensive rice cropping. Soil Science: (in press).

Douglas, G. K. 1984. The meaning of agricultural sustainability. *In* G. K. Douglas [editor] Agricultural sustainability in a changing world order. Westview Press, Boulder, Colorado.

Eicher, C. K. 1988. Food security battles in sub-Saharan Africa. International Service for National Agricultural Research (ISNAR), The Hague, Netherlands.

Etienne, G. 1985. Rural development in Asia: Meeting with peasants. Sage Publications, New Delhi, India.

Goldman, A. 1988. Agricultural changes and variability in high density areas of Imo State, Southeastern Nigeria. International Institute of Tropical Agriculture (IITA), Ibadan, Nigeria.

Hahn, N. 1989. Compound and household farming: A sustainable system of African agriculture. pp. 504-514. *In* Clive A. Edwards, Rattan Lal, Patrick Madden, Robert H. Miller, and Gar House [editors] Sustainable agricultural systems. Soil and Water Conservation Society, Ankeny, Iowa.

Hayami, Y., and V. W. Ruttan. 1985. Agricultural development: An international perspective. Johns Hopkins University Press, Baltimore, Maryland.

Heady, E. O., and D. A. Christiansen. 1984. *In* D. Pimental and C. W. Hall [editors] Food and energy resources. Academic Press, Orlando, Florida.

International Institute of Tropical Agriculture. 1988. IITA strategic plan 1989-2000. IITA,

Ibadan, Nigeria.
Jaeger, J. 1988. Developing policies for responding to climatic change. WMO/TD No. 25. Environment Programme and World Meteorological Organization, United Nations.
Masters, C. 1985. World petroleum resources: A perspective. Open-File Report 85-24B, U.S. Geological Survey, Reston, Virginia.
Okigbo, B. N. 1989. Sustainable agricultural systems in tropical Africa. pp. 323-352. *In* Clive A. Edwards, Rattan Lal, Patrick Madden, Robert H. Miller, and Gar House [editors] Sustainable agricultural systems. Soil and Water Conservation Society, Ankeny, Iowa.
Perkins, D. 1969. Agricultural development in China, 1368 to 1968. Aldine Press, Chicago, Illinois.
Repetto, R. 1985. Paying the price: Pesticide subsidies in developing countries. Research Report No. 2. World Resources Institute, Washington, D.C.
Ruttan, V. W. 1987. Institutional requirements for sustained agricultural development. *In* Ted I. Davis and Isabelle A. Schrimer [editors] Sustainability issues in agricultural development. The World Bank, Washington, D.C.
Schultz, T. W. 1987. The long view in economic policy: The case of agriculture and food. Occasional Paper No. 1. The International Center for Economic Growth, San Francisco, California.
Staatz, J. 1988. Food supply and demand in sub-Saharan Africa. Department of Agricultural Economics, Michigan State University, East Lansing.
Timmer, C. P. 1988. Indonesia: Transition from food importer to food exporter. *In* T. Sicular [editor] Food price policy in Asia: A contemporary study of origins and outcomes. Cornell University Press, Ithaca, New York.

# THE ECONOMIC IMPACT OF SUSTAINABLE AGRICULTURE ON THE AGRICULTURAL CHEMICAL INDUSTRY

C.A.I. Goring

The term "sustainable agriculture" raises many questions. What is sustainable agriculture? How sustainable is it? How must it change as human populations grow? Does it imply partial or complete phasing out of agricultural chemicals, including both fertilizers and pesticides? These are all tough questions, and the answers could have a profound economic impact on the agricultural chemical industry.

Sustainable agriculture is a form of agriculture in which all of the mineral nutrients that are removed from the land during the production of food are returned to the land for use by the next crop. It is probably impossible to construct such a system on a large scale because of the nutrients lost as a result of erosion, leaching, and the dispersion of organic wastes generated by animals and humans. However, agriculture that is sustainable for a long period is possible and is being practiced in a number of different forms.

The shifting type of agriculture that was developed before the advent of modern agriculture, and is still being practiced in some parts of the world, is probably sustainable for a long time, but only at low population levels. As population levels rise, the landscape becomes denuded of forests, the climate changes and rainfall is apt to decrease, erosion becomes catastrophic, and the soils do not have sufficient time between crops to regenerate their natural productivity.

More advanced agricultural systems, such as those in India and China where biological nitrogen fixation is utilized, erosion is controlled, rotation and multiple cropping is practiced, and everything conceivable is done to return to the land the highest possible amount of nutrients removed in the harvest, have been sustainable for a long time and have supported much greater human populations. Similar systems in the United States and many other countries use leguminous crops to supplement the nitrogen needs of nonleguminous crops and involve animal production in conjunction with crop production (Council for Agricultural Science and Technology, 1980).

Recent trends in some countries have been toward crop monocultures and animal production at locations relatively remote from crop production. Farmers specialize and become more knowledgeable and efficient in their production practices. Yields increase. Cash flow increases so as to offset the increasing capital costs of farming. However, some loss of sustainability occurs because of the substitution of inorganic nitrogen fertilizers for legumes and the physical separation of animal waste from crop production, making use of animal waste on cropland uneconomical.

Evolution of these systems was accompanied by increased use of relatively cheap inorganic fertilizers and effective chemicals for pest control and plant growth regulation. Can this trend be reversed and still meet food production needs now and in the future? Food production today roughly matches world needs, but, unfortunately, the distribution system is less than perfect. There is starvation in many parts of the world and an embarrassing surplus in others. Current production levels are being achieved by using most good agricultural land and available water, pesticides, inorganic nitrogen fertilizers, and large quantities of phosphate and potash.

Currently, environmental and organic farming groups are advocating a shift to forms of agricultural production that involve much less use of inorganic fertilizers or pesticides. The consequences of such a shift have been outlined in some detail by the Council for Agricultural Science and Technology (1980). Because supplies of available organic matter are inadequate to meet the nitrogen needs of cash crops, a shift to legumes for hay would have to take place to generate additional nitrogen. Overall crop production would decrease—in some cases dramatically. What was produced would be subject to greater ravages by pests if pesticides were not used. Gradual depletion of the phosphorus and potash reserves of the soil would take place. A substantial amount of marginal land not now being used for food production would eventually have to be planted. Because much of this land slopes more than land currently in use, increased erosion would be expected. Prices of crop commodities and food would increase substantially.

## The Population Growth Issue

How can we meet the needs of an increasing world population and rising demand for food? There is no compelling evidence that the population of the world is stabilizing and will eventually be under control. The social, religious, and economic systems of most countries are not geared to a no-growth concept. Even China, the strongest proponent of zero population growth, seems to be loosening the reins in recent years. Our current world population of about 5 billion is projected to be about 6 billion by the year 2000 (Brown, 1988). The Worldwatch Institute is not optimistic that world population will stabilize before the end of the next century (Brown, 1988). By that time there may be more than 10 billion people. How will they be fed? There is strong doubt that new technologies in pest control and crop production that exclude the use of pesticides and inorganic fertilizers can accomplish that goal.

Ultimately, of course, there must be human population control. All attempts, however admirable, to solve the problems that are being caused by lack of population control—such as rapid depletion of mineral reserves, fossil fuels, and natural fuels, and intolerable pollution—are eventually doomed to fail unless control of human populations is achieved.

In any case, the concept of sustainable agriculture involving decreased use of inorganic fertilizers and pesticides is unlikely to survive increasing population pressure. The best that can be achieved is to minimize the potential hazards of their use.

## The Cyclic Farm Economy

The current interest in reducing the use of agricultural chemicals in the U.S. arises out of the present state of the cyclic farm economy (Council for Agricultural Science and Technology, 1988). When food supplies appear to be short and food prices start to rise precipitously, the political establishment becomes nervous and urges farmers to increase food production. Of course, farmers respond. They increase the use of fertilizers and pesticides to increase crop yields. Overproduction eventually occurs, prices fall, competition becomes fierce, and the less efficient food producers go out of business. The portion of the farm population responsible for most of the food production decreases as it has been doing for decades. It is now in the range of 1 to 2 percent of the U.S. population (Council for Agricultural Science and Technology, 1988). Under pressure from these farmers, the government steps in and provides farm subsidies, at the same time taking acreage out of production. The farmers also try to reduce their costs by decreasing their use of discretionary inputs, such as fertilizers and pesticides. Eventually, food production decreases and food supply may even become short in some cases. Commodity and food prices then rise rapidly, farmers increase their profit, and the cycle starts all over again. The process is not unlike what goes on in other cyclical industries. With each new cycle the peak of food production is higher in response to population growth.

Currently, farmers are suffering financial losses, trying to cut costs, and becoming more efficient. But this time around there is a difference. Environmental groups have advocated the reduction or elimination of pesticides for many years. Organic farming groups have advocated no use of inorganic fertilizers. Some members of these groups seem to have concluded that the time is ripe to minimize or even eliminate both of these production inputs. If these positions become part of the national policy, consumers must be prepared to accept a lower food production capability, substantially decreased quality and increased cost of food, and a reduced amount of food for export to countries that have shortages.

This probably will not occur because of the desires of the vast majority of farmers. Farmers, by and large, have one universal characteristic. They love to grow food—lots of it. They will not voluntarily cut back food production for long by giving up the use of pesticides and inorganic fertilizers. The real pressure will come from the environmental and organic farming groups, and those members of the general public that can be persuaded to support their view.

## Objections to Inorganic Fertilizers and Pesticides

Are there compelling reasons for eliminating the use of inorganic fertilizers and pesticides despite the adverse effect this would have on food production? The objections of the environmental and organic farming groups to these farming practices are based more on a philosophical approach to agriculture than on rational reasons for banning production inputs. As far as sustainability is concerned, phosphate and potash reserves might just as well be stored in soils instead of in mines, especially because they are vital for maintaining or increasing yields per acre. The production of inorganic nitrogen fertilizers does consume fossil fuels, but the amounts involved are very small compared to other uses of fossil fuels (Council for Agricultural Science and Technology, 1980). The production of pesticides requires far less fossil fuel than the production of inorganic nitrogen fertilizers (Council for Agricultural Science and Technology, 1980). The amounts used are going to get even smaller as current chemical pesticides are replaced with more active products and biological pesticides or pest control systems.

Other criticisms of inorganic fertilizers and pesticides are that they are too expensive. But the fact is that farmers have generally found them to be well worth the cost or they wouldn't buy them. Fertilizers and pesticides have also been called addictive, but that is a strange way to talk about the proven benefits of well-established technologies.

The most important attack on inorganic fertilizer and pesticide use is the allegation that they are especially harmful to humans and the environment. In the case of inorganic fertilizers, the perceived hazard is nitrogen. In the soil, ammonium fertilizers convert to nitrates that are easily leached (Council for Agricultural Science and Technology, 1985). Because farmers are continuing to increase application rates of nitrogen fertilizer in search of higher yields, unacceptable nitrate contamination of food and water is becoming a distinct possibility.

## Rebuttal to the Criticisms

Many approaches to this problem can lessen such hazards (Council for Agricultural Science and Technology, 1985). Some of the methods, such as proper placement of the fertilizer, proper timing of application, and efficient irrigation practices, are directed primarily toward achieving maximum possible uptake of nitrogen by the plant. Three methods in particular are also directed toward minimizing nitrate levels in the crop: (1) the selection of optimum fertilizer rates, (2) the selection of crop varieties that utilize nitrogen throughout the growing season, and (3) the use of nitrification inhibitors to maintain ammonium nitrogen in the soil for longer periods.

The initial concern about pesticides was the potential damage to wildlife. The organo-chlorine insecticides that have caused most of the damage have been eliminated for the most part. The vast majority of the remaining pesticides do not cause significant loss of wildlife directly. Those that do are being replaced with more environmentally suitable materials. Wildlife damage is certainly not a problem that warrants a categorical ban on the use of all pesticides. The real problem for wildlife is the expanding destruction of their habitats by humans, not pesticides.

The next issue to receive emphasis was the toxicity of pesticides to humans. Some pesticides are highly toxic and represent a threat to human health unless adequately controlled and properly used. However, the vast majority are not unusually toxic. The Thomson publications (Thomson, 1985-1986, 1986, 1986-1987, 1988) list 617 active ingredients that are acutely toxic to rodents. The LD 50s reported were the lowest figures (highest toxicities) observed. Of those listed, 5.8 percent had LD 50s of 10 milligrams per kilogram or less, 6.8 percent had LD 50s of 10 to 50 milligrams per kilogram, 10.7 percent had LD 50s of 50 to 200 milligrams per kilogram, 22.5 percent had LD 50s of 200 to 1,000 milligrams per kilogram, 21.1 percent had LD 50s of 1,000 to 3,000 milligrams per kilogram, and 33.1

percent had LD50s above 3,000 milligrams per kilogram of body weight. The LD 50s for salt, aspirin, caffeine, nicotine, and vitamin D, respectively, are about 3,000, 1,000, 200, 50, and 10 milligrams per kilogram of body weight (Gleason et al., 1969; Goring, 1981; Ottoboni, 1984; Peters, 1967; Thomson, 1985-1986; Worthing, 1983). In other words, 33.1 percent of pesticides were less toxic than salt, 54.2 percent were less toxic than aspirin, 76.7 percent were less toxic than caffeine, 87.4 percent were less toxic than nicotine, and 94.2 percent were less toxic than vitamin D. A wide variety of poisons occur naturally in plants, animals, and microorganisms (Ames, 1983; Ames and Gold, 1988; Ames et al., 1987; Council for Agricultural Science and Technology, 1979; Goring, 1972; Maclean and Davidson, 1970). Many of them are very toxic in comparison with most pesticides, and some of them are much more toxic than any of the pesticides currently being used (Goring, 1972). But pesticide technology should not be abandoned because of erroneously perceived high acute toxicity, particularly because most active ingredients are formulated to reduce their acute toxicities further.

Reported accidental deaths because of acute toxicity of pesticides are fewer than 30 per year in the U.S. (National Safety Council, 1987). In contrast, accidental motor vehicle deaths are in the range of 45,000 to 50,000 each year.

The estimated number of injuries treated in hospital emergency rooms each year that patients say are related to industrial products is about 10 million (National Electronic Injury Surveillance System, 1987). Pesticides are responsible for about 0.1 to 0.15 percent of these injuries, whereas legal drugs are responsible for about 10 times that number. The incidence rate for occupational injury is lower in the agricultural chemical industry than in 75 percent of all manufacturing and service businesses in the U.S. (National Safety Council, 1987).

All of these data pertaining mostly to acute toxicities hardly suggest a pesticide crisis relative to other technologies that can be solved only by wholesale elimination of pesticides.

What about the chronic toxicity of pesticides? Permissible residue levels are set at about a maximum of 1/100 of the no-effect level in lifetime animal studies. Actual levels are much lower (Council for Agricultural Science and Technology, 1987; National Research Council, 1975). Each person consumes about 40 milligrams of a mixture of pesticides each year (National Research Council, 1975). This mixture has an estimated LD 50 of no less than 100 milligrams per kilogram. It is equivalent in toxicity to about one aspirin pill, a cup of coffee, or the amount of salt people ingest each day (Dempewolff, 1975; Gleason et al., 1969; Goring, 1981; Ottoboni, 1984;

Peters, 1967).

The safety margins for pesticides far exceed those for many of the naturally occurring chemicals that are consumed. Seafood, spices, and vegetables contain a wide variety of naturally occurring toxins, such as goitrogens, estrogens, tumorigenic and carcinogenic agents, lathyrogens, hemagglutinins, cyanogenitic glycosides, stimulants, depressants, antienzymes, and chollinesterase inhibitors, quite often at levels that are mildly toxic (Goring, 1972). Nutritionists generally recommend a well-balanced diet consisting of a wide variety of foods. One of the virtues of following such a recommendation is avoidance of excessive consumption of individual foods at levels that may be toxic because of the naturally occurring toxic chemicals they may contain.

There has also been concern about the potential carcinogenicity of pesticides. It is true that about half of the pesticides tested are regarded as carcinogens. It is also true that about half of the chemicals tested, whether manmade or natural, are also regarded as carcinogens (Ames and Gold, 1988; Ames et al., 1987). It should be noted that the tests are conducted at dosages that are chronically toxic to the animals (Ames and Gold, 1988; Ames et al., 1987; Council for Agricultural Science and Technology, 1978). The more rapid cell division that occurs under these conditions increases the likelihood of genetic errors and cancer. At dosages that are not chronically toxic, no increase in cancer would be expected and none observed if the chemicals are not inherently genotoxic. The chemicals of greatest concern are those that are genotoxic. Few pesticides—and certainly not any of the new ones being developed—are genotoxic.

Ames (1983) has shown that the human diet contains a great variety of natural carcinogens. The contribution of pesticides to human cancer is negligible because the levels in food are so low relative to natural carcinogens (Ames and Gold, 1988; Ames et al., 1987). Using an index that incorporates the relative carcinogenicities and concentrations, Ames and Gold (1988) and Ames et al. (1987) have estimated that the contribution of natural chemicals to human carcinogenicity is at least 10,000 times greater than that of pesticides.

The latest concern is the presence of pesticide residues in groundwater (Council for Agricultural Science and Technology, 1985; Holden, 1986). The concentrations that occur are generally in the low parts-per-billion range. Standards should be set for residues in groundwater, as has been done for residues in food. It is a simple fact that any pesticide used will eventually appear in groundwater if sufficient irrigation water is applied or sufficient rain falls on the soil. The residues that occur can be minimized by confining the use of persistent pesticides that are easily leached to areas where rainfall or irrigation is not high, lowering as much as possible the application rates of all pesticides, and replacing older pesticides with newer ones that are applied at much lower rates. In any case, the current levels of most pesticides in groundwater are generally so low that they constitute a substantially lower risk than the current miniscule risk from pesticide residues in food.

It would appear that the alleged hazards of pesticides and inorganic nitrogen fertilizers are overstated substantially and should not be used to justify the adoption of new production practices that are perceived to be desirable for the long-term future of agriculture. These new practices should be required to stand on their own economic merit in competition with current production practices.

If these alleged hazards are successfully used to justify the wholesale and rapid elimination of pesticides, it would be an economic disaster for the agricultural chemical industry, the farmer, and the consumer. This is not likely to happen. The industry will continue to lose older pesticides and replace them with new and better products. Furthermore, the agricultural chemical industry is investing in seed companies and developing genetically engineered crops resistant to pests. The industry is also developing genetically-engineered organisms designed to control pests. If these products survive the current legal and regulatory battles over genetic engineering, they should appear in the marketplace. It is difficult to predict what fraction of the pesticides being used will eventually be replaced, but the amount certainly could play an important role in agriculture in the future.

## Conclusion

Not all of the companies in the agricultural chemical business will prosper as chemical pesticides continue to evolve and biotechnology makes greater contributions to pest control. Many companies are simply too small to afford the enormous costs of research and development needed to invent and launch products on a worldwide scale. Consolidations and mergers have been taking place and will continue as the weaker companies decide to exit the business. The agricultural chemical industry will eventually have fewer but far larger and stronger companies that will service agriculture worldwide. The demand for effective pesticides and pest control strategies will continue to grow as world population grows and additional food is needed—in spite of the many doubts about the long-term desirability of this technology.

## REFERENCES

Ames, B. N. 1983. Dietary carcinogens and anticarcinogens. Science 221: 1,256-1,264.

Ames, B. N., and L. S. Gold. 1988. Carcinogenic risk estimation: Response to S. S. Epstein and J. B. Swartz. Science 240: 1,045-1,047.

Ames, B. N., R. Magaw, and L. S. Gold. 1987. Ranking possible carcinogenic hazards. Science 236: 271-279.

Brown, L. R. 1988. State of the world. W. W. Norton and Company, New York, New York.

Council for Agricultural Science and Technology. 1978. Identification, classification, and regulation of toxic substances posing a potential occupational carcinogenic risk. Comments from CAST, January 27, 1978. Ames, Iowa.

Council for Agricultural Science and Technology. 1979. Aflatoxin and other mycotoxins: An agricultural perspective. Report No. 80. Ames, Iowa.

Council for Agricultural Science and Technology. 1980. Organic and conventional farming compared. Report No. 84. Ames, Iowa.

Council for Agricultural Science and Technology. 1985. Agriculture and ground water quality. Report No. 103. Ames, Iowa.

Council for Agricultural Science and Technology. 1987. Diet and health. Report No. 111. Ames, Iowa.

Council for Agricultural Science and Technology. 1988. Long-term viability of U.S. agriculture. Report No. 114. Ames, Iowa.

Dempewolff, R. F. 1975. The truth about coffee and your health. Science Digest 77: 30-36.

Gleason, M. N., R. E. Gosselin, H. C. Hodge, and R. P. Smith. 1969. Clinical toxicology of commercial products. Williams and Wilkins Co., Baltimore, Maryland. pp. 209-214.

Goring, C.A.I. 1972. Agricultural chemicals in the environment: A quantative viewpoint. pp. 793-863. *In* C.A.I. Goring and J. W. Hamaker [editors] Organic chemicals in the soil environment. Volume 2. Marcel Dekker Inc., New York, New York.

Goring, C.A.I. 1981. Need for, use, and contribution made by pesticides and the sustainability of the pesticide supply. pp. 1-9. *In* Proceedings, 28th Annual Meeting of the Canadian Pest Management Society.

Holden, P. W. 1986. Pesticides and groundwater quality. National Academy Press, Washington, D.C.

Maclean, G. J., and J. H. Davidson. 1970. Poisonous plants a major cause of livestock disorders. Down to Earth 26(2): 5-11.

National Electronic Injury Surveillance System. 1987. Product summary report. U.S. Consumer Product Safety Commission, National Injury Information Clearinghouse, Washington, D.C.

National Research Council. 1975. Pest control: An assessment of present and alternative technologies. *In* Contemporary pest control practices and prospects: The report of the executive committee. National Academy of Sciences, Washington, D.C.

National Safety Council. 1987. Accident facts, Washington, D.C.

Ottoboni, M. A. 1984. The dose makes the poison. Vincente Books, Berkeley, California.

Peters, J. M. 1967. Factors affecting caffeine toxicity. Journal of Clinical Pharmacology 7: 131-141.

Thomson, W. T. 1985-1986. Agricultural chemicals. Book 1. Insecticides. Thomson Publications, Fresno, California.

Thomson, W. T. 1986. Agricultural chemicals. Book 3. Fumigants, growth regulators, repellents, and rodenticides. Thomson Publications, Fresno, California.

Thomson, W. T. 1986-1987. Agricultural chemicals. Book 2. Herbicides. Thomson Publications, Fresno, California.

Thomson, W. T. 1988. Agricultural chemicals. Book 4. Fungicides. Thomson Publications, Fresno, California.

Worthing, C. R. 1983. The pesticide manual—7th edition. The British Crop Protection Council, Croydon, England.

# COMPOUND AND HOUSEHOLD FARMING: A SUSTAINABLE SYSTEM FOR AFRICAN AGRICULTURE

Natalie D. Hahn

*Our conviction is that, despite errors of the past, it is still possible for Africa to become self-sufficient in food production. But there is a prerequisite: Africa must conceive of an authentic development strategy which takes into account our experiences, our failures, and our successes.*

Seyni Kountche, late President of Niger

*In the primeval pastures, streams ran clear, and clean winds blew. And nomad man came to follow the herds, adapting to the prairie as it was.*

Grassland of Nebraska (Wilson and Wilson)

For most farmers of any continent, the immediate question about modern-day agriculture is not one of sustainability but of survival. Many scientists have become used to the term "farming systems"—which is finally understood by international agencies, donors and, most important, many governments. Gender and nutritional factors are being incorporated into a holistic system. This results in another definition, a new paradigm, and a holistic perspective. Any rethinking about sustainability and a new emphasis is important because it shows a dynamic and evolving shift in perception. But a problem lies in continual intellectual conceptualization without equal energies to ensure testing the technologies and adopting them in some cases, as well as fine channeling research and development for the ultimate impact. There appears to be an over-emphasis of terminology without a revolutionary and substantive change in what is being promoted.

## Low Levels of Adoption of Improved Technologies

The actual transfer and adoption levels of all the research undertaken in sub-Saharan Africa is low. No more than one percent of African rural households use an improved crop variety. There are certainly exceptions, and Zimbabwe provides one exciting example. Eicher (1988) reports that 100 percent of the commercial farms and 80 percent of the communal farms use the improved hybrid maize varieties. In 1986, 50 percent of the surplus in hybrid maize was from the communal farms. In many instances, national and international researchers do not believe that research is ready to leave the experimental area—it must be perfected. Only limited attempts seem to be directed at bridging the research and development stages to ensure the adoption of technologies.

Technology testing is rarely done with a family focus. Women invariably respond differently than male farmers to new technologies. On a conservative basis, an estimated 60 percent of the food in sub-Saharan Africa is produced by women, yet their opinions are often not considered in developing the technologies. The family farm in Nigeria or in Nebraska involves a dependency upon all household members. A 1986 survey in Oyo state, Nigeria, conducted by UNICEF and the International Institute of Tropical Agriculture (IITA) on cassava as a survival crop reported that women undertake 47 percent of the work required in all stages of cassava production and use; children complete 25 percent of the work (particularly in the peeling of cassava), and men provide 28 percent of the labor requirements.

## Post-Harvest Technologies and Utilization

If the post-harvest, nutritional, and garden factors are considered, today's agriculture appears to be at the same stage in terms of sustainability as farming systems were about five years ago. The main concentration is on agronomic production and ecology but little on the post-harvest stages, gender, and nutritional factors. The nutritional situation in Africa has not improved since the early 1970s. According to the Food and Agricultural Organization (FAO), only six of 42 countries had a per-capita food level exceeding 2,400 calories per day in 1985, compared to only one country in 1970 (FAO, 1985a). In sub-Saharan Africa, the average food intake has fallen from 2,109 calories per capita per day in 1970 to 2,093 calories per-capita per day in 1970 to 2,093 calories per capita per day in 1985. Eight countries (Angola, Chad, Ethiopia, Ghana, Guinea, Mozambique, Rwanda, and Sierra Leone) registered levels of less than 2,000 calories per capita per day in 1985 compared to six countries in 1970 (FAO, 1985a). The UNICEF (1988) "State of the World's Children" report states that "undernutrition was a contributing cause in perhaps one-third of all child deaths in the world last year." The report is poignant because each week more than a quarter of a million young children die from infection and malnutrition in the developing world.

The post-harvest stages of processing, storage, marketing, and utilization require more attention in any definition of sustainability. For instance, the varieties improved by the IITA have achieved production goals; yet the bitter (high cyanide) cassava requires more women's labor than any other food crop. In Nigeria, five days for processing and/or fermentation are necessary to decrease HCN to an acceptable level. In many parts of Zaire, the process of making chikwangue (a cassava-based product) takes two weeks. In periods of economic stress or immediate food requirements, these stages are shortened and the food product can be perilous to family members, particularly young children who suffer from malnourishment or low levels of sulphur intake (Rosling, 1987). How can health hazards be assessed in terms of carbohydrate sustainability? What are the dangers to women inhaling fumes as they complete the final stages of frying cassava into gari, as is done in many Nigerian states?

What are the nutritional values of a sustainable system? Can a cassava product containing one percent protein be compared with soybeans that have 40 percent? How can a sustainable system that has the highest potential of nutrient and carbohydrate levels be assessed? Surely the first priority for sustainability must be based on year-round food security at the household level.

## Is Sustainability Achievable Considering the Debt Situation?

How can sustainability be achieved considering the levels of debt of many African countries? Some 30 sub-Saharan countries' debts are more than three times greater than their annual export earnings. The significance of this can be appreciated better by noting that on average these countries must allocate 40 percent of their export earnings to servicing debts. According to the United Nation's Economic Commission for Africa (ECA), debt payments totaled about $30 billion in 1985-1986. The annual cost of servicing Africa's debt is about $12 billion, which is enough to take half the region's total export earnings, according to a study by the Council on

Foreign Relations and the Overseas Development Council (1985).

Important factors in a sustainable system must be national self-sufficiency and cropping systems that are not only biologically and economically sound but also provide sufficient food for family households without an external debt burden.

## Indigenous Knowledge

Farmers have an intellectual capability to use a survival system that incorporates sustainability. As stated concisely by Chambers (1980): "Scientists must recognize there is a parallel farmer system of knowledge to their own, which is complementary, usually valid, and in some respects superior."

Take the example of cassava production and the selection of varieties. Most African households plant between six and eight varieties. Their maturity dates can range from between seven and eight months up to 24 or 36 months. Cassava is a year-round basic survival crop for 160 million people—about 40 percent of the total population in sub-Saharan Africa. In most households both bitter and sweet, and high- and low-cyanide, varieties are planted and consumed. A 1986 survey in Bas-Zaire, Zaire, indicated 20 bitter and 18 sweet cassava types in that region. Southwestern Nigeria has approximately 12 indigenous varieties, of which three are low-cyanide varieties. There is a historical understanding of these crops. The Yoruba names often indicate who brought the cassava to the area or its characteristics. One variety, known as "isunikankiyan," translates: "It is not only yam which can be pounded." The value of the multicropping system with several cassava varieties is an indigenous contribution to both sustainability and survival.

Yet, the vast majority of cassava research funds are applied to the bitter cassava, without considering the work requirements or health hazards to women and children. IITA has completed only limited research on sweet cassava varieties, although sweet cassava requires no processing and fermentation by rural women and can serve as a basis for a multitude of food products and recipes.

Indigenous knowledge can and often does lead to sustainable systems. One example is the compound backyard garden. During periods of war, economic stress, or intensive population pressures the compound farm is a sustainable system.

## The Compound Farm as an Indigenous Survival System

Many characteristics help define a compound farm system. It is physically distinguishable by:

- Continuous cultivation.
- Proximity to the household.
- Diversity of plant species and their density.
- Use of multistoried vegetation.
- Plant boundaries.
- Uses of dried plants for crafts and berries for dyes.
- Presence of livestock.
- Establishment of nurseries with seedlings and experimental plots.
- Presence of post-harvest processing and consumption activities.
- Soil-fertility regeneration activities.
- Supra-seasonal harvesting and early maturity.
- Lack of fallow.
- Lack of burning.
- Use as an experimental area and depository for genetic material.

Compound farm systems are socioeconomically distinguishable by their:

- Land tenure patterns.
- Continuity over time.
- Diversity of labor activities.
- Potential for continuous income generation.
- Contribution to human nutrition.
- Contribution to human welfare through construction, material, fuel, and medicines.
- Use as a depository for capital goods—buildings, livestock, and storage of materials.

A compound farm system is common in certain parts of all humid and sub-humid tropics. However, there have been few systematic studies of the compound farm as a survival system during periods of stress. The overall objective of our 1985 study was to study the composition, structure, and nutritional importance of the compound relative to the other outlying fields and food crops in the farm management system. A second part of the study was to test the objective put forth in the 1970s that the importance and intensity of the compound increases with population density (Lageman, 1974). How is a compound defined? This study distinguished a compound as: "A distinct area adjacent to the house where continuity is physically and socioeconomically safeguarded by continuous cultivation of densely planted, diverse plant species grown to a variety of heights to optimize use of sunlight, shade, and moisture. This environment is manipulated by successive generations of the farm family to maximize output of nutrition, welfare, and income through a variety of production and consumption activities."

The compound is a sub-system of farming systems in the tropics; yet it has been understudied and undervalued. The compound is usually composed of plants labeled by the Food and Agriculture Organization (FAO) as traditional food plants or minor crops.[1]

Research on compound farming has been done by the International Potato Center (CIP). Ninez (1985) defines household gardens for this research at CIP: "Garden cropping is distinguished from arable cropping by the following features: (1) cropping those plants for personal consumption that cannot be collected or supplied by arable farming; (2) small plots; (3) proximity to the house; (4) fencing; (5) mixed or dense planting of a great number of annual, semipermanent, and perennial crops; (6) high intensity land use; (7) land cultivated several times a year; (8) permanence of cultivation; and (9) cultivation with hand implements."

Soemarwoto (1987) explained the virtues of homegardens with stable yields, varied products, continuous or repeated harvests during the year, and low inputs. He states, "[Homegardens] are never static but are capable of responding to new opportunities or adapting to new conditions."

Another example is the garden program of the Asian Vegetable Research and Development Center, which aims for maximum yield of selected nutrients. Designs of the gardens included different crops for three types of gardens—school, home, and market. By the third year of the project, the school garden, 10 meters by 18 meters, provided 142 children with a half-cup of vegetables per day (57 crops). The homegarden yields and nutritional outputs showed the value of a 4-meter by 4-meter garden for a family of five. The market garden aimed to increase net income from a small family farm by more than 30 percent (Gershon et al., 1985).

The compound plants provide an array of basic requirements for the family, including sources of oils and fats, condiments and spices, drugs, structural materials, beverages, animal feed, boundary markers, masticants and stimulants, fuelwood, shade protection of homestead privacy, and facilities for religious and social functions. As explained by Okigbo, the complexity of the compound decreases as one travels from the humid tropics to the drier savanna regions (Okigbo, 1985).

A study of the compound farm in the states of Imo and Anambra involved interviews with people from 49 households. The study was preceded by a literature review. An interdisciplinary team of 19 men and women

[1]The compound plants are called neglected food plants. Many of them are endangered species. Why do these features make the compound plants so important? Consider that the National Academy of Sciences estimates that throughout history about 3,000 plant species have been used for food and at least 150 of them have been commercially cultivated to some extent. However, over the centuries the tendency has been to concentrate on fewer and fewer plant species. Today, most of the people of the world are fed by about 20 crops (National Academy of Sciences, 1975).

completed the exploratory survey in a two-week period. Each compound was measured, and a description of each plant was given. A modeling of the compounds was completed according to (a) population density that ranged from fewer than 300 persons per square kilometer to more than 2,000 persons per square kilometer, (b) ecological zones, and (c) density of plant species and number of livestock. A categorization was made for trees and shrubs, arable crops, plants, and animals.

The general trends and conclusions of the study are as follows:

■ Every farm surveyed had a compound of some type. The range in size was from 0.47 hectare to 1.8 hectares. The average size of the compound was .75 hectare.

■ The size of the compound clustered around 20 percent of the total cultivated land. The trends indicated that when the overall cultivated farmland acreage was the largest, the complementary compound was among the largest of the compounds surveyed.

■ The higher the population density was, the more intense was the compound system. In low-population areas compound farming was the least significant because land is plentiful enough to allow extensive land holdings where bush fallow rotations are practiced. In some of the villages where the population density was the greatest, no land was kept fallow. At the other end of the spectrum was an area near Owerri where only 21 percent of the land was under cultivation.

■ A total of 146 plant species was identified. The number of plant species identified on any compound ranged from 18 to 57. The average was 35 plant species. The majority of farms had a core of about 25 plants.

■ Almost all of the compounds maintained some poultry, but the overall investment in goats was higher. Eighty-five percent of the households in Imo maintained poultry; 94 percent in Anambra had poultry. Also in Anambra, 85 percent of the households had sheep and goats. In Imo, only 60 percent had sheep and goats.

■ In many instances, the compound was managed primarily by women. Commonly, there were three generations on each compound and average household size was 18 persons. Women outnumbered men considerably in the age bracket between 18 and 50.

■ Most of the food nutrients for family consumption came from the compound. Some products were sold to provide cash. In Anambra, between 20 percent and 75 percent of the food items consumed by the household were from the compound.

■ Animal manure was used on their compound by 80 percent of the households. Among the households surveyed in Anambra, 90 percent of the chemicals were used on the outlying fields and only 10 percent were placed on compound plants. Systematic mulching of the compound is done with plant peelings, residues, and bone and human wastes. The high degree of soil fertility is almost by accident rather than by design.

■ There is a delicate and complementary balance between the compound and the outlying fields. The most intensive period of the compound occurs between the first and second season crops in the outlying fields. The timeliness of supplies from the compound minimizes the effects of food shortages during the hungry periods.

## Researching and Developing Compound Systems

Future interest in this project falls into three categories: further research, information and knowledge, and future interventions. These are not discrete categories, but they support time frames for short-, medium-, and long-term interests for a renewed analysis of factors affecting sustainability.

***Further Research.*** Both the physical and socioeconomic environments require further research. Time-series monitoring labor, income, and nutrition during and after project stages would provide a nonformal baseline to distinguish appropriate "levels" for later interventions.

Short-term studies of socioeconomic factors are needed for knowledge about the economics of soil fertility regeneration, labor migration, rural women's labor responsibilities, the importance of alternative income-generating strategies, macro-level market trend analysis, the potential for processing selected tree crops at the village level, and alternative forms of capital investment for rural households.

Studies of the physical components of compound farms would focus primarily on soil fertility regeneration practices and their impacts on the system, plant symbiosis in multistoried regimes, and the potential of tree species for use in later modeling research.

Studies of the potential improvement of tree species would involve collecting samples, providing storage for genetic material, establishing nurseries to provide access, and promoting plant material. Symbiosis between plant species and livestock would study location-specific sources of feeds and plants for maintaining animal health.

***Information and Knowledge.*** The role and importance of compound activities and the compound's diverse species should now be legitimized in southeastern Nigeria.

A series of publications could be developed, translated into local languages, and distributed, dealing with locally used plant species, uses of medicinal plants, locally preferred livestock forage browse, and alternative uses of locally exploited species.

Courses and teaching material should be formulated for extension workers, home economists and nutritionists, vocational students, and students of agronomy and plant and livestock sciences.

***Future Intervention.*** The single goal is to achieve sustainable, intensified, stable farming systems. Models of alternative farming systems that likely integrate trees and arable crops are needed, with emphasis on a simplified system of transporting feasible compound attributes to outer fields for continuous cultivation. Also needed is the ability to incorporate improved crops from the IITA crop improvement programs with adaptable, desirable tree species.

## Conclusion

The compound farm system is an excellent example of sustainable agriculture. It requires further study for improvement and perhaps for replication. For the households managing the compound, the priority is often survival rather than sustainability. The crisis of farming is not only in the developing world but is also evident on farms in America's midwest. My brother and sister-in-law, who farm in northeastern Nebraska, summed it up: "We work off the farm just to keep the farm."

To work with rural farmers, we need to lower scientists' arrogance and heighten their humility and sensibility in understanding that there is an indigenous, intellectual knowledge that must be recognized. Critique of ourselves and our institutions is necessary to ensure that the substance of sustainable programs is sensible and not just "old wine in a new bottle."

The definitions of sustainability should consider post-harvest, nutritional issues, and macro policies to ease the debt burden. There is an ethical obligation to ensure that the technology being promoted increases yields, is environmentally sound, and positively affects each member of the farm family. The UNICEF (1988) "State of the World's Children" discusses the need for the "empowering of individuals" and "families rights to know" about improved technologies. It is our moral obligation to assist in ensuring that right and empowerment.

Timberlake (1985) summed this up well: "Peasants have much to learn from agricultural researchers but so do researchers have much to learn from peasant farmers, and lines must be effectively opened in either direction."

There is a need for better listening and a better understanding of indigenous and sustainable systems, such as the compound that provides options for rural households poor in resources.

## REFERENCES

Brownrigg, L. 1985. Home gardening in international development: What the literature shows. The League for International Food Education, Washington, D.C.

Bruinsma, D. H., W. W. Witsenberg, and W. Wardemann. 1983. Selection of technology for food processing in developing countries. Pudoc, Wageningen, The Netherlands.

Bryant, A. T. 1907. A description of native foodstuffs and their preparation. Privately printed booklet available for reference, London School of Hygiene and Tropical Medicine, and reported by J. Doughty on loss of wild foods in developing countries. Appropriate Technology 5: 3.

Council on Foreign Relations and the Overseas Development Council. 1985. Compact for African development: Report of the committee on African development strategies.

Doughty, J. 1979. Loss of wild foods in developing countries. Intermediate Appropriate Technology 6: 3.

Eicher, C. K. 1988. Food security battles in sub-Saharan Africa. Paper presented at the Seventy World Congress for Rural Sociology, Baogna, Italy.

FAO. 1983. Using vegetables to improve meals in tropical Africa. Rome, Italy.

FAO. 1984. Promoting the consumption of under-exploited plant foods. Rome, Italy.

FAO. 1985a. Food for Africa. Report of a regional workshop organized by Zambia Alliance of Women in collaboration with the Food and Agriculture Organization. Rome, Italy.

FAO. 1985b. Report of the expert consultants on broadening the food base with traditional food plants. Harare, Zimbabwe.

FAO. 1985c. The role of minor crops in nutrition and food security. 8th Session. Committee on Agriculture, Rome, Italy.

FAO. 1986. Traditional food plants. Food and Nutrition 12: 1.

Gershon, J., Yen-Ching Chen, and Jen-Fong Kuo. 1985. The AVRDC garden program, 1983-1984. Asian Vegetable Research and Development Center.

Hahn, N. D. 1986. The African farmer and her husband: Women and farming systems, the IITA experience. Farming Systems Symposium, Kansas State University, Manhattan.

Hennessy, E. F., and O.A.M. Lewis. 1971. Antipellagrenic properties of wild plants used as dietary supplements in Natal, South Africa. Plant Foods for Human Nutrition 2: 75-78.

International Ag-Sieve. 1988. Regenerative agriculture in the third world building on capacity, not on need, 1: 1.

Lageman, J. 1974. Farming systems in relation to increasing population pressure: The case of Eastern Nigeria. International Institute for Tropical Agriculture. Ibadan, Nigeria.

Laumark, S. 1982. Women's contribution to intensive household production in Bangladesh: Vegetable cultivation. Third annual seminar on Maximum Production from Minimum Land, Bangladesh Agricultural Research Institute, India.

LIFE. 1985. Review of household food production systems. LIFE 18: 1.

Moris, J. R. 1985. Indigenous versus introduced solutions to food stress. IFPRI/FAO/AID Workshop on Seasonal Causes of Household Food Insecurity, Policy Implications and Research Needs. IFPRI.

National Academy of Sciences. 1975. Under-exploited tropical plants with promising economic value. Report of an ad hoc panel of the Advisory Committee on Technology Innovation. Board of Science and Technology, Washington, D.C.

New Scientist. 1988. New life in old crops. 160(7): 44-47.

Ninez, V. 1985. Introduction to household gardens and small scale food production. Food and nutrition bulletin 7(3): 1-5.

Okigbo, B. N. 1985. Home gardens in tropical Africa. Paper presented at the International

Conference on Home Gardens, Baundung, Indonesia. (Prepared by editors of Organic Gardening, Rodale Institute, Emmaus, Pennsylvania.)
Pacey, A. 1982. Gardening for better nutrition. Oxfam publication.
Redhead, J. 1984. FAO nutrition consultants reports series no. 74. Rome, Italy.
Rhoades, R., and A. Beggington. 1988. Farmers who experiment—An untapped resource for agricultural research and development. Paper presented at the "International Congress on Plant Physiology," International Potato Center, New Delhi, India.
Rosling, H. 1987. Cassava toxicity and food security. A review of health effects of cyanide exposure from cassava and of ways to prevent the effects. IITA-UNICEF program on household food security and nutrition. International Institute on Tropical Agriculture, Ibadan, Nigeria.
Soemarwoto, O. 1987. Homegardens: A traditional agroforestry system with a promising future. *In* Agroforestry: A decade for development. ICRAF, Nairobi, Africa.
Sommers, P. 1984. Dry season gardening for improving child nutrition. UNICEF publication. New York, New York.
Timberlake, L. 1985. Africa in crisis. International Institute for Environment and Development.
UNICEF. 1985. The UNICEF home gardens handbook for people promoting mixed gardening in the humid tropics. New York, New York.
UNICEF. 1988. The state of the world's children. Oxford University Press, New York, New York.
United Nations University. 1985. Food and nutrition bulletin on household level production 7: 3.
Warren, D. M. 1986. The transformation of international agricultural research and development: Linking scientific and indigenous agricultural systems in the transformation of international agricultural research and development. *In* J. Lin Compton [editor] Some U.S. perspectives. Westview Press, Boulder, Colorado.
Wilson, J. A., and S. Wilson. Grassland of Nebraska. Wide Skies Press, Polk, Nebraska.
WHO/UNICEF. 1985. Gardening for food in the semi-arid tropics: A handbook for programme planners. WHO/UNICEF Joint Nutrition Support Program. UNICEF publication. (Prepared by editors of Organic Gardening, Rodale Institute, Emmaus, Pennsylvania.)

# 30

# SOCIOLOGICAL ASPECTS OF AGRICULTURAL SUSTAINABILITY IN THE UNITED STATES: A NEW YORK STATE CASE STUDY

Frederick H. Buttel, Gilbert W. Gillespie, Jr., and Alison Power

Over the past several years, a silent revolution has occurred in the land grant system, the U.S. Department of Agriculture's Agricultural Research Service (ARS), and other U.S. agricultural research institutions. Whereas these institutions had been ambivalent if not openly hostile toward the symbol, if not the substance, of alternative, sustainable, or organic agriculture for more than a decade (Council on Agricultural Science and Technology, 1978), a number of agricultural research administrators and prominent scientists have now assigned low-input agriculture a high priority in American agricultural research (Joint Council on Food and Agricultural Sciences, 1986) and initiated substantial research and education programs in the area.

As with any other form of technology, socioeconomic factors will be crucial in shaping the nature of the low-input technologies that are developed, the nature and pace of adopting these technologies, and the impacts these technologies will have (Buttel and Youngberg, 1985). Until a few years ago, there was little encouragement to do research on socioeconomic aspects of alternative or low-input agricultural systems. Now, a fairly substantial social science literature exists on these topics (Youngberg, 1984). This literature, however, is dominated by studies comparing so-called alternative or organic farms/farmers with conventional farms/farmers or documenting the social characteristics of alternative-organic farmers (Lockeretz, 1985). Little attention has been devoted to understanding the degree to which rank-and-file American farmers now prefer or can, at some future point, be motivated to prefer lower input, more sustainable agricultural production systems (Buttel et al., 1986; Buttel and Gillespie, 1988).

Here, we explore this topic with a novel methodology that was employed with respect to two samples of farm operators in New York State: (1) a random sample of farm operators, taken from the list maintained by the New York State Crop Reporting Service, which we take to be a sample of conventional farmers, and (2) a companion sample of the membership list of the Natural Organic Farmers Association of New York, the major alternative-organic farming association in the state.

## The Context of Low-input Agriculture

One of the predominant trends in American agriculture in the post-World War II period has been the substitution of petrochemical inputs for land and labor (Conservation Foundation, 1987). Though not without undeniable productivity benefits, this technological change has encountered multiple limits over the past decade. Each of these limits has become a major rationale for low-input systems.

One limit of petrochemical-based agricultural systems has been the problematic potential for long-term productivity increases. Although worries in the 1970s about stagnation of agricultural productivity in the United States (Cochrane, 1979) have proved premature, there is growing evidence that traditional petrochemical-based production systems are reaching a point of diminishing returns (Conservation Foundation, 1987). Low-input agriculture offers a potential mechanism for sustaining (or slightly reducing) output and at the same time significantly reducing input usage, thereby providing new options for an increase in productivity.

Second, agricultural systems based on purchased petrochemical inputs have increasingly encountered environmental limits. These high-input systems tend to be associated with a myriad of environmental problems—soil erosion, pollution, and water resources destruction caused by erosion and runoff, contamination of food and soil, human health problems caused by chemicals, and so on—despite major attempts (e.g., conservation tillage promotion programs, the conservation reserve, pesticide regulation by the Environmental Protection Agency) to bring American agriculture within accepted environmental standards. Agriculture now is the single most important source of water pollution (Clark et al., 1985). Enforcement of pollution standards in agriculture to the degree that such standards are enforced with regard to industrial corporations and municipalities would have a devastating effect on agriculture in many regions of the country. Low-input agriculture is generally more environmentally benign than most conventional, petrochemical-based systems and offers a potentially attractive means for improving the environmental performance of American agriculture and avoiding inevitable confrontations between farmers and environmental regulatory agencies.

The 1980s have also witnessed the most protracted farm crisis since the Great Depression. The farm crisis has had its more immediate origins in the agricultural investment boom of the 1970s, rising real interest rates, the overvalued American dollar (until 1987), declining American agricultural export revenues, and the general global economic crisis that has worn on since 1973—each of which has contributed to farm financial stress (high debt-asset ratios), to a massive decapitalization of U.S. agriculture in the form of plunging land values, and to rising farm bankruptcy rates. The farm crisis, however, also reflects a more enduring, fundamental problem of American agriculture: chronic overproduction. In substantial measure, this problem of overcapacity and overproduction has been exacerbated by the pattern of productivity growth associated with the petrochemical-based trajectory of American agricultural technology; production has tended to outstrip demand and consumption because increased productivity has been so firmly rooted in increased output. Again, low-input agriculture is a plausible though, of course, partial response to overproduction. By stressing input reduction rather than output expansion as the principal goal of research, aggregate output increases should be lower than would otherwise be the case with more conventional technology-development approaches. Low-input agriculture will also tend to appeal to individual farmers who face high debt loads, high real interest rates, reduced credit worthiness, and an inability to finance large input purchases at the beginning of growing seasons.

Finally, the growth of low-input research can be seen as the culmination of a tumultuous decade and a half of political conflict over agricultural research. Beginning with Jim Hightower's (1973) bombshell, *Hard Tomatoes, Hard Times*, and continuing with a plethora of assaults on the public agricultural research system on grounds of social justice, environmental impacts, and the quality of research, the public research system has been put on the defensive (Hadwiger, 1982; Busch and Lacy, 1983; Browne, 1987).

Criticism of the public agricultural research system reached a somewhat surprising climax in the 1980s. After having been criticized by prominent agribusiness firms, foundations, and federal science policymakers for the lack of basic biological (biotechnology) research in 1982 (particularly by the "Winrock Report" [Rockefeller Foundation, 1982]), the crash program of biotechnology research that was implemented by the land grant system shortly thereafter culminated in the bovine growth hormone controversy (see Browne and Hamm, 1988; Buttel, 1986). The first major agricultural biotechnology, bovine growth hormone, met a chilly reception from farmers, particularly in some of the major family-farming dairy states, and attracted a swarm of opponents from environmental and public interest groups. Among the claims of bovine growth hormone opponents were that (1) this output-increasing technology was inappropriate in a time of overproduction in the dairy sector, (2) this technology was being developed more for the benefit of agribusiness firms than for farmers in the states where land grant researchers assisted with development of the technology, and (3) bovine growth hormone technology, because of its ability to increase milk production per cow by 20 to 25 percent, would result in the loss of a large number of American dairy producers, mainly smaller ones.

Low-input agricultural research became an attractive response to the criticisms of public agricultural research and the emergence of farmer opposition to the land grant system. Low-input research, unlike much biotechnology research of a generic nature that tends to be applicable well beyond the borders of a state, represents a return to the kind of applied, locally adapted research that historically has been pivotal in generating loyalty by state farmer groups to their state agricultural experiment station. And low-input research, unlike research that is primarily output-enhancing, became attractive to land grant administrators during the 1980s in addressing criticism that public research was contributing to overproduction and low prices. Finally, low-input research is an attractive means of demonstrating land grant commitment to the technical needs of smaller farmers and those concerned about the environmental impacts of conventional technology.

Despite the social, ecological, and political attractiveness of low-input agriculture, two lingering problems remain. The first—largely beyond our expertise—is whether these systems can perform as well as institutional and public-interest advocates claim they can. Low-input agriculture has been highly touted for its ability to enhance farmer profitability, but the evidence supporting this proposition is by no means conclusive. The second issue is whether there is any substantial constituency for low-input systems, especially if these systems do not live up broadly (across a wide range of commodities and regions) to the claims made for them. Will these systems be attractive to a broad spectrum of American farmers, or will interest in low-input systems be confined largely to their current consituency of smaller, part-time alternative-organic farmers who are critical of and often insulated from the land grant system?

## Development of Hypotheses

There are three plausible points of departure for addressing the matter of the potential constituency for low-input agriculture. The first approach would be to assume that the constituency for low-input, more sustainable agricultural technologies will be limited to those who already use them—that is, to the current (though expanding) group of self-identified alternative agriculturalists or organic farmers. A second approach would be to reason that the number of farmers favoring such systems would tend to be larger than that of self-identified organic farmers, but that these farmers would have similar socioeconomic characteristics. For example, one might expect that interest in low-input alternatives to high-input production systems would be concentrated among small, part-time farm operators.

It can be argued, however, that these two approaches may be misleading, for four reasons. First, there is evidence that a fairly large segment of American farmers employs relatively small amounts of purchased petrochemical inputs[1] and that these farmers' characteristics depart fairly substantially from accepted profiles of alternative-organic operators (Buttel et al., 1986). Although these farmers tend to operate somewhat smaller farms than do high-input-using farmers, these differences, as revealed in a New York State study, are modest, and such farmers differ little in their educational backgrounds, ages, and so on from their high-input-using counterparts.[2] Second, it is plausible that a substantial portion of farmers who use high levels of purchased petrochemical inputs would welcome alternatives to pesticides, chemical fertilizers, and the like for reasons of conserving capital, avoiding environmental problems, and reducing the risks of health problems. Third, there is substantial evidence that, although organic farms tend to be smaller than conventional ones, large organic farms do exist, and there are no inherent diseconomies of scale in low-input agriculture (Lockeretz et al., 1981). Fourth, we would argue that for most rank-and-file conventional operators attempting to deal with the highly varigated management problems of farming the choice between low- and high-input systems as a whole—that is, a choice between conventional and sustainable or alternative practices—may not be a meaningful one. Many farmers may be able to envision and actually implement low-input systems for disease control

[1]Among the reasons that some farmers' current production practices approximate low chemical-input systems is that some commodities (e.g., dairying) tend to involve crop rotations involving legumes, availability of animal manures for fertilization, and other conditions conducive to minimizing the use of purchased chemical inputs.

[2]These data were based on simple dichotomies of whether farmers did or did not use chemical fertilizers or pesticides. The results were reported in the original version of Buttel's (1986) article presented at the 1986 annual meeting of the Southern Association of Agricultural Workers, but they were omitted from the published version for reasons of space. The data also showed that farmers who do not use chemical inputs tended to have significantly higher levels of net worth and total family income than did farmers who used these inputs.

or fertilization but be reluctant to rely on nonchemical, low-input methods for controlling weeds or insects.

The third approach, which we adopted in our study, has three basic working hypotheses: First, there is a considerable constituency for low-input practices among rank-and-file American farmers. Second, farmer support for low-input systems will vary substantially across spheres of crop production (e.g., varietal selection, fertility, weed control, tillage). Third, this farmer constituency for low-input practices is relatively broad-based in the sense that it does not come exclusively or even primarily from the ranks of small, part-time farmers that predominate among alternative-organic agriculturalists.[3]

The first two hypotheses are explored with a methodology that (1) measures farmer preferences for low- versus high-input systems by instructing farmers to assume that each of the two systems would be equally profitable and (2) disaggregates low- and high-input systems into a number of components that reflect major crop production problems or important management decisions. The third hypothesis will be examined by constructing an index of preference for low-input systems and examining the socioeconomic antecedents of this variable through bivariate and multivariate analysis.

## Methods of Data Collection

The data for our study were collected from a random sample of 599 New York farm operators in the late winter and early spring of 1987. The sample was drawn from the list of farmers maintained by the New York State Crop Reporting Service. Each respondent was sent an 18-page mail questionnaire and a follow-up postcard one week later. If necessary, the farm operators in the sample were contacted up to two more times with another cover letter and copy of the survey until a completed questionnaire was received. The response rate, adjusting for those in the sample who said they had not operated a farm since 1982 and for inaccurate addresses, was 57.8 percent.

For comparative purposes, we also report data from a companion sample of the membership list of the Natural Organic Farmers Association of New York that received the same questionnaire at the same time as the

[3]This is not to imply, of course, that all or even a substantial minority of the nation's small, part-time farmers are self-identified alternative-organic farmers. Roughly two-thirds of U.S. farm operators, or about 1.4 million farmers, have annual gross farm sales of less than $40,000. In contrast, it would be reasonable to assume that fewer than five percent of U.S. farm operators, or roughly 100,000 farmers, identify themselves as alternative-organic farmers.

general sample of New York state farm operators. For the organic farmers list, all addresses judged likely to be nonfarm were culled, and two of three of the remaining addresses were randomly selected. The response rate among the organic farmers was 69.9 percent.

Farmers in the two samples were asked to respond to a series of questions about their preferences for crop production practices pertaining to eight production problems or management decisions. The questions, the complete wording of which is reported in Buttel and Gillespie (1988), were preceded by an introductory statement as follows:

"Some farmers have reported that even though they use a particular production practice, they would prefer to get along without this practice. Use of certain chemicals for insect or weed control have been examples. Little is known, however, about farmers' preferences for crop production practices. *For each of the following questions, assume that both practices described would yield about the same net profit*" [emphasis in original]. Thus, farmers were also asked to indicate their preferences on the assumption that their choice would not be based on economic grounds.[4]

A variety of independent variables were included in our study to address the third hypothesis set forth above. Assuming that there is an underlying coherence to farmers' preferences for low-input practices—that is, that there are strong intercorrelations among the items and that they can be summed to construct a reliable index—it is possible to examine the antecedents of preference for low-input practices.

A number of variables that are indicators of farm size and structure were selected for the analysis. These include total farm acres, gross farm sales, net farm income, total family income, net worth, farm assets, and number of hired farm workers. Each was measured with an item that asked for information pertaining to 1986. Total acres operated was measured as the exact number of acres given. Gross farm sales, net farm income, total family income, net worth, and farm assets were all measured with response categories involving a range of dollar amounts. The number of response categories was, respectively, 12, 8, 8, 14, and 12. Respondents were assigned a score from 1 to 14 as appropriate for each variable. Number of full-time hired workers was scored as the exact number (e.g., farmers having four or more workers were assigned a score of four).

In addition, several other variables were included to indicate the degree

[4]This procedure was considered more reasonable than instructing farmers to assume that low-input practices would have superior profitability. Moreover, this approach has the advantage of shedding light on whether low-input, more sustainable agricultural systems will have a substantial constituency if their profitability is comparable only to conventional, high-chemical-input methods.

to which preference for low-input practices has a broad base in the farming community. These variables were part-time farming status, age, education, extension contact, and profit orientation. Each of these variables, with the exception of profit orientation, was measured with a single item. Part-time farming status was measured as a dichotomous variable, with those operators working off the farm in 1986 for wages assigned a score of one and others a score of zero. Age was scored in terms of the exact number of years. Education was scored from one to nine, with the lowest category including respondents who did not finish eighth grade and the highest including those who had received a postgraduate degree. Extension contact was scored from one to four, with those having had no contact with extension over the past two years receiving a score of one and those contacting extension five or more times over the past two years receiving a score of four. Profit orientation was measured with a composite index based on three Likert-type items. For this and other attitudinal indexes, cases were included in the analysis if less than or equal to 25 percent of the constituent items had missing data. Each composite index was constructed by summing each respondent's scores on items for which there were no missing data and dividing the sum by the number of items included.

## Study Results

***Extent of Support for Low-input Production Practices.*** Table 1 presents data on the distribution of responses to the eight items reflecting support for low- versus high-input systems. Data are included for both the general sample of New York farm operators (conventional farmers), and the organic farmers.[5] The conventional farmers are disaggregated into operators of small farms (gross farm sales less than $40,000) and operators of commercial-scale farms (gross sales greater than $40,000). Because these bivariate results are discussed in another media, (Buttel and Gillespie, 1988), we only summarize the patterns in the data here.

First, as preliminary evidence that our method for measuring farmers' preferences for crop production practices has validity, it should be noted that there was a very low incidence of missing data for any of the items. Of the 327 conventional farmers in the sample, 312 provided information

[5]A random sample of farm operators will obviously include some number who are self-identified alternative-organic farmers. Our two mailing lists had one name in common; this individual was arbitrarily assigned to the conventional sample and omitted from the sample of organic farmers. Nonetheless, it is reasonable to assume that 95 to 98 percent of the farmers in the general sample were conventional farmers in that they were not self-identified alternative-organic operators.

**Table 1. Percentage distributions of preferences for crop production practices by type of farm.**

| *Phase of Crop Production** | *Percentage by Type of Farm†* | | | |
|---|---|---|---|---|
| | *Conventional* | | | |
| | *Small* | *Commercial* | *All* | *Organic* |
| Varietal selection | | | | |
| Prefer moderate-yield, low-input varieties | 87.1 | 83.4 | 85.6 | 97.1 |
| No preference | 10.6 | 6.6 | 8.3 | 2.9 |
| Prefer high-yield, high-input varieties | 2.3 | 9.9 | 6.7 | 0.0 |
| Ns | (132) | (181) | (313) | (70) |
| Fertility | | | | |
| Prefer on-farm sources | 66.2 | 65.6 | 65.8 | 100.0 |
| No preference | 12.8 | 13.9 | 13.4 | 0.0 |
| Prefer purchased fertilizers | 21.1 | 20.6 | 20.8 | 0.0 |
| Ns | (133) | (180) | (313) | (71) |
| Weed control | | | | |
| Prefer cultural practices | 59.1 | 35.5 | 45.4 | 98.6 |
| No preference | 10.6 | 13.1 | 12.1 | 1.4 |
| Prefer commercial herbicides | 30.3 | 51.4 | 42.6 | 0.0 |
| Ns | (132) | (183) | (315) | (71) |
| Insect control | | | | |
| Prefer natural controls | 47.0 | 42.3 | 44.3 | 100.0 |
| No preference | 20.5 | 26.4 | 23.9 | 0.0 |
| Prefer commercial insecticides | 32.6 | 31.3 | 31.8 | 0.0 |
| Ns | (132) | (182) | (314) | (71) |
| Disease control | | | | |
| Prefer natural controls | 73.1 | 75.3 | 74.4 | 98.6 |
| No preference | 9.0 | 11.0 | 10.1 | 0.0 |
| Prefer commercial fungicides | 17.9 | 13.7 | 15.5 | 1.4 |
| Ns | (134) | (182) | (316) | (70) |
| Tillage system | | | | |
| Prefer as few operations as possible | 33.6 | 44.5 | 39.8 | 48.6 |
| No preference | 12.7 | 9.9 | 11.1 | 7.1 |
| Prefer as many tillage operations as necessary | 53.7 | 45.6 | 49.1 | 44.3 |
| Ns | (134) | (182) | (316) | (70) |
| Crop mix | | | | |
| Prefer crop rotations | 58.3 | 44.0 | 50.0 | 88.7 |
| No preference | 12.1 | 7.7 | 9.6 | 4.2 |
| Prefer to specialize | 29.5 | 48.4 | 40.4 | 7.0 |
| Ns | (132) | (182) | (314) | (71) |
| Crop production management | | | | |
| Prefer low purchased inputs, high labor | 41.2 | 19.9 | 36.6 | 78.6 |
| No preference | 31.3 | 34.8 | 33.3 | 20.0 |
| Prefer high purchased inputs, low labor | 27.5 | 43.3 | 30.1 | 1.4 |
| Ns | (131) | (178) | (309) | (70) |

*The responses presented in this table are paraphrases of the actual response categories.

†For each practice and type of farm, the three column percentages sum to 100.0, except for rounding errors. The total N is 398.

on gross farm income. Among these 327 cases included in the analysis, the largest number of missing data cases was 15 (with respect to the crop production management item). The items on disease control and tillage system had only 10 missing data cases each. Likewise, relatively few farmers elected the "no preference" response categories, indicating that they did have relatively clear-cut preferences and were readily able to choose one of the response alternatives.[6]

Second, for several aspects of crop production, particularly varietal selection, fertility, and disease control, surprisingly large proportions of conventional farm operators prefer low-input practices over the high-input alternative (85.6, 65.8, and 74.4 percent, respectively). The weed control, insect control, tillage system, and crop production management items each elicited less than 50 percent support. Nonetheless, except for reduced tillage and high-labor-based crop production management, 44 percent or more of the respondents endorsed the low-input alternative, indicating that the potential constituency for most components of low-input agriculture is quite large.

Third, there was a general though modest pattern for operators of small farms to favor low-input practices more than did operators of commercial-scale farms. These differences were greatest in the case of the weed control, crop mix, and crop production management items and to a lesser degree with regard to the tillage system item. There were, however, no significant differences between small and commercial-scale farmers in their preferences with regard to varietal selection, fertility, insect control, and disease control practices.

Finally, with the exception of the tillage system item, the alternative-organic farmers were significantly more likely to choose the low-input alternative than were the conventional farmers as a group. These differences were particularly large with respect to the weed control, insect control, and crop production management items.

***Constructing an Index of Preferences for Low-input Production Practices.*** Table 2 provides a product-moment correlation matrix for the eight crop production practice preference items. For each item, respondents electing the low-input alternative were assigned a score of three; those with no preference were assigned a score of two; and those electing the high-input alternative were assigned a score of one. The results show, as might

[6]The crop production management item was an exception, however. One-third of the conventional sample and one-fifth of the alternative-organic farmers elected the "no preference" category. These data suggest that this item was not as intuitively meaningful to the respondents as were the others.

**Table 2. Product-moment correlation coefficients for the relationships among production practice preference items.**

| *Items* | $X_1$ | $X_2$ | $X_3$ | $X_4$ | $X_5$ | $X_6$ | $X_7$ | $X_8$ |
|---|---|---|---|---|---|---|---|---|
| $X_1$ = Varieties | — | | | | | | | |
| $X_2$ = Fertility | .08* | — | | | | | | |
| $X_3$ = Weed control | .19* | .35* | — | | | | | |
| $X_4$ = Insect control | .18* | .14* | .37* | — | | | | |
| $X_5$ = Disease control | .23* | .19* | .28* | .40* | — | | | |
| $X_6$ = Tillage | −.07 | −.00 | −.07 | −.06 | .01 | — | | |
| $X_7$ = Crop mix | .06 | −.05 | .10* | .05 | .08* | — | | |
| $X_8$ = Crop production management | .08* | .22* | .33* | .11* | .16* | −.07 | .06 | — |

*P < .05.

**Table 3. Principal components factor analysis of crop production preference items.***

| *Item* | *Factor Matrix (Loadings) I* | *Commonality* |
|---|---|---|
| Varietal selection | .434 | .188 |
| Fertility | .573 | .329 |
| Weed control | .788 | .622 |
| Insect control | .705 | .622 |
| Disease control | .646 | 418 |
| Crop mix | .353 | .125 |
| Crop production management | .609 | .371 |
| Eigenvalue | 2.548 | |
| Percent of variance | 36.4 | |

*Because only one factor was extracted, the solution could not be rotated.

be expected, that there are generally positive correlations among these items. The only major exception is the tillage systems item, which correlated inversely with several of the other items. The tillage system item is conceptually ambiguous regarding reduced-input practice preferences because reduced tillage systems for most crops tend to require use of herbicides and possibly insecticides and fungicides as well. Hence, it is not surprising that alternative-organic farmers are ambivalent about minimizing tillage operations. Reduced tillage has tended to be adopted by farmers at least as much for its labor savings as for its ability to conserve soil (Buttel and Swanson, 1986); thus, we eliminated this item from subsequent analysis.

Table 3 gives the results of a principal components factor analysis equation computed for the remaining practice preference items among the sample of conventional farmers. The single-factor solution indicates that the items reflect a single, underlying domain of content, which we label "preference for low-input production practices." These seven items were then subjected to a reliability test, and it was found that exclusion of the crop mix item yielded an index with the maximum Chronbach's alpha coefficient (0.64). The size of the alpha coefficient indicates that the index has acceptable reliability. This index was employed in subsequent bivariate and multivariate analyses.

***Bivariate and Multivariate Results.*** Table 4 provides product-moment correlation coefficients for the relationships between the independent variables and the index of preferences for low-input production practices. The results show, as anticipated, that there were no strong relationships between indicators of farm size and the preference index. The largest such correlation, with respect to number of hired workers (r = −.182) was modest, and the coefficients with respect to total acres, gross farm income, net farm income, farm assets, farm debt, and part-time farming status were all at or around 0.10. Several were nonsignificant. The coefficients for age and education were also nonsignificant. Somewhat larger but still modest coefficients were exhibited for total family income (r = −.143) and profit orientation (r = −.170). Nonetheless, there was a clear pattern for those preferring low-input practices to be drawn broadly from the farming population and to not be disproportionately small, part-time operators or non-profit-oriented.

Table 4 shows that the strongest correlate of preference for low-input

**Table 4. Product-moment correlation coefficients for the relationships between selected independent variables and preference for low-input production practices.**

| *Independent Variables* | *r* | *Significance* |
|---|---|---|
| Total acres | −.086 | .064 |
| Number of full-time hired workers | −.186 | .002 |
| Gross farm income | −.106 | .033 |
| Net farm income | −.112 | .027 |
| Farm assets | −.105 | .035 |
| Farm debt | −.086 | .067 |
| Total family income | −.143 | .007 |
| Total net worth | −.154 | .004 |
| Part-time farming status | .093 | .055 |
| Operator education | .036 | .263 |
| Operator age | −.084 | .071 |
| Profit orientation | −.170 | .001 |
| Extension contract | −.019 | .369 |
| Support for agricultural research | −.210 | .000 |
| Concern with agricultural pollution | .332 | .000 |
| Concern with soil erosion | .172 | .001 |
| Agribusiness cynicism | −.017 | .383 |
| Support for agricultural commodity programs | −.097 | .043 |
| Support for federal action to ameliorate the farm crisis | −.069 | .113 |
| Political liberalism | .011 | .426 |

practices was concern with agricultural pollution (r = −.332). Farmers preferring low-input practices were also more likely to be concerned with soil erosion (r = .177) and to be less favorable toward agricultural research and public agricultural research institutions (r = −.210) than those preferring high-chemical-input systems. Thus, the socioeconomic context within which low-input agriculture has emerged as a significant public agricultural research priority, especially environmental constraints on agriculture and farmer scrutiny of agricultural research, is reflected in farmers' preferences for crop production systems.

The results, however, show that farmers who prefer low-input practices do not differ substantially from their high-input counterparts with respect to agribusiness cynicism, support for federal commodity programs, support for federal action to ameliorate the farm crisis, or political liberalism. Support for low-chemical-input systems is not merely a reaction to farmer concerns about the deteriorating economic conditions that have plagued American agriculture during the 1980s.[7] Neither are these preferences mere reflections of political ideology.

In the multivariate analysis, variables with statistically significant ($p < .05$) bivariate coefficients were retained. First-order, partial correlation coefficients controlling for concern with agricultural pollution and second-order coefficients controlling for concern with agricultural pollution and support for agricultural research were computed (Table 5).

These results were largely consistent with the bivariate data, with one notable exception: the partial coefficients for the farm size indicators (especially full-time hired workers and gross farm income) were larger than their bivariate counterparts when concern with agricultural pollution was controlled. The partial coefficients for full-time hired workers and gross farm income were -.240 and -.215, respectively, when concern with agricultural pollution was controlled. Very similar coefficients were estimated when support for agricultural research was included as an additional control variable. This pattern of relationships indicates that there is a "suppressor effect" of farm size with respect to concern with agricultural pollution.

Table 5 shows, however, that support for federal action to ameliorate the farm crisis, support for agricultural commodity programs, and profit orientation had relatively minor relationships with preference for crop produc-

[7]It should be kept in mind that these data were collected at the height of the nation's farm crisis, and the farmer respondents could, therefore, be expected to have been maximally influenced by farm-crisis reasoning at the time of the survey. The farm crisis, however, was not as severe or longstanding in the northeastern dairy region as it was in the Great Plains, Midwest, and parts of the South.

tion practices when the two control variables were held constant. Concern with soil erosion likewise continued to have a significant relationship with the dependent variable (second-order partial = .169 and third-order partial = .202). The small, first-order partial for support for agricultural research also indicated that its bivariate effect on preference for crop production practices was largely joint with concern with agricultural pollution.

## Ecological Context and Environmental Implications

It is important to interpret these results in a biological context as well as a social one. Although we have emphasized that preferences for low-input strategies are surprisingly high even among commercial-scale conventional farmers, these preferences may be strongly dependent upon the cropping systems used by the sample groups. In the region of this study, a majority of commercial-scale farmers use a forage/dairy cropping system representative of the Great Lakes region (Fick and Power, 1988). For example, 91.3 percent of the commercial-scale conventional respondents milked dairy cows, while only 24.7 and 24.3 percent of the small conventional and alternative farmers, respectively, were involved in dairying ($X^2 = 127.16$; $p < 0.001$). Similarly, the percentages of farmers producing corn for silage were 69.0, 18.8, and 8.5 for large conventional, small conventional, and alternative farms, respectively ($x^2 = 119.94$; $p < 0.001$).

In contrast, only 7.0 and 16.7 percent of the commercial-scale and small

**Table 5. Partial correlation coefficients for the relationship between selected independent variables and preference for low-input production practices.**

| *Independent Variables* | *Second-Order Coefficients** | *Third-Order Coefficients†* |
|---|---|---|
| Number of full-time hired workers | −.240‖ | −.221‖ |
| Gross farm income | −.215‖ | −.194§ |
| Net farm income | −.100 | −.073 |
| Farm assets | −.149‡ | −.127‡ |
| Total family income | −.169§ | −.127‡ |
| Total net worth | −.138‡ | −.117‡ |
| Profit orientation | −.079 | −.046 |
| Support for agricultural research | −.120‡ | — |
| Concern with soil erosion | .169‡ | 202‡ |
| Support for agricultural commodity programs | −.138‡ | −.128‡ |
| Support for federal action to ameliorate the farm crisis | .026 | .026 |

*Controlling for concern with agricultural pollution.
†Controlling for concern with agricultural pollution and support for agricultural research.
‡p < or = .05.
§p < or + .01.
‖p < or + .001.

conventional farmers, respectively, produced vegetables, while 63.4 percent of the alternative farmers did ($x^2 = 101.16$; $p < 0.001$). Because the forage and cash grain systems favored by the conventional farmers tend to have fewer severe pest problems than do vegetable crops, the receptivity to low-input practices for pest control may simply reflect a reduced requirement for pest control in general. This would not be true of the alternative-organic farmers in this sample because a majority of these grow vegetables. To document the influence of cropping system per se, it would be necessary to compare a sample of commercial-scale vegetable producers with the groups included in this study. Nonetheless, we suspect that cropping system may be an important factor in the results described here.

In addition to the influence of cropping system, many crop production practices are related in ecologically meaningful ways that lead to significant intercorrelations. For example, preferences for low-input disease control were correlated with preferences for low-input disease control ($r = 0.23$, Table 2). By definition, low-input varieties are assumed to be more disease-resistant; at the same time, one of the potential low-input methods of disease control is resistant varieties. A similar relationship exists between varieties and insect control.

The highest correlations found among production practice preferences were those among pest management strategies. For example, preferences for low-input methods of insect control were highly correlated with preferences for low-input methods of disease control ($r = 0.40$, Table 2). Although the low-input management strategies for various categories of pests are not necessarily identical, they have in common the characteristic of being information-intensive—that is, they tend to require a greater knowledge of the biology and phenology of the pest than do conventional strategies relying primarily on pesticides. To use modification of planting time as a pest-control strategy, a farmer must have a reasonably complete understanding of the phenology of pest development and attack. A farmer's willingness to become knowledgeable about pest biology is unlikely to be restricted to a single pest category; therefore, preferences for these information-intensive strategies are logically correlated. In a similar sense, concern with agricultural pollution (which is strongly correlated with a preference for low-input practices) is unlikely to be limited to a single class of pesticides. A farmer who chooses low-input strategies for insect control because of environmental concerns will also prefer low-input methods of disease control.

The fact that low-input pathogen and insect control strategies tend to be primarily information-intensive rather than labor-intensive may also explain differences among correlations with crop production management.

For crop production management practices, low-input is explicitly linked with labor-intensive. Although low-input methods of increasing soil fertility and controlling weeds are undoubtedly more labor-intensive than their high-input counterparts, this is not necessarily true for insect and disease control. This lack of congruence is reflected in the strength of the correlations between crop management (i.e., substitution of labor for inputs) and pest management strategies. Preferences for low-input crop management were highly correlated with preferences for low-input methods of weed control (r = 0.33), but much less strongly correlated with low-input control of insects (r = 0.11) or disease (r = 0.16).

A biological analysis may also help to explain the pattern of correlations between crop mix preferences and preferences for other low-input practices. The low-input crop mix preference was crop rotation, which is also a low-input strategy for both weed control and disease control. Although none of the correlations with crop mix was particularly strong, preference for crop rotation was significantly correlated only with preference for low-input weed control and disease control (r = 0.10 and r = 0.08).

As suggested, negative correlations between reduced tillage and other low-input practices also make biological sense. Farmers who opt for low-input methods of pest control, which often use tillage operations as a way of killing pests or disrupting their life cycle, are unlikely to prefer reduced tillage systems, despite the correlation between preferences for low-input practices and concern for soil erosion (Table 4). Farmers thus might be said to exhibit more realism than many researchers who promote reduced tillage systems. In fact, farmers who prefer low-input practices overall tend to be less concerned with soil erosion than with agricultural pollution. This concern is reflected in their nonpreference for reduced tillage.

## Conclusions

This study provides tentative evidence for several arguments about the potential socioeconomic basis for low-input agricultural systems. First, farmer preferences for low-input systems vary considerably, depending upon the phase of crop production. Second, there is considerable interest among farmers in low-input approaches for most phases of crop production; from 40 to 80 percent of conventional farmers expressed interest in low-input practices on the assumption that their profitability will be comparable to conventional practices. Third, although overall levels of interest in low-input methods vary by phase of crop production, preferences for six of the eight items were sufficiently intercorrelated so that one can identify an underlying, unidimensional construct of preference for low-input practices. Fourth, the major antecedent of preference for low-input practices was farmers' level of concern about agricultural pollution, followed by support for agricultural research (which was inversely related to the composite index). Fifth, preference for low-input practices did not bear a major relationship to general or agriculturally-related political ideology, attitudes toward the farm crisis or federal commodity programs, and profit orientation. Sixth, there was no strong pattern for small or part-time farmers to be a disproportionately large constituency for low-input practices or for large farmers to be disinterested in low-input approaches. Finally, the pattern of results seemed sufficiently consistent with established agroecological knowledge to indicate that the production practice preference items have prima facie construct validity. This increased our confidence that the results are valid and may be used as a guide to setting research priorities.

Although there is reason for confidence in the results of this study, the method we employed involves limitations. Further research will be required to determine the degree to which wording of the questions or other methodological factors significantly influenced our results. Also, our dependent variable was an attitudinal measure, and the consistency of these attitudinal measures with future behaviors is unknown. Finally, the results are from one state only—one that because of its agroecological conditions is particularly well-suited to a number of low-input practices. Results in less favorable areas—for example, much of the southeastern region—may be quite different.

## REFERENCES

Browne, W. P. 1987. An emerging opposition? Agricultural interests and federal research policy. pp. 81-90. *In* D. F. Hadwiger and W. P. Browne [editors] Public policy and agricultural technology. St. Martin's Press, New York, New York.

Browne, W. P., and L. G. Hamm. 1988. Political choices, social values, and the economics of biotechnology: A lesson from the dairy industry. Staff Paper 88-33. Department of Agricultural Economics, Michigan State University, East Lansing.

Busch, L., and W. B. Lacy. 1983. Science, agriculture, and the politics of research. Westview Press, Boulder, Colorado.

Buttel, F. H. 1986. Agricultural research and farm structural change: Bovine growth hormone and beyond. Agriculture and Human Values 3: 88-98.

Buttel, F. H., and G. W. Gillespie, Jr. 1988. Preferences for crop production practices among conventional and alternative agriculturalists. American Journal of Alternative Agriculture 3: 11-17.

Buttel, F. H., G. W. Gillespie, Jr., R. Janke, B. Caldwell, and M. Sarrantonio. 1986. Reduced-input agricultural systems: Rationale and prospects. American Journal of Alternative Agriculture 1: 58-64.

Buttel, F. H., and I. G. Youngberg. 1985. Sustainable agricultural research and technology transfer: Sociopolitical opportunities and constraints. pp. 287-297. *In* T. C. Edens et al. [editors] Sustainable agriculture and integrated farming systems. Michigan State University Press, East Lansing.

Buttel, F. H., and L. Swanson. 1986. Soil and water conservation: A farm structural and public policy context. pp. 26-39. *In* S. B. Lovejoy and T. L. Napier [editors] Conserving soil: Insights from socioeconomic research. Soil Conservation Society of America, Ankeny, Iowa.

Clark, E. H. II, J. A. Haverkamp, and W. Chapman. 1985. Eroding soils: The off-farm impacts. The Conservation Foundation, Washington, D.C.

Cochrane, W. W. 1979. The development of American agriculture. University of Minnesota Press, Minneapolis.

Conservation Foundation. 1987. State of the environment. The Conservation Foundation, Washington, D.C.

Council on Agricultural Science and Technology. 1978. Organic and conventional farming compared. Ames, Iowa.

Fick, G. W., and A. G. Power. 1988. Pests and integrated control. *In* C. J. Pearson [editor] Field-crop ecosystems. Elsevier Publications, Cambridge, Massachusetts.

Hadwiger, D. F. 1982. The politics of agricultural research. University of Nebraska Press, Lincoln.

Hightower, J. 1973. Hard tomatoes, hard times. Schenckman, Cambridge, Massachusetts.

Joint Council on Food and Agricultural Sciences. 1986. Five-year plan for the food and agricultural sciences. Washington, D.C.

Lockeretz, W. 1985. U.S. organic farming: What we can and cannot learn from on-farm research. pp. 96-104. *In* T. C. Edens et al. [editors] Sustainable agriculture and integrated farming systems. Michigan State University Press, East Lansing.

Lockeretz, W., G. Shearer, and D. H. Kohl. 1981. Organic farming in the corn belt. Science 211: 540-547.

Rockefeller Foundation. 1982. Science for agriculture. New York, New York.

Youngberg, I. G. 1984. Alternative agriculture in the United States: Ideology, politics, and prospects. pp. 107-135. *In* D. F. Knorr and R. R. Watkins [editors] Alterations in food production. Van Nostrand, New York, New York.

# SOCIOECONOMIC ASPECTS OF MACHINERY REQUIREMENTS FOR ROTATIONAL AGRICULTURE

T. S. Colvin, D. C. Erbach, and W. D. Kemper

Use of rotations is a cornerstone of any sustainable agricultural system. Rotations that contain a close-growing crop, such as a forage, also can be important in developing a soil and water conservation plan.

## Needs for Rotations in Conservation Plans

The conservation compliance provision of the 1985 U.S. Food Security Act (FSA) requires that a farmer, to remain eligible for government price supports on crops, must develop an approved conservation plan on each of the highly erodible fields on his or her farm. The act underscores the nation's commitment to conserving topsoil, essential for long-term food security. It also reflects a growing consciousness that humans are capable of changing their environment. Land stewards play major roles in determining the quantity and quality of groundwater; quality of surface waters; runoff and flooding; life and condition of reservoirs and lakes; availability of gravel beds for spawning fish; dust, pollen, and chemical content of the air; and a host of other factors.

Society has a legitimate interest in the long-term security of food supplies and overall environmental quality. Farm program support payments can be used as a lever to modify crop production practices. Wind and water erosion are among the most obvious detractors from environmental quality. Moreover, when erosion levels are reduced, several of the other environmental factors listed above can be improved. Consequently, conservation programs to reduce water and wind erosion are the primary objectives of the conservation compliance provisions of the Food Security Act and affect all federal farm programs.

The U.S. Department of Agriculture (USDA) assigned the Soil Conservation Service (SCS) the task of identifying fields susceptible to wind and water erosion, of providing guidance on how to reduce erosion to acceptable levels, and of adjudicating cases of questionable compliance. Over a large portion of the United States, conservation tillage systems are the most promising management practices to achieve the required erosion reductions. These systems keep more of the crop residue on the soil surface, where it increases infiltration by protecting the surface from the sealing action of raindrops and by providing food and cover for surface-feeding soil fauna whose burrows drain water into the root zone during intense rainfall. These residues also help stabilize the soil, reducing erosion if runoff does occur. Crop residues on the surface also reduce evaporation from the soil surface, leaving more of the water available for crop use.

Challenges associated with leaving residues on the surface include finding ways to seed and fertilize through the residues. New equipment and management techniques are meeting these challenges. However, weed control and the carry-over of diseases and insects are problems that have caused many farmers to discontinue, or refrain from adopting, conservation tillage. These problems are particularly endemic to fields where only one crop is grown. In these monocropped fields, the populations of pest and disease organisms attacking that crop tend to build up, protected by or sometimes actually carried over to the next season in crop residues left on the surface.

## Benefits of Rotations

***Reducing Populations of Diseases and Pests Adapted to Specific Crops.*** A large portion of the total spectrum of diseases, insects, and nematodes is adapted to specific crop species or closely related species. Rotations with other crops reduce populations of such harmful organisms and decrease the reliance on and costs of chemical disease and pest control.

***Weed Control.*** Several relatively low-cost, broad-spectrum herbicides will kill monocotyledenous weeds in dicotyledenous crops, or dicotyledenous weeds in monocotyledenous crops. In many cases, no herbicides will eliminate monocotyledenous weeds from a monocotyledenous crop (or dicotyledenous weeds out of a dicotyledenous crop) without damaging the crop. Where such herbicides exist, they often are expensive. Consequently, rotations of dicotyledenous and monocotyledenous crops allow use of low-cost herbicides to achieve good weed control.

Other crops inserted into a rotation can control many weeds. For instance, in many portions of the United States, alfalfa will get an earlier start than most weeds; the hay, including the weeds, will be harvested before the weeds have developed seed; and the alfalfa will recover from repeated harvesting more quickly than will most weeds. Consequently, properly managed alfalfa, in which a good stand is maintained, can reduce populations of both annual and perennial weeds.

***Carryover of Legume Nitrogen.*** Legumes can be used in rotations to supply nitrogen to a succeeding crop. When corn or oats follow a soybean crop, the nitrogen application rate should be reduced to make use of the nitrogen made available by the soybeans. Alfalfa accumulates up to 200 kilograms of nitrogen per hectare in its crowns and root systems. Many farmers also kill alfalfa when the last crop of the season is still in the field, which can add from 30 to 100 kilograms per hectare to the amount of nitrogen available to succeeding crops (Robbins and Carter, 1986). This nitrogen becomes available as the organic matter decomposes. Timing of this availability closely matches the needs of corn. Thus, corn is not stressed for nitrogen and takes up most of the nitrogen from the previous alfalfa crop.

However, winter wheat, seeded soon after the alfalfa is killed, is more active during the succeeding cool months than the microorganisms that decompose the alfalfa roots and residues. This results commonly in a nitrogen deficiency during the early season growth and maturation of the wheat before all the nitrogen becomes available; then, large amounts of nitrogen can be left in the soil (Robbins and Carter, 1986). Depending upon growth of the next crop and precipitation or irrigation water management, this leftover nitrogen can benefit the crop or can leach below the crop root zone and move into groundwater.

Annual and biennial clovers, plowed down as green manures, can contribute most of the nitrogen needed by the following crop in addition to improving the physical condition of the soil. Where precipitation is not abundant, these benefits must be balanced against the value of water removed from the soil by the green manure crop. Sometimes, when early season wet conditions prevent early seeding of cash crops, use of water by the preceding legume can be beneficial.

Seeding costs of annual legumes are often high. Rotations that provide the best economic returns are generally those from which the legume develops seed that can be harvested as a cash crop (Power et al., 1983). In southern areas, some vetches and subclovers set seed in the early spring, then die back during the growing season of a summer cash crop, and begin growing from seed again in the fall. Limited grazing of the winter legumes adds a cash return, although it reduces both the amount and uniformity of nitrogen returned to the soil. There may be a need to harvest the green manure crop for forage. This, of course, reduces the nutrients available to the next crop.

***Evaluating Benefits of Rotations.*** The economic values of the potential benefits attendant to good rotations are not well-defined because they depend upon cropping history; soil minerals; populations of weed seeds, insects, and disease organisms; weather; and a host of other factors. Establishing alfalfa in a rotation is costly, and most farmers want to keep the alfalfa for the establishment year and at least two years more. The addition of two or more other crops increases the rotation to five or more years. At least two cycles of a rotation are generally considered essential for its evaluation; consequently, careful evaluation of a rotation is a costly and long-term undertaking. Schumaker et al. (1967) reported on a well-designed, six-year rotation study that ran for two cycles and took about 12 scientific years and 18 technician years. For these reasons, a large portion of our evaluations of benefits of rotations, including crops not widely grown, are qualitative results observed by good farm managers, who have experimented with crops having potentially good markets. Long-term experiments in several countries have observed rotations for many years for a limited number of crops.

## Markets as a Prerequisite for Including New Crops

Markets that provide an economic return—that is, greater than the input costs—are generally a prerequisite for a crop that is successful in a rotation. For crops that have not been grown in the area before, this may require introduction of new processing machinery that is beyond the ability of a farmer to finance, such as canneries, oil seed extraction plants, fiber separation plants, and so on. Assembling the capital, designing the plant, purchasing the machinery, constructing the plant, and organizing farmers to grow the raw material constitute a complex socioeconomic-biological venture that requires extensive organization. Consequently, the United States continues to import jute fiber for rug backing, pulp for paper, and palm oil for soaps, even though U.S. farmers could be growing the raw materials from which these products are made.

USDA attempted to provide the data base needed for production and processing of kenaf to potential manufacturers of kenaf paper and fiber. This venture, in cooperation with a new commercial partner, currently shows excellent promise for eventual success. Nevertheless, it has been subject to criticism because USDA chose to work with one selected company, excluding others. The company, after investing substantial amounts in the venture, also became understandably protective of its ownership rights and prerogatives and now tends to back away from cooperative activities.

In Virginia, a vegetable marketing cooperative was started to provide a market for farmers adding vegetables to their rotation (McPeters, 1986). This required more than 30 growers to join together and build a $500,000 facility for processing and shipping. This cooperative is an example of the cost and social organization that may be required to provide the marketing portion of the infrastructure. Such an organizational structure already exists in the Midwest for cash-grain producers.

## Using Hay, Silage, and Forage Crops Produced in a Rotation

The spectrum of forage, silage, and hay crops that can be fed to animals includes many of the best candidates for interspersing between "cash" crops. The costs of automated feeding systems, manure utilization systems, and fences and enclosures to manage the animals often hinder farmers from availing themselves of this group of components of rotations. When animals are part of an agricultural operation, they tend to need daily attention. The exception is when animals are grazed extensively on open range and may be left on their own for longer periods. Although this may be a substantial factor in a farmer's decision whether to keep animals, another major factor—keeping cattle out of feedlots and alfalfa out of crop rotations—is the economic risk involved in owning a herd of cattle.

Agricultural advisors and farmers may be tempted to see Dick Thompson's well-managed farming and feeding operation near Boone, Iowa, as a model. However, widespread adoption of this model would tend to raise the sale of beef and hogs off the farm and reduce the sale of grains. Hence, a note of caution is in order. Overloading the beef or hog markets has the same effect as overloading grain markets. It drives prices down. Farmers who can sell cattle at these lower prices and still make money may be viable in competing for the market. But the current beef market is so inelastic and prices so near production costs that entry of one farmer in the feeding business generally means the eventual exit of another. Because the individual entering the feeding business will have added initial costs to amortize, that farmer will commonly be among the first ones out.

## Costs and Social Alternatives of Additional Machinery Needed

Assuming that a solid market exists for a crop that could be used in a rotation, the factor most frequently inhibiting use of a crop in a rotation is the cost of additional machinery to seed, cultivate, spray, and harvest the crop. Ball (1987), working in North Dakota, suggests that adding forage to a rotation that previously included only grain crops will increase the overall machinery costs for the farm. As an example, assume that a farmer who has been growing corn and soybeans wishes to add alfalfa to the rotation. The farmer's options to acquire use of the alfalfa harvesting machinery include buying the additional machinery for individual farm use; hiring a commercial contractor who owns and operates the swather, baler, and loader-hauler-stacker machinery; or cooperating with nearby farmers, with each farmer agreeing to care for one crop (i.e., one harvests and cultivates corn, one cultivates and harvests beans, and the third harvests and stacks the alfalfa), with the proceeds to be shared by the farmer owning the land and the farmer doing the harvesting.

The first option will cost a substantial amount per year to amortize the cost of and to maintain equipment. In return, the farmer will have the satisfaction of owning the equipment and a greater ability to harvest at times judged best for yield and quality.

How much the second option will cost depends upon the number of hay harvesters in the area and the demand for their services. Hiring a commercial hay harvester frees considerable amounts of time that may be of great or limited value, depending upon other demands for time during hay harvest seasons.

The third option requires a fair division of the profits for the cooperators to survive as a socioeconomic unit and remain friends. If the farmers trust each other sufficiently, they can lay out their whole budgets and agree to equitable share-cropping formulas for each of the crops. These calculated formulas often require considerable time and experience for development; once they are established by one successful partnership, they often are used by other farmers considering such cooperation.

## Specific Studies of Machine Costs

A number of assumptions must be made for each study of machine requirements. These include the number of days available for operation, work hours per day in the field, ground speed, and organization of the entire operation.

In considering machinery costs for crop rotations, one must look at both the short-term and the long-term aspects. If the rotation used on the farm changes, the machines required for the new crop are needed immediately. The level and type of cost increase will depend on how those services are secured.

**Table 1. Machinery requirements and costs for various rotation systems on 182 hectares (ridges).**

| | | *New Costs ($)* | | | |
|---|---|---|---|---|---|
| *Size* | *Machine* | *Continuous Corn* | *Corn-Soybeans* | *Corn-Soybeans-Oats* | *Corn-Soybeans-Oats-Alfalfa* |
| 8-row | Sprayer | 2,700 | 2,700 | 2,700 | 2,700 |
| 4-row | Planter | 10,000 | 10,000 | 10,000 | 10,000 |
| 4-row | Rotary hoe | 4,500 | 4,500 | 4,500 | 4,500 |
| 4-row | Cultivator | 5,000 | 5,000 | 5,000 | 5,000 |
| | Corn picker-sheller | 15,000 | | | |
| | Tractor | 30,000 | 30,000 | 30,000 | 30,000 |
| | Combine | | 40,000 | 40,000 | 40,000 |
| | Grain drill | | | 8,000 | 8,000 |
| | Baler | | | 6,800 | 6,800 |
| | Mower-conditioner | | | | 6,000 |
| | Rake | | | | 2,000 |
| | Totals | 67,200 | 92,200 | 107,000 | 115,000 |

For example, consider the first option discussed earlier: buying the added equipment necessary. Table 1 shows the cost of machinery required to produce continuous corn on 182 hectares. For this illustration, it was necessary to make assumptions and choices concerning working rates for machines, type of tillage system, and listed costs. Note that items such as wagons, pickup trucks, and the like, are not shown but would obviously be required for a functional farm. Investment in the machines averages $370 per hectare for machines purchased new. Because the only crop is corn, a relatively simple, low-cost picker-sheller is sufficient for harvest.

Adding a second crop, such as soybeans, requires other machinery (Table 1). A farmer making this change would need to trade the picker-sheller for a combine that can handle both beans and corn if the farm is to own all the machinery necessary to produce both corn and soybeans. Keeping all machine values on a new basis, the cost of machinery increases by almost 40 percent, with a per-hectare value of $510 for the 182 hectares. If the rotation on the 182 hectares is expanded to include oats, the farmer needs to buy a grain drill and baler (Table 1), which would increase his or her machinery costs to $590 per hectare.

An even larger machinery set would be required to grow and harvest a four-crop rotation, including alfalfa (Table 1). Machinery costs can be kept to $630 per hectare if the farmer sells alfalfa in the field to a buyer who picks it up and hauls it away.

The machinery investments shown are based on new prices. In actuality, farms have machinery of different ages. Some farms already have machines for crops in addition to those currently in their rotations. Other farmers already use some form of custom or cooperative work for some crops (Dunaway, 1988). Nelson and Kletke (1987) listed custom rates for equipment operations based on an Oklahoma survey, and similar information is available for other states and regions.

Some farmers routinely do custom work for other farmers for cash. Some do custom work on a share basis, particularly forage harvesting, which means that the farmer doing the work receives a portion of the crop as payment. When the person doing hay harvesting has paid no other costs of production, a common charge in the Midwest is 50 percent of the crop. On irrigated land in the West, the harvester may receive only 30 to 35 percent of the crop. Negotiations determining the harvester's share are affected strongly by how many harvesters are available and how much hay has to be harvested. Last-minute deals are affected strongly by weather conditions.

Before making an investment in machinery, a good farm manager would consider the second and third options (custom and cooperative work) for providing the necessary services for the added crop in the rotation. A cost that might be overlooked, however, is the fixed cost of machines that are no longer needed (unless a machine is sold) on the acres devoted to the replacement crop. This cost may be large depending upon the area devoted to crops no longer requiring that machine's services. Some farmers recoup at least part of these costs by doing custom work for neighbors or by leasing additional acreage. In the long term, machines can be resized (within limits) to fit the number of hectares required for each crop in the new rotation. But in the short term, there may be a cost for the additional machines as well as the cost of having the excess capacity of the original equipment.

After the farm has adjusted to the new rotation over time, machinery costs may not differ substantially on a per-hectare basis. Farm records from members of the Iowa Farm Business Association (Duffy and Le Brun, 1987) indicated comparable annual costs for machinery between farms regardless of crop rotations. Farmers who sold at least 95 percent of their crops (cash grain) had an average machinery and power cost of $77 per hectare on 224 hectares. Farms that were identified as cow-calf producers had machine and power costs of $64 per hectare on an average 271 hectares. The cow-calf producers reported lower land values and would need forage, but reported 90 percent as much in total crop sales as the cash-grain producers. When at least three-fourths of the crops produced on the farm were consumed on the farm, for example, with various combinations of hog, beef, and dairy enterprises, the machinery and power costs ranged from $74 to $87 per hectare (Duffy and Le Brun, 1987). This indicates that long-term costs for machinery can be comparable on farms with different levels of forage in their rotations. This does not suggest that a farm adding a new crop will be able to match immediately the costs of farms already producing that crop. After an adjustment period, there seems to be a fairly level machinery cost to grow common midwestern crops.

Information contained in Iowa State University Extension publications designed for farm budgeting (1984 values) showed a drop, or at least no increase, in the cost for crop rotations that included forage as compared with cash grain. Annual machinery costs were $202 per hectare for corn (including drying), compared with $121 per hectare for soybeans (Duffy, 1987), and $129 per hectare per year for harvest machine cost for a 13.5-ton-per-hectare yield of alfalfa hay (Barnhart and Edwards, 1984). In 1985, budget information from the University of Georgia suggested a harvesting machinery cost of $267 per hectare for a 11-ton-per-hectare forage yield. This cost would be somewhat higher than normally expected in the Midwest (Givan, 1985).

Rowshan and Black (1987) reported similar machinery costs between con-inuous corn and a rotation that included corn, navy beans, and sugar beets in Michigan, both under conventional tillage. The farms modeled were larger than average, ranging from 200 to 600 hectares. When conservation tillage was used with the more complex rotation, costs were reduced by more than 10 percent per hectare, compared to continuous corn produced with conventional tillage. These rotations did not include forage but did include sugar beets, which require completely different harvesting equipment than cash grain crops do (Rowshan and Black, 1987). Blobaum (1983) looked at barriers to conversion to organic farming. Among the barriers identified were financial and weed control problems, but machinery or machinery costs were not mentioned.

Farmers already obtain machine services in a number of ways. Table 2 shows the percentage of respondents to surveys, conducted by Farm Progress Publications (1984), Webb Publishing (1984), and HBJ Publications (1987), who owned the types of forage tools listed. It also shows the percentage of respondents to the Farm Progress and HBJ surveys who grew hay. Clearly, more farmers in the Corn Belt grow hay than own all of the equipment necessary to harvest hay. In contrast, more Pennsylvania farmers own conventional balers than report hay production. It is also obvious that in some areas of the Corn Belt, a majority of farmers already are growing hay and must either own the necessary equipment or have made arrangements to secure the services of that equipment. By comparison, 86 percent of the Iowa respondents grew corn in 1983, with 61 percent of the respondents owning combines and corn heads (38 percent of the respondents owned ear-corn pickers).

An alternative to harvesting might be to graze animals on the land. There would be costs associated with obtaining the animals and providing fencing and water, but these might be an effective alternative to some portion of total machine harvest. Turner et al. (1987), working in Kentucky, discussed intensive grazing systems.

In some areas, farm equipment dealers or leasing companies have programs for short-term rental or long-term equipment leasing. The benefits to the farmer depend upon the terms of the arrangements, which can vary widely. Some soil conservation districts in the United States have made arrangements (as an educational program) for farmers to obtain the services of planting equipment, on a short-term basis, to allow farmers to try new conservation tillage planting systems on their own farms before they have to make the investment in new machinery.

Machinery cooperatives have been established or proposed in several areas of the world (Holtkamp, 1988). In France, Federation Nationale des Cooperatives d'Utilisation du Materiel Agricole gathers operating and financial data from 12,000 local farm-equipment-using cooperatives that own more than 100,000 pieces of equipment (Bregand, 1988). One piece of equipment discussed was a four-row, self-propelled forage harvester. As the annual use of that machine went from 100 to 350 hours, the total cost of the machine, excluding fuel, dropped from about 700 francs per hour to 500 francs per hour. This is an example of spreading the costs of machinery over many hectares by using it on several farms. No data were given on timeliness costs associated with unavailability of the machine when it was needed. The timeliness cost would accrue to the individual farm and not to the cooperatives. Over time, high timeliness costs would force individual farmers

**Table 2. Percent of respondents to surveys in selected states having listed equipment or growing forage.**

| *State* | *Mower* | *Mower Conditioner* | *Rake* | *Large Round Baler* | *Small Square Baler* | *Forage Harvester* | *Hay* |
|---|---|---|---|---|---|---|---|
| | | | | % | | | |
| Minnesota | 55 | 23 | 59 | 60 | 20 | 33 | na* |
| Dakotas | 80 | 6 | 65 | 30 | 20 | 31 | na |
| Wisconsin | 45 | 60 | 78 | 5 | 21 | 60 | 87 |
| Illinois | 71 | 23 | 51 | 12 | 17 | 14 | 41 |
| Indiana | 60 | 23 | 49 | 8 | 16 | 13 | 48 |
| Iowa | 69 | 25 | 67 | 14 | 17 | 21 | 58 |
| Ohio | 65 | 40 | 67 | 17 | 63 | 21 | 53 |
| Mississippi | 44 | 50 | 70 | 14 | 65 | 39 | 62 |
| Pennsylvania | 64 | 70 | 85 | 15 | 84 | 55 | 73 |
| Missouri | 71 | 37 | 70 | 39 | 51 | 17 | 74 |
| Kansas | 56 | 29 | 56 | 35 | 44 | 24 | 58 |

*Not available.

to consider the total cost structure of the operation.

In corn-producing areas in the Midwest, local fertilizer outlets, cooperatives, or others commonly supply fertilizer application equipment for a charge, along with the purchase of fertilizer. Equipment that may not work well because of use on many acres can reverse the economies of scale.

## Timeliness Considerations for Owning Machinery

Timeliness costs have been determined for various operations. Colvin et al. (1984) gathered values from several sources. The values normally are reported as a fractional reduction in yield caused by delaying the operation one day. They ranged from almost zero for some noncritical operations to 0.01 for corn planting and soybean harvest. The factor for hay harvest is about 0.005. A factor of 0.01 means that there would be a one percent reduction in the yield caused by a one-day delay in the operation. One of the problems with these values is that they vary depending on the year. In a year that has many days suitable for an operation, a one-day delay may affect the yield very slightly, whereas a delay in the same operation in a year with few good days may have a very high cost. An example would be for two neighbors to share a combine. One might start harvesting in a timely fashion and work at a normal pace until finished. Then, if the second owner were to just begin work, only to have severe weather completely destroy the crop, the second owner would have just experienced a very large timeliness cost. Farmers who are not adverse to risk will accept extra costs associated with excess machine capacity as reasonable.

If livestock is to be added to the farm, as the outlet for a forage crop, machinery costs must be considered because the time needed for livestock care and management may conflict with the time needed for crop production, requiring larger, more costly machines (Guy et al., 1988). On the other hand, if crops are to be grown on contract for vegetable or forage processors, the processor may do the harvesting so that the material flow to the processing plant can be controlled. Even though the processor may charge the costs to the grower, the costs commonly will be relatively low because the processor will generally make heavy and prolonged use of the expensive harvesting equipment.

## Discussion and Conclusions

Rotations can make major contributions to erosion, weed, pest, and disease control. They also may reduce nitrogen fertilizer needs. These contributions can enable many farmers to avoid or reduce the application and materials costs of herbicides, insecticides, nematicides, fungicides, and nitrate fertilizer. By helping to keep infiltration rates high, rotations can also reduce the need for terracing, although some form of conservation tillage commonly would be required to completely avoid that need on steep lands.

The initial cost of additional equipment needed for additional crops is perceived by many monocropping farmers as a major disincentive to adoption of rotations. These initial costs can be reduced or avoided by cooperative farming with neighbors, whereby each cooperator manages a specific crop and the landowner and crop manager share the product. Hiring equipment owners to plant, spray, and harvest the additional crops is another means of avoiding the initial machinery costs inherent in expanding from monocropping into rotations. Hiring custom operators to do this work may release some of the farmer's time, which may be used to do custom work for other farmers with the equipment originally used in the monocrop system. Adjustments of this type can help defray or avoid the costs of machinery for the additional crops. However, such adjustments often require some time, reduce the farmer's independence, and affect the timeliness of some operations. Long-term machinery costs of rotation farming can be as low or lower than for monocropping if the farmers are willing to cooperate.

The primary factor determining the long-term economic viability of rotations is the market for the added crops. Broad-scale increases in forage production and subsequent animal production will reduce prices for forage and the animals drastically because American meat consumption does not increase as rapidly as prices for the farmer's product decrease.

Assuming availability of a continuing market for the additional crops that can be used in rotations, farmers still need help to understand that perceived increased costs of machinery can be avoided. Socioeconomic systems for avoiding such increases in machinery costs should developed, evaluated, and demonstrated, along with the improved agronomic systems, to achieve farmer adoption.

## REFERENCES

Ball, W. S. 1987. Crop rotations for North Dakota. Publication EB-48. North Dakota Cooperative Extension Service, Bismarck.

Barnhart, S., and W. Edwards. 1984. Estimated costs of pasture and hay production. Publication AG-96. Iowa Cooperative Extension Service, Ames.

Blobaum, R. 1983. Barriers to conversion to organic farming practices in the midwestern United States. pp. 263-278. *In* W. Lockeretz [editor] Environmentally sound agriculture. Praeger Publishers, New York, New York.

Bregand, J. 1988. A data bank for using costs of farm equipment. Paper 88.147. pp. 53-54. *In* Book of abstracts. International Conference on Agricultural Engineering (Agricultural Engineering 88, March 2-6, 1988, Paris). National Center for Farm Machinery, Agricultural Engineering, Water and Forestry. Antony, France.

Colvin, T. S., C. A. Hamlett, and K. L. McConnell. 1984. Timeliness costs due to farm dispersion. Paper 84-1022. American Society of Agricultural Engineers, St. Joseph, Michigan.

Duffy, M. 1987. Estimated costs of crop production in Iowa, 1988. Publication FM-1712. Iowa Cooperative Extension Service, Ames.

Duffy, M., and J. Le Brun. 1987. Iowa farm costs and returns. Publication FM-1789. Iowa Cooperative Extension Service, Ames.

Dunaway, B. 1988. Swap, share, rent, and custom hire—Farm tested ways to cut your machinery costs. Farming 31(4): 2-5.

Farm Progress Publications. 1984. Farm equipment surveys of Iowa, Indiana, Illinois, and Wisconsin. Lombard, Illinois.

Givan, W. 1985. Economics of alfalfa production. *In* J. T. Johnson [editor] Alfalfa production in Georgia. Bulletin 898. Georgia Cooperative Extension Service, Athens.

Guy, M., M. Lund, and M. Duffy. 1988. Labor constraints on farm size and alternative farming systems. Internal report. Agricultural Economics Department, Iowa State University, Ames.

HBJ Publications. 1987. Harvesting equipment survey of Ohio, Michigan, Pennsylvania, Missouri, and Kansas. Cleveland, Ohio.

Holtkamp, R. 1988. Evaluation of different systems of multifarm use of agricultural machinery in developing countries. Paper 88.131. pp. 148-149. *In* Book of abstracts. International Conference on Agricultural Engineering (Agricultural Engineering 88, March 2-6, 1988, Paris). National Center for Farm Machinery, Agricultural Engineering, Water and Forestry, Antony, France.

McPeters, L. 1986. Marketing, then production: Organizing a marketing co-op. *In* Adapt 100. Successful Farming, Des Moines, Iowa. pp. 155.

Nelson, T. R., and D. D. Kletke. 1987. Oklahoma farm and ranch custom rates, 1986-87. Publication 140. Oklahoma Cooperative Extension Service, Stillwater.

Power, J. F., R. F. Follett, and G. E. Carlson. 1983. Legumes in conservation tillage systems: A research perspective. Journal of Soil and Water Conservation 38: 217-218.

Robbins, C. W., and D. L. Carter. 1986. Nitrate-nitrogen leaching below the root zone during and following alfalfa. Journal of Environmental Quality 9: 447-450.

Rowshan, S., and J. R. Black. 1987. Impact of farm size on machinery requirements and costs for conventional and conservation tillage systems on multiple crop farms. Paper 87-1045. American Society of Agricultural Engineers, St. Joseph, Michigan.

Schumaker, G. A., C. W. Robinson, W. D. Kemper, H. M. Golus, and M. Amemiya. 1967. Soil productivity in western Colorado improved with fertilizers and alfalfa. Technical Bulletin 91. Colorado Agricultural Experiment Station, Fort Collins.

Turner, L. W., C. W. Absher, J. K. Evans, and S. G. McNeill. 1987. Planning fencing systems for intensive grazing management. Paper 87-4084. American Society of Agricultural Engineers, St. Joseph, Michigan.

Webb Publishing Company. 1984. Farm equipment market profile of Minnesota, South Dakota, and North Dakota. St. Paul, Minnesota.

# IMPROVED ECOLOGICAL IMPACTS OF SUSTAINABLE AGRICULTURE

# LOWER INPUT EFFECTS ON SOIL PRODUCTIVITY AND NUTRIENT CYCLING

Fred P. Miller and William E. Larson

The weathering of various geologic materials under a given climatic regimen results in development of a biological system, the whole of which ultimately yields soil. The genesis of soil through this solar-energy-powered system results in increased entropy—that is, increased disorder through continual weathering, decay, and transformations within and between the spheres (i.e., lithosphere, biosphere, hydrosphere, and atmosphere) comprising the soil system.

The elemental, molecular, and mineral fractions and components of these spheres each has a characteristic cycle. None of these cycles is closed. They all leak. Thus, there must be repositories or sinks accumulating the leakages from these weathering and soil-forming processes—witness the salts of the oceans and gases (e.g., $N_2$) of the atmosphere. Some of the soluble mineral fraction released via the weathering process and not lost from the soil system is tied up in both the living biomass and decayed organic matter. Thus, the biotic system within the soil, as well as the plant community growing from the soil, is nurtured by nutrients recycled from decaying plant residue and soil organic matter, along with nutrients released through solubilization of the mineral fraction.

## Nutrient Withdrawals and Yield Response

Even under the comparatively stable virgin prairie and forest ecosystems of temperate North America, leakages from the nutrient cycles resulted in losses from these nutrient-rich soil-plant ecosystems via leaching, volatilization, erosion, and what little intact biomass may have been lost through small harvest or carried away by the wind or smoke when burned. Gains and losses of nutrients in natural ecosystems are roughly in balance so that continued biological growth or net fixation of carbon depends upon the cycling of nutrients between the biomass and the organic and inorganic stores (White, 1979). Removing or harvesting portions of the biomass from ecosystems without replacing the nutrients contained in the harvested biomass fraction ultimately depletes one or more of the nutrient stores. This nutrient removal disrupts the balance between the nutrient stores and reduces the subsequent biological yield.

Except for desert soils that have inherently low organic matter contents, placing new land under cultivation without the use of fertilizers results in a gradual but pronounced decline in crop yield, usually within a few years (Thorne and Thorne, 1978). Experiments (e.g., Rothamsted Experiment Station, England; Morrow Plots, Illinois) exceeding 100 years in duration also clearly show that reduced yields will result from continuous nutrient withdrawal from the natural nutrient pools. Tables 1 and 2 from these loca-

**Table 1. Corn yield comparison by treatment and rotation, Morrow Plots, Illinois, 1893-1982 (Morrow Plots, 1984).**

| | *Continuous Corn* | | *Corn, Oats, Hay* | |
|---|---|---|---|---|
| *Date* | *No Treatment* | *LNPK(55)** | *No Treatment* | *LNPK(55)** |
| | | *bushels/acre* | | |
| 1893 | 21.7 | | 34.1 | |
| 1899 | 50.1 | | 53.5 | |
| 1907 | 29.0 | | 80.5 | |
| 1913 | 19.4 | | 33.8 | |
| 1919 | 24.0 | | 52.2 | |
| 1925 | 19.1 | | 42.0 | |
| 1931 | 24.8 | | 45.4 | |
| 1937 | 43.1 | | 67.4 | |
| 1943 | 16.4 | | 52.0 | |
| 1949 | 20.0 | | 72.7 | |
| 1955 | 35.9 | 85.9 | 63.1 | 102.4 |
| 1961 | 46.2 | 104.0 | 72.7 | 126.4 |
| 1967 | 44.9 | 131.5 | 80.4 | 148.7 |
| 1973 | 44.0 | 128.7 | 99.3 | 164.1 |
| 1979 | 35.2 | 112.5 | 85.5 | 170.6 |
| 1982 | 28.4 | 149.3 | 97.9 | 215.2 |

*LNPK treatment: In 1955 and 1963, lime was applied at the rate of 2.5 tons/acre and 3 tons/acre, respectively; in 1955, N at the rate of 200 pounds/acre as urea was applied on corn; in 1955, triple superphosphate was applied at the rate of 150 pounds/acre of $P_2O_5$ on corn and 40 pounds/acre/year in 1956-1966. In 1955, potassium chloride was applied at the rate of 100 pounds/acre of $K_2O$ and 30 pounds/acre/year in 1956-1966. Starting in 1967, LPK was applied to maintain soil-test levels of pH $\geq$ 6.5, $P_1 \geq 40$, and K $>$ 300, with N applied to corn as urea at the rate of 200 pounds/acre/year.

**Table 2. Comparison of wheat yields under various nutrient supplements and rotations, 1852-1967, Broadbalk, Rothamsted, England (Rothamsted Experimental Station, 1969).**

| *Treatment** | *N Rate (lbs/acre)* | *Continuous Wheat 1852-1925* | *Six Fallow Cycles 1935-1964* | *Whole Period 1852-1967* |
|---|---|---|---|---|
| | | | *bushes/acre*† | |
| None | None | 12.5 | 21.1 | 14.4 |
| FYM (14 tons/acre/year) | ca. 200 | 36.2 | 42.4 | 36.4 |
| NPK NaMg salts | 86 | 32.9 | 36.8 | 33.0 |
| NPK NaMg salts | 129 | 37.5 | 41.3 | 37.5 |

*FYM = farmyard manure was not analyzed but was assumed that 14 tons/acre would contribute 200 lbs. N, 30 lbs. P, and 140 lbs. K. The "mineral manures" PKNaMg provided nearly the same amount of P and a little more than half of the K contained in the FYM.

†Original data reported in cwt/acre; conversion for wheat using 60 pounds/bushel.

tions illustrate the reduced yields for corn *(Zea mays L.)* and wheat *(Triticum aestivum L.)* when the crop must rely totally on the native soil nutrient pool, compared with nutrient supplement treatments and crop rotations.

Corn yields on the Morrow Plots remained relatively low compared to those on plots supplemented by lime, nitrogen (N), phosphorus (P), and potassium (K) beginning in 1955. Figure 1 illustrates the general trends for corn yields under two nutrient regimens (none and MLP, which consisted of manure at two tons per acre per year, plus lime and P) and three cropping systems in the Morrow Plots. The yield upturn (Figure 1) of the three no-treatment systems after the 1930s was a result of improved cultivars and increased plant populations. Both data sets (Morrow Plots and Rothamsted) also show the benefits to corn and wheat when following a rotated crop or fallow period. The Rothamsted data illustrate as well the response of wheat to manure and mineral salt supplements, indicating that plant response and yields can be sustained under either system.

In essence, these experiments show that continued cultivation and harvest without recycling or supplementing nutrients result in reduced yields and nutrient deficiency symptoms. Supplementing the crop, either continuously grown or in rotation, with nutrients, regardless of source (e.g., organic, mineral), increases yields dramatically.

## Nutrient Supplementation

Although there is no denying that the aforementioned observations and data clearly substantiate the need to supplement nutrients in the soil-crop ecosystem if sustained or increased yields are to be produced, there

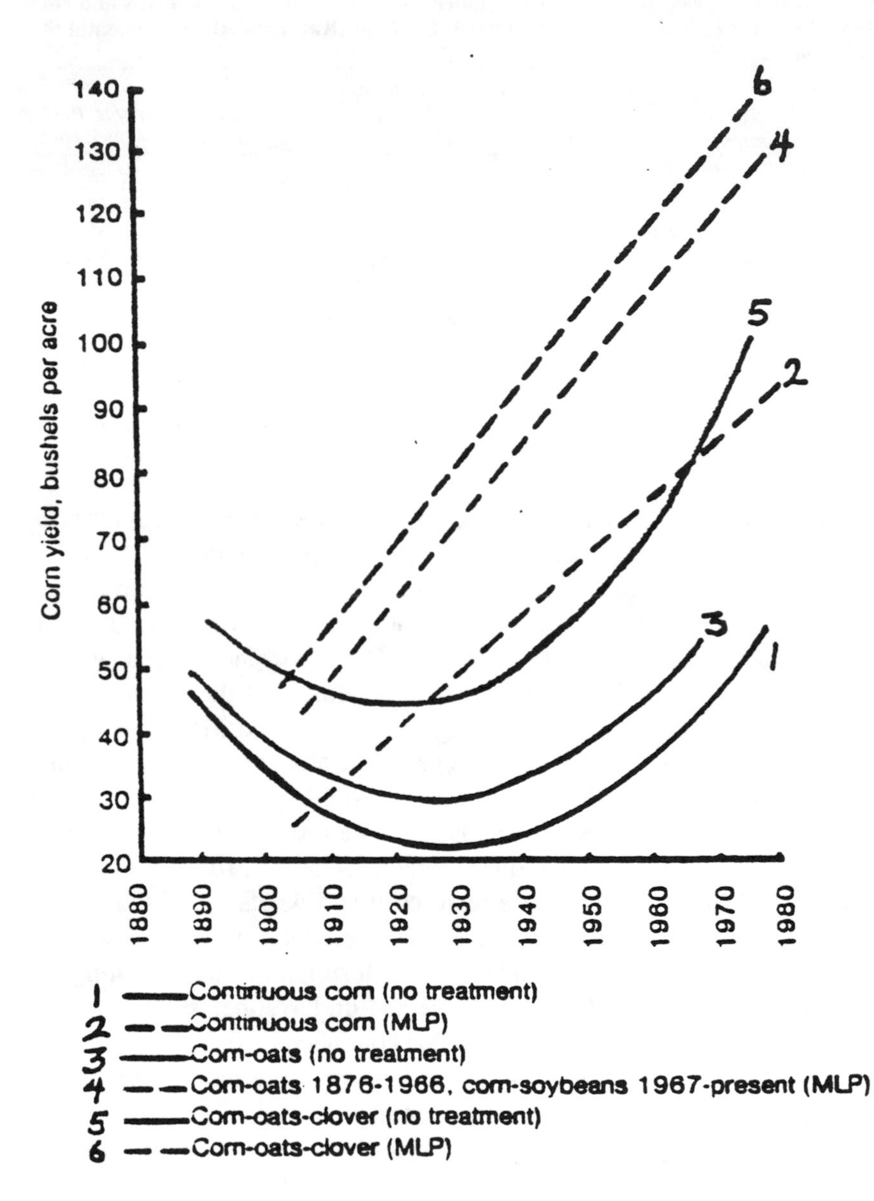


**Figure 1. Corn yields on Morrow Plants with different cropping systems and soil treatments, 1888-1978 (Morrow Plots, 1984). MLP treatment. Note. Manure applied at 2 tons per acre per year through 1908, an amount equal to the amount of dry matter removed in the crop, 1909-1966; lime applied in 1904 at a rate of 0.85 tons per acre, 5 tons per acre in 1919, 3 tons per acre in 1943, and 2 tons per acre in 1949; P applied (ground rock phosphate) on two subplots at rates totaling 13,200 pounds per acre from 1904 through 1919, and 3,300 pounds per acre (steamed bonemeal) from 1904 through 1919, with two subplots averaged. No P added after 1919.**

are differences of opinion about the most desirable nature and source (e.g., mineral, organic, legumes, etc.) of nutrients as well as quantities to be added. Poincelot (1986) stated that sustainable agriculturalists think in terms of supplementation of nutrients, whereas organic agriculturalists think in terms of replacement of nutrients. Poincelot (1986) concedes that if U.S. agriculture were to shift to organic practices, crop production would decline enough to forfeit the production now consumed by our current export demands and food costs would rise. Barrons (1988) points out that if U.S. farmers were still harvesting the same annual yields per acre in the 1980-1985 period that they did in the 1938-1940 period, an additional 418 million acres of cropland, about twice the current cropland area, would be necessary to produce the same volume of crops produced in 1980-1985.

As the scarcity of labor and, therefore, its costs have increased, the use of on-farm and recycled resources (e.g., manures, legume-hay, labor) has given way to resources purchased off the farm. Table 3 shows the substitution of energy and other purchased resources for labor in response to the technological advances in agriculture. The result has been a dramatic increase in production per unit of land rather than through expanded land area. Researchers in Kansas and Iowa assessed the substitution of N for land and found that a ton of N fertilizer replaced 24.3 and 16.1 acres of land, respectively, at the irrigated Kansas and dryland Iowa locations (Carlson, 1987). In developing countries, the substitution ratio (fertilizer to land) is even more dramatic. Results show that a ton of fertilizer applied to rice *(Oryza sativa L.)* substituted for 29 and 54 acres of land in Thailand and Peru, respectively. With wheat, a ton of fertilizer substituted for 11 acres in Argentina and 24 acres in Chili (Carlson, 1987). These yield increases require nutrient supplementation far beyond what the natural soil system can provide, at least for the macronutrients. Therefore, nutrient additions through fertilizers have become a primary production investment under today's production systems, offsetting the need to use additional land and eliminating the potential environmental impacts (e.g. erosion) from using this additional land.

A farmer's choice of nutrient sources is often predicated upon the availability of the nutrient source, its cost, and compatibility with the farmer's production system. Reliance on manure or legumes to supply nutrients in a predominantly grain-producing area, while a scientifically acceptable scenario, is precluded by the lack of adequate resources (e.g., manure, animal demand for hay) to support such a system. Thus, one's philosophical persuasion regarding a farming system is often compromised by economics and resource availability.

Singer and Munns (1987) listed the requisites for selecting a nutrient or

Table 3. Indexes of total farm input and major input subgroups (1977 = 100) (U.S. Department of Agriculture, 1987).

| | *Total Inputs* | | | | | | | | | |
|---|---|---|---|---|---|---|---|---|---|---|
| *Year* | *All* | *Non-purchased** | *Purchased†* | *Farm Labor‡* | *Farm Real Estate§* | *Mechanical Power and Machinery≈* | *Agricultural Chemicals#* | *Feed, Seed and Livestock Purchases*** | *Taxes and Interest††* | *Miscellaneous‡‡* |
| 1920 | 95 | 198 | 37 | 485 | 105 | 27 | 5 | 23 | 62 | 65 |
| 1930 | 99 | 195 | 43 | 463 | 104 | 34 | 6 | 27 | 76 | 60 |
| 1940 | 97 | 175 | 50 | 417 | 107 | 36 | 9 | 39 | 74 | 57 |
| 1950 | 101 | 166 | 60 | 309 | 109 | 72 | 19 | 58 | 83 | 63 |
| 1960 | 98 | 131 | 74 | 206 | 103 | 83 | 32 | 77 | 95 | 77 |
| 1970 | 97 | 107 | 88 | 126 | 105 | 85 | 75 | 96 | 102 | 89 |
| 1980 | 103 | 98 | 107 | 92 | 103 | 101 | 123 | 114 | 100 | 96 |
| 1981 | 102 | 97 | 107 | 90 | 103 | 98 | 129 | 108 | 99 | 108 |
| 1982 | 99 | 95 | 103 | 87 | 103 | 94 | 118 | 106 | 99 | 114 |
| 1983 | 95 | 91 | 96 | 79 | 101 | 89 | 105 | 106 | 99 | 110 |
| 1984 | 96 | 89 | 103 | 80 | 99 | 88 | 120 | 106 | 95 | 122 |

*Includes operator and unpaid family labor, and operator-owned real estate and other capital inputs.
†Includes all inputs other than nonpurchased inputs.
‡Includes hired, operator, and unpaid family labor.
§Includes all land in farms, service buildings, grazing fees, and repairs on service buildings.
≈Includes interest and depreciation on mechanical power and machinery repairs, licenses, and fuel.
#Includes fertilizer, lime, and pesticides.
**Includes nonfarm value of feed, seed, and livestock purchases.
††Includes real estate and personal property taxes, and interest on livestock and crop inventory.
‡‡Includes things such as insurances, telephone, veterinary fees, containers, and binding materials.
§§Preliminary.

fertilizer material; the fertilizer material must:

- Contain the required nutrients.
- Release these nutrients at the right time.
- Be obtained at the right price, thus recognizing the necessity for profitability.
- Be convenient to use—that is, be of minimum bulk and suitable for the user's application equipment.
- Have acceptable side effects.

Nutrient supplementation, therefore, must be balanced against (1) the nutrient needs of the crop and removal through harvest, (2) the nutrient resources available in the soil system and (3) those nutrients added through crop residues, symbiotic N fixation, manures and other organic sources, and commercial fertilizers.

## Nutrient Balance in the Soil Ecosystem—Inputs

The balance of nutrients in soil ecosystems, whether natural or agricultural, can be described by the following equation (Follett et al., 1987):

$$RN_{tn} = \sum^{tn} (AP_t + AR_{\Delta t} - RM_{\Delta t} - L_{\Delta t})$$

where RN is the soil inorganic and organic nutrients remaining at time (tn), AP is the soil inorganic and organic nutrients present at time t, AR is the inorganic and organic nutrients added or returned to the soil during the time interval Δt, RM is the plant nutrients removed with the harvested product during the time interval Δt, L is the inorganic and organic nutrients lost during the time interval Δt, t is the beginning time, tn is the ending time, and Δt is the time interval between t and tn.

The equation simply states that if nutrients removed are greater than additions, the reservoir of nutrients remaining within the total pool will decline.

***The Native Nutrient Pool.*** Soils of the United States contained considerable amounts of organic carbon (C) and N at the time of modern human intervention. For example, many of the highly fertile mineral soils developed under grass (mollisols) in the Corn Belt contained as much as 15,000 pounds of N per acre. This amount accumulated over thousands of years from symbiotic N fixation by native legumes, from free-fixing microorganisms, from N contained in precipitation, and from geologic materials. Following the breaking of sod or clearing of trees for cultivation, the organic N declined rapidly because of (a) mineralization and uptake by crops removed from the land, (b) mineralization and leaching or denitrification, or (c) removal by erosion. In many mineral soils developed under grass in the Corn Belt and northern Great Plains, the C and N contents today are about one-half to two-thirds of the amount contained in the virgin state (Haas et al., 1957). Thus, a Typic Hapludoll having 15,000 pounds of organic N per acre in its native state may have lost about 7,500 pounds per acre. If the soil had been cultivated for 150 years, an average of 50 pounds of N was removed each year.

The organic matter and organic N content of most cultivated soils probably have now reached an equilibrium under current management systems. Under some conservation tillage systems, the organic C and N contents of the soil appear to be increasing slightly, but will not reach the levels contained in the soil in its native state. Larson et al. (1972) evaluated changes in organic C and N over 11 years on a Typic Hapludoll under a continuous corn-moldboard plow system in which differing amounts of corn stover and alfalfa *(Medicago sativa L.)* were applied annually. They concluded that 4,500 pounds per acre per year of residue were required to maintain organic C at the initial level of 1.81 percent.

The nutrients supplied by a soil to a crop within a year vary widely. For many high N-demanding crops, such as corn, the amount made available may range from 0 to 100 pounds or more per acre per year. More N will be made available from soils high in organic matter and in years following a legume in the rotation. If the cropping history and the characteristics of the soil and climate are known, experienced soil scientists usually can make reasonable estimates of available N. In dry areas, a nitrate soil test at the start of the season is desirable to determine the amount of carryover N available. This can then be taken into account in determining fertilizer applications.

A soil test for P and K is usually a good indicator of the availability of these two elements.

***Crop Residues.*** Cultivated crop residue is an important source of organic material in the soil and contains significant quantities of nutrients. An estimated 429 million tons of crop residue is produced each year in the United States (USDA, 1978), returning to the soil about 3.3, 0.4, and 3.5 million tons of N, P, and K, respectively. About 40 percent of this residue is produced in the Corn Belt. Nationally, corn, wheat, and soybeans *(Glycine max L. Merr.)* produce the greatest amounts of residue (Larson et al., 1978). Crop residue production is usually estimated by multiplying grain yield by average straw/grain ratios (harvest index) (Gupta et al., 1979). Using these ratios, a 150 bushel-per-acre corn grain yield (harvest index of 1.0) produces 8,400 pounds of residue, and a 60-bushel winter wheat crop (harvest index of 1.7) produces 6,120 pounds of residue per acre.

Crop residues, if returned to the soil, are a source of available nutrients or, if removed from the soil, can be a drain on the nutrient supply. Holt (1979) gives the percentages of N, P, and K in crop residue as a percentage of those in residue plus grain. Examples of the N, P, and K percentages for corn are 43, 41, and 78; for wheat, 29, 15, and 70; and for soybeans, 38, 36, and 48 percent, respectively. Hence, if a 150-bushel-per-acre crop of corn grain contains 135, 25, and 30 pounds of N, P, and K in the grain, respectively, corresponding amounts in the residue would be 100, 17, and 110 pounds per acre. Contents of N, P, and K in all crop residue returned to the land represent about 31, 18, and 72 percent, respectively, of the average amounts applied to all crops as fertilizer in the United States. Some crops, such as corn, that take up large amounts of nutrients can sometimes increase the available P and K in surface horizons by root withdrawal from deep horizons and deposition on or within the surface horizon.

***Animal Manures.*** Follett et al. (1987) estimated that 174 million tons of animal manure are produced annually in the United States; this manure contains 891,642 tons, 649,028 tons, and 1,242,942 tons of N, P, and K, respectively. These quantities are 8.4, 16.5, and 25.8 percent of the amounts applied as N, P, and K in mineral fertilizer, respectively. Follett et al.'s (1987) calculations are based on estimates of feces and urine produced by the major domestic animal populations. About 90 percent of the approximately 174 million tons of manure generated annually under both confined and unconfined conditions is estimated to be used on land. Manure from animals represents about 22 percent of all organic wastes; crop residue comprises about 54 percent (Follett et al. 1987). Although all of the other organic wastes (garbage, sewage sludges, etc.) represent 24 percent of the total, they represent only 1.0 percent of that applied to land. These wastes appear to offer a potential resource for use on land.

***Commercial Fertilizers.*** Commercial fertilizers are the major source of nutrients added to cropland. In 1977, a total of 10.6 million tons of N, 2.44 million tons of P, and 4.81 million tons of K were applied to cropland. Average N, P, and K rates applied to corn in 1984 were estimated to be 138, 28.6, and 72.3 pounds per acre, respectively. For wheat, the N, P, and K rates were 62.5, 16.1, and 38.4 pounds per acre, respectively (USDA, 1985). Estimated corn yield increases from fertilizer range from 20 to 50 percent (Walsh, 1985).

***Fixed Nitrogen.*** It has been estimated that total annual biologically fixed N from U.S. agriculture amounts to 7.94 million tons (Follet et al., 1987).

Because most of this biologically fixed N is immobilized in the legume plant herbage and ultimately harvested, only a relatively small portion is returned to cropland via residue, primarily in the form of seed legumes and alfalfa. Of the estimated 3.3 million tons of N returned to U.S. cropland annually as crop residue, seed legume residue accounts for about 793,649 tons, or 24 percent of this amount. Soybeans account for 96 percent of residue N from seed legumes. Follett et al. (1987) estimated that 48 percent of the 793,649 tons of N contained in seed legume residue is due to biological $N_2$ fixation. Therefore, about 385,800 tons of N per year, or about 12 percent of the N estimated to be returned by all crop residue in the United States, result from biological $N_2$ fixation by seed legumes. Likewise, Follett et al. (1987) estimated that alfalfa also returns about 385,800 tons of symbiotically fixed N annually to U.S. cropland. Thus, the annual biologically fixed N contribution from seed legumes and alfalfa to U.S. cropland accounts for approximately 24 percent of the total N contribution from all crop residue.

## Nutrient Balance in the Soil Ecosystem—Removals

***Erosion.*** Soil erosion is a serious factor in soil degradation, including nutrient loss. Follett et al. (1987) estimated that the total additions of N, P, and K from fertilizer, manure, crop residue, and N from biological fixation, compared to the total amounts of N, P, and K in eroded sediments, are about 1.5:1.0, 1.9:1.0, and 0.2:1.0, respectively. The available nutrients in sediments are a small percentage of that applied as fertillizer, however.

***Leaching.*** Some nutrients were leached out of the soil root-zone under cultivated conditions even before the advent of modern fertilizers and, indeed, even in the soil's virgin state. Earlier, it was estimated that perhaps an average of 50 pounds of N per acre were removed each year from the mollisols of the western Corn Belt during their 150 years of cultivation. The amount mineralized from the native organic matter during the early years of cultivation was probably on the order of 100 pounds of N per acre. Prior to 1940, corn yields averaged 40 bushels per acre, which contained approximately 50 pounds of N per acre. Most excess mineralized N over that removed in the corn grain was probably leached from the root zone, although some may have been denitrified.

Since the advent of modern fertilizers, the input of nutrients to cropland has increased dramatically, as have the amounts removed through increased yields from harvested crops. No generalizations can be made about the amounts leached, although a hypothetical case can be made to illustrate

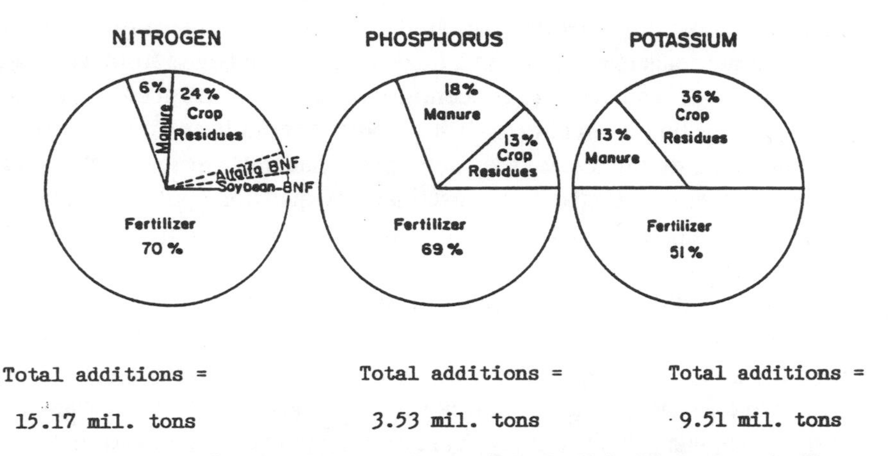


**Figure 2. Additions of N, P, and K to cropland soils in the United States from fertilizers and by return of crop residue and manures (Follett et al., 1987).**

the magnitude. If four inches of water containing 10 milligrams per liter of N is leached annually, the total N leached is nine pounds per acre. If the four inches of water averages 40 milligrams per liter, the total N leached is 36 pounds per acre.

***Denitrification.*** A major loss of N occurs as conversion of nitrate to gaseous forms, such as nitrous oxide and denitrogen. Under wet, poorly aerated soil conditions, where anaerobic conditions in the soil medium occur either in part or wholly, losses of N can be considerable. It is not possible to generalize about the magnitude of the losses because of the difficulty in measurement and the spatial variabilities over landscapes.

***Shifts in Nutrient Use.*** The additions of N, P, and K to cropland soils in the United States are summarized in figure 2 (Follett et al., 1987). Off-farm nutrient sources are by far the major component of today's nutrient budget. Shifting to other on-farm nutrient sources will require major farm management changes.

Today's nutrient cycles are more geographically dispersed than those of yesteryear, when on-farm nutrient cycling was nearly closed. The three primary nutrients may be manufactured or mined, processed, and shipped from sources far from the cropland upon which they are finally applied. Likewise, the harvested biomass carrying large quantities of nutrients off the farm can be dispersed over long distances to municipalities or feedlots, rendering the recycling of these nutrients impossible, at least back to the land from which they were harvested. The wastes resulting from the consumption of the biomass can now present waste management problems to these municipalities, feedlots, or other consumers of the biomass. The harvested nutrients may end up in a landfill or be discharged from a sewage treatment plant in such dilute concentrations and so far from cropland that practical recycling is precluded. Thus, today's agriculture and infrastructures have greatly broadened the geographic sphere of nutrient cycles. Can these systems be sustained? The sustainability question applies to the adequacy of the nutrient resources as well as the cropping systems within which they are used.

Edwards (1988) defines low-input/sustainable agriculture as "integrated systems of agricultural production that are less dependent on high inputs of energy and synthetic chemicals, and more dependent on intensive management than conventional monocultural systems. These lower input sustainable systems maintain or only slightly decrease productivity, maintain or increase net income for the farmer, and are ecologically desirable and protective of the environment."

For nutrient management under this definition of lower input/sustainable agriculture, Edwards (1988) argues that synthetic fertilizer use can be compensated for mainly by use of rotations, particularly those involving legumes, and by the use of organic manures and on future developments in nutrient sources and management. More than a half century of experience and literature from land grant university agricultural experiment stations and other sources scientifically verifies the soundness of agronomic practices such as crop rotations, use of legumes in rotations, use of manure, and application of nutrient supplements from various sources. Even under today's production technology, crop yields are generally higher under rotations than in continuous monoculture, regardless of the nutrient source (Sahs and Lesoing, 1985; see also Figure 1). Shifting to other nutrient sources, such as manure, legumes, and crop residue, is not without economic and environmental costs, however.

If a goal of sustainable agriculture is to reduce off-farm inputs (fertilizers), N, P, and K inputs obviously must be increased from manures, biological N fixation, and other sources, such as wastes. Because fertilizers now comprise about two-thirds of the total N, P, and K additions to U.S. cropland, major additional amounts of nutrients from manures and biological N fixation clearly must be achieved, or other sources of nutrients must be found.

Additional nutrients from manure are not likely without major additions of livestock. More efficient use of animal manure is possible by improved collection, storage, and distribution on land. An obstacle to improved distribution is the concentration of livestock in large rearing facilities. And there is both an economic and ecological price to pay for handling, collecting, storing, and distributing this bulky material, including soil compaction from heavy spreading equipment.

Increased use of domestic wastes, such as sewage wastes and garbage, on land has some potential. Currently, about 8.4 million tons of sewage sludge are produced nationally. This sludge contains about 328,000, 210,000, and 34,000 tons of N, P, and K, respectively. About 70 pounds of sewage sludge is produced per person annually. The N, P, and K in sewage sludge is about 3, 9, and 0.7 percent of that used in fertilizer, respectively. Major obstacles to widespread use of sewage sludge include its contamination with potentially toxic chemicals and heavy metals as well as the aforementioned costs of handling, storage, transportation, and distribution.

Likewise, domestic garbage contains large amounts of nutrients. An estimated 150 million tons of municipal solid wastes are produced annually in the United States. Of this, about 80 percent is organic.

Currently, about 5 percent of the N taken up by harvested crops comes from symbiotic fixation, chiefly by alfalfa and soybeans (Follett et al., 1987). The United States grows about 70 million acres of soybeans and 26 million acres of alfalfa. Assuming fixation efficiency remains constant, the acreage of these crops will have to be increased markedly to have a significant impact on total N availability to crops. Some improvement in fixation efficiency can be expected from research.

## Nutrient Efficiency Through More Intense Management

Today's fertilizer inputs undoubtedly could be reduced to some extent without greatly impacting yields. Unneeded nutrients are sometimes applied when soil test results clearly indicate that such supplements are unnecessary. Another component of reducing fertilizer use could be accommodated by increasing management intensity—that is, being more realistic in setting yield goals and calibrating application equipment. Studies in Nebraska and Iowa (Schepers and Martin, 1986; Padgitt, 1986) indicate that about half of the farmers in these states overestimate their yield goal by at least 20 to 25 percent.[1] Likewise, many farmers, in planning their nutrient budgets, do not take full credit for the available N stores or pools contributed by the soil organic matter, irrigation water, some crop residue or cover crops, legumes, manure, or carryover from previous fertilization. As Hallberg (1987) pointed out, it is easy to see why the overestimate of yield goals, coupled with inadequate credits for N in the other pools, can result in excessive leaching losses of nitrate-N.

[1]It must not be inferred from this statement that farmers overfertilize by 20 to 25 percent, only that they tend to overestimate their yield goals by this amount.

Farmers manage their resources as they do because they must make decisions based on the probability of a given outcome—an unknown probability determined by the vagaries of weather, insects and disease pressures, weed competition, rooting zone stresses, and market fluctuations. Many farmers apply high rates of fertilizer (particularly P and K) as much to offset the effects of poor growing seasons as to take advantage of good years. It can also be argued that because many, if not most, farmers do follow recommended N application rates, they are taking credit for the nutrient stores in soil organic matter, crop residues, and other nutrient "pools" because such credits are built into these recommendations. Nevertheless, simple management alertness can result in more nutrient efficiencies in current production systems, over and above the benefits of using rotations.

Regardless of how a farmer manages nutrient supplements, accountability must be measured in terms of profit and sustainability. Holt (1988) proposed a graphic model (Figure 3) that can be used to visualize the planning challenge facing the farmer. Because the variable cost curve is not linear, the point of maximum profit will occur at different yield levels, depending upon a commodity's price. Figure 3 depicts three different price-income scenarios. Once again, this model, like all economic models, shows that the point of maximum profit is below the point of maximum yield. As Holt (1988) pointed out, this simple two-dimensional graph cannot fully portray the real world production system. "The total-cost curve is an extremely complex, multidimensional response surface. The farmer cannot predict with certainty the exact point of maximum return or minimum loss because there is uncertainty associated with the responses of crops to inputs and the prices that can be expected" (Holt, 1988). Subsequent research into production and marketing systems can reduce this uncertainty with the aid of computer models and expert systems through which farmers can obtain risk management strategies (Holt, 1988).

Using Edwards' (1988) definition of lower input/sustainable agriculture, the concept is not to reduce nutrient inputs to the production system except when necessary to attain nutrient efficiencies and reduce nutrient losses, but rather to replace synthetically produced and energy-intensive nutrient sources with onfarm resources when possible with economic advantage (legumes, manures) and intensive management (rotations, accounting for contributions from all nutrient pools). It is important to distinguish between Edwards' concept of lower input/sustainable agriculture and other concepts in which yields are also reduced significantly with lower inputs. As Holt (1988) pointed out, those who say lower input systems are appropriate for all land are flying in the face of economic reality. Likewise, the environmental impact of well-managed nutrient additions should not

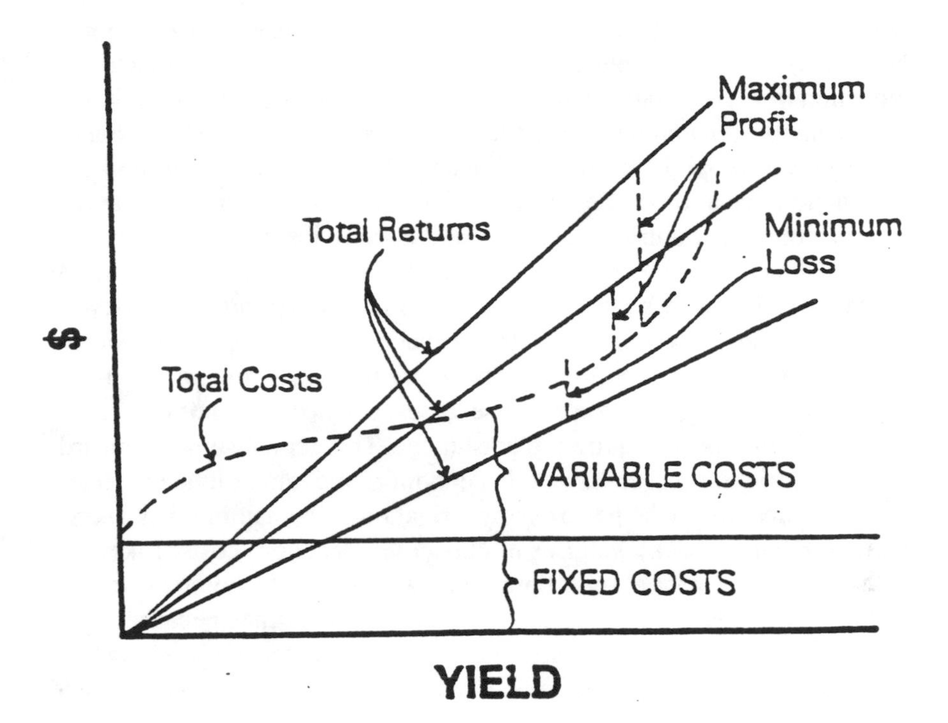


**Figure 3. A graphic model of costs and profits associated with productivity goals (Holt, 1988).**

be greatly different, regardless of the nutrient source, because each nutrient source has its own characteristic cycle with leakage pathways. Referring to Singer and Munns' (1987) concept of selecting nutrient materials, the ecological impact of some on-farm nutrient sources (e.g., manures) could be more negative than commercial fertilizers. This is because of the potential soil compaction under heavier loads (to obtain the same nutrient input) and the potential for runoff to streams, which contributes not only nutrients but also biochemical oxygen demand (BOD) loads and microbial contamination.

## Considering Several Nutrient Supplement Scenarios

There is a finite point for each cost/return scenario in which maximum profit (or minimum loss) occurs (Table 3). Table 4 provides several scenarios for N use. Even under the most optimistic low-input scenario, in which manure, legumes, increased nutrient management efficiencies, wastes, and nutrient reductions are used, the fertilizer requirement remains the greatest nutrient component. Scenarios 3 and 4 in table 4 illustrate that although N reductions can take place yield reductions will also occur—thus the reduction in N contribution from crop residue. We contend that, although N substitutions for commercial fertilizers can take place in the United States, moving from scenario 1 to scenario 3 in table 4 will not be without economic and environmental costs. Certainly, some specific farm enterprises can rely on noncommercial nutrient sources, but the scenarios in table 4 reflect the national situation, which is not conducive to major shifts in nutrient alternatives. In our view, to sustain current and projected production levels, commercial fertilizer will remain the primary nutrient supplement source.

Regardless of nutrient source, the nutrient cycles remain open and vulnerable to leakages or losses. These losses, coupled with nutrient removals in farm products, require replenishment. This replenishment, regardless of nutrient source, has both economic and environmental costs. Spreading manure, whether by spray irrigation or by hauling, has energy and equipment costs as well as soil impact (compaction) costs. Likewise, manure has the potential runoff problem, such as BOD loading of streams and infectious agents, as well as the added cost of storage because application can occur only during specific "windows" prior to and during the growing season. Growing legumes in rotation or as a cover crop is a sound

**Table 4. Several scenarios for nitrogen utilization and sources in U.S. agricultural production projected to the year 2000.**

| | *N Use Scenarios Based on 1988 Comparison* | | | | |
|---|---|---|---|---|---|
| *N Source* | *1980s* | *1* | *2* | *3* | *4* |
| | | | % | | |
| Fertilizer | 71 | 51 | 56 | 48 | 32 |
| Manure* | 5 | 8 | 8 | 8 | 8 |
| Nonlegume crop residue | 18 | 18 | 18 | 16 | 16 |
| Biological N fixation† | 6 | 8 | 8 | 8 | 12 |
| Wastes‡ | tr | 5 | tr | tr | 5 |
| Efficiency§ | — | 10 | 10 | 10 | 12 |
| Reduction≈ | — | 0 | 0 | 10 | 15 |
| Total use t × $10^6$ | 15.17 | 15.17 | 17.17 | 13.65 | 12.90 |

*Manure—better on-farm storage, spreading, incorporation, improved credit for N in determining amount of use.

†Biological N fixation—greater use of legumes in rotation, greater use of legume cover and green manure crops, improved rhizobia.

‡Wastes—sewage sludge, domestic garbage, composts.

§Efficiency—better soil tests, N use based on realistic yield goals, timing, placement, less erosion, less leaching, and denitrification.

≈Reduction—10% reduction assumes maximum economic yield now or 95% of optimum, future 85% of optimum (scenario 3); 15% reduction reduces yield below economic maximum (scenario 4).

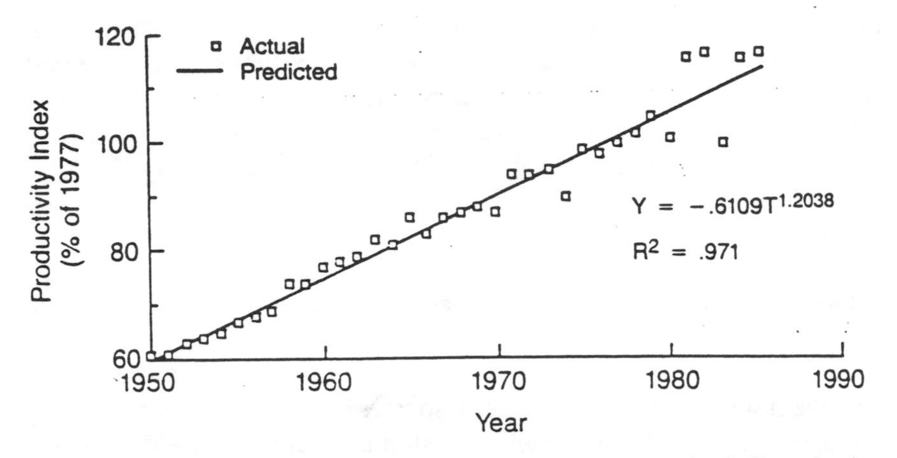


**Figure 4. Actual and predicted productivity index (output per unit of all production inputs) for U.S. agriculture (Tweeten, 1987).**

agronomic practice, but it presents an economic dilemma for cash grain farmers who have no market for the forage. The more farmers who incorporate legume forages into their rotations, the higher is the probability that supply/demand forces will require farmers to consider other nutrient options. This is because the land must be forfeited from cash grain production to produce forages that most likely will have limited marketability—a situation now faced by millions of cash grain farmers. Incentives through government programs could increase the amount of forage legume acreage. Figure 2 shows the relatively small contribution of N from seed legumes and alfalfa; a large acreage increase would be necessary to increase this N source significantly.

To sustain current levels of agricultural productivity will require major nutrient infusions into the production process. Certainly, increased efficiencies and on-farm sources can be captured to reduce these off-farm nutrient sources. But unless additional land is brought into production to offset the reduced production resulting from lower nutrient inputs, there appears to be no alternative to the concept of nutrient supplementation as a major production factor.

Is current agricultural technology sustainable? Tweeten (1987) analyzed yield trends in U.S. agriculture and concluded that, although the rate of yield increases has declined, absolute yields are increasing (Figure 4). He pointed out that successive technological revolutions bring ever-higher rates of productivity growth even though the growth rate slows as the revolution matures. As pointed out in the Council for Agricultural Science and Technology report on the long-term viability of U.S. agriculture (CAST, 1988), "The U.S. is between revolutions, a difficult time for prediction, and of interest is whether the next decades will be characterized by diminishing productivity rates of the past revolution or by accelerating rates of the incipient revolution featuring the new biotechnology and information system."

## Considering the Alternatives

From the standpoint of nutrient management in agriculture and its sustainability, several alternatives seem possible.

First are the inorganic nutrient sources and energy requirements to produce them sustainable? Both of these resources have finite limits. Economics and innovation will determine when we shift to other sources—for example, constructing an ammonia plant next to a nuclear power plant in a developing country and using the off-peak power to hydrolyze water for a hydrogen source rather than using fossil fuel or, more likely, shifting from natural gas to coal.

Second are the efficiencies that can and must be captured in the current use of nutrients, including accounting for existing, on-farm nutrient pools (soil mineral and organic matter, legumes, manure, crop residue, carryover nutrient stocks, irrigation, concentrations, etc.). These increased efficiencies will require more intensive management, driven by economics and by environmental regulation.

Third, the sustainability of agriculture in countries in the developed and developing worlds is not predicated on one set of management options. The variation in soils, ecosystems, and farming practices allows for a variety of nutrient management scenarios. Certain soil and climatic conditions can apparently accommodate almost total reliance on inorganic nutrient supplementations under a given farming system. But some other soils must be nurtured with high organic returns, regardless of nutrient source. One can make the argument for many ecosystems (e.g., marginal, erosion-prone land; soils with rooting zone constraints, both physical and chemical; soils with low moisture-holding capacity) that sustainability can occur only with high levels of inputs. Much of the agriculture now practiced on marginal ecosystems in Africa and elsewhere is not sustainable because of low inputs. Likewise, agricultural systems in large parts of Florida and California could not be sustained under low-input systems. An important component of all agricultural systems is climate, which may not be matched with ideal soils and landscapes. To tap this climatic resource, high inputs must be made in certain ecosystems. Assuming that environmental impacts can be minimized in these ecosystems, economics, not conference-derived

derived definitions, will eventually dictate whether such agricultural systems are sustainable.

## Conclusions

More than 100 years of systematic evaluation of nutrient supplementation under a variety of nutrient sources and cropping systems (Tables 1 and 2, Figure 1) has shown that various management systems are sustainable and high-yielding. The maximum economic yield concept (Figure 3) will have a different point on the production/yield axis each year, depending upon the price of a commodity and nutrient resource. The level of management required to adjust this production response to costs will also vary, from year to year and from cropping system to cropping system. The ecological price for operating these systems will vary depending upon the weather and ecosystem in which the cropping system is deployed. Many sustainable systems can be devised. Neither exceeding the resource requirements nor undershooting the resource investments for the desired yield is economically or ecologically sustainable.

World population pressures, to say nothing of the potential markets for biomass in the future to fuel chemical feedstock needs, will not allow us to return to the on-farm recycling and resource dependence of the first 99 centuries of agricultural history. It would appear that, at least for the foreseeable future, a viable fertilizer industry will be needed. It may well become an industry that will supply nutrients as the final component in a total nutrient management system after all other nutrient sources in the total nutrient pool have been tapped. But this concept is no different from the one most agronomists and land grant experiment stations have espoused for years.

## REFERENCES

Barrons, K. C. 1988. Body building for soils. Science of Food and Agriculture, CAST 6(2): 22-26.

Carlson, C. W. 1987. Technology unlocks cropland productivity. pp. 309-314. *In* W. Whyte [editor] Our American land, 1987 yearbook of agriculture. U.S. Department of Agriculture, Washington, D.C.

Council for Agricultural Science and Technology. 1988. Long-term viability of U.S. agriculture. Report No. 114. CAST, Ames, Iowa.

Edwards, C. A. 1988. The concept of integrated systems in lower input/sustainable agriculture. American Journal of Alternative Agriculture 2(4): 148-152.

Follett, R. F., S. C. Gupta, and P. G. Hunt. 1987. Conservation practices: Relation to the management of plant nutrients for crop production. *In* Soil fertility and organic matter as critical components of production systems. Special Publication No. 19. Soil Science Society of America, Madison, Wisconsin.

Gupta, S. C., C. A. Onstad, and W. E. Larson. 1979. Predicting the effect of tillage and crop residue management on soil erosion. Journal of Soil and Water Conservation 34: 7-9.

Haas, H. J., C. E. Evans, and E. F. Miles. 1957. Nitrogen and carbon changes on Great Plains soils as influenced by cropping and soil treatments. Technical Bulletin 1164. U.S. Department of Agriculture, Washington, D.C.

Hallberg, G. R. 1987. Agricultural chemicals in ground water: Extent and implications. American Journal of Alternative Agriculture 2(1): 3-15.

Holt, D. 1988. Agricultural production systems research. Phi Kappa Phi Journal 68(3): 14-18.

Holt, R. F. 1979. Crop residue, soil erosion, and plant nutrient relationships. Journal of Soil and Water Conservation 34: 26-28.

Larson, W. E. 1978. Residues for soil conservation. pp. 1-15. *In* Crop residue management systems. American Society of Agronomy, Crop Science Society of America, Soil Science Society of America, Madison, Wisconsin.

Larson, W. E., C. E. Clapp, W. H. Pierre, and Y. B. Morachan. 1972. Effects of increasing amounts of organic residues on continuous corn: II. Organic carbon, nitrogen, phosphorus, and sulfur. Agronomy Journal 64: 204-208.

Morrow Plots-A Century of Learning. 1984. Bulletin No. 775. Illinois Agricultural Experiment Station, Urbana.

Padgitt, S. 1986. Agriculture and groundwater quality as a social issue: Assessing farming practices and potential for change. pp. 134-144. *In* Agricultural impacts on ground water. National Water Well Association, Worthington, Ohio.

Poincelot, R. P. 1986. Toward a more sustainable agriculture. AVI Publishing Co., Inc., Westport, Connecticut.

Rothamsted Experimental Station Report for 1968, Part 2. 1969. Harpenden, England.

Sahs, W. W., and G. Lesoing. 1985. Crop rotations and manure versus agricultural chemicals in dryland grain production. Journal of Soil and Water Conservation 40(6): 511-516.

Schepers, J. S., and D. L. Martin. 1986. Public perception of groundwater quality and producers' dilemma. pp. 349-411. *In* Agricultural impacts on ground water. National Water Well Association, Worthington, Ohio.

Singer, M. J., and D. N. Munns. 1987. Soils, an introduction. Macmillan Publishing Co., New York, New York.

Thorne, W., and A. C. Thorne. 1978. Land resources and quality of life. *In* J. S. Bethel and M. A. Massengale [editors] Renewable resource management for forestry and agriculture. University of Washington Press, Seattle.

Tweeten, L. 1987. Agricultural technology—the potential socioeconomic impact. Bovine Practitioner 22: 4-14.

U.S. Department of Agriculture. 1978. Improving soils with organic wastes. Report to Congress in response to Section 1461 of the Food and Agriculture Act of 1977 (P.L. 95-113). Washington, D.C.

U.S. Department of Agriculture. 1985. Inputs, outlook and situation report. IOS-7. Economic Research Service, Washington, D.C.

U.S. Department of Agriculture, 1987. 1987 fact book of U.S. agriculture. Miscellaneous Publication No. 1063. Washington, D.C.

Walsh, L. M. 1985. Plant nutrients and food production. *In* Proceedings, plant nutrient use and the environment symposium, Kansas City, Missouri, October 21-23. The Fertilizer Institute, Washington, D.C.

White, R. E. 1979. Introduction to the principles and practices of soil science. John Wiley and Sons, New York, New York.

# SOIL EROSION AND A SUSTAINABLE AGRICULTURE

John M. Laflen, Rattan Lal, and Samir A. El-Swaify

A sustainable agriculture is an agriculture maintained to meet humankind's needs. There is little argument that humankind will require land upon which to produce sustenance and that soil erosion is a major threat to the available agricultural land—hence, a major threat to a sustainable agriculture. In the United States, this threat led to a federal agency for the control of soil erosion—the Soil Conservation Service of the U.S. Department of Agriculture. Other regions of the world also have programs to deal with the soil erosion threat to a sustainable agriculture.

In some regions of the world, it is too late to create a sustainable agriculture. The collapse of the Mayan civilization in South America was caused at least in part by soil erosion (Deevey et al., 1979). Loudermilk (1953) attributed agriculture's decline in areas where civilization began to siltation in irrigation canals caused by soil erosion in the watersheds contributing the irrigation water. In regions of Africa, soil erosion is contributing to present-day starvation. In the United States, considerable land, particularly in the Southeast, has been lost to crop production because of soil erosion.

A sustainable agriculture is one that can provide for humankind's needs. These needs in the future, as in the past, will not be static. An agriculture that is sustainable today will likely not be relevant tomorrow. An agriculture that provided for the draft horse in the early 20th century in the United States might have been considered sustainable in its day. Today, however, the system would hardly be relevant and certainly would be incapable of providing the agricultural commodities required near the end of the 20th century. But the agriculture of the early 20th century required an adequate supply of good land, and the agriculture of the late 20th century also requires an adequate supply of good land, as will the agriculture of the 25th century in all likelihood.

What an adequate supply of land will be is anyone's guess. Our guess is that the amount of land required will not decline, because of increases in the world's population, an improvement of diet hoped for within the world's population, and an increased reliance on products grown on the land for use as energy and as raw materials for industrial processes. We take as a given, then, the need to maintain the land resources of the world if we are to have a sustainable agriculture. We believe this is the first and foremost requirement for a sustainable agriculture. Without it, there can be no sustainable agriculture.

## Allowable Erosion Rates for a Sustainable Agriculture

Smith (1941) first expressed the concept of an allowable erosion rate for a sustainable agriculture. Since then, there has been considerable debate about just what an allowable erosion rate should be so that agriculture can be sustained. It is unlikely that our comments will end the debate.

Several scientists have expressed the view that geologic rates of soil erosion are not harmful (Kellogg, 1948; Bennett and Chapline, 1928; and Nikiforoff, 1942). Apparently, rates that are very low are considered natural and, hence, not undesirable. McCormack and Larson (1981) concluded that "soil productivity is directly tied to the overall thickness of the rooting zone." Other studies using modeling techniques have shown the same results.

Although the results of such studies indicate little cause for immediate alarm when erosion rates are moderate, the time frames usually evaluated are so short as to raise concern about a sustainable agriculture that will be producing to meet the world's needs several centuries from now. Perhaps these are not time frames we should worry about. They certainly render time frames meaningless. But examples exist the world over where agriculture cannot be maintained because of what took place centuries ago.

In our view, a sustainable agriculture is one where topsoil removal is very slow, near what might be called a "geologic rate," and at the rate at which topsoil is formed from subsoil and near the rate at which bedrock and other subsoil materials are transformed into a satisfactory material in which a crop root can thrive. Although these rates generally are too low to measure, some estimates have been made. Menard (1961) estimated some past geologic rates for Appalachia as 6.2 centimeters per 1,000 years, for the Mississippi drainage basin as 4.6 centimeters per 1,000 years, and for the Himalayas-Indian Plains-Indian Ocean as 21 centimeters per 1,000 years

Owens and Watson (1979) estimated soil formation rates of 11 and 4.1 millimeters per 1,000 years for two watersheds in Rhodesia, based on rates of granite weathering. Soil loss rates of 1 metric ton per hectare would result in the loss of less than 10 millimeters of soil in a thousand years. On many soils, this loss would be considered inconsequential, only somewhat greater than the rate at which bedrock and other materials form subsoil.

Sustainable agriculture cannot exist in the presence of intense gullying, but such gullying seldom occurs unless soil erosion is a major concern. And intense gullying can occur from very rare storms in areas where severe erosion would not normally be the case.

Our viewpoint, then, is that soil erosion standards currently used are considerably higher than would support a sustainable agriculture, if we consider that agriculture in five or more centuries will require at least as much good land as it does presently. Who knows what wonders will come along as a result of mother nature and technology? Our bet is that, whatever it is, agriculture will always need a vast supply of arable land to provide for the earth's needs.

## Soil Erosion Processes

Soil erosion by water is a process of detachment and transport by raindrops and flowing water. The energy of raindrops detaches and transports soil short distances to nearby rills. Detachment, transport, and deposition of soil may occur in these rills and in progressively larger drainage channels as runoff water accumulates from many smaller channels and moves downslope toward reservoirs or oceans.

Raindrop detachment and transport is termed "interrill erosion." Generally, the rate of interrill erosion is proportional to the square of the rainfall intensity, but this relationship apparently depends upon soil properties (Meyer, 1981). Interrill erosion is controlled by reducing the velocity of raindrops reaching the surface and increasing the resistance of soil to detachment by raindrops. The velocity of raindrops reaching the surface can be reduced by keeping the soil covered with a mulch of residue and by maintaining a canopy that intercepts rainfall. Surface resistance may be increased by using agronomic practices that strongly aggregate the surface soil and by using management practices that seldom disturb the soil.

Soil detachment in small rills and concentrated flow channels that transport excess rainfall overland is caused by the hydraulic shear of the flowing water, by undercutting of channel banks, and by headcutting. Soil detachment in channels can be reduced by reducing surface runoff rates and volumes or by introducing channels that are nonerodible. Nonerodible channels are those that do not erode because the hydraulic forces are less than those required to detach channel material. Grassed waterways are examples of channels designed to be nonerodible on the basis of their slope, cross-section shape, and channel material.

Soil erosion by water can be controlled through a combination of conservation and cropping and management practices that result in lower rates and amounts of surface runoff, reduce the energy of rainfall as it strikes the ground, increase the resistance of soil to detachment, reduce the hydraulic shear of runoff water, and discharge excess surface runoff through nonerodible rills and channels.

For example, contouring can reduce the hydraulic shear of runoff water if the runoff follows rills around the hill rather than flowing directly downhill. Practices that leave the residue on the soil surface reduce the velocity of raindrops striking the soil surface. These same practices counteract the hydraulic shear of surface runoff. When the soil is not tilled, the surface compacts, increasing the soil's resistance to detachment. After a sod is tilled, the soil is highly resistant to detachment. Long-term no-till and sod-based rotations may also reduce surface runoff rates and amounts. Cropping and management practices can usually be implemented at low cost and frequently to the economic benefit of those who apply them. Because there is often an economic incentive, practices can also be applied quickly over large areas.

In some cases, additional practices are required to prevent soil erosion and land degradation. Surface runoff frequently causes small channels to form. Eventually, this can result in large gullies within fields. In these cases, nonerodible channels to carry runoff water are required.

Grassed waterways can be designed for many topographies to transport runoff water in a nonerodible channel to field drains. In addition, temporary storage of surface runoff in terraces, tied ridges, or small reservoirs may reduce runoff rates to acceptable levels.

## Sustainable Agricultural Practices in the United States

A number of crop production systems used in the United States might result in a very low erosion rate and could be considered sustainable. Conservation tillage systems are extremely effective in controlling erosion (Moldenhauer et al., 1983). The emphasis here is on only two of these systems, but they can be used with many variations over an extensive area on some major crops in areas where erosion is a problem.

***Ridge Tillage in the Corn Belt.*** A ridge tillage system for continuous corn has been in place on an experimental farm in the loessial area of

**Table 1. Runoff and sediment yield from the Treynor watersheds.**

| | *Ridged Watershed* | | *Conventional Watershed* | |
|---|---|---|---|---|
| *Year* | *Runoff (mm)* | *Soil Loss (t/ha)* | *Runoff (mm)* | *Soil Loss (t/ha)* |
| 1973 | 27 | 0.2 | 75 | 1.1 |
| 1974 | 2 | 0 | 14 | 0.7 |
| 1975 | 3 | 0 | 21 | 1.8 |
| 1976 | 10 | 2.5 | 4 | 0 |
| 1977 | 8 | 0.2 | 104 | 18.4 |
| 1978 | 40 | 1.1 | 81 | 9.3 |
| 1979 | 76 | 0.2 | 102 | 4.3 |
| 1980 | 50 | 4.5 | 116 | 51.8 |
| Average | 27 | 1.1 | 65 | 10.9 |

western Iowa since the mid-1970s. The ridge-tilled watershed is one of four watersheds, all similar in size and topography (from 30 to 60 hectares). These intensively studied lands, known as the Treynor watersheds, are steeply sloping. Soil loss is measured from each watershed.

A ridge-planted watershed and a conventionally tilled watershed are managed quite similarly, except for the differences in tillage. The conventionally tilled watershed is farmed on the contour when feasible. The ridge-tilled watershed is also farmed on the contour, with most rows draining directly into a grassed waterway that transports runoff to an untreated gully. Table 1 provides unpublished data on runoff and soil loss from these watersheds.

The remarkable performance of ridge-tilled watersheds on such an erosion-prone area demonstrates the potential for such a system to reduce erosion to that of a sustainable system. Many farmers have advocated and used ridge-tillage systems as a means of reducing costs. Because the crop is planted on a tall ridge, the system is acceptable in areas that have high rainfall and cool springs, when quick germination is needed because of short growing seasons. At planting, the top of the ridge is shaved off, leaving a surface exposed to the sun. As a result, germination tends to be quick. In addition, aggressive cultivation controls many weeds and grasses, and chemical usage is usually no more than that needed for conventional tillage—and often much less when herbicides are banded. It is a system that works well for continuous row-cropping in climates and soils similar to those of the U.S. Corn Belt.

Erosion is greatly reduced because much of the surface is covered by crop residue and a mulch of loose soil and decaying organic matter. Thus, interrill erosion and runoff volumes are reduced. Rows are on the contour, and they are tall enough not to overtop. At locations where they might overtop, a grassed waterway is installed to transport runoff water. The contoured rows are mostly of low slope, so the ability of the flow to detach and transport sediment is minimized.

In areas of moderate erosion potential, where a ridge tillage system will produce adequately, soil erosion rates can be held to acceptable rates with a continuous row crop using ridge tillage on the contour in conjunction with grassed waterways.

***Stubble Mulch in Wheat-Fallow in the Great Plains.*** Wind and water erosion are serious problems in the Great Plains of the United States. Fallowing, a system in which cropping is forfeited for a time so water can be accumulated for the succeeding crop, is used on more than 10 million hectares west of the Mississippi River. Fallowing conserves water for crop production, controls weeds, and stabilizes yields, but it can contribute greatly to soil erosion (both wind and water) problems. In particular, using a bare fallow system in the 1930s contributed greatly to the "dust bowl" occurring during the early parts of that decade. A requirement for the successful use of fallowing is to control weeds and grasses that use precious water. During this period, the potential for soil erosion is great unless the surface is protected from rainfall by a mulch that also enhances infiltration.

Stubble mulching involves undercutting weeds using a sweep that runs beneath the soil surface. Weeds are controlled, but relatively little residue is buried. The subsurface sweep can have a number of different designs, depending on conditions. The implement generally operates between 6 and 15 centimeters beneath the soil surface. Sweeps typically are more than 60 centimeters wide and can range up to about two meters in width. They are equipped with coulters to cut through residue. Several operations may be required to satisfactorily control weeds.

An additional tool used to control weeds and grass in a stubble mulch system is a rodweeder. The rodweeder consists of a square rod that revolves backward beneath the ground. It is a secondary tillage tool, operated about 5 centimeters below the ground surface. If operated properly, it reduces residue cover by about 10 percent.

As shown in table 2, soil erosion in a stubble mulch system is considerably less than that in a bare-fallow system, but more than that in a no-till system. The stubble mulch system is also intermediate in terms of runoff volumes. The results are from a rainfall simulation study, but they do show the potential of systems that leave a considerable amount of residue on the soil surface to control water erosion. Residues and canopies are also extremely effective in controlling wind erosion.

The stubble mulch system is more acceptable to farmers than a no-till

**Table 2. Comparisons of surface cover, soil erosion, and runoff from a rainfall simulation study for bare-fallow, stubble-mulch, and no-till wheat-fallow systems (Dickey et al., 1983).**

| *Period* | *System* | *Cover (%)* | *Runoff (cm)* | *Soil Loss (kg/ha)* |
|---|---|---|---|---|
| Fallow after harvest (October) | bare fallow | 62 | 0.9 | 662 |
| | stubble mulch | 91 | 1.5 | 803 |
| | no-till | 91 | 1.1 | 718 |
| Fallow after tillage (May) | bare fallow | 4 | 3.6 | 9,401 |
| | stubble mulch | 92 | 0.9 | 208 |
| | no-till | 96 | 0.1 | 17 |
| Wheat 10-cm tall (October) | bare fallow | 26 | 3.5 | 7,246 |
| | stubble mulch | 38 | 2.4 | 2,576 |
| | no-till | 85 | 0.5 | 550 |
| Wheat 45-cm tall (May) | bare fallow | 78 | 4.3 | 2,094 |
| | stubble mulch | 83 | 2.9 | 836 |
| | no-till | 88 | 1.6 | 337 |

fallow system because of the reduced dependency on chemicals for weed control. The expense of chemicals and their uncertainty make them less desirable than stubble mulching.

## Sustainable Agricultural Practices in the Tropics

Accelerated soil erosion is a severe threat to sustainable agriculture in tropical Africa (Fournier, 1967; Lal, 1976, 1984; Roose, 1977, 1987; FAO/UNEP, 1978; Walling, 1984). Severe erosion in the forest region occurs when the protective vegetation is removed for intensive cultivation of row crops. Erosion rates in the forest region reportedly are as high as 15 tons per hectare per year. The sub-humid region is characterized by a dry season of four to six months. In this region erosion rates under the worst conditions can be as high as 20 tons per hectare per year. The sub-Sahelian or semiarid region of West Africa is the most severely affected by water erosion; the mean erosion rate by water may be 25 tons per hectare per year. The Sahel suffers from both wind and water erosion. The combined mean erosion rates in this region may be as high as 35 tons per hectare per year. Considering the shallow effective rooting depth, these rates are excessive and can have severe adverse effects on crop yields (Lal, 1987).

Sustainable management technologies exist. If successfully applied, improved techniques of soil surface management can drastically reduce the soil erosion risk. But technological options differ among soils and regions.

***Humid/Forest Region of Africa.*** Severe soil erosion occurs on land where forest vegetation has just been removed. Erosion is more severe on land cleared mechanically than manually. Lal (1981) reported that if machine clearing is absolutely essential, clearing with a shear blade is better than clearing with a dozer blade, tree pusher, or root rake. The data in table 3 show that soil erosion is the most severe when the forest is removed with the tree-pusher root rake than with the shear blade or by manual clearing. The soil compaction and erosion risk associated with shear-blade clearing can be avoided by seeding an appropriate cover crop soon after the removal of tree cover.

Once the land has been cleared, soil erosion can be controlled by adopting mulch farming techniques based on no-till or conservation tillage. Lal (1976) observed that soil erosion under a maize-cowpea rotation is controlled by no-till and mulch farming systems on slopes up to 15 percent (Table 4). Mulch creates a favorable soil temperature and moisture regime and encourages the activity of earthworms and other soil fauna. Biochannels made by earthworms and other soil animals enhance water infiltration and reduce runoff. Beneficial effects of mulch farming on soil and water conservation are also reported for pineapple in Ivory Coast (Roose, 1977) and coffee in Kenya (Othieno and Laycock, 1977). The effectiveness of a no-till system on soil and water conservation can be greatly enhanced by growing legume covers in the rotation. A planted cover grown over three or four years is an important component of the improved cropping system. Quick growing

**Table 3. Effects of land clearing methods on runoff and soil erosion (Lal, 1981).**

| *Land Clearing Method* | *Runoff (mm)* | *Soil Erosion (t/ha)* |
|---|---|---|
| Forest control | 1 | 0.01 |
| Traditional | 7 | 0.02 |
| Manual | 16 | 0.4 |
| Shear blade | 105 | 4 |
| Tree-pusher root rake | 107 | 16 |

**Table 4. Soil and water losses from no-till and moldboard plowing for a maize-cowpea rotation (Lal, 1976).**

| *Slope (%)* | *Runoff (mm)* | | *Soil Erosion (t/ha)* | |
|---|---|---|---|---|
| | *No-till* | *Plowed* | *No-till* | *Plowed* |
| 1 | 11 | 55 | 0 | 1 |
| 5 | 12 | 159 | 0.2 | 8 |
| 10 | 20 | 52 | 0.1 | 4 |
| 15 | 21 | 90 | 0.1 | 24 |

**Table 5. Alley cropping effects on runoff, erosion, and grain yield of maize and cowpea (Lal, 1986).**

| Treatment | *Runoff (mm)* | *Erosion (t/ha)* | *Grain Yield (t/ha)* | |
|---|---|---|---|---|
| | | | *Maize* | *Cowpea* |
| Plowed | 232 | 14.9 | 3.6 | 0.58 |
| No-till | 6 | 0.03 | 4.0 | 1.19 |
| Leucaena-4m | 10 | 0.2 | 3.7 | 0.58 |
| Leucaena-2m | 13 | 0.1 | 3.8 | 0.45 |
| Gliricidia-4m | 38 | 1.7 | 3.6 | 0.67 |
| Gliricidia-2m | 20 | 3.3 | 3.3 | 0.50 |

grass and legume covers provide large quantities of mulch, improve soil structure and organic matter content, and alleviate soil compaction.

Agroforestry is another innovation that conserves soil and water resources. It involves growing deep-rooted, perennial leguminous shrubs and trees in association with food crop annuals or animals. Alley cropping is a special case of agroforestry; it is the practice of growing food crops between rows of specially planted woody shrubs or trees. The woody shrubs are regularly pruned during the cropping season to prevent shading, provide mulch, and reduce water use by the perennials. Lal (1986) reported that three-year-old contour hedges of *Leucaena leucocephala,* planted at two- or four-meter intervals, effectively reduce runoff and erosion (Table 5). This system of management controls erosion while enabling satisfactory yields of corn and other nitrophilic crops. With this system, about 15 percent of the land goes out of production. The practice is labor-intensive.

In addition to soil and crop management techniques, terraces and diversion channels can also be used as back-up systems. These engineering structures, however, must be properly constructed and maintained to avoid the severe land degradation that can result from their failure. Water runoff and soil erosion in steep terrain, however, can be significantly reduced with adequately designed and properly constructed and maintained engineering systems. The data in table 6, from terraced and unterraced watersheds, indicate that terraces are more effective in reducing soil erosion than in reducing water runoff.

***Semiarid/Savanna in Africa.*** Alfisols and entisols of semiarid and arid Africa are structurally inert soils. They do not swell or shrink, and there is little activity by soil fauna. Some form of mechanical tillage is needed to improve water infiltration through the compacted, crusted surfaces. Charreau (1972) and Nicou (1974) have shown, from their studies in the Sahel, that soil structure is greatly improved by deep plowing and soil inversion. Plowing increases total porosity, but it brings about only transient improvements, and soil structure is easily degraded during the cropping phase.

Ridge-cropping has proved to be a stable system for soil and water conservation in structurally unstable soils of the semiarid tropics (IITA, 1981). Tied-ridging or ridging with the addition of cross ties in the furrows is an improvement over the simple, traditional ridge-furrow system. This practice is designed to hold surplus water and allow more time for infiltration.

Although the beneficial effects of mulch farming in arid and semiarid Africa are widely recognized, procurement of mulch material is a severe constraint. Some tree species (*Acacia albida*) and perennial shrubs can be grown on plot boundaries and in contiguous areas to produce biomass and reduce the risks of soil erosion.

***Legume Intercrops in Hawaii.*** Intercropping with legumes is an excellent practice for controlling soil erosion and sustaining crop production in the tropics (El-Swaify et al., 1988). Where rainfall is excessive in the tropics, cropping management practices that leave the soil bare for part of the season may permit excessive erosion and runoff, eventually resulting in an infertile soil with poor characteristics for crop production. Practices that provide for a continuous cover, particularly if the cover is a legume, may result in a soil with superior characteristics and with a good fertility level.

**Table 6. Runoff and soil loss from terraced and unterraced watersheds in Ibadan, Nigeria, on July 6, 1981 (Lal, 1981).**

| *Watershed Management* | *Runoff (mm)* | *Soil Erosion (t/ha)* |
|---|---|---|
| Terraced | 18.1 | 0.7 |
| Unterraced | 18.8 | 2.3 |

**Table 7. Effects of legume intercrops and nitrogen fertilization on yield of maize.**

| *Nitrogen Applied (kg/ha)* | *Yield of Maize* | |
|---|---|---|
| | *No Intercrop (t/ha)* | *Legume Intercrop (t/ha)* |
| First crop | | |
| 190 | 9.2 | 8.8 |
| 380 | 9.9 | 9.4 |
| 570 | 10.1 | 9.8 |
| Second crop | | |
| 95 | 5.7 | 6.2 |
| 190 | 9.0 | 8.9 |
| 285 | 9.8 | 9.4 |

**Table 8. Yield of cassava (*Ceiba cultivar*) roots and intercropped groundnuts.**

| | *Yields (t/ha)* | |
|---|---|---|
| | *Cassava Roots* | *Groundnuts* |
| First crop | | |
| No legume | 36.3 | — |
| Legume-stylosanthes | 25.5 | — |
| Second crop | | |
| No legume | 37.6 | — |
| Residual stylosantes | 75.2 | — |
| Groundnut intercrop | 32.7 | 2.1 |

Although intercropping may result in increased competition for limiting plant growth resources, such as water, nutrients, and light, the advantages of an intercrop can result in superior yields in some cases.

In a study on an aridisol soil in Hawaii, El-Swaify et al. (1988) demonstrated that an intercrop leguminous ground cover planted between maize rows did not have a significant effect on the yield of irrigated maize (Table 7). Two maize crops were grown consecutively, and apparently there was litle benefit derived from the legumes for the second crop. Soil loss and runoff volumes were also lower where an intercrop was grown between the maize rows.

In another study dealing with intercropping of legumes and groundnuts cassava was produced on the same site several years later. Table 8 summarizes the results. Succeeding crops of cassava were produced, the first with and without intercropping of legumes, and the second with both legumes and groundnuts. In this case, the legume intercrop provided competition for cassava. Other data from the same study, including another cultivar, were not as conclusive, nor did top yields show competition for resources.

The results in table 8 indicate the importance of the residual nitrogen provided by a prior intercrop. In fact, root yields of cassava after legume intercropping were quite similar to those grown at high nitrogen fertilization rates. Groundnuts grown as an intercrop reduced the yield of cassava somewhat, but the groundnuts could provide additional income.

Legume intercrops, when used wisely, show promise for maintaining yields and providing nitrogen for succeeding crops—important in resource-poor areas of the world. And erosion is lowered because of the more continous protection of the soil surface from rill and interrill erosion.

## Conclusion

Soil erosion is a major threat to a sustainable agriculture everywhere. Yet, carefully selected agricultural and conservation systems can sustain the land for much-needed agricultural production.

Agricultural and conservation systems must concentrate on complete systems. A system that controls interrill erosion through use of a canopy may fail because the excess surface runoff that will inevitably occur is not discharged through a nonerodible channel. Likewise, systems must control rill and interrill erosion to provide for an adequate depth of topsoil to serve as a storage place for resources necessary for crop growth.

Rapid progress has been made in developing sustainable management technologies, but many of these are awaiting the adaptive research and demonstration needed before the technologies can be delivered to the farmer. Many of these technologies promise conservation benefits and economic benefits alike. This seems to be true whether the practice is used in a developed or in a developing country.

Although cropping systems and conservation practices must control soil erosion wherever it occurs, specific regions require specific systems and practices. For that reason, those interested in a sustainable agriculture must understand the fundamental erosion processes and the principles of erosion control and develop technology suitable for their specific climate, culture, cropping, soils, and topography that will sustain agricultural production indefinitely.

## REFERENCES

Bennett, H. H., and W. R. Chapline. 1928. Soil erosion a national menace. Circular 331. U.S. Department of Agriculture, Washington, D.C.

Charreau, C. 1972. Problemes poses par l'utilisation agricole des sols tropicaux par des cultures annuellles. Agronomy Tropicale 27: 905-929.

Deevey, E. S., D. S. Rice, P. M. Rice, H. H. Vaughan, M. Brenner, and M. S. Flannery. 1979. Mayan urbanism: Impact on a tropical karst environment. Science 206: 298-306.

Dickey, E. C., C. R. Fenster, J. M. Laflen, and R. H. Mickelson. 1983. Effects of tillage on soil erosion in a wheat-fallow rotation. Transactions, American Society of Agricultural Engineers 26: 814-820.

El-Swaify, S. A., A. Lo, R. Joy, L. Shinshiro, and R. S. Yost. 1988. Achieving conservation-effectiveness in the tropics using legume intercrops. Soil Technology 1: 1-12.

Food and Agriculture Organization/United Nations Environmental Programme. 1978. Methodology for assessment of soil degradation. Food and Agriculture Organization, Rome, Italy.

Fournier, F. 1967. Research on soil erosion and soil conservation in Africa. African Soils 12: 53-96.

International Institute for Tropical Agriculture. 1981. Role of tied ridges in maize production in Upper Volta. pp. 259-270. *In* IITA Research Highlights. Ibadan, Nigeria.

Kellogg, C. E. 1948. Modern soil science. American Scientist 36: 515-525.

Lal, R. 1976. Soil erosion problems on an Alfisol in western Nigeria and their control. IITA Monograph 1. International Institute for Tropical Agriculture, Ibadan, Nigeria.

Lal, R. 1981. Deforestation and hydrological problems. pp. 131-140. *In* R. Lal and E. W. Russell [editors] Tropical agricultural hydrology. John Wiley & Sons, Chichester, United Kingdom.

Lal, R. 1984. Soil erosion from tropical arable lands and its control. Advanced Agronomy 37: 183-248.

Lal, R. 1986. Soil surface management in the tropics for intensive land use and high and sustained production. Advanced Soil Science 5: 1-108.

Lal, R. 1987. Effects of soil erosion on crop productivity. CRC Critical Reviews in Plant Science 5: 303-367.

Loudermilk, W. C. 1953. Conquest of the land through seven thousand years. Agriculture Information Bulletin 99. U.S. Department of Agriculture, Washington, D.C.

McCormack, D. E., and W. E. Larson. 1981. A values dilemma: Standards for soil quality tomorrow. *In* Walter E. Jeske [editor] Economics, ethics, ecology: Roots of productive conservation. Soil Conservation Society of America, Ankeny, Iowa.

Menard, H. W. 1961. Some rates of regional erosion. Journal of Geology 69:154-161.

Meyer, L. D. 1981. How rain intensity affects interrill erosion. Transactions, American Society of Agricultural Engineers 24: 1,472-1,475.

Moldenhauer, W. C., A. W. Langdale, W. Frye, D. K. McCool, R. I. Papendick, D. E. Smika, and D. W. Fryrear. 1983. Conservation tillage for erosion control. Journal of Soil and Water Conservation 38: 144-151.

Nicou, R. 1974. Contribution on the study and improvement of the porosity of sand and sandy clay soils in the dry tropical zone. Agronomy of Tropics 30: 325-343.

Nikiforoff, C. C. 1942. Soil dynamics. American Scientist 30: 36-50.

Othieno, C. O., and D. H. Laycock. 1977. Factors affecting soil erosion under tea field. Tropical Agriculture 54: 323-331.

Owens, L. B., and J. P. Watson. 1979. Rates of weathering and soil formation on granite in Rhodesia. Soil Science Society of America Journal 43: 160-166.

Roose, E. J. 1977. Adoption of soil conservation techniques to the ecological and socio-economic conditions of West Africa. Agronomy Tropics 32: 132-140.

Roose, E. J. 1987. Reseau erosion. Bulletin 7. ORSTOM, Paris, France.

Smith, D. D. 1941. Interpretation of soil conservation data for field use. Agricultural Engineering 22: 173-175.

Walling, D. E. 1984. The sediment yield of African rivers. International Association of Hydrologic Scientists Bulletin 144: 265-283.

# SUSTAINABLE AGRICULTURE AND WATER QUALITY

T. J. Logan

The impact of agricultural practices on water quality has been of concern in developed countries for at least 25 years. Accelerated erosion with its associated nutrients, direct discharge of manure to streams, leaching of nutrients and pesticides to groundwater, and bacterial contamination of surface water and groundwater are some of the recorded effects of modern agriculture on the environment. Agriculture has responded to these problems with a number of approaches. Best management practices (BMPs) were developed in the 1970s as a voluntary approach to controlling nonpoint-source pollution. Many of these practices, such as conservation tillage, deal almost exclusively with erosion control but have been promoted as a means of controlling nonpoint-source phosphorus (Forster et al., 1985). There is, however, some question of whether large-scale adoption of conservation tillage will increase the use and leaching potential of pesticides as well as nitrate leaching (Logan et al., 1987). Another major effort that has received limited federal funding has been manure handling and storage. The physical facilities have had significant local impact on surface water quality, but the larger question of how to manage land application of livestock waste to minimize nutrient losses has not been addressed satisfactorily.

The question in the 1960s and 1970s of surface water contamination from sediment, nutrients, and pesticides has given way in the 1980s to public concern for groundwater contamination by pesticides and nitrates. Although surface water contamination still is far more extensive than groundwater contamination, localized groundwater problems have occurred with nitrates and some more mobile and persistent pesticides, such as EDB and aldicarb.

The current debate over the concept of sustainable agriculture in the United States has a strong environmental focus. Sustainability of agricultural practices in the United States may well hinge on whether they are environmentally acceptable—or perceived by the public to be so. Those who define sustainable agriculture narrowly as low-input or no-input farming clearly envision positive impacts on water quality. Although this is most obvious for pesticide contamination, it is not immediately clear that nitrate contamination would be prevented. If maintenance of water quality is to be a primary requisite of any sustainable agricultural system, that system will have to incorporate principles that prevent or minimize contamination of surface and groundwater.

## The Water Quality Perspective

A remarkable examination of the basic foundations of modern agriculture in the United States has taken place during the last decade. This critical inspection of farming practices and their impact on farm economics, social problems, and the environment is driven by a series of public beliefs, developing since the publication of Rachel Carson's *Silent Spring* in 1962, that something is not right down on the farm. Rising government support of agriculture, an apparent inability to curb production, declining farm exports, and widespread bankruptcies of marginal producers, together with a national concern for environmental problems, have prompted a call for change. The spectacular rise in U.S. farm production since the 1940s has been accomplished with some expansion in farm acreage, although farmland acreage has been essentially constant over the last 15 years. But that expansion is attributable primarily to improvements in the genetic yield potential and a dependence on power traction, synthetic fertilizers, and chemical pest control (Pimental et al., 1973). The labor requirements of U.S. grain production have been reduced to the point where one farmer produces food for every 98 members of the population (USDA, 1987). U.S. farmers have streamlined their production practices to the point where many can supplement their income with off-farm employment.

For several decades, a small but growing number of U.S. farmers—and their counterparts around the world—have been practicing various forms of what is today called low-input farming, regenerative agriculture, or sustainable agriculture. At the extreme of this philosophy are the organic farmers who eschew any use of synthetic chemicals but rely instead on green manures and livestock wastes for nutrients, and tillage and biological agents for weed and insect control. More moderate proponents advocate the use of crop rotations instead of monoculture, pasture legumes as a source of nitrogen and to improve soil properties, recycling of livestock wastes, maintenance of mulch cover, and reduction of pesticide use through integrated pest management.

Water quality problems suggest that one of the major constraints to sustainability of modern U.S. agriculture will be its ability to meet society's growing expectations for a clean environment. Until recently, agricultural production has been relatively unaffected by state and federal regulations. There are no mandated controls on erosion and sedimentation, fertilizer use, and livestock waste handling (except for operations larger than 1,000 animal units). Pesticide use, regulated through the Federal Insecticide, Fungicide and Rodenticide Act (FIFRA), limits compounds that can be used, restricts use of some compounds to specific crops, and requires that some applicators be certified. Pesticides are, in fact, the most regulated of all potential agricultural pollutants. Sustainable agricultural systems in the United States, then, will be those that can meet future environmental criteria. These criteria include reductions in erosion and sedimentation, phosphorus loadings to lakes, and nitrate and pesticide contamination of surface water and groundwater.

The ecological attributes of the various low-input or regenerative systems being proposed also have the potential to remedy many of the environmental problems facing production agriculture.

## Major U.S. Crop Production Systems

The major crop production systems in the United States are described here in terms of those components of the system most related to water quality: tillage, fertilizer management, pest management, livestock waste management, and irrigation management. The primary focus is on the major feed grains—corn, soybeans, and wheat. These crops represent a majority of U.S. cropland acreage and use large percentages of the nitrogen and phosphorus fertilizers and pesticides applied to farm fields.

***Tillage.*** Mannering et al. (1987) recently described the various tillage systems used in the United States for grain production. These include conventional tillage, no-till, ridge-till, strip-till, mulch-till, and reduced till. They point out that although conventional tillage refers to the combined primary and secondary operations normally performed for a given crop and geographical area, many researchers have used this term to mean inversion tillage, as with a moldboard plow, or to imply an operation or operations that result in a bare soil surface. Conventional tillage often is used as the standard in experiments to assess other tillage systems.

Mannering et al. (1987) provided the following descriptions of conservation tillage systems in use in various parts of the United States:

*No-till or Slot Planting.* The soil is left undisturbed prior to planting. Planting is completed in a narrow seedbed about 2 to 8 centimeters wide. Weed control is accomplished primarily with herbicides.

*Ridge-till (Includes No-till on Ridges).* The soil is left undisturbed prior to planting. About one-third of the soil surface is tilled at planting with sweeps or row cleaners. Planting is completed on ridges usually 10 to 15 centimeters higher than the row middles. Weed control is accomplished with a combination of herbicides and cultivation. Cultivation is used to rebuild ridges.

*Strip-till.* The soil is left undisturbed prior to planting. About one-third of the soil surface is tilled at planting time. Tillage in the row may be done by a rototiller, in-row chisel, row cleaners, and so on. Weed control is accomplished with a combination of herbicides and cultivation.

*Mulch-till.* The total surface is disturbed by tillage prior to planting. Tillage tools, such as chisels, field cultivators, disks, sweeps, or blades, are used. A combination of herbicides and cultivation are used to control weeds.

*Reduced-till.* This system consists of any other tillage and planting system not covered above that meets the Soil Conservation Service's (SCS) criterion for conservation tillage of 30 percent surface residue cover after planting.

Tillage methods are classified as conservation tillage systems if they maintain 30 percent surface residue cover after planting (Mannering et al., 1987). This figure, considered minimal for protecting the soil from rainfall erosion, is used by SCS to estimate conservation tillage use in a county. Recently, direct measurements of soil residue cover (Dickey et al., 1987) have indicated that the current subjective practice of local conservation officials' estimating the use of conservation tillage in a county is resulting in overestimation of the extent of these practices and their impacts on erosion.

The primary impact of conservation tillage systems is to maintain as much residue cover on the soil surface as possible. Residue cover at planting is essentially zero with moldboard plowing and secondary tillage, 50 to 70 percent with chisel plowing, 30 to 60 percent for strip-till, and 50 to 90 percent for no-till (all in corn stalk residue). The corresponding values for soybean residue are 30 to 50 percent residue cover less than that for corn. The environmental impacts associated with tillage are erosion and sedimentation, loss of sediment-bound phosphorus and pesticides, and runoff and leaching losses of nitrate and pesticides.

***Fertilizer Management.*** Crops require nitrogen and phosphorus—the two most important environmental plant nutrients—in large amounts.

Together with potassium, nitrogen and phosphorus represent the bulk of the chemical fertilizer applied by producers in the United States. Total fertilizer use increased from 34.1 million tons in 1969 to 42.8 million tons in 1985 (USDA, 1984, 1987). Nitrogen fertilizer use increased 32 percent during this period, while phosphorus use declined by 21 percent. Management of nitrogen and phosphorus is distinctly different as a consequence of the differing soil chemical and microbiological behavior of these two elements.

Nitrogen exists in many chemical forms in the soil and is transformed rapidly by microbiological processes from one form to another. Most nitrogen fertilizer applied to crops in the United States is in the form of ammonia ($NH_3$). Major nitrogen fertilizer forms include anhydrous ammonia ($NH_3$), urea ($NH_2CONH_2$), and mono- and diammonium phosphate [$NH_4H_2PO_4$ and $(NH_4)_2HPO_4$)]. In the soil, these ammonia forms are nitrified rapidly to nitrate ($NO_3^-$), the form most readily used by plants. Nitrate is a highly soluble ion and is not retained by soil minerals as a consequence of its net negative charge and its inability to form strong chemical bonds with mineral cations, such as aluminum, iron, and calcium. As a result, soil nitrate not readily taken up by plants is susceptible to leaching losses to groundwater or in tile flow, in which it is returned to the surface water system (Gilliam et al., 1985).

Because of its highly dynamic nature in soil, the available nitrogen supply of a given soil is difficult to predict based on soil tests. Nitrate measured in a soil today could be denitrified or immobilized by the time the crop needs it. Mineralization of soil organic nitrogen is variable, and its measurement does not lend itself to routine soil analysis. Only in relatively dry areas where water is managed through irrigation and drainage is a residual nitrate soil test meaningful for predicting soil nitrogen supply to crops. In lieu of soil testing, the usual approach to determining nitrogen fertilizer rates is to conduct nitrogen rate field studies.

In addition to determining nitrogen application rates accurately, the method and timing of nitrogen fertilizer application are important considerations in optimizing nitrogen fertilizer use efficiency. The implications for nitrate contamination of surface water and groundwater are obvious because nitrate taken up by crops cannot be lost by runoff or leaching.

Unlike nitrogen, phosphorus in the soil is relatively stable, exists primarily in the form of orthophosphate ($PO_4^{3-}$), and is tightly bound to soil particles by adsorption to mineral surfaces or by precipitation of insoluble phosphate minerals (Logan, 1982; Nelson and Logan, 1983). Native soils are relatively low in total and plant-available phosphorus. In natural ecosystems, net biomass production is low, and although the pool of available phosphorus is small, its turnover is sufficient for biomass production (Jordan, 1985). In high-yield crop production systems, net phosphorus removal may be as high as 50 kilograms per hectare per year (Nelson and Logan, 1983), and external phosphorus sources are required to supplement native soil phosphorus.

Phosphorus fertilizer requirements for crops usually are based upon soil tests and field trials that determine sufficiency levels for each crop. Soils low in available phosphorus receive large applications of phosphorus fertilizer that are incorporated uniformly into the root zone. After the sufficiency level has been attained, annual applications are made—often as a band next to the seed—to replace phosphorus removed by the crop. All commercial phosphorus fertilizers are some form of soluble orthophosphate that dissolves readily in soil solution and reacts rapidly by adsorption or precipitation with the soil constituents. More than 90 percent of the phosphorus in soil (Logan, 1982) is in the form of particulate (soil-bound) phosphorus and is, therefore, susceptible to environmental loss only by surface runoff and erosion.

***Pest Management.*** Modern crop production systems rely almost entirely on chemical pest control—primarily weed control. Between 1964 and 1985, farm use of pesticides increased almost 170 percent (The Conservation Foundation, 1986). Herbicides accounted for most of this increase. By 1985, 88 percent of the pesticides used by U.S. farmers were herbicides. Some 1,500 active pesticide ingredients are registered for use by the U.S. Environmental Protection Agency (EPA) (The Conservation Foundation, 1986). Corn, cotton, soybeans, and wheat received 90 percent of all herbicide and insecticide applications to U.S. crops in 1982; corn alone accounted for 54 percent of all herbicide and 43 percent of all insecticides applied that year.

A major change in pesticide use in recent years has been a shift from persistent, chlorinated hydrocarbon insecticides, such as DDT, to compounds, such as the organophosphates that are less persistent but have higher acute mammalian toxicities. Modern herbicides, in general, are short-lived but tend to have fairly high water solubilities and low soil-water partition coefficients that make them susceptible to groundwater leaching. Some experimental compounds are not only short-lived but are applied at rates of grams per hectare rather than kilograms per hectare.

***Livestock Waste Management.*** Livestock operations in the United States today include dairy, hogs, poultry, cow-calf rangeland, beef feedlots, and sheep on range (Table 1). Numbers have been relatively constant from 1969 to 1987 for beef, dairy, and hogs, have declined for chickens and sheep, and have increased for poultry. Tablc 2 shows cstimates of manure and manure nitrogen and phosphorus production in the United States from current livestock production figures and standard values for manure and nutrients produced annually by each livestock unit. Beef, dairy, and hogs account for most of the manure and manure nutrients. At least half of the beef and all of the dairy and hogs are completely or partially confined, as is the poultry. Confined animal operations are growing in size. The small, integrated, animal/grain farm has given way to large, specialized grain and livestock operations (Walter et al., 1987). The three general areas of livestock concentration are: (1) New York, Pennsylvania, and Vermont; (2) Wisconsin, Iowa, southern Minnesota, northern Illinois, eastern South Dakota, and eastern Nebraska; and (3) southern California and New Mexico (Gilbertson et al., 1979). To this list can be added the growing number of large poultry operations (more than one million birds) in the Delmarva peninsula and the southeast.

This distribution of livestock means that the grain nutrients normally returned to the same land as manure are now concentrated in localized areas. As Walter et al. (1987) point out, this system presently is economically viable because "inexpensive" chemical fertilizer is substituted for manure nutrients in grain production. Chemical fertilizer is inexpensive when the costs of collecting, storing, hauling, and spreading manure are considered. Costs of transporting nutrients in manure back to the areas of grain production are high, and in the absence of stronger regulations regarding manure disposal, producers are not likely to spread their wastes any farther from their central operations than is absolutely necessary. At present, state and federal regulations and cost-sharing programs deal primarily with manure collection and storage, and these regulations and programs do not consider adequately the problem of having land that is adequate for completely receiving the manure nutrients.

**Table 1. Livestock production in the United States for 1969 and 1985-1987 (USDA, 1984, 1987).**

| | *1969* | *1985-1987* | |
|---|---|---|---|
| | *millions* | | |
| Beef cattle | 110.0 | 102.5 | (1987) |
| Dairy | 16.5 | 16.9 | (1986) |
| Hogs | 57.0 | 52.3 | (1985) |
| Sheep | 21.4 | 10.0 | (1986) |
| Chickens | 422.0 | 368.5 | (1985) |
| Turkeys | 106.7 | 185.4 | (1985) |

**Table 2. Estimated manure and manure nitrogen and phosphorus produced annually in the United States for current (1985-1987) livestock production.**

| | *Livestock Numbers (millions)** | *Manure†* | *Nitrogen†* | *Phosphorus†* |
|---|---|---|---|---|
| | | *million tons/year* | | |
| Beef cattle | 102.5 | 78.9 | 3.1 | 1.08 |
| Dairy | 16.5 | 31.2 | 1.0 | 0.17 |
| Hogs | 57.0 | 12.0 | 0.9 | 0.21 |
| Sheep | 21.4 | 3.9 | 0.17 | 0.04 |
| Chickens | 422.0 | 3.4 | 0.18 | 0.18 |
| Turkeys | 106.7 | 2.7 | 0.05 | 0.02 |

*U.S. Department of Agriculture (USDA, 1987).
†Data from USDA and U.S. Environmental Protection Agency.

***Irrigation Management.*** In 1954, there were 73.1 million hectares of irrigated cropland in the United States. By 1986, this had increased to 110.5 million hectares (USDA, 1987). The serious drought of 1988 and generally dry conditions throughout the continental United States have stimulated an increasing interest in irrigation. Continued low fuel costs for pumping water have done nothing to dampen enthusiasm for irrigation. Irrigation commonly is practiced in the western states but in recent years has pushed into the central plains and parts of the southeast. Flood irrigation and furrow irrigation remain the most common practices in the West, with center pivot systems increasingly used for grain production in the central plains. In the eastern United States, there has been considerable interest recently in subsurface irrigation, using subsurface drainage systems to deliver water to the root zone during dry periods. In addition, use of drip irrigation systems for high-value tree and vegetable crops is expanding.

Environmental problems associated with cropland irrigation are ancient, as attested to by the archeological evidence of failed irrigation systems. Excess soluble salts—salinity—remains the most pervasive water quality problem associated with irrigation, affecting about one-third of all irrigated land (Yaron, 1981). In the United States, about 28 percent of irrigated land suffers from depressed yields due to salinity (Yaron, 1981). Salinity affects plant growth by reducing water uptake as a consequence of the high osmotic potential of high salt concentrations in water. Individual ions found in saline waters, such as sodium, chloride, and boron, may also be toxic to plants and animals. Recently, high selenium levels in irrigation return flows in the central valley of California have resulted in toxicities to migratory birds using wetlands in the area receiving irrigation drainage (Bureau, 1985).

Skogerboe and Walker (1981) indicated that water quality problems associated with irrigation differ for surface and subsurface return flows. Surface runoff from excess irrigation usually has increased concentrations of sediment and sediment-associated chemicals: slightly higher salt levels; increased but variable levels of pesticides, fertilizer nutrients, and bacteria; and increased amounts of crop residues and other debris. Drainage water that has moved through the soil profile will have much higher salt concentrations, little sediment and sediment-associated chemicals, and generally increased nitrate and soluble pesticides.

The primary cause of water quality problems with irrigation of agricultural land is poor or no water management (Skogerboe and Walker, 1981). The major emphasis in modern irrigation technology has been on water delivery hardware (Skogerboe and Walker, 1981), with minimal attention to on-farm water management. Relatively cheap irrigation water in the western United States also has led to a tendency to overirrigate, with resultant increases in the volume of irrigation tailwater and associated pollutant loads. From a water quality standpoint, the major focus historically has been on salinity and sodicity (excessive sodium concentrations) and their effects on crop yields. Other water quality concerns, such as bacterial, nitrate, and pesticide contamination, are of relatively recent origin and have not been addressed adequately. Chemigation, the delivery of fertilizer and pesticides in irrigation water, is an efficient and growing practice. When used in conjunction with low-volume drip irrigation systems, it offers the promise of greatly reducing chemical losses in drainage. Of concern with these systems, however, is the potential for back siphoning of chemicals from the irrigation delivery pipe into the well from which the water is removed. This problem can be remedied easily by installing check valves on the well pump. Amendments to the Safe Drinking Water Act of 1986 require groundwater protection for all chemigation systems.

## Sustainable Agriculture and Water Quality

***Surface Water Quality.*** *Soil Erosion and Sediment Loss.* Soil erosion and the subsequent transport of sediment from the land surface to a receiving water body is a function of complex soil, hydrological, and climatological processes. The emphasis in this discussion will be on those processes acting on the land surface and controlling sediment transport to the edge-of-field—the geographical dimension within which agricultural practices will have maximum impact.

Soil erosion and sediment transport within the field (rill and interrill erosion) are a function of raindrop impact that causes soil particle detachment and the energy of overland flow that contributes to detachment and carries suspended sediment downslope. These processes depend highly on the nature of the soil surface at the time of rainfall and, specifically, on the extent to which the surface is covered by plant residues or growing plants. Soil surface cover—either by material directly on the soil surface or as plant canopy—is inversely proportional to soil erosion. Therefore, any practice that promotes the maintenance of surface cover throughout the year will decrease erosion. In addition, practices that break the slope length, such as terraces and diversion ditches, reduce erosion and sediment transport by reducing the velocity and energy of runoff.

The universal soil loss equation (USLE) was developed by USDA (Wischmeier and Smith, 1978) as an empirical tool for predicting the relative effects of various crop and soil management practices on soil erosion. Although the USLE is of limited use in predicting actual erosion rates, it gives reliable predictions of relative long-term soil loss and is useful in selecting those practices that would reduce long-term erosion to sustainable levels.

Of greater controversy than the estimation of erosion rates with various agricultural production systems have been the attempts to define levels of annual erosion that permit sustained crop production (Hall et al., 1985). In the United States, SCS has used the concept of soil loss tolerance (T value) as a general guide or objective in conservation planning (Hall et al., 1985). The T value varies from 2 to 11 tons per hectare per year and depends primarily upon existing rooting depth. The T value concept does not attempt to limit allowable soil loss to the absolute rate of soil regeneration—on the order of 0.5 ton per hectare per year—but is based on the assumption that desirable topsoil (primarily the A horizon) properties can be regenerated more rapidly. It also implicitly (Logan, 1982) accepts that some deeper soils will be permitted to erode in the long-term to the point at which their soil loss tolerance values must be reduced to lower values. The concept permits erosion of better, deeper soils at higher rates than poorer, shallower soils. The long-term consequence of this approach is to reduce the overall productivity of the soil resource.

The concept of the T value is compatible with current thinking on the sustainability of agricultural systems if one takes a short-term view (less than 50 years). Modern, high-input farming with residue conservation is capable of maintaining near optimum crop yields at annual erosion rates of 2 to 11 tons per hectare. Soil physical degradation and loss of nutrients at these rates of erosion usually are not great enough to significantly reduce crop yields. A drive through the drought-stricken Corn Belt in 1988, however, points out dramatically that existing erosion conditions can seriously impact crop production during years of extreme stress. Will agricultural production systems be sustainable with occasional years such as 1988? How often will farmers have to have drought before erosion does indeed limit production to the point where the system is not sustainable?

If one takes the long-term view, the current T-value approach to conservation planning is not sustainable unless one assumes that, during the same period, technological advances will be made to offset productivity losses due to erosion. These advances may include more efficient and cost-effective irrigation systems or land reclamation techniques. Expanded use of irrigation will be in direct conflict with increasing nonfarm water uses, however, and land reclamation can help to improve existing rooting depth but can do little to increase it.

Are there existing soil and crop management practices that can achieve the 0.5 to 1 tons per hectare per year annual soil losses that would be required for true sustainability from the standpoint of soil erosion alone? Application of the USLE to a range of rainfall, soil erodibility, and slope conditions suggests that combinations of conservation tillage, small grains and forages in the rotation, and use of conservation practices can achieve these levels of soil erosion, excluding marginal lands of slopes greater than 8 to 10 percent. The challenge is to develop systems that combine these elements and still meet the other requirements of a sustainable system—namely, acceptability by the farmer and economic viability. There also must be economic inducements, either positive or negative, that will force farmers to take the long-term rather than the short-term view. Presently, the short-term loss of productive soil capacity from existing erosion rates is not great enough for farmers to adopt conservation measures on purely economic grounds.

*Nutrient Losses.* The two nutrients most associated with pollution from agriculture are phosphate ($H_2PO_4^-$) and nitrate ($NO_3^-$). These two chemical forms are quite different in their chemical and biological behavior, and their fates and transport in soils and waters are also markedly different. Although both are anions, phosphate is held tightly to soil by strong chemical bonds, while nitrate has no affinity for soil particles. Phosphorus exists in relatively few chemical forms in soil and water—primarily as orthophosphate adsorbed to soil surfaces; precipitated as iron, aluminum, or calcium phosphates; and as dissolved orthophosphate. Nitrate can exist in several ionic ($NO_3^-$, $NO_2^-$, $NH_4^+$) and gaseous ($N_2$, NO, $NO_2$, $N_2O$, $NH_3$) forms. Because of the differences in soil reactivity of nitrate and phosphate, phosphate pollution is associated primarily with surface runoff and erosion, and nitrate contamination is associated with tile and irrigation drainage and percolation to groundwater. It is important to note that most tile and irrigation return flow eventually becomes part of surface waters.

Plants require nitrogen and phosphorus together with potassium, which is not generally regarded as a water pollutant, in large quantities. In natural systems with little net export of nutrients in the form of harvested biomass, nitrogen and phosphorus needs are met by native supplies of these two elements through recycling and, in the case of nitrogen, atmospheric fixation. In crop production today, the net export of nutrients from the land ultimately must be counteracted to sustain productivity of the system.

Before the advent of synthetic chemical fertilizers, exported nitrogen was replenished by using leguminous forage crops in rotation with cereal grains or pasture grasses and by the application of manures. Phosphorus was supplied by manure (or sewage wastes in some countries) or with applications of rock phosphate, guano, seaweed, or other biological material. With the simultaneous advent of chemical fertilizers and high-yielding, improved crop varieties, crop production systems became highly dependent upon external nutrient sources for sustained productivity. The nutrient-supplying capacity of the soil itself has declined in importance.

Three trends in the United States in the last 30 years have led to the problems of water contamination with phosphate and nitrate: (1) application of phosphorus fertilizer in excess of crop needs; (2) the growing hectares of continuous corn and its associated high use of nitrogenous fertilizer; and (3) the growth of large, confined-livestock operations with the associated concentration of manure in land areas too small for optimum use of the manure nutrients.

*Phosphorus.* Table 3 summarizes annual phosphorus discharges in streams draining watersheds of different land use (Nelson and Logan, 1983). The data show, as expected, that most of the phosphorus in surface water is in the form of particulate phosphorus. A much lower percentage is in the form of dissolved phosphorus. Where row crops are the dominant farming system, particulate phosphorus generally is more than 90 percent of total phosphorus (Nelson et al., 1980). High dissolved phosphorus loads usually are associated with surface-applied manure (Armstrong et al., 1974). The values for agricultural watersheds (Table 3) are much higher than phosphorus loads from forested watersheds, where total phosphorus loads rarely exceed 0.2 kilogram per hectare per year. Likens et al. (1977) reported that total phosphorus loss from the forested Hubbard Brook watershed in New Hampshire was 0.02 kilogram per hectare per year.

Table 4 gives data on annual phosphorus losses from agricultural land in tile drainage. In contrast to the data for surface runoff, a greater percentage of total phosphorus from tile flow is dissolved. Studies on fine-textured soils (the Ohio data in table 4) indicate that sediment loads in tile flow may exceed 100 kilograms per hectare per year and can contribute significant amounts of particulate phosphorus. Generally, however, phosphorus loads

**Table 3. Annual phosphorus discharges in stream flow from selected agricultural watersheds in the eastern United States.**

| *Watershed* | | | *Phosphorus Transported*[i] | |
|---|---|---|---|---|
| *Location* | *Size (ha)* | *Land Use* | *Dissolved P* | *Particulate P* |
| | | | (kg/ha/yr) | |
| Ohio | | | | |
| Maumee River Basin[a]* | $1.64\times10^6$ | Mixed† | 0.29 | 1.53 |
| Portage River Basin[a]* | $1.11\times10^5$ | Mixed | 0.30 | 0.84 |
| Plot 111[a]* | 3.2 | Soybeans | 0.13 | 1.09 |
| Ohio[b] | 123 | Pasture and forest | 0.07 | — |
| Indiana[c]* | $5\times10^3$ | Mixed | 0.15 | 1.90 |
| Michigan | | | | |
| Average of plots[d]* | 0.8 | Row crops | 0.71 | — |
| Mill Creek[e] | — | Mixed | 0.2 | 0.2 |
| Illinois | | | | |
| Kaskaskia River Basin[f] | $1.3\times10^4$ | Mixed | 0.1 | — |
| Wisconsin | | | | |
| Tributary to Lake Kegonsa[g] | — | — | 0.11 | 0.46 |
| Dairy farming[h] | 546 | Mixed | 0.58 | 0.77 |
| Iowa[i] | | | | |
| Watershed 2* | 3.3 | Corn | 0.09 | — |
| Connecticut[j] | $8.5\times10^2$ | Forested | — | 0.22 |
| Arkansas[k] | $3.1\times10^5$ | Mixed | — | 2.3 |
| Maryland | | | | |
| Potomac River Basin[l] | $2.8\times10^4$ | Mixed | — | 0.27 |
| North Carolina | | | | |
| Pigeon River[m] | $3.5\times10^4$ | Mixed | — | 0.17 |
| Watershed 2[n]* | 1.5 | Rotation | 0.27 | — |
| Oklahoma | | | | |
| Watershed C3[o]* | 17.9 | Cotton | 1.1 | 5.6 |
| Maine | | | | |
| Stetson River[p] | $7.4\times10^3$ | — | — | 0.04 |
| Agricultural watersheds[e]* | — | Mixed | 0.6-0.9 | 0.3-0.4 |

Sources: Nelson and Logan, 1983; [a]Logan and Stiefel, 1979; [b]Taylor et al., 1971; [c]Nelson et al., 1980; [d]Ellis et al., 1978; [e]PLUARG, 1978; [f]Harmeson et al., 1971; [g]Armstrong et al., 1974 (in Sawyer, 1947); [h]Armstrong et al., 1974 (in Zitter, 1968); [i]Alberts et al., 1978; [j]Frink, 1967; [k]Armstrong et al., 1974 (in Gearheart, 1969); [l]Jaworski and Hetling, 1970; [m]Keup, 1968; [n]Kilmer et al., 1974; [o]Menzel, 1978; [p]Armstrong et al., 1974 (in MacKenthum et al., 1968).

*Average data.

†Predominantly cropland with some forest and pasture.

**Table 4. Annual phosphorus losses from soil in tile drainage water (Nelson and Logan, 1983).**

| Location | Crop | Treatment | Phosphorus Leached* Dissolved $P_i$ | Particulate P |
|---|---|---|---|---|
| | | | (kg/ha/yr) | |
| Ohio[a] | | | | |
| Castalia | Corn-oats | Fall plow | 0.58 | 0.47 |
| Dellinger farm† | Corn | 224 kg N/ha | 0.31 | — |
| Hoytville plots† | Soybeans | — | 0.13 | 0.33 |
| Rohrs farm† | Soybeans | — | 0.07 | 0.50 |
| Iowa plots[a] | Corn-oats-soybeans | — | 0.003 | 0.015 |
| Minnesota[a] | | | | |
| Plots | Corn | 112 kg N/ha | 0.09 | 0.11 |
| Large system† | Corn-soybeans | 220 kg N/ha | 0.02 | 0.04 |
| Indiana field†[b] | Soybeans | — | 0.027 | 0.08 |
| Idaho[c] | Mixed | — | 0.13 | — |
| Ontario[d] | Corn | Fertilized | 0.26 | — |
| | Alfalfa | — | 0.10 | — |
| | Bluegrass sod | — | 0.01 | — |

Sources: [a]Logan et al., 1980; [b]Lake and Morrison, 1977; [c]Carter et al., 1971; [d]Bolton et al., 1970.
*Dissolved inorganic P; particulate P.
†Averaged data.

in tile drainage are lower than those in surface runoff. This suggests that tile drainage systems should decrease the overall loading of phosphorus from agricultural land.

The higher phosphorus losses in surface runoff from agricultural land are due to increasingly higher levels of total and available soil phosphorus as a result of excessive fertilization (Logan and Forster, 1982) combined with soil erosion and transport of phosphorus-bearing sediment. Thus, any crop production system that is to meet locally mandated phosphorus load limits, such as those in the Great Lakes Basin or in Chesapeake Bay, must address both erosion control and phosphorus fertility management.

Erosion control, such as that provided by conservation tillage, will significantly reduce total phosphorus loads in surface runoff. Logan and Adams (1981) calculated that the reduction of total phosphorus in runoff with conservation tillage was 90 percent as effective as the corresponding reduction in soil loss. The less-than-100-percent effectiveness of conservation tillage in reducing total phosphorus loads was attributed to the dissolved phosphorus fraction that is relatively unaffected by conservation tillage and to the fact that as sediment load decreases with erosion control, there is selective enrichment of the sediment with finer particles. These particles are primarily reactive clay minerals and organic matter that have higher phosphorus contents than the coarser silts and sands.

Table 5 summarizes phosphorus losses in surface runoff with conservation tillage systems as a percentage of losses with conventional tillage (Baker and Johnson, 1983). Conservation tillage can reduce particulate phosphorus loads up to 95 percent, with the greatest reductions coming from no-till. Dissolved phosphorus loads, on the other hand, generally increased greatly with conservation tillage compared to conventional tillage. The reasons for this are twofold. As indicated previously, reduced soil loss results in runoff sediment that is higher in clay and organic matter and enriched in phosphorus. The phosphorus bound to this sediment is also more labile and will maintain a higher equilibrium dissolved phosphorus concentration than sediment that is more coarse-grained (Logan, 1982; Nelson and Logan, 1983). The other reason is that conservation tillage systems result in soils that are relatively undisturbed. With surface application of phosphorus fertilizer, this results in a build-up of labile phosphorus at the surface and increased dissolved phosphorus. In spite of the higher dissolved phosphorus loads with conservation tillage, however, the overall reduction of total phosphorus with conservation tillage is as high as 89 percent (Table 5), but generally in the range of 20 to 70 percent. In a few cases, conservation tillage increased total phosphorus loads. These instances usually were due to runoff occurring soon after surface application of phosphorus fertilizer.

The other approach to reducing phosphate load in surface runoff that offers some chance of success is phosphorus fertility management. Although this approach will not produce as dramatic a reduction in total phosphorus runoff loads as will erosion control, the impact on bioavailable phosphorus is greater (Logan and Forster, 1982).

Phosphorus fertility management has two components: (1) maintaining plant-available soil phosphorus levels at the sufficiency level for the crops grown, and (2) placement of phosphorus fertilizer below the soil surface where it is less likely to be removed as eroded sediment or to be desorbed into surface runoff water. Plant-available sufficiency levels have been established through extensive field correlation research for all major crops (Kamprath and Watson, 1980). These levels are based on several commonly used soil tests. Soil test recommendations have a dual objective. The first is to bring all soils in crop production up to the sufficiency level by relatively large build-up applications of phosphorus fertilizer. These are commonly and more effectively surface-applied and then incorporated uniformly into the top 15 to 20 centimeters of soil. The second objective is to maintain the sufficiency level by regular maintenance applications of phosphorus

**Table 5. Runoff, erosion, and nutrient losses from conservation tillage (Baker and Johnson, 1983).**

| | | Soil | | | | Nitrogen | | | Phosphorus | | | |
|---|---|---|---|---|---|---|---|---|---|---|---|---|
| *Practice* | *Study** | *Texture* | *Slope* | *Runoff* | *Erosion* | *Solution* | *Sediment* | *Total* | *Solution* | *Sediment* | *Total* | *Comments* |
| | | | (%) | | | | | | | | | |
| Till-plant[a] | N,W | sil | 10-15 | 65 | 38 | 68 | 54 | 55 | 180 | 112† | 130 | Continuous corn |
| No-till (ridge) | | | | 58 | 11 | 44 | 19 | 20 | 230 | 36 | 58 | Continuous corn |
| No-till[b] | N,W | sil | 9 | 9 | 1 | — | — | — | — | — | — | Continuous corn |
| No-till[c] | N,P | sil | 5 | 51 | 1 | 70 | 6 | 10 | 1,400 | 6 | 16 | Beans-beans |
| No-till | | | | 38 | 1 | 190 | 6 | 21 | 1,600 | 5 | 17 | Beans-wheat |
| No-till | | | | 106 | 12 | 180 | 40 | 52 | 450 | 36 | 39 | Beans-corn |
| No-till | | | | 80 | 3 | 410 | 14 | 47 | 1,650 | 13 | 25 | Corn-beans |
| Till-plant[d] | N,P | sicl | 6 | 71 | 58 | — | — | — | — | — | — | Corn |
| Till-plant[e] | S,P | sil | 8-12 | 86 | 33 | 2,100 | 41 | 92 | 2,250 | 46 | 47 | Continuous corn‡ |
| Chisel | | | | 49 | 5 | 1,900 | 9 | 52 | 1,950 | 10 | 11 | Continuous corn‡ |
| Disk | | | | 85 | 15 | 1,050 | 18 | 42 | 1,850 | 16 | 17 | Continuous corn‡ |
| No-till (coulter) | | | | 74 | 8 | 3,950 | 10 | 99 | 100,000 | 24 | 55 | Continuous corn‡ |
| Till-plant[f] | S,P | sil | 5-12 | 83 | 77 | 200 | 68 | 70 | 315 | 165† | 170 | Continuous corn |
| Chisel | | | | 96 | 62 | 205 | 59 | 62 | 335 | 130 | 135 | Continuous corn |
| Disk | | | | 84 | 31 | 215 | 38 | 41 | 390 | 84 | 93 | Continuous corn |
| No-till (ridge) | | | | 77 | 15 | 280 | 28 | 33 | 585 | 67 | 82 | Continuous corn |
| No-till (coulter) | | | | 75 | 8 | 270 | 14 | 19 | 625 | 32 | 50 | Continuous corn |
| Disk-chisel[g] | S,P | sil | 5 | 72 | 27 | 260 | 31 | 34 | 100 | 34 | 35 | Corn |
| Coulter-chisel | | | | 65 | 24 | 120 | 27 | 29 | 83 | 30 | 31 | Corn |
| Chisel | | | | 70 | 39 | 140 | 40 | 41 | 83 | 42 | 42 | Corn |
| Disk | | | | 70 | 20 | 240 | 26 | 29 | 75 | 26 | 27 | Corn |
| No-till | | | | 90 | 17 | 120 | 21 | 22 | 100 | 21 | 22 | Corn |
| Chisel-disk[h] | S,P | sil | 11 | 87 | 61 | — | — | — | — | — | — | Row-crop |
| No-till | | | | 109 | 36 | — | — | — | — | — | — | Row-crop |
| Chisel-disk | S,P | sl | 5 | 69 | 45 | — | — | — | — | — | — | Row-crop |
| No-till | | | | 85 | 28 | — | — | — | — | — | — | Row-crop |

Sources: [a]Johnson et al., 1979; [b]Harold et al., 1970; [c]McDowell and McGregor, 1980; [d]Onstad, 1972; [e]Romkens et al., 1973; [f]Barisas et al., 1978; Laflen et al., 1978; [g]Siemens and Oschwald, 1978; [h]Laflen and Colvin, 1981.
Note: Losses are expressed as a percentage of those for conventional tillage.
*N indicates natural precipitation; S, simulated rainfall; W, watershed; and P, plot.
†Phosphorus lost with sediment was as available P, for other studies as total P.
‡Fertilizer treatment.

fertilizer, calculated on the basis of crop removal rates. These applications may be broadcast or conveniently applied as starter fertilizer near the seed at planting.

By the late 1970s, soil test levels in the major crop production regions of the United States, and particularly in the Corn Belt (Logan, 1977), had risen to the point at which few production soils were testing below the sufficiency level. Logan (1977) examined state soil test records for Ohio and Michigan and found that most of the soils tested were at the sufficiency level but that the percentage of samples testing well above the level of crop response had risen rapidly. Some evidence, based on current fertilizer use statistics (USDA, 1987) and recent soil test summary data, indicates that phosphorus fertilizer use has leveled off in response to extension attempts to reduce unnecessary applications. There is still the opportunity, however, to reduce phosphorus loads in runoff by using residual plant-available phosphorus in soils testing very high in phosphorus. In a recent research demonstration study on farm fields in northwest Ohio, Kroetz and Logan (1987) showed that, on soils testing at or above the sufficiency level, yields of corn and soybeans with no phosphorus fertilizer applied were equal to those with starter fertilizer applied. The appeal of this approach to reducing phosphorus losses in runoff is that it requires little change in the farmer's management and will reduce his or her variable production costs. It is sustainable and easily monitored by annual soil and plant tissue analyses.

The second approach to phosphorus fertility management—placement of phosphorus fertilizer—works on the assumption that most phosphorus lost in surface runoff is removed by surface soil erosion and by the desorption of phosphate ($H_2PO_4^-$) in water interacting with the near-surface soil. Sharpley et al. (1978) estimated that this depth of soil interaction is less than 1 centimeter. Therefore, phosphorus placed at least 1 to 2 centimeters below the soil surface should be well-protected from normal runoff and rill-interrill erosion processes. Starter fertilizer applied to soils at the sufficiency level, commonly placed with the seed at a depth of 5 to 7 centimeters, should be effective in reducing runoff phosphorus losses.

*Nitrogen.* Nitrogen is a much more difficult nutrient to manage than phosphorus because of its highly dynamic nature and the dependence of most nitrogen transformations on microbiological processes. The most environmentally important form of nitrogen is nitrate ($NO_3^-$). Nitrate is highly water-soluble and is not retained by soil. It is, therefore, highly susceptible to movement in water and particularly movement in tile flow and deeper percolation. Surface runoff losses of nitrate can be particularly significant, however, when rainfall occurs soon after surface fertilizer application. Organic nitrogen in sediment and exchangeable $NH_4^+$ on eroded soil

**Table 6. Exampless of nitrogen loss to surface waters from forest, pasture, and cultivated lands (Gilliam et al., 1985).**

| | | *Nitrogen Loss* | | | |
|---|---|---|---|---|---|
| *Location* | *Crop* | *Surface Drainage* | *Subsurface Drainage* | *Total* | *Range Measured* |
| | | *kg/ha/year* | | | |
| New Hampshire[a] | Forest | 1.8 | | 1.8 | |
| West Virginia[b] | Forest | 0.8 | | 0.8 | |
| Oklahoma[c] | Pasture | 6 | | 6 | |
| North Carolina[d] | Pasture | | | 8 | 3-12 |
| Great Britain[e] | Pasture | | 33* | | 11.55* |
| Iowa[f] | Corn | | 38 | | 27-48 |
| Minnesota[g] | Corn | | 56* | | 19-120* |
| Oklahoma[h] | Cotton | 13 | | | |
| California[i] | Citrus | | 64 | | |
| North Carolina[j] | Corn | 25 | 21 | 46 | 45-48 |
| Canada[k] | Mixed crops | | 34*† | | 4-64* |
| Canada[k] | Mixed crops | | 145‡ | | 37-245 |
| Texas[l] | Mixed crops | 8 | | | |
| The Netherlands[m] | Mixed crops | | 30 | | 0-60 |

Sources: [a]Borman et al., 1968; [b]Aubertin & Patric, 1974; [c]Olness et al., 1975; [d]Kilmer et al., 1974; [e]Hood, 1976; [f]Baker & Johnson, 1981; [g]Gast et al., 1978; [h]Olness et al., 1975; [i]Bingham et al., 1971; [j]Gambrell et al., 1975; [k]Miller, 1979; [l]Kissel et al., 1976; [m]Kolenbrander, 1969.
*These values include losses from fields fertilized at rates higher than those recommended.
†Mineral soils.
‡Organic soils.

particles can contribute nitrate through subsequent mineralization and nitrification reactions. Table 6 summarizes nitrogen losses in runoff and tile drainage.

Like phosphorus, the water quality problems associated with nitrogen in production agriculture are a result of inefficiencies in nitrogen use by crops and of inadequacies in methods of predicting the dynamics of nitrogen in the soil-plant-water system. This is particularly true of crops such as corn, potatoes, cotton, and sugar beets, which have high nitrogen requirements that must be supplied externally (they do not fix nitrogen as do the leguminous crops). Although the absolute nitrogen demands of crops for optimum yield have been well established through extensive field trials, environmental effects on nitrogen supply to crops are less predictable and subject to seasonal environmental change. Denitrification, volatilization, and immobilization—all processes that reduce the supply of nitrogen to the crop—are highly variable and difficult to measure and predict in the time frames that are important to crop management.

The approach that has been taken to handle the lack of predictability of nitrogen supply to crops is to base nitrogen fertilizer requirements on field response trials that integrate, over some period of study (a few years to several decades), the effects of soil and environmental variables on nitrogen supply. Research has shown that, except for highly permeable soils where leaching losses are difficult to avoid under any circumstances, nitrogen fertilizer applications at recommended rates result in nitrogen losses in runoff and tile drainage that are in the range found for soybeans when no fertilizer is used, and that these concentrations rarely result in flow-weighted mean concentrations that exceed the drinking water standard of 10 milligrams of nitrate per liter (Logan et al., 1980; Logan, 1987; Baker and Johnson, 1983). Applications much above the recommended rate result in greatly elevated concentrations of nitrate in tile drainage (Logan et al., 1980; Baker and Johnson, 1983).

Besides carefully following nitrogen rate recommendations, several other approaches can be taken to increase nitrogen use efficiency and decrease nitrogen losses in runoff and tile drainage. One approach is to maximize nitrogen fertilizer uptake by the crop by strategic placement of the fertilizer near the plant roots and by timing nitrogen fertilizer applications to coincide with periods of greatest crop demand. Baker and Johnson (1983) summarized the relative effects of various fertilizer application methods on water quality, and their table is reproduced here (Table 7). These methods require more labor and may be more risky than using a single, large application of fertilizer. For these methods to be acceptable to the farmer, and therefore truly sustainable, they must be integrated into the overall crop management system. Also, a water quality benefit will be realized only if the overall rate of fertilizer application is reduced when these more efficient practices are used.

The other approach is to substitute leguminous green manure crops and forage legumes, such as clovers, vetches, and alfalfa, for chemical fertilizer. The assumption here is that the legume nitrogen will be mineralized slowly and will be used more efficiently than single large doses of chemical nitrogenous fertilizer. In addition, some deep-rooted legumes, such as alfalfa, have been shown to produce little nitrate movement to tile drains (Logan and Schwab, 1976; Armstrong et al., 1974). Armstrong et al. (1974) reviewed the literature on field studies of nutrient movement in runoff and percolation (lysimeter and tile drainage). They showed that, in general, losses in percolation were higher with fertilized crops and were particularly low for well-established perennial grass and legume stands. Logan and Schwab (1976) found that nitrate losses in tile drainage from an established alfalfa stand fertilized with dairy manure were much lower (1.3 kilograms nitrogen per hectare per year) than corresponding losses from two corn fields (30.8 and

**Table 7. Effects of timing and placement of chemical applications on runoff losses or concentrations (Baker and Johnson, 1983).**

| Practice | Study* | Soil Texture | Soil Slope | Chemical 1 Solution | Chemical 1 Sediment | Chemical 1 Total | Chemical 2 Solution | Chemical 2 Sediment | Chemical 2 Total | Comments |
|---|---|---|---|---|---|---|---|---|---|---|
| | | | | (%) | | | | | | |
| | | | | Atrazine | | | | | | |
| Timing[a] | S,P | sl | 6 | — | — | 44 | | | | Loss 96 hours after application relative to 1 hour |
| | | | | 2,4-D† | | | | | | |
| Timing[b] | S,P | sl | 5-7 | — | — | 75 | | | | Loss 48 hours after application relative to 1 hour |
| | | | | — | — | 108 | | | | Loss 96 hours after application relative to 1 hour |
| | | | | Nitrogen | | | | | | |
| Timing[c] | S,P | sil | 13 | 121 | — | — | | | | Concentration 120 hours after application relative to 24 hours; sod and fallow |
| | | | | Carbofuran | | | | | | |
| Placement[d] | N,P | sil | 9 | — | — | 57 | | | | Loss for in-furrow application relative to broadcast application |
| | | | | Alachlor | | | Atrazine | | | |
| Placement[e] | S,P | sl | 5 | 79 | 82 | — | 83 | 123 | — | Concentration for application below corn residue relative to above it |
| | | | | Picloram | | | | | | |
| Placement[f] | N,P | c | 3 | 28 | — | — | — | — | — | Concentration for subsurface application relative to surface application |
| | | | | Alachlor | | | Atrazine | | | |
| Placement[g] | SP | sl | 7 | 23 | 29 | 24 | 34 | 47 | 36 | Losses for disk incorporation relative to surface application |
| | | | | 300 | 235 | 290 | 415 | 355 | 405 | Losses for surface application on traffic-compacted soil relative to surface application on uncompacted soil |
| | | | | Nitrogen | | | Phosphorus | | | |
| Placement[h] | S,P | l | 7 | 51 | — | — | 23 | — | — | Concentration for disk incorporation relative to surface application; fallow |
| | | | | 20 | — | — | 13 | — | — | Concentration for plow-down relative to surface application; fallow |
| | | | | Nitrogen | | | Phosphorus | | | |
| Placement[i] | S,P | sl | 5 | 40 | — | — | 14 | — | — | Concentration for point injection relative to surface application; fallow |
| | | | | 90 | — | — | 98 | — | — | Concentration for application below corn residue relative to above it |

Sources: [a]White et al., 1967; [b]Barnett et al., 1967; [c]Moe et al., 1968; [d]Caro et al., 1973; [e]Baker et al., 1982; [f]Bovey et al., 1978; [g]Baker and Laflen, 1979; [h]Timmons et al., 1973; [j]Baker and Laflen, 1982.

*N indicates natural precipitation; S, rainfall simulation; and P, plot.

†A mine salt.

45.6 kilograms nitrogen per hectare per year). I am aware of no field experiments that have monitored nutrient movement from legume stands killed by incorporation or by herbicide as a nutrient source for a subsequent grain crop.

Yet the question of nutrient release and movement from this type of source is important to address in determining if rotation-based crop production systems will affect water quality any differently than monoculture systems.

Compared to the observed differences in tile drainage, Armstrong et al. (1974) found little difference between crops, including forage legumes and grasses, in the loss of nitrate in surface runoff. This is not unexpected for living stands where nutrients are actively recycled. But it is not clear what nitrate levels in runoff would be for no-till systems in which the forage stand is killed with herbicide and there is no incorporation. Hall et al. (1984) indicate that runoff from no-till birdsfoot trefoil and crownvetch was less than 0.25 centimeter in a growing season compared to 1 centimeter from no-till corn and 9.7 centimeters from plowed corn stalk residue. Soil loss was 32.2, 1.1, 0.04, and 0.04 tons per hectare for plowing, no-till in corn stalk residue, no-till birdsfoot trefoil, and no-till crownvetch, respectively. Although no data on nutrient losses were reported, the runoff data suggest that runoff would be reduced sufficiently with no-till forages to preclude any significant nitrogen losses in runoff.

*Pesticide Losses.* Pesticide losses in runoff and tile drainage have been studied for several years, although the literature is far less extensive than that for sediment and for nutrients. In considering the impacts of pesticide losses on water quality, it is important to keep several points in mind. First, the concentration of pesticides in water usually is much lower than that of nitrate or dissolved orthophosphate—usually micrograms per liter. Second, the public sees the water quality impacts of these concentrations in terms of human health risk and not in terms of ecosystem damage. Third, the human health risk is not acute toxicity, because the concentrations are always well below the $LC_{50}$ for humans, but chronic toxicity is of concern. Chronic toxicity of currently used pesticides is difficult to determine. As a result, drinking water standards for many of the commonly used pesticides have yet to be developed.

Much of the data on losses of pesticides from agricultural land (Weber et al., 1980) is reported in terms of loads, and a common reference is the percentage of the applied compound that is lost. These data are significant in determining the impact of specific practices on pesticide loss but are not helpful in determining human exposure and health risk. Of more significance is the frequency of concentrations in an impacted water body, and this can be presented as a concentration exceedency curve (Baker, 1988).

As threshold concentrations for health effects are determined, the concentration exceedency curve can be used to determine the health risk.

Health risk from pesticides can be reduced in three basic ways: (1) reduce or eliminate the use of pesticides in crop production systems, (2) reduce the losses of applied pesticides in runoff or percolation, and (3) develop nontoxic compounds. Of these, the third has always been an explicit goal in the development of new compounds, but evidence suggests that until recently (and perhaps still) the toxicity screening of new pesticides has not been adequate to guarantee acceptable risk protection.

The first approach, controlling the amounts and kinds of pesticides used, is the subject of extensive research and development and much debate in the agricultural community. At one end of the spectrum is organic farming, which has as a basic tenet complete independence from chemical pest control. This is achieved by a combination of cultural practices, including cultivation for weed control and use of rotations and beneficial organisms and plants for insect and disease control. These systems have been successful, but the labor requirements are high and present economic conditions preclude their widespread adoption.

An intermediate approach to reduce pesticide use is integrated pest management, which has as its aim elimination of unnecessary routine pesticide use and substitutes strategic application of chemicals combined with other techniques as needed on the basis of weed, insect, and disease pressure. This practice is a powerful tool for pest management but remains to be fully adopted by U.S. farmers.

One apparent conflict of environmental quality goals that has arisen recently is the use of conservation tillage for erosion and phosphorus control. It has been suggested that conservation tillage, in addition to being more dependent on pesticides than are plow-based systems, also increases the risk of percolation losses to tile drains and groundwater by increasing infiltration and decreasing runoff.

At an EPA-sponsored workshop to deal specifically with this issue (Logan et al., 1987), several papers considered in detail the actual use of pesticides in conservation tillage systems as well as the effects of conservation tillage on surface and subsurface hydrology. Fawcett (1987) summarized current statistics on the use of pesticides with conservation tillage in the United States and concluded that there was little difference in the amount and type of pesticides used in conservation tillage and in plow-based systems. In modern crop production systems, tillage is not a substitute for herbicides. In separate papers on hydrologic effects of conservation tillage, Onstad and Voorhees (1987) and Baker (1987) suggested that the most extreme of the conservation tillage systems, no-till, evidenced increased, decreased, or no effect on runoff, as compared to conventionally plowed systems. Considering that most conservation tillage in the United States involves some tillage, the workshop participants concluded that widespread adoption of conservation tillage by U.S. farmers would have a minor effect on pesticide movement in runoff or percolation.

The third method of reducing pesticide losses in runoff and percolation is through formulation, timing, and placement of the chemical. Baker and Johnson (1983) have summarized the effectiveness of some of these approaches (Table 7). Subsurface injection or soil incorporation appeared to be the most promising practices to reduce runoff losses of pesticides. For the more persistent chemicals, such as atrazine, reduced runoff losses may result in greater movement in percolation.

*Manure Management.* The water quality problems associated with manure disposal are primarily from nutrient losses, although increased biological oxygen demand and pathogens in manure runoff can be of significant local importance. Even if manure is collected and stored so as to avoid direct runoff to streams, the problem of manure nutrients in excess of crop needs is present in areas of large, confined-livestock operations.

Manure contains significant amounts of nitrogen and phosphorus (Gilbertson et al., 1979). Table 2 indicates that 5 and 1.5 million tons of nitrogen and phosphorus, respectively, were produced in manure in the U.S. in 1985-1987. This compares to 10.4 and 1.8 million tons of nitrogen and phosphorus, respectfully, used in chemical fertilizer in the United States in 1985 (USDA, 1987). Considering that feed grains—corn and wheat, primarily—and hay are major consumers of fertilizer, much of the nutrient content in manure originally came from fertilizer. If these manure nutrients were efficiently recycled for crop production, overall fertilizer consumption would decrease. The problem is that with the trend toward large, confined-livestock operations, manure is regionally concentrated and there is little economic incentive to redistribute manure nutrients to the areas of grain and feed production.

Table 8 shows that, even at agronomic rates of application (not given but assumed to be those that supply nitrogen needs of the crop), nitrogen and phosphorus concentrations in runoff increase significantly with surface application without incorporation. Walter et al. (1987) indicated, however, that incorporation of manure as shallow as 3 centimeters can reduce nutrient losses by as much as 80 percent. They also indicated that manure solids act as a surface mulch to reduce runoff and increase infiltration. This favors reduction of erosion and total phosphorus losses, but could contribute to increased nitrate leaching. With the exception of runoff of liquid manure in rainfall, shortly after manure application (a hydraulic problem), Walter

**Table 8. Estimated concentrations of total nitrogen, total phosphorus, and chemical oxygen demand dissolved in runoff from land with and without livestock or poultry manure surface-applied at agronomic rates (Gilbertson et al., 1979).**

| | *Rainfall Runoff* | | | | | | *Snowmelt Runoff.* | | |
|---|---|---|---|---|---|---|---|---|---|
| | *Total Nitrogen* | | *Total Phosphorus* | | *COD* | | *Total Nitrogen* | *Total Phosphorus* | *COD* |
| *Cropping Condition* | *Manure With* | *Manure Without* | *Manure With* | *Manure Without* | *Manure With* | *Manure Without* | *With Manure* | *With Manure* | *With Manure* |
| | *milligrams/liter* | | | | | | | | |
| Grass | 11.9 | 3.2 | 3.0 | 0.44 | 360 | 50 | 36 | 8.7 | 370 |
| Small grain | 16.0 | 3.2 | 4.0 | 0.40 | 170 | 20 | 25 | 5.0 | 270 |
| Row crop | 7.1 | 3.0 | 1.7 | 0.40 | 88 | 55 | 12.2 | 1.9 | 170 |
| Rough plow | 13.2 | 3.0 | 1.7 | 0.20 | 88 | 55 | 12.2 | 1.9 | 170 |

et al. (1987) view runoff of manure nutrients as a less serious long-term water quality problem than nitrate leaching.

***Groundwater Quality.*** Groundwater quality, including that attributed to agricultural practices, has become one of the dominant environmental issues of the 1980s, although contamination of groundwater by nitrates and pesticides has been recognized in the United States and elsewhere for decades. Agricultural groundwater contamination is highly localized and, in this regard, is a more limited problem than surface water contamination. The potential impact of groundwater contamination on human health is greater, however, because about 50 percent of the U.S. population relies on groundwater for drinking, much of it untreated (EPA, 1987).

Research on the effects of agriculture on groundwater contamination is limited. Canter (1987) summarized the current literature on agricultural groundwater research—63 papers, 34 on nitrate and 29 on pesticides. Of the literature on pesticides, the majority dealt with aldicarb (Temik$^{R}$), a systemic insecticide and nematicide. This compound is used primarily for potato production. Perhaps the most serious instance of pesticide groundwater contamination is that by ethylene dibromide (EDB), a soil fumigant. It has been detected in wells in 16 counties in California, Florida, Georgia, and Hawaii at concentrations typically in the range of 0.05 to 5 micrograms per liter (EPA, 1986). EDB is a potent cancer-causing substance and has since been banned by EPA.

*Nitrate.* Unlike surface water, in which options are available for regulating pollutant transport by introducing erosion control and runoff reduction, controlling water percolation to groundwater is difficult. Surface and tile drainage will intercept infiltrating and percolating water, but tradeoffs between surface and groundwater contamination must be considered. Irrigation management can reduce excess percolation, but again there is a tradeoff, with the potential for salinity to be considered. Besides water management, the most effective means of controlling groundwater contamination is source control. Figure 1 shows clearly the direct correlation between nitrate levels in groundwater in the Big Springs watershed in Iowa and nitrogen fertilizer use in the same period. Manure nitrogen application also increased slightly. This area is particularly susceptible to groundwater contamination because it is underlain by karst—limestone bedrock with extensive water-conducting solution channels (Libra et al., 1987). Nitrogen source controls include more precise estimations of application rates for corn, nitrogen application rates that are in line with realistic yield goals, full credit for nitrogen supplied by manure and legumes, and an adequate land base for manure disposal. Practices that attempt to improve nitrogen use efficiency by crops, such as split applications, fertilizer placement, and use of nitrification inhibitors, have a lower potential to reduce nitrate leaching in soil and have not been widely adopted.

*Pesticides.* Evidence to date from private well surveys suggests that pesticide contamination is not widespread but is highly regional and involves relatively few compounds. In 1985, 17 pesticides had been detected in 23 states, with normal agricultural practices (EPA, 1986). Aldicarb, EDB,

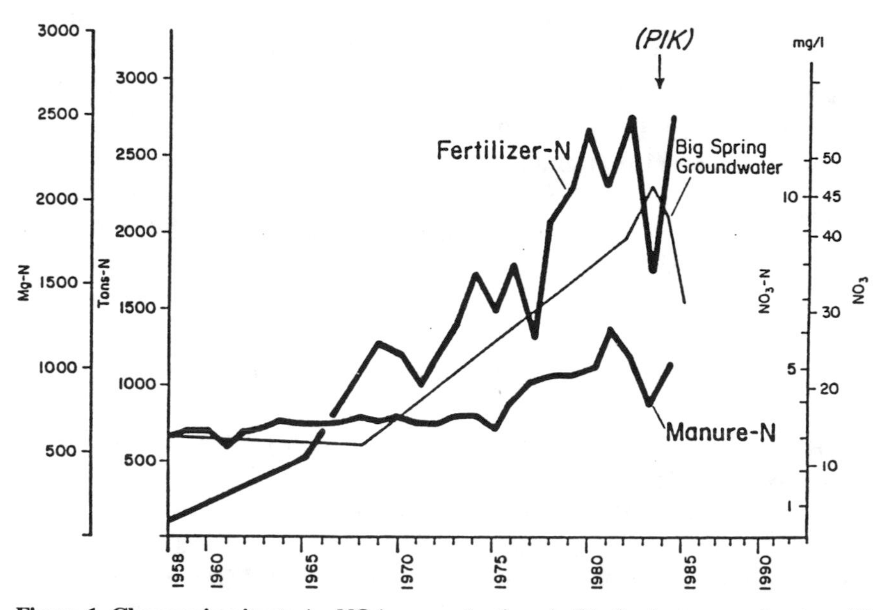


**Figure 1. Changes in nitrate (as $NO_3$) concentrations in Big Springs groundwater with estimates of nitrogen fertilizer and manure-nitrogen use (Libra et al., 1987).**

DBCP, and atrazine were detected most often. These compounds all have relatively high water solubility, have a low affinity for soil, and are relatively long-lived.

Source control is the only effective means of reducing the potential for pesticide groundwater contamination. Source controls include restricting use of the more mobile compounds in areas of high groundwater vulnerability; using integrated pest management or alternative pest control measures instead of prescription pesticide application wherever possible; and good housekeeping measures, such as proper rinsing of spray tanks and disposing of containers. Of these approaches, the first and last are likely to be more effective in the short run. Farmers have yet to adopt integrated pest management strategies to any large extent, and alternative pest control is highly pest- and crop-specific.

### Systems Management

One of the deficiencies of current agricultural research has been the lack of a systems approach. Separate components of the soil-crop management system have been studied in detail and have provided invaluable data on the function and effectiveness of various crop production practices. However, there have been few, if any, studies of entire crop production systems in which the interactions of various components can be observed. This is particularly true with studies of the impacts of crop production systems on water quality. If U.S. production agriculture will have regulatory constraints in the future—and this is probable—sustainable agricultural systems will be those that can meet these constraints. These systems can be developed on the basis of available knowledge of the individual components of the system, but these systems must be evaluated under realistic field conditions.

Field research is expensive even when the experimental design is simple—one or two variables, as in a fertilizer rate study. And such research becomes prohibitive if it proposes to measure the myriad variables that define a system. Computer modeling can reduce the amount of data that must be collected, but only comprehensive field studies of crop production systems will answer satisfactorily the questions that have been discussed. Because of their great cost, comprehensive field studies cannot be replicated everywhere. But a few, each representing major agroecosystem regions of the United States, should be established at universities, experiment stations, and federal laboratories. Crop production systems with a range of tillage, pest management, and nutrient management options should be established and monitored for ecosystem function, crop production, and water quality.

The data gathered can be used to develop or validate system models. The data produced should be made available to the research community in computer form, and the sites themselves made available to researchers for specific system studies. By intensifying research at a few locations, we may begin to quantify the complex processes by which agricultural ecosystems function and have a more sound basis for selecting sustainable agricultural systems.

## Conclusions

- Current crop production systems in the United States are intensive and rely heavily on chemical fertilizers and pesticides. Water quality impacts associated with these practices include accelerated erosion and nutrient and pesticide losses.
- Large, confined-livestock operations have resulted in a concentration of manure nutrients in locations far removed from the areas of feed and feed grain production. This results in inadequate land in the manure-producing area for proper manure nutrient disposal.
- A number of alternative management practices have the potential to decrease agricultural water pollution. These, however, have yet to be developed into functioning systems that are acceptable to the farmer or are economically viable. In the absence of specific agricultural water quality regulations, these systems are not likely to be adopted.
- The current public concern for groundwater contamination is detracting from the fact that surface waters usually are more contaminated by nutrients and pesticides than is groundwater. Groundwater contamination is highly localized and involves areas vulnerable to chemical transport in conjunction with agricultural enterprises, such as confined livestock operations and potato production, that are sources of nitrate or pesticides.

## Research Needs

- Establish long-term agroecosystem research sites in each of the major U.S. cropping system regions, with water quality assessment as a major objective.
- Perform a comprehensive assessment of environmental risk from present and alternative agricultural practices.
- Develop crop production systems that simultaneously address crop production, erosion control, nutrient management, and pest management.

## REFERENCES

Alberts, E. E., G. E. Shuman, and R. E. Burwell. 1978. Seasonal runoff losses of nitrogen and phosphorus from Missouri Valley loess watersheds. Journal of Environmental Quality 7: 401-406.

Armstrong, D. E., K. W. Lee, P. D. Uttomark, D. R. Keeney, and R. F. Harris. 1974. Pollution of the Great Lakes by nutrients from agricultural land. International Refer Group on Great Lakes Pollution from Land Use Activities. Volume 1. International Joint Commission, Windsor, Ontario.

Aubertin, G. M., and J. H. Patric. 1974. Water quality after clearcutting a small watershed in West Virginia. Journal of Environmental Quality 3: 243-249.

Baker, D. B. 1988. Sediment, nutrient and pesticide transport in selected lower Great Lakes tributaries. EPA-905/4-88-001. Great Lakes National Program Office. U.S. Environmental Protection Agency, Region V, Chicago, Illinois.

Baker, J. L. 1987. Hydrologic effects of conservation tillage and their importance relative to water quality. pp. 113-124. *In* T. J. Logan, J. M. Davidson, J. L. Baker, and M. R. Overcash [editors] Effects of conservation tillage on groundwater quality. Lewis Publications, Chelsea, Michigan.

Baker, J. L., and H. P. Johnson. 1981. Nitrate-nitrogen in tile drainage as affected by fertilization. Journal of Environmental Quality 10: 519-522.

Baker, J. L., and H. P. Johnson. 1983. Evaluating the effectiveness of BMPs from field studies. pp. 281-304. *In* F. W. Schaller and G. W. Bailey [editors] Agricultural management and water quality. Iowa State University Press, Ames.

Baker, J. L., and J. M. Laflen. 1979. Runoff losses of surface-applied herbicides as affected by wheel tracks and incorporation. Journal of Environmental Quality 8: 602-607.

Baker, J. L., and J. M. Laflen. 1982. Effects of corn residue and fertilizer management on soluble nutrient runoff losses. Transactions, American Society of Agricultural Engineers 25: 344-348.

Baker, J. L., J. M. Laflen, and R. O. Hartwig. 1982. Effects of corn residue and herbicide placement on herbicide runoff losses. Transactions, American Society of Agricultural Engineers 25: 340-343.

Barisas, S. G., J. L. Baker, H. P. Johnson, and J. M. Laflen. 1978. Effect of tillage systems on runoff losses of nutrients, a rainfall simulator study. Transactions, American Society of Agricultural Engineers 21: 893-897.

Barnett, A. P., E. W. Hauser, A. W. White, and J. H. Holladay. 1967. Loss of 2,4-D in washoff from cultivated fallow land. Weeds 15: 133-137.

Bingham, F. T., S. Davis, and E. Shade. 1971. Water relations, salt balance, and nitrate leaching losses of a 960-acre citrus watershed. Soil Science 112: 410-417.

Bolton, E. F., J. W. Aglesworth, and F. R. Hore. 1970. Nutrient losses through tile lines under three cropping systems and two fertility levels on a Brookston clay soil. Canadian Journal of Soil Science 50: 275-279.

Borman, F. H., G. E. Likens, D. W. Fisher, and R. S. Pierce. 1968. Nutrient losses accelerated by clear-cutting of a forest ecosystem. Science 159: 882-884.

Bovey, R. W., C. Richardson, E. Burnett, M. G. Merkle, and R. E. Meyer. 1978. Loss of spray and pelleted picloram in surface runoff water. Journal of Environmental Quality 7: 178-180.

Bureau, R. G. 1985. Environmental chemistry of selenium. California Agriculture 39 (July-August): 16-18.

Canter, L. W. 1987. Nitrates and pesticides in ground water: An analysis of a computer-based literature search. pp. 153-174. *In* D. M. Fairchild [editor] Ground water quality and agricultural practices. Lewis Publishers, Chelsea, Michigan.

Caro, J. H., H. P. Freeman, D. E. Glotfelty, B. C. Turner, and W. M. Edwards. 1973. Dissipation of soil-incorporated carbofuran in the field. Journal of Agricultural Food

Chemicals 21: 1,010-1,015.

Carson, R. 1962. Silent spring. Houghton Mifflin, Boston, Massachusetts.

Carter, D. L., J. A. Bondurant, and C. W. Robbins. 1971. Water soluble NO3-N, PO4-P, and total salt balances on a large irrigation tract. Soil Science Society of America Proceedings 35: 331-335.

Dickey, E. C., P. J. Jasa, B. J. Dolesh, L. A. Brown, and S. K. Rockwell. 1987. Conservation tillage: Perceived and actual use. Journal of Soil and Water Conservation 42: 431-434.

Ellis, B. G., A. E. Erickson, and A. R. Wolcott. 1978. Nitrate and phosphorus runoff losses from small watersheds in the Great Lakes basin. EPA-600/3-78-028. U.S. Environmental Protection Agency, Athens, Georgia.

Fawcett, R. S. 1987. Overview of pest management for conservation tillage. pp. 19-37. *In* T. J. Logan, J. M. Davidson, J. L. Baker, and M. R. Overcash [editors] Effects of conservation tillage on groundwater quality. Lewis Publications, Chelsea, Michigan.

Forster, D. L., T. J. Logan, S. M. Yaksich, and J. R. Adams. 1985. An accelerated implementation program for reducing the diffuse source phosphorus load to Lake Erie. Journal of Soil and Water Conservation 40: 136-141.

Frink, C. R. 1967. Nutrient-budget rational analysis of eutrophication in a Connecticut lake. Environmental Science and Technology 1: 425-428.

Gambrell, R. P., J. W. Gilliam, and S. B. Weed. 1975. Nitrogen losses from soils of the North Carolina coastal plain. Journal of Environmental Quality 4: 317-323.

Gast, R. G., W. W. Nelson, and G. W. Randall. 1978. Nitrate accumulation in soils and loss in tile drainage following nitrogen applications to continuous corn. Journal of Environmental Quality 7: 258-261.

Gilbertson, C. B., F. A. Norstadt, A. C. Mathers, R. F. Holt, A. P. Barnett, T. M. McCalla, C. A. Onstad, and R. A. Young. 1979. Animal waste utilization on cropland and pastureland. EPA-600/2-70-59. U.S. Environmental Protection Agency, Ada, Oklahoma.

Gilliam, J. W., T. J. Logan, and F. E. Broadbent. 1985. Fertilizer use in relation to the environment. pp. 561-588. *In* O. P. Engelstad [editor] Fertilizer technology and use (3rd edition). Soil Science Society of America, Madison, Wisconsin.

Hall, G. F., T. J. Logan, and K. K. Young. 1985. Criteria for determining tolerable erosion rates. pp. 173-188. *In* R. F. Follett and B. A. Stewart [editors] Soil erosion and crop productivity. American Society of Agronomy, Madison, Wisconsin.

Hall, J. K., N. L. Hartwig, and L. D. Hoffman. 1984. Cyanazine losses in runoff from no-tillage corn in "living" and dead mulch vs. unmulched, conventional tillage. Journal of Environmental Quality 13: 105-110.

Harmeson, R. H., F. W. Sollo, and T. E. Larson. 1971. The nitrate situation in Illinois. Journal of American Water Works Association 63: 303-310.

Harrold, L. L., G. B. Triplett, Jr., and W. M. Edwards. 1970. No-tillage corn—Characteristics of the system. Agricultural Engineering 51: 128-131.

Hood, A.E.M. 1976. The high nitrogen trial on grassland at Jealott's Hill. Stikstof (English edition) 83: 395-404.

Jaworski, N. A., and L. J. Hetling. 1970. Relative contributions of nutrients to the Potomac River Basin from various sources. *In* Relationship of agriculture to soil and water pollution. Cornell University, Ithaca, New York.

Johnson, H. P., J. L. Baker, W. D. Schrader, and J. M. Laflen. 1979. Tillage system effects on sediment and nutrients in runoff from small watersheds. Transactions, American Society of Agricultural Engineers 22: 1,110-1,114.

Jordan, C. F. 1985. Nutrient cycling in tropical forest ecosystems. John Wiley and Sons, New York, New York. pp. 77-80.

Kamprath, E. J., and M. E. Watson. 1980. Conventional soil and tissue tests for assessing the phosphorus status of soils. pp. 433-469. *In* F. E. Khasawneh, E. C. Sample, and E. J. Kamprath [editors] The role of phosphorus in agriculture. American Society of Agronomy, Madison, Wisconsin.

Keup, E. 1968. Phosphorus in flowing waters. Water Resources 2: 373-386.
Kilmer, V. J., J. W. Gilliam, J. F. Lutz, R. T. Joyce, and C. K. Eklunk. 1974. Nutrient losses from fertilized grassed watersheds in western North Carolina. Journal of Environmental Quality 3: 214-219.
Kissel, D. E., C. W. Richardson, and E. Burnett. 1976. Losses of nitrogen in surface runoff in the blackland prairie of Texas. Journal of Environmental Quality 5: 288-292.
Kolenbrander, G. J. 1969. Nitrate content and nitrogen loss in drainwater. Netherlands Journal of Agricultural Science 17: 246-255.
Kroetz, M. E., and T. J. Logan. 1987. Phosphorus fertilizer rate demonstrations. Northwest Ohio (1982-84). Ohio Cooperative Extension Service, Wooster.
Laflen, J. M., J. L. Baker, R. O. Hartwig, W. F. Buchele, and H. P. Johnson. 1978. Soil and water loss from conservation tillage systems. Transactions, American Society of Agricultural Engineers 21: 881-885.
Laflen, J. M., and T. S. Colvin. 1981. Effect of crop residue on soil loss from continuous row cropping. Transactions, American Society of Agricultural Engineers 24: 605-609.
Lake, J., and J. Morrison. 1977. Environmental impact of land use on water quality. EPA-905/9-77-007-B. U.S. Environmental Protection Agency, Region V, Chicago, Illinois.
Libra, R. D., G. R. Hallberg, and B. E. Hoyer. 1987. Impacts of agricultural chemicals on ground water quality in Iowa. pp. 185-215. *In* D. M. Fairchild [editor] Ground water quality and agricultural practices. Lewis Publishers, Chelsea, Michigan.
Likens, G. E., F. H. Bormann, R. S. Pierce, J. S. Eaton, and N. M. Johnson. 1977. Biogeochemistry of a forested ecosystem. Springer-Verlag, New York, New York.
Logan, T. J. 1977. Levels of plant available phosphorus in agricultural soils in the Lake Erie Basin. Lake Erie Wastewater Management Study. Technical Report Series. U.S. Army Corps of Engineers, Buffalo District, Buffalo, New York.
Logan, T. J. 1982. Mechanisms for release of sediment-bound phosphate to water and the effects of agricultural land management on fluvial transport of particulate and dissolved phosphate. 9: 511-530. *In* P. G. Sly [editor] Sediment/freshwater interaction. Junk Publishers, The Hague, The Netherlands.
Logan, T. J. 1987. Tile drainage water quality: A long term study in NW Ohio. C-53-64. *In* Proceedings, Third International Workshop on Land Drainage, Ohio State University, Columbus.
Logan, T. J., and D. L. Forster. 1982. Alternative management options for the control of diffuse phosphorus loads to Lake Erie. Lake Erie Wastewater Management Study. Technical Report Series. U.S. Army Corps of Engineers, Buffalo, New York.
Logan, T. J., and G. O. Schwab. 1976. Chemical characterization of tile drainage in western Ohio. Journal of Soil and Water Conservation 31: 24-27.
Logan, T. J., G. W. Randall, and D. R. Timmons. 1980. Nutrient content of tile drainage from cropland in the North Central region. North Central Research Publication No. 268. Ohio Agricultural Research and Development Center, Wooster.
Logan, T. J., J. M. Davidson, J. L. Baker, and M. R. Overcash, editors. 1987. Effects of conservation tillage on groundwater quality. Lewis Publishers, Chelsea, Michigan.
Logan, T. J., and J. R. Adams. 1981. The effects of reduced tillage on phosphate transport from agricultural land. Lake Erie Wastewater Management Study. Technical Report Series. U.S. Army Corps of Engineers, Buffalo, New York.
Logan, T. J., and R. C. Stiefel. 1979. The Maumee river basin pilot watershed study. Volume 1. Watershed characteristics and pollutant loadings. EPA-905/9-79-005-A. U.S. Environmental Protection Agency, Region V, Chicago, Illinois.
Mannering, J. V., D. L. Schertz, and B. A. Julian. 1987. Overview of conservation tillage. pp. 3-17. *In* T. J. Logan, J. M. Davidson, J. L. Baker, and M. R. Overcash [editors] Effects of conservation tillage on groundwater quality. Lewis Publications, Chelsea, Michigan.
McDowell, L. L., and K. C. McGregor. 1980. Nitrogen and phosphorus losses from no-

till soybeans. Transactions, American Society of Agricultural Engineers 23: 643-648.
Menzel, R. G., E. D. Rhoades, A. E. Olness, and S. J. Smith. 1878. Variability of annual nutrient and sediment discharges in runoff from Oklahoma cropland and rangeland. Journal of Environmental Quality 7: 203-20.
Miller, M. H. 1979. Contribution of nitrogen and phosphorus to subsurface drainage water from intensively cropped mineral and organic soils in Ontario. Journal of Environmental Quality 8: 42-48.
Moe, P. G., J. V. Mannering, and C. B. Johnson. 1968. A comparison of nitrogen losses from urea and ammonium nitrate in surface runoff water. Soil Science 105: 428-433.
Nelson, D. W., D. B. Beasley, S. Amin, and E. J. Monke. 1980. Water quality: Sediment and nutrient loadings from cropland. *In* Proceedings, water quality management trade-offs. Point Source vs. Diffuse Source Pollution. EPA-905/9-80-009. U.S. Environmental Protection Agency, Region V, Chicago, Illinois.
Nelson, D. W., and T. J. Logan. 1983. Chemical processes and transport of phosphorus. pp. 65-91. *In* F. W. Schaller and G. W. Bailey [editors] Agricultural management of water quality. Iowa State University Press, Ames.
Olness, A., S. J. Smith, E. D. Rhoades, and R. G. Menzel. 1975. Nutrient and sediment discharge from agricultural watersheds in Oklahoma. Journal of Environmental Quality 4: 331-336.
Onstad, C. A. 1972. Soil and water losses as affected by tillage practices. Transactions, American Society of Agricultural Engineers 15: 287-289.
Onstad, C. A., and W. B. Voorhees. 1987. Hydrologic soil parameters affected by tillage. pp. 95-112. *In* T. J. Logan, J. M. Davidson, J. L. Baker, and M. R. Overcash [editors] Effects of conservation tillage on groundwater quality. Lewis Publishers, Chelsea, Michigan. pp. 95-112.
Pimental, D. et al. 1973. Food production and the energy crisis. Science 182: 443-449.
Romkens, M.J.M., D. W. Nelson, and J. V. Mannering. 1973. Nitrogen and phosphorus composition of surface runoff as affected by tillage method. Journal of Environmental Quality 2: 292-295.
Sharpley, A. N., J. K. Syers, and R. W. Tillman. 1978. An improved soil-sampling procedure for the prediction of dissolved inorganic phosphate concentrations in surface runoff from pasture. Journal of Environmental Quality 7: 455-456.
Siemens, J. C., and W. R. Oschwald. 1978. Corn-soybean tillage systems: Erosion control, effects on crop production cost. Transactions, American Society of Agricultural Engineers 21: 293-302.
Skogerboe, G. V., and W. R. Walker. 1981. Impact of irrigation on the quality of groundwater and river flows. pp. 121-157. *In* D. Yaron [editor] Salinity in irrigation and water resources. Marcel Dekker, Inc., New York, New York.
Taylor, A. W., W. M. Edwards, and E. C. Simpson. 1971. Nutrients in streams draining woodland and farmland near Coshocton, Ohio. Water Resources Research 7: 81-89.
The Conservation Foundation. 1986. Agriculture and the environment in a changing world economy. Washington, D.C.
Timmons, D. R., R. E. Burwell, and R. F. Holt. 1973. Nitrogen and phosphorus losses in surface runoff from agricultural land as influenced by placement of broadcast fertilizer. Water Resources Research 9: 658-667.
U.S. Department of Agriculture. 1984. Agricultural Statistics. Washington, D.C.
U.S. Department of Agriculture. 1987. Agricultural Statistics. Washington, D.C.
U.S. Environmental Protection Agency. 1986. Pesticides in ground water: Background document. EPA 440/6-86-002. Office of Groundwater Protection, Washington, D.C.
U.S. Environmental Protection Agency. 1987. Improved protection of water resources from long-term and cumulative pollution. EPA 440/6-87-013. Office of Groundwater Protection, Washington, D.C.
Walter, M. F., T. L. Richard, P. D. Robillard, and R. Muck. 1987. Manure management

with conservation tillage. pp. 253-270. *In* T. J. Logan, J. M. Davidson, J. L. Baker, and M. R. Overcash [editors] Effects of conservation tillage on groundwater quality. Lewis Publications, Chelsea, Michigan.

Weber, J. B., P. J. Shea, and H. J. Strek. 1980. An evaluation of nonpoint sources of pesticide pollution in runoff. pp. 69-68. *In* M. R. Overcash and J. M. Davidson [editors] Environmental impact of nonpoint source pollution. Ann Arbor Science Publications, Ann Arbor, Michigan.

White, A. W., A. P. Barnett, B. G. Wright, and J. H. Holladay. 1967. Atrazine losses from fallow land caused by runoff and erosion. Environmental Science Technology 1: 740-744.

Wischmeier, W. H., and D. D. Smith. 1978. Predicting rainfall erosion losses—A guide to conservation planning. Agricultural Handbook No. 537. U.S. Department of Agriculture, Washington, D.C.

Yaron, D. 1981. The salinity problem in irrigation—An introductory review. pp. 1-20. *In* D. Yaron [editor] Salinity in irrigation and water resources. Marcel Dekker, Inc., New York, New York.

# SOIL MICROBIOLOGICAL INPUTS FOR SUSTAINABLE AGRICULTURAL SYSTEMS

R. H. Miller

The soil, the soil rhizophere (soil immediately around a plant root), and the rhizoplane (the root surface, including the mucigel and adhering root debris) are marvelously complex and scientifically interesting ecosystems. The number of microorganisms found in these environments is impressive (Tables 1 and 2). Even more important is the influence, both positive and negative, that some of these microorganisms have on plant growth and development. Detrimental effects include nutrient immobilization, plant diseases, and the microbial production of phytotoxic substances. Important as these effects are, they will not be discussed further here. Instead, the focus will be the exciting challenge of how to enhance the positive influence of soil, rhizosphere, and rhizoplane microorganisms on plants for a more sustainable agriculture.

Some ways in which soil microorganisms positively influence plant growth and development have been known and appreciated since the inception of soil microbiology in the late nineteenth and early twentieth centuries. Others have more recent origin and recognition. Table 3 summarizes some of the beneficial activities of soil microorganisms. Many of these activities are particularly important and significant in the plant rhizosphere and rhizoplane regions, where microbial populations and activity are high.

All of the reactions listed in table 3 occur normally to some degree in plant-soil systems. Some can be enhanced by adding more soil organic residues and waste materials. But what is particularly intriguing for sustainable agricultural systems is the methodology by which these beneficial activities can be selectively enhanced by soil, seed, or seedling inoculation. Nontoxic, environmentally neutral microorganisms might be used to enhance nutrient use efficiency; stimulate plant growth; reduce insect, weed, and disease pressures; and so forth.

## Current Status of Microbial Products for Plant Inoculation

How successful are we in using beneficial soil microorganisms in agriculture? Some people would say that these products are already available in the marketplace. But this answer is only partly true and largely false. Certainly, there are attractively packaged and actively marketed "microbial fertilizers" or "microbial activators" available for purchase. Microbial fertilizers constitute the various living cultures that are said to contain strains of soil bacteria, fungi, actinomycetes, and algae, alone or in combination. Microbial activators are products, often of ill-defined chemical composition, claimed to increase the number and activity of beneficial soil microorganisms. The claims made for these products are always impressive and generally mirror the potential beneficial effects of microorganisms summarized in table 3. However, with the exception of specific legume inoculants, some ectomycorrhizal products used in forest nurseries, and a new bacterial fungicide for cotton (Brosten, 1988), none of these products fulfill the claims made for them, and the reasons for their failures are evident (Miller, 1979, 1979a,b).

Microbial products (microbial fertilizers) are generally applied by soil,

**Table 1. Number of soil microorganisms in cultivated temperate soils.**

| *Group* | *Abundance (no./g)* | *Live biomass (kg/ha)* |
|---|---|---|
| Bacteria | $10^4$-$10^9$ | 300-3,000 |
| Actinomycetes | $10^5$-$10^8$ | |
| Fungi | $2 \times 10^4$-$10^6$ | 500-5,000 |
| Algae | $10^2$-$5 \times 10^4$ | 7-300 |
| Protozoa | $10^4$-$10^5$ | 50-200 |

**Table 2. Extent of rhizosphere of 18-day-old blue lupin seedlings (no. $\times 10^6$/g dry soil) (Papavizas and Davey, 1961).**

| *Distance From Root (mm)* | *Bacteria* | *Streptomycetes* | *Fungi* |
|---|---|---|---|
| Rhizoplane | | | |
| 0 | 159 | 47 | .36 |
| Rhizosphere | | | |
| 0-3 | 49 | 16 | .18 |
| 3-6 | 38 | 11 | .17 |
| 9-12 | 37 | 10 | .12 |
| 15-18 | 34 | 10 | .12 |
| Soil | | | |
| 80 | 27 | 9 | .09 |

seed, or seedling inoculation, with or without some carrier for the microorganisms—for example, peat, organic compost, or stickers. Regardless of the method, the number of cells reaching the soil from commercial products is very small (100 or fewer cells per gram of soil). When compared to the existing numbers of soil or rhizosphere microorganisms (Tables 1 and 2), these added cells are unlikely to have a beneficial impact on the plant unless multiplication occurs. Considerable scientific data support the concept that multiplication does not occur. In actuality, the population of introduced microorganisms will decline and be eliminated in a very short time, often days or weeks. This die-back should be considered a normal event in line with fundamental ecological principles.

Agricultural or horticultural soils (without sterilization) contain a complex community of microorganisms that have proven themselves, by an extensive period of selection, competition, and adaptation, to best occupy a particular niche. The microorganisms occupying these niches do so because they have a favorable combination of nutritional, biochemical, or morphological attributes that have given an advantage over other microorganisms that could potentially occupy these same niches. Once this stability is achieved and all of the recognized niches are filled, this community of micro-

**Table 3. Beneficial activities of microorganisms in the soil, rhizosphere, or rhizoplane.**

Decomposition of plant residues, manures, and organic wastes
- Humus synthesis
- Mineralization of organic N, S, and P
- Improved soil aggregation

Increase in the availability of plant nutrients—for example, P, Mn, Fe, Zn, Cu
- Symbiotic mycorrhizal associations
- Production of organic chelating agents
- Oxidation—reduction reactions

Biological nitrogen fixation
- Free-living bacteria and bluegreen algae
- Associative microorganisms
- Symbiotic—Legume and nonlegume

Plant growth promotion: changes in seed germination, floral development, root and shoot biomass
- Production of plant growth hormones
- Protection against root pathogens and pseudopathogens
- Enhanced nutrient use efficiency

Control of soil nematodes and insects

Biological control of weeds—for example, biological herbicides

Biodegradation of synthetic pesticides or industrial contaminants

Enhanced drought tolerance of plants

organisms will remain remarkably stable (quantitatively and qualitatively) against the intrusion of foreign microorganisms or moderate perturbations of the soil habitat (Alexander, 1971). Given these ecological principles, microbiological products can be successful only when they are introduced in sterile or partially sterilized soil or plant growth media, when the desired plant growth response early in seedling development will influence subsequent plant development, or when a microorganism introduced is more soil- or rhizosphere-competent than the natural population. The former can occur in some greenhouse culture of flowers, ornamentals, or vegetables. The latter two are addressed later.

Finally, currently marketed microbial products are likely to be ineffective because of problems in handling, storage, and applying the microorganisms to plants. Inoculants of beneficial microorganisms are notoriously unstable (Schroth and Weinhold 1986). Good initial field responses are followed by erratic or even negative responses in subsequent studies. Likewise, the formulation of inoculum, method of application, and storage of the product are all critical to the success of a biological product. If research scientists continue to have problems with product stability, the companies currently producing or distributing mircobial fertilizers most likely will have even greater problems delivering a quality product to the producer. Companies currently marketing the microbial products of questionable merit are not backed by strong research and development efforts to support their products. Many times they are not supported by research at all. And their product labels do not give ample recognition of the importance of application techniques and proper storage.

The biological activators marketed at present are as suspect as the microbial fertilizers. The approach itself has some scientific basis, and Katznelson (1940a,b) was able to show the influence of soil amendments or foliar application of chemicals in altering the rhizosphere. These attempts were largely empirical, however; and there has been little progress in concepts or scientific basis since these pioneering studies almost 50 years ago. Because of the enormous variety of crops and soils that must be addressed by a single formulation, a product designed to stimulate particular groups of beneficial microorganisms cannot be expected to work. The chance of success for current products seems remote.

## The Future for Microbiological Products

Regardless of current pessimism about successfully introducing beneficial microorganisms into the soil or rhizosphere, the potential is so great that increased research and product development in this area seem well worth the effort. Rather recent research has been successful in demonstrating this potential. In some instances, the development of marketable biological products seems within reach. Table 4 presents a partial listing of some demonstrated plant responses associated with the introduction of beneficial rhizosphere microorganisms. The implications of these introductions for reducing plant disease, for enhancing or changing plant growth, for increasing nutrient use efficiency, and for biological control of weeds are certainly important for more sustainable, less chemically intensive agricultural systems.

Additional successes with beneficial microorganisms may also be achievable because of the considerable progress in understanding what physiological, environmental, and genetic factors influence rhizosphere competence for microorganisms. Some mechanisms responsible for influencing root colonization and enhancing rhizosphere competence include:

- Niche in soil, rhizosphere, or rhizoplane empty.
- Early microbial activity before cell die-back.
- Enhanced cellulase production.
- Antibiotic tolerance or production.
- Siderophore influence on Fe nutrition.
- Unique physiological attributes.
- Altered plant genetics.
- Tolerance to fungicides or other chemicals.
- Foliar treatments.

The concept of niche occupancy discussed earlier is used to designate the unique function of an organism in its habitat (Miller, 1979a,b). The function performed by a species or strain of microorganism in the soil or rhizosphere is dependent upon the biochemical, nutritional, and someimes morphological properties of the microorganism itself. Gause's (1934) principle states that, as a general rule, only one species occupies any one specific niche in a habitat. In the soil, all niches in the rhizosphere or rhizoplane of normal nonsterilized soil are generally occupied by adapted organisms; that is why it is so difficult for introduced beneficial microorganisms to survive.

In sterilized or partially sterilized soils, or in artificial growth media used in commercial floriculture or vegetable production in greenhouses, introduction of beneficial microorganisms can often be achieved. Many of the successful examples listed in table 4 are in this category. For agronomic crops, soils (except soils that have been fumigated) are seldom sterilized. In field soils, the successful introduction of beneficial microorganisms is much more difficult or rare. For example, the nodulation of legumes by rhizobium inoculants is now successful only when background populations of indigenous

**Table 4. Microorganisms that have shown beneficial plant responses when used as soil or plant rhizosphere introductions.**

| *Microorganism* | *Effect* | *Reference* |
|---|---|---|
| *Trichoderma harzianum* | Shorter germination time for pepper seed, hastening of flowering of periwinkle, increase in number of chrysanthemum blooms. Increase in height or dry weight of periwinkle, pepper, tomato, and cucumber | 1 |
| | Increased dry weight of radish | 3 |
| | Increased tobacco seed emergence in $CH_3Br$-treated plant beds, as well as increase in plant dry matter | 11 |
| | Increased growth and earlier flowering in allyssum, pepper, marigold, periwinkle, and petunia in commercial production | 4,5 |
| | Faster rooting of carnation and chrysanthemum cuttings | 5, 35 |
| *Trichoderma spp.* | Reduced damping-off caused by *Pythium* or *Rhizoctonia* solani on pea and radish | 16, 28 |
| *Azospirillum brasilense* | Enhanced uptake of $NO_3-$, $K+$, and $H_2PO_4-$. Improved growth and dry weight of corn and sorghum | 29 |
| Fluorescent *Pseudomonas spp.* | Suppression of "take-all" of wheat | 45 |
| *Pseudomonas fluorescens* | Increased growth of carnation, sunflower vinca, and zinnia growth from seeds or cuttings | 46 |
| | Reduced damping-off of cotton seedlings | 7, 18, 19 |
| | Increased yield of potato and sugar beet | 25, 43 |
| | Increased yield of radish | 24 |
| | Changed serogroup distribution of *Bradyrhizobium* strains on soybean root systems | 13 |
| *Pseudomonas putida* | Increased yield of sugar beet | 42 |
| | Reduced *Fusarium* wilt of carnation | 47 |
| *Pseudomonas spp.* | Reduced incidence of *Fusarium* wilt of flax, cucumber, and radish | 37 |
| | Reduced in stand growth and seed production on downy brome (cheat grass) | * |
| *Pseudomonas syringae pv. Tabaci* | Enhanced alfalfa growth, plant nitrogen nodulation, and dintrogen fixation | 27 |
| Binucleate *Rhizoctonia* | Controlled Rhizoctonia root rot of snap bean | 9 |
| | Suppressed brown patch disease of creeping bentgrass | 8 |
| *Agrobacteria tumefaciens* | Controlled crown gall in grapes and some avirulent strains other plant species | 22 |
| *Alcaligenes spp.* | Reduced *Fusarium* wilt of carnations | 47 |
| *Bacillus subtilis* | Increased germination and growth of cabbage | 6 |

*A. Kennedy, Washington State University (unpublished).

rhizobia are very low—about 10 to 100 cells per gram of soil (Singleton and Tavares, 1986). An empty niche occurs only when a new leguminous plant is introduced into a region where the crop has never been grown. This occurred when soybeans were first introduced in the United States in the early twentieth century.

Some benefits may be associated with introduced microorganisms even though the microorganisms survive only a few weeks. This early benefit, before microbial die-back, may be sufficient to enhance crop growth and development. Examples include small grains (Rovira, 1963), in which early plant stimulation may enhance yields, depending upon later climatic factors or supression of early seedling diseases by *Pythium* or *Rhizoctonia* (Howell and Stipanovic, 1979, 1980).

One exciting breakthrough in our understanding of rhizosphere competence was made by Baker et al., (1986). These researchers concluded that mutants of a beneficial fungus, *Trichoderma harzianum*, were more successful in root colonization because of enhanced cellulase activity. The mutants, according to the researchers, use the remnants of primary cell walls of the root mucigel better and thus are better prepared to occupy this common niche than are other microorganisms. Primary cell walls are very high in cellulose.

The influence of antibiotic production or antibiotic tolerance on rhizosphere competence has received little attention (Kloepper and Schroth, 1981; Schroth and Weinhold, 1986) but should be of considerable importance. Conversely, Howell and Stipanovic (1979, 1980) have shown that production of the antibiotics pyrrolnitrin and pyoluteorin by a beneficial strain of *Pseudomonas fluorescens* was the principal reason why the pathogens *Rhizoctonia* and *Pythium spp.* are controlled in the rhizosphere of cotton.

In recent years, microbial metabolites capable of chelating iron (siderophores) have been shown to be the mechanism by which introduced rhizosphere bacteria influence plant growth or suppress disease-producing microorganisms (Loper and Schroth, 1986; Yuen and Schroth, 1986a,b). Moreover, the influence of siderophore production on rhizosphere competence, although not demonstrated, seems equally logical.

Different physiological properties of a microorganism should also affect a microorganism's rhizosphere competence and its ability to colonize the rhizosphere or rhizoplane. Included are generation time, ability to use different substrates, and the capacity to grow and survive within a range of soil temperature and moisture conditions and pH (Schroth and Hancock, 1985). Certainly, this remains a fertile area for ecological research.

One novel approach to enhance the establishment of a beneficial microorganism is to alter the plant genetically. Keyser and Cregan (1987) have demonstrated that indigenous and dominant isolates of serogroup 123 of *Bradyrhizobium japonicum* can be excluded from nodule occupancy by genetic manipulation of soybeans. Although I am unaware of attempts to use altered plant genetics to improved rhizosphere colonization of non-symbiotic microorganisms, the potential is there. For example, genetics could easily alter the spectrum of root exudates and rhizosphere pH.

Chemical seed treatment offers another possibility to introduce a microorganism on the root system if the desired microorganism is resistant to the chemical. Mendez-Castro and Alexander (1983) were able to develop a mancozeb-resistant *Pseudomonas* strain and establish this strain on corn roots by soil amendment of the fungicide mancozeb.

The final approach to altering plant rhizosphere populations involves the use of foliarly applied chemicals. Changes have been documented by use of foliarly applied streptomycen, urea (Davey and Papavizas, 1961; Horst and Herr, 1962), and plant growth regulators (Gupta, 1971; Sethunathan, 1970). Persistent problems with the isolation and quantitative and qualitative characterization of soil and rhizosphere microorganisms will render this approach much too empirical for rapid progress, however.

## REFERENCES

Ahmad, J. S., and R. Baker. 1987. Competitive saprophytic ability and cellulolytic activity of rhizosphere—Competant mutants of Trichoderma harzianum. Phytopathology 77: 358-362.

Alexander, M. 1971. Microbial ecology. John Wiley & Sons, Inc., New York, New York.

Baker, R. 1988. Trichoderma spp. as plant-growth stimulants. CRC Critical Reviews in Biotechnology 7: 97-106.

Baker, R., T. Paulitz, M. H. Windham, and Y. Elad. 1986. Enhancement of growth of ornamentals by a biological control agent. Colorado Greenhouse Growers Association Research Bulletin 431: 1.

Baker, R., Y. Elad, and I. Chet. 1984. The controlled experiment in the scientific method with special emphasis on biological control. Phytopathology 74: 1,019.

Broadbent, B., K. F. Baker, N. Franks, and J. Holland. 1977. Effect of Bacillus spp. on increased growth of seedlings in steamed and in nontreated soil. Phytopathology 67: 1,027-1,034.

Brosten, D. 1988. In-furrow biological fungicide set to protect seedlings from disease. Agrichemical Age 32: 10-12.

Burpee, J. L., and I. G. Gaulty. 1984. Suppression of brown patch disease of creeping bentgrass by isolates of nonpathogenic Rhizoctonia. spp. Phytopathology 74: 692-694.

Cardoso, J. E., and E. Echandi. 1987. Biological control of Rhizoctonia root rot of snap bean with binucleate Rhizoctonia-like fungi. Plant Disease 71: 167-170.

Chang, Ya-Chun, Yih-Chang, R. Baker, O. Kleifeld, and I. Chet. 1986. Increased growth of plants in the presence of the biological control agent Trichoderma harzianum. Plant Disease 70: 145-148.

Cole, J. S., and Z. Zvenyika. 1986. Integrating Trichoderma harzianum and triadimenol for the control of tobacco sore shin in Zimbabwe. Bulletin Information Coresta #68.

Davey, C. B., and G. C. Papavizas. 1961. Translocation of streptomycin from coleus leaves

and its effect on rhizosphere bacteria. Science 134: 1,368-1,369.
Fuhrmann, J., and A. G. Wollum. 1988. Nodulation competition among Bradyrhizobium japonicum strains as influenced by rhizosphere bacteria and iron availability. Biology and Fertility of Soils 6 (in press).
Gause, G. F. 1934. The struggle for existence. Williams and Wilkins, Baltimore, Maryland.
Gupta, P. C. 1971. Foliar spray of gibberellic acid and its influence on rhizosphere and rhizoplane mycoflora. Plant and Soil 34: 233.
Harman, G. E., I. Chet, and R. Baker. 1980. Trichoderma hamatum on seedling disease induced in radish and pea by Pythium spp., or Rhizoctonia solani. Phytopathology 70: 1,167-1,172.
Horst, R. K., and L. J. Herr. 1962. Effects of foliar urea treatment on numbers of actinomycetes antagonistic to Fusarium roseum f. cerealis in the rhizosphere of corn seedlings. Phytopathology 52: 423-427.
Howell, C. R., and R. D. Stipanovic. 1979. Control of Rhizoctonia solani on cotton seedlings with Pseudomonas fluorescens and with an antibiotic produced by the bacterium. Phytopathology 69: 480-482.
Howell, C. R., and R. D. Stipanovic. 1980. Suppression of Pythium ultimum-induced damping off of cotton seedlings by Pseudomonas fluorescens and its antibiotic, pyluteorin. Phytopathology 70: 712-715.
Katznelson, H. 1940a. Survival of Azotobacter in soil. Soil Science 49: 21-35.
Katznelson, H. 1940b. Survival of microorganisms introduced into soil. Soil Science 49: 283-293.
Kerr, A. 1980. Biological control of crown gall through production of agrocin 84. Plant Disease 64: 24-30.
Keyser, H. H., and P. B. Cregan. 1987. Nodulation and competition for nodulation of selected soybean genotypes among Bradyrhizobium japonicum serogroup 123 isolates. Applied Environmental Microbiology 53: 2,631-2,635.
Kloepper, J. W., and M. N. Schroth. 1981. Relationship of invitro antibiosis of plant growth promoting rhizobacteria to plant growth and the displacement of root microflora. Phytopathology 71: 1,020-1,024.
Kloepper, J. W., and M. N. Schroth. 1978. Plant growth promoting rhizobacteria on radishes. pp. 879-882. *In* Proceedings, international conference on plant pathogenic bacteria, Volume 2. INRA, Angers, France.
Knight, T. J., and P. J. Langston-Unkeefer. 1988. Enhancement of symbiotic dinitrogen fixation by a toxin-releasing plant pathogen. Science 241: 951-954.
Lifschitz, R., MN. T. Windham, and R. Baker. 1986. Mechanism of biological control of preemergence damping-off of pea by seed treatment with Trichoderma spp. Phytopathology 76: 720-725.
Lin, W., Y. Okon, and R.W.F. Hardy. 1983. Enhanced mineral uptake by Zea mays and Sorghum bicolor roots inoculated with Azospirillum brasilense. Applied Environmental Microbiology 45: 1,775-1,779.
Loper, J. E., and M. N. Schroth. 1986. Importance of siderophores in microbial interactions. pp. 85-98. *In* T. R. Swinburne [editor] Iron, siderophores, and plant diseases. Plenum Publishing Co., London, England.
Mendez-Castro, F. A., and M. Alexander. 1983. Method for establishing a bacterial inoculum on corn roots. Applied Environmental Microbiology 45: 248-254.
Miller, R. H. 1979a. Ecological factors which influence the success of microbial fertilizers or activators. Developments in Industrial Microbiology 20: 335-342.
Miller, R. H. 1979b. Microbial soil amendments from the biologist's point of view. New Farm 1: 25-28.
Papavizas, G. C., and C. B. Davey. 1961. Extent and nature of the rhizosphere of lupinus. Plant and Soil 14: 215-234.
Paulitz, T., M. Windham, and R. Baker. 1985. The effects of Trichoderma harzianum on

rooting of chrysanthemum cuttings. Phytopathology 75: 1,333.
Rovira, A. D. 1963. Microbial inoculation of plants I. Establishment of free-living nitrogen fixing bacteria in the rhizosphere and their effects on maize, tomato and wheat. Plant and Soil 19: 304-314.
Scher, F. M., and R. Baker. 1980. Mechanism of biological control in a Fusarium-suppressive soil. Phytopathology 70: 412-417.
Schroth, M. N., and A. R. Weinhold. 1986. Root-colonizing bacteria and plant health. Horticulture Science 21: 1,295-1,298.
Schroth, M. N., and J. G. Hancock. 1985. Soil antagonists in IPM systems. pp. 415-431. *In* Biological Control in Agricultural IPM Systems. Academic Press, Inc., Orlando, Florida.
Sethunathan, N. 1970. Foliar sprays of growth regulators and rhizosphere effect in Cajanus cajan. I. Quantitative changes. Plant and Soil 33: 62-70.
Singleton, P. W., and J. W. Tavares. 1986. Inoculation response of legumes in relation to the number and effectiveness of indigenous rhizobium populations. Applied Environmental Microbiology 51: 1,013-1,018.
Suslow, T. V., and M. N. Schroth. 1982. Rhizobacteria of sugar beets: Effects of seed application and root colonization on yield. Phytopathology 72: 199-206.
Suslow, T. V., and M. N. Schroth. 1982. Role of deleterious rhizobacteria as minor pathogens in reducing crop growth. Phytopathology 72: 111-115.
Weller, D. M., and R. J. Cook. 1983. Suppression of take-all of wheat by seed treatment with fluorescent Pseudomonas. Phytopathology 73: 463-469.
Yuen, G. Y., and M. N. Schroth. 1986a. Inhibition of Fusarium oxysporum f. sp. dianthi by iron competition with an Alcaligenes sp. Phytopathology 76: 171-176.
Yuen, G. Y., and M. N. Schroth. 1986b. Interactions of Pseudomonas fluorescens strain E-6 with ornamental plants and its effect on the composition of root-colonizing microflora. Phytopathology 76: 176-180.
Yuen, G. Y., M. N. Schroth, and A. H. McCain. 1985. Reduction of Fusarium wilt of carnation with suppressive soils and antagonistic bacteria. Plant Disease 69: 1,071-1,075.

# ROLE OF SUSTAINABLE AGRICULTURE IN RURAL LANDSCAPES

G. W. Barrett, N. Rodenhouse, and P. J. Bohlen

Landscape ecology is a rapidly expanding field of study (Forman, 1983; Forman and Godron, 1986) that considers the development and maintenance of spatial heterogeneity on a regional or global basis. Landscape ecology focuses on (1) the spatial and temporal interactions among habitats via exchanges of organisms and materials across the landscape, (2) the influence of heterogeneity on biotic and abiotic processes, and (3) the management of heterogeneity (Risser et al., 1984). Agriculture has had a profound impact on the development and maintenance of landscape heterogeneity in the midwestern United States. Thus far, however, the role of agroecosystems in maintaining the stability and sustainability of landscape systems (Lowrance et al., 1986) is poorly understood. The potential role of agriculture in rural landscapes becomes clear only when natural and agricultural systems are compared and contrasted.

Natural ecosystems are unsubsidized, solar-powered systems. These systems are composed of abiotic and biotic components that interact in ordered, regulated ways (Figure 1). Agroecosystems are also solar-powered, but they are increasingly driven by external subsidies of energy and nutrients (Figure 2). Agroecosystems have also been increasingly modified by chemical pesticides, mechanical technology, selected genetic inputs, and socioeconomic goals (Risser, 1985). Consequently, the capacity of these systems for natural feedback and regulation has been greatly reduced. Agroecosystems probably cannot be made sustainable without taking maximum advantage of natural ecosystem feedbacks and regulation (Altieri, 1987). Management of agroecosystems will thus require that we under-

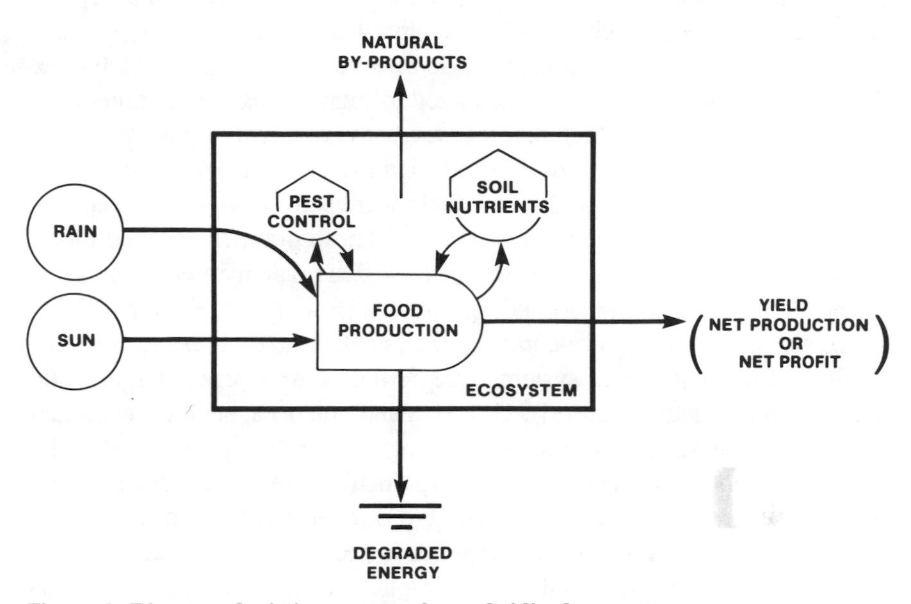


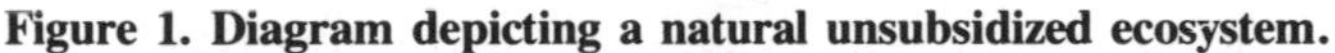
**Figure 1. Diagram depicting a natural unsubsidized ecosystem.**

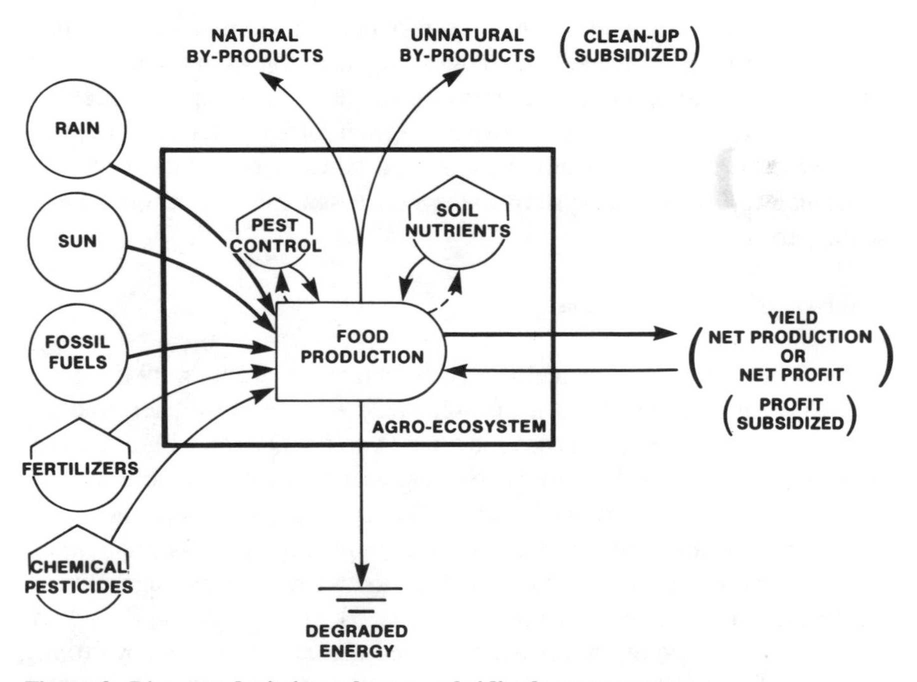


**Figure 2. Diagram depicting a human-subsidized agroecosystem.**

stand the theory and behavior of natural ecosystems and how the processes of these natural systems—for example, nutrient recycling and species regulation (Figure 1)—are modified by agricultural objectives.

Agroecosystem management seems to be diverging along two pathways. One management strategy is characterized by highly subsidized, large-scale agricultural systems in which natural processes are largely ignored (Figure 2). Pesticides are used to control weeds and insects, fertilizers are applied to maintain soil fertility, and fossil fuels increasingly supply the energy needed to achieve management objectives. These practices reduce biotic diversity as well as landscape heterogeneity. This management strategy is often referred to as conventional agriculture (Edens and Koenig, 1981).

The alternative strategy encourages the establishment of less subsidized, smaller scale agricultural systems integrating natural processes into the management strategy (Figure 1). Concepts and practices, such as integrated pest management, optimal rate and timing of nutrient cycling within the system, and increased energy efficiency (particularly through reduced tillage and fertilizer inputs) (Pimental et al., 1983), exemplify this management strategy. These agricultural systems are designed to reduce wasteful inputs, minimize nutrient outputs, and increase biotic and landscape diversity. This management strategy is often described as sustainable agriculture (Lowrance et al., 1986).

Landscape patterns and processes can be determined by cropping practices, landscape heterogeneity, and agroecosystem inputs and outputs. We analyzed the development of existing agricultural landscapes to identify changes needed to achieve a sustainable agriculture. By contrasting present landscape trends with our view of a sustainable rural landscape, we developed future agroecosystem management schemes based on a landscape perspective.

## Analysis of Landscape Data

The U.S. Census of Agriculture (U.S. Bureau of Census, 1945, 1982) and Ohio Agricultural Statistics (1942, 1983) were used to determine the types and area coverage of crops grown in Ohio for 1940 and 1982. Census of Agriculture data for these two years were not entirely comparable because of changes in the definition of a farm. The data were sufficient, however, for us to discern overall crop and landscape patterns for 1940 and 1982.

We determined crop and landscape diversity for the Ohio agricultural landscape using the Shannon-Wiener index: $H' = -\Sigma \, Pi \log Pi$, where Pi is the percentage of total acres harvested that is represented by crop i. This index is often used to measure diversity within and between ecological communities (Pielou, 1975); it has also been used to measure crop diversity in tropical swidden agriculture (Eden, 1987). We used it here to measure the diversity of crops and to describe the agricultural landscape in Ohio for the years 1940 and 1982. Crop diversity was computed for crops with 10,000 or more acres (equal to or greater than 4,447 hectares) harvested. Landscape diversity was based on six categories of agricultural land use: annual cropland, hay, pasture, woodland, fallow land, and orchards. Each category was represented as a proportion (Pi) of land in farms.

The Shannon-Wiener index measures the number of categories present (i.e., crop richness) and the acreage of each crop harvested (i.e., crop apportionment). To integrate these two components of diversity, we also measured apportionment using the evenness index (Pielou, 1966): $J' = H' / \log S$, where $H'$ is the Shannon-Wiener index and S is the number of crop types or landscape categories. We also examined changes in agricultural inputs of fertilizer (Steiner, 1987; Tennessee Valley Authority, 1982) and pesticides (U.S. Bureau of Census, 1964, 1982) and outputs of crop yield and soil loss in 1940 and 1982.

Soil Conservation Service aerial photographs were used to quantify changes in landscape elements for Butler County, Ohio, in 1938 and 1983. Thirty-five randomly selected photos were used for landscape analysis; photos containing urbanized areas were avoided. A three-kilometer east-west transect was centered on each photo and the number of major landscape elements recorded (e.g., crop fields, wooded areas, pasture fields, and uncultivated corridors, such as streams and wooded fence rows). The length of each wooded area was measured where it crossed the transect. Data for the two years were compared using a Student's t-test.

## Analytical Results

***Landscape Structure.*** Crop diversity declined from 0.80 in 1940 to 0.60 in 1982 (Table 1). Evenness of distribution among crops, however, was nearly the same in both years. The decline in diversity was due mainly to a decline in the number of crops. For example, fewer hectares were planted to small grains or harvested for hay seed (e.g., clovers and timothy) in 1982 than in 1940 (Table 1). Among the five major crops—corn, soybeans, wheat, hay, and oats—evenness declined from 0.94 in 1940 to 0.82 in 1982. This decline was due to a greater proportion of cropland devoted to corn and soybeans in 1982 than in 1940.

Landscape diversity declined from 0.61 in 1940 to 0.48 in 1982. Because the number of landscape categories was the same in both years, this decline was due exclusively to a decline in evenness ($J' = 0.79$ in 1940 and 0.62

in 1982). Annual crops increased from 37 percent to 65 percent of the farm landscape, whereas pasture and hay declined from 33 percent to 11 percent and 16 percent to 9 percent, respectively. As a result of these shifts to annual grain crops, a greater percentage of acres in farms was harvested in 1982 (67 percent) than in 1940 (45 percent) (Table 2).

The aerial photo analysis for Butler County revealed no significant differences ($p > 0.05$) in the number of major landscape elements between 1938 and 1983. There were, however, significantly more, and larger wooded areas in 1983 than in 1938, even though woodland reported as a percentage of total land in farms did not change (12 percent). The increase in woodland may have been on land not reported earlier as land in farms, however.

***Agroecosystem Inputs and Outputs.*** Fertilizer inputs, particularly nitrogen and potassium, increased greatly from 1940 to 1982, outstripping increases

**Table 1. Relative importance (percentage of total acres harvested), and diversity and evenness indices for agricultural crops with more than 10,000 acres (4,047 ha) harvested in Ohio for 1940 and 1982.**

| *Crop* | *Relative Importance* ($p_i$) *1940* | *1982* |
|---|---|---|
| Corn | .312 | .392 |
| Hay | .252 | .118 |
| Wheat | .190 | .113 |
| Oats | .098 | .031 |
| Soybeans | .055 | .338 |
| Red clover seed | .029 | — |
| Orchards | .015 | .002 |
| Potatoes | .009 | .001 |
| Vegetable crops* | .009 | .005 |
| Rye | .007 | — |
| Timothy seed | .007 | — |
| Sugar beets | .004 | — |
| Alsike clover seed | .004 | — |
| Tobacco | .003 | .001 |
| Barley | .003 | — |
| Buckwheat | .002 | — |
| Alfalfa seed | .002 | — |
| Sweet clover seed | .001 | — |
| Crop diversity ($H' = \Sigma\ p_i \log p_i$) | .80 | .60 |
| Evenness ($J' = H'/\log S$) | .64 | .63 |

*Includes all vegetables grown for fresh market and for processing.

**Table 2. Agricultural land use in Ohio for 1940 and 1982.**

| | *1940* | *1982* |
|---|---|---|
| | (thousand acres) | |
| Land in farms | 21,908 | 15,404 |
| Cropland harvested | 9,772 | 10,396 |
| Acres harvested/acres in farms | .45 | .67 |

**Table 3. Crop yield and fertilizer use in Ohio for 1940 and 1982.**

| | *1940* | *1982* | *Increase (x)* |
|---|---|---|---|
| Yield (bu/acre) | | | |
| Corn | 38 | 117 | 3.1 |
| Soybeans | 16 | 37 | 2.4 |
| Wheat | 22 | 44 | 2.0 |
| Oats | 44 | 70 | 1.6 |
| Hay (ton/acre) | 1.5 | 2.5 | 1.7 |
| Consumption (thousand tons) | | | |
| Total fertilizer | 36 | 2,250 | 6.1 |
| Total N | 9 | 451 | 50.1 |
| Total $P_2O$ | 48 | 252 | 5.3 |
| Total $K_2O$ | 23 | 389 | 16.9 |
| Total plant nutrients | 80 | 1,092 | 13.7 |

in the yields of the five major crops (Table 3). Data on pesticide use were not available for 1940. From 1964 to 1982, however, the number of acres treated for insects, diseases, and weeds increased 62 percent.

An additional output of great importance to agricultural productivity (Crosson and Stout, 1983; Larson et al., 1983) and to the interactions between landscape elements (Lowrance et al., 1983; 1988) occurs via soil erosion. Quantitative data on soil losses per hectare first became available in the 1970s. Qualitative comparisons, however, between the 1930s and 1970s can be made by considering changes in factors influencing rates of soil erosion. Major factors potentially influencing rates of soil erosion include the percentage of cropland in particular crops and farming practices. We assessed the relative rates of erosion in the 1930s by considering these factors.

The large percentage increase in row crops and decrease in hay crops no doubt led to an increase in the amount of erosion on a per-farm basis because row crops have the highest rates of soil loss of all major midwestern crops (Crosson and Stout, 1983). Considering row crops only, technological changes in farming practices since the 1930s (U.S. Department of Agriculture, 1981) would also tend to increase erosion per hectare. Such changes include (1) increased soil compaction (National Agricultural Lands Study, 1980) because of more frequent and intense tillage; (2) reductions in soil fauna, particularly earthworms that improve soils (Hendrix et al., 1986), as a result of the use of pesticides, inorganic fertilizers, and tillage; and (3) declines in the water-holding capacity of soils (Dick et al., 1986) because of the degradation of soil organic matter (which was brought about in part by reduced application of manure and shorter rotations without hay crops).

Farming practices that would tend to reduce soil erosion include use of conservation tillage. However, such practices have been less widely accepted in Ohio, compared to all other Corn Belt states; only nine percent of the cropland in Ohio received some form of conservation tillage treatment (U.S. Department of Agriculture, 1981).

In summary, rates of soil erosion for row crops in Ohio have probably increased since the 1930s where conservation tillage is not used, and rates of soil erosion per farm probably have also increased because a greater percentage of farmland is producing row crops.

## Implications for Pattern and Process in Rural Landscapes.

Changes in agriculture since the 1930s have had extensive impacts on patterns and processes in rural agricultural landscapes. The data presented here indicate a trajectory of change in rural landscapes that includes (1) less diversity in crops produced, (2) less diversity in landscape elements (i.e., larger habitat types), (3) increased external inputs of pesticides and fertilizers, and (4) higher outputs of crop yields as well as soil losses. The questions now are these: How have these changes altered the functioning of agricultural landscapes, and how must the trajectory of change be shifted to create more sustainable rural landscapes?

Less crop diversity can slow soil development and contribute to crop losses by pests (arthropods and weeds) and diseases. Hay crops, which contribute significantly to soil development (Frye et al., 1985; Power, 1987), have been dropped from almost all crop rotations in Ohio. The result is soil degradation that affects landscape-level processes. For example, such degradation reduces soil moisture-holding capacity, which allows greater leaching and surface runoff of soil, nutrients, and pesticides (Frye et al., 1985). Less diversity may also result in greater crop losses to pests (Power, 1987) in which the success of regional (i.e., field to field) dispersal of pests is enhanced (Gould and Stinner, 1984), the rate of adaptation to crops by pests is accelerated (Kogan, 1981), and the associated resistance of crops to pests is reduced (Root, 1973).

Reduced landscape diversity also alters patterns of microclimate and crop-pest interactions. Microclimate is altered by changes in a suite of correlated meterological variables that are each influenced by the heterogeneity of landscape; these variables include air temperature, the pattern and speed of winds, and evaporation (Rosenberg et al., 1983). The meterological variables influence plant stress and, hence, the susceptibility of crops to pests (Mattson and Haack, 1987). Furthermore, less landscape diversity may mean the elimination of overwintering habitats for beneficial insects, the enhanced movement of pests between crop fields, and the creation of ecological traps for beneficial species (Best, 1986).

Increased use of pesticides and fertilizers has overloaded agroecosystems with these materials. Little regard has been given to how these agrochemicals might affect internal agroecosystem processes. Toxic effects on resident fauna, such as earthworms (Pimentel and Edwards, 1982); the degradation of soil quality (Hendrix et al., 1986); and the export of agrochemicals into ground and surface waters (Hallberg, 1987) have resulted. Agricultural inputs are having dramatic, detrimental effects on adjacent and downstream ecosystems (Baker, 1985; Pimentel et al., 1987).

Increased yield output of crops is the greatest benefit of conventional agriculture; yet yield increases have not kept pace of late with those of energy and material inputs. Part of the slowdown in agricultural yield increases may be due to the detrimental effects of soil loss on soil quality (Frye et al., 1985). Additional agroecosystem outputs that are not beneficial to the functioning of rural landscapes include soil, nutrients, pesticides, and pests (Canter, 1986).

### Goals for Creating Sustainable Rural Landscapes.

Clearly the trajectory of change in agriculture must be shifted from its present course. Analyses of the trends outlined here indicate that three goals are needed for the development of sustainable rural landscapes: (1) reduction in inputs of fertilizers and pesticides; (2) optimization of internal ecosystem regulation necessary for the retention and recycling of nutrients, for the creation of favorable microclimates for crop growth, and for the control of pests; and (3) reduction in the export of soil and nutrients from agroecosystems. Long-term crop yields will be maintained only by implementing programs that take maximum advantage of the natural ecological processes operating within and among landscape elements (Figure 3). This approach must encompass a larger view—a more holistic landscape perspective—of the agricultural arena (Waddell and Bower, 1988).

Focus on landscape patterns and processes will be required to achieve these goals. Retention and recycling of nutrients will be enhanced by (1) maintaining continuous vegetative cover on croplands, (2) adjusting the timing and amount of nutrient inputs to crop requirements, and (3) employing tillage selectively as needed to stimulate mineralization or to control pest populations. Obviously, holding soil in place and building soil quality (e.g., organic matter content, beneficial microflora and fauna) must also be of primary concern.

Regulation of crop pests using reduced inputs of pesticides will be promoted by greater landscape diversity within and among field crops (Kemp and Barrett, 1989). Favorable habitats for the overwintering and growth of predator populations must be provided and located so that entire crop fields are included within the dispersal distances of predators. Novel approaches to avoiding the pesticide paradox (i.e., pests must be killed to protect crops, yet killing a pest reduces the effectiveness of the pesticide via strong directional selection) must be used. For example, combinations of control techniques (e.g., enhancing populations of generalist predators, rotating crops, managing the timing of planting, and juxtaposing crop types to slow the movement of pests) should be used to control insect pests in an attempt to minimize directional selection (Schultz, 1983).

Vigorous crop plants are less susceptible to pests and diseases. Thus, the creation of favorable microclimates for crop growth must also be a goal when considering sustainable landscapes. Factors enhancing favorable microclimates on cropland include within-field diversity of vegetation structure,

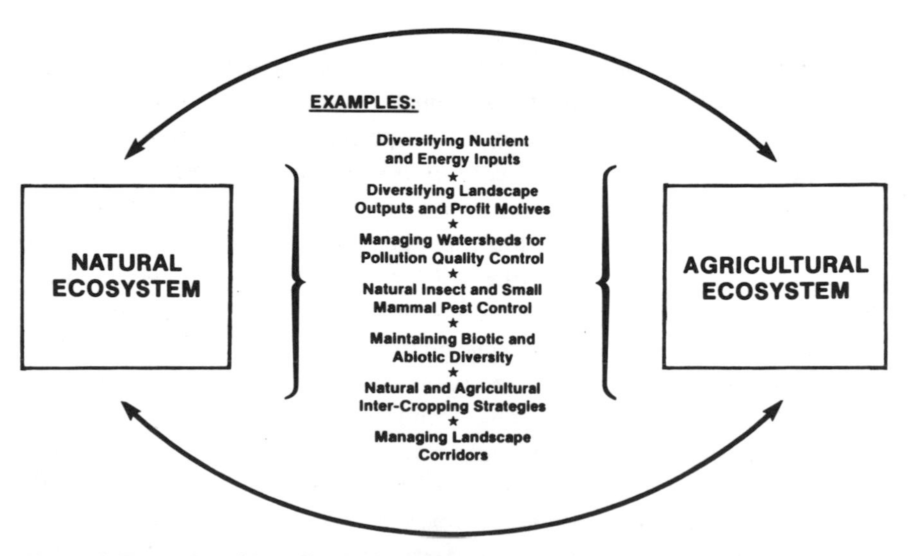


**Figure 3. Examples of benefits derived from managing natural and agricultural ecosystems based on a landscape perspective.**

such as stripcropping (Allen et al., 1976); fencerows and shelterbelts (Rosenberg et al., 1983); field size and shape; and the juxtaposition of different types of landscape elements (Forman and Godron, 1986).

In summary, sustainable rural landscapes of the future will differ substantially from those of today. They will have greater crop and landscape diversity, tighter cycling of nutrients, and lower inputs of energy and materials.

### Managing Sustainable Rural Landscapes.

Management in conventional agriculture is focused at the individual field level. Sustainable landscapes can best be created by management at the watershed or regional scale because the flux of materials and organisms at these scales significantly influences management decisions for individual fields (Gould and Stinner, 1984). Management at such a scale will require a more flexible administrative approach. For example, cross-compliance of federal conservation and farm programs (Myers, 1988; Steiner, 1987) could be used to manipulate landscape patterns and, hence, to promote beneficial (both ecological and economic) agricultural practices.

### Research Needs.

Research that integrates landscape ecology and agroecosystem ecology is just beginning. Particularly in need of research are the interactions of landscape elements. Questions to be addressed include: What is the spatial scale of influence of a particular landscape element above ground and below ground? Are there combinations of landscape elements that create synergisms favorable for crop growth and pest control? How do farm management practices influence the interactions between landscape elements? Examination of these questions must include the integration of hierarchical levels, from organisms to ecosystems, and be carried out on a variety of spatial scales (Addicott et al., 1987; Urban et al., 1987). Long-term research funding (Barrett, 1985; Callahan, 1984) is needed, especially in the area of applied ecology integrating agroecosystem and landscape ecology (Barrett, 1984; 1987), because changes in many ecosystem and landscape parameters are detectable only on time scales measured in years.

### Conclusions.

The greatest challenge to modern agriculture is to create sustainable agricultural systems. These systems will be created by taking maximum advantage of natural ecological processes to retain and recycle nutrients, to enhance pest control, and to create microclimates favorable for crop growth. Because these ecological processes function, in part, at the landscape level, creation of a sustainable agriculture will depend upon understanding and managing the ecology of rural landscapes.

## REFERENCES


Addicott, J. F., J. M. Aho, M. F. Anotlin, D. K. Padilla, J. S. Richardson, and D. A. Soluk. 1987. Ecological neighborhoods: Scaling environmental patterns. Oikos 49(3): 340-346.

Allen, L. H., Jr., T. R. Sinclair, and E. R. Lemon. 1976. Radiation and microclimate relationships in multiple cropping systems. pp. 171-199. *In* R. I. Papendick, P. A. Sanchez, and G. B. Triplett [editors] Multiple cropping. Madison, Wisconsin.

Altieri, M. A. 1987. Agroecology: The scientific basis of alternative agriculture. Westview Press, Boulder, Colorado.

Baker, D. B. 1985. Regional water quality impacts of intensive row-crop agriculture: A Lake Erie basin case study. Journal of Soil and Water Conservation 40(1): 125-132.

Barrett, G. W. 1984. Applied ecology: An integrative paradigm for the 1980s. Environmental Conservation 11(4): 319-322.

Barrett, G. W. 1985. A problem-solving approach to resource management. BioScience 35(7): 423-427.

Barrett, G. W. 1987. Applied ecology at Miami University: An integrative approach. Ecological Society of America Bulletin 68(2): 154-155.

Best, L. B. 1986. Conservation tillage: Ecological traps for nesting birds? Wildlife Society Bulletin 14(3): 308-317.

Callahan, J. T. 1984. Long-term ecological research. BioScience 34(6): 363-367.

Canter, L. W. 1986. Environmental impacts of agricultural production activities. Lewis Publishers, Inc., Chelsea, Michigan.

Crosson, P. R., and A. T. Stout. 1983. Productivity effects of cropland erosion in the United States. Resources for the Future, Washington, D.C.

Dick, W. A., D. M. Van Doren, Jr., G. B. Triplett, Jr., and J. E. Henry. 1986. Influence of long-term tillage and rotation combinations on crop yields and selected soil parameters. II. Results obtained for a typic fragiudalf soil. Research Bulletin 1181. OARDC, Ohio State University, Wooster.

Eden, M. J. 1987. Crop diversity in tropical swidden cultivation: Comparative data from Colombia and Papua New Guinea. Agricultural Ecosystems and Environment 20(2): 127-136.

Edens, T. C., and H. E. Koenig. 1981. Agroecosystem management in a resource limited world. BioScience 30(9): 697-701.

Forman, R.T.T. 1983. An ecology of the landscape. BioScience 33(9): 535.

Forman, R.T.T., and M. Godron. 1986. Landscape ecology. John Wiley & Sons, New York, New York.

Frye, W. W., O. L. Bennett, and G. J. Buntley. 1985. Restoration of crop productivity on eroded or degraded soils. pp. 335-356. *In* R. F. Follett and B. A. Stewart [editors] Soil erosion and crop productivity. American Society of Agronomy, Madison, Wisconsin.

Gould, F., and R. E. Stinner. 1984. Insects in heterogeneous habitats. pp. 427-449. *In* C. B. Huffaker and R. L. Rabb [editors] Ecological entomology. John Wiley & Sons, New York, New York.

Hallberg, G. R. 1987. Agricultural chemicals in groundwater: Extent and implications. American Journal of Alternative Agriculture (2(1): 3-15.

Hendrix, P. F., R. W. Parmelee, D. A. Crossley, Jr., D. C. Coleman, E. P. Odum, and P. M. Groffman. 1986. Detritus food webs in conventional and no-tillage agroecosystems.

BioScience 36(6): 374-380.
Kemp, J. C., and G. W. Barrett. 1989. Spatial patterning: Impact of uncultivated corridors on arthropod populations within soybean agroecosystems. Ecology 70(1): 114-128.
Kogan, M. 1981. Dynamics of insect adaptations to soybean: Impact of integrated pest management. Environment Entomology 10(3): 363-371.
Larson, W. E., F. J. Pierce, and R. H. Dowdy. 1983. The threat of soil erosion to long-term crop productivity. Science 219(4584): 458-465.
Lowrance, R. R., R. F. Hendrix, and E. P. Odum. 1986. A hierarchical approach to sustainable agriculture. American Journal of Alternative Agriculture 1(4): 169-173.
Lowrance, R. R., R. L. Todd, and L. E. Asmussen. 1983. Waterborne nutrient budgets for the riparian zone of an agricultural watershed. Agriculture, Ecosystems, and Environment 10: 371-384.
Lowrance, R. R., S. McIntrye, and C. Lance. 1988. Erosion and deposition in a field/forest system estimated using cesium-137 activity. Journal of Soil and Water Conservation 43(2): 195-199.
Mattson, W. J., and R. A. Haack. 1987. The role of drought stress in provoking outbreaks of phytophagous insects. *In* P. Barbosa and J. C. Schultz [editor] Insect outbreaks. Academic Press, Inc., New York, New York.
Myers, P. C. 1988. Conservation at the crossroads. Journal of Soil and Water Conservation 43(1): 10-13.
National Agricultural Lands Study. 1980. Soil degradation: Effects on agricultural productivity. Interim Report Number 4, U.S. Government Printing Office, Washington, D.C.
Ohio Agricultural Statistics Service. 1942. Ohio Agricultural Statistics 1940 and 1941. Bulletin 472. Ohio Agricultural Experiment Station, Wooster.
Ohio Agricultural Statistics Service. 1983. 1982 Ohio Agricultural Statistics. Columbus.
Pielou, E. C. 1966. Species-diversity and pattern-diversity in the study of ecological succession. Journal of Theoretical Biology 10(2): 370-383.
Pielou, E. C. 1975. Ecological diversity. John Wiley & Sons, New York, New York.
Pimental, D., and C. Edwards. 1982. Pesticides and ecosystems. BioScience 32(7): 595-600.
Pimental, D., G. Berardi, and S. Fast. 1983. Energy efficiency of farming systems: Organic and conventional agriculture. Agriculture, Ecosystems, and Environment 9(1983): 359-372.
Pimental, D., J. Allen, A. Beers, L. Guinand, R. Linder, P. McLaughlin, B. Meer, D. Musonda, D. Perdue, S. Poisson, S. Siebert, K. Stoner, R. Salazar, and A. Hawkins. 1987. World agriculture and soil erosion. BioScience 37(4): 277-283.
Power, J. F. 1987. Legumes: Their potential role in agricultural production. American Journal of Alternative Agriculture 2(2): 69-73.
Risch, S. J., D. Andow, and M. A. Altieri. 1983. Agroecosystem diversity and pest control: Data, tentative conclusions, and new research directions. Environmental Entolomology 12(3): 625-629.
Risser, P. G. 1985. Toward a holistic management perspective. BioScience 35(7): 414-418.
Risser, P. G., J. R. Karr, and R. T. Forman. 1984. Landscape ecology: Directions and approaches. Special Publication Number 2. Illinois Natural History Survey, Champaign, Illinois.
Root, R. B. 1973. Organization of a plant-arthropod association in simple and diverse habitats: The fauna of collards (Brassica oleracea). Ecological Monographs 43(1): 95-124.
Rosenberg, N. J., B. L. Blad, and S. B. Verma. 1983. Microclimate: The biological environment. John Wiley & Sons, New York, New York.
Schultz, J. C. 1983. Impact of variable plant defensive chemistry on susceptibility of insects to natural enemies. American Chemical Society Symposium Series 208: 37-54.
Shannon, C. E., and W. Weaver. 1964. The mathematical theory of communication. University of Illinois Press, Urbana.
Steiner, F. 1987. Soil conservation policy in the United States. Environmental Management 11(2): 209-223.

Tennessee Valley Authority. 1959. Fertilizer summary data. Knoxville, Tennessee.

Tennessee Valley Authority. 1982. Fertilizer summary data. Muscle Shoals, Alabama.

Urban, D. L., R. V. O'Neill, and H. H. Shugart, Jr. 1987. Landscape ecology. BioScience 37(2): 119-127.

U.S. Bureau of Census. 1945. U.S. Census of Agriculture, 1945. U.S. Government Printing Office, Washington, D.C.

U.S. Bureau of Census. 1964. U.S. Census of Agriculture, 1964. U.S. Government Printing Office, Washington, D.C.

U.S. Bureau of Census, 1982. U.S. Census of Agriculture, 1982. U.S. Government Printing office, Washington, D.C.

U.S. Department of Agriculture. 1981. Soil, water, and related resources in the United States: Analysis of resource trends. Soil and Water Resources Conservation Act, 1980 Appraisal, Part II. U.S. Government Printing Office, Washington, D.C.

Waddell, T. E., and B. T. Bower. 1988. Agriculture and the environment: What do we really mean? Journal of Soil and Water Conservation 43(3): 241-242.

# SOIL BIOTA AS COMPONENTS OF SUSTAINABLE AGROECOSYSTEMS

P. F. Hendrix, D. A. Crossley, Jr., J. M. Blair, and D. C. Coleman

Soil biota refers to a highly diverse assemblage of organisms that spend at least part of their life cycle in or on the soil. As their numerical diversity suggests, soil biota undertake a wide range of functions, including the detrimental activities of pathogens, parasites, and other agricultural pests. However, the majority of soil organisms are free-living and participate directly or indirectly in the decomposition and mineralization of plant and animal residues. In natural ecosystems decomposing organic matter is a principal source of nutrients for plant growth. Thus, soil biota can be viewed as regulators of nutrient availability to primary producers. In addition to influencing soil processes, soil biota play an important role in affecting soil structure. For example, biochemical exudates from microbes and physical activities of soil animals (burrowing, mixing of organic and mineral particles) are major factors in the formation of soil organic matter and in the maintenance of soil aggregate and pore structure.

Despite the wealth of literature on the soil biota in natural ecosystems, surprisingly little is known about the biology and ecology of free-living organisms in agricultural soils (Edwards, 1984, Crossley et al., 1988). In particular, there has been little consideration of the role they might play in the development of sustainable agricultural practices.

## Sustainability

Any discussion of sustainable agriculture must deal with the inevitable problem of defining sustainability. Definitions and controversy abound.

However, from ecological and economic perspectives, two basic criteria must be met by any system that is to be sustainable in the long-term (Dover and Talbot, 1987):

1. Minimal utilization of nonrenewable resources, especially fossil energy. This implies the need for increased use of renewable resources, such as organic fertilizers, and greater resource use efficiency, such as increased recycling of nutrients. Activities of soil organisms may contribute substantially to these areas.

2. Minimal degradation of environmental systems, both on and off the farm. This includes off-farm environmental "life-support systems" (Odum, 1983), such as air and water resources, and on-farm soil resources, especially soil fertility and the biological processes that mediate it.

Although these criteria do not define the sociological or political factors of sustainable farming systems, they recognize the ultimate ecological constraints on agriculture and emphasize the importance of ecosystem processes in the design and function of sustainable agroecosystems. A wide range of management options is left open, including organic, regenerative, low-input (or reduced-input) systems, as well as appropriate high-technology systems, such as those that use engineered organisms or their products. These options fall within the broader concept of "integrated" sustainable systems (Edwards, 1989).

With respect to the first criterion, current agricultural practices in developed countries are highly dependent upon fossil energy inputs. Fertilizer, tillage, and pesticide inputs account for as much as 33, 17, and 10 percent, respectively, of total energy use in U.S. agriculture (Lockeretz, 1983; Pimental, 1984; Sprague and Triplett, 1986). Interestingly, these inputs represent the managerial factors that appear most likely to benefit from enhanced soil biotic activities, such as nutrient cycling, soil conditioning, and biological pest control. The present ability to utilize these biotic activities to effectively replace or supplement high energy inputs is limited. Nonetheless, several possibilities for soil biotic management already exist, and many others appear to be fruitful areas for research.

## Soil Biota in Natural Ecosystems

Several recent treatises and symposium proceedings summarize the state of knowledge about soil ecology (Swift et al., 1979; Lebrun et al., 1982; Petersen and Luxton, 1982; Cooley, 1985; Fitter et al., 1985; Spence, 1985; Mitchell and Nakas, 1986; Eisenbeis and Wichard, 1987; Anderson 1988).

Taxonomically, soil biota include hundreds of thousands of species representing all five biotic kingdoms, at least 11 animal phyla, and all known

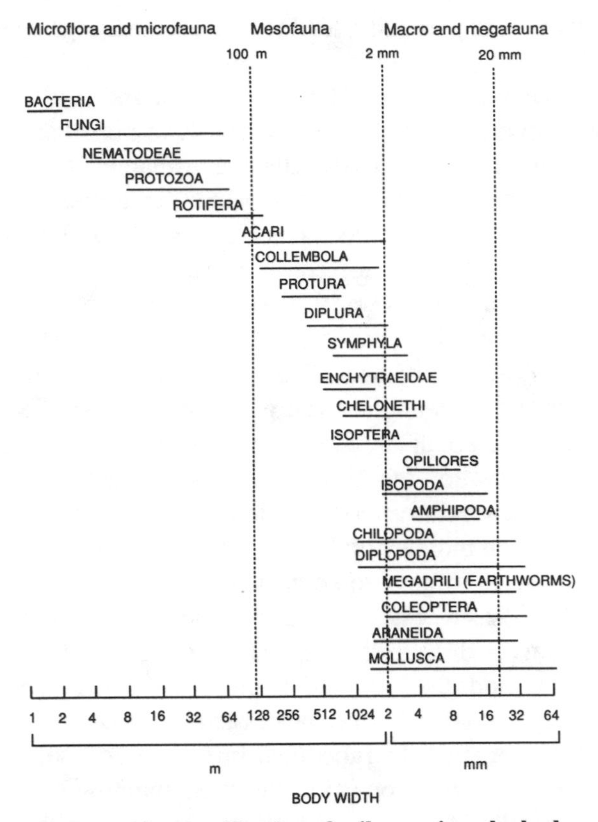


**Figure 1. Size classification of soil organisms by body width (Swift et al., 1979).**

types of microorganisms (Swift et al., 1979; Lynch, 1983). Morphologically, they range from less than 1 micrometer (bacteria) to several centimeters in diameter (snails) and up to 1.5 meters in length (the giant Australian earthworm, *Megascolides australis* (Lee, 1985). Figure 1 presents a classification of the soil biota according to body width, suggesting the scale at which they are involved in soil processes.

Soil organisms may inhabit surface layers of soil and litter or deeper soil horizons, or move throughout the soil profile. Among earthworms these habits are termed epigeic, endogeic, or anecic, respectively. Although many microorganisms are motile over short distances, small organisms and propagules (spores, eggs) are also transported throughout the soil by larger soil fauna, inoculating other areas. The vertical and horizontal distribution of the soil biota is generally limited by temperature, water, and soil texture. Within these constraints the quality, quantity, and distribution of organic matter inputs are major factors controlling soil biotic processes.

Microbial activities, in particular, tend to increase near organically enriched substrates.

The abundance and activity of the soil biota in undisturbed ecosystems vary widely. Although complete enumerations of all organisms in a soil at a given time are rare, a few estimates are available. Ryszkowski (1985) reported nonprotozoan faunal biomass values (in grams dry weight per square meter) of 3.5 to 45.0 for deciduous forests, 9.0 to 12.0 for deciduous-coniferous forests, 5.4 to 9.0 for coniferous forests, and 0.8 to 32.2 for grasslands. Protozoan biomass was 0.20 to 0.24 grams dry weight per square meter in a meadow soil. In contrast, estimates of microbial biomass ranged between about 100 and more than 5,000 grams dry weight per square meter in coniferous forest soils and alluvial grassland soils studied by Persson et al. (1980) and Kaszubiak and Kaczmarek (1985). These values illustrate the consistent observation that soil biomass is dominated by bacteria and fungi, although their relative contributions vary.

In general, the total biomass of the soil biota constitutes a relatively small fraction (1 to 8 percent) of total soil organic matter (Doran and Smith, 1987). In fact, Zlotin (1985) suggests that total organic standing stocks in any terrestrial ecosystem are described by the ratio $n \times 1{,}000 : n \times 100 : n \times 10 : n \times 1$ for humus, living and dead vegetation, microorganisms, and soil fauna, respectively. Only the n-coefficient changes across ecosystems. It is generally recognized, however, that the functional importance of soil organisms in ecosystems is not directly proportional to their standing biomass. Rather, the activities of soil biota may regulate system performance (primary production) through their effects on nutrient cycling and soil structure (Table 1).

## Functional Groups

Fungi and bacteria (including actinomycetes), the soil *microflora,* are the dominant soil organisms in terrestrial ecosystems, both in numbers and in biomass. Saprophytic microflora are the primary decomposers that attack complex organic materials and convert them into simpler molecules and into new microbial biomass and byproducts. This process, or more accurately suite of processes because of the array of enzymes involved, initiates the cycling of nutrients by making them available to plants, to other microflora, to soil animals that feed upon the microbial biomass and organic residues, or to loss through leaching. Soil microbes involved in nitrogen cycling may also directly affect nutrient cycling through the processes of nitrogen fixation and denitrification.

Soil *microfauna* consist primarily of protozoans, nematodes, and some small mites (Acarina) and springtails (Collembola). Rotifers and tardigrades

**Table 1. Influences of soil biota on soil processes in ecosystems.**

| | *Nutrient Cycling* | *Soil Structure* |
|---|---|---|
| Microflora | Catabolize organic matter<br>Mineralize and immobilize nutrients | Produce organic compounds that bind aggregates<br>Hyphae entangle particles onto aggregates |
| Microfauna | Regulate bacterial and fungal populations<br>Alter nutrient turnover | May affect aggregate structure through interactions with microflora |
| Mesofauna | Regulate fungal and microfaunal populations<br>Alter nutrient turnover<br>Fragment plant residues | Produce fecal pellets<br>Create biopores<br>Promote humification |
| Macrofauna | Fragment plant residues<br>Stimulate microbial activity | Mix organic and mineral particles<br>Redistribute organic matter and microorganisms<br>Create biopores<br>Promote humification<br>Produce fecal pellets |

(water bears) also may be abundant at times. These small animals feed upon the microflora and organic particles. Their abundance often fluctuates dramatically in response to food availability and to wetting and drying cycles in soil and litter. Their feeding activities can regulate population densities and activity of the microflora.

Soil *mesofauna* are composed of mites, collembola and other small soil insects, and enchytraeids (or potworms). These animals display a wide variety of feeding habits, including microbivory, saprobivory, omnivory, and predation. As a result, the functional roles of these animals in ecosystem processes are numerous. Although the mesofauna can attack and fragment plant residues, they are considered to be more important in regulating microbial populations and in reworking the feces of larger fauna (Swift et al., 1979).

The large and more conspicuous soil animals are *macrofauna* and *megafauna*, including amphipods, isopods (pillbugs), centipedes, and millipedes, adult and larval insects, earthworms, and mollusks (snails and slugs). Vertebrate scavengers also may be considered as megafauna. These animals are the principal agents of fragmentation and redistribution of plant residues in soil. Their activities enhance decomposition by stimulating and increasing substrate surface areas for microbial activity. The macrofauna and megafauna also affect soil structure by physically mixing the soil and through the formation of soil pores. Fecal pellets produced by mesofauna, macrofauna, and megafauna can persist in the soil and may be important in the formation of stable aggregates (Spence, 1985).

## Detritus Food Webs

The cascading of energy and nutrients from plant residue to microflora to microfauna to meso-, macro-, and megafauna during decomposition can be visualized as a detritus food web (Persson et al., 1980; Ingham et al., 1985; Hendrix et al., 1986; Hunt et al., 1987). Figure 2 illustrates a complex detritus food web from a grassland ecosystem. Using a simulation model of this food web, Hunt et al. (1987) estimated that soil animals account directly for 37 percent of nitrogen mineralization over an annual cycle. The bacterial-based subsystem was more important than the fungal-based one because bacteriophagous nematodes and protozoa were responsible for a large fraction of the nitrogen cycled through the soil fauna. Similar results were obtained in experimental studies by Parker et al. (1984).

The functioning of detritus food webs is driven by inputs of organic matter. Energy, in the form of available carbon substrates, often limits the activity of soil organisms. Given inputs of these substrates (plant or animal residues),

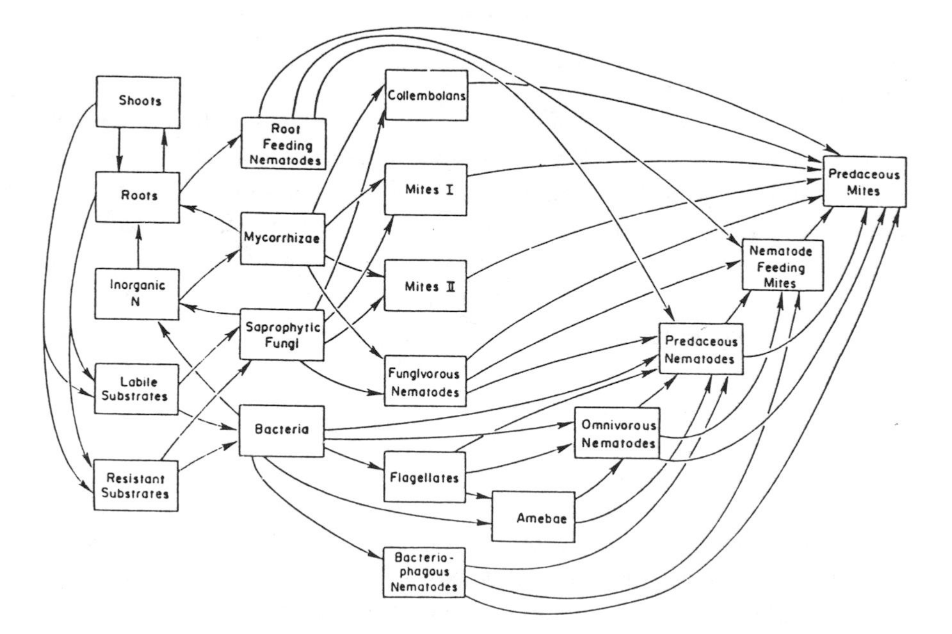


**Figure 2. Conceptual model of a detritus food web from a shortgrass prairie (Hunt et al., 1987).**

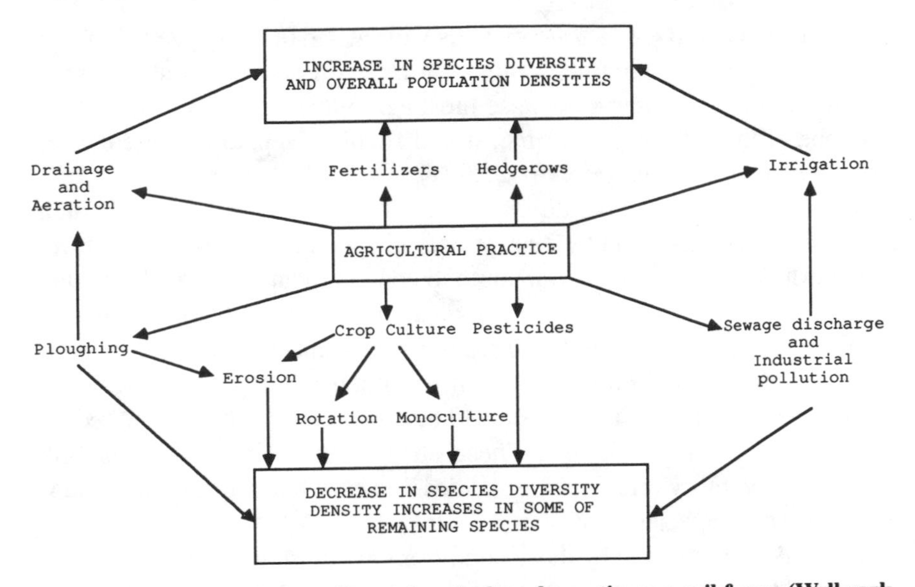


**Figure 3. Diagram showing effects of agricultural practices on soil fauna (Wallwork, 1976; modified from Edwards and Lofty, 1969).**

microbes utilize the energy, carbon, and available mineral nutrients to produce new microbial biomass. Turnover of the nutrient pool contained in microbial biomass during the decomposition and mineralization of organic inputs is a key element in soil fertility because it regulates the availability of nutrients to plants. Feeding on microbes by soil fauna can accelerate or slow microbial turnover and, thus, indirectly affect plant growth and nutrient content (Andren and Lagerlof, 1983; Coleman et al., 1984). Detritus food web manipulation may be an important means of managing soil fertility in sustainable agroecosystems.

## Effects of Conventional Agriculture on Soil Biota

Cultivation of undisturbed soil brings about a number of changes in community structure and activity of indigenous soil biota. These effects have been thoroughly reviewed by Edwards and Lofty (1969), Wallwork (1976), Marshall (1977), Madge (1981), Andren and Lagerlof (1983), Edwards (1983, 1984), Lynch (1983), Curry (1986), Kaszubiak and Kaczmarek (1985), Ryszkowski (1985), and Anderson (1988). Major effects of various agricultural practices on soil organisms are illustrated in figure 3.

Apart from the introduction of toxic chemicals (pesticides and industrial pollutants), changes in abundance and activity of soil biota can be related largely to changes in the factors previously discussed that regulate soil biotic activity, such as temperature, water, soil type, and organic matter inputs and distribution. However, because most agricultural practices alter more than one of these factors at a time, it is difficult to separate single cause-and-effect relationships (Hendrix, 1987).

Plowing alters soil temperature and water regimes by reducing the mulch effect of surface litter and by disrupting soil aggregate and pore structure. These effects, in turn, may contribute to soil degradation through organic matter depletion and soil erosion. The quantity and quality of organic matter inputs to the soil are often reduced when native plant communities are converted to monoculture cropping systems. Similarly, the principal effect of most herbicides on soil fauna appears to be exerted indirectly through reductions in organic matter inputs. These changes in habitat conditions and resource availability often reduce species diversity but may increase abundance of some species.

Soil organisms favored by these conditions are those with a short generation time, small size, rapid dispersal, low degree of food and habitat specialization, and high metabolic activity (Andren and Lagerlof, 1983; Ryszkowski, 1985). These changes in species composition may alter the trophic structure of detritus food webs and change nutrient cycling dynamics in agroecosystems (Hendrix et al., 1986; Holland and Coleman, 1987).

On the other hand, some agricultural practices increase abundance and diversity of soil organisms by enhancing habitat conditions or resource availability. For example, irrigation and/or drainage may optimize soil water content, and plowing may loosen compacted soils and improve soil aeration. Similarly, hedgerows and riparian strips near agroecosystems increase species diversity by providing refuges for pests as well as beneficial species. Organic fertilizers (animal or green manures) increase both quantity and quality of available organic substrates for soil organisms. Inorganic fertilizers also tend to enhance organic inputs by stimulating plant production. Interestingly, crop rotations themselves appear not to greatly affect the soil biota except for plant-specific organisms. Rather, effects of crop rotations appear to be exerted primarily through quantity or quality of organic inputs (Wallwork, 1976; Andren and Lagerlof, 1983).

Effects of management intensity on the abundance of some soil organisms are illustrated in table 2 with data from our research site on the Georgia Piedmont. All systems are adjacent to one another on a sandy clay loam bottomland soil (Typic Rhodudult) except the low organic matter plowed system that is on a highly degraded, upland sandy clay loam (Typic Hapludult) (see Groffman et al., 1986; Langdale et al., 1987, for site descriptions).

**Table 2. Abundance of soil fauna in ecosystems on the Georgia Piedmont. Values are numbers of organisms per square meter to a depth of 5 cm for microarthropods and to 15 cm for earthworms.**

| | *Forest* | *Meadow* | *No-till** | *Plowed* (High Soil Organic Matter)* | *Plowed (Low Soil Organic Matter)* |
|---|---|---|---|---|---|
| Prostigmata | 96,272 | 51,376 | 63,859 | 25,978 | 7,551 |
| Mesostigmata | 6,023 | 510 | 6,799 | 2,645 | 612 |
| Oribatid | 78,378 | 8,163 | 33,272 | 5,097 | 357 |
| Astigmata | 0 | 0 | 97 | 3,486 | 1,378 |
| Collembola | 21,230 | 1,173 | 12,487 | 7,727 | 23,267 |
| Others | 6,824 | 663 | 2,594 | 1,070 | 0 |
| Microarthropod Totals | 208,727 | 61,885 | 119,108 | 46,003 | 33,165 |
| Earthworms | ND† | 190 | 967 | 149 | 127 |

*Microarthropod data from House and Parmelee (1985); earthworm data from Parmelee (1987).
†ND=not determined.

Abundance of the dominant microarthropods (prostigmatid and oribatid mites) generally follows the organic matter content of the soils [forest greater than no-till greater than meadow greater than plowed (high organic matter) greater than plowed (low organic matter)]. Both groups consist primarily of fungivores, possibly reflecting the relative importance of fungi as primary decomposers in these systems. Organism numbers and organic matter content are higher in the no-till cropping system than in the grass meadow. This is probably because of higher primary productivity and return of plant residues to the soil in the more intensively managed, double-cropped, no-till agroecosystem. As observed in other studies, astigmatid mites show higher abundance and earthworms lower abundance in plowed soils (Andren and Lagerlof, 1983; Edwards, 1983). The high numbers of Collembola at the highly degraded site are unexplained.

## Soil Biota in Sustainable Agroecosystems

Our premise is that, as in natural ecosystems, the activities of soil biota can be important in nutrient cycling and maintenance of soil conditions in agroecosystems, thereby contributing to agricultural sustainability. Does it follow that the conversion of conventional agroecosystems into sustainable ones requires less intensive management to reverse the effects of conventional practices and to allow systems to approach more natural conditions?

It has been argued that to be sustainable, agroecosystems should more closely resemble natural ecosystems in certain respects (Gliessman, 1987; Altieri, 1987). For example, increased species richness (polycultures, hedgerows, greater tolerance of weeds) may decrease risks of production failure by providing alternate crops and by promoting natural predators of pests. However, high species diversity can be very difficult to manage in practice. Thus, species richness for its own sake may be counter-productive (Dover and Talbot, 1987). Nutrient cycles in undisturbed ecosystems tend to be more closed and less "leaky" than in agroecosystems. However, an important characteristic of agroecosystems is that they export large amounts of nutrients in crop biomass and, therefore, require large inputs, regardless of the amount of internal recycling. By design, these systems have large nutrient through-flows that will necessarily alter their biological and physicochemical characteristics relative to natural ecosystems. The challenge is to minimize nutrient losses from these systems and maximize the efficiency of internal nutrient recycling.

Therefore, rather than less management, sustainable agroecosystems will probably require more and better-informed management of all ecosystem components, including soil biota. The primary goal of management should be to optimize internal natural processes, using exogenous inputs within the constraints of sustainability as discussed previously. This is consistent with the concept of integrated management (Edwards, 1989).

## Possibilities for Managing the Soil Biota

In theory, at least, probably enough is known about soil biota to manage some of their activities to enhance crop growth or soil fertility (Elliott and Coleman, 1988). As discussed previously, soil organisms are inadvertently managed when crops and soils are managed. However, considerable basic and applied research is needed to increase the capabilities in soil biotic "husbandry." Specific cases of primarily microbial management are reviewed by Lynch (1983, 1987) and by Miller (1989).

Herein, management is considered in more general terms to include both microflora and soil fauna. Possibilities for management include direct manipulation of target groups of organisms and indirect manipulation of food resources and/or habitat conditions.

***Direct Methods.*** Direct methods attempt to alter the abundance or activity of specific groups of organisms. Inoculation of seeds or roots with rhizobia, mycorrhizae, and Trichoderma are examples of direct manipulation of microflora to enhance plant performance (Lynch, 1987). An intriguing tripartite association of microbes has been described in which a fungus converts cellulose from wheat straw into simple sugars that are used by a nonsymbiotic nitrogen-fixing bacterium. Another bacterium produces polysaccharides that help maintain the anaerobic conditions necessary for nitrogenase activity. The authors report gains of 12 grams of nitrogen used per kilogram of straw which, when combined with nitrogen present in the straw, could amount to 106 kilograms of nitrogen per hectare. Other such microbial associations and their limitations are discussed in Lynch (1987).

Introductions of predatory organisms to control pests have been used in agriculture for more than a century. Many of these introductions have proven successful for controlling a variety of insect and weed pests (Pimental et al., 1986). Introductions of earthworms have been used to increase soil structure and fertility (see examples in Edwards and Lofty 1977; Lee, 1985a). Earthworms have also been introduced successfully in a number of instances for "zoological conditioning" of soils in reclaimed lands (Zlotin, 1985; Coleman, 1985a). However, the physiological and life-history conditions of particular genera and species have to be considered. Lavelle et al. (1987) have shown considerable transformation and utilization of humic materials in tropical agroecosystems that have high populations of a geophagous Glossoscolecid earthworm.

It may be possible to manage soil nutrient availability by manipulating detritus food webs. For example, altering the number of predatory mites to increase or decrease abundances of their fungivorous prey (mites, Collembola, or nematodes) could slow or accelerate fungal decomposition of plant residues. Although food web manipulations have not been attempted in practice, they are currently an exciting area for modeling and experimental research.

A major problem to overcome in the use of inoculations and introductions is ensuring the establishment of the introduced organisms. Competition from a diverse indigenous soil biota may overwhelm introduced organisms. Additionally, limited availability of food resources may result in extinction or emigration. It may be necessary to add food supplies or organic amendments along with inocula to aid establishment (Chen and Avnimelech, 1986).

Another method of direct manipulation is the use of inhibitors to reduce activity or abundance of selected organisms. Pesticides have long been used for this purpose, both above and below the ground. Nitrification inhibitors (NServe) have also been used to manage soil microbial activities. Biocides with varying degrees of selectivity have also been used to reduce particular groups of organisms in ecological research (Coleman, 1985b).

Problems traditionally involved with using inhibitors include detrimental effects on nontarget organisms, residual toxicity in the soil, high costs, and "nonsustainability" (in terms of the criteria herein). Nonetheless, highly selective and potent compounds arc currently being developed for pest control, and some may prove useful for manipulating free-living soil organisms.

***Indirect Methods.*** Indirect methods can be used to manage soil biotic processes by manipulating the factors that control biotic activity (habitat structure, microclimate, nutrients, and energy resources) rather than the organisms themselves. Most agricultural practices alter these factors, as discussed previously, but are not commonly viewed in terms of managing soil biota.

In theory, distinctions can be made between habitat and resource quality manipulations, but in practice they often confound one another. For example, surface mulching with organic residues alters habitat structure, soil temperature, and water regimes, while simultaneously affecting the availability of organic matter to the decomposer community.

Some physical and chemical manipulations alter habitat conditions directly and may be useful for stimulating or suppressing biotic activity. Soil tillage is an effective means of manipulating habitat conditions. Plowing and plant residue incorporation select for communities of opportunistic organisms adapted to pulses of rapid biotic activity and consequent rapid residue decomposition and nutrient turnover. Reduced tillage (with surface placement of residues) creates a relatively more stable environment and encourages development of more diverse decomposer communities and slower nutrient turnover. Available evidence suggests that conditions in no-till systems favor a higher ratio of fungi to bacteria, whereas in conventionally tilled systems bacterial decomposers may predominate (Hendrix et al., 1986; Holland and Coleman, 1987). The water content of soil or surface residues can be controlled by irrigation to adjust soil biotic activity. Below-ground drip irrigation and above-ground sprinklers might be used in concert to manage plant growth and surface residue decomposition rates simultaneously. Soil pH is commonly adjusted to optimize conditions for plant growth, but pH also affects soil organisms and might be manipulated to select for certain biotic groups—i.e., fungal activity under acid conditions.

Resource manipulations control inputs of nutrients and energy to soil organisms. Managing the quantity, quality, and placement of organic fertilizers and residues is common practice in organic farming and will undoubtedly be a cornerstone of sustainable agricultural management. Because of the importance of both quantity and quality of carbon substrates in controlling soil biotic activity, some researchers have suggested managing carbon inputs as a means of managing soil nutrient availability. For exam-

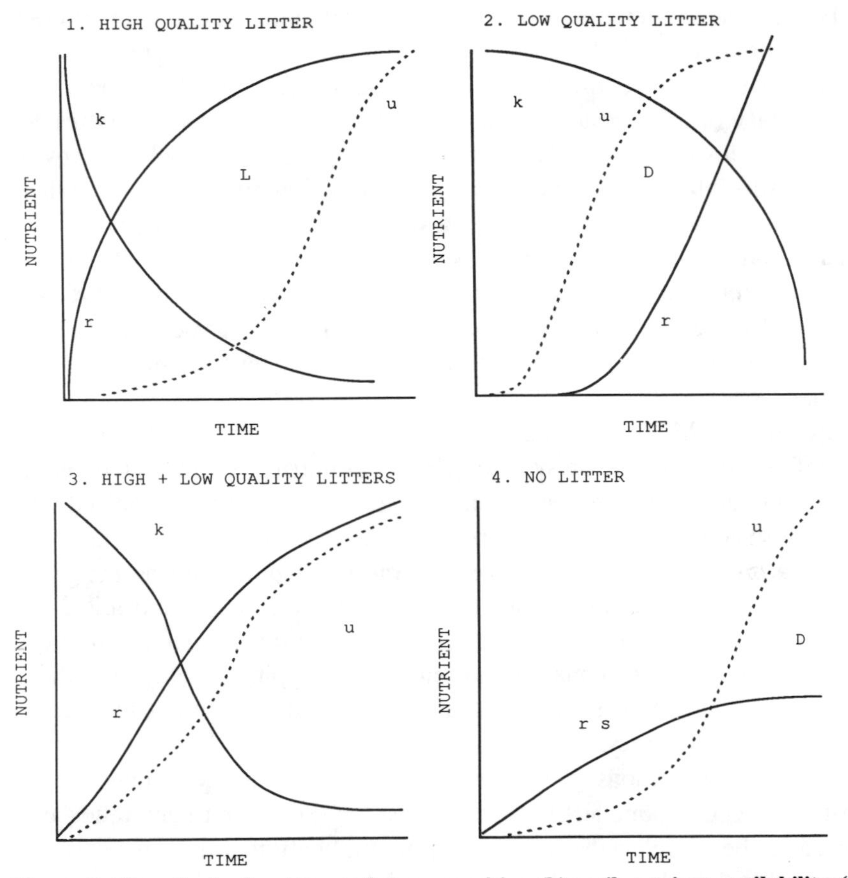


**Figure 4. Hypothesized patterns of decomposition (k), soil nutrient availability (r), and plant uptake of nutrients (u) in systems subject to inputs of plant litter of various qualities; (rs) is release of nutrients from soil. Area L represents potential for nutrient loss by leaching, and area D represents potential nutrient deficit (Swift, 1987).**

ple, high carbon:nitrogen ratio materials stimulate immobilization of other nutrients into microbial biomass, removing them from the soil solution. Nutrients in higher quality substrates (low carbon:nitrogen ratio) are more rapidly mineralized, resulting in less immobilization and greater nutrient availability.

A key goal of soil biotic management is to manipulate the processes of residue decomposition, nutrient immobilization, and mineralization so that nutrient release is synchronized with plant growth (Brussaard et al., 1988; McGill and Myers, 1987; Sanchez et al., 1989; Swift, 1987). The rationale is that timing of increased nutrient availability to coincide with plant demand will then increase nutrient use efficiency and reduce leaching losses of soluble nutrients.

Four scenarios in figure 4 represent inputs of plant residue with different carbon:nitrogen ratios. High quality litter (panel 1), such as legume residue, decomposes relatively quickly (k), releasing nutrients (r) out of phase with plant uptake (u). The result is high potential for loss (L). Low quality residue (panel 2) (for example, wheat straw) decomposes too slowly to provide nutrients for plant uptake and, in fact, may stimulate microbial immobilization of nutrients. The net result is a nutrient deficit (D) for plants. The ideal situation (panel 3) is, of course, one in which nutrients are released from decomposing residues in concert with plant demand. Nutrient release from soil mineral and organic pools (panel 4) may also be synchronized with initial stages of plant growth but may not meet total plant demand in many soils.

McGill and Myers (1987) present an interesting series of data showing the influence of climate on synchronization of nutrient supply and demand. They conclude that cropping systems are generally well-synchronized with soil biotic activity in humid temperate and tropical life-zones where soil temperature and water regimes promote mineralization during the growing seasons. Asynchrony appears to be a problem in semiarid tropical and subtropical agroecosystems because characteristic seasonal wet and dry periods increase the potential for nutrient loss. Cropping systems and residue management techniques have to be adapted to such conditions to increase nutrient use efficiency.

A number of methods exist for adjusting resource quality of inputs. Composts and aged organic fertilizers may have low carbon:nitrogen ratios and, therefore, begin releasing nutrients upon application. Fresh manures can be mixed with high carbon:nitrogen material (such as straw) to slow nutrient release. Inorganic fertilizers applied to low-quality plant residues may be immobilized into microbial biomass during residue decomposition and then released later during remineralization. Manipulating soil conditions or organisms to speed or slow this process could help synchronize nutrient supply and demand. Green manure crops might even be designed to achieve a desired resource quality by interplanting legume and nonlegume seeds in certain proportions.

An important problem that must be dealt with in soil biotic management strategies is the relative unpredictability of system performance. Decomposing organic matter and soil organisms may not behave as predictably as conventional systems managed with well-defined and prescribed chemical formulations. This unpredictability results from: (1) the inherent variability of the factors that drive soil biotic activities, especially temperature and moisture regimes; and (2) limited understanding of the mechanisms at work in these physically, chemically, and biologically diverse systems. Laboratory and modeling studies suggest that soil biotic processes are predictable based upon knowledge of climatic inputs. In principle, these results should also apply to field conditions, but further research is needed to increase the predictive capabilities. Better understanding of regulatory mechanisms and trophic relationships within detritus food webs should improve the ability to manage soil communities and the processes they mediate.

## Conclusions

There is a diverse array of organisms comprising the soil biota and the habitats and microhabitats in which they reside. For the purposes of agroecosystem management, the shoot/litter/root/microbial/faunal/soil system should be considered as a series of interacting components, all of which play roles in nutrient immobilization or mineralization at various times. Achieving an effective synchrony of biotic and chemical interactions (principally in terms of nitrogen and phosphorus cycling) will require careful thought and management.

The soil biota can play an important role in regulating nutrient cycling processes and maintaining soil structure in sustainable agroecosystems. Key areas in which the soil biota can influence sustainability include internal recycling of nutrients and altering soil structure, including affecting soil aggregates and porosity and the formation and distribution of soil organic matter. Practitioners of low-input sustainable agriculture should focus their attention on biotic management, particularly of detrital food webs, to better attain their goals of successful, optimal, long-term sustainability.

## REFERENCES


Altieri, M. A. 1987. Agroecology: The scientific basis of alternative agriculture. Westview Press, Boulder, Colorado.

Anderson, J. M. 1988. The role of soil fauna in agricultural systems. pp. 59-112. *In* J. R. Wilson [editor] Adances in nitrogen cycling in agricultural ecosystems. C.A.B. International, Wallingford, England.

Andren, O. and J. Lagerlof. 1983. Soil fauna (microarthropods, enchytraeids, nematodes) in Swedish agricultural cropping systems. Acta Agri. Scand. 33: 33-52.

Brussaard, L., J. A. van Veen, M.J. Kooistra, and G. Lebbink. 1988. The Dutch programme on soil ecology of arable farming systems. I. Objectives, approach and some preliminary results. Ecological Bulletin 39.

Chen, Y., and Y. Avnimelech. 1986. The role of organic matter in modern agriculture. Martinus Nijhoff Publishers.

Coleman, D. C. 1985a. Role of soil fauna in pedogenesis and nutrient cycling in natural and human-managed ecosystems. INTECOL Bulletin 12: 19-28.

Coleman, D. C. 1985b. Through a ped darkly: An ecological assessment of root-soil-microbial-faunal interactions. pp. 1-21. *In* A. H. Fitter, D. Atkinson, D. J. Read, and M. B. Usher [editors] Ecological interactions in soil. Blackwell Scientific Publications, Palo Alto, California.

Coleman, D. C., R. V. Anderson, C.V. Cole, J.F. McClellan, L. E. Woods, J.A. Trofymow, and E. T. Elliott. 1984. Roles of protozoa and nematodes in nutrient cycling. pp. 17-28. *In* Microbial-plant interactions. American Society of Agronomy, Madison, Wisconsin.

Cooley, J. 1985. Soil ecology and management. INTECOL Bulletin 12: 1-132.

Crossley, D. A., Jr., D. C. Coleman, and P.F. Hendrix. 1988. The biota of agricultural soils: Importance of the fauna. Agriculture Ecosystems Environment (in press).

Curry, J. P. 1986. Effects of management on soil decomposers and decomposition processes in grassland and croplands. pp. 349-398. *In* M. J. Mitchell and J.P. Nukas [editors] Microfloral and faunal interactions in natural and agro ecosystems. Martinus Nijhoff Publishers.

Doran, J. W., and M. S. Smith. 1987. Organic matter management and utilization of soil and fertilizers nutrients. pp. 53-71. *In* R. F. Follett, J.W.B. Stewart, and C. V. Cole [editors] Soil fertility and organic matter as critical components of production systems. Soil Science Society of America and American Society of Agronomy, Madison, Wisconsin.

Dover, M. J., and L. M. Talbot. 1987. To feed the earth: Agro-ecology for sustainable development. World Resources Institute, Wasshington, D.C..

Edwards, C. A. 1983. Earthworm ecology in cultivated soils. pp. 126-137. *In* J. E. Satchell [editor] Earthworm ecology. Chapman and Hall, New York, New York.

Edwards, C. A. 1984. Changes in agricultural practice and their impact on soil organisms. Rothamsted Experiment Station, Harpenden. American Journal of Alternative Agriculture 2: 148-152.

Edwards, C. A. 1989. The importance of integration in sustainable agricultural systems. *In* C. A. Edwards, R. Lal, P. Madden, R. H. Miller, and G. House [editors] Sustainable agricultural systems. Soil and Water Conservation Society, Ankeny, Iowa.

Edwards, C. A., and J. R. Lofty. 1969. The influence of agricultural practice on soil microarthropod populations. pp. 237-247. *In* J. G. Sheals [editor] The soil ecosystems. systematics association, London, England.

Edwards, C. A., and J. R. Lofty. 1977. Biology of earthworms. Chapman and Hall, London, England.

Eisenbeis, G., and W. Wichard. 1987. Atlas on the biology of soil arthropods. Springer-Verlag, New York, New York.

Elliott, E. T., and D. C. Coleman. 1988. Let the soil work for us. Ecological Bulletins (Copenhagen) 39: 23-32.

Fitter, A. H., D. Atkinson, D. J. Read, and M. B. Usher. 1985. Ecological interactions in soil Volume 4: Plants, microbes and animals. Blackwell Scientific Publications, Palo Alto, California.

Gliessman, S. R. 1987. Species interactions and community ecology in low external-input agriculture. American Journal of Alternative Agriculture 2: 160-165.

Groffman, P. M., G. J. House, P. F. Hendrix, D. E. Scott, and D. A. Crossley, Jr. 1986. Nitrogen cycling as affected by interactions of components in a Georgia Piedmont agroecosystems. Ecology 67:80-87.

Heal, O. W., and J. Dighton. 1985. Ecological interactions in soil Volume 4. Resource quality and trophic structure in the soil system. Blackwell Scientific Publications, Palo Alto, California.

Hendrix, P. F. 1987. Strategies for research and management in reduced-input agroecosystems. American Journal of Alternative Agriculture 2: 166-172.

Hendrix, P. F., R. W. Parmelee, D. A. Crossley, Jr., D. C. Coleman, E. P. Odum, and P. M. Groffman. 1986. Detritus food webs in conventional and notillage agroecosystems. Bio Science 36: 374-380.

Holland, E. A., and D. C. Coleman. 1987. Litter placement effects on microbial communities and organic matter dynamics. Ecology 68: 425-433.

House, G. J., and R. W. Parmelee. 1985. Comparison of soil arthropods and earthworms from conventional and no-tillage agroecosystems. Soil and Tillage Research 5: 351-360.

Hunt, H. W., D. C. Coleman, E. R. Ingham, E. T. Elliott, J. C. Moore, S. L. Rose, C.P.P.

Reid, and C. R. Morley. 1987. The detrital food web in a shortgrass prairie. Biology and Fertility of Soils 3: 57-68.

Ingham, R. E., J. A. Trofymow, E. R. Ingham, and D. C. Coleman. 1985. Interactions of bacteria, fungi and their nematode grazers: Effects on nutrient cycling and plant growth. Ecological Monographs 55: 119-140.

Kaszubiak, H., and W. Kaczmarek. 1985. Differentiation of bacterial biomass in croplands and grasslands. INTECOL Bulletin 12: 29-37.

Knorr, D. 1983. Sustainable food systems. AVI Publishing Co., Inc., Westport, Connecticut.

Langdale, G. W., R. R. Bruce, and A. W. Thomas. 1987. Restoration of eroded Southern Piedmont land in conservation tillage systems. pp. 142-143. *In* J. F. Power [editor] The role of legumes in conservation tillage system. Soil Conservation Society of America, Ankeny, Iowa.

Lebrun, P., H. M. Andre, A. De Medts, C. Gregoire-Wibo, and G. Wauthy. 1982. New trends in soil biology. Proceedings, VIII International Colloquium of Soil Ecology.

Lee, J. A., S. McNeill, and I. H. Rorison. 1983. Nitrogen as an ecological factor. Blackwell Scientific Publications, Palo Alto, California.

Lee, K. E. 1985. Earthworms: Their ecology and relationships to soils and land use. Academic Press.

Levelle, P., I. Barois, I. Cruz, C. Fragoso, A. Hernandez, A. Pineda, and R. Rangel. 1987. Adaptive strategies of Pontoscolex corethrurus (Glossoscolecidae, Oligochaeta), a perergrine geophagous earthworm of the humid tropics. Biology and Fertility of Soils 5: 188-194.

Lockeretz, W. 1983. Energy in U.S. agricultural production. pp. 77-103. *In* D. Knorr [editor] Sustainable food systems. AVI Publishing Company, Inc., Westport, Connecticut.

Lynch, J. M. 1983. Soil biotechnology: Microbiological factors in crop production. Blackwell Scientific Publications, Palo Alto, California.

Lynch, J. M. 1987. Soil biology: Accomplishments and potential. Soil Science Society of America Journal 51: 1,409-1,412.

Madge, D. S. 1981. Influence of agricultural practice of soil invertebrate animals. pp. 79-98. *In* B. Stonehouse [editor] Biological husbandry. Butterworths, London, England.

Marshall, V. G. 1977. Effects of manures and fertilizers on soil fauna: A review. (Special publication no. 3). Commonwealth Bureau of Soils, Harpenden, United Kingdom.

Miller, R. H. 1989. Soil microbiological inputs for sustainable agricultural systems. pp. 614-623. *In* C. A. Edwards, R. Lal, P. Madden, R. H. Miller, and G. House [editors] Sustainable agricultural systems. Soil and Water Conservation Society, Ankeny, Iowa.

Mitchell, M. J., and J. P. Nakas. 1986. Microfloral and faunal interactions in natural and agro-ecosystems. Martinus Nijhoff Publishers.

McGill, W. B., and R.J.K. Myers. 1987. Controls on dynamics of soil and fertilizers nitrogen. pp. 73-99. *In* R. F. Follett, J.W.B. Stewart, and C. V. Cole [editors] Soil fertility and organic matter as critical components of production systems. Soil Science of America and American Society of Agronomy, Madison, Wisconsin.

Odum, E. P. 1983. Basic ecology. Saunders College Publishing.

Parker, L. W., P.F. Santos, J. Phillips, and W. G. Whitford. 1984. Carbon and nitrogen dynamics during the decomposition of litter and roots of Chihauahuan desert annual, Lepidium lasiocarpum. Ecological Monographs 54: 339-360.

Persson, T., E. Baath, M. Clarholm, H. Lundkvist, B. E. Soderstrom, and B. Sohlenius. 1980. Trophic structure, biomass dynamics and carbon metabolism of soil organisms in a scots pine forest. Ecological Bulletin: 419-459.

Petersen, H., and M. Luxton. 1982. A comparative analysis of soil fauna populations and their role in decomposition processes. Oikos 39: 287-388.

Pimentel, D. 1984. Energy flow in agroecosystems. pp. 121-132. *In* R. Lowrance, B. R. Stinner, and G. J. Horse [editors] Agricultural ecosystems. John Wiley and Sons, Inc.,

Somerset, New Jersey.

Pimentel, D., C. Glenister, S. Fast, and D. Gallahan. 1986. Environmental risks associated with the use of biological and cultural pest controls. *In* D. Pimentel [editor] Some aspects of integrated pest management. Cornell University Press, Ithaca, New York.

Ryszkowski, L. 1985. Impoverishment of soil fauna due to agriculture. INTECOL Bulletin 12: 7-17.

Sanchez, P. A., C. A. Palm, L. T. Szott, E. Cuevas, and R. Lal. 1989. Organic input management in tropical agroecosystems. *In* D. C. Coleman, J. M. Oades, and G. Uehara [editors] Organic matter in tropical soils.

Spence, J. R. 1985. Oil, toil & soil: An introduction to the symposium on faunal influences on soil structure. Quaestiones Entomologicae 21(4): 371.3-371.4.

Sprague, M. A., and G. B. Triplett. 1986. No-tillage and surface-tillage agriculture: The tillage revolution. John Wiley & Sons, Inc., Somerset, New Jersey.

Swift, M. J., editor. 1987. Tropical soil biology and fertility: Interregional research planing workshop. Biology International Special Issue 13. International Union of Biological Sciences.

Swift, M. J., O. W. Heal, and J. M. Anderson. 1979. Decomposition in terrestrial ecosystems. University of California Press, Berkeley.

Wallwork, J. A. 1976. The distribution and diversity of soil fauna. Academic Press, London, England.

Zlotin, R. I. 1985. Role of soil fauna in the formation of soil properties. INTECOL Bulletin 12: 39-47.

# AGRICULTURE AND HUMAN HEALTH

Katherine L. Clancy

Things that tend to be damaging to the environment also tend to be damaging to human health (Satin, 1988). The interfaces of human health and sustainable agriculture are multifaceted. At a minimum, they include biological health effects, worker safety, dietary quality (including water quality), and environmental hazards. They also can encompass recreational needs, stress reduction, human scale activities, and other concepts that fit within the parameters of the World Health Organization's definition of health as a "state of complete physical, mental, and social well-being." If sustainable agriculture is to gain adequate public support, its proponents must respond to needs and concerns in all of these areas, if not every need or concern in every part of the country. What follows is a review of the literature since 1985 on the health hazards of conventional agriculture (Clancy, 1986) and policy issues raised by an array of private and government organizations in response to the dangers.

## Health Risks

The four major agricultural health risks to farmers and consumers (excluding farm accidents) derive from pesticides, antibiotics, nitrates, and altered nutrient levels. Although no calculated risk estimates are available, the order listed is probably that of risk priorities. The available literature and number of different policy issues are certainly greater for pesticides than for the other three risks.

***Pesticides.*** Pesticide use poses significant health hazards to farm workers, farmers, and consumers through three pathways of exposure: acute and chronic work-related exposure and chronic, and low-level exposure through consumption of residues in food. Acute poisoning of farm workers is clearly the primary danger. The numbers of poisonings in the United States is unknown; the most recent (July 1988) review was still citing an earlier estimate from California of 110,000 per year, and 200 deaths (Council on Scientific Affairs, 1988). This number is much higher in developing countries (Davies and Lee, 1987).

The health effects of chronic exposure to pesticide compounds by farm workers and farmers have been chronicled and reviewed in a number of widely dispersed sources (Coye, 1984; Davies and Lee, 1987; Moses, 1986). They include neurotoxicity (Coye, 1984), infertility (Davies and Lee, 1987), dermatologic lesions (Council on Scientific Affairs, 1988; Coye, 1984), and immune system incompetence (Olson, 1986). But the potential carcinogenicity of pesticides has been the major topic of review, in large part because cancer registries are available. The report of the Council on Scientific Affairs (1988), of the American Medical Association, is the most recent and provides a fairly thorough review of literature since 1979. The conclusions drawn from this report are as follows:

- A number of pesticide compounds have been found, through a combination of animal, human, and in vitro tests reviewed by both the Environmental Protection Agency (EPA) and the International Agency for Research on Cancer, to be probably carcinogenic in humans. Two of these compounds, vinyl chloride and arsenic, definitely cause cancer in humans.
- A number of recent epidemiologic studies suggest that farmers have an increased risk of developing certain cancers (particularly non-Hodgkins lymphoma and leukemia), and pesticides appear to play a role in some studies.
- Factors in the agricultural environment other than pesticides may be responsible for the elevated risk.
- There is apparently a difference in risks for different cancers in dairy- and beef-producing areas, as compared to corn-, hog-, and chicken-producing areas.
- The data from the wide array of studies are not consistent.
- Testing of compounds with animals, humans, and in vitro has been inadequate.
- There is an urgent need to continue surveillance and assessment of the delayed human health effects of pesticide use.

For some reason, the American Medical Association's review did not include studies of farm workers, which might have shed a stronger light on some of the increased cancer risks.

The third path is of low-level, chronic exposure to pesticides through proximity to agricultural fields, use in the home, and, of course, consumption of residues on foods. Very little information is available on the health effects of chronic, low-level exposure. Schwartz and Lo Gerfo's (1988) recent analysis of California birth records suggests the possible role of pesticide exposure in the development of birth defects in children born in areas where pesticide use is heavy. Another recent study of rural residents documents a relation of serum pesticide levels to consumption of eggs from home-raised hens and root vegetables but not ground-vine vegetables (Stehr-Green et al., 1988).

There is so little research on the risks to the general public of chronic exposure that confirmatory studies have not been done and the use of different models makes it difficult to compare predictions. For example, the National Research Council/Institute of Medicine report on diet and cancer reviewed some epidemiological evidence of pesticide workers and one farm worker study, as well as bioassays for some organochlorine compounds, some organophosphates, and two carbamates. Based on this modest review, the report concluded that the amounts of residues present in the average U.S. diet do not make a major contribution to the overall risks of cancer in humans (National Research Council, 1982). The Board on Agriculture, National Academy of Science panel used a totally different model that identified those foods that pose the greatest risk to consumers based on worst-case assumptions (National Research Council, 1987). Obviously, these two studies are not comparable.

No study has calculated total predicted risk from food and other avenues, such as household and garden use and water. Apparently, a much larger number of wells have tested positive for pesticides than had originally been reported by EPA in 1987. The latest count notes that well water in 34 states is contaminated with 73 different pesticides (Anderson, 1988). It has been reported that one-third of U.S. counties have the potential for groundwater contamination (Gianessi, 1987). Cancer risk estimates have not been calculated for this exposure.

Another avenue of exposure is breast milk. A valid national sample has not been taken to test for nursing infants' intake of pesticides, so it is not possible to measure the extent of breast milk contamination. A recent article about risks in East Africa contains a review of data from other countries (Kanja et al., 1986). Concerns about the inequitable body burdens of residues borne by children have been raised because, among other things, tolerance-setting by EPA does not consider these differences (Mott and Snyder, 1987).

The latter point is only one of a very long list of policy issues regarding pesticide regulation that has surfaced in the last couple of years. Many of the issues appear in the book *Pesticide Alert* (Mott and Snyder, 1987), and have been corroborated by reports of government and consumer protection organizations (Mongomery, 1987; U.S. General Accounting Office, 1986a, 1986b). According to Mott and Snyder, the brief against EPA tolerance-setting procedures is based upon five facts: (1) EPA has established tolerances without necessary health and safety data; (2) EPA relies on outdated assumptions about what constitutes an average diet, particularly in terms of low levels of fruit and vegetable consumption; (3) tolerances are rarely revised when EPA receives new scientific data about the risks of a pesticide; (4) ingredients in pesticides that may leave hazardous residues in food, such as inert ingredients, are not considered in tolerance-setting; and (5) EPA's tolerances allow carcinogenic residues to occur in foods, even though there is strong debate about the existence of threshold levels.

Other regulatory issues not discussed in *Pesticide Alert* include those raised in the National Research Council report regarding the inconsistency in handling residues on fresh and processed foods and the fact that the synergistic effects of compounds are not considered in setting tolerances (National Research Council, 1987). Yarbrough et al. (1982) have reported that when pesticides are given in pairs, the effects appear to be generally additive. Other researchers have also discovered that teratogenic effects can go undetected in standard bioassays (Karlock et al., 1987). Finally, the adequacy of the Food and Drug Administration's (FDA) activities has been questioned. FDA is responsible for monitoring residue levels and implementing seizure of violative lots. The system has many problems—inadequate financing being one of them. Among other problems are that the laboratory methods that FDA uses can detect only 40 percent of the chemicals used and cannot detect, by routine methods, 40 percent of pesticides that have a moderate-to-high health hazard (U.S. General Accounting Office, 1986b).

Many other policy issues could be raised. But the question that should be asked is why producers and researchers using alternative farming practices should pay attention to these issues if they are already supplying residue-free or reduced-residue products. The first reason is obvious: to provide a rationale, besides economics, for the promotion and adoption of alternative farming methods. The second reason is to help define the research agenda for developing appropriate chemicals for use in integrated pest management and other alternative farming programs. The third reason is to gauge the progress of the research agenda in sustainable agriculture vis-a-vis public concerns.

Virtually all public concerns about pesticide residues are focused on fruits and vegetables. The Nutri-Clean system is used only for fresh produce. Supermarket chains are attracting customers by providing organically grown produce, not organically grown grains (although there are a number of organic grain products in the market). The great majority of food safety reports released in the last two to three years are on pesticides. Furthermore, according to calculations done for the National Research Council report, 6 of the top 7, and 9 of the 15 foods with the greatest oncogenic risk are produce items (National Research Council, 1987). In the meantime, the public is being urged, in campaigns all over the country, to increase the intake of fruits and vegetables, either as a way of reducing the risk of cancer and heart disease or as a way to support state and locally grown products. The demand that is likely to grow from these messages poses a challenge to both conventional and nonconventional growers; yet a review of the research agenda for sustainable agriculture reflects this public concern in only a minute way. Research is inadequate in several ways:

- Reports on the long-term effects of pesticide exposure through fruit and vegetable production are heavily underrepresented in the cancer literature.
- Only a small number of ongoing research projects are devoted to the development of pesticide-free or reduced vegetable and fruit crops relative to research on grains.
- Few of the low-input sustainable agriculture financial research grants went to fruit and vegetable projects.

It is important to recognize the disparity in acreages devoted to grain versus fruit and vegetable production (the minor crops), the dollar value of these crops, and the present skewedness in geographic location of major producing areas, as well as the much greater contribution to consumers' caloric intake of grains versus fruits and vegetables. But a contradiction is emerging. The public is beginning to show an interest in sustainable agriculture through emphasis on the pesticide/safety problems of produce. However, at this time many of these crops are difficult to grow organically, so increased demands encourage growers to cut corners and conventional brokers to commit fraud (Zind, 1988). Thus, it is difficult to believe that these crops have been given sufficient research attention.

The emphasis of organic agriculture on soil preservation and decreased fertilizer use, which does lead to a decrease in nitrate levels in water, may not be a reflection of the greatest taxpayer interest. However, consumers have come to prefer unblemished fruit and vegetable products. Research is just starting to measure the trade-offs between cost and appearance and safety. So far, appearance is still winning (Sachs et al., 1987). Feenstra's review (1988) of the number of pesticide applications used on tomatoes and oranges solely to meet cosmetic standards is a useful reference, and it may be worthwhile to review grading standards. But it will take a lot of education to bring public pressure to change grading systems reflecting taste and safety rather than size and color. On the other hand, organic growers are working hard to meet appearance standards, so the question may become moot.

***Antibiotics.*** Controversy about the human health effects of prophylactic antibiotic treatments in animal production has been intense for 20 years (National Research Council, 1988). The last significant data appeared in 1984 (Holmberg et al., 1984a, 1984b), and a review was published by Cohen and Tauxe (1986). In brief, the concern is that the subtherapeutic use of antibiotics in animal feeds may cause bacteria in animals to become resistant to antibiotics. This resistance might be transferred to bacteria in humans, thus making the antibiotics ineffective in treating human bacterial infections (Frappaolo, 1986). Confirming the presence of disease caused in this way in humans has been difficult. In the meantime, food-borne illness is becoming an increasing problem. FDA's recent figures estimate 33 million cases and 9,000 human deaths in the U.S. each year (Young, 1987). Over half of this illness is traceable to bacteria from meat and dairy products (Brady and Katz, 1988).

Other animal drugs also pose human health problems. Residues of these have been found with some frequency in milk products (Brady and Katz, 1988) and in meat. Despite these findings, FDA has not made many changes in its drug review process and has frequently dismissed many of the criticisms raised as uninformed.

The alternatives to using antibiotics are similar to those used to decrease other hazards. They consist of better management, reduced confinement of animals, and alternative therapies. Meat produced under alternative systems is on the market, and steadily increasing demand for these products seems likely. Government policies that require less use of aversive substances will clearly encourage the adoption of an integrated set of alternative animal production systems.

***Nitrates.*** Vogtmann's treatment of the "nitrate story" is quite comprehensive (Vogtmann and Biedermann, 1985) and predominates in this review. Nitrates are a human health hazard because, under various conditions, they convert to nitrites, which can be converted in vivo to two different compounds: nitrosamines, which are carcinogenic, and methemoglobin. Methemoglobin is a threat to infants, and the U.S. drinking water standards for nitrates are based on levels that minimize risk to the infant age group

(Blodgett and Clark, 1986). There is also preliminary evidence that nitrates can cause birth defects (Dorsch et al., 1984).

Hallberg (1987) has been in the forefront in calling attention to the hazards to human health from increasing groundwater contamination from nitrates. One must not forget, however, that water is the second most important source of nitrates. Vegetables are the primary source, contributing 70 percent of the total intake of nitrates in average human diets (Vogtmann and Biedermann, 1985). This reinforces the need for development of alternative farm production systems. Studies show that between 25 and 85 percent of the nitrogen in plants is derived from added mineral-nitrogen fertilizers. Plants use nitrate as an internal nitrogen reserve for protein synthesis. It can remain in the vacuoles, and will do so if plants are provided with more nitrogen than they can use. A number of experiments have demonstrated that the use of compost and other slower-releasing fertilizers results in much lower nitrate levels in plants than do mineral-nitrogen fertilizers (Vogtmann and Biedermann, 1985).

Also, researchers have found that lower dry matter and vitamin C levels result under conventional practices. Corroboration of the latter finding is being pursued by researchers at the U.S. Department of Agriculture Beltsville labs and other places (Kendall, 1988). Vogtmann and Biedermann (1985) point out that vegetables produced during periods of low light intensity (such as green house production) and root crops can store excessively high nitrate concentrations, which surpass the maximum suggested levels in European countries. According to a study reported in 1982, between 2.7 percent and 5.8 percent of rural drinking water supplies exceeded the nitrate standard of 10 milligrams per liter. In smaller, localized studies, 20 to 30 percent of the wells exceeded the standard (Blodgett and Clark, 1986). Increasing vegetable consumption could lead to nitrate uptakes that begin to approach average daily intake values from the World Health Organization of 255 milligrams of nitrate per day (Vogtmann and Biedermann, 1985).

Therefore, there are good reasons for encouraging the use of alternative methods in the production of vegetable crops. This is underscored by the health hazard that arises from the formation of nitroso compounds from fungicides, particularly the dithiocarbamates (Hallberg, 1987; Vogtmann and Biedermann (1985).

However, nitrate levels in food and water supplies are not of great concern, except in a few midwest states. As pesticide contamination of water supplies becomes more visible, it will raise general questions about water quality in all states and lead to greater questioning of conventional farming practices.

***Nutrient Quality.*** The final issue on the agriculture safety agenda is nutrient quality. Few recent papers provide useful information (Termine et al., 1987). Only research that holds cultivar, weather, and other factors constant and manipulates only inputs will be able to support claims of nutrient differences linked to production methods (Clancy, 1986).

## Agriculture and Health

Food grown organically in local areas is fresher, is transported shorter distances, and has a longer shelf life. Consumers in various parts of the country are going to great lengths to purchase local organic food and are willing to pay more for it. The reasons are not well-researched yet, but they include concern for personal safety, wildlife, and farmers (Sachs et al., 1987; Yarbrough and Yarbrough, 1985), and some less easily measured factors, such as attitudes toward technology, general destruction of the environment, and feelings that "food is more than a commodity to be produced, sold, and consumed" (Callicott, 1988). The latter point is an inarticulate statement of the idea of sustainability—one that arises not just from a concern about the environment but about many other phenomena that are seen to compromise health in both its natural and human manifestations. Not surprisingly, most of the definitions of sustainable agriculture incorporate a notion of health: "both of soils and people," as Wendell Berry (1985) has said; ecologically sound or "healthy, whole, and in good condition," stated by Gips (1988); or another Wendell Berry (1985) statement that, "an agriculture that is whole nourishes the whole person body and soul."

The latter description is close to the World Health Organization's (WHO) definition of health, and it seems appropriate to conclude by calling attention to a new health campaign that is attempting to make the definition of health real and not just a group of unattainable concepts. The idea is to change the measurement of progress and development from economic growth to health. The conceptual framework of this new public health movement is being called "healthy public policy" (Satin, 1988), as opposed to "public health policy" that isolates public health from the activities in all sectors that generate public health problems. It seems obvious how such a concept might be understood in the context of agriculture; for example, farm policies that reward farmers for using high chemical inputs are "unhealthy" farm policies and those that have as their ends low pesticide, lower fertilizer inputs, are "healthy" farm policies—with the idea being that health is a powerful concept and a powerful word. Although all farmers may not be interested in the new idea, consumers and taxpayers are likely to find it compelling and worthy of support.

Healthy public policy embraces many ideas, including prevention, wellness, less emphasis on health care and technological medicine, and others. It also has a strong ecological underpinning. The policy focus moves away from single linear goals, such as profit, production, and total numbers, to relationships between people and the environment. In short, this approach requires producers to think about the effects of their inputs and to recognize that the effects move through the entire plant, animal, and human ecosystem.

Slowly but surely, agriculture and health will be joined at the farm, consumer, and policy levels. Apparently, sustainable agriculture practitioners and researchers are ahead of their conventional colleagues in awareness of this need. But they have not thought broadly enough yet. Attention must be turned to fruit and vegetable crops, along with grains and meat. Research in sustainable agriculture must recognize the need for information about social and health concerns and provide support for doing these studies. Finally, proponents of "systems" research must realize that the system is much bigger than the farm and its immediate environment. Classroom and extension education should stress this fact and see this new knowledge redound to the benefit of both farmers and consumers.

## REFERENCES

Anderson, J. 1988. Well-water imperiled. Syracuse Post-Standard, September 13. pp. A-9.

Berry, W. 1985. Preface. *In* M. Fukuoka. The one straw revolution. Bantam Books, Toronto, Canada.

Blodgett, J., and E. A. Clark. 1986. Fertilizers, nitrates, and groundwater: An overview. Colloquium on Agrichemical Management to Protect Water Quality. National Research Council, Washington, D.C. pp. 1-15.

Brady, M. S., and S. E. Katz. 1988. Antibiotic/antimicrobial residues in milk. Journal of Food Protection 51(1): 8-11.

Callicott, J. B. 1988. Agroecology in context. pp. 39-43. *In* P. Allen and D. van Dusen [editors] Global perspectives on agroecology and sustainable agricultural systems. Agroecology Program, University of California, Santa Cruz.

Clancy, K. 1986. The role of sustainable agriculture in improving the safety and quality of the food supply. American Journal of Alternative Agriculture 1: 11-18.

Cohen, M., and R. Tauxe. 1986. Drug-resistant salmonella in the United States: An epidemiological perspective. Science 334: 964-967.

Council on Scientific Affairs. 1988. Cancer risk of pesticides in agricultural workers. JAMA 260: 959-966.

Coye, M. 1984. Occupational health and safety of agricultural workers in the United States. pp. 183-206. *In* L. Busch and W. B. Lacy [editors] Food security in the United States. Westview Press, Boulder, Colorado. pp. 183-206.

Davies, J. E., and J. A. Lee. 1987. Changing profiles in human effects of pesticides. *In* R. Greenhalgh and T. R. Roberts [editors] Pesticide science and biotechnology. Sixth International Congress of Pesticide Chemistry. Blackwell Scientific Publications, Oxford, England.

Dorsch, M. M., R. K. Scragg, A. J. McMichael, P. A. Baghurst, and K. F. Dyer. 1984. Congenital malformations and maternal drinking water supply in rural South Australia: A case control study. American Journal of Epidemology 119: 473-486.

Feenstra, G. 1988. Who chooses your food? A study of the effect of cosmetic standards on the quality of fresh and processed produce. CALPIRG Project Report, Davis, California.

Frappaolo, P. 1986. Risks to human health from the use of antibiotics in animal feeds. pp. 100-111. *In* W. Moats [editor] Agricultural uses of antibiotics. American Chemical Society, Washington, D.C.

Gianessi, L. 1987. Text of press conference on national pesticide usage. Resources for the Future, Washington, D.C.

Gips, T. 1988. What is a sustainable agriculture? *In* P. Allen and D. van Dusen [editors] Global perspectives on agroecology and sustainable agricultural systems. Agroecology Program, University of California, Santa Cruz.

Hallberg, G. 1987. Agricultural chemicals in ground water: Extent and implications. American Journal of Alternative Agriculture 2(1): 3-15.

Holmberg, S. D., J. G. Wells, and M. L. Cohen. 1984a. Animal-to-man transmission of antimicrobial-resistant Salmonella: Investigations of U.S. outbreak, 1971-1983. Science 225: 833-835.

Holmberg, S. D., M. T. Osterholm, K. A. Senger, and M. L. Cohen. 1984b. Drug resistant Salmonella from animals fed antimicrobials. New England Journal of Medicine 331: 617-622.

Kanja, L., J. U. Skane, J. Nafstad, C. Maitai, and P. Lokken. 1986. Organochlorine pesticides in human milk from different areas of Kenya, 1983-1985. Journal of Toxicology and Environmental Health 19: 449-464.

Karlock, R., J. Rogen, L. E. Gray, and N. Chernoff. 1987. Post-natal alternations in development resulting from prenatal exposure to pesticides. *In* R. Greenhalgh and T. R. Roberts [editors] Pesticide science and biotechnology. Sixth International Congress of Pesticide Chemistry. Blackwell Scientific Publication, Oxford, England.

Kendall, Dave. 1988. It's organic—but is it better? The New Farm 10: 28-32.

Mongomery, A. 1987. America's pesticide permeated food. Nutrition Action 14: 1, 4-7.

Moses, M. 1986. Epidemiologic studies of cancer in humans related to farming, agricultural work, or pesticide application or manufacturing. Abstracts. National Farm Workers Health Group, San Francisco, California.

Mott, L., and K. Snyder. 1987. Pesticide alert—A guide to pesticides in fruits and vegetables. Sierra Club Books, San Francisco, California.

National Research Council. 1982. Diet, nutrition and cancer. National Academy Press, Washington, D.C.

National Research Council. 1987. Regulating pesticides in food: The Delaney paradox. National Academy Press, Washington, D.C.

National Research Council. 1988. News report. Antibiotics in feed. March. p. 9.

Olson, L. 1986. The immune system and pesticides. Journal of Pesticide Reform (Summer).

Sachs, C., D. Blair, and C. Richter. 1987. Consumer pesticide concerns: A 1965 and 1984 comparison. The Journal of Consumer Affairs 21(1): 96-107.

Satin, M. 1988. Shifting the framework from "growth" to "health." New Options 50 (July 25).

Schwartz, David, and James P. Lo Gerfo. 1988. Congenital limb reduction defects in the agricultural setting. American Journal of Public Health 78(6): 654-657.

Stehr-Green, P. A., J. Farrar, V. Burse, W. Royce, and J. Wohlleb. 1988. A survey of measured levels and dietary sources of selected organochlorine pesticide residues and metabolites in human sera from a rural population. American Journal of Public Health 78(7): 828-830.

Termine, E., D. Lairon, B. Tauper-Letage, S. Gautier, R. Lafont, and H. Lafont. 1987. Yield and content in nitrates, minerals and ascorbic acid of leeks and turnips grown under mineral or organic nitrogen fertilization. Plant Foods for Human Nutrition 37: 321-332.

U.S. General Accounting Office. 1986a. Pesticides—Better sampling and enforcement needed

on imported food. U.S. GAO, Washington, D.C.
U.S. General Accounting Office. 1986b. Pesticides—Need to enhance FDA's ability to protect the public from illegal residues. U.S. GAO, Washington, D.C.
Vogtmann, H., and R. Biedermann. 1985. The nitrate story—No end in sight. Nutrition and Health 3: 203-216.
Yarbrough, J., J. Chambers, and K. Robinson. 1982. Alterations in liver structure and function resulting from chronic insecticide exposure. *In* J. Chambers and J. Yarbrough [editors] Effects of chronic exposures to pesticides on animal systems. Raven Press, New York, New York.
Yarbrough, P., and F. Yarbrough. 1985. Pesticides and related environmental issues—A study of the opinions and behaviors of New York adults. Cornell University, Ithaca, New York.
Young, F. 1987. Food safety and FDA's action plan, Phase II. Food Technology (November): 116-124.
Zind, T. 1988. California company fails to document organic-grown claim. The Packer (July 2).

# A MATTER OF COMMITMENT

Stephen Viederman

If a definition of the word "presumptuous" were needed, here it is: a New Yorker, born and bred, trained as a historian of sixteenth-century England, no less, who had not been on a farm until two years ago (and then, not for very long), talking to some of the most eminent specialists in the field of agriculture, on their subject. However, perhaps there are some advantages in having an outsider, who shares values with you, looking at sustainable low-input agriculture from the outside.

It is incumbent upon me to deliver to you "good news." Recalling that "a pessimist is a well-informed optimist," let me proceed.

The "good news" is the environmental assaults that have been inflicted upon us. North America and other parts of the world have experienced an unprecedented level of environmental awareness and concern. Thus, if we can find a good side to the serious droughts here and elsewhere, the prospect of global warming, ocean and beach pollution, devastating floods in Bangladesh, and PCB fires in Canada—to highlight a few recent events, it is that the media have taken upon themselves the responsibility to report and analyze what we are doing to our world and to help us as a society begin thinking about what we need to do to create a more sustainable society in the future.

Clearly, these assaults are extremely serious and complex. But we have within our capacity, in varying degrees, the knowledge and the means to ameliorate them and to lessen their impacts. Modern society, especially the so-called developed, industrialized countries, has probably done more to destroy the environment than any society in history. The "good news" is that we now perhaps are developing the will to act to do something about it.

Journalist David Wier has suggested that "we borrow our world from our grandchildren." The question we are left with, then, is: What will we do to reasonably ensure that we return this borrowed world in a form that leaves our grandchildren at least the same, if not better, chances than we have had to make a go of it?

At a U.S. Department of Agriculture Conference on low-input, sustainable agriculture in September 1988, one USDA speaker implied that a concern for environmental matters in agriculture, especially in the so-called less developed countries, is a luxury. The "good news" is that this view is less widely held today than it was a decade ago, both at policymaking levels and among populations as a whole, and both in developed and in developing countries. The World Commission on Environment and Development's report, "Our Common Future," was almost universally acclaimed, except by the Reagan Administration, and has been the subject of intensive discussion. The report makes it eminently clear that we are not talking about luxuries when we talk about environmental and developmental issues. Environment and sustained development are one and the same. In effect, the question is: Do we want to pay now, or do we want to pay later, or have our children make the payments? But there can be no denying that there will be payments to be made.

The development of sustainable agriculture reflects a growing awareness of the social and environmental problems of agriculture and of the need to seek solutions to these problems in our lifetime. This, too, is part of the "good news." The USDA conference probably could not have been held five years ago. The size and scope of that meeting is another indication of how far we have come in this movement. Without wishing to suggest complacency with regard to the acceptance of sustainable agriculture, the fact is that we have come a long way.

But before we become self-satisfied with our "wisdom" in moving ahead now with sustainable agriculture, we must make certain that it does not simply become another technological fix to the problems in agriculture—problems that will reappear in other guises or create other difficulties as we find the need to deal with the unanticipated consequences of actions that we take now and in the near future. Americans have an historical propensity to believe that technology will solve our ills. A New York Times' article of August 16, 1988, reporting various technological approaches to minimizing the impacts of global warming, rather than approaches to arresting the problems at their source, is a recent example of this tendency. So, too, is a recent suggestion that a significant investment be made in biotechnology and genetic-engineering techniques to create vegetation that will release more oxygen into the atmosphere to stave off the effects of ozone depletion. Efforts to achieve greater fuel efficiency are within our reach and are more likely to have a long-term beneficial effect. A technological fix will not satisfy the needs of creating a sustainable society—a society that gives equal weight to the needs of present and future generations.

## The Three Es

Wes Jackson of the Land Institute has suggested that we have to look beyond the problems *in* agriculture to the problems *of* agriculture. In doing this, he argues that we have to look at ecology with a big "E." Rather than tinkering with nature, we must be certain that we are working with nature, avoiding seemingly simple solutions to complex problems within nature's systems. We must understand agriculture in its own terms, using nature as our standard. In seeking fast results to deal with real problems, we have to avoid incomplete results, whose long-term consequences we do not understand.

Baltimore journalist H. L. Mencken, is alleged to have said that "for every human problem there is a solution that is simple, neat, and wrong!" Yes, we need to include ecology with a big "E" as our base. But we must also include economics, also with a big "E." Farmers need to know the profitability of sustainable systems, as compared with conventional systems. This information is important to encourage the transition from conventional agriculture. Big "E" economics, however, would go beyond what is important to the farmer today. Big "E" economics would deal with the real costs of agriculture to the society as a whole. What is the burden to present and future generations of soil erosion? Of nonpoint source pollution? Of the ill-effects of conventional farming on the health of farmers, farm workers, and consumers, caused by high doses of agricultural chemicals? On habitat and wildlife? Of irrigation subsidies? Of overpumping of aquifers?

Big "E" economics also would focus attention on the costs to agriculture of environmental assaults generated outside of agriculture by society at large. Who pays the estimated $5 billion each year as a result of damage to crops such as soybeans, peanuts, cotton, and winter wheat resulting from air pollution, especially ozone? What are the costs of tree damage? Until we begin to develop the answers to these and a myriad of other big "E" economic questions, we cannot determine the true costs of agriculture.

In addition to issues of ecology and economics, we are confronted with yet another "E"—the issue of equity, which involves issues of social organization, ethics, and governance.

A concern for equity requires, for example, that we examine the growing emphasis on biotechnology in agriculture, especially as it is being touted by many as the key to a low-input system. Experience with another technological innovation in agriculture, the Green Revolution, provides us with a convenient historical analogy. Yes, indeed, Green Revolution technologies provided food in quantities previously believed to be out of reach, and that was positive—but at what costs to the landless farmers pushed off their lands, unable to pay for the high cost of the inputs needed? And at what costs to the environment that present and future generations will have to deal with? Biotechnology, or any other technology, in and of itself is neither good nor evil. It is a question of how the society chooses to govern its use, how the society calculates its costs and benefits, in whose hands the technology is placed, and who does not have access to it. One organic farmer at the USDA meeting observed that she did not want to become beholden to the biotechnology/genetic engineering companies as generations of farmers in the past had become beholden to the energy and chemical companies. We have to learn to live with nature.

A concern for social organization will also require that we consider the positive and negative impacts of sustainable agriculture on the community. Sustainable agriculture will have positive impacts on rural communities because it could help arrest the decline of the family farm, which is at the heart of the rural community. But there may be unanticipated negative consequences, as well. For example, what happens to the people and the incomes that are derived from the sale of chemicals and fertilizers and of the other goods and services that a sustainable agriculture might make less necessary but that are an important part of the sustainability of the community now? Thus, social organization and governance are as important to the development of a sustainable agricultural system as are issues more traditionally considered agricultural. The "good news" is that I think there is a growing recognition of the importance of the three "Es"—Ecology, Economics, and Equity—as essential elements of a sustainable agriculture.[1]

## Making Commitments

I think there is a growing commitment on the part of many to make agriculture not an alternative, but the conventional agriculture of the future. The ultimate commitment is of the farmers, because they are the ones who

[1] I subsequently learned of Herman Daly's book *Economics, Ecology and Ethics,* whose title I seem to have reinvented.

are taking the greatest risk in making the transition. Our responsibility, as scientists, funders, and government officials, is to meet the farmers' needs and to better understand the barriers they face as they look to the future. Why is it, for example, that all too often a low-input farm is an island in a sea of conventional farms, even when the low-input farmer is doing well financially?

Part of the commitment to farmers, particularly from scientists and researchers, will require a change in the traditional relationship in setting agendas for research and Extension. The land grant institutions in this country for years said, in effect, "Listen, we are speaking." This appeared to work well when we lived under the delusion that land and nature were limitless. But the fact that limits are imposed by nature is now abundantly clear, and the attitudes of scientists and researchers must change along with this new perception. The needs and nature of a sustainable agriculture are very different from those of a chemically based agriculture. Now the emphasis in setting agendas must be, "Speak—we are listening," or "Let us sit down and talk together."

This change in relationships also will require a change in the way we share information. An extension service geared to the provision of information, a one-way flow, will not serve today's needs. Today and in our sustainable agriculture future, the emphasis must be more on the exploration of problems and the range of possible solutions. The change will require new forms of extension, perhaps with farmers and field days playing larger roles, and perhaps different types of people as extension agents.

The executive and legislative branches of governments, at state and national levels, must be encouraged to make clear a commitment to sustainable agriculture, not with words alone, but also with money and action. At the USDA conference it was suggested that if the U.S. government's investment of $84 million per year in biotechnology in agriculture is a drop in the bucket, then the $3.9 million for sustainable, low-input agriculture does not even represent a vapor. This must be just the beginning. The branches of government, once they have made a commitment to sustainability in agriculture, must ensure, to the degree possible, complementary, between the low-input program and other research and demonstration programs. It makes no sense to fund contradictory activities. In the United States, particular attention must be paid to the Agricultural Research Service's program.

Legislators have an obligation to ensure that farm and agricultural legislation contribute to a sustainable agriculture that meets the needs of farmers, the communities in which they live, consumers, and the natural environment that supports all life. What are the unintended consequences of legislation on the move to sustainability? Do federal commodity programs, for example, encourage a focus on monocultures? How can federal and state policies regulating pesticides contribute to a sustainable agriculture system? What are the implications of agricultural trade policies? Such a comprehensive approach should be the focus of the debate on the 1990 Farm Bill.

Many private, voluntary organizations, large and small, have made a major commitment to sustainable agriculture. They have carried much of the burden to this point in time, often at great cost to themselves. The time has now come to enlist the support of the more traditional organizations that have for years ostensibly spoken in the name of farmers. In this category I would include industrial farming and agribusiness, as well as the farm bureaus. If we are concerned with future generations, we cannot look simply at the bottom line of the profit statement today. The corporations that own the large farms and allege to serve the needs and interests of farmers are run by people who, just like you and I, have a stake in the future, directly and through their children. A positive profit statement will not ensure the future unless environmental and community balance sheets are given their proper place in the calculation of profit and loss.

Universities and research institutes, especially the land grant institutions in the United States, have a special responsibility. They were established to serve the needs of farmers. In recent years, however, criticism has been directed to some of them for their failure to do so. Fortunately, there are hopeful signs of growing commitment on the part of many individual scientists to greater attention to sustainable agriculture. However, there are fewer signs of institutional commitment to make the necessary changes for sustainability to become a central and permanent thrust of the institutions' concerns.

We all know that it is easy to entice people, in and out of the universities, to take one's money for some specific purpose. But we must avoid marginalization and cooption: we must ensure commitment. At the Jessie Smith Noyes Foundation, in every grant cycle we see any number of proposals for work that seem to be dusted-off versions of something that the individual has been doing for years, supposedly made attractive to us by adding the words "...and sustainable agriculture." We call this the "...and found God syndrome." But the foundation is in the business of effecting change, not simply giving away money. Thus, the challenge that we face is to reasonably ensure that the grantees, and especially the land grant institutions, are serious about change. To put it crassly, where is the recipient institution's money, up front?

At the USDA conference, one speaker suggested that because universities are short of funds, funding support for low-input sustainable agriculture is difficult. But we are all short of funds, in our personal and professional lives. As a result, we are always faced with making choices that reflect our values. These same institutions that are strapped for funds, making work on sustainable agriculture difficult, do not have the same difficulty when it comes to funding biotechnology.

What do we mean by commitment at universities and in research institutions? Released time of a professor or the waiving of overhead is not commitment; it is accounting. Setting up a center may be useful, but funding it adequately from the institution's budget is an indication of real commitment. Making it clear in word and deed to young faculty members that they will not be disadvantaged, and might even be in a better position in their quest for tenured positions if they work on topics related to sustainability, is another example of commitment. So, too, is the willingness to fill vacancies that occur on the faculty with people who have an interest in sustainable agriculture, not just people in molecular biology and biotechnology.

The foundation community is also showing its commitment. Foundation support for sustainable agriculture is not sufficient to meet the demand, but it is significant. Since 1985, the Jessie Smith Noyes Foundation has made grants totalling \$2.4 million, representing 20 percent of our total funding. If related grants on groundwater were added, the total would exceed \$3.0 million. For a small foundation, that is significant. Other foundations, such as the Northwest Area Foundation and the Joyce Foundation, also have made significant contributions. The foundation community must continue to play its traditional role in encouraging and supporting new and, hopefully on occasion, outrageous ideas. The foundations must foster experimental projects and institutions. We must work together in this important endeavor, to help those responsible for doing the work, and those who are to be the ultimate beneficiaries of the work, to define their own agendas and then to help support them in carrying out the agendas.

In closing, I should note the obvious: Foundations are not the font of all wisdom, although we would sometimes like to think we are and we all too often act as if we were. Part of our commitment must also be to listen. And part of your commitment—you the farmers, the private organizations, the scientists and researchers—is to speak. We recognize that there is a power relationship between the fund giver and the fund receiver that inevitably makes the relationship uneven. Yet, we need to create a dialogue that is as open as possible. We need to be prepared, each one of us, to praise and to criticize in a constructive manner.

In moving toward a sustainable, low-input agriculture, we share a common goal. It is nothing less than an effort to save our planet and make it more livable for ourselves and future generations. The "good news" is that there is no reason that all of us, working together with deep commitment, cannot make that dream a reality. We can all hope that in the not-too-distant future, conferences on "conventional" agriculture will be under the auspices of history departments, leaving the agriculture schools to deal with the real agriculture, a sustainable agriculture.

# SUSTAINABLE AGRICULTURAL SYSTEMS: A CONCLUDING VIEW

Francille M. Firebaugh

To have an ecologically and economically healthy world, we have to make changes in agriculture. The question and ecological philosophy of sustainable agricultural systems must have meaning and substance and not just be a new name.

Sustainable agriculture systems are complex, and work related to sustainable agriculture should include many aspects: the consumer's perspective, the total food system from production to consumption, the social implications of agriculture, the role of women in agriculture.

Dankelman and Davidson (1988) suggest basic requirements of sustainable agriculture:

- Equitable access by all farmers to fertile land, credit, and agricultural information.
- The maintenance and support of independent agriculture over which farmers, both women and men, have control.
- The development of cultivation, food processing, and food storage methods that ease the intense demands on women's labor.
- A high degree of species diversification to maintain flexible cropping patterns.
- The conservation of fertile soils in which organic matter is recycled (to avoid dependence on imported nutrients).
- An appropriate use of water and fuel resources.

These requirements may not engender full agreement, just as the requirements for on-farm and off-farm inputs for sustainable agriculture are not agreed upon. Input descriptions include: "low input," "lower input," "minimal input," and "appropriate input." I believe that with wide environmental diversity, such as soil and climatic conditions, and the varying capacities of market systems, there will be a range of appropriate inputs and methods for ecologically and economically sound systems of food and fiber production. The U.S. Department of Agriculture's term "range of opportunities" suggests the ecumenical nature of sustainable agriculture,

The complex nature of sustainable agriculture leads to the need for interrated farming systems. Increased knowledge about inputs and their interaction is necessary before fully integrated farming systems can be designed. My own work in family systems suggests that research about systems requires a new look at methodologies and, indeed, requires rather sophisticated methodologies.

In describing sustainable agriculture in the United States, J. F. Parr and colleagues, in this book, propose that the ultimate goals of farmers in sustainable agriculture are to (1) maintain or improve the natural resource base, (2) protect the environment, (3) ensure profitability, (4) conserve energy, (5) increase productivity, (6) improve food quality and safety, and (7) create more viable socioeconomic infrastructure for farms and rural communities.

The human side of sustainable agriculture is mentioned in the CGIAR Technical Advisory Committee definition of sustainable agriculture as involving "successful management of resources for agriculture to satisfy changing human needs while maintaining or enhancing the natural resource base and avoiding environmental degradation." The definition is worthy of continued consideration.

The economic aspects of sustainable agriculture for the United States and developing countries are of obvious concern. A number of challenges face profitability assessment methods, and the world of scholarship and academia does not encourage the interdisciplinary and collaborative work that has to be done. Further, cooperation among researchers, farmers, and extension must be strengthened and rewarded.

Some recurring characteristics of the search for sustainable agriculture are that we need interdisciplinary and collaborative approaches (as I have just mentioned); that a holistic, systems, or whole-farm view is necessary; that sustainable agriculture must be site-specific, with local adaptation; and that a stable, appropriate relation between agricultural production and consumption is needed. If production and consumption are to be related satisfactorily in an environment of continued population growth, close attention to appropriate inputs will be essential. It will likewise be essential to convert some of the fragile and marginal land now in crop production to more appropriate uses.

The body of current and increasing indigenous knowledge about sustainable agriculture is a resource we must identify and use. We can learn from centuries-old practices, just as we benefit from germplasm that has survived over time.

Education—primary, secondary, and higher education, as well as informal education—should take an open view of sustainable agriculture.

Let's take a look at some questions in the area of sustainable agriculture—questions that require answers.

1. Will the human and social aspects of sustainable agriculture be considered? Will sustainable food systems be important issues?

2. Will surface and groundwater quality and aquifer depletion receive attention?

3. Will there be multiple definitions of sustainable agriculture?

4. Can more highly productive land successfully use higher fertilizer and pesticide inputs than poorer land?

5. Will more attention be given to fruits and vegetables, and will there be increased emphasis on a healthy food supply?

6. Will perennial crops, with their potential for sustaining or increasing production with limited inputs, receive emphasis?

7. Will the increase in U.S crop yields more nearly parallel the off-farm inputs?

8. Will plant and animal biotechnology research yield contributions to sustainable agriculture?

9. Will there be an increase in systems research, including on-farm research?

10. Will there be a broadened base of public support?

11. Will U.S. "assistance" to developing countries be innovative, less arrogant, and more sensitive to differing environments, indigenous knowledge, and cultural values?

12. Will the concept of a "fully integrated systems approach" be accepted? Will the social, biological, engineering, and life sciences contribute to the approach?

13. Will government policies reflect an openness to and support for sustainable agriculture approaches?

14. Will there be increased participation by land grant universities, USDA, and the mainstream press?

## REFERENCE

Dankelman, I., and J. Davidson. 1988. Women and environment in the Third World. Earthscan Publications, Ltd., London, England.

# CONTRIBUTORS

**I. P. Abrol**
ICAR
New Delhi, India

**Randolph Barker**
Agricultural Economics Department
Cornell University
Ithaca, New York

**G. W. Barrett**
Department of Zoology
Miami University
Oxford, Ohio

**Charles M. Benbrook**
Board on Agriculture
National Research Council
Washington, D.C.

**J. M. Blair**
Department of Entomology
Ohio State University
Columbus

**P. J. Bohlen**
Department of Zoology
Miami University
Oxford, Ohio

**N. C. Brady**
Science and Technology
U.S. Agency for International Development
Washington, D.C.

**H.C.P. Brown**
Department of Zoology
College of Biological Science
University of Guelph
Guelph, Ontario

**Frederick H. Buttel**
Department of Rural Sociology
Cornell University
Ithaca, New York

**Richard F. Celeste**
Governor, State of Ohio
Columbus

**Duane Chapman**
Agricultural Economics Department
Cornell University
Ithaca, New York

**Katherine L. Clancy**
Department of Nutrition and Food Management
Syracuse University
Syracuse, New York

**M. D. Clegg**
Department of Agronomy
University of Nebraska
Lincoln

**D. C. Coleman**
Institute of Ecology and Department of Entomology
University of Georgia
Athens

**T. S. Colvin**
Agricultural Research Service
U.S. Department of Agriculture
Ames, Iowa

**D. A. Crossley, Jr.**
Institute of Ecology and Department of Entomology
University of Georgia
Athens

**Carlos Magno C. da Rocha**
Researches of Cerrados Agricultural Center
EMBRAPA
Planaltina, D.F. Brazil

**Thomas L. Dobbs**
Department of Agricultural Economics
South Dakota State University
Brookings

**D. J. Eckert**
Department of Agronomy
Ohio State University
Columbus

**Clive A. Edwards**
Department of Entomology
Ohio State University
Columbus

**W. M. Edwards**
Agricultural Research Service
U.S. Department of Agriculture
Coshocton, Ohio

**Samir A. El-Swaify**
Department of Agronomy and Soil Science
University of Hawaii
Honolulu

**A. El Titi**
State Institute for Plant Protection
Stuttgart, Federal Republic of Germany

**D. C. Erbach**
Agricultural Research Service
U.S. Department of Agriculture
Ames, Iowa

**N. R. Fausey**
Agricultural Research Service
U.S. Department of Agriculture
Columbus, Ohio

**Francille Firebaugh**
Cornell University
Ithaca, New York

**C. A. Francis**
Department of Agronomy
University of Nebraska
Lincoln

**Gilbert W. Gillespie, Jr.**
Department of Rural Sociology
Cornell University
Ithaca, New York

**Stephen R. Gliessman**
Agroecology Program
University of California
Santa Cruz

**Robert M. Goodman**
Calgne, Inc.
Davis, California

**C.A.I. Goring**
Reno, Nevada

**Natalie D. Hahn**
International Institute of Tropical Agriculture
Ibadan, Nigeria

**Chun ru Han**
Agricultural University
Beijing, China

**Richard R. Harwood**
Winrock International Institute for Agricultural Development
Morrilton, Arkansas

**Holly Hauptli**
Calgne, Inc.
Davis, California

**P. F. Hendrix**
Institute of Ecology and Department of Entomology
University of Georgia
Athens

**Garfield J. House**
Department of Entomology
North Carolina State University
Raleigh

**Rhonda R. Janke**
Rodale Research Center
Emmaus, Pennsylvania

**David Katz**
AgExcess
Davis, California

**W. D. Kemper**
Agricultural Research Service
U.S. Department of Agriculture
Beltsville, Maryland

**Larry D. King**
Soil Science Department
North Carolina State University
Raleigh

**John M. Laflen**
National Soil Erosion Research Laboratory
Agricultural Research Service
U.S. Department of Agriculture
West Lafayette, Indiana

**Rattan Lal**
Agronomy Department
Ohio State University
Columbus

**H. Landes**
The Estate of Lautenbach
Oedheim, Federal Republic of Germany

**William E. Larson**
Soil Science Department
University of Minnesota
St. Paul

**T. J. Logan**
Agronomy Department
Ohio State University
Columbus

**John M. Luna**
Department of Entomology
Virginia Polytechnic Institute and State University
Blacksburg

**Shi ming Luo**
Agroecology Laboratory
South China Agricultural University
Gang Zhou, China

**J. Patrick Madden**
Madden Associates
Glendale, California

**R. E. Meyer**
U.S. Agency for International Development
Rosslyn, Virginia

**Fred P. Miller**
Agronomy Department
Ohio State University
Columbus

**R. H. Miller**
Department of Soil Science
North Carolina State University
Raleigh

**Bill Murphy**
Department of Plant and Soil Science
University of Vermont
Burlington

**Paul F. O'Connell**
Cooperative State Research Service
U.S. Department of Agriculture
Washington, D.C.

**Bede N. Okigbo**
Department of Crop and Soil Sciences
Michigan State University
East Lansing

**R. I. Papendick**
Agricultural Research Service
U.S. Department of Agriculture
Pullman, Washington

**Charles F. Parker**
Department of Animal Science
Ohio State University
Columbus

**J. F. Parr**
Agricultural Research Service
U.S. Department of Agriculture
Beltsville, Maryland

**Jose Roberto Peres**
Researches of Cerrados Agricultural Center
EMBRAPA
Planaltina, D.F. Brazil

**Donald L. Plucknett**
Consultative Group on International Agricultural Research
Washington, D.C.

**Alison Power**
Section of Ecology and Systematics
Program on Science, Technology and Science
Cornell University
Ithaca, New York

**N. S. Randhawa**
ICAR
New Delhi, India

**Emilie E. Regnier**
Agronomy Department
Ohio State University
Columbus

**Robert Rodale**
Rodale Press, Inc.
Emmaus, Pennsylvania

**N. Rodenhouse**
Department of Zoology
Miami University
Oxford, Ohio

**Jose E. Silva**
Researches of Cerrados Agricultural Center
EMBRAPA
Planaltina, D.F. Brazil

**Benjamin R. Stinner**
Department of Entomology
Ohio Agricultural Research and Development Center
Ohio State University
Wooster

**Bruce R. Thomas**
Calgne, Inc.
Davis, California

**V. G. Thomas**
Department of Zoology
University of Guelph
Guelph, Ontario

**P. Vereijken**
Centre for Agrobiological Research
Wageningen, The Netherlands

**Stephen Viederman**
Jessie Smith Noyes Foundation, Inc.
New York City, New York

**Hugo Villachica**
Tropical Crops Research Program
INIAA
Lima, Peru

**I. G. Youngberg**
Institute for Alternative Agriculture
Greenbelt, Maryland

# INDEX